Introduction to Health and Safety at Work

Introduction to Health and Safety at Work

Fourth Edition

The Handbook for the NEBOSH National General Certificate

Phil Hughes MBE, MSc, CFIOSH, former Chairman NEBOSH 1995–2001

Ed Ferrett PhD, BSc (Hons Eng), CEng, MIMechE, MIET, CMIOSH, Vice Chairman NEBOSH 1999–2008

endorsed by

nebosh

AMSTERDAM • BOSTON • HEIDELBERG • LONDON • NEW YORK • OXFORD
PARIS • SAN DIEGO • SAN FRANCISCO • SINGAPORE • SYDNEY • TOKYO

ELSEVIER

Butterworth-Heinemann is an imprint of Elsevier

Butterworth-Heinemann is an imprint of Elsevier
The Boulevard, Langford Lane, Kidlington, Oxford, OX5 1GB
30 Corporate Road, Burlington, MA 01803

First edition 2003
Reprinted 2003 (twice)
Second edition 2005
Reprinted 2006
Third edition 2007
Fourth edition 2009
Reprinted 2010

British Library Cataloguing in Publication Data
A catalogue record for this book is available from the British Library

Library of Congress Cataloging in Publication Data
A catalog record for this book is available from the Library of Congress

ISBN 978-1-85617-668-2

For information on all Butterworth-Heinemann publications
visit our web site at www.elsevierdirect.com

Typeset by Macmillan Publishing Solutions
www.macmillansolutions.com

Printed and bound in China

10 11 12 13 14   6 5 4 3 2

Contents

Contents

Contents

Preface to the fourth edition

Introduction to Health and Safety at Work has quickly established itself, in its first six years, as the foremost text for students taking the NEBOSH National General Certificate. It is also of great value to those studying for level 3 N/SVQ, the IOSH Managing Safely Award and similar management and supervisor focused learning. It has become a significant work of reference for managers with health and safety responsibilities and it is therefore a matter of primary importance that it should be kept up-to-date, as far as is possible, with new legislation and recent developments. It has now been endorsed by NEBOSH as a recommended text book for the National General Certificate course.

This fourth edition has been produced in order to update health and safety legislation, with particular regard to changes in legislation relating to corporate manslaughter and corporate homicide, the Health and Safety (Offences) Act, changes to the Health and Safety (Information for Employees) Regulations and the introduction of environmental permitting. More information has been given on the Registration, Evaluation, Authorization and Chemicals Regulations (REACH), the introduction of a new regime for domestic gas safety and future changes to the Control of Pesticides Regulations. Finally, the merger of the Health and Safety Commission and the Health and Safety Executive into a single unified body called the Health and Safety Executive is also covered.

The qualification is now divided into three distinct units each of which is assessed separately. This development offers the opportunity for additional and more flexible course formats and students may now study parallel courses (in, say, fire and construction) without repeating the management unit. Students who decide to take individual units will, on passing, receive a Unit Certificate.

Since the first edition of this book was published, there has been a change in the style of examination questions and NEBOSH has introduced a commendably thorough system for question paper preparation to ensure that no candidates are disadvantaged by question ambiguity. NEBOSH is anxious to dispel the myths surrounding their examinations and have introduced regional meetings for course providers to introduce changes to the syllabuses and to answer any queries so that their students get the best possible preparation for the assessment tasks. The NEBOSH website is also a very useful channel of communication with course providers and students.

Many questions now cover the contents of more than one chapter. It is recommended that students have sight of the published examiner's reports, which are available from NEBOSH. These reports not only provide an excellent guide on the expected answers to the questions but also indicate areas of student misunderstanding.

Since 1 September 2004, there have been changes in the National Qualifications Framework (NQF) that are relevant to the NEBOSH General Certificate. However, up to level 3 the NQF is unchanged and therefore its effect on the Certificate will be minimal. The Certificate syllabus is designed to provide the underpinning knowledge for the 'Health and Safety for People at Work' standard, an approved level 3 vocational standard developed by the Employment National Training Organisation (ENTO). The syllabus is unitized to make it as compatible as possible with short courses based on the eight units of the ENTO standard. The NEBOSH National General Certificate is accredited as a level 3 award by Ofqual. More detailed information on these eight units is available in the guide to the NEBOSH National General Certificate. The fourth edition, which continues to follow closely the eight units, will therefore be particularly useful to people involved with the 'Health and Safety for People at Work' level 3 ENTO standard.

As mentioned earlier, the Certificate syllabus has been updated to include much of the new legislation. However, it is the policy of NEBOSH to examine new relevant legislation 6 months after its introduction whether it is specifically mentioned in the syllabus document or not.

Finally, one of the objectives of the book is to provide a handbook for the use of any person who has health and safety as part of his/her responsibilities. We thought that it would be useful, therefore, to add a few useful topics which are outside the syllabus. These include more on the International Labour Organization, managing occupational road safety (Chapter 9), fast-track settlement of compensation claims following the Woolfe reforms (Chapter 8) and the effects of alcohol and drugs on occupational health and safety (Chapter 15). A sample question on the effects of alcohol has also been included.

We hope that you find this new edition to be useful.

Phil Hughes
Ed Ferrett

Acknowledgements

Throughout the book, definitions used by the relevant legislation and the Health and Safety Executive and advice published in Approved Codes of Practice or various Health and Safety Commission/Executive publications have been utilized. Most of the references produced at the end of each Act or Regulation summary in Chapter 17 are drawn from the HSE Books range of publications.

At the end of each chapter, there are some examination questions taken from recent NEBOSH National General Certificate papers. Some of the questions may include topics which are covered in more than one chapter. The answers to these questions are to be found within the preceding chapter of the book. NEBOSH publishes a very full examiners' report after each public examination which gives further information on each question. Most accredited NEBOSH training centres will have copies of these reports and further copies may be purchased directly from NEBOSH. We have been asked about the allocation of marks to each question shown throughout the book. However, as the marks awarded to each part of a question can vary, only general guidance can be given.

All one- or two-part questions are 8-mark questions with a minimum of 2 marks awarded to each part.

Questions with three or more parts are 20-mark questions with a minimum of 4 marks for each part.

The authors would like to thank NEBOSH for giving them permission to use these questions.

The authors' grateful thanks go to Liz Hughes and Jill Ferrett for proof reading and patience and their administrative help during the preparation of this edition. The authors are particularly grateful to Liz for the excellent study guide that she has written for all NEBOSH students, which is included at the end of this book and for the section on report writing (Section 7.9). Liz gained an honours degree in psychology at the University of Warwick, later going on to complete a Master's degree at the same university. She taught psychology in further and higher education, where most of her students were either returning to education after a gap of many years, or were taking a course to augment their existing professional skills. She went on to qualify as a social worker specializing in mental health, and later moved into the voluntary sector where she managed development for a number of years. Liz then helped to set up and manage training for the National Schizophrenia Fellowship (now called Rethink) in the Midlands.

We would also like to acknowledge the contribution made by Hannah Ferrett for the help that she gave during the research for the book and with some of the word processing. The advice given on the specimen practical application and risk assessments (Appendices 5.3 and 5.4) by John Tremelling, Health and Safety Consultant from Penzance, is also gratefully acknowledged.

We would like to thank Teresa Budworth, the Chief Executive of NEBOSH, for her support during this fourth edition and various HSE staff for their generous help and advice. Finally we would like to thank Stephen Vickers, the immediate past Chief Executive of NEBOSH for his encouragement at the beginning of the project and Doris Funke and all the production team at Elsevier who have worked hard to translate our dream into reality.

About the authors

Phil Hughes MBE is a well-known UK safety professional with over 30 years' worldwide experience as Head of Environment, Health and Safety at two large multinationals, Courtaulds and Fisons. Phil started work in health and safety in the Factory Inspectorate at the Derby District in 1969 and moved to Courtaulds in 1974. He joined IOSH in that year and became Chairman of the Midland Branch, then National Treasurer and was President in 1990–1991. Phil has been very active in the NEBOSH Board for over 10 years and served as Chairman from 1995 to 2001. He is also a Professional Member of the American Society of Safety Engineers and has lectured widely throughout the world. Phil received the RoSPA Distinguished Service Award in May 2001 and became a Director and Trustee of RoSPA in 2003. He received an MBE in the New Year Honours List 2005 for services to Health and Safety.

Ed Ferrett is an experienced health and safety consultant who has practised for over 20 years. With a PhD and honours degree in mechanical engineering from Nottingham University, Ed spent 30 years in higher and further education, retiring as the Head of the Faculty of Technology of Cornwall College in 1993. Since then he has been an independent consultant to several public and private sector organizations including construction businesses and the Regional Health and Safety Adviser for the Government Office (West Midlands), and was Chair of West of Cornwall Primary Care NHS Trust for 6 years until 2006.

Ed has been a member of the NEBOSH Board since 1995 and was Vice Chair from 1999 to 2008. He has delivered many health and safety courses and is a lecturer in NEBOSH courses at the Cornwall Business School and for other course providers. He has recently been appointed as the External Examiner for the MSc course in Health and Safety at a UK University. Ed is a Chartered Engineer and a Member of IOSH.

List of principal abbreviations

Most abbreviations are defined within the text. Abbreviations are not always used if it is not appropriate within the particular context of the sentence. The most commonly used ones are as follows:

ACL   Approved carriage list
ACM   Asbestos-containing material
ACOP   Approved Code of Practice
ACPO   Association of Chief Police Officers in Scotland
AIB   Asbestos Insulation Board
ALARP   As low as reasonably practicable
APAU   Accident Prevention Advisory Unit, now Operations Unit
ARCA   Asbestos Removal Contractors Association
BA   Breathing apparatus
BAT   Best available techniques
BRE   Building Research Establishment
BSI   British Standards Institution
CAR   Control of Asbestos Regulations
CBI   Confederation of British Industry
CD   Consultative document
CDM   Construction (Design and Management) Regulations
CECA   The Civil Engineering Contractors Association
CEN   Comite Europeen de Normalisation
CENELEC   Comite Europeen de Normalisation Electrotechnique
CHIP   Chemicals (Hazard Information and Packaging) Regulations
CIB   Chartered Institute of Building
CIRA   Construction Industry Research and Information Association
CLAW   Control of Lead at Work Regulations
CONIAC   Construction Industry Advisory Committee
COPFS   Crown Office and Procurator Fiscal Service
CORGI   Council for Registered Gas Installers
COSHH   Control of Substances Hazardous to Health Regulations
COSLA   Convention of Scottish Local Authorities
dB(A)   Decibel (A-weighted)
dB(C)   Decibel (C-weighted)
DSE   Display screen equipment

DSEAR	Dangerous Substances and Explosive Atmospheres Regulations
DWP	Department for Work and Pensions
E&W	England and Wales
EAV	Exposure action value
EC	European Community
ELV	Exposure limit value
EMAS	Employment Medical Advisory Service
EPA	Environmental Protection Act 1990
EU	European Union
FSA	Financial Services Authority
FSB	Federation of Small Businesses
HAV	Hand–arm vibration
HGV	Heavy goods vehicle
HIE	Highlands and Islands Enterprise
HOPE	Healthcare, Occupational and Primary for Employees
HSAC	Health and Safety Advice Centre
HSCER	Health and Safety (Consultation with Employers) Regulations
HSE	Health and Safety Executive
HSL	Health and Safety Laboratory
HSW Act	Health and Safety at Work etc. Act 1974
HWL	Healthy Working Lives
IAC	Industry Advisory Committee
ILO	International Labour Office
IOSH	Institution of Occupational Safety and Health
LBRO	Local Better Regulation Office
LEAL	Lower exposure action level
LOLER	Lifting Operations and Lifting Equipment Regulations
LPG	Liquefied petroleum gas
MCG	The Major Contractors Group
MEL	Maximum exposure limit
MHOR	Manual Handling Operations Regulations
MHSW	Management of Health and Safety at Work Regulations
MORR	Management of Occupational Road Risk
MoT	Ministry of Transport (still used for vehicle tests)
NAWR	Control of Noise at Work Regulations
NEBOSH	National Examination Board in Occupational Safety and Health
NVQ	National Vocational Qualification
OHSAS	Occupational Health and Safety Assessment Series
OSH	Occupational Safety and Health
PF	Procurator Fiscal
PHASS	The Partnership on Health and Safety in Scotland
POOSH	Scotland Professional Organisations in Occupational Safety & Health
PPE	Personal protective equipment
ppm	Parts per million
PUWER	The Provision and Use of Work Equipment Regulations
RCD	Residual current device
REACH	Registration Evaluation and Authorization of Chemicals
RES	Representative(s) of employee safety
RIDDOR	The Reporting of Injuries, Diseases and Dangerous Occurrences Regulations
RoSPA	Royal Society for the Prevention of Accidents

RPE	Respiratory protective equipment
RRFSO	Regulatory Reform Fire Safety Order
RTA	Road traffic accident
SaHW	Safe and Healthy Working
SBSA	Scottish Building Standards Agency
ScotPHO	Scottish Public Health Observatory
SCVO	Scottish Council for Voluntary Organisations
SE	Scottish Executive
SEPA	Scottish Environment Protection Agency
SHAW	Scotland's Health at Work
SPL	Sound pressure level
STEL	Short-term exposure limit
STUC	Scottish Trades Union Congress
SWL	Safe working load
SWP	Safe working pressure
TLV	Threshold limit value
TUC	Trades Union Congress
TWA	Time-weighted average
UEAL	Upper exposure action level
UK	United Kingdom
VAWR	Vibration at Work Regulations
WAHR	Work at Height Regulations
WBV	Whole body vibration
WEL	Workplace exposure limit
WHO	World Health Organization
WRULD	Work-related upper limb disorder

Illustrations credits

Figure 1.5	HSE © Crown copyright material is reproduced with the permission of the Controller of HMSO and Queen's Printer for Scotland.
Figure 1.12	From HSG65 *Successful Health and Safety Management* (HSE Books 1997) ISBN 0717612767. © Crown copyright material is reproduced with the permission of the Controller of HMSO and Queen's Printer for Scotland.
Figure 3.7	HSE INDG232 (rev) page 1 HSE Web.
Figure 4.3	From HSG57 *Seating at Work* (HSE Books 1998) ISBN 0717612317. © Crown copyright material is reproduced with the permission of the Controller of HMSO and Queen's Printer for Scotland.
Figure 4.7	From HSG48 *Reducing Error and Influencing Behaviour* (HSE Books 1999) ISBN 0717624528. © Crown copyright material is reproduced with the permission of the Controller of HMSO and Queen's Printer for Scotland.
Figure 5.1	From HSG149 *Backs for the Future: Safe Manual Handling in Construction* (HSE Books 2000) ISBN 0717611221. © Crown copyright material is reproduced with the permission of the Controller of HMSO and Queen's Printer for Scotland.
Figure 6.1	Reproduced with permission from *The Argus*, Brighton.
Figure 6.4	Courtesy of Stocksigns.
Figure 6.7	Courtesy of Stocksigns.
Figure 6.15	From HSG150 (rev 1) *Health and Safety in Construction* (HSE Books 2006) ISBN 0717661822. © Crown copyright material is reproduced with the permission of the Controller of HMSO and Queen's Printer for Scotland.
Figure 6.18	Adapted from *PUWER 2008. Provision and Use of Work Equipment Regulations 1998: Open Learning Guidance* (HSE Books 1999) ISBN 9780717662852. © Crown copyright material is reproduced with the permission of the Controller of HMSO and Queen's Printer for Scotland.
Figure 6.19	Cover of INDG98 *Permit-to-Work Systems* (HSE 1998) ISBN 0717613313. © Crown copyright material is reproduced with the permission of the Controller of HMSO and Queen's Printer for Scotland.
Figure 7.1	From HSG65 *Successful Health and Safety Management* (HSE Books 1997) ISBN 0717612767. © Crown copyright material is reproduced with the permission of the Controller of HMSO and Queen's Printer for Scotland.

Figure 7.2	From *Guide to Measuring Health and Safety Performance* (HSE 2001). © Crown copyright material is reproduced with the permission of the Controller of HMSO and Queen's Printer for Scotland.
Figure 8.5	From BI 510 *Accident Book* (HSE Books 2003) ISBN 0717626032. © Crown copyright material is reproduced with the permission of the Controller of HMSO and Queen's Printer for Scotland.
Figure 9.1	From HSG155 *Slips and Trips* (HSE Books 1996) ISBN 0717611450. © Crown copyright material is reproduced with the permission of the Controller of HMSO and Queen's Printer for Scotland.
Figure 9.4(a)	From HSG76 *Health and Safety in Retail and Wholesale Warehouses* (HSE Books 1992) ISBN 0118857312. © Crown copyright material is reproduced with the permission of the Controller of HMSO and the Queen's Printer for Scotland.
Figure 9.6	From HSG6 *Safety in Working with Lift Trucks* (HSE Books 2000) ISBN 0717617815. © Crown copyright material is reproduced with the permission of the Controller of HMSO and Queen's Printer for Scotland.
Figure 10.4	From L23 *Manual Handling Operations – Guidance on Regulations* (HSE Books 2004) ISBN 071762823X. © Crown copyright material is reproduced with the permission of the Controller of HMSO and Queen's Printer for Scotland.
Figure 10.5	From *Manual Handling in the Health Services* (HSE Books 1998) ISBN 0717612481. © Crown copyright material is reproduced with the permission of the Controller of HMSO and Queen's Printer for Scotland.
Figure 10.6	From HSG115 *Manual Handling Solutions You Can Handle* (HSE Books 1994) ISBN 0717606937. © Crown copyright material is reproduced with the permission of the Controller of HMSO and Queen's Printer for Scotland.
Figure 10.8(a)–(c)	From HSG115 *Manual Handling Solutions You Can Handle* (HSE Books 1994) ISBN 0717606937. © Crown copyright material is reproduced with the permission of the Controller of HMSO and Queen's Printer for Scotland.
Figure 10.9	From HSG149 *Backs for the Future: Safe Manual Handling in Construction* (HSE Books 2000) ISBN 0717611221. © Crown copyright material is reproduced with the permission of the Controller of HMSO and Queen's Printer for Scotland.
Figure 10.10	From HSG76 *Health and Safety in Retail and Wholesale Warehouses* (HSE Books 1992) ISBN 0118857312. Crown copyright material is reproduced with the permission of the Controller of HMSO and the Queen's Printer for Scotland.
Figure 10.13	From HSG150 (rev 1) *Health and Safety in Construction* (HSE Books 2001) ISBN 0717621065. © Crown copyright material is reproduced with the permission of the Controller of HMSO and Queen's Printer for Scotland.
Figure 11.4	Courtesy of Draper.
Figure 11.6	Courtesy of Draper. Speedy catalogue 2004, page 23.
Figure 11.11	Reprinted from *Safety with Machinery* Second Edition, John Ridley and Dick Pearce, pages 26–34, 2005, with permission from Elsevier.
Figure 11.15	Reprinted from *Safety with Machinery* Second Edition, John Ridley and Dick Pearce, pages 26–34, 2005, with permission from Elsevier.
Figure 11.16	Courtesy of Allen-Bradley Guardmaster brand from Rockwell Automation.
Figure 11.17	Reprinted from *Safety with Machinery* Second Edition, John Ridley and Dick Pearce, page 74, 2005, with permission from Elsevier.

Figure 11.18	Reprinted from *Safety with Machinery* Second Edition, John Ridley and Dick Pearce, pages 72 and 73, 2005, with permission from Elsevier.
Figure 11.19	Courtesy of Allen-Bradley Guardmaster brand from Rockwell Automation.
Figure 11.20	Reprinted from *Safety with Machinery* Second Edition, John Ridley and Dick Pearce, page 90, 2005, with permission from Elsevier.
Figure 11.21	Courtesy of Canon.
Figure 11.22	Courtesy of Fellowes.
Figure 11.23	Courtesy of Draper.
Figure 11.24	Courtesy of Draper.
Figure 11.25	Courtesy of Atco-Qualcast.
Figure 11.28	Picture supplied courtesy of STIHL GB.
Figure 11.29	Courtesy of Pakawaste.
Figure 11.30	Courtesy of Winget.
Figure 11.31	From L114 *Safe Use of Woodworking Machinery* (HSE Books 1998) ISBN 0717616304. © Crown copyright material is reproduced with the permission of the Controller of HMSO and Queen's Printer for Scotland.
Figure 12.2	Courtesy of Stocksigns.
Figure 12.7 (a)–(c)	Courtesy of DeWalt.
Figure 12.9 (a), (b)	From *Essentials of Health and Safety* (HSE Books 1999) ISBN 071760716X. © Crown copyright material is reproduced with the permission of the Controller of HMSO and Queen's Printer for Scotland.
Figure 12.10 (a), (b)	From *Essentials of Health and Safety* (HSE Books 1999) ISBN 071760716X. © Crown copyright material is reproduced with the permission of the Controller of HMSO and Queen's Printer for Scotland.
Figure 13.9	Courtesy of Armagard.
Figure 13.10	Courtesy of NEBOSH.
Figure 14.3	Reprinted from *Anatomy and Physiology in Health and Illness*, ninth edition, Waugh and Grant, pages 240 and 248, 2002, with permission from Elsevier.
Figure 14.4	Reprinted from *Anatomy and Physiology in Health and Illness*, ninth edition, Waugh and Grant, page 9, 2002, with permission from Elsevier.
Figure 14.5	Reprinted from *Anatomy and Physiology in Health and Illness*, ninth edition, Waugh and Grant, page 8, 2002, with permission from Elsevier.
Figure 14.6	Reprinted from *Anatomy and Physiology in Health and Illness*, ninth edition, Waugh and Grant, page 340, 2002, with permission from Elsevier.
Figure 14.7	Reprinted from *Anatomy and Physiology in Health and Illness*, ninth edition, Waugh and Grant, page 363, 2002, with permission from Elsevier.
Figure 14.9	Courtesy of Draeger Safety UK Limited.
Figure 14.13	From HSG53 *The Selection, Use and Maintenance of Respiratory Protective Equipment* (HSE Books 1998) ISBN 0717615375. © Crown copyright material is reproduced with the permission of the Controller of HMSO and Queen's Printer for Scotland.

Figure 14.14	Courtesy of Draper.
Figure 15.1	From HSG121 *A Pain in Your Workplace* (HSE Books 1994) ISBN 0717606686. © Crown copyright material is reproduced with the permission of the Controller of HMSO and Queen's Printer for Scotland.
Figure 15.3	From INDG175 (rev 1) *Health Risks from Hand-Arm Vibration* (HSE Books 1998) ISBN 0717615537. © Crown copyright material is reproduced with the permission of the Controller of HMSO and Queen's Printer for Scotland.
Figure 15.4	From HSG170 *Vibration Solutions* (HSE Books 1997) ISBN 0717609545. © Crown copyright material is reproduced with the permission of the Controller of HMSO and Queen's Printer for Scotland.
Figure 15.9	Reprinted from *Anatomy and Physiology in Health and Illness*, ninth edition, Waugh and Grant, page 195, 2002, with permission from Elsevier.
Figure 15.12	Heat Stress Card published by the Occupational Safety & Health Administration, USA.
Figure 15.15	From INDG69 (rev) *Violence at Work: A Guide for Employers* (HSE Books 2000) ISBN 0717612716. © Crown copyright material is reproduced with the permission of the Controller of HMSO and Queen's Printer for Scotland.
Figure 16.1	From HSG151 *Protecting the Public – Your Next Move* (HSE Books 1997) ISBN 0717611485. © Crown copyright material is reproduced with the permission of the Controller of HMSO and Queen's Printer for Scotland.
Figure 16.3	From HSG185 *Health and Safety in Excavations* (HSE Books 1999) ISBN 0717615634. © Crown copyright material is reproduced with the permission of the Controller of HMSO and Queen's Printer for Scotland.
Figure 16.7	From CIS49 (rev 1) *General Access Scaffolds and Ladders*. Construction Information Sheet No. 49 (revision) (HSE Books 2003). © Crown copyright material is reproduced with the permission of the Controller of HMSO and Queen's Printer for Scotland.
Figure 16.9	From www.brattsladders.com.
Figure 16.10	From HSG149 *Backs for the Future: Safe Manual Handling in Construction* (HSE Books 2000) ISBN 0717611221. © Crown copyright material is reproduced with the permission of the Controller of HMSO and Queen's Printer for Scotland.
Figure 16.13	From HSG185 *Health and Safety in Excavations* (HSE Books 1999) ISBN 0717615634. © Crown copyright material is reproduced with the permission of the Controller of HMSO and Queen's Printer for Scotland.
Figure 17.1	From INDG350 *The Idiot's Guide to CHIP: Chemicals (Hazard Information and Packaging for Supply) Regulations 2002* (HSE Books 2002) ISBN 0717623335. © Crown copyright material is reproduced with the permission of the Controller of HMSO and Queen's Printer for Scotland.
Figure 17.8	From L23 *Manual Handling Operations – Guidance on Regulations* (HSE Books 2004) ISBN 071762823X. © Crown copyright material is reproduced with the permission of the Controller of HMSO and Queen's Printer for Scotland.
Figure 18.6	From HSG65 *Successful Health and Safety Management* (HSE Books 1997) ISBN 0717612767. © Crown copyright material is reproduced with the permission of the Controller of HMSO and Queen's Printer for Scotland.
Figure 20.1	Courtesy of Robert Kirkham.

Health and safety foundations

1

After reading this chapter you should be able to:

1. outline the scope and nature of occupational health and safety
2. explain briefly the moral, legal and financial reasons for promoting good standards of health and safety
3. outline the legal framework for the Regulation of health and safety
4. describe the roles and powers of enforcement agencies, the judiciary and external agencies
5. identify the nature and key sources of health and safety information.

1.1 Introduction

Occupational health and safety is relevant to all branches of industry, business and commerce including traditional industries, information technology companies, the National Health Service, care homes, schools, universities, leisure facilities and offices.

The purpose of this chapter is to introduce the foundations on which appropriate health and safety management systems may be built. Occupational health and safety affects all aspects of work. In a low hazard organization, health and safety may be supervised by a single competent manager. In a high hazard manufacturing plant, many different specialists, such as engineers (electrical, mechanical and civil), lawyers, medical doctors and nurses, trainers, work planners and supervisors, may be required to assist the professional health and safety practitioner in ensuring that there are satisfactory health and safety standards within the organization.

There are many obstacles to the achievement of good standards. The pressure of production or performance targets, financial constraints and the complexity of the organization are typical examples of such obstacles. However, there are some powerful incentives for organizations to strive for high health and safety standards. These incentives are moral, legal and economic.

Corporate responsibility, a term used extensively in the 21st century world of work, covers a wide range of issues. It includes the effects that an organization's business has on the environment, human rights and Third World poverty. Health and safety in the workplace is an important corporate responsibility issue.

Corporate responsibility has various definitions. However, broadly speaking, it covers the ways in which organizations manage their core business to add social, environmental and economic value in order to produce a positive, sustainable impact on both society and the business itself. Terms such as 'corporate social responsibility', 'socially responsible business' and 'corporate citizenship' all refer to this concept.

The UK Health and Safety Executive's (HSE) mission is to ensure that the risks to health and safety of workers are properly controlled. In terms of corporate responsibility, it is working to encourage organizations to:

➤ improve health and safety management systems to reduce injuries and ill health;

➤ demonstrate the importance of health and safety issues at board level;

➤ report publicly on health and safety issues within their organization, including their performance against targets.

The HSE believes that effective management of health and safety:

➤ is vital to employee well-being;

➤ has a role to play in enhancing the reputation of businesses and helping them achieve high-performance teams;

➤ is financially beneficial to business.

1.2 Some basic definitions

Before a detailed discussion of health and safety issues can take place, some basic occupational health and safety definitions are required.

Health – The protection of the bodies and minds of people from illness resulting from the materials, processes or procedures used in the workplace.

Safety – The protection of people from physical injury. The borderline between health and safety is ill defined and the two words are normally used together to indicate concern for the physical and mental well-being of the individual at the place of work.

Welfare – The provision of facilities to maintain the health and well-being of individuals at the workplace. Welfare facilities include washing and sanitation arrangements, the provision of drinking water, heating, lighting, accommodation for clothing, seating (when required by the work activity or for rest), eating and rest rooms. First-aid arrangements are also considered as welfare facilities.

Occupational or work-related ill health – This is concerned with those illnesses or physical and mental disorders that are either caused or triggered by workplace activities. Such conditions may be induced by the particular work activity of the individual or by activities of others in the workplace. The time interval between exposure and the onset of the illness may be short (e.g. asthma attacks) or long (e.g. deafness or cancer).

Environmental protection – These are the arrangements to cover those activities in the workplace which affect the environment (in the form of flora, fauna, water, air and soil) and, possibly, the health and safety of employees and others. Such activities include waste and effluent disposal and atmospheric pollution.

Accident – This is defined by the Health and Safety Executive (HSE) as 'any unplanned event that results in injury or ill health of people, or damage or loss to property, plant, materials or the environment or a loss of a business opportunity'. Other authorities define an accident more narrowly by excluding events that do not involve injury or ill health. This book will always use the HSE definition.

Near miss – This is any incident that could have resulted in an accident. Knowledge of near misses is very important as research has shown that, approximately, for every 10 'near miss' events at a particular location in the workplace, a minor accident will occur.

Dangerous occurrence – This is a 'near miss' which could have led to serious injury or loss of life. Dangerous occurrences are defined in the Reporting of Injuries, Diseases and Dangerous Occurrences Regulations 1995 (often known as RIDDOR) and are always reportable to the enforcement authorities. Examples include the collapse of a scaffold or a crane or the failure of any passenger carrying equipment.

Hazard and risk – A **hazard** is the *potential* of a substance, person, activity or process to cause harm. Hazards take many forms including, for example, chemicals, electricity and working from a ladder. A hazard can be ranked relative to other hazards or to a possible level of danger.

A **risk** is the *likelihood* of a substance, activity or process to cause harm. A risk can be reduced and the hazard controlled by good management.

It is very important to distinguish between a *hazard* and a *risk* – the two terms are often confused and activities such as construction work are frequently called *high risk* when they are *high hazard*. Although the hazard will continue to be high, the risks will be reduced as controls are implemented. The level of risk remaining when controls have been adopted is known as the *residual risk*. There should only be *high residual risk* where there is poor health and safety management and inadequate control measures.

1.3 The legal framework for health and safety

1.3.1 Sub-divisions of law

There are two sub-divisions of the law that apply to health and safety issues: criminal law and civil law.

Figure 1.1 At work.

Criminal law

Criminal law consists of rules of behaviour laid down by the Government or the State and, normally, enacted by Parliament through Acts of Parliament. These rules or Acts are imposed on the people for the protection of the people. Criminal law is enforced by several different Government Agencies who may prosecute individuals and organizations for contravening criminal laws. It is important to note that, except for very rare cases, only these agencies are able to decide whether to prosecute an individual or not.

An individual or organization who breaks criminal law is deemed to have committed an offence or crime and, if he/she is prosecuted, the court will determine whether he/she is guilty or not. If the individual is found guilty, the court could sentence him/her to a fine or imprisonment. Owing to this possible loss of liberty, the level of proof required by a Criminal Court is very high and is known as proof 'beyond reasonable doubt', which is as near certainty as possible. Although the prime object of a Criminal Court is the allocation of punishment, the court can award compensation to the victim or injured party. One example of criminal law is the Road Traffic Act, which is enforced by the police. However, the police are not the only criminal law enforcement agency. The Health and Safety at Work (HSW) etc. Act is another example of criminal law and this is enforced either by the HSE or Local Authority Environmental Health Officers (EHOs). Other agencies which enforce criminal law include the Fire Authority, the Environment Agency, Trading Standards and Customs and Excise.

There is one important difference between procedures for criminal cases in general and criminal cases involving health and safety. The prosecution in a criminal case has to prove the guilt of the accused beyond reasonable doubt. Although this obligation is not totally removed in health and safety cases, section 40 of the HSW Act 1974 transferred, where there is a duty to do something 'so far as is reasonably practicable' or 'so far as is practicable' or 'use the best practicable means', the onus of proof to the accused to show that there was no better way to discharge his/her duty under the Act. However, when this burden of proof is placed on the accused, they only need to satisfy the court on the balance of probabilities that what they are trying to prove has been done.

Civil law

Civil law concerns disputes between individuals or individuals and companies. An individual sues another individual or company to address a civil wrong or tort (or delict in Scotland). The individual who brings the complaint to court is known as the claimant or plaintiff (pursuer in Scotland), and the individual or company who is being sued is known as the defendant (defender in Scotland).

The Civil Court is concerned with liability and the extent of that liability rather than guilt or non-guilt. Therefore, the level of proof required is based on the 'balance of probability', which is a lower level of certainty than that of 'beyond reasonable doubt' required by the Criminal Court. If a defendant is found to be liable, the court would normally order him/her to pay compensation and possibly costs to the plaintiff. However, the lower the balance of probability, the lower the level of compensation awarded. In extreme cases, where the balance of probability is just over 50%, the plaintiff may 'win' the case but lose financially because costs may not be awarded and the level of compensation is low. The level of compensation may also be reduced through the defence of **contributory negligence**, which is discussed later under Section 1.8. For cases involving health and safety, civil disputes usually follow accidents or illnesses and concern negligence or a breach of statutory duty. The vast majority of cases are settled 'out of court'. Although actions are often between individuals, where the defendant is an employee who was acting in the course of his/her employment during the alleged incident, the defence of the action is transferred to his/her employer – this is known as **vicarious liability**. The civil action then becomes one between the individual and the employer.

The Employers' Liability (Compulsory Insurance) Act makes it a legal requirement for all employers to have employers' liability insurance. This ensures that any employee, who successfully sues his/her employer following an accident, is assured of receiving compensation irrespective of the financial position of the employer.

There is a maximum penalty of up to £2500 for every day without appropriate cover for employers who do not have such insurance. In addition, one or more copies of the current certificate must be displayed at each place of business and be 'reasonably protected' from being defaced or damaged. Recently, the rules requiring an employer to display the certificate have changed, so that the requirement will be satisfied if the certificate is made available in electronic format and is reasonably accessible to relevant employees.

1.4 The legal system in England and Wales

The description that follows applies to England and Wales (and with a few minor differences to Northern Ireland). Only the court functions concerning health and safety are mentioned. Figure 1.2 shows the court hierarchy in schematic form.

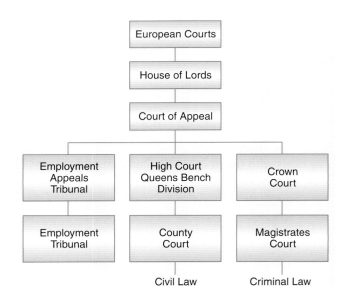

Figure 1.2 The court system in England and Wales for health and safety showing the principal courts.

1.4.1 Criminal law

Magistrates Courts

Most criminal cases begin and end in the Magistrates Courts. Health and safety cases are brought before the court by enforcement officers (Health and Safety Executive or Local Authority Environmental Health Officers) and they are tried by a bench of three lay magistrates (known as Justices of the Peace) or a single district judge. The lay magistrates are members of the public, usually with little previous experience of the law, whereas the district judge is legally qualified.

The Magistrates Court has limited powers with a maximum fine of £5000 for employees to £20 000 for employers or for those who ignore prohibition notices. Magistrates are also able to imprison for up to 6 months. The vast majority of health and safety criminal cases are dealt with in the Magistrates Court. See Table 17.1 for details of new penalties under the Health and Safety Offences Act 2008.

Crown Court

The Crown Court hears the more serious cases, which are passed to them from the Magistrates Court – normally because the sentences available to the magistrates are felt to be too lenient. Cases are heard by a judge and jury, although some cases are heard by a judge alone. The penalties available to the Crown Court are an unlimited fine and up to 2 years' imprisonment for breaches of enforcement notices. The Crown Court also hears appeals from the Magistrates Court.

Appeals from the Crown Court are made to the Court of Appeal (Criminal Division) who may then give leave to appeal to the most senior court in the country – the House of Lords. The most senior judge at the Court of Appeal is the Lord Chief Justice.

1.4.2 Civil law

County Court

The lowest court in civil law is the County Court, which only deals with minor cases (for compensation claims of up to £50 000 if the High Court agrees). Cases are normally heard by a judge sitting alone. For personal injury claims of less than £5000, a small claims court is also available.

High Court

Many health and safety civil cases are heard in the High Court (Queens Bench Division) before a judge only. It deals with compensation claims in excess of £50 000 and acts as an appeal court for the County Court.

Appeals from the High Court are made to the Court of Appeal (Civil Division). The House of Lords receives appeals from the Court of Appeal or, on matters of law interpretation, directly from the High Court. The most senior judge at the Court of Appeal is the Master of the Rolls.

The judges in the House of Lords are called Law Lords and are sometimes called upon to make judgements on points of law, which are then binding on lower courts. Such judgements form the basis of common law, which is covered later.

Other courts – Employment Tribunals

These were established in 1964 and primarily deal with employment and conditions of service issues, such as unfair dismissal. However, they also deal with appeals over health and safety enforcement notices, disputes between recognized safety representatives and their employers and cases of unfair dismissal involving health and safety issues. There are usually three members who sit on a tribunal. These members are appointed and are often not legally qualified. Appeals from the tribunal may be made to the Employment Appeal Tribunal or, in the case of enforcement notices, to the High Court. Appeals from tribunals can only deal with the clarification of points of law.

1.5 The legal system in Scotland

Scotland has both criminal and civil courts but prosecutions are initiated by the procurator-fiscal rather than the Health and Safety Executive. The lowest Criminal Court is called the District Court and deals with minor offences. The Sheriff Court has a similar role to that of the Magistrates Court (for criminal cases) and the County Court (for civil cases), although it can deal with more serious cases involving a sheriff and jury.

The High Court of Judiciary, in which a judge and jury sit, has a similar role to the Crown Court and appeals are made to the Court of Criminal Appeal. The Outer House of the Court of Session deals with civil cases in a similar way to the English High Court. The Inner House of the Court of Session is the appeal court for civil cases.

For both appeal courts, the House of Lords is the final court of appeal. There are Employment Tribunals in Scotland with the same role as those in England and Wales.

1.6 European Courts

There are two European Courts – the European Court of Justice and the European Court of Human Rights.

The European Court of Justice, based in Luxembourg, is the highest court in the European Union (EU). It deals primarily with community law and its interpretation. It is normally concerned with breaches of community law by member states and cases may be brought by other member states or institutions. Its decisions are binding on all member states. There is currently no right of appeal.

The European Court of Human Rights, based in Strasbourg, is not directly related to the EU – it covers most of the countries in Europe including the EU member states. As its title suggests, it deals with human rights and fundamental freedoms. With the introduction of the Human Rights Act 1998 in October 2000, many of the human rights cases are now heard in the UK.

1.7 Sources of law (England and Wales)

There are two sources of law – common law and statute law.

1.7.1 Common law

Common law dates from the 11th century when William I set up Royal Courts to apply a uniform (common) system of law across the whole of England. Prior to that time, there was a variation in law, or the interpretation of the same law, from one town or community to another. Common law is based on judgements made by courts (or strictly judges in courts). In general, courts are bound by earlier judgements on any particular point of law – this is known as 'precedent'. Lower courts must follow the judgements of higher courts. Hence judgements made by the Law Lords in the House of Lords form the basis of most of the common law currently in use.

In health and safety, the legal definition of negligence, duty of care and terms such as 'practicable' and 'as far as is reasonably practicable' are all based on legal judgements and form part of the common law. Common law also provides the foundation for most civil claims made on health and safety issues.

1.7.2 Statute law

Statute law is a law which has been laid down by Parliament as Acts of Parliament and other legislation. In health and safety, an Act of Parliament, the HSW Act 1974, lays down a general legal framework. Specific health and safety duties are, however, covered by Regulations or

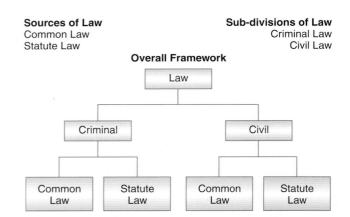

Figure 1.3 Sub-divisions and sources of law.

Statutory Instruments – these are also examples of statute law. If there is a conflict between statute and common law, statute law takes precedence. However, as with common law, judges interpret statute law usually when it is new or ambiguous. Although for health and safety, statute law is primarily the basis of criminal law, there is a tort of breach of statutory duty which can be used when a person is seeking compensation following an accident or illness. Breaches of the HSW Act 1974 cannot be used for civil action but breaches of most of the Regulations produced by the Act may give rise to civil actions.

1.7.3 The relationship between the sub-divisions and sources of law

The two sub-divisions of law may use either of the two sources of law (as shown in Figure 1.3). For example, murder is a common law crime. In terms of health and safety, however, criminal law is only based on statute law, whereas civil law may be based on either common law or statute law.

In summary, criminal law seeks to protect everyone in a society whereas civil law seeks to protect and recompense any individual citizen.

1.8 Common law torts and duties

1.8.1 Negligence

The only tort (civil wrong) of real significance in health and safety is negligence. Negligence is the lack of reasonable care or conduct which results in injury, damage (or financial loss) of or to another. Whether the Act or omission was reasonable is usually decided as a result of a court action or out-of-court settlement.

There have been two important judgements that have defined the legal meaning of negligence. In 1856, negligence was judged to involve actions or omissions and the need for reasonable and prudent behaviour. In 1932, Lord Atkin said the following:

> *You must take reasonable care to avoid Acts or omissions which you reasonably foresee would be likely to injure your neighbour. Who then, in law is my neighbour? The answer seems to be persons who are so closely and directly affected by my Act that I ought reasonably to have them in contemplation as being so affected when I am directing my mind to the Acts or omissions which are called in question.*

It can be seen, therefore, that for negligence to be established, it must be reasonable and foreseeable that the injury could result from the Act or omission. In practice, the Court may need to decide whether the injured party is the neighbour of the perpetrator. A collapsing scaffold could easily injure a member of the public who could be considered a neighbour to the scaffold erector.

An employee who is suing his/her employer for negligence needs to establish the following three criteria:

1. a duty was owed to the employee by the employer as the incident took place during the course of his/her employment;
2. there was a breach of that duty because the event was foreseeable and all reasonable precautions had not been taken;
3. the breach resulted in the specific injury, disease, damage and/or loss suffered.

These tests should also be used by anyone affected by the employer's undertaking (such as contractors and members of the public) who is suing the employer for negligence.

If the employer is unable to defend against the three criteria, two further partial defences are available. It could be argued that the employee was fully aware of the risks that were taken by not complying with safety instructions (known as *volenti non fit injuria* or 'the risk was willingly accepted'). This defence is unlikely to be totally successful because courts have ruled that employees have not accepted the risk voluntarily as economic necessity forces them to work.

The second possible defence is that of 'contributory negligence' where the employee is deemed to have contributed to the negligent Act. This partial defence, if successful, can significantly reduce the level of compensation (by up to 80% in some cases).

Any negligence claim must be made within a set time period (currently 3 years).

1.8.2 Duty of care

Several judgements have established that employers owe a duty of care to each of their employees. This duty cannot be assigned to others, even if a consultant is employed to advise on health and safety matters or if the employees are sub-contracted to work with another employer. This duty may be sub-divided into five groups. Employers must:

1. provide a safe place of work, including access and egress;
2. provide safe plant and equipment;
3. provide a safe system of work;
4. provide safe and competent fellow employees;
5. provide adequate levels of supervision, information, instruction and training.

Employer duties under common law are often mirrored in statute law. This, in effect, makes them both common law and statutory duties.

These duties apply even if the employee is working at a third party premises or if he/she has been hired by the employer to work for another employer.

The requirements of a safe workplace, including the maintenance of floors and the provision of walkways and safe stairways, for example, are also contained in the Workplace (Health, Safety and Welfare) Regulations.

The requirement to provide competent fellow employees includes the provision of adequate supervision, instruction and training. As mentioned earlier, employers are responsible for the actions of their employees (**vicarious liability**) provided that the action in question took place during the normal course of their employment.

1.9 Levels of statutory duty

There are three principal levels of statutory duty which form a hierarchy of duties. These levels are used extensively in health and safety statutory (criminal) law but have been defined by judges under common law. The three levels of duty are absolute, practicable and reasonably practicable.

1.9.1 Absolute duty

This is the highest level of duty and, often, occurs when the risk of injury is so high that injury is inevitable unless safety precautions are taken. It is a rare requirement regarding physical safeguards, although it was more common before 1992 when certain sections of the Factories Act 1961 were still in force. The duty is absolute and the employer has no choice but to undertake the duty. The verbs used in the Regulations are 'must' and 'shall'.

An example of this is Regulation 11(1a) of the Provision and Use of Work Equipment Regulations which requires that every employer shall ensure that effective measures are taken to prevent access to any dangerous part of machinery or rotating stock bar.

For certain health and safety Regulations, such as The Electricity at Work Regulations and The Control of Substances Hazardous to Health Regulations, an absolute duty may be defended using the argument that 'all reasonable precautions and all due diligence' were taken to comply with the law.

Many of the health and safety management requirements contained in health and safety law place an absolute duty on the employer. Examples of this include the need for written safety policies and risk assessments when employee numbers rise above a basic threshold.

1.9.2 Practicable

This level of duty is more often used than the absolute duty as far as the provision of safeguards is concerned and, in many ways, has the same effect.

A duty that 'the employer ensures, so far as is practicable, that any control measure is maintained in an efficient state' means that if the duty is technically possible or feasible then it must be done irrespective of any difficulty, inconvenience or cost. Examples of this duty may be found in the Provision and Use of Work Equipment Regulations [Regulation 11(2) (a and b)].

1.9.3 Reasonably practicable

This is the most common level of duty in health and safety law and was defined by Judge Asquith in Edwards v. the National Coal Board (1949) as follows:

> *Reasonably practicable' is a narrower term than 'physically possible', and seems to me to imply that computation must be made by the owner in which the quantum of risk is placed on one scale and the sacrifice involved in the measures necessary for averting the risk (whether in time, money or trouble) is placed in the other, and that, if it be shown that there is a gross disproportion between them – the risk being insignificant in relation to the sacrifice – the defendants discharge the onus on them.*

In other words, if the risk of injury is very small compared to the cost, time and effort required to reduce it, no action is necessary. It is important to note that money, time and trouble must 'grossly outweigh', not balance (Figure 1.4), the risk. This duty requires judgement on the part of the employer (or his/her adviser) and clearly needs

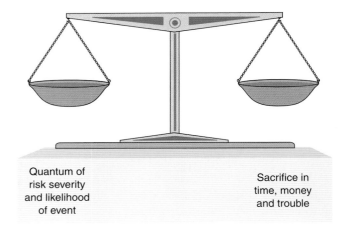

Quantum of risk severity and likelihood of event

Sacrifice in time, money and trouble

Figure 1.4 Diagrammatic view of 'reasonably practicable'.

a risk assessment to be undertaken with conclusions noted. Continual monitoring is also required to ensure that risks do not increase. There are numerous examples of this level of duty, including the HSW Act [e.g. section 2(3)], the Manual Handling Operations Regulations [see Chapter 17 (Section 17.23.3)] and the Control of Substances Hazardous to Health Regulations [see Chapter 17 (Section 17.11.6)]. If organizations follow 'good practice', defined by the HSE as 'those standards which satisfy the law', then they are likely to satisfy the 'reasonably practicable' test. Approved Codes of Practice and Guidance published by the HSE describe and define many of these standards.

The term 'suitable and sufficient' is used to define the scope and extent required for health and safety risk assessment and may be interpreted in a similar way to 'reasonably practicable'. More information is given on this definition in Chapter 5.

1.10 The influence of the European Union on health and safety

As Britain is part of the European Union, much of the health and safety law originates in Europe. Proposals from the European Commission may be agreed by member states. The member states are then responsible for making them part of their national law.

In Britain itself and in much of Europe, health and safety law is based on the principle of risk assessment described in Chapter 5. The main role of the EU in health and safety is to harmonize workplace and legal standards and remove barriers to trade across member states. A Directive from the EU is legally binding on each member state and must be incorporated into the national law of each member state. Directives set out specific minimum aims which must be covered within the national law. Some states incorporate Directives more speedily than others.

Directives are proposed by the European Commission, comprising commissioners, who are citizens of each of the member states. The proposed Directives are sent to the European Parliament which is directly elected from the member states. The European Parliament may accept, amend or reject the proposed Directives. The proposed Directives are then passed to the Council of Ministers, who may accept the proposals on a qualified majority vote, unless the European Parliament has rejected them, in which case they can only be accepted on a unanimous vote of the Council. The Council of Ministers consists of one senior government minister from each of the member states.

The powers of the EU in health and safety law are derived from the Treaty of Rome 1957 and the Single European Act 1986. For health and safety, the Single European Act added two additional Articles – 100A and 118A. The 1997 Treaty of Amsterdam renumbered them as 95A and 138A, respectively. Article 95A is concerned with health and safety standards of equipment and plant and its Directives are implemented in the UK by the Department for Business Enterprise and Regulatory Reform.

Article 138A is concerned with minimum standards of health and safety in employment and its Directives are implemented by the Health and Safety Commission/Executive.

The objective of the Single European Act 1986 is to produce a 'level playing field' for all member states so that goods and services can move freely around the EU without any one state having an unfair advantage over another. The harmonization of health and safety requirements across the EU is one example of the 'level playing field'.

The first introduction of an EU Directive into UK health and safety law occurred on 1 January 1993 when a Framework Directive on Health and Safety Management and five daughter Directives were introduced using powers contained in the HSW Act 1974. These Directives, known as the European Six Pack, covered the following areas:

➤ management of health and safety at work;
➤ workplace;
➤ provision and use of work equipment;
➤ manual handling;
➤ personal protective equipment;
➤ display screen equipment.

Since 1993, several other EU Directives have been introduced into UK law. Summaries of the more common UK Regulations are given in Chapter 17.

1.11 The Health and Safety at Work Act 1974 (HSW Act)

1.11.1 Background to the Act

The HSW Act resulted from the findings of the Robens Report, published in 1972. Earlier legislation had tended to relate to specific industries or workplaces. This resulted in over 5 million workers being unprotected by any health and safety legislation. Contractors and members of the public were generally ignored. The law was more concerned with the requirement for plant and equipment to be

Figure 1.5 Health and safety law poster – must be displayed or brochure given to employees.

Figure 1.6 HSW Act.

safe rather than the development of parallel arrangements for raising the health and safety awareness of employees.

A further serious problem was the difficulty that legislation had in keeping pace with developments in technology. For example, following a court ruling in 1955 which, in effect, banned the use of grinding wheels throughout industry, it took 15 years to produce the Abrasive Wheels Regulations 1970 to address the problem raised by the 1955 court judgement (John Summers and Sons Ltd. v. Frost). In summary, health and safety legislation before 1974 tended to be reactive rather than proactive.

Lord Robens was asked, in 1970, to review the provision made for the health and safety of people at work. His report produced conclusions and recommendations upon which the HSW Act 1974 was based. The principal recommendations were as follows:

➤ There should be a single Act that covers all workers and that Act should contain general duties which should 'influence attitudes'.

➤ The Act should cover all those affected by the employer's undertaking such as contractors, visitors, students and members of the public.

➤ There should be an emphasis on health and safety management and the development of safe systems of work. This would involve the encouragement of employee participation in accident prevention. (This was developed many years later into the concept of the health and safety culture.)

➤ Enforcement should be targeted at 'self-regulation' by the employer rather than reliance on prosecution in the courts.

These recommendations led directly to the introduction of the HSW Act in 1974.

1.11.2 An overview of the Act

Health and Safety Executive (HSE)

The HSW Act established the Health and Safety Commission and gave it the responsibility to draft new Regulations and to enforce them either through its executive arm, known as the HSE, or through the Local Authority Environmental Health Officer (EHO).

In April 2008, the Health and Safety Commission and HSE merged into a single unified body called the HSE. The aim of the merger is to increase the outside, or non-executive, input to its work. This should improve communication, accountability and the oversight of longer term strategy. The merger means that the size of the Board of the new Executive should be no more than 12 members. The Board of the new Executive has assumed responsibility for running all aspects of the organization, including setting the overall strategic direction, financial and performance management and prioritization of resources.

The merger does not change the fundamental day-to-day operations of the HSE but should lead to closer working throughout the organization. The HSE retains its independence, reflecting the interests of employers, employees and Local Authorities and is committed to maintaining its service delivery.

Although occupational health and safety in Scotland is a reserved issue (it was not devolved to the Scottish Executive), there are some differences in some areas of activity. A summary of the concordat between the HSE and the Scottish Executive is given in Chapter 17 (Section 17.2.3) and covers several of these differences.

The prime function of the HSE remains the same – to monitor, review and enforce health and safety law and to produce codes of practice and guidance. However, the HSE also undertakes many other activities, such as the compilations of health and safety statistics, leading national health and safety campaigns, investigations into accidents or complaints, visiting and advising employers and the production of a very useful website.

Regulations

The HSW Act is an **Enabling Act** which allows the Secretary of State to make further laws (known as Regulations) without the need to pass another Act of Parliament. Regulations are laws, approved by Parliament. These are usually made under the HSW Act, following proposals from the HSE. This applies to Regulations based on EU Directives as well as 'home-grown' ones. It is a criminal offence to breach a Regulation and any breaches may result in enforcement action as explained later in this chapter (Section 1.11.4).

The HSW Act, and general duties in the Management Regulations, aim to help employers to set goals, but leave them free to decide how to control hazards and risks which they identify. Guidance and Approved Codes of Practice give advice, but employers are free to take other routes to achieving their health and safety goals, so long as they do what is reasonably practicable. But some hazards are so great, or the proper control measures so expensive, that employers cannot be given discretion in deciding what to do about them. Regulations identify these hazards and risks and set out specific action that must be taken. Often these requirements are absolute – employers have no choice but to follow them and there is no qualifying phrase of 'reasonably practicable' included.

Some Regulations apply across all organizations – the Manual Handling Regulations would be an example. These apply wherever things are moved by hand or bodily force. Equally, the Display Screen Equipment Regulations apply wherever visual display units are used at work. Other Regulations apply to hazards unique to specific industries, such as mining or construction.

Wherever possible, the HSE will set out the Regulations as goals and describe what must be achieved, but not how it must be done.

Sometimes it is necessary to be *prescriptive*, and to spell out what needs to be done in detail, as some standards are absolute. For example, all mines should have two exits; contact with live electrical conductors should be avoided. Sometimes European law requires prescription.

Some activities or substances are so dangerous that they have to be licensed, for example explosives and asbestos removal. Large, complex installations or operations require 'safety cases', which are large-scale risk assessments subject to scrutiny by the Regulator. An example would be the recently privatized railway companies. They are required to produce safety cases for their operations.

Approved Code of Practice (ACOP)

An ACOP is produced for most sets of Regulations by the HSE and attempts to give more details on the requirements of the Regulations. It also attempts to give the level of compliance needed to satisfy the Regulations. ACOPs have a special legal status (sometimes referred to as quasi-legal). The relationship of an ACOP to a Regulation is similar to the relationship of the Highway Code to the Road Traffic Act. A person is never prosecuted for contravening the Highway Code but can be prosecuted for contravening the Road Traffic Act. If a company is prosecuted for a breach of health and safety law and it is proved that it has not followed the relevant provisions of the ACOP, a court can find it at fault, unless the company can show that it has complied with the law in some other way.

As most health and safety prosecutions take place in a Magistrates Court, it is likely that the lay magistrates will consult the relevant ACOP as well as the Regulations when dealing with a particular case. Therefore, in practice, an employer must have a good reason for not adhering to an ACOP.

Codes of Practice generally are only directly legally binding if:

➤ the Regulations or Act indicates that they are so, for example The Safety Signs and Signals Regulations Schedule 2 specify British Standard Codes of Practice for alternative hand signals;

➤ they are referred to in an Enforcement Notice.

Guidance

Guidance, which has no formal legal standing, comes in two forms – legal and best practice. The Legal Guidance series of booklets is issued by the HSE to cover the technical aspects of Health and Safety Regulations. These booklets generally include the Regulations and the ACOP, where one has been produced.

Best practice guidance is normally published in the HSG series of publications by the HSE. Examples of best practice guidance books include *Health and Safety in Construction* HSG150 and *Lighting at Work* HSG38.

An example of the relationship between these three forms of requirement/advice can be shown using a common problem found throughout industry and commerce – minimum temperatures in the workplace.

Regulation 7 of the Workplace (Health, Safety and Welfare) Regulations states '(1) During working hours, the temperature in all workplaces inside buildings shall be reasonable'.

The ACOP states 'The temperature in workrooms should normally be at least 16 degrees Celsius unless much of the work involves severe physical effort in which case the temperature should be at least 13 degrees Celsius. ....'

It would, therefore, be expected that employers would not allow their workforce to work at temperatures below those given in the ACOP unless it was **reasonable** for them to do so (if, for example, the workplace was a refrigerated storage unit).

Best practice guidance to cover this example is given in HSG194 (*Thermal Comfort in the Workplace*) in which possible solutions to the maintenance of employee welfare in low-temperature environments is given.

1.11.3 *General duties and key sections of the Act*

A summary of the HSW Act is given in Chapter 17 (Section 17.4) – only an outline is given here.

Section 2 Duties of employers to employees

To ensure, so far as is reasonably practicable, the health, safety and welfare of all employees. In particular:

➤ safe plant and systems of work;
➤ safe use, handling, transport and storage of substances and articles;
➤ provision of information, instruction, training and supervision;
➤ safe place of work, access and egress;
➤ safe working environment with adequate welfare facilities;
➤ a written safety policy together with organizational and other arrangements (if there are five or more employees);

➤ consultation with safety representatives and formation of safety committees where there are recognized trade unions.

Section 3 Duties of employers to others affected by their undertaking

A duty to safeguard those not in their employment but affected by the undertaking. This includes members of the public, contractors, patients, customers, visitors and students.

Section 4 Duties of landlords, owners and those in control of premises

A duty to ensure that means of access and egress are safe for those using the premises. Those in control of non-domestic premises also have a duty to ensure, so far as is reasonably practicable, that the premises, the means of access and exit, and any plant (such as boilers and air conditioning units) or substances are safe and without risks to health.

Section 6 Duties of suppliers

Persons who design, manufacture, import or supply any article or substance for use at work must ensure, so far as is reasonably practicable, that it is safe and without risk to health. Articles must be safe when they are set, cleaned, used and maintained. Substances must be without risk to health when they are used, handled, stored or transported. This requires that information must be supplied on the safe use of the articles and substances. There may be a need to guarantee the required level of safety by undertaking tests and examinations.

Section 7 Duties of employees

Two main duties are:

➤ to take reasonable care for the health and safety of themselves and others affected by their acts or omissions;
➤ to co-operate with the employer and others to enable them to fulfil their legal obligations.

Section 8

No person is to misuse or interfere with safety provisions (sometimes known as the 'horseplay section').

Section 9

Employees cannot be charged for health and safety requirements such as personal protective equipment.

Section 37 Personal liability of directors

Where an offence is committed by a corporate body with the consent or connivance of, or is attributable to any

Figure 1.7 Employees at work.

neglect of, a director or other senior officer of the body, both the corporate body and the person are liable to prosecution.

1.11.4 Enforcement of the Act

Powers of inspectors

Inspectors under the Act work either for the HSE or the Local Authority. Local Authorities are responsible for retail and service outlets such as shops (retail and wholesale), restaurants, garages, offices, residential homes, entertainment and hotels. The HSE is responsible for all other work premises including the Local Authorities themselves. Both groups of inspectors have the same powers. The detailed powers of inspectors are given in Chapter 17 (Section 17.4.6). In summary, an inspector has the right to:

➤ enter premises at any reasonable time, accompanied by a police officer, if necessary;
➤ examine, investigate and require the premises to be left undisturbed;
➤ take samples, photographs and, if necessary, dismantle and remove equipment or substances;
➤ require the production of books or other relevant documents and information;
➤ seize, destroy or render harmless any substance or article;
➤ issue enforcement notices and initiate prosecutions.

Figure 1.8 The inspector inspects.

An inspector may issue a formal caution when an offence has been committed, but it is deemed that the public interest does not require a prosecution. Formal cautions are not normally considered if the offender has already had a formal caution.

Enforcement notices

There are two types of enforcement notice.

Improvement Notice – This identifies a specific breach of the law and specifies a date by which the situation is to be remedied. An appeal must be made to the Employment Tribunal within 21 days. The notice is then suspended until the appeal is either heard or withdrawn. There are five main grounds for an appeal:

1. The inspector interpreted the law incorrectly.
2. The inspector exceeded his/her powers.
3. The breach of the law is admitted but the proposed solution is not practicable or reasonably practicable.
4. The time allowed to comply is too short.
5. The breach of the law is admitted but the breach is so insignificant that the notice should be cancelled.

Prohibition Notice – This is used to halt an activity which the inspector feels could lead to a serious personal injury. The notice will identify which legal requirement is being or is likely to be contravened. The notice takes effect as soon as it is issued. As with the improvement notice, an appeal may be made to the Employment Tribunal but, in this case, the notice remains in place during the appeal process.

There are two forms of prohibition notice:

➤ an immediate prohibition notice – this stops the work activity immediately until the specified risk is reduced;
➤ a deferred prohibition notice – this stops the work activity within a specified time limit.

Penalties

In 2008, the Health and Safety (Offences) Act raised the maximum penalties available to the courts in respect of certain health and safety offences by amending section 33 of the HSW Act. The main changes were:

➤ £20 000 fines in lower courts for nearly all summary offences, unlimited fines in higher courts;
➤ imprisonment for nearly all offences – up to 12 months in Magistrates Courts (when permitted, currently 6 months) and 2 years in the Crown Court.

Magistrates Court (or lower court)

The Magistrates Court deals with summary offences.

For most health and safety offences, the maximum penalties are a fine of £20 000 and/or up to 6 months' imprisonment. In the future, this period of imprisonment will be increased to 12 months when the changes introduced by the Health and Safety Offences Act are permitted in Magistrates Courts.

Crown Court (or higher court)

The Crown Court deals with indictable offences and some summary cases referred to it by the lower court which may feel that their powers are too limited.

Fines are unlimited in the Crown Court for all health and safety offences. Possible imprisonment for up to 2 years may also be imposed by the court.

The courts also have the power to disqualify directors who are convicted of health and safety offences up to 5 years at the lower court and 15 years at the higher court.

Imprisonment for health and safety offences has been rare. From 1996 until 2005, only five people were sent to prison for health and safety offences. With the changes in the new Act this may increase.

The HSE believes that by extending the £20 000 maximum fine to the lower courts and making imprisonment an option, more cases will be resolved in the lower courts and justice will be faster, less costly and more efficient.

Summary of the actions available to an inspector

Following a visit by an inspector to premises, the following actions are available to an inspector:

➤ take no action;
➤ give verbal advice;
➤ give written advice;
➤ formal caution;
➤ serve an improvement notice;
➤ serve a prohibition notice;
➤ prosecute.

In any particular situation, more than one of these actions may be taken.

The decision to prosecute will be made either when there is sufficient evidence to provide a realistic possibility of conviction or when prosecution is in the public interest.

Work-related deaths

Work-related deaths are investigated by the police jointly with the HSE or Local Authority to ascertain whether a charge of manslaughter (culpable homicide in Scotland) or corporate manslaughter is appropriate. An inquest by a coroner may be made into any workplace death whether due to an accident or industrial disease. Before the inquest, the coroner will decide whether to order a post-mortem. At the inquest, the coroner sits with a jury and decides which witnesses will be called to give evidence. Decisions are usually made on the balance of probability rather than 'beyond reasonable doubt'. The appropriate enforcement agency, such as the HSE, may attend the inquest either to give evidence or to decide whether the evidence presented warrants a prosecution.

If the police decide not to pursue a manslaughter charge, the HSE or Local Authority will continue the investigation under health and safety law.

The Corporate Manslaughter and Corporate Homicide Act 2007

The Act creates a new statutory offence of corporate manslaughter which will replace the common law offence of manslaughter by gross negligence where corporations and similar entities are concerned. In Scotland, the new offence will be called 'corporate homicide'. The Act introduces a new offence but does not introduce any new health and safety duties. An organization will have committed the new offence if it:

➤ owes a duty of care to another person in defined circumstances;
➤ there is a management failure by its senior managers; and
➤ is judged that its actions or inaction amount to a gross breach of that duty resulting in a person's death;.

The health and safety duties relevant to the Act are:

➤ employer and occupier duties including the provision of safe systems of work and training on any equipment used;
➤ duties connected with:
  • the supply of goods and services to customers;
  • the operation of any activity on a commercial basis;
  • any construction and maintenance work;
  • the use or storage of plant, vehicles or any other item.

A breach of duty will be gross if the management of health and safety at the organization falls far below that which would reasonably be expected. It is, therefore, a judgement on the health and safety culture of the organization (as described in Chapter 4) at the time of the incident. The emphasis of the Act is on the responsibility of the senior management of the organization rather than a single individual. A conviction is only likely if the management and organization of health and safety by senior management is a substantial element in the breach of the duty of care.

On conviction, the offence will be punishable by an unlimited fine and the courts will be able to make:

➤ a remedial order requiring the organization to take steps to remedy the management failure concerned;
➤ a publicity order requiring the organization to publicize details of its conviction and fine.

The publicity order will probably require an organization convicted of corporate manslaughter to publicize:

➤ the fact that it has been convicted;
➤ the particulars of the offence;
➤ the amount of any fine;
➤ the terms of any remedial order.

It is important to note that the Act does not create a new individual liability. Individuals may still be charged with the existing offence of manslaughter by gross negligence and/or a breach of section 37 of the HSW Act. Crown immunity will not apply to the offence, although a number of public bodies and functions will be exempt from it (such as military operations and the response of emergency services).

The police will investigate the death with the technical support of the relevant enforcement agency (often but not always the HSE).

The health and safety poster

The Health and Safety (Information for Employees) Regulations require that the approved poster entitled 'Health and Safety Law – what you should know' is displayed or the approved leaflet is distributed. This information tells employees in general terms about the requirements of health and safety law. Employers must also inform employees of the local address of the enforcing authority (either the HSE or the Local Authority) and the Employment Medical Advisory Service (EMAS). This should be marked on the poster or supplied with the leaflet.

The previous text focused on the modern framework of general duties of the employer and employee, supplemented by some basic information on health and safety management and risk assessment. It included two additional boxes – one for details of trade union or other safety representatives and another for competent persons appointed to assist with health and safety and their responsibilities.

The Health and Safety Information (Amendment) Regulations allows the HSE to approve and publish new posters and leaflets which do not need organizations to update or add enforcing authority and Employment Medical Advisory Service (EMAS) contact information. However, the poster will still need to be displayed and provide employees with basic health and safety information. The HSE is concerned that the poster needs to be understood by employees:

➤ who have visual and/or learning difficulties;
➤ who have poor English reading skills;
➤ who work in an environment where the risk of being denied employment rights is high.

The new poster (Figure 1.5) is a simplified version of the previous one and contains a single Incident Contact Centre number for the HSE helpline. The leaflets that employers can give to workers are replaced with pocket cards.

1.12 The Management of Health and Safety at Work Regulations 1999

As mentioned earlier, on 1 January 1993, following an EC Directive, the Management of Health and Safety at Work Regulations became law in the UK. These Regulations were updated in 1999 and are described in detail in Chapter 17 (Section 17.22). In many ways, the Regulations were not introducing concepts or replacing the 1974 Act – they simply reinforced or amended the requirements of the HSW Act. Some of the duties of employers and employees were re-defined.

1.12.1 Employers' duties

Employers must:

➤ undertake suitable and sufficient written risk assessments when there are five or more employees;
➤ put in place effective arrangements for the planning, organization, control, monitoring and review of health and safety measures in the workplace (including health surveillance). Such arrangements should be recorded if there are five or more employees;
➤ employ (to be preferred) or contract competent persons to help them comply with health and safety duties;
➤ develop suitable emergency procedures. Ensure that employees and others are aware of these procedures and can apply them;
➤ provide health and safety information to employees and others, such as other employers, the self-employed and their employees who are sharing the same workplace and parents of child employees or those on work experience;
➤ co-operate in health and safety matters with other employers who share the same workplace;
➤ provide employees with adequate and relevant health and safety training;
➤ provide temporary workers and their contract agency with appropriate health and safety information;
➤ protect new and expectant mothers and young persons from particular risks;
➤ under certain circumstances, as outlined in Regulation 6, provide health surveillance for employees.

The information that should be supplied by employers under the Regulations is:

➤ risks identified by any risk assessments including those notified to the employer by other employers sharing the same workplace;

➤ the preventative and protective measures that are in place;
➤ the emergency arrangements and procedures and the names of those responsible for the implementation of the procedures.

Finally, it is important to note that the Regulations outline the principles of prevention which employers and the self-employed need to apply so that health and safety risks are addressed and controlled. These principles are discussed in detail in Chapter 6.

1.12.2 Employees' duties

Employees must:

➤ use any equipment or substance in accordance with any training or instruction given by the employer;
➤ report to the employer any serious or imminent danger;
➤ report any shortcomings in the employer's protective health and safety arrangements.

1.13 Role and function of external agencies

The HSE and the Local Authorities (a term used to cover county, district and unitary councils) are all external agencies that have a direct role in the monitoring and enforcement of health and safety standards. There are, however, three other external agencies which have a regulatory influence on health and safety standards in the workplace (see Figure 1.9).

1.13.1 Fire and Rescue Authority

The Fire and Rescue Authority is situated within a single or group of local authorities and is normally associated with rescue, fire fighting and the offering of general advice. It has also been given powers to enforce fire precautions within places of work under fire safety law. The powers of the Authority are very similar to those of the HSE on health and safety matters. The Authority issues Alteration Notices to workplaces and conducts routine and random fire inspections (often to examine fire risk assessments). It should be consulted during the planning stage of proposed building alterations when such alterations may affect the fire safety of the building (e.g. means of escape). The Fire and Rescue Authority can issue both Enforcement and Prohibition Notices. The Authority can prosecute for offences against fire safety law.

Figure 1.9 Diagram showing all external agencies impacting on workplace.

1.13.2 The Environment Agency (Scottish Environmental Protection Agency)

The Environment Agency was established in 1995 and was given the duty to protect and improve the environment. It is the regulatory body for environmental matters and has an influence on health and safety issues.

It is responsible for authorizing and regulating emissions from industry.

Other duties and functions of the Agency include:

➤ ensuring effective controls of the most polluting industries;
➤ monitoring radioactive releases from nuclear sites;
➤ ensuring that discharges to controlled waters are at acceptable levels;
➤ setting standards and issuing permits for collecting, transporting, processing and disposal of waste (including radioactive waste);

➤ enforcement of the Producer Responsibility Obligations (Packaging Waste) Regulations. These resulted from an EU Directive which seeks to minimize packaging and recycle at least 50% of it;
➤ enforcement of the Waste Electrical and Electronic Equipment (WEEE) Directive and its associated Directives.

The Agency may prosecute in the criminal courts for the infringement of environmental law – in one case a fine of £4 millon was imposed.

1.13.3 Insurance companies

Insurance companies play an important role in the improvement of health and safety standards. Since 1969, it has been a legal requirement for employers to insure their employees against liability for injury or disease arising out of their employment. This is called employers' liability insurance

Figure 1.10 Good standards prevent harm and save money.

and was covered earlier in this chapter. Certain public sector organizations are exempted from this requirement because any compensation is paid from public funds. Other forms of insurance include fire insurance and public liability insurance (to protect members of the public).

Premiums for all these types of insurance are related to levels of risk which is related to standards of health and safety. In recent years, there has been a considerable increase in the number and size of compensation claims and this has placed further pressure on insurance companies.

Insurance companies are becoming effective health and safety Regulators by weighing the premium offered to an organization according to its safety and/or fire precaution record (Figure 1.10).

1.14 Sources of information on health and safety

When anybody, whether a health and safety professional, a manager or an employee, is confronted with a health and safety problem, they will need to consult various items of published information to ascertain the scale of the problem and its possible remedies. The sources of this information may be internal to the organization and/or external to it.

Internal sources which should be available within the organization include:

➤ accident and ill-health records and investigation reports;
➤ absentee records;
➤ inspection and audit reports undertaken by the organization and by external organizations such as the HSE;

➤ maintenance, risk assessment (including COSHH) and training records;
➤ documents which provide information to workers;
➤ any equipment examination or test reports.

External sources, which are available outside the organization, are numerous and include:

➤ health and safety legislation;
➤ HSE publications, such as Approved Codes of Practice, guidance documents, leaflets, journals, books and their website;
➤ International (e.g. International Labour Office, ILO), European and British Standards;
➤ health and safety magazines and journals;
➤ information published by trade associations, employer organizations and trade unions;
➤ specialist technical and legal publications;
➤ information and data from manufacturers and suppliers;
➤ the internet and encyclopaedias.

Many of these sources of information will be referred to throughout this book.

1.15 Moral, legal and financial arguments for health and safety management

1.15.1 Moral arguments

The moral arguments are supported by the occupational accident and disease rates.

Accident rates

Accidents at work can lead to serious injury and even death. Although accident rates are discussed in greater detail in later chapters, some trends are shown in Tables 1.1–1.4. A major accident is a serious accident typically involving a fracture of a limb or a 24-hour stay in a hospital. An 'over 3-day accident' is an accident which leads to more than 3 days off work. Statistics are collected on all people who are injured at places of work not just employees.

It is important to note that since 1995 suicides and trespassers on the railways have been included in the HSE figures – this has led to a significant increase in the figures. Table 1.1 details accidents involving all people at a place of work over a 4-year period. In 2007/08, there were 358 fatal injuries to members of the public and about 73% (263) of these were due to acts of suicide or trespass on the railways.

Table 1.1 Accidents involving all people at a place of work

Injury	2004/05	2005/06	2006/07	2007/08
Death	593	618	662	587
Major	46 018	46 085	47 511	47 326
Over 3-day	122 922	120 268	115 799	109 912

Table 1.2 Accidents for different groups of people for 2007/08

	Deaths	Major	Over 3 day
Total	587	47 326	109 912
Employees	179 (30%)	27 976	108 795
Self–employed	50 (9%)	1187	1117
Members of the public	358 (61%)	18 163	n/a

Table 1.3 Accidents involving employees in the workplace

Injury	2004/05	2005/06	2006/07	2007/08
Death	172	164	191	179
Major	30 451	28 914	28 544	27 976
Over 3-day	121 779	119 045	114 653	108 795

Table 1.2 shows, for the year 2007/08, the breakdown in accidents between employees, self-employed and members of the public. Table 1.3 shows the figures for employees only. It shows that there has been a decline in fatalities and major accidents since 2006/07.

Table 1.4 gives an indication of accidents in different employment sectors for 2007/08.

These figures indicate that there is a need for health and safety awareness even in occupations which many would consider very low hazard, such as the health services and hotels. In fact over 70% (423) of all deaths occur in the service sector and manufacturing is considerably safer than construction and agriculture. These latter two industries accounted for almost half of all fatal injuries to workers in 2007/08.

In 2007/08, slips and trips accounted for 38% of major accidents whereas 41% of over 3-day injuries were caused by handling, lifting and carrying. These two percentages have hardly changed over the last few years.

Although there has been a decrease in fatalities since 2006/07, there is still a very strong moral case for improvement in health and safety performance.

Table 1.4 Accidents (per 100 000 employees) to all employees in various employment sectors (2007/08)		
Sector	Deaths	Major
Agriculture	8.8	231.8
Construction	4.3	302.9
Transport	2.6	381.5
Manufacturing	1.2	180.8
Retail and wholesale	0.5	242.9
Hotel and catering	0	65.8
Health services	0	77.7

Table 1.5 Self-reported work-related illness for 2007/08	
Type of illness	Prevalence (thousands)
Musculoskeletal disorders of the back	241
Musculoskeletal disorders of the neck or upper limbs	213
Musculoskeletal disorders of the lower limbs	85
Musculoskeletal disorders in total	**539**
Stress, depression or anxiety	442
Breathing or lung problems	36
Infectious diseases	40
Other (e.g. skin or heart problems)	204
Total	1261

Disease rates

Work-related ill-health and occupational disease can lead to absence from work and, in some cases, to death. Such occurrences may also lead to costs to the State (the Industrial Injuries Scheme) and to individual employers (sick pay and, possibly, compensation payments).

In 2007/08, there were an estimated 2.1 million people in the UK suffering from work-related illness, of whom 563 000 were new cases in that year. This led to 28 million working days lost compared to 6 million due to workplace injury.

Over the last 3 years, 5700 cases have been assessed for industrial injuries disablement benefit. The largest groups were vibration white finger, carpal tunnel syndrome and respiratory diseases.

In 2007/08, 8.8 million working days were lost due to musculoskeletal disorders causing each sufferer to have 21 days off work on average and 13.5 million working days were lost due to stress, depression and anxiety causing each sufferer to have 31 days off work on average. Recent research has shown that one in five people who are on sickness leave from work for 6 weeks will stay off work and leave paid employment.

Ill-health statistics are estimated either from self-reported cases (Table 1.5) or from data supplied by hospital specialists and occupational physicians. The number of self-reported cases will always be much higher than those seeking advice from hospital-based specialists (Table 1.6).

1.15.2 Legal arguments

The legal arguments concerning the employer's duty of care in criminal and civil law have been covered earlier.

Some statistics on legal enforcement indicate the legal consequences resulting from breaches in health and safety law. There have been some very high compensation awards for health and safety cases in the Civil Courts and fines in excess of £100 000 in the Criminal Courts. Table 1.7 shows the number of enforcement notices served over a 3-year period by the HSE. Most notices are served in the manufacturing sector followed by construction and agriculture. Local Authorities serve 40% of all improvement notices and 20% of all prohibition notices.

Table 1.6 Proportion (%) of cases and certified days lost reported by General Practitioners for 2006/07

Type of illness	Diagnosis	Certified days lost
Musculoskeletal disorders	53	43
Mental ill health (stress, anxiety)	29	49
Dermatitis and other skin disorders	10	2
Other diagnosis including infections	4	4
Respiratory disease	3	2
Hearing loss	1	0

Table 1.7 Number of enforcement notices issued over a 3-year period by the HSE

Year	Improvement notice	Prohibition notice
2005/06	3922	2654
2006/07	5165	3109
2007/08	4528	3212

Table 1.8 shows the number of prosecutions by the HSE over the same 3-year period and it can be seen that approximately 80% of those prosecuted are convicted. HSE present 80% of the prosecutions and the remainder are presented by Local Authority EHOs. Most of these prosecutions are for infringements of the Construction Regulations (including the Work at Height Regulations) and the Provision and Use of Work Equipment Regulations.

There are clear legal reasons for sound health and safety management systems.

Table 1.8 Number of prosecutions by the HSE over a 3-year period

Year	Offences prosecuted	Convictions
2005/06	1056	840
2006/07	1051	852
2007/08	1028	839

1.15.3 Financial arguments

Costs of accidents

Any accident or incidence of ill health will cause both direct and indirect costs and incur an insured and an uninsured cost. It is important that all of these costs are taken into account when the full cost of an accident is calculated. In a study undertaken by the HSE, it was shown that indirect costs or hidden costs could be 36 times greater than direct costs of an accident. In other words, the direct costs of an accident or disease represent the tip of the iceberg when compared to the overall costs (Figure 1.11).

Direct costs

These are costs which are directly related to the accident and may be insured or uninsured.

Insured direct costs normally include:

➤ claims on employers and public liability insurance;
➤ damage to buildings, equipment or vehicles;
➤ any attributable production and/or general business loss;
➤ the absence of employees.

Uninsured direct costs include:

➤ fines resulting from prosecution by the enforcement authority;
➤ sick pay;
➤ some damage to product, equipment, vehicles or process not directly attributable to the accident (e.g. caused by replacement staff);
➤ increases in insurance premiums resulting from the accident;
➤ any compensation not covered by the insurance policy due to an excess agreed between the employer and the insurance company;
➤ legal representation following any compensation claim.

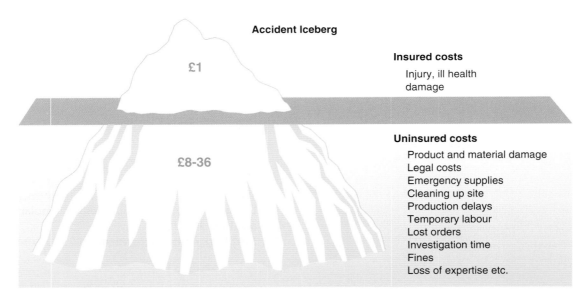

Accident Iceberg

Insured costs

Injury, ill health
damage

£1

£8-36

Uninsured costs

Product and material damage
Legal costs
Emergency supplies
Cleaning up site
Production delays
Temporary labour
Lost orders
Investigation time
Fines
Loss of expertise etc.

Figure 1.11 Insured and uninsured costs.

Indirect costs

These are costs which may not be directly attributable to the accident but may result from a series of accidents. Again these may be insured or uninsured.

Insured indirect costs include:

- a cumulative business loss;
- product or process liability claims;
- recruitment of replacement staff.

Uninsured indirect costs include:

- loss of goodwill and a poor corporate image;
- accident investigation time and any subsequent remedial action required;
- production delays;
- extra overtime payments;
- lost time for other employees, such as a first aider, who tend to the needs of the injured person;
- the recruitment and training of replacement staff;
- additional administration time incurred;
- first-aid provision and training;
- lower employee morale possibly leading to reduced productivity.

Some of these items, such as business loss, may be uninsurable or too prohibitively expensive to insure. Therefore, insurance policies can never cover all of the costs of an accident or disease because either some items are not covered by the policy or the insurance excess is greater than the particular item cost.

1.16 The framework for health and safety management

Most of the key elements required for effective health and safety management are very similar to those required for good quality, finance and general business management. Commercially successful organizations usually have good health and safety management systems in place. The principles of good and effective management provide a sound basis for the improvement of health and safety performance (see Figure 1.12).

HSE, in HSG65, *Successful Health and Safety Management*, have identified six key elements involved in a successful health and safety management system. The following chapters will describe and discuss this framework in detail. The six elements are:

1. **Policy** – A clear health and safety policy contributes to business efficiency and continuous improvement throughout the operation. The demonstration of senior management involvement provides evidence to all stakeholders that responsibilities to people and the environment are taken seriously. The policy should state the intentions of the organization in terms of clear aims, objectives and targets.

2. **Organizing** – A well-defined health and safety organization offering a shared understanding of the organization's values and beliefs at all levels of the

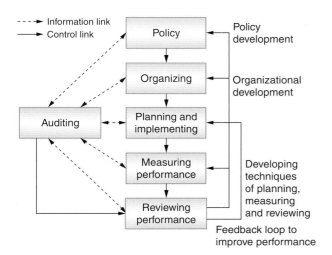

Figure 1.12 Key elements of successful health and safety management.

organization is an essential component of a positive health and safety culture. An effective organization will be noted for good staff involvement and participation; high-quality communications; the promotion of competency; and the empowerment and commitment of all employees to make informed contributions.

3. **Planning and implementing** – A clear health and safety plan involves the setting and implementation of performance standards, targets and procedures through an effective heath and safety management system. The plan is based on risk assessment methods to decide on priorities and set objectives for the effective control or elimination of hazards and the reduction of risks. Measuring success requires the establishment of practical plans and performance targets against which achievements can be identified.

4. **Measuring performance** – This includes both active (sometimes called proactive) and reactive monitoring to see how effectively the health and safety management system is working. Active monitoring involves looking at the premises, plant and substances along with the people, procedures and systems. Reactive monitoring discovers through investigation of accidents and incidents why controls have failed. It is also important to measure the organization against its own long-term goals and objectives.

5. **Reviewing performance** – The results of monitoring and independent audits should be systematically reviewed to evaluate the performance of the management system against the objectives and targets established by the health and safety policy. It is at the review stage that the objectives and targets set in the health and safety policy may be changed. Changes in the health and safety environment in the organization, such as an accident, should also trigger a performance review – this is discussed in more detail in Chapter 2. Performance reviews are not only required by the HSW Act but are part of any organization's commitment to continuous improvement. Comparisons should be made with internal performance indicators and the external performance indicators of similar organizations with exemplary practices and high standards.

6. **Auditing** – An independent and structured audit of all parts of the health and safety management system reinforces the review process. Such audits may be internal or external (the differences are discussed in Chapters 7 and 18). The audit assesses compliance with the health and safety management arrangements and procedures. If the audit is to be really effective, it must assess both the compliance with stated procedures and the performance in the workplace. It will identify weaknesses in the health and safety policy and procedures and identify unrealistic or inadequate standards and targets.

The conclusions from an audit of an organization's health and safety performance should be included in the annual report for discussion at Board meetings. This is considered best corporate practice.

More detail on HSG65 and other health and safety management systems is given in Chapter 18.

1.17 Sources of reference

Sources of reference

The Management of Health and Safety at Work ACOP L21, HSE Books ISBN 978 0 7176 2488 1.
Successful Health and Safety Management HSG65, HSE Books ISBN 978 0 7176 1276 5.
Health & Safety Executive 'Ready Reckoner' website at www.hse.gov.uk/costs.

Relevant statutory provisions

The Health and Safety at Work etc. Act 1974 – sections 2, 20–25, 33 and 39–40.
The Management of Health and Safety at Work Regulations 1999.

1.18 Practice NEBOSH questions for Chapter 1

1. **Explain**, using an example in **EACH** case, the meaning of the following terms:
   (i) 'hazard'
   (ii) 'risk'.

2. (i) **Outline** the main functions of:
      (i) criminal law
      (ii) civil law.
   (ii) **Explain** the principal differences between common law and statute law.

3. **Outline** the **four** key differences between civil law and criminal law.

4. An employer has common law duty of care for the health, safety and welfare of their employees. **Giving** an example in **EACH** case, **identify** what the employer must provide in order to fulfil this common law duty of care.

5. (i) **Define** the term 'negligence'.
   (ii) **Outline** the three standard conditions that must be met for an injured employee to prove a case of negligence against his/her employer following an accident at work.
   (iii) **Outline** the general defences available to an employer in a case of alleged negligence brought by an employee.
   (iv) **State** the circumstances in which an employer may be held vicariously liable for the negligence of an employee.

6. An employer has general and specific duties under section 2 of the Health and Safety at Work etc. Act 1974 that are qualified by the phrase 'so far as is reasonably practicable'.
   (i) **Explain**, using a practical example, the meaning of the term 'so far as is reasonably practicable'.
   (ii) **Describe** the general duty of the employer under section 2(1).
   (iii) **Giving** a workplace example of **EACH**, **state** the **FIVE** specific duties of the employer under section 2(2).
   (iv) Name the courts that can hear a prosecution for breaches of section 2 of the Health and Safety at Work etc. Act 1974 **AND state** the penalties that can be imposed if the prosecution is successful.

7. **Outline** the general duties placed on employees by:
   (i) the Health and Safety at Work etc. Act 1974
   (ii) the Management of Health and Safety at Work Regulations 1999.

8. With respect to section 6 of the Health and Safety at Work etc. Act 1974, **outline** the general duties of designers, manufacturers and suppliers of articles and substances for use at work to ensure that they are safe and without risk.

9. During a routine visit, a health and safety enforcement officer has discovered an unguarded lift shaft, left by a contractor, whilst working on an employer's premises.
   (i) **State** the powers given to the enforcement officer under the Health and Safety at Work etc. Act 1974.
   (ii) **Outline** the breaches of the Health and Safety at Work etc. Act 1974 by:
   (a) the employer
   (b) the person (contractor) carrying out repairs on the lift.

10. In order to meet a production deadline, a supervisor instructed an employee to operate a machine which they both knew to be defective.
    (i) **Giving** reasons in **EACH** case, **identify** possible breaches of the Health and Safety at Work etc. Act 1974 in relation to this scenario.

11. With reference to the Management of Health and Safety at Work Regulations 1999:
    (i) **outline** the information that an employer must provide to their employees
    (ii) **identify FOUR** classes of persons, other than their own employees, to whom an employer must provide health and safety information
    (iii) **identify** the specific circumstances when health and safety training should be given to employees.

12. **Explain** the meaning, legal status and roles of:
    (i) health and safety Regulations
    (ii) HSC Approved Codes of Practice
    (iii) HSE guidance.

13. **Outline**, with an example of **EACH**, the differences between health and safety Regulations and HSC Approved Codes of Practice.

14. **Explain** the differences between *HSC Approved Codes of Practice* and *HSE guidance*, giving an example of **EACH**.

15. **List** the powers given to inspectors appointed under the Health and Safety at Work etc. Act 1974.

16. **(i) Outline FOUR** powers available to an inspector when investigating a workplace accident.
(ii) Identify the two types of enforcement notice that may be served by an inspector, stating the conditions that must be satisfied before each type of notice is served.

17. **(i) Explain**, using a relevant example, the circumstances under which a health and safety inspector may serve an improvement notice.
(ii) Identify the time period within which an appeal may be lodged against an improvement notice **AND state** the effect that the appeal will have on the notice.
(iii) State the penalties for contravening the requirements of an improvement notice when heard **BOTH** summarily **AND** on indictment.

18. **(i) Explain**, giving an example in each case, the circumstances under which a health and safety inspector may serve:
    (i) an improvement notice
    (ii) a prohibition notice.
(ii) State the effect on each type of enforcement notice of appealing against it.

19. **Outline** ways in which the Health and Safety Executive can influence the health and safety performance of an organization.

20. **Outline** the sources of published information that may be consulted when dealing with a health and safety problem at work.

21. **(i) Outline** the purpose of employer liability insurance.
(ii) Outline SIX costs of a workplace accident that might be uninsured.

22. **State FOUR** possible direct **AND FOUR** possible indirect costs to an organization following a serious accident at work.

23. Replacement or repair of damaged equipment is a cost that an organization may incur following an accident at work.
List EIGHT other possible costs to the organization following a workplace accident.

24. **List EIGHT** possible costs to an organization when employees are absent due to work-related ill health.

25. **Outline** reasons for maintaining good standards of health and safety within an organization.

26. **(i) Draw** a flowchart to show the relationships between the six elements of the health and safety management model in HSE's *Successful Health and Safety Management* (HSG65).
(ii) Outline the part that **EACH** element of the HSG65 model plays within the health and safety management system.

27. **(i) Draw** a flowchart to identify the main components of the health and safety management system described in the HSE publication *Successful Health and Safety Management* (HSG65).
(ii) Outline TWO components of the health and safety management system identified in (a).

28. **Outline** the main components of a health and safety management system.

29. **Outline** the economic benefits that an organization may obtain by implementing a successful health and safety management system.

Policy

2

2.1 Introduction

Every organization should have a clear policy for the management of health and safety so that everybody associated with the organization is aware of its health and safety aims and objectives and how they are to be achieved. For a policy to be effective, it must be honoured in the spirit as well as the letter. A good health and safety policy will also enhance the performance of the organization in areas other than health and safety, help with the personal development of the workforce and reduce financial losses.

2.2 Legal requirements

Section 2(3) of the Health and Safety at Work (HSW) Act and the Employers' Health and Safety Policy Statements (Exception) Regulations 1975 require employers, with five or more employees, to prepare and review on a regular basis a written health and safety policy together with the necessary organization and arrangements to carry it out

and to bring the policy and any revision of it to the notice of their employees.

This does not mean that organizations with four or less employees do not need to have a health and safety policy – it simply means that it does not have to be written down. The number of employees is the maximum number at any one time whether they are full time, part time or seasonal.

This obligation on employers was introduced for the first time by the HSW Act and is related to the employers' reliance on the Act on self-regulation to improve health and safety standards rather than on enforcement alone. A good health and safety policy involves the development, monitoring and review of the standards needed to address and reduce the risks to health and safety produced by the organization.

Figure 2.1 Well-presented policy documents.

The law requires that the written health and safety policy should include the following three sections:

➤ a health and safety policy statement of intent which includes the health and safety aims and objectives of the organization;
➤ the health and safety organization detailing the people with specific health and safety responsibilities and their duties;
➤ the health and safety arrangements in place in terms of systems and procedures.

The Management of Health and Safety at Work Regulations also requires the employer to 'make and give effect to such arrangements as are appropriate, having regard to the nature of their activities and the size of their undertaking, for the effective planning, organization, control, monitoring and review of the preventative and protective measures'. It further requires that these arrangements must be recorded when there are five or more employees.

When an inspector from the Health Service Executive (HSE) or Local Authority visits an establishment, it is very likely that they will wish to see the health and safety policy as an initial indication of the management attitude and commitment to health and safety. There have been instances of prosecutions being made due to the absence of a written health and safety policy. (Such cases are, however, usually brought before the courts because of additional concern.)

2.3 Key elements of a health and safety policy

2.3.1 Policy statement of intent

The health and safety policy statement of intent is often referred to as the health and safety policy statement or simply (and incorrectly) as the health and safety policy. It should contain the aims (which are not measurable) and objectives (which are measurable) of the organization or company. Aims will probably remain unchanged during policy revisions whereas objectives will be reviewed and modified or changed every year. The statement should be written in clear and simple language so that it is easily understandable. It should also be fairly brief and broken down into a series of smaller statements or bullet points.

The statement should be signed and dated by the most senior person in the organization. This will demonstrate management commitment to health and safety and give authority to the policy. It will indicate where ultimate responsibility lies and the frequency with which the policy statement is reviewed.

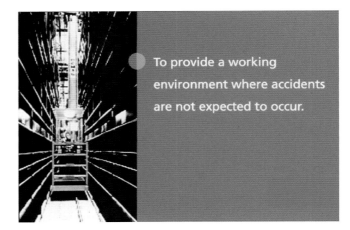

To provide a working environment where accidents are not expected to occur.

Figure 2.2 Part of a policy commitment.

The most senior manager is normally the Chief Executive Officer (CEO) or the Managing Director. It is the responsibility of the CEO, under the HSW Act, to ensure that the health and safety policy is developed and communicated to all employees in the organization. They will need to ensure the following:

➤ key functions of health and safety management, such as monitoring and audit, accident investigation and training, are included in the organizational structure;
➤ adequate resources are available to manage health and safety effectively;
➤ the production of various health and safety arrangements in terms of rules and procedures;
➤ arrangements for the welfare of employees;
➤ the regular review and, if necessary, updating of the health and safety policy.

The policy statement should be written by the organization and not by external consultants, as it needs to address the specific health and safety issues and hazards within the organization. In large organizations, it may be necessary to have health and safety policies for each department and/or site with an overarching general policy incorporating the individual policies. Such an approach is often used by local authorities and multinational companies.

The following points should be included or considered when a health and safety policy statement is being drafted:

➤ the aims, which should cover health and safety, welfare and relevant environmental issues;
➤ the position of the senior person in the organization or company who is responsible for health and safety (normally the chief executive);

- the names of the Health and Safety Adviser and any safety representatives or other competent health and safety persons;
- a commitment to the basic requirements of the HSW Act (access, egress, risk assessments, safe plant and systems of work, use, handling, transport and handling of articles and substances, information, training and supervision);
- a commitment to the additional requirements of the Management of Health and Safety at Work Regulations (risk assessment, emergency procedures, health surveillance and employment of competent persons);
- duties towards the wider general public and others (contractors, customers, students, etc.);
- the principal hazards in the organization;
- specific policies of the organization (e.g. smoking policy, violence to staff, etc.);
- a commitment to employee consultation possibly using a safety committee or plant council;
- duties of employees (particularly those defined in the Management of Health and Safety at Work Regulations);
- specific health and safety performance targets for the immediate and long-term future;
- a commitment to provide the necessary resources to achieve the objectives outlined in the policy statement.

Health and safety **performance targets** are an important part of the statement of intent because:

- they indicate that there is management commitment to improve health and safety performance;
- they motivate the workforce with tangible goals resulting, perhaps, in individual or collective rewards;
- they offer evidence during the monitoring, review and audit phases of the management system.

The type of target chosen depends very much on the areas which need the greatest improvement in the organization. The following list, which is not exhaustive, shows common health and safety performance targets:

- a specific reduction in the number of accidents, incidents (not involving injury) and cases of work-related ill health (perhaps to zero);
- a reduction in the level of sickness absence;
- a specific increase in the number of employees trained in health and safety;
- an increase in the reporting of minor accidents and 'near miss' incidents;
- a reduction in the number of civil claims;
- no enforcement notices from the HSE or Local Authority;
- a specific improvement in health and safety audit scores;

- the achievement of a nationally recognized health and safety management standard such as OHSAS 18001 (see Chapter 18 – 18.3.3).

The policy statement of intent should be posted on prominent notice boards throughout the workplace and brought to the attention of all employees at induction and refresher training sessions. It can also be communicated to the workforce during team briefing sessions, at 'toolbox' talks which are conducted at the workplace or directly by email, intranet, newsletters or booklets. It should be a permanent item on the agenda for health and safety committee meetings where it and its related targets should be reviewed at each meeting.

2.3.2 Organization of health and safety

This section of the policy defines the names, positions and duties of those within the organization or company who have a specific responsibility for health and safety. Therefore, it identifies those health and safety responsibilities and the reporting lines through the management structure. This section will include the following groups together with their associated responsibilities:

- directors and senior managers (responsible for setting policy, objectives and targets);
- supervisors (responsible for checking day-to-day compliance with the policy);
- health and safety advisers (responsible for giving advice during accident investigations and on compliance issues);
- other specialist, such as an occupational nurse, chemical analyst and an electrician (responsible for giving specialist advice on particular health and safety issues);
- health and safety representatives (responsible for representing employees during consultation meetings on health and safety issues with the employer);
- employees (responsible for taking reasonable care of the health and safety of themselves and others who may be affected by their acts or omissions);
- fire marshals (responsible for the safe evacuation of the building in an emergency);
- first aiders (responsible for administering first aid to injured persons).

For smaller organizations, some of the specialists mentioned above may well be employed on a consultancy basis.

For the health and safety organization to work successfully, it must be supported from the top (preferably at Board level) and some financial resource made available.

It is also important that certain key functions are included in the organization structure. These include:

➤ accident investigation and reporting;
➤ health and safety training and information;
➤ health and safety monitoring and audit;
➤ health surveillance;
➤ monitoring of plant and equipment, their maintenance and risk assessment;
➤ liaison with external agencies;
➤ management and/or employee safety committees – the management committee will monitor day-to-day problems and any concerns of the employee health and safety committee.

The role of the health and safety adviser is to provide specialist information to managers in the organization and to monitor the effectiveness of health and safety

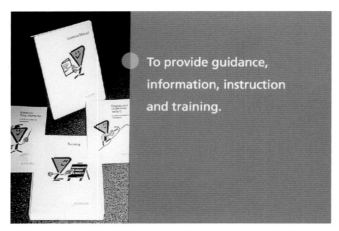

To provide guidance, information, instruction and training.

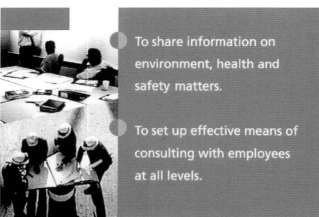

To share information on environment, health and safety matters.

To set up effective means of consulting with employees at all levels.

Figure 2.3 Good information, training and working with employees is essential.

procedures. The adviser is not 'responsible' for health and safety or its implementation; that is the role of the line managers.

Finally the job descriptions, which define the duties of each person in the health and safety organizational structure, must not contain responsibility overlaps or blur chains of command. Each individual must be clear about his/her responsibilities and the limits of those responsibilities.

2.3.3 Arrangements for Health and Safety

The arrangements section of the health and safety policy gives details of the specific systems and procedures used to assist in the implementation of the policy statement. This will include health and safety rules and procedures and the provision of facilities such as a first-aid room and wash rooms. It is common for risk assessments (including COSHH, manual handling and personal protective equipment (PPE) assessments) to be included in the arrangements section, particularly for those hazards referred to in the policy statement. It is important that arrangements for fire and other emergencies and for information, instruction, training and supervision are also covered. Local codes of practice (e.g. for fork-lift truck drivers) should be included.

The following list covers the more common items normally included in the arrangements section of the health and safety policy:

➤ employee health and safety code of practice;
➤ accident and illness reporting and investigation procedures;
➤ emergency procedures, first aid;
➤ fire drill procedure;
➤ procedures for undertaking risk assessments;
➤ control of exposure to specific hazards (noise, vibration, radiation, manual handling, hazardous substances, etc.);
➤ machinery safety (including safe systems of work, lifting and pressure equipment);
➤ electrical equipment (maintenance and testing);
➤ maintenance procedures;
➤ permits to work procedures;
➤ use of PPE;
➤ monitoring procedures including health and safety inspections and audits;
➤ procedures for the control and safety of contractors and visitors;
➤ provision of welfare facilities;
➤ training procedures and arrangements;
➤ catering and food hygiene procedures;
➤ arrangements for consultation with employees;

- terms of reference and constitution of the safety committee;
- procedures and arrangements for waste disposal.

The three sections of the health and safety policy are usually kept together in a health and safety manual and copies distributed around the organization.

2.4 Review of health and safety policy

It is important that the health and safety policy is monitored and reviewed on a regular basis. For this to be successful, a series of benchmarks needs to be established. Such benchmarks, or examples of good practice, are defined by comparison with the health and safety performance of other parts of the organization or the national performance of the occupational group of the organization. The HSE publish an annual report, statistics and a bulletin, all of which may be used for this purpose. Typical benchmarks include accident rates per employee and accident or disease causation.

There are several reasons to review the health and safety policy. The more important reasons are:

- significant organizational changes may have taken place;
- there have been changes in key personnel;
- there have been changes in legislation and/or guidance;
- new work methods have been introduced;
- there have been alterations to working arrangement and/or processes
- there have been changes following consultation with employees;
- the monitoring of risk assessments or accident/incident investigations indicates that the health and safety policy is no longer totally effective;
- information from manufacturers has been received;
- advice from an insurance company has been received;
- the findings of an external health and safety audit;
- enforcement action has been taken by the HSE or Local Authority;
- A sufficient period of time has elapsed since the previous review.

A positive promotion of health and safety performance will achieve far more than simply preventing accidents and ill health. It will:

- support the overall development of personnel;
- improve communication and consultation throughout the organization;
- minimize financial losses due to accidents and ill health and other incidents;
- directly involve senior managers in all levels of the organization;

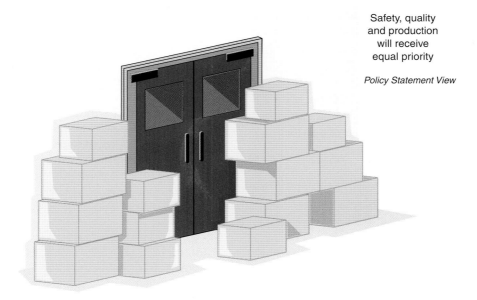

Safety, quality
and production
will receive
equal priority

Policy Statement View

Figure 2.4 Sound policy exists but not put into practice – blocked fire exit.

improve supervision, particularly for young persons and those on occupational training courses;

improve production processes;

improve the public image of the organization or company.

It is apparent, however, that some health and safety policies appear to be less than successful. There are many reasons for this. The most common are:

the statements in the policy and the health and safety priorities not understood by or properly communicated to the workforce;

minimal resources made available for the implementation of the policy;

too much emphasis on rules for employees and too little on management policy;

a lack of parity with other activities of the organization (such as finance and quality control) due to mistaken concerns about the costs of health and safety and the effect of those costs on overall performance. This attitude produces a poor health and safety culture;

lack of senior management involvement in health and safety, particularly at board level;

employee concerns that their health and safety issues are not being addressed or that they are not receiving adequate health and safety information. This can lead to low morale among the workforce and, possibly, high absenteeism;

high labour turnover;

inadequate or no PPE;

unsafe and poorly maintained machinery and equipment;

a lack of health and safety monitoring procedures.

In summary, a successful health and safety policy is likely to lead to a successful organization or company. A checklist for assessing any health and safety policy has been produced by the HSE and has been reproduced in Appendix 2.1.

2.5 Sources of reference

The Management of Health and Safety at Work ACOP L21, HSE Books 2000 ISBN 978 0 7176 2488 1.

Successful Health and Safety Management HSG65, HSE Books 1997 ISBN 978 0 7176 1276 5.

An Introduction to Health and Safety (INDG259 Rev 1) HSE Books 2003 ISBN 978 0 7176 2685 4.

Relevant statutory provisions
The Health and Safety at Work etc. Act 1974 – section 2(3).

The Management of Health and Safety at Work Regulations 1999 – Regulation 5.

2.6 Practice NEBOSH questions for Chapter 2

1. **(i) Outline** the legal requirements whereby an employer must prepare a written health and safety policy.
   (ii) Identify the **THREE** main sections of a health and safety policy document and **explain** the purpose and general content of **EACH** section.

2. **(i) State** the legal requirements whereby employers must prepare a written statement of their health and safety policy.
   (ii) Outline the various methods for communicating the contents of a health and safety policy to a workforce.

3. **Explain** why a health and safety policy should be signed by the most senior person in an organization, such as a Managing Director or Chief Executive.

4. **(i) Outline** the typical responsibilities of a managing director in relation to the health and safety policy.
   (ii) Outline the possible breaches of the Health and Safety at Work etc. Act 1974 if a managing director neglects his/her legal responsibilities in relation to the health and safety policy.

5. **(i) Identify** the typical content of the 'statement of intent' section of an organization's health and safety policy document.
   (ii) Outline the factors that may indicate that health and safety standards within an organization do not reflect the objectives within the 'statement of intent'.

6. With respect to the 'statement of intent' section of a health and safety policy:
   (i) Explain its purpose.
   (ii) Outline the issues that may be addressed in this section of the health and safety policy.

7. **Identify** the purposes of **EACH** of the following sections of a health and safety policy document:
   (i) 'statement of intent'
   (ii) 'organization'
   (iii) 'arrangements'

8. **Identify SIX** categories of persons who may be shown in the 'organization' section of a health and safety policy document **AND state** their likely general or specific health and safety responsibilities.

9. **Outline** the issues that are typically included in the arrangements section of a health and safety document.

10. **Outline** the key areas that should be addressed in the 'arrangements' section of a health and safety policy document.

11. **(i) Explain** why it is important for an organization to set targets in terms of its health and safety performance.
(ii) Outline SIX types of target that an organization might typically set in relation to health and safety.

12. **Outline** the circumstances that would require a health and safety policy to be reviewed.

13. Outline **FOUR** external **AND FOUR** internal influences that might initiate a health and safety policy review.

Appendix 2.1 Health and Safety Policy checklist

The following checklist is intended as an aid to the writing and review of a health and safety policy. It is derived from HSE Information.

General policy and organization

➤ Does the statement express a commitment to health and safety and are your obligations towards your employees made clear?

➤ Does it say which senior manager is responsible for seeing that it is implemented and for keeping it under review, and how this will be done?

➤ Is it signed and dated by you or a partner or senior director?

➤ Have the views of managers and supervisors, safety representatives and the safety committee been taken into account?

➤ Were the duties set out in the statement discussed with the people concerned in advance, and accepted by them, and do they understand how their performance is to be assessed and what resources they have at their disposal?

➤ Does the statement make clear that co-operation on the part of all employees is vital to the success of your health and safety policy?

➤ Does it say how employees are to be involved in health and safety matters, for example by being consulted, by taking part in inspections and by sitting on a safety committee?

➤ Does it show clearly how the duties for health and safety are allocated and are the responsibilities at different levels described?

➤ Does it say who is responsible for the following matters (including deputies where appropriate)?

  ● reporting investigations and recording accidents
  ● fire precautions, fire drill and evacuation procedures
  ● first aid
  ● safety inspections
  ● the training programme
  ● ensuring that legal requirements are met, for example regular testing of lifts and notifying accidents to the health and safety inspector.

Arrangements that need to be considered

➤ keeping the workplace, including staircases, floors, ways in and out, washrooms, etc. in a safe and clean condition by cleaning, maintenance and repair;

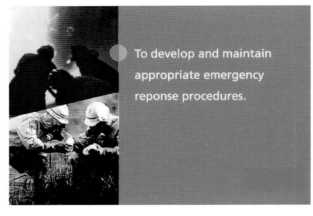

To develop and maintain appropriate emergency reponse procedures.

Figure 2.5 Emergency procedures.

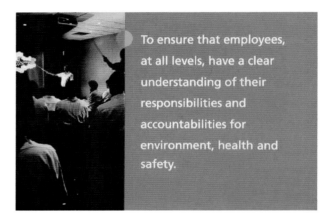

To ensure that employees, at all levels, have a clear understanding of their responsibilities and accountabilities for environment, health and safety.

Figure 2.6 Responsibilities.

➤ the requirements of the Work at Height Regulations;
➤ any suitable and sufficient risk assessments.

Plant and substances

➤ maintenance of equipment such as tools, ladders, etc. (are they in a safe condition?);

➤ maintenance and proper use of safety equipment such as helmets, boots, goggles, respirators, etc.;

➤ maintenance and proper use of plant, machinery and guards;

➤ regular testing and maintenance of lifts, hoists, cranes, pressure systems, boilers and other dangerous machinery, emergency repair work, and safe methods of carrying out these functions;

➤ maintenance of electrical installations and equipment;

➤ safe storage, handling and, where applicable, packaging, labelling and transport of flammable and/or hazardous substances;

Figure 2.7 Fork-lift truck.

- controls of work involving harmful substances such as lead and asbestos;
- the introduction of new plant, equipment or substances into the workplace by examination, testing and consultation with the workforce;
- exposure to non-ionizing and ionizing radiation.

Other hazards

- noise problems – wearing of hearing protection, and control of noise at source;
- vibration problems – hand–arm and whole-body control techniques and personal protection;
- preventing unnecessary or unauthorized entry into hazardous areas;
- lifting of heavy or awkward loads;
- protecting the safety of employees against assault when handling or transporting the employer's money or valuables;
- special hazards to employees when working on unfamiliar sites, including discussion with site manager where necessary;
- control of works transport, for example fork-lift trucks, by restricting use to experienced and authorized

operators or operators under instruction (which should deal fully with safety aspects).
- driving on public roads while at work.

Emergencies

- ensuring that fire exits are marked, unlocked and free from obstruction;
- maintenance and testing of fire-fighting equipment, fire drills and evacuation procedures;
- first aid, including name and location of person responsible for first aid and deputy, and location of first-aid box.

Communication

- giving employees information about the general duties under the HSW Act and specific legal requirements relating to their work;
- giving employees necessary information about substances, plant, machinery and equipment with which they come into contact;
- discussing with contractors, before they come on site, how they plan to do their job, whether they need any equipment from your organization to help them, whether they can operate in a segregated area or only when part of the plant is shut down and, if not, what hazards they may create for your employees and vice versa.

Training

- training employees, supervisors and managers to enable them to work safely and to carry out their health and safety responsibilities efficiently.

Supervising

- supervising employees so far as necessary for their safety – especially young workers, new employees and employees carrying out unfamiliar tasks.

Keeping check

- regular inspections and checks of the workplace, machinery appliances and working methods.

Organizing for health and safety

3

3.1 Introduction

This chapter is about managers in businesses, or other organizations, setting out clear responsibilities and lines of communications for everyone in the enterprise. The chapter also covers the legal responsibilities that exist between people who control premises and those who use them, and between contractors and those who hire them; and the duties of suppliers, manufacturers and designers of articles and substances for use at work. Chapter 2 is concerned with policy, which is an essential first step. The policy will only remain as words on paper, however good the intentions, until there is an effective organization set up to implement and monitor its requirements.

The policy sets the direction for health and safety within the enterprise and forms the written intentions of the principals or directors of the business. The organization needs to be clearly communicated and people need to know what they are responsible for in the day-to-day operations. A vague statement that 'everyone is responsible for health and safety' is misleading and fudges the real issues.

Everyone is responsible (Figure 3.1), but management in particular. There is no equality of responsibility under law between those who provide direction and create policy and those who are employed to follow. Principals, or employers in terms of the Health and Safety at work (HSW) Act, have substantially more responsibility than employees.

Some policies are written so that most of the wording concerns strict requirements laid on employees and only a few vague words cover managers' responsibilities. Generally, such policies do not meet the HSW Act or the Management of Health and Safety at Work Regulations, which require an effective policy with a robust organization and arrangements to be set up.

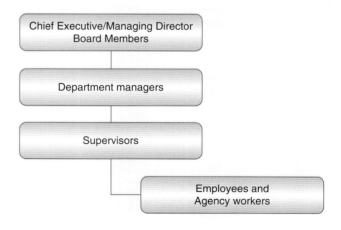

Figure 3.1 Everyone from senior managers down has health and safety responsibilities.

3.2 Control

Like all management functions, establishing control and maintaining it day in day out is crucial to effective health and safety management. Managers, particularly at senior levels, must take proactive responsibility for controlling issues that could lead to ill health, injury or loss. A nominated senior manager at the top of the organization needs to oversee policy implementation and monitoring. The nominated person will need to report regularly to the most senior management team and will be a director or principal of the organization.

Health and safety responsibilities will need to be assigned to line managers and expertise must be available, either inside or outside the enterprise, to help them achieve the requirements of the HSW Act and the Regulations made under the Act. The purpose of the health and safety organization is to harness the collective enthusiasm, skills and effort of the entire workforce with managers taking key responsibility and providing clear direction. The prevention of accidents and ill health through management systems of control becomes the focus rather than looking for individuals to blame after the incident occurs.

The control arrangements should be part of the written health and safety policy. Performance standards will need to be agreed and objectives set which link the outputs required to specific tasks and activities for which individuals are responsible. For example, the objective could be to carry out a workplace inspection once a week to an agreed checklist and rectify faults within three working days. The periodic, say annual, audit would check to see if this was being achieved and if not the reasons for non-compliance with the objective.

People should be held accountable for achieving the agreed objectives through existing or normal procedures such as:

➤ job descriptions, which include health and safety responsibilities;
➤ performance appraisal systems, which look at individual contributions;
➤ arrangements for dealing with poor performance;
➤ where justified, the use of disciplinary procedures.

Such arrangements are only effective if health and safety issues achieve the same degree of importance as other key management concerns and a good performance is considered to be an essential part of career and personal development.

3.3 Employers' responsibilities

Employers have duties under both criminal and civil law. The civil law duties are covered in Chapter 1. The general duties of employers under the HSW Act relate to:

➤ the health, safety and welfare at work of employees and others whether part-time, casual, temporary or homeworkers, on work experience, government training schemes or on site as contractors – that is anyone working under their control or direction;
➤ the health and safety of anyone who visits or uses the workplace and anyone who is allowed to use the organization's equipment;
➤ the health and safety of those affected by the work activity, for example neighbours, and the general public.

3.3.1 What employers must do

Employers must protect the health and safety of employees and others who might be affected by their work activities. Health and safety is about sensible, proportionate actions that protect people – not unnecessary bureaucracy and paperwork.

This 10-point list shows some of the key actions required by law that apply to nearly every organization:

1. Take out Employers' Liability Compulsory Insurance and display the certificate.
2. Make sure the business has someone competent to help meet their health and safety duties. This does not have to be an external consultant.
3. Decide how health and safety is going to be managed. This is the health and safety policy.
4. Decide what could harm anyone involved in the business activities and what precautions should be taken. This is the risk assessment.
5. Managers must act on the findings of the risk assessment by putting sensible controls in place to prevent accidents and ill health and making sure they are followed.
6. Provide basic welfare facilities, such as toilets, washing facilities and drinking water.
7. Provide free health and safety training for workers.
8. Consult workers on health and safety.
9. Display the health and safety law poster or give workers a leaflet with the information.
10. Report certain work-related accidents, diseases and dangerous occurrences.

Failure to comply with these requirements can have serious consequences – for both organizations and individuals. Sanctions include fines, imprisonment and disqualification.

Under the Corporate Manslaughter and Corporate Homicide Act 2007 an offence will be committed where failings by an organization's senior management are a substantial element in any gross breach of the duty of care owed to the organization's employees or members of the public that results in death. The maximum penalty is an unlimited fine and the court can additionally make a publicity order requiring the organization to publish details of its conviction and fine.

Under the HSW Act 1974, employers have a duty to protect the health, safety and welfare of their employees, including homeworkers.

Employers are required to do a risk assessment of the work activities carried out by homeworkers. It may be necessary for employers to visit the home of the worker to carry out a risk assessment, although homeworkers can also help in identifying the hazards for their employers themselves. See also chapters 1 and 7.

3.3.2 Visitors and general public

Organizations have a duty to ensure the health and safety of the public while on their premises, even if the individuals concerned, like children, are not supposed to be there. Two cases reported in the Autumn of 2008 have highlighted how far this liability can extend.

Case study

Two companies pleaded guilty to charges of breaching the HSW Act 1974 following the drowning of a 9-year-old girl, who was playing with other children in the company car park when they strayed onto nearby reservoirs. At the time of the accident, in 2004, the main gates to the factory were off their hinges because work was being carried out on the site and a second gate, which led to the reservoir, was only secured with a nylon rope.

In a separate case, a construction company has been found guilty of failing to prevent unauthorized persons, including children, from gaining access to an area where construction material and equipment were stored. A child was seriously injured by falling paving stones while playing on a partly built housing estate where materials were being stored during construction work.

Misunderstandings over who is responsible for monitoring and protecting contractors and sub-contractors – leading to the failure to carry out responsibilities – is often at the heart of high-profile health and safety cases. But situations such as these could be **avoided** with a clearer understanding of employers' and contractors' responsibilities.

Visitors to a site whether authorized or not are often more at risk than employees because:

➤ they are unfamiliar with the workplace processes, the hazards and associated risks they present;
➤ they may not have the appropriate personal protective equipment (PPE);
➤ they will have a lack of knowledge of the site or premises layout;
➤ walkways are often inadequate, unsigned or poorly lit;
➤ they are not familiar with the emergency procedures or means of escape;
➤ they may be particularly vulnerable if they suffer from a disability or are very young.

Many of these problems with visitors can be overcome by, for example:

➤ visitors signing in and being provided with a site escort;
➤ providing appropriate PPE and identity badges;
➤ providing simple induction procedures with a short video and information on site rules, hazards and emergency procedures;
➤ clear marking of walkways and areas where unauthorized people are not permitted.

3.3.3 Night working

Employers should ascertain whether they employ people who would be classified as night workers. If so, they should check:

➤ how much working time night workers normally work;
➤ if night workers work more than eight hours per day on average, whether the number of hours can be reduced and if any exceptions apply;
➤ how to conduct a health assessment and how often health checks should be carried out;
➤ that proper records of night workers are maintained, including details of health assessments;
➤ that night workers are not involved in work which is particularly hazardous.

3.3.4 Temporary workers Directive adopted by EU Parliament

The European Parliament has adopted a Directive on temporary agency work which enables temporary workers to be treated equally, from day one, with those of the employer company.

However, following an agreement reached May 2008 between the social partners in the UK, agency workers will get the same pay and conditions as permanent staff after being employed for 12 weeks. This is the final stage in the process and member states have a maximum of 3 years to transpose the Directive.

The agreement has been made between the Government, the CBI and the TUC, and the following points have been decided upon:

➤ After 12 weeks in a given job there will be entitlement to equal treatment.
➤ Equal treatment will be defined to mean at least the basic working and employment conditions that would apply to the workers concerned if they had been recruited directly by that undertaking to occupy the same job. This will not cover occupational social security schemes.
➤ The Government will consult regarding the implementation of the Directive, in particular how disputes will be resolved regarding the definition of equal treatment and how both sides of the industry can reach appropriate agreements on the treatment of agency workers.
➤ The new arrangements will be reviewed at an appropriate point 'in the light of experience'.

Other duties of employers are covered in Chapter 1 and the summary of the HSW Act in Chapter 17.

3.4 Employees' and agency workers' responsibilities

They have specific responsibilities under the HSW Act, which are:

➤ to take reasonable care for the health and safety of themselves and of other persons who may be affected by their acts or omissions at work. This involves the same wide group that the employer has to cover, not just the people on the next desk or bench;
➤ to cooperate with employers in assisting them to fulfil their statutory duties;
➤ not to interfere with deliberately or misuse anything provided, in accordance with health and safety legislation, to further health and safety at work. See also chapters 1 and 17.

3.5 Organizational health and safety responsibilities – directors

In addition to the legal responsibilities on management, there are many specific responsibilities imposed by each organization's health and safety policy. Section 3.6 gives a typical summary of the health and safety responsibilities and accountability of each level of the line organization. More details are given in Appendix 3.1. Many organizations will not fit this exact structure but most will have those who direct, those who manage or supervise and those who have no line responsibility, but have responsibilities to themselves and fellow workers.

Because of the special role and importance of directors, these are covered here in detail.

The HSE and the Institute of Directors have published INDG417, which replaces INDG243. The guide is based on a plan, i.e. to deliver, monitor and review the management concept and the following information is closely based on this guide.

Effective health and safety performance comes from the top; members of the board have both collective and individual responsibility for health and safety. Directors and boards need to examine their own behaviours, both individually and collectively, against the guidance given, and, where they see that they fall short of the standards it sets them, to change what they do to become more effective leaders in health and safety.

Why directors and board members need to act:

➤ Protecting the health and safety of employees or members of the public who may be affected by their activities is an essential part of risk management and must be led by the board.
➤ Failure to include health and safety as a key business risk in board decisions can have catastrophic results. Many high-profile safety cases over the years have been rooted in failures of leadership.
➤ Health and safety law places duties on organizations and employers, and directors can be personally liable when these duties are breached: members of the board have both collective and individual responsibility for health and safety.

3.5.1 Plan the direction of health and safety

The board should set the direction for effective health and safety management. Board members need to establish a health and safety policy that is much more than a document – it should be an integral part of the organization's culture, its values and performance standards.

All board members should take the lead in ensuring the communication of health and safety duties and benefits throughout the organization. Executive directors must develop policies to avoid health and safety problems and must respond quickly where difficulties arise or new risks are introduced; non-executives must make sure that health and safety is properly addressed.

Core actions

To agree a policy, boards will need to ensure they are aware of the significant risks faced by their organization. The policy should set out the board's own role and that of individual board members in leading the health and safety of its organization. It should require the board to:

➤ 'own' and understand the key issues involved;
➤ decide how best to communicate, promote and champion health and safety.

The health and safety policy is a 'living' document and it should evolve over time, for example, in the light of major organizational changes such as restructuring or a significant acquisition.

Good practice

➤ Health and safety should appear regularly on the agenda for board meetings.
➤ The Chief Executive can give the clearest visibility of leadership, but some boards find it useful to name one of their members as the health and safety 'champion'.
➤ The presence on the Board of a Health and Safety Director can be a strong signal that the issue is being taken seriously and that its strategic importance is understood.
➤ Setting targets helps define what the board is seeking to achieve.
➤ A non-executive director can act as a scrutinizer – ensuring that the processes to support boards facing significant health and safety risks are robust.

3.5.2 Deliver health and safety

Delivery depends on an effective management system to ensure, so far as is reasonably practicable, the health and safety of employees, customers and members of the public.

Organizations should aim to protect people by introducing management systems and practices that ensure risks are dealt with sensibly, responsibly and proportionately.

Core actions

To take responsibility and 'ownership' of health and safety, members of the board must ensure that:

➤ health and safety arrangements are adequately resourced;
➤ they obtain competent health and safety advice;
➤ risk assessments are carried out;
➤ employees or their representatives are involved in decisions that affect their health and safety.

The board should consider the health and safety implications of introducing new processes, new working practices or new personnel, dedicating adequate resources to the task and seeking advice where necessary.

Boardroom decisions must be made in the context of the organization's health and safety policy; it is important to 'design-in' health and safety when implementing change.

Good practice

➤ Leadership is more effective if visible – board members can reinforce health and safety policy by being seen on the 'shop floor', following all safety measures themselves and addressing any breaches immediately.
➤ Consider health and safety when deciding senior management appointments.
➤ Having procurement standards for goods, equipment and services can help prevent the introduction of expensive health and safety hazards.
➤ The health and safety arrangements of partners, key suppliers and contractors should be assessed: their performance could adversely affect director's own performance.
➤ Setting up a separate risk management or health and safety committee as a subset of the board, chaired by a senior executive, can make sure the key issues are addressed and guard against time and effort being wasted on trivial risks and unnecessary bureaucracy.
➤ Providing health and safety training to some or all of the board can promote understanding and knowledge of the key issues in the organization.
➤ Supporting worker involvement in health and safety, above the legal duty to consult worker representatives, can improve participation and help prove senior management commitment.

3.5.3 Monitor health and safety

Monitoring and reporting are vital parts of a health and safety system. Management systems must allow the board to receive both specific (e.g. incident-led) and routine reports on the performance of health and safety policy.

Much day-to-day health and safety information need be reported only at the time of a formal review. But only a strong system of monitoring can ensure that the formal review can proceed as planned – and that relevant events in the interim are brought to the board's attention.

Core actions

The board should ensure that:

➤ appropriate weight is given to reporting both preventive information (such as progress of training and maintenance programmes) and incident data (such as accident and sickness absence rates);
➤ periodic audits of the effectiveness of management structures and risk controls for health and safety are carried out;
➤ the impact of changes such as the introduction of new procedures, work processes or products, or any major health and safety failure, is reported as soon as possible to the board;
➤ there are procedures to implement new and changed legal requirements and to consider other external developments and events.

Good practice

➤ Effective monitoring of sickness absence and workplace health can alert the board to underlying problems that could seriously damage performance or result in accidents and long-term illness.
➤ The collection of workplace health and safety data can allow the board to benchmark the organization's performance against others in its sector.
➤ Appraisals of senior managers can include an assessment of their contribution to health and safety performance.
➤ Boards can receive regular reports on the health and safety performance and actions of contractors.
➤ Some organizations have found they win greater support for health and safety by involving workers in monitoring.

3.5.4 Review health and safety

A formal boardroom review of health and safety performance is essential. It allows the board to establish whether the essential health and safety principles – strong and active leadership, worker involvement, and assessment and review – have been embedded in the organization. It tells senior managers whether their system is effective in managing risk and protecting people.

Core actions

The board should review health and safety performance at least once a year. The review process should:

➤ examine whether the health and safety policy reflects the organization's current priorities, plans and targets;
➤ examine whether risk management and other health and safety systems have been effectively report to the board;
➤ report health and safety shortcomings, and the effect of all relevant board and management decisions;
➤ decide actions to address any weaknesses and a system to monitor their implementation;
➤ consider immediate reviews in the light of major shortcomings or events.

Good practice

➤ Performance on health and safety and well-being is increasingly being recorded in organizations' annual reports to investors and stakeholders.
➤ Board members can make extra 'shop floor' visits to gather information for the formal review.
➤ Good health and safety performance can be celebrated at central and local level.

3.6 Typical managers' organizational responsibilities

A summary of the organizational responsibilities for health and safety for typical line managers is given here. A more detailed list of the responsibilities is given in Appendix 3.1.

3.6.1 Managing Directors/Chief Executives

Managing Directors/Chief Executives are responsible for the health, safety and welfare of all those who work or visit the organization. In particular, they:

1. are responsible and accountable for health and safety performance within the organization;
2. must ensure that adequate resources are available for the health and safety requirements within the organization;

3. establish, implement and maintain a formal, written health and safety programme for the organization that encompasses all areas of significant health and safety risk;
4. approve, introduce and monitor all site health and safety policies, rules and procedures;
5. review annually the effectiveness and, if necessary, require revision of the health and safety programme.

3.6.2 Departmental managers

The principal departmental managers may report to the Site Manager, Managing Director or Chief Executive. In particular, they:

1. are responsible and accountable for the health and safety performance of their department;
2. must ensure that any machinery, equipment or vehicles used within the department are maintained, correctly guarded and meet agreed health and safety standards. Copies of records of all maintenance, statutory and insurance inspections must be kept by the Departmental Manager;
3. develop a training plan that includes specific job instructions for new or transferred employees and follow up on the training by supervisors. Copies of records of all training must be kept by the Departmental Manager;
4. personally investigate all lost workday cases and dangerous occurrences and report to their line manager. Progress any required corrective action.

3.6.3 Supervisors

The supervisors are responsible to and report to their Departmental Manager. In particular, they:

1. are responsible and accountable for their team's health and safety performance;
2. enforce all safe systems of work procedures that have been issued by the Departmental Manager;
3. instruct employees in relevant health and safety rules, make records of this instruction and enforce all health and safety rules and procedures;
4. enforce PPE requirements, make spot checks to determine that protective equipment is being used and periodically appraise condition of equipment. Record any infringements of the PPE policy.

3.7.1 Competent person

One or more competent persons must be appointed to help managers comply with their duties under health and safety law. The essential point is that managers should have access to expertise to help them fulfil the legal requirements. However, they will always remain as advisers and do not assume responsibility in law for health and safety matters. This responsibility always remains with line managers and cannot be delegated to an adviser whether inside or outside the organization (Figure 3.2). The appointee could be:

➤ the employer themselves if they are sure they know enough about what to do. This may be appropriate in a small, low-hazard business;
➤ one or more employees, as long as they have sufficient time and other resources to do the task properly;
➤ someone from outside the organization who has sufficient expertise to help.

The HSE have produced a leaflet entitled *Getting specialist help with health and safety* INDG420. This gives simple guidance for those looking for specialist health and safety help.

When there is an employee with the ability to do the job, it is better for them to be appointed than use outside specialists. Many health and safety issues can be tackled by people with an understanding of current best practice

Figure 3.2 Safety practitioner at the front line.

and an ability to judge and solve problems. Some specialist help is needed long term, other help for a one-off short period. There is a wide range of specialists available for different types of health and safety problem. For example:

➤ engineers for specialist ventilation or chemical processes;
➤ occupational hygienists for assessment and practical advice on exposure to chemical (dust, gases, fumes, etc.), biological (viruses, fungi, etc.) and physical (noise, vibration, etc.) agents;
➤ occupational health professionals for medical examinations and diagnosis of work-related disease, pre-employment and sickness advice, health education;
➤ ergonomists for advice on suitability of equipment, comfort, physical work environment, work organization;
➤ physiotherapists for treatment and prevention of musculoskeletal disorders;
➤ radiation protection advisers for advice on compliance with the Ionising Radiation Regulations 1999;
➤ health and safety practitioners for general advice on implementation of legislation, health and safety management, risk assessment, control measures and monitoring performance.

3.7.2 Health and Safety Practitioner

Status and **competence** are essential to the role of health and safety practitioners and other advisers. They must be able to advise management and employees or their representatives with authority and independence. They need to be able to advise on:

➤ creating and developing health and safety policies. These will be for existing activities in addition to new acquisitions or processes;
➤ the promotion of a positive health and safety culture. This includes helping managers to ensure that an effective health and safety policy is implemented;
➤ health and safety planning. This will include goal-setting, deciding priorities and establishing adequate systems and performance standards. Short- and long-term objectives need to be realistic;
➤ day-to-day implementation and monitoring of policy and plans. This will include accident and incident investigation, reporting and analysis;
➤ performance reviews and audit of the whole health and safety management system.

To do this properly, health and safety practitioners need to:

➤ have proper training and be suitably qualified – for example NEBOSH Diploma or competence-based IOSH membership, relevant degree and, where appropriate, a Chartered Safety and Health Practitioner NEBOSH certificate in small to medium-sized low-hazard premises, like offices, call centres, warehouses and retail stores;
➤ keep up-to-date information systems on such topics as civil and criminal law, health and safety management and technical advances;
➤ know how to interpret the law as it applies to their own organization;
➤ actively participate in the establishment of organizational arrangements, systems and risk control standards relating to hardware and human performance. Health and safety practitioners will need to work with management on matters such as legal and technical standards;
➤ undertake the development and maintenance of procedures for reporting, investigating, recording and analysing accidents and incidents;
➤ develop and maintain procedures to ensure that senior managers get a true picture of how well health and safety is being managed (where a benchmarking role may be especially valuable). This will include monitoring, review and auditing;
➤ be able to present their advice independently and effectively.

3.7.3 Relationships within the organization

Health and safety practitioners:

➤ support the provision of authoritative and independent advice;
➤ report directly to directors or senior managers on matters of policy and have the authority to stop work if it contravenes agreed standards and puts people at risk of injury;
➤ are responsible for professional standards and systems. They may also have line management responsibility for other health and safety practitioners, in a large group of companies or on a large and/or high-hazard site.

3.7.4 Relationships outside the organization

Health and safety practitioners also have a function outside their own organization. They provide the point of liaison with a number of other agencies including the following:

➤ environmental health officers and licensing officials;
➤ architects and consultants;
➤ HSE and the fire and rescue authorities;
➤ the police;
➤ HM coroner or the procurator fiscal;

➤ local authorities;
➤ insurance companies;
➤ contractors;
➤ clients and customers;
➤ the public;
➤ equipment suppliers;
➤ the media;
➤ general practitioners;
➤ IOSH and occupational health specialists and services.

3.8   Persons in control of premises

Section 4 of the HSW Act requires that 'Persons in control of **non-domestic** premises' take such steps as are reasonable in their position to ensure that there are no risks to the health and safety of people who are not employees but use the premises. This duty extends to:

➤ people entering the premises to work;
➤ people entering the premises to use machinery or equipment, for example a launderette;
➤ access to and exit from the premises;
➤ corridors, stairs, lifts and storage areas.

Those in control of premises are required to take a range of steps depending on the likely use of the premises and the extent of their control and knowledge of the actual use of the premises.

3.9   Self-employed

The duties of the self-employed under the HSW Act are fairly limited. The Revitalising Health and Safety Strategy document expressed concern about whether the HSW Act sufficiently covers the area, in view of the huge growth in the use of contractors and sub-contractors throughout the UK. However, no further legislation has as yet been introduced. Under the HSW Act the self-employed are:

➤ responsible for their own health and safety;
➤ responsible to ensure that others who may be affected are not exposed to risks to their health and safety.

These responsibilities are extended by the Management of Health and Safety at Work Regulations, which requires self-employed people to:

➤ carry out risk assessment;
➤ cooperate with other people who work in the premises and, where necessary, in the appointment of a health and safety coordinator;

➤ provide comprehensible information to other peoples' employees working in their undertaking.

3.10   The supply chain

3.10.1 Introduction

Market leaders in every industry are increasing their grip on the chain of supply. They do so by monitoring rather than managing, and also by working more closely with suppliers. The result of this may be that suppliers or contractors are absorbed into the culture of the dominant firm, while avoiding the costs and liabilities of actual management. Powerful procurement departments emerge to define and impose the necessary quality standards and guard the lists of preferred suppliers.

The trend in many manufacturing businesses is to involve suppliers in a greater part of the manufacturing process so that much of the final production is the assembly of pre-fabricated subassemblies. This is particularly true of the automotive and aircraft industries. This is good practice as it:

➤ involves the supplier in the design process;
➤ reduces the number of items being managed within the business;
➤ reduces the number of suppliers;
➤ improves quality management by placing the onus on suppliers to deliver fully checked and tested components and systems.

In retail, suppliers are even given access to daily sales and forecasts of demand which would normally be considered as highly confidential information. In the process, the freedom of local operating managers to pick and choose suppliers is reduced. Even though the responsibility to do so is often retained, it is strongly qualified by centrally imposed rules and lists, and assistance or oversight (Figure 3.3).

Suppliers have to be:

➤ trusted;
➤ treated with fairness in a partnership;
➤ given full information to meet the demands being placed on them.

Under these conditions, suppliers and contractors looking for business with major firms need greater flexibility and wider competence than earlier. This often implies increased size and perhaps mergers, though in principle bids could be, and perhaps are, made by loose partnerships of smaller firms organized to secure such business.

Figure 3.3 NEBOSH is in control here (former premises).

3.10.2 Advantages of good supply chain management

Reduction of waste

This is an important objective of any business and involves not only waste of materials but also that of time. Examples of waste are:

➤ unwanted materials due to over ordering, damage or incorrect specifications;

➤ extraneous activities like double handling, for example between manufacturer, builders merchant and the site;

➤ re-working and re-fitting due to poor quality, design, storage or manufacture;

➤ waste of time such as waiting for supplies due to excessive time from ordering to delivery or early delivery long before they are needed.

Faster reaction

A well-managed supply chain should be able to respond rapidly to changing requirements (Figure 3.4). Winter conditions require very different materials than during warmer or drier seasons. A contractor may have to modify plans rapidly and suppliers may need to ramp up or change production at short notice.

Reduction in accidents

A closer relationship between client, designers, principal contractors and suppliers of services and products can result not only in a safer finished product but, in construction, a safer method of erection. If more products are pre-assembled in ideal factory conditions and then fixed in place on site, it is often safer than utilizing a full assembly

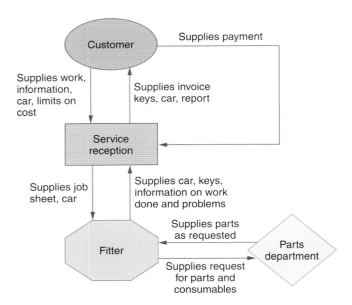

Figure 3.4 Typical supply chain.

approach in poor weather conditions on temporary work platforms. Examples are made-up roof trusses and pre-fabricated doors and windows already fitted to their frames.

3.10.3 Legislation

The HSW Act section 6 places a duty on everyone in the supply chain, from the designer to the final installer, of articles of plant or equipment for use at work or any article of fairground equipment to:

➤ ensure that the article will be safe and without risk to health at all times when it is being set, used, cleaned or maintained;

➤ carry out any necessary testing and examination to ensure that it will be safe;

➤ provide adequate information about its safe setting, use, cleaning, maintenance, dismantling and disposal.

There is an obligation on designers or manufacturers to do any research necessary to prove safety in use. Erectors or installers have special responsibilities to make sure, when handed over, that the plant or equipment is safe to use.

Similar duties are placed on manufacturers and suppliers of substances for use at work to ensure that the substance is safe when properly used, handled, processed, stored or transported and to provide adequate information and do any necessary research, testing or examining.

Where articles or substances are imported, the suppliers' obligations outlined above attach to the importer, whether they are a separate importing business or the user personally who does the importing.

Often items are obtained through hire purchase, leasing or other financing arrangements with the ownership of the items being vested with the financing organization. Where the financing organization's only function is to provide the money to pay for the goods, the supplier's obligations do not attach to them.

3.10.4 Information for customers

The quality movement has drawn attention to the need to ensure that there are processes in place which ensure quality, rather than just inspecting and removing defects when it is too late. In much the same way, organizations need to manage health and safety rather than acting when it is too late.

Customers need information and specifications from the manufacturer or supplier – especially where there is a potential risk involved for them. When deciding what the supplier needs to pass on, careful thought is required about the health and safety factors associated with any product or service.

This means focusing on four key questions and then framing the information supplied so that it deals with each one. The questions are as follows:

➤ Are there any inherent dangers in the product or service being passed on – what could go wrong?
➤ What can the manufacturer or supplier do while working on the product or service to reduce the chance of anything going wrong later?
➤ What can be done at the point of handover to limit the chances of anything going wrong?
➤ What steps should customers take to reduce the chances of something going wrong? What precisely would they need to know?

Depending on what is being provided to customers, the customer information may need to comply with the following legislation:

➤ Supply of Machinery (Safety) Regulations 1992 from December 2009 Supply of Machinery (Safety) Regulations 2008;
➤ Provision and Use of Work Equipment Regulations 1998;
➤ Control of Substances Hazardous to Health Regulations 2002 and 2005 amendments;
➤ Chemicals (Hazard Information and Packaging for Supply) Regulations 2002 and 2005 amendments.

This list is not exhaustive. See Chapter 17 for summaries.

Figure 3.5 Inadequate chair – take care when buying second-hand.

3.10.5 Buying problems

Examples of problems that may arise when purchasing include:

➤ second-hand equipment which does not conform to current safety standards such as an office chair which does not provide adequate back support or have five feet/castors;
➤ starting to use new substances which do not have safety data sheets;
➤ machinery which, while well guarded for operators, may pose risks for a maintenance engineer.

A risk assessment should be done on any new product, taking into account the likely life expectancy (delivery, installation, use, cleaning, maintenance, disposal, etc.). The supplier should be able to provide the information needed to do this. This will help the purchaser make an informed decision on the total costs because the risks will have been identified as will the precautions needed to control those risks. A risk assessment will still be needed for a CE-marked product. The CE marking signifies the manufacturer's declaration that the product conforms to relevant European Directives. Declarations from reputable manufacturers will normally be reliable. However, purchasers should be alert to fake or inadequate declarations and technical standards which may affect the health and safety of the product despite the CE marking. The risk assessment is still necessary to consider how and where the product will be used, what effect it might have on existing operations and what training will be required.

Employers have some key duties when buying plant and equipment:

➤ They must ensure that work equipment is safe and suitable for its purpose and complies with the relevant legislation. This applies equally to equipment which is adapted to be used in ways for which it was not originally designed.
➤ When selecting work equipment, they must consider existing working conditions and health and safety issues.
➤ They must provide adequate health and safety information, instructions, training and supervision for operators. Manufacturers and suppliers are required by law to provide information that will enable safe use of the equipment, substances, etc. and without risk to health.

Some of the issues that will need to be considered when buying in product or plant include:

➤ ergonomics – risk of work-related upper limb disorders (WRULD);
➤ manual handling needs;
➤ access/egress;
➤ storage, for example of chemicals;
➤ risk to contractors when decommissioning old plant or installing new plant;
➤ hazardous materials – provision of extraction equipment or PPE;
➤ waste disposal;
➤ safe systems of work;
➤ training;
➤ machinery guarding;
➤ emissions from equipment/plant, such as noise, heat or vibration.

3.11 Contractors

3.11.1 Introduction

The use of contractors is increasing as many companies turn to outside resources to supplement their own staff

and expertise. A contractor is anyone who is brought in to work and is not an employee. People on work experience or on labour only contracts or temps are considered to be employees. Contractors are used for maintenance, repairs, installation, construction, demolition, computer work, cleaning, security, health and safety and many other tasks. Sometimes there are several contractors on site at any one time. Clients need to think about how their work may affect each other and how they interact with the normal site occupier.

3.11.2 Legal considerations

The HSW Act applies to all work activities See chapters 1 and 17. It requires employers to ensure, so far as is reasonably practicable, the health and safety of:

➤ their employees;
➤ other people at work on their site, including contractors;
➤ members of the public who may be affected by their work.

All parties to a contract have specific responsibilities under health and safety law, and these cannot be passed on to someone else:

➤ Employers are responsible for protecting people from harm caused by work activities. This includes the responsibility not to harm contractors and subcontractors on site.
➤ Employees and contractors have to take care not to endanger themselves, their colleagues or others affected by their work.
➤ Contractors also have to comply with the HSW Act and other health and safety legislation. Clearly, when contractors are engaged, the activities of different employers do interact. So cooperation and communication are needed to make sure all parties can meet their obligations.
➤ Employees have to cooperate with their employer on health and safety matters, and not do anything that puts them or others at risk.
➤ Employees must be trained and clearly instructed in their duties.
➤ Self-employed people must not put themselves in danger, or others who may be affected by what they do.
➤ Suppliers of chemicals, machinery and equipment have to make sure their products or imports are safe, and provide information on this.

The Management of Health and Safety at Work Regulations apply to everyone at work and encourage employers to take a more systematic approach to dealing with health and safety by:

➤ assessing the risks which affect employees and any-one who might be affected by the site occupier's work, including contractors;

➤ setting up emergency procedures;

➤ providing training;

➤ cooperating with others on health and safety matters, for example contractors who share the site with an occupier;

➤ providing temporary workers, such as contractors, with health and safety information.

The principles of cooperation, coordination and communication between organizations underpin the Management of Health and Safety at Work Regulations and the CDM Regulations, explained next. See Section 3.12 on joint occupation of premises. For more information on the Management of Health and Safety at Work Regulations, see chapters 1 and 17.

3.11.3 Construction Design and Management (CDM2007) Regulations

Businesses often engage contractors for construction projects at one time or another to build plant, convert or extend premises and demolish buildings. The CDM 2007 Regulations apply to all construction projects. Larger projects which are notifiable (see details in Chapters 16 and 17) have more extensive requirements.

All projects require the following:

➤ non-domestic clients to: check the competence of all their appointees; ensure there are suitable manage-ment arrangements for the project; allow sufficient time and resources for all stages; provide pre-con-struction information to designers and contractors;

➤ designers to: eliminate hazards and reduce risks during design; and provide information about remaining risks;

➤ contractors to: plan, manage and monitor their own work and that of employees; check the competence of all their appointees and employees; train their own employees; provide information to their employees; comply with the requirements for health and safety on site detailed in Part 4 of the Regulations and other Regulations such as the Work at Height Regulations; and ensure there are adequate welfare facilities for their employees;

➤ everyone to: assure their own competence; cooperate with others and coordinate work so as to ensure the health and safety of construction workers and others who may be affected by the work; report obvious risks;

take account of the general principles of prevention in planning or carrying out construction work; and comply with the requirements in Schedule 3, Part 4 of CDM 2007 and other Regulations for any work under their control.

For even small projects, clients should ensure that contractors provide:

(a) information regarding the contractor's health and safety policy;

(b) information on the contractor's health and safety organization detailing the responsibilities of individuals;

(c) information on the contractor's procedures and standards of safe working;

(d) the method statements for the project in hand;

(e) details on how the contractor will audit and imple-ment its health and safety procedure;

(f) do they have procedures for investigating incidents and learning the lessons from them.

Smaller contractors may need some guidance to help them produce suitable method statements. While they do not need to be lengthy, they should set out those features essential to safe working, for example access arrange-ments, PPE, control of chemical risks, etc.

Copies of relevant risk assessments for the work to be undertaken should be requested. These need not be very detailed but should indicate the risk and the control methods to be used.

The client, designer, CDM coordinator, principal contractor and other contractors all have specific roles under CDM 2007 Regulations. For more information see Chapters 16 and 17.

3.11.4 Contractor selection

The selection of the right contractor for a particular job is probably the most important element in ensuring that the risks to the health and safety of everybody involved in the activity and people in the vicinity, are reduced as far as possible. Ideally, selection should be made from a list of approved contractors who have demonstrated that they are able to meet the client's requirements (Figure 3.6).

The selection of a contractor has to be a balanced judgement with a number of factors taken into account. Fortunately, a contractor who works well and meets the client's requirements in terms of the quality and timeli-ness of the work is also likely to have a better than aver-age health and safety performance. Cost, of course, will have to be part of the judgement but may not provide any indication of which contractor is likely to give the best performance in health and safety terms. In deciding

Figure 3.6 Contractors at work.

which contractor should be chosen for a task, the following should be considered:

➤ Do they have an adequate health and safety policy?
➤ Can they demonstrate that the person responsible for the work is competent?
➤ Can they demonstrate that competent safety advice will be available?
➤ Do they monitor the level of accidents at their work site?
➤ Do they have a system to assess the hazards of a job and implement appropriate control measures?
➤ Will they produce a method statement which sets out how they will deal with all significant risks?
➤ Do they have guidance on health and safety arrangements and procedures to be followed?
➤ Do they have effective monitoring arrangements?
➤ Do they use trained and skilled staff who are qualified where appropriate? (Judgement will be required, as many construction workers have had little or no training except training on the job.) Can the company demonstrate that the employees or other workers used for the job have had the appropriate training and are properly experienced and, where appropriate, qualified?
➤ Can they produce good references indicating satisfactory performance?

3.11.5 Contractor authorization

Contractors, their employees, sub-contractors and their employees, should not be allowed to commence work on any client's site without authorization signed by the company contact. The authorization should clearly define

the range of work that the contractor can carry out and set down any special requirements, for example protective clothing, fire exits to be left clear, and isolation arrangements.

Permits will be required for operations such as hot work. All contractors should keep a copy of their authorization at the place of work. A second copy of the authorization should be kept at the site and be available for inspection.

The company contact signing the authorization will be responsible for all aspects of the work of the contractor. The contact will need to check as a minimum the following:

➤ that the correct contractor for the work has been selected;
➤ that the contractor has made appropriate arrangements for supervision of staff;
➤ that the contractor has received and signed for a copy of the contractor's safety rules;
➤ that the contractor is clear what is required, the limits of the work and any special precautions that need to be taken;
➤ that the contractor's personnel are properly qualified for the work to be undertaken.

The company contact should check whether sub-contractors will be used. They will also require authorization, if deemed acceptable. It will be the responsibility of the company contact to ensure that sub-contractors are properly supervised.

Appropriate supervision will depend on a number of factors, including the risk associated with the job, experience of the contractor and the amount of supervision the contractor will provide. The responsibility for ensuring there is proper supervision lies with the person signing the contractor's authorization.

The company contact will be responsible for ensuring that there is adequate and clear communication between different contractors and company personnel where this is appropriate.

3.11.6 Safety rules for contractors

In the conditions of contract, there should be a stipulation that the contractor and all of their employees adhere to the contractor's safety rules. Contractors' safety rules should contain as a minimum the following points:

➤ **Health and safety** – that the contractor operates to at least the minimum legal standard and conforms to accepted industry good practice;

- **Supervision** – that the contractor provides a good standard of supervision of their own employees;
- **Sub-contractors** – that they may not use sub-contractors without prior written agreement from the organization;
- **Authorization** – that each employee must carry an authorization card issued by the organization at all times while on site.

3.11.7 Example of rules for contractors

Contractors engaged by the organization to carry out work on its premises will:

- familiarize themselves with so much of the organization's health and safety policy as affects them and will ensure that appropriate parts of the policy are communicated to their employees, and any sub-contractors and employees of sub-contractors who will do work on the premises;
- cooperate with the organization in its fulfilment of its health and safety duties to contractors and take the necessary steps to ensure the like cooperation of their employees;
- comply with their legal and moral health, safety and food hygiene duties;
- ensure the carrying out of their work on the organization's premises in such a manner as not to put either themselves or any other persons on or about the premises at risk;
- ensure that where they wish to avail themselves of the organization's first-aid arrangements/facilities while on the premises, written agreement to this effect is obtained prior to first commencement of work on the premises;
- supply a copy of its statement of policy, organization and arrangements for health and safety written for the purposes of compliance with The Management of Health and Safety at Work Regulations and section 2(3) of the HSW Act where applicable and requested by the organization;
- abide by all relevant provisions of the organization's safety policy, including compliance with health and safety rules and CDM 2007;
- ensure that on arrival at the premises, they and any other persons who are to do work under the contract report to reception or their designated organization contact.

Without prejudice to the requirements stated above, contractors, sub-contractors and employees of contractors and sub-contractors will, to the extent that such matters are within their control, ensure:

- the safe handling, storage and disposal of materials brought onto the premises;
- that the organization is informed of any hazardous substances brought onto the premises and that the relevant parts of the Control of Substances Hazardous to Health Regulations in relation thereto are complied with;
- that fire prevention and fire precaution measures are taken in the use of equipment which could cause fires;
- that steps are taken to minimize noise and vibration produced by their equipment and activities;
- that scaffolds, ladders and other such means of access, where required, are erected and used in accordance with Work at Height Regulations and good working practice;
- that any welding or burning equipment brought onto the premises is in safe operating condition and used in accordance with all safety requirements;
- that any lifting equipment brought onto the premises is adequate for the task and has been properly tested/certified;
- that any plant and equipment brought onto the premises is in safe condition and used/operated by competent persons;
- that for vehicles brought onto the premises, any speed, condition or parking restrictions are observed;
- that compliance is made with the relevant requirements of the Electricity at Work Regulations 1989;
- that connection(s) to the organization's electricity supply is from a point specified by its management and is by proper connectors and cables;
- that they are familiar with emergency procedures existing on the premises;
- that welfare facilities provided by the organization are treated with care and respect;
- that access to restricted parts of the premises is observed and the requirements of food safety legislation are complied with;
- that any major or lost-time accident or dangerous occurrence on the organization's premises is reported as soon as possible to their site contact;
- that where any doubt exists regarding health and safety requirements, advice is sought from the site contact.

The foregoing requirements *do not* exempt contractors from their statutory duties in relation to health and safety, but are intended to assist them in attaining a high standard of compliance with those duties.

3.12 Joint occupation of premises

The Management of Health and Safety at Work Regulations specifically states that where two or more employers share a workplace – whether on a temporary or a permanent basis – each employer shall:

➤ cooperate with other employers;
➤ take reasonable steps to coordinate between other employers to comply with legal requirements;
➤ take reasonable steps to inform other employers where there are risks to health and safety.

All employers and self-employed people involved should satisfy themselves that the arrangements adopted are adequate. Where a particular employer controls the premises, the other employers should help to assess the shared risks and coordinate any necessary control procedures. Where there is no controlling employer, the organizations present should agree joint arrangements to meet regulatory obligations, such as appointing a health and safety coordinator.

3.13 Consultation with the workforce

3.13.1 General

It is important to gain the cooperation of all employees if a successful health and safety culture is to become established. This cooperation is best achieved by consultation. Joint consultation can help businesses be more efficient and effective by reducing the number of accidents and work-related ill health and also to motivate staff by making them aware of health and safety problems.

There are two pieces of legislation that cover health and safety consultation with employees and both are summarized in Chapter 17. They are the Safety Representatives and Safety Committees Regulations 1977 and the Health and Safety (Consultation with Employees) Regulations 1996. These Regulations require employers to consult with their employees on health and safety matters and make provision for both trade-union-appointed safety representatives and representatives of employee safety (RES) elected by the workforce (Figure 3.7).

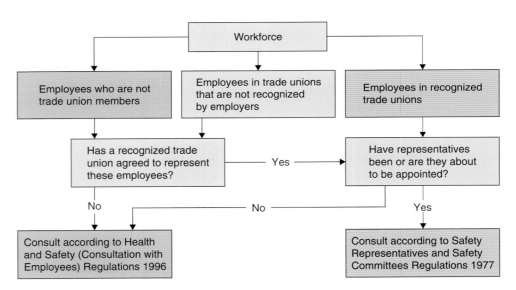

Figure 3.7 The law on consulting employees about health and safety in your workplace. References to the Regulations are colour-coded to help find the parts that are most relevant to a particular organization: ☐ – for workplaces where the Safety Representatives and Safety Committees Regulations 1977 apply; ☐ – for workplaces where the Health and Safety (Consultation with Employees) Regulations 1996 apply. *Source*: HSE INDG232(rev1).

Employers must consult their workforce on a range of health and safety issues, including:

➤ any measures at the workplace which may substantially affect their health and safety, for example changes in systems of work, types of equipment or chemicals being used;
➤ arrangements for getting competent people to help them comply with health and safety requirements;
➤ the information that must be given to employees about risks to health and safety and any preventative measures;
➤ planning and organizing health and safety training;
➤ the health and safety consequences of using new technology or substances.

Also, in workplaces in which the employer recognizes a trade union:

➤ the employer must consult with trade-union-appointed safety representatives on health and safety matters affecting the employees they represent.

Trade-union-appointed safety representatives may:

➤ investigate possible dangers at work, the causes of accidents there and general complaints by employees on health and safety issues and take these up with the employer;
➤ carry out inspections of the workplace;
➤ represent employees in discussions with the HSE inspectors and receive information from them;
➤ attend safety committee meetings.

In workplaces in which trade unions are not recognized:

➤ employees must be consulted on health and safety, either directly or through their elected representatives.

Elected RES may:

➤ take up with employers concerns about possible risks and dangerous events in the workplace that may affect the employees they represent;
➤ take up with employers general matters affecting the health and safety of the employees they represent;
➤ represent employees who elected them in consultations with health and safety inspectors.

Finally, an employer should ensure that all representatives receive reasonable training in health and safety at the employer's expense, are allowed time during working hours to perform their duties, and provide other facilities and assistance that the representatives should reasonably require.

3.13.2 The Safety Representatives and Safety Committees Regulations 1977

These Regulations apply only to those organizations that have recognized trade unions for collective bargaining purposes. The recognized trade union may appoint safety representatives from among the employees and notify the employer in writing. In some workplaces, safety representatives have agreed to represent the entire workforce.

The Regulations give safety representative several functions not duties (a failure to carry out a function is not a breach of law as it would be if they were legal duties). Functions include:

(a) representing employees in consultation with the employer;
(b) investigating potential hazards and dangerous occurrences;
(c) being involved with risk assessment procedures;
(d) investigating the causes of accidents, cases of work-related diseases or ill health and dangerous occurrences;
(e) investigating employee complaints relating to health, safety and welfare;
(f) making representations to the employer on health, safety and welfare matters;
(g) carrying out inspections of the workplace. They must also be allowed to inspect the workplace at least once a quarter or sooner if there has been a substantial change in the conditions of work;
(h) representing employees at the workplace in consultation with enforcing inspectors;
(i) receiving information from health and safety inspectors;
(j) attending safety committee meetings.

Representatives must be allowed time off with pay to fulfil these functions and to undergo health and safety training. They must have access to suitable facilities and assistance to carry out their functions.

Safety committees

If two or more safety representatives have requested in writing that a safety committee be set up, the employer has 3 months to comply.

In medium to large organizations the easiest and often the most effective method of consultation is the health and safety committee. It will realize its full potential if its recommendations are seen to be implemented and both management and employee concerns are freely discussed. It will not be so successful if it is seen as a talking shop.

The committee should have stated objectives which mirror the objectives in the organization's health and safety policy statement, and its own terms of reference. Terms of reference should include the following:

➤ the study of accident and notifiable disease statistics to enable reports to be made of recommended remedial actions;
➤ the examination of health and safety audits and statutory inspection reports;
➤ the consideration of reports from the external enforcement agency;
➤ the review of new legislation, Approved Codes of Practice and guidance and their effect on the organization;
➤ the monitoring and review of all health and safety training and instruction activities in the organization;
➤ the monitoring and review of health and safety publicity and communication throughout the organization;
➤ development of safe systems of work and safety procedures;
➤ reviewing risk assessments;
➤ considering reports from safety representatives;
➤ continuous monitoring of arrangements for health and safety and revising them whenever necessary.

There are no fixed rules on the composition of the health and safety committee except that it should be representative of the whole organization. It should have representation from the workforce and the management including at least one senior manager (other than the Health and Safety Advisor). Managers and representatives should agree who should chair the meetings, how often meetings should be held and what they hope to achieve.

Accident and ill-health investigations

Properly investigated accidents and ill health can reveal weaknesses which need to be remedied. A joint investigation with safety representatives is more likely to inspire confidence with workers so that they cooperate fully with the investigation, as in many cases those involved may be concerned about being blamed for the accident.

Safety representatives are entitled to contact enforcing authority inspectors. If this is just for information, they can be contacted directly. If it is a formal complaint against the employer, the inspector will need to know if the employer has been informed. The inspectors can be contacted anonymously. They will keep the person's identity secret in such circumstances.

Training, facilities and assistance

Safety representatives are legally entitled to paid time off for training, which is usually freely available from their trade union or the TUC.

Training courses topics often include:

➤ the role and functions of the safety representative;
➤ health and safety legislation;
➤ how to identify and minimize hazards;
➤ how to carry out a workplace inspection and accident investigation;
➤ employer's health and safety arrangements, including emergency procedures, risk assessments and health and safety policies;
➤ further information on training courses.

The employer is also required to provide facilities and assistance for safety representatives. Depending on the circumstances, these could include:

➤ noticeboard;
➤ telephone;
➤ lockable filing cabinet;
➤ access to an office to meet workers in private;
➤ camera;
➤ key health and safety information;
➤ access to specialist assistance and support in understanding technical issues.

3.13.3 The Health and Safety (Consultation with Employees) Regulations 1996

In non-unionized workplaces where there are no safety representatives or in a workplace that has trade union recognition but either the trade union has not appointed a safety representative or they do not represent the whole workforce, the Health and Safety (Consultation with Employees) Regulations will apply.

Employers must consult either on an individual basis (e.g. very small companies) or through RES elected by the workforce. See Table 3.1 for a comparison of functions with union-appointed safety representatives.

The guidance to these Regulations emphasizes the difference between informing and consulting. Consultation involves listening to the opinion of employees on a particular issue and taking it into account before a decision is made. Informing employees means providing information on health and safety issues such as risks, control systems and safe systems of work.

The functions of these RES are to:

➤ represent the interest of workers on health and safety matters to the employer;
➤ approach the employer regarding potential hazards and dangerous occurrences at the workplace;
➤ approach the employer regarding general matters affecting the health and safety of the people they represent;

Table 3.1 A comparison of the functions of health and safety representatives

Safety Representatives and Safety Committees Regulations 1977	Health and Safety (Consultation with Employees) Regulations 1996
Representatives	
Appointed in writing by a trade union recognized for collective bargaining purposes.	Elected by the workforce, where the employer has decided not to consult directly.
Known as	
Safety representatives.	Representatives of employee safety.
Functions	
Investigate potential hazards and dangerous occurrences at the workplace, and complaints by an employee relating to health, safety and welfare at work, and examine causes of workplace accidents.	
Representation to the employer on the above investigations, and on general matters affecting the health and safety of employees they represent.	Representation to the employer on: ● potential hazards and dangerous occurrences ● general matters affecting the health and safety of employees they represent ● specific matters on which the employer must consult.
Inspect the workplace.	
Represent employees in dealings with health and safety inspectors.	Represent employees in dealings with health and safety inspectors.
Receive certain information from inspectors.	
Attend health and safety committee meetings.	

Source: HSE INDG232(rev) page 4 HSE Web.

➤ to speak for the people they represent in consultation with inspectors.

The employer must consult RES on the following:

➤ the introduction of any measure or change which may substantially affect employees' health and safety;

➤ the arrangements for the appointment of competent persons to assist in following health and safety law;

➤ any information resulting from risk assessments or their resultant control measures which could affect the health, safety and welfare of employees;

➤ the planning and organization of any health and safety training required by legislation;

➤ the health and safety consequences to employees of the introduction of new technologies into the work place.

However, the employer is not expected to disclose information if:

➤ it violates a legal prohibition;
➤ it could endanger national security;
➤ it relates specifically to an individual without his/her consent;
➤ it could harm substantially the business of the employer or infringe commercial security;
➤ it was obtained in connection with legal proceedings.

3.14 Sources of reference

The Management of Health and Safety at Work ACOP L21, HSE Books 2000 ISBN 978 0 7176 2488 1.
Successful Health and Safety Management HSG65, HSE Books 1997 ISBN 978 0 7176 1276 5.
Health & Safety Executive 'Ready Reckoner' website at www.hse.gov.uk/costs

Relevant statutory provisions

The Health and Safety at Work etc. Act 1974, sections 2–4, 6–9, 36 and 37.
The Management of Health and Safety at Work Regulations 1999, 7–9, 11, 12, 14, 15; the Safety Representatives and Safety Committees Regulations 1977.
The Health and Safety (Consultation with Employees) Regulations 1996.

3.15 Practice NEBOSH questions for Chapter 3

1. **Outline** the factors that will determine the level of supervision that a new employee should receive during their initial period of employment within an organization.

2. (i) **Explain** the meaning of the term 'competent person'.
   (ii) **Outline** the organizational factors that may cause a person to work unsafely even though they are competent.

3. **Outline** the practical means by which a manager could involve employees in the improvement of health and safety in the workplace.

4. (i) **Give TWO** reasons why visitors to a workplace might be at greater risk of injury than an employee.
   (ii) **Outline** measures to be taken to ensure the health and safety of visitors to the workplace.

5. **List** the factors that should be considered when assessing the health and safety competence of a contractor.

6. **Outline** the checks that could be made in assessing the health and safety competence of a contractor.

7. A contractor has been engaged by a manufacturing company to undertake extensive maintenance work on the interior walls of a factory workshop.
   (i) **State** the legal duties that the manufacturing company owes the contractor's employees under the Health and Safety at Work etc. Act 1974.
   (ii) **Outline** the information relevant to health and safety that should be provided before work commences by:
       (a) the manufacturing company to the contractor and
       (b) the contractor to the manufacturing company.
   (iii) **Describe** additional procedural measures that the manufacturing company should take to help ensure the health and safety of their own and the contractor's employees.

8. (i) **State** the circumstances under which an employer must establish a health and safety committee.
   (ii) **Give SIX** reasons why a health and safety committee may prove to be ineffective in practice.

9. **Outline** ways to help ensure the effectiveness of a safety committee.

10. **Outline** the benefits to an organization of having a health and safety committee.

11. In relation to the Safety Representatives and Safety Committees Regulations 1977, **outline**:
    (i) the rights and functions of a trade union appointed safety representative
    (ii) the facilities that an employer may need to provide to safety representatives.

12. With respect to the Safety Representatives and Safety Committees Regulations 1977, **state** when a safety representative is legally entitled to inspect the workplace.

13. By reference to the Safety Representatives and Safety Committees Regulations 1977:
    (i) **explain** the circumstances under which an employer must form a health and safety committee
    (ii) **identify TWO** reasons why a safety representative may have cause to complain to an Employment

Tribunal **AND state** the period of time within which such a complaint must be made.

14. In relation to the Health and Safety (Consultation with Employees) Regulations 1996, **identify**:
(i) the health and safety matters on which employers have a duty to consult their employees
(ii) four types of information that an employer is not obliged to disclose to an employee representative.

15. With reference to the Health and Safety (Consultation with Employees) Regulations 1996:
(i) **explain** the difference between consulting and informing
(ii) **outline** the health and safety matters on which employers must consult their employees

(iii) outline the entitlements of representatives of employee safety who have been elected under the Regulations.

16. With reference to the Health and Safety (Consultation with Employees) Regulations:
(i) **Outline** the ways in which an employer can consult with the workforce:
(ii) **State** the criteria that would help determine the appropriate number of representatives of employee safety in the workplace.

17. In relation to the Health and Safety (Consultation with Employees) Regulations 1996, **outline** the types of facility that an employer may need to provide to representatives of employee safety.

Appendix 3.1 Detailed health and safety responsibilities

Managing Directors/Chief Executives

1. are responsible and accountable for health and safety performance at the organization;
2. develop a strong, positive health and safety culture throughout the company and communicate it to all managers. This should ensure that all managers have a clear understanding of their health and safety responsibilities;
3. provide guidance and leadership on health and safety matters to their management team;
4. establish minimum acceptable health and safety standards within the organization;
5. ensure that adequate resources are available for the health and safety requirements within the organization and authorize any necessary major health and safety expenditures;
6. evaluate, approve and authorize health and safety related projects developed by the organization's health and safety advisers;
7. review and approve health and safety policies, procedures and programmes developed by the organization's managers;
8. ensure that a working knowledge of the areas of health and safety that are regulated by various governmental agencies, particularly the HSE, Local Authorities and the Environment Agency, are maintained;
9. ensure that health and safety is included as an agenda topic at all formal senior management meetings;
10. review and act upon major recommendations submitted by outside loss prevention consultants and insurance companies;
11. ensure that health and safety is included in any tours such as fire inspections of the organization's sites and note any observed acts or conditions that fall short of or exceed agreed health and safety standards;
12. ensure that all fatalities, major property losses, serious lost workday injuries and dangerous occurrences are investigated;
13. establish, implement and maintain a formal, written health and safety programme for the organization that encompasses all areas of significant health and safety risk;
14. establish controls to ensure uniform adherence to the health and safety programme across the organization. These controls should include both corrective and follow-up actions;

15. attend the health and safety committee meetings at the organization;
16. review, on a regular basis, all health and safety activity reports and performance statistics;
17. review health and safety reports submitted by outside agencies and determine that any agreed actions have been taken;
18. review annually the effectiveness of, and, if necessary, require revision of the site health and safety programme;
19. appraise the performance of the health and safety advisers and provide guidance or training where necessary;
20. monitor the progress of managers and others towards achieving their individual health and safety objectives;

Departmental Managers

1. are responsible and accountable for the health and safety performance of their department;
2. contact each supervisor frequently (daily) to monitor the health and safety standards in the department;
3. hold departmental health and safety meetings for supervisors and employee representatives at least once a month;
4. ensure that any machinery, equipment or vehicles used within the department are maintained and correctly guarded and meet agreed health and safety standards. Copies of records of all maintenance, statutory and insurance inspections must be kept by the Departmental Manager;
5. ensure that all fire and other emergency equipment is properly maintained on a regular basis with all faults rectified promptly and that all departmental staff are aware of fire and emergency procedures;
6. ensure that there is adequate first-aid cover on all shifts and all first-aid boxes are adequately stocked;
7. ensure that safe systems of work procedures are in place for all jobs and that copies of all procedures are submitted to the site Managing Director for approval;
8. review all job procedures on a regular basis and require each supervisor to check that the procedures are being used correctly;
9. approve and review, annually, all departmental health and safety risk assessments, rules and procedures, maintain strict enforcement and develop plans to ensure employee instruction and re-instruction;
10. ensure that all health and safety documents (such as the organization's health and safety manual, risk

assessments, rules and procedures) are easily accessible to all departmental staff;

11. establish acceptable housekeeping standards, defining specific areas of responsibility, and assign areas to supervisors; make a weekly spot check across the department, hold a formal inspection with supervisors at least monthly and submit written reports of the inspections to the health and safety adviser with deadlines for any required actions;

12. authorize purchases of tools and equipment necessary to attain compliance with the organization's specifications and relevant statutory Regulations;

13. develop a training plan that includes specific job instructions for new or transferred employees and follow up on the training by supervisors. Copies of records of all training must be kept by the Departmental Manager;

14. review the health and safety performance of their department each quarter and submit a report to the Managing Director/Chief Executive;

15. personally investigate all lost workday cases and dangerous occurrences and report to the Managing Director/Chief Executive. Progress any required corrective action;

16. adopt standards for assigning PPE to employees, insist on strict enforcement and make spot field checks to determine compliance;

17. evaluate the health and safety performances of supervisors;

18. develop in each supervisor strong health and safety attitudes and a clear understanding of their specific duties and responsibilities;

19. instil, by action, example and training, a positive health and safety culture among all departmental staff;

20. instruct supervisors in site procedures for the care and treatment of sick or injured employees;

21. ensure that the names of any absentees, written warnings and all accident reports are submitted to the Human Resources Manager.

Supervisors

1. are responsible and accountable for their team's health and safety performance;

2. conduct informal health and safety meetings with their employees at least monthly;

3. enforce all safe systems of work procedures that have been issued by the Departmental Manager;

4. report to the Departmental Manager any weaknesses in the safe system of work procedures or any actions taken to revise such procedures. These weaknesses may be revealed by either health and safety risk assessments or observations;

5. report any jobs that are not covered by safe systems of work procedures to the Departmental Manager;

6. review any unsafe acts and conditions and either eliminate them or report them to the Departmental Manager;

7. instruct employees in relevant health and safety rules, make records of this instruction and enforce all health and safety rules and procedures;

8. make daily inspections of assigned work areas and take immediate steps to correct any unsafe or unsatisfactory conditions, report to the Departmental Manager those conditions that cannot be immediately corrected and instruct employees on housekeeping standards;

9. instruct employees that tools/equipment are to be inspected before each use and make spot checks of tools'/equipment's condition;

10. instruct each new employee personally on job health and safety requirements in assigned work areas;

11. provide on-the-job instruction on safe and efficient performance of assigned jobs for all employees in the work area;

12. report any apparent employee health problems to the Departmental Manager;

13. enforce PPE requirements; make spot checks to determine that protective equipment is being used and periodically appraise condition of equipment. Record any infringements of the PPE policy;

14. in the case of a serious injury, ensure that the injured employee receives prompt medical attention, isolate the area and/or the equipment as necessary and immediately report the incident to the Departmental Manager. In case of a dangerous occurrence, the Supervisor should take immediate steps to correct any unsafe condition and, if necessary, isolate the area and/or the equipment. As soon as possible, details of the incident and any action taken should be reported to the Departmental Manager;

15. investigate all accidents, serious incidents and cases of ill health involving employees in assigned work areas. Immediately after an accident, complete accident report form and submit it to the Departmental Manager for onward submission to the health and safety adviser. A preliminary investigation report and any recommendations for preventing a recurrence should be included on the accident report form;

16. check for any changes in operating practices, procedures and other conditions at the start of each shift/day

and before relieving the 'on duty' Supervisor (if applicable). A note should be made of any health and safety related incidents that have occurred since their last working period;

17. at the start of each shift/day, make an immediate check to determine any absentees. Report any absentees to the Departmental Manager;

18. make daily spot checks and take necessary corrective action regarding housekeeping, unsafe acts or practices, unsafe conditions, job procedures and adherence to health and safety rules;

19. attend all scheduled and assigned health and safety training meetings;

20. act on all employee health and safety complaints and suggestions;

21. maintain, in their assigned area, health and safety signs and notice boards in a clean and legible condition.

Employees and agency workers

1. are responsible for their own health and safety;

2. ensure that their actions will not jeopardize the safety or health of other employees;

3. obey any safety rules, particularly regarding the use of PPE or other safety equipment;

4. learn and follow the operating procedures and health and safety rules and procedures for the safe performance of the assigned job;

5. must correct, or report to their Supervisor, any observed unsafe practices and conditions;

6. maintain a healthy and safe place to work and cooperate with managers in the implementation of health and safety matters;

7. make suggestions to improve any aspect of health and safety;

8. maintain an active interest in health and safety;

9. follow the established procedures if accidents occur by reporting any accident to the Supervisor;

10. report any absence from the company caused by illness or an accident.

Appendix 3.2 Checklist for supply chain health and safety management

This checklist is taken from the HSE leaflet INDG268 *Working Together: Guidance on Health and Safety for Contractors and Suppliers 2002*. It is a reminder of the topics that might need to be discussed with people with whom individual contractors may be working.

It is not intended to be exhaustive and not all questions will apply at any one time, but it should help people to get started.

1. **Responsibilities**
   - ➤ What are the hazards of the job?
   - ➤ Who is to assess particular risks?
   - ➤ Who will coordinate action?
   - ➤ Who will monitor progress?

2. **The job**
   - ➤ Where is it to be done?
   - ➤ Who with?
   - ➤ Who is in charge?
   - ➤ How is the job to be done?
   - ➤ What other work will be going on at the same time?
   - ➤ How long will it take?
   - ➤ What time of day or night?
   - ➤ Do you need any permit to do the work?

3. **The hazards and risk assessments**
   Site and location

   Consider the means of getting into and out of the site and the particular place of work – are they safe? – and

   - ➤ Will any risks arise from environmental conditions?
   - ➤ Will you be remote from facilities and assistance?
   - ➤ What about physical/structural conditions?
   - ➤ What arrangements are there for security?

   Substances

   - ➤ What supplier's information is available?
   - ➤ Is there likely to be any microbiological risk?
   - ➤ What are the storage arrangements?
   - ➤ What are the physical conditions at the point of use? Check ventilation, temperature, electrical installations, etc.
   - ➤ Will you encounter substances that are not supplied, but produced in the work, for example fumes from hot work during dismantling plant? Check how much, how often, for how long, method of work, etc.
   - ➤ What are the control measures? For example, consider preventing exposure, providing engineering controls, using personal protection (in that order of choice).
   - ➤ Is any monitoring required?
   - ➤ Is health surveillance necessary, for example for work with sensitizers? (Refer to health and safety data sheet.)

Plant and equipment

➤ What are the supplier/hirer/manufacturer's instructions?
➤ Are any certificates of examination and test needed?
➤ What arrangements have been made for inspection and maintenance?
➤ What arrangements are there for shared use?
➤ Are the electrics safe to use? Check the condition of power sockets, plugs, leads and equipment. (Don't use damaged items until they have been repaired.)
➤ What assessments have been made of noise levels?

4. **People**

➤ Is information, instruction and training given, as appropriate?
➤ What are the supervision arrangements?
➤ Are members of the public/inexperienced people involved?
➤ Have any disabilities/medical conditions been considered?

5. **Emergencies**

➤ What arrangements are there for warning systems in case of fire and other emergencies?
➤ What arrangements have been made for fire/emergency drills?
➤ What provision has been made for first-aid and fire-fighting equipment?
➤ Do you know where your nearest fire exits are?
➤ What are the accident reporting arrangements?
➤ Are the necessary arrangements made for availability of rescue equipment and rescuers?

6. **Welfare**

Who will provide:

➤ shelter
➤ food and drinks
➤ washing facilities
➤ toilets (male and female)
➤ clothes changing/drying facilities?

There may be other pressing requirements which make it essential to re-think health and safety as the work progresses.

Promoting a positive health and safety culture

4

After reading this chapter you should be able to:

1. describe the concept of health and safety culture and its significance in the management of health and safety in an organization
2. identify indicators which could be used to assess the effectiveness of an organization's health and safety culture and recognize factors that could cause its deterioration
3. identify the factors which influence safety related behaviour at work
4. identify methods which could be used to improve the health and safety culture of an organization
5. outline the internal and external influences on an organization's health and safety standards.

4.1 Introduction

In 1972, the Robens report recognized that the introduction of health and safety management systems was essential if the ideal of self-regulation of health and safety by industry was to be realized. It further recognized that a more active involvement of the workforce in such systems was essential if self-regulation was to work. Self-regulation and the implicit need for health and safety management systems and employee involvement were incorporated into the Health and Safety at Work (HSW) Act.

Since the introduction of the HSW, health and safety standards have improved considerably but there have been some catastrophic failures. One of the worst was the fire on the off-shore oil platform, Piper Alpha, in 1988 when 167 people died. At the subsequent enquiry, the concept of a safety culture was defined by the Director General of the Helth and Safety Executive (HSE) at that time, J. R. Rimington. This definition has remained as one of the key points for a successful health and safety management system.

4.2 Definition of a health and safety culture

The health and safety culture of an organization may be described as the development stage of the organization

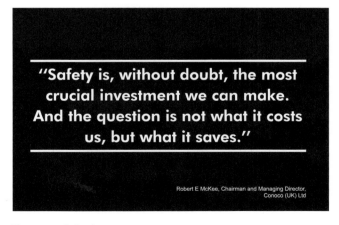

"Safety is, without doubt, the most crucial investment we can make. And the question is not what it costs us, but what it saves."

Robert E McKee, Chairman and Managing Director, Conoco (UK) Ltd

Figure 4.1 Safety investment.

in health and safety management at a particular time. HSG65 gives the following definition of a health and safety culture:

> The safety culture of an organization is the product of individual and group values, attitudes, perceptions, competencies and patterns of behaviour that determine the commitment to, and the style and proficiency of, an organization's health and safety management.
>
> Organizations with a positive safety culture are characterized by communications founded on mutual trust, by shared perceptions of the importance of safety and by confidence in the efficacy of preventive measures.

There is concern among some health and safety professionals that many health and safety cultures are developed and driven by senior managers with very little input from the workforce. Others argue that this arrangement is sensible because the legal duties are placed on the employer. A positive health and safety culture needs the involvement of the whole workforce just as a successful quality system does. There must be a joint commitment in terms of attitudes and values. The workforce must believe that the safety measures put in place will be effective and followed even when financial and performance targets may be affected.

4.3   Safety culture and safety performance

4.3.1 The relationship between health and safety culture and health and safety performance

The following elements are the important components of a positive health and safety culture:

- leadership and commitment to health and safety throughout and at all levels of the organization;
- acceptance that high standards of health and safety are achievable as part of a long-term strategy formulated by the organization;
- a detailed assessment of health and safety risks in the organization and the development of appropriate control and monitoring systems;
- a health and safety policy statement outlining short- and long-term health and safety objectives. Such a policy should also include codes of practice and required health and safety standards;
- relevant employee training programmes and communication and consultation procedures;
- systems for monitoring equipment, processes and procedures and the prompt rectification of any defects;
- the prompt investigation of all incidents and accidents and reports made detailing any necessary remedial actions.

If the organization adheres to these elements, then a basis for a good performance in health and safety will have been established. However, to achieve this level of performance, sufficient financial and human resources must be made available for the health and safety function at all levels of the organization.

All managers, supervisors and members of the governing body (e.g. directors) should receive training in health and safety and be made familiar during training sessions with the health and safety targets of the organization. The depth of training undertaken will depend on the level of competence required of the particular manager. Managers should be accountable for health and safety within their departments and be rewarded for significant improvements in health and safety performance. They should also be expected to discipline employees within their departments who infringe health and safety policies or procedures.

4.3.2 Important indicators of a health and safety culture

There are several outputs or indicators of the state of the health and safety culture of an organization. The most important are the numbers of accidents, near misses and occupational ill-health cases occurring within the organization.

Although the number of accidents may give a general indication of the health and safety culture, a more detailed examination of accidents and accident statistics is normally required. A calculation of the rate of accidents enables health and safety performance to be compared between years and organizations.

The simplest measure of accident rate is called the incident rate and is defined as:

$$\frac{\text{Total number of accidents}}{\text{Number of persons employed}} \times 1000$$

or the total number of accidents per 1000 employees.

A similar measure (per 100 000) is used by the HSE in its annual report on national accident statistics and enables comparisons to be made within an organization between time periods when employee numbers may change. It also allows comparisons to be made with the national occupational or industrial group relevant to the organization.

There are four main problems with this measure which must be borne in mind when it is used. These are:

➢ there may be a considerable variation over a time period in the ratio of part-time to full-time employees;
➢ the measure does not differentiate between major and minor accidents and takes no account of other incidents, such as those involving damage but no injury (although it is possible to calculate an incidence rate for a particular type or cause of accident);
➢ there may be significant variations in work activity during the periods being compared;
➢ under-reporting of accidents will affect the accuracy of the data.

Subject to the above limitations, an organization with a high accident incidence rate is likely to have a negative or poor health and safety culture.

There are other indications of a poor health and safety culture or climate. These include:

➢ a high sickness, ill health and absentee rate among the workforce;
➢ the perception of a blame culture;
➢ high staff turnover leading to a loss of momentum in making health and safety improvements;
➢ no resources (in terms of budget, people or facilities) made available for the effective management of health and safety;
➢ a lack of compliance with relevant health and safety law and the safety rules and procedures of the organization;
➢ poor selection procedures and management of contractors;

➢ poor levels of communication, cooperation and control;
➢ a weak health and safety management structure;
➢ either a lack or poor levels of health and safety competence;
➢ high insurance premiums.

In summary, a poor health and safety performance within an organization is an indication of a negative health and safety culture.

4.3.3 Factors affecting a health and safety culture

The most important factor affecting the culture is the commitment to health and safety from the top of an organization. This commitment may be shown in many different ways. It needs to have a formal aspect in terms of an organizational structure, job descriptions and a health and safety policy, but it also needs to be apparent during crises or other stressful times. The health and safety procedures may be circumvented or simply forgotten when production or other performance targets are threatened.

Structural reorganization or changes in market conditions will produce feelings of uncertainty among the workforce which, in turn, will affect the health and safety culture.

Poor levels of supervision, health and safety information and training are very significant factors in reducing health and safety awareness and, therefore, the culture.

Finally, the degree of consultation and involvement with the workforce in health and safety matters is crucial for a positive health and safety culture. Most of these factors may be summed up as human factors.

4.4 Human factors and their influence on safety performance

4.4.1 Human factors

Over the years, there have been several studies undertaken to examine the link between various accident types, graded in terms of their severity, and near misses. One of the most interesting was conducted in the USA by H. W. Heinrich in 1950. He looked at over 300 accidents/incidents and produced the ratios.

This study indicated that for every 10 near misses, there will be an accident. Although the accuracy of this study may be debated and other studies have produced different ratios, it is clear that if near misses are continually ignored, an accident will result. Further, the HSE Accident

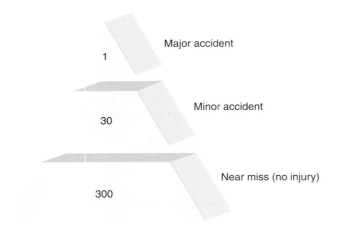

Figure 4.2 Heinrich's accidents/incidents ratios.

Prevention Unit has suggested that 90% of all accidents are due to human error and 70% of all accidents could have been avoided by earlier (proactive) action by management. It is clear from many research projects that the major factors in most accidents are human factors.

The HSE has defined human factors as, 'environmental, organizational and job factors, and human and individual characteristics which influence behaviour at work in a way which can affect health and safety'.

In simple terms in addition to the environment, the health and safety of people at work are influenced by:

- the organization
- the job
- personal factors.

These are known as human factors as they each have a human involvement. The personal factors which differentiate one person from another are only one part of those factors – and not always the most important.

Each of these elements will be considered in turn.

The organization

The organization is the company or corporate body and has the major influence on health and safety. It must have its own positive health and safety culture and produce an environment in which it:

- manages health and safety throughout the organization, including the setting and publication of a health and safety policy and the establishment of a health and safety organizational structure;
- measures the health and safety performance of the organization at all levels and in all departments. The performance of individuals should also be measured.

There should be clear health and safety targets and standards and an effective reporting procedure for accidents and other incidents so that remedial actions may be taken;
- motivates managers within the organization to improve health and safety performance in the workplace in a proactive rather than reactive manner.

The HSE has recommended that an organization needs to provide the following elements within its management system:

- a clear and evident commitment from the most senior manager downwards, which provides a climate for safety in which management's objectives and the need for appropriate standards are communicated and in which constructive exchange of information at all levels is positively encouraged;
- an analytical and imaginative approach identifying possible routes to human factor failure. This may well require access to specialist advice;
- procedures and standards for all aspects of critical work and mechanisms for reviewing them;
- effective monitoring systems to check the implementation of the procedures and standards;
- incident investigation and the effective use of information drawn from such investigations;
- adequate and effective supervision with the power to remedy deficiencies when found.

It is important to recognize that there are often reasons for these elements not being present, resulting in weak management of health and safety. The most common reason is that individuals within the management organization do not understand their roles – or their roles have never been properly explained to them. The higher a person is within the structure the less likely it is that he has received any health and safety training. Such training at board level is rare.

Objectives and priorities may vary across and between different levels in the structure, leading to disputes which affect attitudes to health and safety. For example, a warehouse manager may be pressured to block walkways so that a large order can be stored prior to dispatch.

Motivations can also vary across the organization, which may cause health and safety to be compromised. The production controller will require that components of a product are produced as near simultaneously as possible so that their final assembly is performed as quickly as possible. However, the health and safety adviser will not want to see safe systems of work compromised.

In an attempt to address some of these problems, the Health and Safety Commission (HSC) produced guidance in 2001 on the safety duties of company directors. Each director and the Board, acting collectively, will be expected to provide health and safety leadership in the organization. The Board will need to ensure that all its decisions reflect its health and safety intentions and that it engages the workforce actively in the improvement of health and safety. The Board will also be expected to keep itself informed of changes in health and safety risks. (See Chapter 3 for more details on directors' responsibilities.)

The following simple checklist may be used to check any organizational health and safety management structure.

Does the structure have:

➤ an effective health and safety management system?
➤ a positive health and safety culture?
➤ arrangements for the setting and monitoring of standards?
➤ adequate supervision?
➤ effective incident reporting and analysis?
➤ learning from experience?
➤ clearly visible health and safety leadership?
➤ suitable team structures?
➤ efficient communication systems and practices?
➤ adequate staffing levels?
➤ suitable work patterns?

HSG48 Reducing Error and Influencing Behaviour gives the following causes for failures in organizational and management structures:

➤ poor work planning leading to high work pressure
➤ lack of safety systems and barriers
➤ inadequate responses to previous incidents
➤ management based on one-way communications
➤ deficient coordination and responsibilities
➤ poor management of health and safety
➤ poor health and safety culture.

Organizational factors play a significant role in the health and safety of the workplace. However, this role is often forgotten when health and safety is being reviewed after an accident or when a new process or piece of equipment is introduced.

The job

Jobs may be highly dangerous or present only negligible risk of injury. Health and safety is an important element

during the design stage of the job and any equipment, machinery or procedures associated with the job. Method study helps to design the job in the most cost-effective way and ergonomics helps to design the job with health and safety in mind. Ergonomics is the science of matching equipment, machines and processes to people rather than the other way round. An ergonomically designed machine will ensure that control levers, dials, meters and switches are sited in a convenient and comfortable position for the machine operator. Similarly, an ergonomically designed workstation will be designed for the comfort and health of the operator. Chairs, for example, will be designed to support the back properly throughout the working day (Figure 4.3).

Physically matching the job and any associated equipment to the person will ensure that the possibility of human error is minimized. It is also important to ensure that there is mental matching of the person's information and decision-making requirements. A person must be capable, either through past experience or through specific training, of performing the job with the minimum potential for human error.

The major considerations in the design of the job, which would be undertaken by a specialist, have been listed by the HSE as follows:

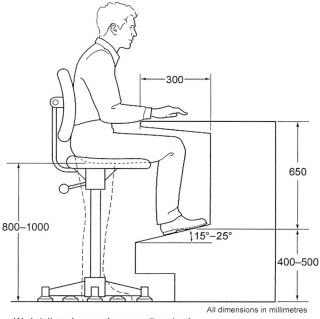

All dimensions in millimetres

Workstation where workers can sit or stand

Figure 4.3 Well-designed workstation for sitting or standing.

➤ the identification and detailed analysis of the critical tasks expected of individuals and the appraisal of any likely errors associated with those tasks;

➤ evaluation of the required operator decision making and the optimum (best) balance between the human and automatic contributions to safety actions (with the emphasis on automatic whenever possible);

➤ application of ergonomic principles to the design of man–machine interfaces, including displays of plant and process information, control devices and panel layout;

➤ design and presentation of procedures and operating instructions in the simplest terms possible;

➤ organization and control of the working environment, including the workspace, access for maintenance, lighting, noise and heating conditions;

➤ provision of the correct tools and equipment;

➤ scheduling of work patterns, including shift organization, control of fatigue and stress and arrangements for emergency operations;

➤ efficient communications, both immediate and over a period of time.

For some jobs, particularly those with a high risk of injury, a job safety analysis should be undertaken to check that all necessary safeguards are in place. All jobs should carry a job description and a safe system of work for the particular job. The operator should have sight of the job description and be trained in the safe system of work before commencing the job. More information on both these latter items is given in Chapter 6.

The following simple checklist may be used to check that the principal health and safety considerations of the job have been taken into account:

➤ Have the critical parts of the job been identified and analysed?

➤ Have the employee's decision-making needs been evaluated?

➤ Has the best balance between human and automatic systems been evaluated?

➤ Have ergonomic principles been applied to the design of equipment displays, including displays of plant and process information, control information and panel layouts?

➤ Has the design and presentation of procedures and instructions been considered?

➤ Has the guidance available for the design and control of the working environment, including the workspace, access for maintenance, lighting, noise and heating conditions, been considered?

➤ Have the correct tools and equipment been provided?

➤ Have the work patterns and shift organization been scheduled to minimize their impact on health and safety?

➤ Has consideration been given to the achievement of efficient communications and shift handover?

HSG48 gives the following causes for failures in job health and safety (Figure 4.4):

➤ illogical design of equipment and instruments;

➤ constant disturbances and interruptions;

➤ missing or unclear instructions;

➤ poorly maintained equipment;

➤ high workload;

➤ noisy and unpleasant working conditions.

It is important that health and safety monitoring of the job is a continuous process. Some problems do not become apparent until the job is started. Other problems do not surface until there is a change of operator or a change in some aspect of the job.

It is very important to gain feedback from the operator on any difficulties experienced because there could be a health and safety issue requiring further investigation.

Personal factors

Personal factors, which affect health and safety, may be defined as any condition or characteristic of an individual which could cause or influence him/her to act in an unsafe manner. They may be physical, mental or psychological in nature. Personal factors, therefore, include issues such as attitude, motivation, training and human error and their interaction with the physical, mental and perceptual capability of the individual.

Figure 4.4 Poor working conditions.

These factors have a significant effect on health and safety. Some of them, normally involving the personality of the individual, are unchangeable but others, involving skills, attitude, perception and motivation can be changed, modified or improved by suitable training or other measures. In summary, the person needs to be matched to the job.

Studies have shown that the most common personal factors which contribute to accidents are low skill and competence levels, tiredness, boredom, low morale and individual medical problems.

It is difficult to separate all the physical, mental or psychological factors because they are interlinked. However, the three most common factors are psychological factors – attitude, motivation and perception.

Attitude is the tendency to behave in a particular way in a certain situation. Attitudes are influenced by the prevailing health and safety culture within the organization, the commitment of the management, the experience of the individual and the influence of the peer group. Peer group pressure is a particularly important factor among young people and health and safety training must be designed with this in mind by using examples or case studies that are relevant to them. Behaviour may be changed by training, the formulation and enforcement of safety rules and meaningful consultation – attitude change often follows.

Motivation is the driving force behind the way a person acts or the way in which people are stimulated to act. Involvement in the decision-making process in a meaningful way will improve motivation as will the use of incentive schemes. However, there are other important influences on motivation such as recognition and promotion opportunities, job security and job satisfaction. Self-interest, in all its forms, is a significant motivator and personal factor.

Perception is the way in which people interpret the environment or the way in which a person believes or understands a situation (Figure 4.6). In health and safety, the perception of hazards is an important concern. Many accidents occur because people do not perceive that there is a risk. There are many common examples of this, including the use of personal protective equipment (PPE, such as hard hats) and guards on drilling machines and the washing of hands before meals. It is important to understand

Setting goals leads to vigorous activity if:

The goal is realistic.

A serious commitment is made, especially if it is made publicly.

Feedback is received.

Figure 4.5 Motivation and activity.

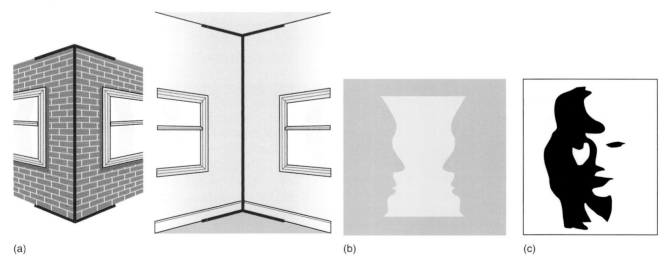

(a) (b) (c)

Figure 4.6 Visual perception. (a) Are the lines of the same length? (b) Faces or vase? (c) Faces or saxophone player?.

that when perception leads to an increased health and safety risk, it is not always caused by a conscious decision of the individual concerned. The stroboscopic effect caused by the rotation of a drill at certain speeds under fluorescent lighting will make the drill appear stationary. It is a well-known phenomenon, especially among illusionists, that people will often see what they expect to see rather than reality. Routine or repetitive tasks will reduce attention levels, leading to the possibility of accidents.

Other personal factors which can affect health and safety include physical stature, age, experience, health, hearing, intelligence, language, skills, level of competence and qualifications.

Memory is an important personal factor, as it is influenced by training and experience. The efficiency of memory varies considerably between people and during the lifetime of an individual. The overall health of a person can affect memory as can personal crises. Owing to these possible problems with memory, important safety instructions should be available in written as well as verbal form.

Finally, it must be recognised that some employees do not follow safety procedures either due to peer pressure or a wilful disregard of those procedures.

The following checklist given in HSG48 may be used to check that the relevant personal factors have been covered:

➤ Has the job specification been drawn up and included age, physique, skill, qualifications, experience, aptitude, knowledge, intelligence and personality?
➤ Have the skills and aptitudes been matched to the job requirements?
➤ Have the personnel selection policies and procedures been set up to select appropriate individuals?
➤ Has an effective training system been implemented?
➤ Have the needs of special groups of employees been considered?
➤ Have the monitoring procedures been developed for the personal safety performance of safety critical staff?
➤ Have fitness for work and health surveillance been provided where it is needed?
➤ Have counselling and support for ill health and stress been provided?

Personal factors are the attributes that employees bring to their jobs and may be strengths or weaknesses. Negative personal factors cannot always be neutralized by improved job design. It is, therefore, important to ensure that personnel selection procedures should match people to the job. This will reduce the possibility of accidents or other incidents.

Human errors and violations

Human failures in health and safety are classified either as errors or violations. An error is an unintentional deviation from an accepted standard, whereas a violation is a deliberate deviation from the standard (Figure 4.7).

4.5.1 Human errors

Human errors fall into three groups – slips, lapses and mistakes, which can be further sub-divided into rule-based and knowledge-based mistakes.

Slips and lapses
These are very similar in that they are caused by a momentary memory loss often due to lack of attention or loss of concentration. They are not related to levels of training, experience or motivation and they can usually be reduced by re-designing the job or equipment or minimizing distractions.

Slips are failures to carry out the correct actions of a task. Examples include the use of the incorrect switch, reading the wrong dial or selecting the incorrect component

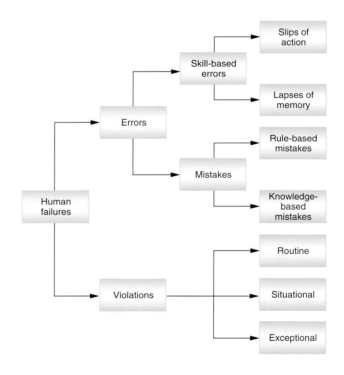

Figure 4.7 Types of human failure.

for an assembly. A slip also describes an action taken too early or too late within a given working procedure.

Lapses are failures to carry out particular actions which may form part of a working procedure. A fork-lift truck driver leaving the keys in the ignition lock of his truck is an example of a lapse as is the failure to replace the petrol cap on a car after filling it with petrol. Lapses may be reduced by re-designing equipment so that, for example, an audible horn indicates the omission of a task. They may also be reduced significantly by the use of detailed checklists.

Mistakes

Mistakes occur when an incorrect action takes place but the person involved believes the action to be correct. A mistake involves an incorrect judgement. There are two types of mistake – rule-based and knowledge-based.

Rule-based mistakes occur when a rule or procedure is remembered or applied incorrectly. These mistakes usually happen when, due to an error, the rule that is normally used no longer applies. For example, a particular job requires the counting of items into groups of ten followed by the adding together of the groups so that the total number of items may be calculated. If one of the groups is miscounted, the final total will be incorrect even though the rule has been followed.

Knowledge-based mistakes occur when well-tried methods or calculation rules are used inappropriately. For example, the depth of the foundations required for a particular building was calculated using a formula. The formula, which assumed a clay soil, was used to calculate the foundation depth in a sandy soil. The resultant building was unsafe.

The HSE has suggested the following points to consider when the potential source of human errors is to be identified:

➤ What human errors can occur with each task? There are formal methods available to help with this task.
➤ What influences are there on performance? Typical influences include time pressure, design of controls, displays and procedures, training and experience, fatigue and levels of supervision.
➤ What are the consequences of the identified errors? What are the significant errors?
➤ Are there opportunities for detecting each error and recovering it?
➤ Are there any relationships between the identified errors? Could the same error be made on more than one item of equipment due, for example, to the incorrect calibration of an instrument?

Errors and mistakes can be reduced by the use of instruction, training and relevant information. However, communication can also be a problem, particularly at shift handover times. Environmental and organizational factors involving workplace stress will also affect error levels.

The following steps are suggested to reduce the likelihood of human error:

➤ Examine and reduce the workplace stressors (e.g. noise, poor lighting) which increase the frequency of errors.
➤ Examine and reduce any social or organizational stressors (e.g. insufficient staffing levels, peer pressure).
➤ Design plant and equipment to reduce error possibilities – poorly designed displays, ambiguous instructions.
➤ Ensure that there are effective training arrangements.
➤ Simplify any complicated or complex procedures.
➤ Ensure that there is adequate supervision, particularly for inexperienced or young trainees.
➤ Check that job procedures, instructions and manuals are kept up to date and are clear.
➤ Include the possibility of human error when undertaking the risk assessment.
➤ Isolate the human error element of any accident or incident and introduce measures to reduce the risk of a repeat.
➤ Monitor the effectiveness of any measures taken to reduce errors.

4.5.2 Violations

There are three categories of violation – routine, situational and exceptional.

Routine violation occurs when the breaking of a safety rule or procedure is the normal way of working. It becomes routine not to use the recommended procedures for tasks. An example of this is the regular speeding of fork-lift trucks in a warehouse so that orders can be fulfilled on time.

There are many reasons given for routine violations; for example:

➤ taking short cuts to save time and energy;
➤ a belief that the rules are unworkable or too restrictive;
➤ lack of knowledge of the procedures;
➤ perception that the rules are no longer applied;
➤ poor supervision and a lack of enforcement of the rules;
➤ new workers thinking that routine violations are the norm and not realizing that this was not the safe way of working.

Finally, it must be recognized that there are some situations where peer pressure or simply a wilful disregard for procedures or other peoples' safety may result in routine violations. Routine violations can be reduced by regular monitoring, ensuring that the rules are actually necessary, or re-designing the job.

The following features are very common in many workplaces and often lead to routine violations:

➤ poor working posture due to poor ergonomic design of the workstation or equipment;

➤ equipment difficult to use and/or slow in response;

➤ equipment difficult to maintain or pressure on time available for maintenance;

➤ procedures unduly complicated and difficult to understand;

➤ unreliable instrumentation and/or warning systems;

➤ high levels of noise and other poor aspects to the environment (fumes, dusts, humidity);

➤ associated PPE either inappropriate, difficult and uncomfortable to wear or ineffective due to lack of maintenance.

Situational violations occur when particular job pressures at particular times make rule compliance difficult. They may happen when the correct equipment is not available or weather conditions are adverse. A common example is the use of a ladder rather than a scaffold for working at height to replace window frames in a building. Situational violations may be reduced by improving job design, the working environment and supervision.

Exceptional violations rarely happen and usually occur when a safety rule is broken to perform a new task. A good example are the violations which can occur during the operations of emergency procedures such as for fires or explosions. These violations should be addressed in risk assessments and during training sessions for emergencies (e.g. fire training).

Everybody is capable of making errors. It is one of the objectives of a positive health and safety culture to reduce them and their consequences as much as possible.

4.6 The development of a positive health and safety culture

No single section or department of an organization can develop a positive health and safety culture on its own. There needs to be commitment by the management, the promotion of health and safety standards, effective communication within the organization, cooperation from and with the workforce and an effective and developing training programme. Each of these topics will be examined in turn to show their effect on improving the health and safety culture in the organization.

4.6.1 Commitment of management

As mentioned earlier, there needs to be a commitment from the very top of the organization and this commitment will, in turn, produce higher levels of motivation and commitment throughout the organization. Probably, the best indication of this concern for health and safety is shown by the status given to health and safety and the amount of resources (money, time and people) allocated to health and safety. The management of health and safety should form an essential part of a manager's responsibility and they should be held to account for their performance on health and safety issues. Specialist expertise should be made available when required (e.g. for noise assessment), either from within the workforce or by the employment of external contractors or consultants. Health and safety should be discussed on a regular basis at management meetings at all levels of the organization. If the organization employs sufficient people to make direct consultation with all employees difficult, there should be a health and safety committee at which there is employee representation. In addition, there should be recognized routes for anybody within the organization to receive health and safety information or have their health and safety concerns addressed.

The health and safety culture is enhanced considerably when senior managers appear regularly at all levels of an organization whether it be the shop floor, the hospital ward or the general office and are willing to discuss health and safety issues with staff. A visible management is very important for a positive health and safety culture.

Finally, the positive results of a management commitment to health and safety will be the active involvement of all employees in health and safety, the continuing improvement in health and safety standards and the subsequent reduction in accident and occupational ill-health rates. This will lead, ultimately, to a reduction in the number and size of compensation claims.

HSG48 makes some interesting suggestions to managers on the improvements that may be made to health and safety which will be seen by the workforce as a clear indication of their commitment. The suggestions are:

➤ review the status of the health and safety committees and health and safety practitioners. Ensure that any recommendations are acted upon or implemented;

ensure that senior managers receive regular reports on health and safety performance and act on them;

ensure that any appropriate health and safety actions are taken quickly and are seen to have been taken;

any action plans should be developed in consultation with employees, based on a shared perception of hazards and risks, be workable and continually reviewed.

4.6.2 The promotion of health and safety standards

For a positive health and safety culture to be developed, everyone within the organization needs to understand the standards of health and safety expected by the organization and the role of the individual in achieving and maintaining those standards. Such standards are required to control and minimize health and safety risks.

Standards should clearly identify the actions required of people to promote health and safety. They should also specify the competencies needed by employees and should form the basis for measuring health and safety performance.

Health and safety standards cover all aspects of the organization. Typical examples include:

the design and selection of premises;

the design and selection of plant and substances (including those used on site by contractors);

the recruitment of employees and contractors;

the control of work activities, including issues such as risk assessment;

competence, maintenance and supervision;

emergency planning and training;

the transportation of the product and its subsequent maintenance and servicing.

Having established relevant health and safety standards, it is important that they are actively promoted within the organization by all levels of management. The most effective method of promotion is by leadership and example. There are many ways to do this such as:

the involvement of managers in workplace inspections and accident investigations;

the use of PPE (e.g. goggles and hard hats) by all managers and their visitors in designated areas;

ensuring that employees attend specialist refresher training courses when required (e.g. first aid and fork-lift truck driving);

full cooperation with fire drills and other emergency training exercises;

comprehensive accident reporting and prompt follow-up on recommended remedial actions.

The benefit of good standards of health and safety will be shown directly in less lost production, accidents and compensation claims, and lower insurance premiums. It may also be shown in higher product quality and better resource allocation.

An important and central necessity for the promotion of high health and safety standards is health and safety competence. What is meant by 'competence'?

Competence

The word 'competence' is often used in health and safety literature. One definition, made during a civil case in 1962, stated that a competent person is:

> *a person with practical and theoretical knowledge as well as sufficient experience of the particular machinery, plant or procedure involved to enable them to identify defects or weaknesses during plant and machinery examinations, and to assess their importance in relation to the strength and function of that plant and machinery.*

This definition concentrates on a manufacturing rather than service industry requirement of a competent person.

Regulation 7 of the Management of Health and Safety at Work Regulations requires that 'every employer shall employ one or more competent persons to assist him in undertaking the measures he needs to take to comply with the requirements and prohibitions imposed upon him by or under the relevant statutory provisions'. In other words, competent persons are required to assist the employer in meeting his obligations under health and safety law. This may mean a health and safety adviser in addition to, say, an electrical engineer, an occupational nurse and a noise assessment specialist. The number and range of competent persons will depend on the nature of the business of the organization.

It is recommended that competent employees are used for advice on health and safety matters rather than external specialists (consultants). It is recognized, however, that if employees, competent in health and safety, are not available in the organization, then an external service may be enlisted to help. The key is that management and employees need access to health and safety expertise.

The regulations do not define 'competence' but do offer advice to employers when appointing health and safety advisers who should have:

➤ a knowledge and understanding of the work involved, the principles of risk assessment and prevention and current health and safety applications;

➤ the capacity to apply this to the task required by the employer in the form of problem and solution identification, monitoring and evaluating the effectiveness of solutions and the promotion and communication of health and safety and welfare advances and practices.

Such competence does not necessarily depend on the possession of particular skills or qualifications. It may only be required to understand relevant current best practice, be aware of one's own limitations in terms of experience and knowledge and be willing to supplement existing experience and knowledge. However, in more complex or technical situations, membership of a relevant professional body and/or the possession of an appropriate qualification in health and safety may be necessary. It is important that any competent person employed to help with health and safety has evidence of relevant knowledge, skills and experience for the tasks involved. The appointment of a competent person as an adviser to an employer does not absolve the employer from his responsibilities under the HSW Act and other relevant statutory provision.

Finally, it is worth noting that the requirement to employ competent workers is not restricted to those having a health and safety function but covers the whole workforce.

Competent workers must have sufficient training, experience, knowledge and other qualities to enable them to properly undertake the duties assigned to them.

Effective communication

Many problems in health and safety arise due to poor communication. It is not just a problem between management and workforce – it is often a problem the other way or indeed at the same level within an organization. It arises from ambiguities or, even, accidental distortion of a message.

The Health and Safety (Information for Employees) Regulations require that the latest version of the approved health and safety poster be displayed prominently in the workplace. Additional information on these regulations is given in Chapter 1.

There are three basic methods of communication in health and safety – verbal, written and graphic.

Verbal communication is the most common. It is communication by speech or word of mouth. Verbal communication should only be used for relatively simple pieces of information or instruction. It is most commonly used in the workplace, during training sessions or at meetings.

There are several potential problems associated with verbal communication. The speaker needs to prepare the communication carefully so that there is no confusion about the message. It is very important that the recipient is encouraged to indicate their understanding of the communication. There have been many cases of accidents occurring because a verbal instruction has not been understood. There are several barriers to this understanding from the point of view of the recipient, including language and dialect, the use of technical language and abbreviations, background noise and distractions, hearing problems, ambiguities in the message, mental weaknesses and learning disabilities, and lack of interest and attention.

Having described some of the limitations of verbal communication, it does have some merits. It is less formal, enables an exchange of information to take place quickly and the message to be conveyed as near to the workplace as possible. Training or instructions that are delivered in this way are called toolbox talks and can be very effective.

Written communication takes many forms from the simple memo to the detailed report.

A memo should contain one simple message and be written in straightforward and clear language. The title should accurately describe the contents of the memo. In recent years, e-mails have largely replaced memos, as it has become a much quicker method to ensure that the message gets to all concerned (although a recent report has suggested that many people are becoming overwhelmed by the number of e-mails which they receive!). The advantage of memos and e-mails is that there is a record of the message after it has been delivered. The disadvantage is that they can be ambiguous or difficult to understand or, indeed, lost within the system.

Reports are more substantial documents and cover a topic in greater detail. The report should contain a detailed account of the topic and any conclusions or recommendations. The main problem with reports is that they are often not read properly due to the time constraints on managers. It is important that all reports have a summary attached so that the reader can decide whether it needs to be read in detail (see Chapter 7 – 7.9).

The most common way in which written communication is used in the workplace is the **notice board**. For a notice board to be effective, it needs to be well positioned within the workplace and there needs to be a regular review of the notices to ensure that they are up to date and relevant.

In addition to the health and safety poster, mentioned earlier, the following types of health and safety information could be displayed on a workplace notice board:

➤ a copy of the Employer's Liability Insurance Certificate;
➤ details of first-aid arrangements;
➤ emergency evacuation and fire procedures;
➤ minutes of the last health and safety committee meeting;
➤ details of health and safety targets and performance against them;
➤ health and safety posters and campaign details.

There are many other examples of written communications in health and safety, such as employee handbooks, company codes of practice, minutes of safety committee meetings and health and safety procedures.

Graphic communication is communication by the use of drawings, photographs or DVDs. It is used to impart either health and safety information (e.g. fire exits) or health and safety propaganda. The most common forms of health and safety propaganda are the poster and the DVD. Both can be used very effectively as training aids, as they can retain interest and impart a simple message. Their main limitation is that they can become out of date fairly quickly or, in the case of posters, become largely ignored.

There are many sources of health and safety information which may need to be consulted before an accurate communication can be made. These include regulations, judgements, Approved Codes of Practice, guidance, British and European standards, periodicals, case studies and HSE publications.

4.8 Health and safety training

The provision of information and training for employees will develop their awareness and understanding of the specific hazards and risks associated with their jobs and working environment. It will inform them of the control measures that are in place and any related safe procedures that must be followed. Apart from satisfying legal obligations, several benefits will accrue to the employer by the provision of sound information and training to employees. These benefits include:

➤ a reduction in accident severity and frequency;
➤ a reduction in injury and ill-health related absence;
➤ a reduction in compensation claims and, possibly, insurance premiums;
➤ an improvement in the health and safety culture of the organization;
➤ improved staff morale and retention.

Health and safety training is a very important part of the health and safety culture and it is also a legal requirement, under the Management of Health and Safety at Work Regulations and other regulations, for an employer to provide such training. Training is required on recruitment, at induction or on being exposed to new or increased risks due to:

➤ being transferred to another job or given a change in responsibilities;
➤ the introduction of new work equipment or a change of use in existing work equipment;
➤ the introduction of new technology;
➤ the introduction of a new system of work or the revision of an existing system of work;
➤ an increase in the employment of more vulnerable employees (young or disabled persons);
➤ particular training required by the organization's insurance company (e.g. specific fire and emergency training).

Additional training may well be needed following a single or series of accidents or near misses, the introduction of new legislation, the issuing of an enforcement notice or as a result of a risk assessment or safety audit (Figure 4.8).

It is important during the development of a training course that the target audience is taken into account. If the audience are young persons, the chosen approach must be capable of retaining their interest and any illustrative examples used must be within their experience. The trainer must also be aware of external influences, such as peer pressures, and use them to advantage. For example, if everybody wears PPE then it will be seen as the thing to do. Levels of literacy and numeracy are other important factors.

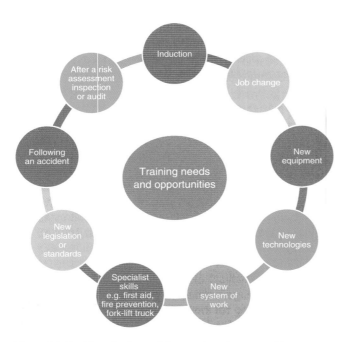

Figure 4.8 Health and safety training needs and opportunities.

The way in which the training session is presented by the use of DVDs, powerpoint slides, case studies, lectures or small discussion groups needs to be related to the material to be covered and the backgrounds of the trainees. Supplementary information in the form of copies of slides and additional background reading is often useful. The environment used for the training sessions is also important in terms of room layout and size, lighting and heating.

Attempts should be made to measure the effectiveness of the training by course evaluation forms issued at the time of the session, by a subsequent refresher session and by checking for improvements in health and safety performance (such as a reduction in specific accidents).

There are several different types of training; these include induction, job specific, supervisory and management, and specialist. Informal sessions held at the place of work are known as 'toolbox talks'. Such sessions should only be used to cover a limited number of issues. They can become a useful route for employee consultation.

4.8.1 Induction training

Induction training should always be provided to new employees, trainees and contractors. While such training

covers items such as pay, conditions and quality, it must also include health and safety. It is useful if the employee signs a record to the effect that training has been received. This record may be required as evidence should there be a subsequent legal claim against the organization.

Most induction training programmes would include the following topics:

- ➤ the health and safety policy of the organization including a summary of the organization and arrangements including employee consultation;
- ➤ a brief summary of the health and safety management system including the name of the employee's direct supervisor, safety representative and source of health and safety information;
- ➤ the employee responsibility for health and safety including any general health and safety rules (e.g. smoking prohibitions);
- ➤ the accident reporting procedure of the organization, the location of the accident book and the location of the nearest first aider;
- ➤ the fire and other emergency procedures including the location of the assembly point;
- ➤ the hazards that are specific to the workplace;
- ➤ a summary of any relevant risk assessments and safe systems of work;
- ➤ the location of welfare, canteen facilities and rest rooms;
- ➤ procedures for reporting defects or possible hazards and the name of the responsible person to whom the report should be made;
- ➤ details of the possible disciplinary measures that may be enacted for non-compliance with health and safety rules and procedures.

Additional items which are specific to the organization may need to be included such as:

- ➤ internal transport routes and pedestrian walkways (e.g. fork-lift truck operations);
- ➤ the correct use of PPE and maintenance procedures;
- ➤ manual handling techniques and procedures;
- ➤ details of any hazardous substances in use and any procedures relating to them (e.g. health surveillance).

There should be some form of follow-up with each new employee after 3 months to check that the important messages have been retained. This is sometimes called a refresher course, although it is often better done on a one-to-one basis.

It is very important to stress that the content of the induction course should be subject to constant review and updated following an accident investigation, new legislation, changes in the findings of a risk assessment or the introduction of new plant or processes.

4.8.2 Job-specific training

Job-specific training ensures that employees undertake their job in a safe manner. Such training, therefore, is a form of skill training and is often best done 'on the job' – sometimes known as 'toolbox training'. Details of the safe system of work or, in more hazardous jobs, a permit to work system, should be covered. In addition to normal safety procedures, emergency procedures and the correct use of PPE also need to be included. The results of risk assessments are very useful in the development of this type of training. It is important that any common causes of human errors (e.g. discovered as a result of an accident investigation), any standard safety checks or maintenance requirements, are addressed.

It is common for this type of training to follow an operational procedure in the form of a checklist which the employee can sign on completion of the training. The new employee will still need close supervision for some time after the training has been completed.

4.8.3 Supervisory and management training

Supervisory and management health and safety training follows similar topics to those contained in an induction training course but will be covered in more depth. There will also be a more detailed treatment of health and safety law. There has been considerable research over the years into the failures of managers that have resulted in accidents and other dangerous incidents. These failures have included:

➤ lack of health and safety awareness, enforcement and promotion (in some cases, there has been encouragement to circumvent health and safety rules);
➤ lack of consistent supervision of and communication with employees;
➤ lack of understanding of the extent of the responsibility of the supervisor.

It is important that all levels of management, including the Board, receive health and safety training. This will not only keep everybody informed of health and safety legal requirements, accident prevention techniques and changes in the law, but also encourage everybody to monitor health and safety standards during visits or tours of the organization.

4.8.4 Specialist training

Specialist health and safety training is normally needed for activities that are not related to a specific job but more to an activity. Examples include first-aid, fire prevention, fork-lift truck driving, overhead crane operation, scaffold inspection and statutory health and safety inspections. These training courses are often provided by specialist organizations and successful participants are awarded certificates. Details of two of these courses will be given here by way of illustration.

Fire prevention training courses include the causes of fire and fire spread, fire and smoke alarm systems, emergency lighting, the selection and use of fire extinguishers and sprinkler systems, evacuation procedures, high-risk operations and good housekeeping principles.

A fork-lift truck drivers' course would include the general use of the controls, loading and unloading procedures, driving up or down an incline, speed limits, pedestrian awareness (particularly in areas where pedestrians and vehicles are not segregated), security of the vehicle when not in use, daily safety checks and defect reporting, refuelling and/or battery charging and emergency procedures.

Details of other types of specialist training appear elsewhere in the text.

Training is a vital part of any health and safety programme and needs to be constantly reviewed and updated. Many health and safety Regulations require specific training (e.g. manual handling, PPE and display screens). Additional training courses may be needed when there is a major reorganization, a series of similar accidents or incidents, or a change in equipment or a process. Finally, the methods used to deliver training must be continually monitored to ensure that they are effective.

A list of typical legislation that requires some form of health and safety training is given in Appendix 4.1. The list has been compiled to cover the needs of most small- to medium-sized organizations.

4.9 Internal influences

There are many influences on health and safety standards, some are positive and others negative (Figure 4.9). No business, particularly small businesses, is totally divorced from their suppliers, customers and neighbours. This section considers the internal influences on a business, including management commitment, production demands, communication, competence and employee relations.

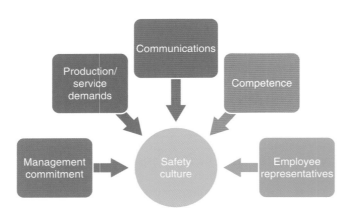

Figure 4.9 INTERNAL influences on safety culture.

4.9.1 Management commitment

Managers, particularly senior managers, can give powerful messages to the workforce by what they do for health and safety. Managers can achieve the level of health and safety performance that they demonstrate they want to achieve. Employees soon get the negative message if directors disregard safety rules and ignore written policies to get urgent production to the customer or to avoid personal inconvenience. It is what they do that counts not what they say. In Chapter 3, managers' organizational roles were listed showing the ideal level of involvement for senior managers. Depending on the size and geography of the organization, senior managers should be personally involved in:

➤ health and safety inspections or audits;
➤ meetings of the central health and safety or joint consultation committees;
➤ involvement in the investigation of accidents, ill health and incidents. The more serious the incident the more senior the manager who takes an active part in the investigation.

4.9.2 Production/service demands

Managers need to balance the demands placed by customers with the action required to protect the health and safety of their employees. How this is achieved has a strong influence on the standards adopted by the organization. The delivery driver operating to near impossible delivery schedules and the manager agreeing to the strident demands of a large customer regardless of the risks to employees involved are typical dilemmas facing workers and managers alike. The only way to deal with these issues is to be well organized and agree with the client/employees ahead of the crisis as to how they should be prioritized.

Rules and procedures should be intelligible, sensible and reasonable. They should be designed to be followed under normal production or service delivery conditions. If following the rules involves very long delays or impossible production schedules, they should be revised rather than ignored by workers and managers alike, until it is too late and an accident occurs. Sometimes the safety rules are simply used in a court of law, in an attempt to defend the company concerned, following an accident.

Managers who plan the impossible schedule or ignore a safety rule to achieve production or service demands are held responsible for the outcomes. It is acceptable to balance the cost of the action against the level of risk being addressed but it is never acceptable to ignore safety rules or standards simply to get the work done. Courts are not impressed by managers who put profit considerations ahead of safety requirements.

4.9.3 Communication

Communications were covered in depth earlier in the chapter and clearly will have significant influence on health and safety issues. This will include:

➤ poorly communicated procedures that will not be understood or followed;
➤ poor verbal communications which will be misunderstood and will demonstrate a lack of interest by senior managers;
➤ missing or incorrect signs which may cause accidents rather than prevent them;
➤ managers who are nervous about face-to-face discussions with the workforce on health and safety issues, which will have a negative effect.

Managers and supervisors should plan to have regular discussions to learn about the problems faced by employees and discuss possible solutions. Some meetings, like that of the safety committee, are specifically planned for safety matters, but this should be reinforced by discussing health and safety issues at all routine management meetings. Regular one-to-one talks should also take place in the workplace, preferably to a planned theme or safety topic, to get specific messages across and get feedback from employees.

4.9.4 Competence

Competent people, who know what they are doing and have the necessary skills to do the task correctly and safely, will make the organization effective. Competence can be bought in through recruitment or consultancy but it is often much more effective to develop it among employees. It demonstrates commitment to health and safety and a sense of security for the workforce. The loyalty that it creates in the workforce can be a significant benefit to safety standards. Earlier in this chapter, there is more detail on how to achieve competence.

4.9.5 Employee representation

Given the resources and freedom to fulfil their function effectively, enthusiastic, competent employee safety representatives can make a major contribution to good health and safety standards. They can provide the essential bridge between managers and employees. People are more willing to accept the restrictions that some precautions bring if they are consulted and feel involved, either directly in small workforces or through their safety representatives. See Chapter 3 (Section 3.13) for the details.

4.10 External influences

The role of external organizations is set out in Chapter 1. Here the influence they can have is briefly discussed, including societal expectations, legislation and enforcement, insurance companies, trade unions, economics, commercial stakeholders (Figure 4.10).

4.10.1 Societal expectations

Societal expectations are not static and tend to rise over time, particularly in a wealthy nation like the UK. For example, the standards of safety accepted in a motor car 50 years ago would be considered to be totally inadequate at the beginning of the 21st century. We expect safe, quiet, comfortable cars that do not break down and which retain their appearance for many thousands of miles. Industry should strive to deliver these same high standards for the health and safety of employees or service providers. The question is whether societal expectations are as great an influence on workplace safety standards as they are on product safety standards. Society can influence standards through:

➤ people only working for good employers. This is effective in times of low unemployment;
➤ national and local news media highlighting good and bad employment practices;

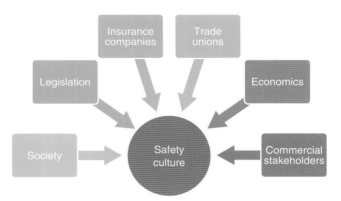

Figure 4.10 EXTERNAL influences on safety culture.

➤ schools teaching good standards of health and safety;
➤ the purchase of fashionable and desirable safety equipment, such as trendy crash helmets for mountain bikes;
➤ buying products only from responsible companies. The difficulty of defining what is responsible has been partly overcome through ethical investment criteria but this is possibly not widely enough understood to be a major influence;
➤ watching TV and other programmes which improve safety knowledge and encourage safe behaviour from an early age.

4.10.2 Legislation and enforcement

Good comprehensible legislation should have a positive effect on health and safety standards. Taken together, legislation and enforcement can affect standards by:

➤ providing a level to which every employer has to conform;
➤ insisting on minimum standards which also enhances peoples' ability to operate and perform well;
➤ providing a tough, visible threat of getting shut down or a heavy fine;
➤ stifling development by being too prescriptive; for example, woodworking machines did not develop quickly in the 20th century, partly because the Regulations were so detailed in their requirements that new designs were not feasible;
➤ providing well presented and easily read guidance for specific industries at reasonable cost or free.

On the other hand, a weak enforcement regime can have a powerful negative effect on standards.

4.10.3 Insurance companies

Insurance companies can influence health and safety standards mainly through financial incentives. Employers' liability insurance is a legal requirement in the UK and therefore all employers have to obtain this type of insurance cover. Insurance companies can influence standards through:

➤ discounting premiums to those in the safest sectors or best individual companies;
➤ insisting on risk reduction improvements to remain insured. This is not very effective where competition for business is fierce;
➤ encouraging risk reduction improvements by bundling services into the insurance premium;
➤ providing guidance on standards at reasonable cost or free.

4.10.4 Trade unions

Trade unions can influence standards by:

➤ providing training and education for members;
➤ providing guidance and advice cheaply or free to members;
➤ influencing governments to regulate, enhance enforcement activities and provide guidance;
➤ influencing employers to provide high standards for their members. This is sometimes confused with financial improvements with health and safety getting a lower priority;
➤ encouraging members to work for safer employers;
➤ helping members to get proper compensation for injury and ill health if it is caused through their work.

4.10.5 Economics

Economics can play a major role in influencing health and safety standards. The following ways are the most common:

➤ lack of orders and/or money can cause employers to try to ignore health and safety requirements;
➤ if employers were really aware of the actual and potential cost of accidents and fires, they would be more concerned about prevention. The HSE believes that the ratio between insured and uninsured costs of accidents is between 1:8 and 1:36 (see *The Costs of Accidents at Work*, HSE Books, HSG96;
➤ perversely, when the economy is booming activity increases and, particularly in the building industry, accidents can sharply increase. The pressures to perform and deliver for customers can be safety averse;

➤ businesses that are only managed on short-term performance indicators seldom see the advantage of the long-term gains that are possible with a happy, safe and fit workforce.

The cost of accidents and ill health, in both human and financial terms, needs to be visible throughout the organization so that all levels of employee are encouraged to take preventative measures.

4.10.6 Commercial stakeholders

A lot can be done by commercial stakeholders' influence standards. This includes:

➤ insisting on proper arrangements for health and safety management at supplier companies before they tender for work or contracts;
➤ checking on suppliers to see if the workplace standards are satisfactory;
➤ encouraging ethical investments;
➤ considering ethical standards as well as financial when banks provide funding;
➤ providing high-quality information for customers;
➤ insisting on high standards to obtain detailed planning permission (where this is possible);
➤ providing low-cost guidance and advice.

4.11 Sources of reference

The Management of Health and Safety at Work (ACOP) L21, HSE Books 2000 ISBN 978 0 7176 2488 1.
Successful Health and Safety Management HSG65, HSE Books, 1997 ISBN 978 0 7176 1276 5.
Reducing Error and Influencing Behaviour HSG 48, HSE Books, 1999, ISBN 978 0 7176 2452 2.

Relevant statutory provisions
The Health and Safety at Work etc. Act 1974.
The Management of Health and Safety at Work Regulations 1999.
The Health and Safety Information for Employees Regulations 1989.

4.12 Practice NEBOSH questions for Chapter 4

1. **Describe FIVE** components of a positive health and safety culture.

2. **(i) Define** the term 'accident incidence rate'.
(ii) Outline how information on accidents could be used to promote health and safety in the workplace.

3. **Outline** the personal factors that might place an individual at a greater risk of harm while at work.

4. **(i) Explain**, using an example, the meaning of the term 'attitude'.
(ii) Outline THREE influences on the attitude towards health and safety of employees within an organization.
(iii) Outline the ways in which employers might motivate their employees to comply with health and safety procedures.

5. **(i) Explain** the meaning of the term 'motivation'.
(ii) Other than lack of motivation, **outline SIX** reasons why employees may fail to comply with safety procedures at work.
(iii) Outline ways in which employers may motivate their employees to comply with health and safety procedures.

6. **(i) Explain** the meaning of the term 'perception'.
(ii) Outline the factors relating to the individual that may influence a person's perception of an occupational risk.
(iii) Outline ways in which employees' perceptions of hazards in the workplace might be improved.

7. With reference to HSG48 'Reducing error and influencing behaviour', **using** practical examples, **define** the terms 'error' and 'violation'.

8. **(i) Giving** a practical example, **explain** the meaning of the term 'human error'.
(ii) Outline *individual (or personal)* factors that may contribute to human errors occurring at work.

9. **Describe**, using practical examples, **FOUR** types of human error that can lead to accidents in the workplace.

10. **(i) Outline** ways of reducing the likelihood of human error in the workplace.
(ii) Give FOUR reasons why the seriousness of a hazard may be underestimated by someone exposed to it.
(iii) Outline ways in which managers can motivate employees to work safely.

11. **Outline** the ways in which the health and safety culture of an organization might be improved.

12. **Outline** the factors that might cause the safety culture within an organization to decline.

13. **Outline** the practical means by which a manager could involve employees in the improvement of health and safety in the workplace.

14. Following a significant increase in accidents, a health and safety campaign is to be launched within an organization to encourage safer working by employees.
(i) Outline how the organization might ensure that the nature of the campaign is effectively communicated to, and understood by, employees.
(ii) Explain why it is important to use a variety of methods to communicate health and safety information in the workplace.
(iii) Other than poor communications, **describe** the organizational factors that could limit the effectiveness of the campaign.

15. **Give** reasons why a verbal instruction may not be clearly understood by an employee.

16. Non-compliance with safety procedures by employees has been identified as one of the possible causes of a serious accident at work. **Outline** reasons why the safety procedures may not have been followed.

17. A new process was introduced into a workplace. Operators, supervisors and managers have received information and training on a safe system of work associated with the process.
(i) Outline how provision of *information* and *training* for the employee contributes to controlling and reducing risks.
(ii) Identify FOUR benefits to an employer of providing information and training.

18. **Outline FOUR** advantages **AND FOUR** disadvantages of using 'propaganda' posters to communicate health and safety information to the workforce.

19. **(i) Identify FOUR** types of health and safety information that might usefully be displayed on a notice board within a workplace.
(ii) Explain how the effectiveness of notice boards as a means of communicating health and safety information to the workforce can be maximized.

20. **Identify** a range of methods that an employer can use to provide health and safety information directly to individual employees.

21. **(i)** **Identify** the **TWO** means by which employers may provide information to employees in order to comply with the Health and Safety Information for Employees Regulations 1989.
    (ii) **Outline** the categories of information provided to employees by the means identified in (a).

22. **Explain** how induction training programmes for new employees can help to reduce the number of accidents in the workplace.

23. A contractor has been engaged to undertake building maintenance work in a busy warehouse. **Outline** the issues that should be covered in an induction programme for the contractor's employees.

24. **Outline** the content of an induction training programme for new employees designed to reduce the risk of accidents.

25. An independent audit of an organization has concluded that employees have received insufficient health and safety training.

    (i) **Describe** the factors that should be considered when developing an extensive programme of health and safety training within an organization.
    (ii) **Outline** the various measures that might be used to assess the effectiveness of such training.
    (iii) **Give FOUR** reasons why it is important for an employer to keep a record of the training provided to each employee.

26. **Outline** reasons why an employee might require additional health and safety training at a later stage of employment within an organization.

27. **Describe TWO** internal **AND TWO** external influences on the health and safety culture of an organization.

28. **Identify** *external* influences that may affect health and safety management within an organization.

Appendix 4.1 List of typical legislation requiring health and safety training

Legislation

1. Management of Health and Safety at Work Regulations
2. Regulatory Reform (Fire Safety) Order 2004
3. Provision and Use of Work Equipment Regulations
4. Control of Substances Hazardous to Health Regulations
5. Health and Safety (First Aid) Regulations
6. Health and Safety (Display Screen Equipment) Regulations
7. Manual Handling Operations Regulations
8. Control of Noise at Work Regulations
9. Personal Protective Equipment at Work Regulations
10. Health and Safety (Safety Signs and Signals) Regulations
11. Control of Asbestos Regulations
12. Confined Spaces Regulations
13. Health and Safety (Consultation with Employees) Regulations
14. Control of Vibration at Work Regulations.

Risk assessment

5.1 Introduction

Risk assessment is an essential part of the planning stage of any health and safety management system. HSE, in the publication HSG65 *Successful Health and Safety Management*, states that the aim of the planning process is to minimize risks.

Risk assessment methods are used to decide on priorities and to set objectives for eliminating hazards and reducing risks. Wherever possible, risks are eliminated through selection and design of facilities, equipment and processes. If risks cannot be eliminated, they are minimized by the use of physical controls or, as a last resort, through systems of work and personal protective equipment.

5.2 Legal aspects of risk assessment

The general duties of employers to their employees in section 2 of the HSW Act 1974 imply the need for risk assessment. This duty was also extended by section 3 of the Act to anybody else affected by activities of the employer – contractors, visitors, customers or members of the public. However, the Management of Health and Safety at Work Regulations are much more specific concerning the need for risk assessment. The following requirements are laid down in those Regulations:

the risk assessment shall be 'suitable and sufficient' and cover both employees and non-employees affected by the employer's undertaking (e.g. contractors, members of the public, students, patients, customers); every self-employed person shall make a 'suitable and sufficient' assessment of the risks to which they or those affected by the undertaking may be exposed;

any risk assessment shall be reviewed if there is reason to suspect that it is no longer valid or if a significant change has taken place;

where there are five or more employees, the significant findings of the assessment shall be recorded and any specially at risk group of employees identified. (This does not mean that employers with four or less employees need not undertake risk assessments.)

The term 'suitable and sufficient' is important as it defines the limits to the risk assessment process. A suitable and sufficient risk assessment should:

➤ identify the significant risks and ignore the trivial ones;
➤ identify and prioritize the measures required to comply with any relevant statutory provisions;
➤ remain appropriate to the nature of the work and valid over a reasonable period of time;
➤ identify the risk arising from or in connection with the work. The level of detail should be proportionate to the risk.

The significant findings that should be recorded include a detailed statement of the hazards and risks, the preventative, protective or control measures in place and any further measures required to reduce the risks present (Figure 5.1).

When assessing risks under the Management of Health and Safety at Work Regulations, reference to other Regulations may be necessary even if there is no specific requirement for a risk assessment in those Regulations. For example, reference to the legal requirements of the Provision and Use of Work Equipment Regulations will be necessary when risks from the operation of machinery are being considered. However, there is no need to repeat a risk assessment if it is already covered by other Regulations (e.g. a risk assessment considering personal protective equipment is required under the COSHH Regulations so there is no need to undertake a separate risk assessment under the Personal Protective Equipment Regulations).

Apart from the duty under the Management of Health and Safety at Work Regulations to undertake a health and safety assessment of the risks to any person (employees, contractors or members of the public), who may be affected by the activities of the organization, the following Regulations require a specific risk assessment to be made:

➤ Ionising Radiation Regulations;
➤ Control of Asbestos Regulations;
➤ the Control of Noise at Work Regulations;
➤ Manual Handling Operations Regulations;
➤ Health and Safety (Display Screen Equipment) Regulations;
➤ the Personal Protective Equipment at Work Regulations;
➤ the Confined Spaces Regulations;
➤ Work at Height Regulations;
➤ the Regulatory Reform (Fire Safety) Order;
➤ the Control of Vibration at Work Regulations;
➤ Control of Lead at Work Regulations;
➤ Control of Substances Hazardous to Health Regulations.

A detailed comparison of the risk assessments required for most of these and more specialist Regulations was given in the HSE Guide to Risk Assessment Requirements, INDG218.

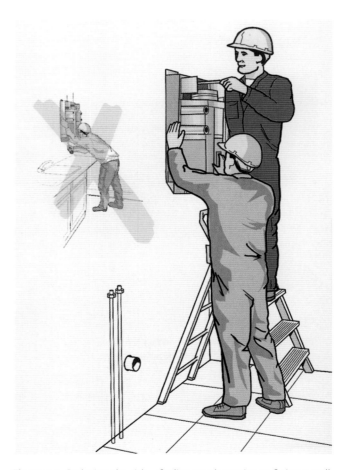

Figure 5.1 Reducing the risk – finding an alternative to fitting a wall-mounted boiler.

5.3 Forms of risk assessment

There are two basic forms of risk assessment.

A quantitative risk assessment attempts to measure the risk by relating the probability of the risk occurring to the possible severity of the outcome and then giving the risk a numerical value. This method of risk assessment is used in situations where a malfunction could be very serious (e.g. aircraft design and maintenance or the petrochemical industry).

The more common form of risk assessment is the qualitative assessment, which is based purely on personal judgement and is normally defined as high, medium or low. Qualitative risk assessments are usually satisfactory as the definition (high, medium or low) is normally used to determine the time frame in which further action is to be taken.

The term 'generic' risk assessment is sometimes used and describes a risk assessment which covers similar activities or work equipment in different departments, sites or companies. Such assessments are often produced by specialist bodies, such as trade associations. If used, they must be appropriate to the particular job and they will need to be extended to cover additional hazards or risks.

5.4 Some definitions

Some basic definitions were introduced in Chapter 1 and those relevant to risk assessment are reproduced here.

5.4.1 Hazard and risk

A hazard is the *potential* of a substance, activity or process to cause harm. Hazards take many forms including, for example, chemicals, electricity and the use of a ladder. A hazard can be ranked relative to other hazards or to a possible level of danger.

A risk is the *likelihood* of a substance, activity or process to cause harm. Risk (or strictly the level of risk) is also linked to the severity of its consequences. A risk can be reduced and the hazard controlled by good management.

It is very important to distinguish between a hazard and a risk – the two terms are often confused and activities often called high risk are in fact high hazard. There should only be high residual risk where there is poor health and safety management and inadequate control measures.

Electricity is an example of a high hazard as it has the potential to kill a person. The risk associated with electricity – the likelihood of being killed on coming into contact with an electrical device – is, hopefully, low.

5.4.2 Occupational or work-related ill health

This is concerned with those illnesses or physical and mental disorders that are either caused or triggered by workplace activities. Such conditions may be induced by the particular work activity of the individual or by activities of others in the workplace. The time interval between exposure and the onset of the illness may be short (e.g. asthma attacks) or long (e.g. deafness or cancer).

5.4.3 Accident

This is defined by the HSE as 'any unplanned event that results in injury or ill health of people, or damage or loss to property, plant, materials or the environment or a loss of a business opportunity'. Other authorities define an accident more narrowly by excluding events that do not involve injury or ill health. This book will always use the HSE definition (Figure 5.2).

5.4.4 Near miss

This is any incident that could have resulted in an accident. Knowledge of near misses is very important as research has shown that, approximately, for every 10 'near miss' events at a particular location in the workplace, a minor accident will occur.

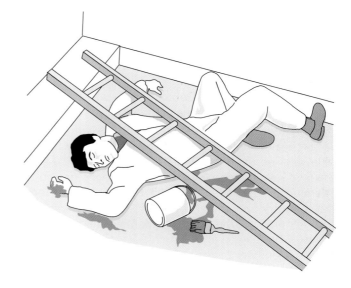

Figure 5.2 Accident at work.

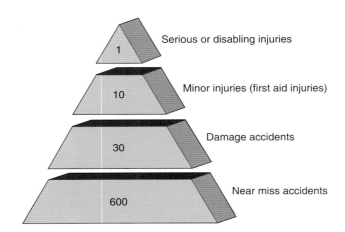

Figure 5.3 F. E. Bird's well-known accident triangle.

5.4.5 *Dangerous occurrence*

This is a 'near miss' which could have led to serious injury or loss of life. Dangerous occurrences are defined in the Reporting of Injuries, Diseases and Dangerous Occurrences Regulations 1995 (often known as RIDDOR) and are always reportable to the Enforcement Authorities. Examples include the collapse of a scaffold or a crane or the failure of any passenger-carrying equipment.

In 1969, F. E. Bird collected a large quantity of accident data and produced a well-known triangle (Figure 5.3).

It can be seen that damage and near miss accidents occur much more frequently than injury accidents and are, therefore, a good indicator of risks. The study also shows that most accidents are predictable and avoidable.

5.5   The objectives of risk assessment

The main objective of risk assessment is to determine the measures required by the organization to comply with relevant health and safety legislation and, thereby, reduce the level of occupational injuries and ill health. The purpose is to help the employer or self-employed person to determine the measures required to comply with their legal statutory duty under the HSW Act 1974 or its associated Regulations. The risk assessment will need to cover all those who may be at risk, such as customers, contractors and members of the public. In the case of shared workplaces, an overall risk assessment may be needed in partnership with other employers.

In Chapter 1, the moral, legal and financial arguments for health and safety management were discussed in detail. The important distinction between the direct and indirect costs of accidents is reiterated here.

Any accident or incidence of ill health will cause both direct and indirect costs and incur an insured and an uninsured cost. It is important that all of these costs are taken into account when the full cost of an accident is calculated. In a study undertaken by the HSE, it was shown that indirect costs or hidden costs could be 36 times greater than direct costs of an accident. In other words, the direct costs of an accident or disease represent the tip of the iceberg when compared with the overall costs (*The Cost of Accidents at Work* HSG96).

Direct costs are costs that are directly related to the accident. They may be insured (claims on employers' and public liability insurance, damage to buildings, equipment or vehicles) or uninsured (fines, sick pay, damage to product, equipment or process).

Indirect costs may be insured (business loss, product or process liability) or uninsured (loss of goodwill, extra overtime payments, accident investigation time, production delays).

There are many reasons for the seriousness of a hazard not to be obvious to the person exposed to it. It may be that the hazard is not visible (radiation, certain gases and biological agents) or have no short-term effect (work-related upper limb disorders). The common reasons include lack of attention, lack of experience, the wearing of PPE, sensory impairment and inadequate information, instruction and training.

5.6   Accident categories

There are several categories of accident, all of which will be dealt with in more detail in later chapters. The principal categories are as follows:

- contact with moving machinery or material being machined;
- struck by moving, flying or falling object;
- hit by a moving vehicle;
- struck against something fixed or stationary;
- injured while handling, lifting or carrying;
- slips, trips and falls on the same level;
- falls from a height;
- trapped by something collapsing;
- drowned or asphyxiated;
- exposed to, or in contact with, a harmful substance;
- exposed to fire;

➤ exposed to an explosion;
➤ contact with electricity or an electrical discharge;
➤ injured by an animal;
➤ physically assaulted by a person;
➤ other kind of accident.

5.7 Health risks

Risk assessment is not only concerned with injuries in the workplace but also needs to consider the possibility of occupational ill health. Health risks fall into the following four categories:

➤ chemical (e.g. paint solvents, exhaust fumes);
➤ biological (e.g. bacteria, pathogens);
➤ physical (e.g. noise, vibrations);
➤ psychological (e.g. occupational stress).

There are two possible health effects of occupational ill health.

They may be **acute**, which means that they occur soon after the exposure and are often of short duration, although in some cases emergency admission to hospital may be required.

They may be **chronic**, which means that the health effects develop with time. It may take several years for the associated disease to develop and the effects may be slight (mild asthma) or severe (cancer).

Health risks are discussed in more detail in Chapter 14 (Section 14.4).

5.8 The management of risk assessment

Risk assessment is part of the planning and implementation stage of the health and safety management system recommended by the HSE in its publication HSG65. All aspects of the organization, including health and safety management, need to be covered by the risk assessment process. This will involve the assessment of risk in areas such as maintenance procedures, training programmes and supervisory arrangements. A general risk assessment of the organization should reveal the significant hazards present and the general control measures that are in place. Such a risk assessment should be completed first and then followed by more specific risk assessments that examine individual work activities.

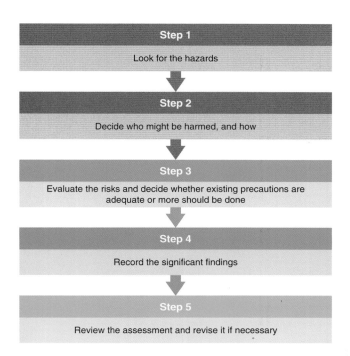

Figure 5.4 Five steps to risk assessment.

HSE has produced a free leaflet entitled *Five Steps to Risk Assessment* INDG163 (Figure 5.4). It gives practical advice on assessing risks and recording the findings and is aimed at small- and medium-sized companies in the service and manufacturing sectors. The five steps are:

➤ look for the hazards
➤ decide who might be harmed, and how
➤ evaluate the risks and decide whether existing precautions are adequate or more should be done
➤ record the significant findings
➤ review the assessment and revise it if necessary.

Each of these steps will be examined in turn in the following section.

Finally, it is important that the risk assessment team is selected on the basis of its competence to assess risks in the particular areas under examination in the organization. The Team Leader or Manager should have health and safety experience and relevant training in risk assessment. It is sensible to involve the appropriate line manager, who has responsibility for the area or activity being assessed, as a team member. Other members of the team will be selected on the basis of their experience, their technical and/or design knowledge and any relevant standards or Regulations relating to the activity or process. At least one team member must have communication and report writing skills.

A positive attitude and commitment to the risk assessment task are also important factors. It is likely that team members will require some basic training in risk assessment.

5.9 The risk assessment process

The HSE approach to risk assessment (five steps) will be used to discuss the process of risk assessment. It is, however, easier to divide the process into six elements:

➤ hazard identification;
➤ persons at risk;
➤ evaluation of risk level;
➤ risk controls (existing and additional);
➤ record of risk assessment findings;
➤ monitoring and review.

Each element will be discussed in turn.

5.9.1 Hazard identification

Hazard identification is the crucial first step of risk assessment. Only significant hazards, which could result in serious harm to people, should be identified. Trivial hazards should be ignored.

A tour of the area under consideration by the risk assessment team is an essential part of hazard identification as is consultation with the relevant section of the workforce.

A review of accident, incident and ill-health records will also help with the identification. Other sources of information include safety inspection, survey and audit reports, job or task analysis reports, manufacturers' handbooks or data sheets and Approved Codes of Practice and other forms of guidance.

Hazards will vary from workplace to workplace but the checklist in Appendix 5.1 shows the common hazards that are significant in many workplaces. Many questions in the NEBOSH examinations involve several common hazards found in most workplaces.

It is important that unsafe conditions are not confused with hazards, during hazard identification. Unsafe conditions should be rectified as soon as possible after observation. Examples of unsafe conditions include missing machine guards, faulty warning systems and oil spillage on the workplace floor.

5.9.2 Persons at risk

Employees and contractors who work full time at the workplace are the most obvious groups at risk and it will be a necessary check that they are competent to perform their particular tasks. However, there may be other groups who spend time in or around the workplace. These include young workers, trainees, new and expectant mothers, cleaners, contractor and maintenance workers and members of the public. Members of the public will include visitors, patients, students or customers as well as passers-by.

The risk assessment must include any additional controls required due to the vulnerability of any of these groups, perhaps caused by inexperience or disability. It must also give an indication of the numbers of people from the different groups who come into contact with the hazard and the frequency of these contacts.

5.9.3 Evaluation of risk level

During most risk assessment it will be noted that some of the risks posed by the hazard have already been addressed or controlled. The purpose of the risk assessment, therefore, is to reduce the remaining risk. This is called the **residual risk**.

The goal of risk assessment is to reduce all residual risks to as low a level as reasonably practicable. In a relatively complex workplace, this will take time so that a system of ranking risk is required – the higher the risk level the sooner it must be addressed and controlled.

For most situations, a **qualitative** risk assessment will be perfectly adequate. (This is certainly the case for NEBOSH Certificate candidates and is suitable for use during the practical assessment.) During the risk assessment, a judgement is made as to whether the risk level is high, medium or low in terms of the risk of somebody being injured. This designation defines a timetable for remedial actions to be taken thereby reducing the risk. High-risk activities should normally be addressed in days, medium risks in weeks and low risks in months or in some cases no action will be required. It will usually be necessary for risk assessors to receive some training in risk level designation.

A **quantitative** risk assessment attempts to quantify the risk level in terms of the likelihood of an incident and its subsequent severity. Clearly the higher the likelihood and severity, the higher the risk will be. The likelihood depends on such factors as the control measures in place, the frequency of exposure to the hazard and the category of person exposed to the hazard. The severity will depend on the magnitude of the hazard (voltage, toxicity, etc.). HSE suggests in HSG65 a simple 3×3 matrix to determine risk levels.

Likelihood of occurrence	Likelihood level
Harm is certain or near certain to occur	High 3
Harm will often occur	Medium 2
Harm will seldom occur	Low 1

Severity of harm	Severity level
Death or major injury (as defined by RIDDOR)	Major 3
3-day injury or illness (as defined by RIDDOR)	Serious 2
All other injuries or illnesses	Slight 1

Risk = Severity × Likelihood

Likelihood	Severity		
	Slight 1	Serious 2	Major 3
Low 1	Low 1	Low 2	Medium 3
Medium 2	Low 2	Medium 4	High 6
High 3	Medium 3	High 6	High 9

Thus:

6–9 High risk
3–4 Medium risk
1–2 Low risk

It is possible to apply such methods to organizational risk or to the risk that the management system for health and safety will not deliver in the way in which it was expected or required. Such risks will add to the activity or occupational risk level. In simple terms, poor supervision of an activity will increase its overall level of risk. A risk management matrix has been developed which combines these two risk levels, as shown below.

		Occupational risk level		
		Low	Medium	High
Organizational risk level	Low	L	L	M
	Medium	L	M	H
	High	M	H	Unsatisfactory

L Low risk   **M** Medium risk   **H** High risk

Whichever type of risk evaluation method is used, the level of risk simply enables a timetable of risk reduction to an acceptable and tolerable level to be formulated. The legal duty requires that all risks should be reduced to as low as is reasonably practicable.

5.9.4 Risk control measures

The next stage in the risk assessment is the control of the risk. In established workplaces, some control of risk will be in place already. The effectiveness of these controls needs to be assessed so that an estimate of the residual risk may be made. Many hazards have had specific acts, Regulations or other recognized standards developed to reduce associated risks. Examples of such hazards are fire, electricity, lead and asbestos. The relevant legislation and any accompanying Approved Codes of Practice or guidance should be consulted first and any recommendations implemented. Advice on control measures may also be available from trade associations, trade unions or employers' organizations.

Where there are existing preventative measures in place, it is important to check that they are working properly and that everybody affected has a clear understanding of the measures. It may be necessary to strengthen existing procedures, for example by the introduction of a permit to work system. More details on the principles of control are contained in Chapter 6.

Hierarchy of risk control

When assessing the adequacy of existing controls or introducing new controls, a hierarchy of risk controls should be considered. The Management of Health and Safety at Work Regulations 1999 Schedule 1 specifies the general principles of prevention which are set out in the European Council Directive. These principles are:

1. avoiding risks;
2. evaluating the risks which cannot be avoided;
3. combating the risks at source;
4. adapting the work to the individual, especially as regards the design of the workplace, the choice of work equipment and the choice of working and production methods, with a view, in particular, to alleviating monotonous work and work at a predetermined work rate and to reducing their effects on health;
5. adapting to technical progress;
6. replacing the dangerous by the non-dangerous or the less dangerous;
7. developing a coherent overall prevention policy which covers technology, organization of work, working

conditions, social relationships and the influence of factors relating to the working environment;

8. giving collective protective measures priority over individual protective measures;

9. giving appropriate instruction to employees.

These principles are not exactly a hierarchy but must be considered alongside the usual hierarchy of risk control, which is as follows:

➤ elimination;
➤ substitution;
➤ changing work methods/patterns;
➤ reduced or limited time exposure;
➤ engineering controls (e.g. isolation, insulation and ventilation);
➤ good housekeeping;
➤ safe systems of work;
➤ training and information;
➤ personal protective equipment;
➤ welfare;
➤ monitoring and supervision;
➤ review.

See Chapter 6 (Section 6.3) for more information on the risk control hierarchy.

Prioritization of risk control

The prioritization of the implementation of risk control measures will depend on the risk rating (high, medium and low) but the timescale in which the measures are introduced will not always follow the ratings. It may be convenient to deal with a low-level risk at the same time as a high-level risk or before a medium-level risk. It may also be that work on a high-risk control system is delayed due to a late delivery of an essential component – this should not halt the overall risk reduction work. It is important to maintain a continuous programme of risk improvement rather than slavishly following a predetermined priority list.

5.9.5 Record of risk assessment findings

It is very useful to keep a written record of the risk assessment even if there are less than five employees in the organization. For an assessment to be 'suitable and sufficient', only the significant hazards and conclusions need be recorded. The record should also include details of the groups of people affected by the hazards and the existing control measures and their effectiveness. The conclusions should identify any new controls required and a review date. The HSE booklet *Five Steps to Risk Assessment* provides a very useful guide and examples of the detail required for most risk assessments.

There are many possible layouts which can be used for the risk assessment record. Examples are given in Appendices 5.2, 5.3 and 5.4. It should be noted that in Appendices 5.3 and 5.4, the initial qualitative risk level at the time of the risk assessment is given – the residual risk level when all the additional controls have been implemented will be 'low'. This should mean that an annual review will be sufficient.

The written record provides excellent evidence to a health and safety inspector of compliance with the law. It is also useful evidence if the organization should become involved in a civil action.

The record should be accessible to employees and a copy kept with the safety manual containing the safety policy and arrangements.

5.9.6 Monitoring and review

As mentioned earlier, the risk controls should be reviewed periodically. This is equally true for the risk assessment as a whole. Review and revision may be necessary when conditions change as a result of the introduction of new machinery, processes or hazards. There may be new information on hazardous substances or new legislation. There could also be changes in the workforce, for example the introduction of trainees. The risk assessment needs to be revised only if significant changes have taken place since the last assessment was done. An accident or incident or a series of minor ones provides a good reason for a review of the risk assessment. This is known as the post-accident risk assessment.

5.9.7 Cost–benefit analysis

In recent years, risk assessment has been accompanied by a cost–benefit analysis that attempts to evaluate the costs and benefits of risk control and reduction. The costs could include capital investment, maintenance and training and produce benefits such as reduced insurance premiums, higher productivity and better product quality. The payback period for most risk reduction projects (other than the most simple) has been shown to be between 2 and 5 years. Although the benefits are often difficult to quantify, cost–benefit analysis does help to justify the level of expenditure on a risk reduction project.

5.10 Special cases

There are several groups of persons who require an additional risk assessment due to their being more 'at risk' than other groups. Three such groups will be considered – young persons, expectant and nursing mothers and disabled workers.

5.10.1 *Young persons*

There were 21 fatalities of young people at work in 2002/03 and more recent figures have not shown much of an improvement.

A risk assessment involving young people needs to consider the particular vulnerability of young persons in the workplace. Young workers clearly have a lack of experience and awareness of risks in the workplace, a tendency to be subject to peer pressure and a willingness to work hard. Many young workers will be trainees or on unpaid work experience. Young people are not fully developed and are more vulnerable to physical, biological and chemical hazards than adults.

An amendment to the HSW Act enables trainees on government sponsored training schemes to be treated as employees as far as health and safety is concerned. The Management of Health and Safety at Work Regulations defines a young person as anybody under the age of 18 years and stipulates that a special risk assessment must be completed which takes into account their immaturity and inexperience. The assessment must be completed before the young person starts work. If the young person is of school age (16 years or less), the parents or guardian of the child should be notified of the outcome of the risk assessment and details of any safeguards which will be used to protect the health and safety of the child.

The following key elements should be covered by the risk assessment:

➤ details of the work activity, including any equipment or hazardous substances;
➤ details of any prohibited equipment or processes;
➤ details of health and safety training to be provided;
➤ details of supervision arrangements.

The extent of the risks identified in the risk assessment will determine whether employers should restrict the work of the people they employ. Except in special circumstances, young people should not be employed to do work which:

➤ is beyond their physical or psychological capacity;
➤ exposes them to substances chronically harmful to human health, for example toxic or carcinogenic substances, or effects likely to be passed on genetically or likely to harm the unborn child;
➤ exposes them to radiation;

➤ involves a risk of accidents which they are unlikely to recognize because of, for example, their lack of experience, training or insufficient attention to safety;
➤ involves a risk to their health from extreme heat, noise or vibration.

These restrictions will not apply in **special circumstances** where young people *over* the minimum school leaving age are doing work necessary for their training, under proper supervision by a competent person, and providing the risks are reduced to the lowest level, so far as is reasonably practicable. Under no circumstances can children of compulsory school age do work involving these risks, whether they are employed or under training such as work experience.

Induction training is important for young workers and such training should include site rules, restricted areas, prohibited machines and processes, fire precautions, emergency procedures, welfare arrangements and details of any further training related to their particular job. At induction, they should be introduced to their mentor and given close supervision, particularly during the first few weeks of their employment.

If work-experience trainees are not at school on any days during the placement, then they should not work more than 8 hours a day or 40 hours a week (Working Time Directive). They should also have a 20-minute rest within a 4.5-hour working period.

A guide is available to employers who organize site visits for young people. The guide (www.safevisits.org.uk) gives practical advice and a checklist that can be used during the organization of the visit.

5.10.2 *Expectant and nursing mothers*

The Management of Health and Safety at Work Regulations 1999 incorporates the Pregnant Workers Directive from the EU. If any type of work could present a particular risk to expectant or nursing mothers, the risk assessment must include an assessment of such risks. Should these risks be unavoidable, then the woman's working conditions or hours must be altered to avoid the risks. The alternatives for her are to be offered other work or be suspended from work on full pay. The woman must notify the employer in writing that she is pregnant, or has given birth within the previous 6 months and/or is breastfeeding.

Pregnant workers should not be exposed to chemicals, such as pesticides and lead, or to biological hazards, such as hepatitis. Female agricultural workers, veterinaries or farmers' wives who are pregnant, should not assist with lambing so that any possible contact with ovine chlamydia is avoided.

Other work activities that may present a particular risk to pregnant women at work are radiography, involving possible exposure to ionizing radiation, and shop work when long periods of standing are required during shelf filling or stock-taking operations.

Typical factors which might affect such women are:

➤ manual handling;
➤ chemical or biological agents;
➤ ionizing radiation;
➤ passive smoking;
➤ lack of rest room facilities;
➤ temperature variations;
➤ ergonomic issues related to prolonged standing, sitting or the need for awkward body movement;
➤ issues associated with the use and wearing of personal protective equipment;
➤ working excessive hours;
➤ stress and violence to staff.

Detailed guidance is available in *New and Expectant Mothers at Work*, HSG122, HSE Books.

5.10.3 *Workers with a disability*

Organizations have been encouraged for many years to employ workers with disabilities and to ensure that their premises provide suitable access for such people. From a health and safety point of view, it is important that workers with a disability are covered by special risk assessments so that appropriate controls are in place to protect them. For example, employees with a hearing problem will need to be warned when the fire alarm sounds or a fork-lift truck approaches. Special vibrating signals or flashing lights may be used. Similarly workers in wheelchairs will require a clear, wheelchair-friendly route to a fire exit and onwards to the assembly point. Safe systems of work and welfare facilities need to be suitable for any workers with disabilities.

The final stage of the goods and services provisions in Part III of the Disability Discrimination Act came into force on 1 October 2004. The new duties apply to service providers where physical features make access to their services impossible or unreasonable for people with a disability. The Act requires equal opportunities for employment and access to workplaces to be extended to all people with disabilities.

5.10.4 *Lone workers*

People who work alone, like those in small workshops, remote areas of a large site, social workers, sales personnel or mobile maintenance staff, should not be at more risk than other employees (Figure 5.5). It is important to consider whether the risks of the job can be properly controlled by one person. Other considerations in the risk assessment include:

➤ does the particular workplace present a special risk to someone working alone?
➤ is there safe egress and exit from the workplace?
➤ can all the equipment and substances be safely handled by one person?
➤ is violence from others a risk?
➤ would women and young persons be specially at risk?
➤ is the worker medically fit and suitable for working alone?
➤ are special training and supervision required?
➤ has the worker access to first aid?

For details of further precautions for lone workers, see Section 6.6 of Chapter 6.

5.11 Sources of reference

Successful Health and Safety Management HSG65, HSE Books, 1997 ISBN 978 0 7176 1276 5.
The Management of Health and Safety at Work (ACOP) L21, HSE Books 2000 ISBN 978 0 7176 2488 1.
Five Steps to Risk Assessment (INDG163), HSE Books ISBN 978 0 7176 6189 3.

Figure 5.5 A lone worker – special arrangements are required.

Five Steps to Risk Assessment – Case Studies (HSG183), HSE Books ISBN 0 7176 1580 4 (no longer published but see HSE website for example risk assessments).

Young People at Work HSG165, HSE Books ISBN 0 7176 1889 7 (no longer published).

Relevant statutory provisions

The Management of Health and Safety at Work Regulations 1999

5.12 Practice NEBOSH questions for Chapter 5

1. **(i) Explain**, using examples, the meaning of the following terms:
   (a) hazard
   (b) risk.
   (ii) Outline the key stages of a general risk assessment.
   (iii) Give THREE reasons why the seriousness of a hazard may not be obvious to someone exposed to it.
   (iv) Outline the logical steps to take in managing risks at work.

2. **(i) Identify** the particular requirements of Regulation 3 of the Management of Health and Safety at Work Regulations 1999 in relation to an employer's duty to carry out risk assessments.
   (ii) Outline the factors that should be considered when selecting individuals to assist in carrying out risk assessments in the workplace.

3. A factory manager intends to introduce a new work process for which a risk assessment is required under Regulation 3 of the Management of Health and Safety at Work Regulations 1999.
   (i) Outline the factors that should be used in carrying out the risk assessment, **identifying** the issues that would need to be considered at **EACH** stage.
   (ii) Explain the criteria that must be met for the assessment to be deemed 'suitable and sufficient'.
   (iii) Identify the various circumstances that may require a review of the risk assessment at a later date.

4. In relation to a risk assessment undertaken to comply with the Management of Health and Safety at Work Regulations 1999:
   (i) Outline TWO 'significant findings' that should be recorded.
   (ii) Identify THREE groups of workers who might be especially at risk.

5. With respect to undertaking general risk assessments on activities within a workplace:
   (i) Outline the **FIVE** key stages of the risk assessment process, **identifying** the issues that would need to be considered at **EACH** stage.
   (ii) Identify FOUR items of information from within and **FOUR** items of information from outside the organization that may be useful when assessing the activities.
   (iii) State the legal requirements for recording workplace risk assessments.

6. **(i) Identify** the 'five' steps involved in the assessment of risk from workplace activities [as described in HSE's *Five Steps to Risk Assessment* (INDG163)].
   (ii) Explain the criteria that should be applied to help develop an action plan to prioritize the control of health and safety risks in the workplace.

7. **Outline** the hazards that might be encountered in a busy hotel kitchen.

8. **Identify SIX** hazards that might be considered when assessing the risk to health and safety of a multi-storey car park attendant.

9. **Outline** the hazards that might be encountered by a gardener employed by a local authority parks department.

10. **Outline** the content of a training course for staff who are required to assist in carrying out risk assessments.

11. An employer has agreed to accept a young person on a work experience placement for 1 week. **Outline** the factors that the employer should consider prior to the placement.

12. **(i) Define** the meaning of the term 'young person' as used in health and safety legislation.
    (ii) Outline the factors to be taken into account when undertaking a risk assessment on young persons who are to be employed in the workplace.

13. **(i) Identify FOUR** 'personal' factors that may place young persons at a greater risk from workplace hazards.
    (ii) Outline FOUR measures that could be taken to minimize the risks to young persons in the workplace.

14. **Outline** the factors that may increase risks to pregnant employees.

15. (i) **Identify** work activities that may present a particular risk to pregnant women at work, giving an example of **EACH** type of activity.
(ii) **Outline** the actions that an employer may take when a risk to a new or expectant mother cannot be avoided.

16. An employee who works on a production line has notified her employer that she is pregnant. **Outline** the factors that the employer should consider when undertaking a specific risk assessment in relation to this employee.

17. **Outline** the issues to be considered to ensure the health and safety of disabled workers in the workplace.

18. **Identify** the factors to be considered to ensure the health and safety of persons who are required to work on their own away from the workplace.

19. **Outline** the issues that should be considered to ensure the health and safety of cleaners employed in a school out of normal working hours.

Appendix 5.1 Hazard checklist

The following checklist may be helpful.

1. **Equipment/mechanical**
   entanglement
   friction/abrasion
   cutting
   shearing
   stabbing/puncturing
   impact
   crushing
   drawing-in
   air or high-pressure fluid injection
   ejection of parts
   pressure/vacuum
   display screen equipment
   hand tools

2. **Transport**
   works vehicles
   mechanical handling
   people/vehicle interface

3. **Access**
   slips, trips and falls
   falling or moving objects
   obstruction or projection
   working at height
   confined spaces
   excavations

4. **Handling/lifting**
   manual handling
   mechanical handling

5. **Electricity**
   fixed installation
   portable tools and equipment

6. **Chemicals**
   dust/fume/gas
   toxic
   irritant
   sensitizing
   corrosive
   carcinogenic
   nuisance

7. **Fire and Explosion**
   flammable materials/gases/liquids
   explosion
   means of escape/alarms/detection

8. **Particles and dust**
   inhalation
   ingestion
   abrasion of skin or eye

9. **Radiation**
   ionizing
   non-ionizing

10. **Biological**
    bacterial
    viral
    fungal

11. **Environmental**
    noise
    vibration
    light
    humidity
    ventilation
    temperature
    overcrowding

12. **The individual**
    individual not suited to work
    long hours
    high work rate
    violence to staff
    unsafe behaviour of individual
    stress
    pregnant/nursing women
    young people

13. **Other factors to consider**
    poor maintenance
    lack of supervision
    lack of training
    lack of information
    inadequate instruction
    unsafe systems

Appendix 5.2 Example of a risk assessment record

General Health and Safety Risk Assessment	No.
Firm/Company	Department
Contact Name	Nature of Business
Telephone Number	

Principal Hazards

Risks to employees and members of the public could arise due to the following hazards:

1. hazardous substances
2. electricity
3. fire
4. dangerous occurrences or other emergency incidents
5.
6.

Persons at Risk
Employees, contractors, and members of the public.

Main Legal Requirements

1. Health and Safety at Work Act 1974 – sections 2 and 3
2. Management of Health and Safety at Work Regulations 1999
3. The Noise at Work Regulations 1989
4. Common Law Duty of Care
5. ?

Significant Risks

➤ acute and chronic health problems caused by the use or release of hazardous substances
➤ injuries to employees and members of the public due to equipment failure such as electric shock
➤ injuries to employees and members of the public from slips, trips and falls
➤ injuries to employees and members of the public caused by fire
➤ ?

Consequences
Fractures, bruising, smoke inhalation and burns, acute and chronic health problems and death.

(Continued)

Existing Control Measures

Possible examples include:

1. All sub-contractors are vetted prior to appointment.
2. All hazardous and harmful materials are identified and the risks to people assessed. COSHH assessments are provided and the appropriate controls are implemented. Health surveillance is provided as necessary.
3. Fire risk assessment has been produced. Fire procedures are in place and all employees are trained to deal with fire emergencies. A carbon dioxide fire extinguisher is available at every work site.
4. A minimum of flammable substances are used on the premises, no more than a half day's supply at a time. Kept in fire resistant store.
5. No smoking is allowed on the premises.
6. Manual handling is kept to a minimum. Where there is a risk of injury manual handling assessments are carried out.
7. Method statements are used for complex and/or hazardous jobs and are followed at all times.
8. All accidents on or around the site are reported and investigated by management. Any changes found necessary are quickly implemented. All accidents, reportable under RIDDOR 1995, are reported to the HSE on form F2508.
9. At least 1 qualified First Aider is available during working hours.
10. ?

Residual Risk, i.e. after controls are in place.

Severity .... **Likelihood** .... **Residual Risk** ....

Information

Details of various relevant HSE and trade publications.

Comments from Line Manager   **Comments from the Risk Assessor**

Signed.......................... **Date** ....   **Signed** .... **Date** ....

Review Date

Likelihood	Severity		
	Slight 1	Serious 2	Major 3
Low 1	Low 1	Low 2	Medium 3
Medium 2	Low 2	Medium 4	High 6
High 3	Medium 3	High 6	High 9

Appendix 5.3 Risk assessment example 2: Hairdressing salon

Name of Company: His and Hers Hairdressers
Name of Assessor: A.R. Smith

Date of Assessment: 11 January 2008
Date of Review: 12 July 2008

Hazards	Persons affected	Risks	Initial risk level	Existing controls	Additional controls	Action by whom?	Action by when?	Done
Hairdressing products and chemicals Various bleaching and cleansing products in particular: Lightening (bleach) product Hydrogen peroxide Oxidative colourants	Staff and customers	Eye and/or skin irritation Possible allergic reaction	Medium	• COSHH assessment completed • Non-latex gloves are provided for staff when using products • Customers are protected with single use towels • Only non-dusty bleaches used • Staff report any allergies at induction • Store-room and salon well ventilated • Products stored as per manufacturer's recommendations • Staff are specifically trained in the correct use of products • Staff check whether customers have allergies to any products	• Needs to be reviewed • Eye baths to be purchased for treatment of eye splashes • Repeat allergy checks every 3 months and records kept • Storage in salon kept to 1 day requirement • Refresher training every 3 months and records kept • Records kept of customer allergies and simple patch tests introduced	Manager Owner Manager Manager Manager Staff Manager	12/2/08 12/2/08 14/2/08 then every 3 months 18/1/08 11/4/08 25/4/08	8/2/08 24/1/08 12/2/08 16/1/08 8/4/08 7/5/08
Sharp instruments	Staff and customers	Cuts, grazes and blood-borne infections	Medium	• All sharp instruments sterilized after use • Sterilizing liquid changed daily • Sharps box available for disposable blades • First aid box available	• Regular recorded checks that sterilization procedures correctly followed • Contents of first aid box checked weekly	Manager Manager	18/1/08 then monthly 18/1/08 then weekly	18/1/08 18/1/08
Fire	Staff and customers	Smoke inhalation and burns	Low	• Fire risk assessment completed	• No flammable products will be displayed in the windows or near heater	Manager	18/1/08	18/1/08

Hazard	Who might be harmed	What is the harm	Risk level	Control measures in place	Further action needed	Action by whom	Action by when	Done
Electricity	Staff and customers	Electrical shocks and burns. Also a risk of fire	Low	• Any damaged cables, plugs or electrical equipment is reported to the manager • All portable electrical equipment and thermostats checked every 6 months by a competent person • All electrical equipment purchased from a reliable supplier • Staff shown how to use and store hairdryers, etc. safely and isolate electrical supply	• An appropriate fire extinguisher (carbon dioxide) should be available • Make a visual check of electrical equipment, cables and sockets every month and record findings • All internal wiring should be checked by a qualified electrician	Owner Manager Owner	25/1/08 25/1/08 then every month 25/4/08	23/1/08 25/1/08 3/4/08
Standing for long periods	Staff	Back pain and pain in neck, shoulders, legs and feet – musculoskeletal injuries	Medium	• Staff given regular breaks • Stools available for staff for use while trimming hair • Customer chairs are adjustable in height	• Develop a formal rota system for breaks	Manager	25/1/08	21/1/08
Wet hand work	Staff	Skin sensitization, dry skin, dermatitis	Medium	• Staff are trained to wash and dry hands thoroughly between hair washes • Non-latex gloves are provided for staff • Moisturizing hand cream is provided for staff	• Ensure that a range of glove sizes are available for staff • Staff will always wear gloves for all wet work	Manager Manager	25/1/08 25/1/08	31/1/08 1/2/08
Slips and trips	Staff and customers	Bruising, lacerations and possible fractures	Low	• Cut hair is swept up regularly • Staff must wear slip-resistant footwear • No trailing leads on floor • Any spills are cleaned immediately • Door mat provided at shop entrance	• Organize repair of worn floor covering	Manager	8/2/08	5/2/08

Appendix 5.4 Risk assessment example 3: Office cleaning

Name of Company: Apex Cleaning Company
Name of Assessor: T W James

Date of Assessment: 14 May 2008
Date of Review: 14 November 2008

Hazards	Persons affected	Risks	Initial risk level	Existing controls	Additional controls	Action by whom?	Action by when?	Done
Machine cleaning of floors	Staff and others	Injury to ankles due to incorrect use of machinery	Low	• Machine supplied for the job is suitable • Machine is maintained regularly and examined by a competent person • Cleaners trained in the safe use of the machine	• Maintenance and inspections to be documented • Training to be documented	Manager Manager	30/5/08 30/5/08	21/5/08 28/5/08
Electrical	Staff	Electric shock and burns	Low	• Staff trained in visual inspection of plugs, cables and switches before use on each shift • Staff inform manager of any defects • All portable electrical equipment is regularly tested by a competent person • Staff trained not to splash water near machines or wall sockets	• Training to be documented • Defect forms and a procedure to be developed and communicated to staff • All electrical equipment to be listed and results of portable appliance testing (PAT) tests recorded • Training to be documented	Manager Manager Manager Manager	30/5/08 13/6/08 20/6/08 30/5/08	28/5/08 18/6/08 18/6/08 28/5/08
Lone working	Staff	Accident, illness or attack by intruder	Medium	• Staff sign in and out at either end of shift with security staff • Security staff visit staff regularly • All staff issued with mobile telephone number of manager for emergency use	• Check that staff are aware that security staff are trained first aiders • Check on regularity of these visits • If manager cannot respond, staff to be told to ring emergency services	Manager Manager Manager	22/5/08 22/5/08 22/5/08	20/5/08 20/5/08 20/5/08
Work at height	Staff	Bruising, sprains, lacerations and fractures	Low	• Staff issued with long-handled equipment for work at high level • Staff told not to stand on chairs or stepladders	• Staff to be trained in safe system of work for cleaning stairways, escalators and external windows	Manager	11/6/2008	3/6/08

Hazard	Who at risk	Harm	Risk level	Existing controls	Further action	By whom	Target date	Date completed
Contact with bleach and other cleaning chemicals	Staff	Skin irritation. Possible allergic reaction and eye injuries from splashes	Low	• Staff trained in safe use and storage of cleaning chemicals • Staff induction questionnaire includes details of any allergies and skin problems • Impervious rubber gloves are issued for use with any chemical cleaners • Equipment used with chemicals is regularly serviced and/or cleaned	• COSSH assessment to be reviewed • Records of the staff training to be kept and updated • Check whether products marked 'harmful' or 'irritant' can be substituted with milder alternatives • Staff to report any health problems	Manager Manager Manager Staff	27/6/08 30/5/08 20/6/08 Ongoing	18/6/08 28/5/08 25/6/08
Musculoskeletal disorders and injuries	Staff	Musculoskeletal injuries to the back, neck, shoulders, legs and feet	Medium	• Staff trained in correct lifting technique • Long-handled equipment used to prevent stooping • Each floor is provided with all necessary cleaning equipment • Staff trained not to overfill buckets	• Check that staff are not lifting heavy objects or furniture or unduly stretching while cleaning • Check whether more up-to-date equipment is available (long-handled wringers and buckets on wheels)	Manager Manager	28/5/08 28/5/08	26/5/08 28/5/08
Slips, trips and falls	Staff and others	Bruising, sprains, lacerations and possible fractures	Medium	• Warning signs placed on wet floors • Staff use nearest electrical socket to reduce trip hazards • Staff told to wear slip-resistant footwear • Wet floor work restricted to less busy times at client offices • Client encouraged in a good housekeeping policy	• Introduce a wet and dry mops cleaning system for floors	Manager	28/5/08	26/5/08
Fire	Staff and others	Smoke inhalation and burns	Low	• Staff trained in client's emergency fire procedures including assembly point	• On each floor, a carbon dioxide fire extinguisher is available for cleaning staff	Manager	6/6/08	2/6/08

Principles of control

6

After reading this chapter you should be able to:

1. describe the general principles of control and a basic hierarchy of risk reduction measures that encompass technical, behavioural and procedural controls
2. describe what factors should be considered when developing and implementing a safe system of work for general work activities and explain the key elements of a safe system applied to the particular situations of working in confined spaces and lone working
3. explain the role and function of a permit-to-work system
4. explain the need for emergency procedures and the arrangements for contacting emergency services
5. describe the requirements for, and effective provision of, first aid in the workplace.

6.1 Introduction

The control of risks is essential to secure and maintain a healthy and safe workplace which complies with the relevant legal requirements. Hazard identification and risk assessment are covered in Chapter 5 and these together with appropriate risk control measures form the core of the HSG65 'implementing and planning' section of the management model. Chapter 1 covers this in more detail.

In industry today safety is controlled through a combination of **engineered measures** such as the provision of safety protection (e.g. guarding and warning systems), and **operational measures** in training, safe work practices, operating procedures and method statements, along with **management supervision**.

These measures (collectively) are commonly known in health and safety terms as **control measures**. Some of these more common measures will be explained in more detail later.

This chapter concerns the principles that should be adopted when deciding on suitable measures to eliminate or control both acute and chronic risks to the health and safety of people at work (Figure 6.1). The principles of control can be applied to both health risks and safety risks, although health risks have some distinctive features that require a special approach.

Chapters 9–16 deal with specific workplace hazards and controls, subject by subject. The principles of prevention now enshrined in the Management of Health and Safety at Work (MHSW) Regulations need to be used jointly with the hierarchy of control methods which give the preferred order of approach to risk control.

When risks have been analysed and assessed, decisions can be made about workplace precautions.

All final decisions about risk control methods must take into account the relevant legal requirements, which establish minimum levels of risk prevention or control. Some of the duties imposed by the HSW Act and the relevant statutory provisions are **absolute** and must be complied with. Many requirements are, however, qualified

Figure 6.1 When controls break down.

by the words **so far as is reasonably practicable**, or **so far as is practicable**. These require an assessment of cost, along with information about relative costs, effectiveness and reliability of different control measures. Further guidance on the meaning of these three expressions is provided in Chapter 1.

6.2 Principles of prevention

The MHSW Regulations Schedule 1 specifies the general principles of prevention which are set out in Article 6(2) of the European Council Directive 89/391/EEC. For the first time the principles have been enshrined directly in Regulations which state, at Regulation 4, that *Where an employer implements any preventative measures he shall do so on the basis of the principles specified in Schedule 1.* These principles are:

1. **Avoiding risks**
   This means, for example, trying to stop doing the task or using different processes or doing the work in a different, safer way.

2. **Evaluating the risks which cannot be avoided**
   This requires a risk assessment to be carried out.

3. **Combating the risks at source**
   This means that risks, such as a dusty work atmosphere, are controlled by removing the cause of the dust rather than providing special protection; or that slippery floors are treated or replaced rather than putting up a sign.

4. **Adapting the work to the individual**
   This involves the design of the workplace, the choice of work equipment and the choice of working and production methods, with a view, in particular, to alleviating monotonous work and work at a predetermined work rate and to reducing their effect on health.

   This will involve consulting those who will be affected when workplaces, methods of work and safety procedures are designed. The control individuals have over their work should be increased, and time spent working at predetermined speeds and in monotonous work should be reduced where it is reasonable to do so.

5. **Adapting to technical progress**
   It is important to take advantage of technological and technical progress, which often gives designers and employers the chance to improve both safety and working methods. With the Internet and other international information sources available, very wide knowledge, going beyond what is happening in the UK or Europe, will be expected by the enforcing authorities and the courts.

6. **Replacing the dangerous by the non-dangerous or the less dangerous**
   This involves substituting, for example, equipment or substances with non-hazardous or less hazardous substances.

7. **Developing a coherent overall prevention policy**
   This covers technology, organization of work, working conditions, social relationships and the influence of factors relating to the working environment.
   Health and safety policies should be prepared and applied by reference to these principles.

8. **Giving collective protective measures priority over individual protective measures**
   This means giving priority to control measures which make the workplace safe for everyone working there so giving the greatest benefit, for example removing hazardous dust by exhaust ventilation rather than providing a filtering respirator to an individual worker. This is sometimes known as a 'Safe Place' approach to controlling risks.

9. **Giving appropriate instruction to employees**
   This involves making sure that employees are fully aware of company policy, safety procedures, good practice, official guidance, any test results and legal requirements. This is sometimes known as a 'Safe Person' approach to controlling risks where the focus is on individuals. A properly set-up health and safety

management system should cover and balance both a Safe Place and Safe Person approach.

6.3 General control measures

6.3.1 Hierarchy of control

When assessing the adequacy of existing controls or introducing new controls, a hierarchy of control should be considered. The principles of prevention in the MHSW Regulations 1999 is not a hierarchy but a list of prevention principles which must all be considered when controlling risks. However, there is a preferred hierarchy of control following these principles and those found in HSG65. The 2006 NEBOSH General Certificate syllabus uses basically the same hierarchy with a few minor differences. These are shown in Table 6.1 alongside the HSG65 hierarchy of risk control principles.

The hierarchy reflects that risk elimination and risk control by the use of physical engineering controls and safeguards can be more reliably maintained than those which rely solely on people. These concepts are now written into the Control of Substances Hazardous to Health (COSHH) Regulations and the Management of Health and Safety at Work (MHSW) Regulations.

Where a range of control measures are available, it will be necessary to weigh up the relative costs of each against the degree of control each provides, both in the short and

Table 6.1 Hierarchy of control

2006 NEBOSH General Certificate	HSG65 Summary of Risk Control Principles
Avoidance of risks	**Eliminate risk** by substituting the dangerous by the inherently less dangerous, for example: ● Use less hazardous substances. ● Substitute a type of machine which is better guarded to make the same product. ● Avoid the use of certain processes.
Elimination of hazards or substitution for something less hazardous	
Reducing or limiting the duration of exposure to the hazard	**Combat risks** at source by engineering controls and giving collective protective measures priority, for example: ● Separate the operator from the risk of exposure to a known hazardous substance by enclosing the process. ● Protect the dangerous parts of a machine by guarding. ● Design process machinery and work activities to minimize the release of, or to suppress or contain, airborne hazards. ● Design machinery which is remotely operated and to which materials are fed automatically, thus separating the operator from danger areas.
Isolation/segregation	
Engineering controls	
Safe systems of work	**Minimize risk** by: ● designing suitable systems of working ● using personal protective clothing and equipment; this should only be used as a last resort.
Training and information	
PPE	
Welfare	
Monitoring and supervision	

long term. Some control measures, such as eliminating a risk by choosing a safer alternative substance or machine, provide a high degree of control and are reliable. Physical safeguards such as guarding a machine or enclosing a hazardous process need to be maintained. In making decisions about risk control, it will therefore be necessary to consider the degree of control and the reliability of the control measures along with the costs of both providing and maintaining the measure.

6.3.2 Avoidance of risks by elimination or substitution

The best and most effective way of reducing risks is by avoiding a hazard and its associated risks. For example avoid working at height by installing a permanent working platform with stair access; avoid entry into a confined space by, for example, using a sump pump in a pit which is removed by a lanyard for maintenance; eliminate the fire risks from tar boilers by using bitumen which can be applied cold.

Substitution describes the use of a less hazardous form of a substance or process. There are many examples of substitution such as the use of water-based rather than oil-based paints; the use of asbestos substitutes and the use of compressed air as a power source rather than electricity to reduce both electrical and fire risks; and the use of mechanical excavators instead of hand digging.

In some cases it is possible to change the method of working so that risks are reduced. For example use rods to clear drains instead of strong chemicals; use a long-handled water hose brush to clean windows instead of ladders. Sometimes the pattern of work can be changed so that people can do things in a more natural way, for example when placing components for packing consider whether people are right- or left-handed; encourage people in offices to take breaks from computer screens by getting up to photocopy, fetch files or print documents.

Care must be taken to consider any additional hazards which may be involved and thereby introduce additional risks, as a result of a substitution.

6.3.3 Reduced time exposure

This involves reducing the time during the working day that the employee is exposed to the hazard, by giving the employee either other work or rest periods. It is normally only suitable for the control of health hazards associated with, for example, noise, vibration, excessive heat or cold, display screens and hazardous substances. However, it is important to note that for many hazards, there are short-term

exposure limits as well as normal workplace exposure limits (WELs) over an 8-hour period (see Chapter 14). Short-term limits must not be exceeded during the reduced time exposure intervals.

It cannot be argued that a short time of exposure to a dangerous part of a machine is acceptable. However, it is possible to consider short bouts of intensive work with rest periods when employees are engaged in heavy labour such as manual digging when machines are not permitted due to the confines of the space or buried services.

6.3.4 Isolation/Segregation

Controlling risks by isolating them or segregating people and the hazard is an effective control measure and used in many instances; for example separating vehicles and pedestrians on factory sites, providing separate walkways for the public on road repairs, providing warm rooms on sites or noise refuges in noisy processes.

The principle of isolation is usually followed with the storage of highly flammable liquids or gases which are put into open, air ventilated compounds away from other hazards such as sources of ignition or from people who may be at risk from fire or explosion.

6.3.5 Engineering controls

This describes the control of risks by means of engineering design rather than a reliance on preventative actions by the employee. There are several ways of achieving such controls:

1. Control the risks at the source (e.g. the use of more efficient dust filters or the purchase of less noisy equipment).
2. Control the risk of exposure by:
   ➤ **isolating** the equipment by the use of an enclosure, a barrier or guard;
   ➤ **insulating** any electrical or temperature hazard;
   ➤ **ventilating** away any hazardous fumes or gases either naturally or by the use of extractor fans and hoods (Figure 6.2).

6.3.6 Safe systems of work

Operating procedures or safe systems of work are probably the most common form of control measure used in industry today and may be the most economical and, in some cases, the only practical way of managing a particular risk. They should allow for methodical execution of tasks. The development of safe operating procedures should address the hazards that have been identified in the risk assessment. The system of work describes the safe

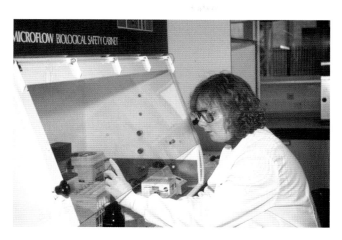

Figure 6.2 Proper control of gases and vapours in a laboratory.

method of performing the job or activity. A safe system of work is a requirement of the Health and Safety at Work Act and is dealt with in detail later.

If the risks involved in the task are high or medium, the details of the system should be in writing and should be communicated to the employee formally in a training session. Details of systems for low-risk activities may be conveyed verbally. There should be records that the employee (or contractor) has been trained or instructed in the safe system of work and that they understand it and will abide by it.

6.3.7 Training

Training helps people acquire the skills, knowledge and attitudes to make them competent in the health and safety aspects of their work. There are generally two types of safety training:

➤ specific safety training (or on the job training) which aims at tasks where training is needed due to the specific nature of such tasks. This is usually a job for supervisors, who by virtue of their authority and close daily contact, are in a position to convert safety generalities to the everyday safe practice procedures that apply to individual tasks, machines, tools and processes;

➤ planned training, such as general safety training, induction training, management training, skill training or refresher courses that are planned by the organization, and relate to managing risk through policy, legislative or organizational requirements that are common to all employees.

Before any employee can work safely, they must be shown safe procedures for completing their tasks. The purpose of safety training should be to improve the safety awareness of employees and show them how to perform their jobs employing acceptable safe behaviour.

See Chapter 4 for more detail on health and safety training.

6.3.8 Information

Organizations need to ensure that they have effective arrangements for identifying and receiving relevant health and safety information from outside the organization including:

➤ ensuring that pertinent health and safety information is communicated to all people in the organization who need it;

➤ ensuring that relevant information is communicated to people outside the organization who require it;

➤ encouraging feedback and suggestions from employees on health and safety matters.

Anyone who is affected by what is happening in the workplace will need to be given safety information. This does not only apply to staff. It can also apply to visitors, members of the public and contractors.

Information to be provided for people in a workplace include:

➤ who is at risk and why;
➤ how to carry out specific tasks safely;
➤ correct operation of equipment;
➤ emergency action;
➤ accident and hazard reporting procedures;
➤ the safety responsibilities of individual people.

Information can be provided in a variety of ways. These include safety signs, posters, newsletters, memos, emails, personal briefings, meetings, toolbox talks, formal training, written safe systems of work and written health and safety arrangements.

For more details see Chapters 1, 4 and 9.

6.3.9 Safety signs

A summary of the Safety Signs and Signals Regulations is given in Chapter 17. All general health and safety signs used in the workplace must include a pictorial symbol categorized by shape, colour and graphic image (Figure 6.3).

All workplaces need to display safety signs of some kind but deciding what is required can be confusing. Here are the basic requirements for the majority of small premises or sites like small construction sites, canteens, shops, small workshop units and offices. This does not cover any signs which food hygiene law may require.

Most requirements are covered by the Health and Safety (Safety Signs and Signals) Regulations 1996. These require signs wherever a risk has not been controlled by other means. For example if a wet area of floor is cordoned off, a warning sign will not be needed, because the barrier will keep people out of the danger area. Signs are not needed where the sign would not reduce the risk or the risk is insignificant.

The following signs are typical of some of the ones most likely to be needed in these premises. Others may be necessary, depending on the hazards and risks present.

(i) Overhead obstacles, construction site and Prohibition notices (Figures 6.4 and 6.5)

(ii) Wet floors

These need to be used wherever a slippery area is not cordoned off. Lightweight stands holding double-sided signs are readily available (Figure 6.6).

(iii) Chemical storage

Where hazardous cleaning chemicals are stored, apart from keeping the store locked, a suitable warning notice should be posted if it is considered this would help to reduce the risk of injury (Figure 6.7).

(iv) Fire safety signs

The Regulations apply in relation to general fire precautions. The guidance under the Fire Safety Order requires signs to comply with BS5499-4 and 5 and the Safety Signs and Signals Regulations.

Since 24 December 1998 the older, text-only 'fire exit' signs should have been supplemented or replaced with pictogram signs. Fire safety signs complying with BS5499-4 and 5 already contain a pictogram and do not require changing (Figure 6.8).

(v) Fire action signs

These and other fire safety signs, such as fire extinguisher location signs, will be needed (Figure 6.9).

(vi) First aid

Signs showing the location of first-aid facilities will be needed. Advice on the action to take in the case of electric shock is no longer a legal requirement but is recommended (Figure 6.10).

Prohibition
A red circular band with diagonal crossbar on a white background, the symbol within the circle to be black denoting a safety sign that indicates that a certain behaviour is prohibited.

Warning
A yellow triangle with black border and symbol within the yellow area denoting a safety sign that gives warning of a hazard.

Mandatory
A blue circle with white symbol denoting a sign that indicates that a specific course of action must be taken.

Safe condition
A green oblong or square with symbol or text in white denoting a safety sign providing information about safe conditions.

Fire equipment
A red oblong or square with symbol in white denoting a safety sign that indicates the location of fire-fighting equipment.

Figure 6.3 Colour categories and shapes of signs.

Overhead load

Safety helmet must be worn

Safety harness must be worn

Not drinkable

No access for pedestrians

Figure 6.4 Examples of warning, mandatory and prohibition signs.

Figure 6.5 Falling object and construction site entrance signs.

Figure 6.9 Examples of fire action signs.

Figure 6.6 Wet floor signs.

Figure 6.10 Examples of first-aid signs.

Toxic   Corrosive

Figure 6.7 Examples of chemical warning signs.

Fire exit

Figure 6.8 Examples of fire safety signs.

Figure 6.11 LPG sign.

(vii) Gas pipes and LPG cylinder stores

LPG cylinder stores should have the sign shown in Figure 6.11.

(viii) No smoking

Areas substantially enclosed should have the sign shown in Figure 6.12 under smokefree legislation. See Chapters 14 and 17 for more details.

Figure 6.12 Smokefree – no-smoking sign.

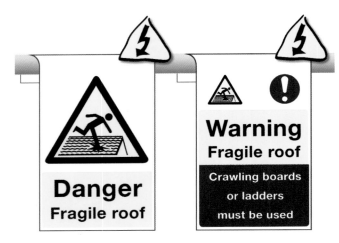

Figure 6.13 Fragile roof sign.

(ix) Fragile roofs

Signs should be erected at roof access points and at the top of outside walls where ladders may be placed (Figure 6.13).

(x) Obstacles or dangerous locations

For example low head height, tripping hazard, etc. – alternating yellow and black stripes.

Figure 6.14 PPE used for loading a textile dye vessel.

(xi) Other signs and posters

- Health and safety law – What you should know (there is a legal requirement to display this poster or distribute equivalent leaflet).
- Certificate of Employer's Liability Insurance (there is a legal requirement to display this).
- Scalds and burns are common in kitchens. A poster showing recommended action is advisable, for example 'First Aid for Burns'.

(xii) Sign checklist

Existing signs should be checked to ensure that:

➤ they are correct and up to date;
➤ they carry the correct warning symbol where appropriate;
➤ they are relevant to the hazard;
➤ they are easily understood;
➤ they are suitably located and not obscured;
➤ they are clean, durable and weatherproof where necessary;
➤ illuminated signs have regular lamp checks;
➤ they are used when required (e.g. 'Caution wet floor' signs);
➤ they are obeyed and effective.

6.3.10 Personal protective equipment

Personal protective equipment (PPE) should only be used as a last resort. There are many reasons for this (Figure 6.14). The most important limitations are that PPE:

➤ only protects the person wearing the equipment, not others nearby;

> relies on people wearing the equipment at all times;
> must be used properly;
> must be replaced when it no longer offers the correct level of protection. This last point is particularly relevant when respiratory protection is used.

The benefits of PPE are:

> it gives immediate protection to allow a job to continue while engineering controls are put in place;
> in an emergency it can be the only practicable way of effecting rescue or shutting down plant in hazardous atmospheres;
> it can be used to carry out work in confined spaces where alternatives are impracticable. But it should never be used to allow people to work in dangerous atmospheres, which are, for example, enriched with oxygen or potentially explosive.

See Chapter 14 for more details on PPE.

6.3.11 Welfare

Welfare facilities include general workplace ventilation, lighting and heating and the provision of drinking water, sanitation and washing facilities. There is also a requirement to provide eating and rest rooms. Risk control may be enhanced by the provision of eye washing and shower facilities for use after certain accidents (Figure 6.15).

Good housekeeping is a very cheap and effective means of controlling risks. It involves keeping the workplace clean and tidy at all times and maintaining good storage systems for hazardous substances and other potentially dangerous items. The risks most likely to be influenced by good housekeeping are fire and slips, trips and falls.

See Chapter 15 for more information on the work environment.

6.3.12 Monitoring and supervision

All risk control measures, whether they rely on engineered or human behavioural controls, must be monitored for their effectiveness with supervision to ensure that they have been applied correctly. Competent people who have a sound knowledge of the equipment or process should undertake monitoring. Checklists are useful to ensure that no significant factor is forgotten. Any statutory inspection or insurance company reports should be checked to see whether any areas of concern were highlighted and if any recommendations were implemented. Details of any accidents, illnesses or other incidents will give an indication on the effectiveness of the risk control measures. Any

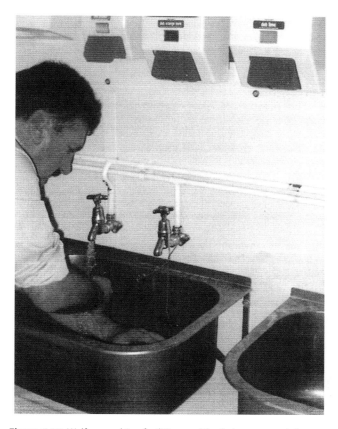

Figure 6.15 Welfare washing facilities: washbasin large enough for people to wash their forearms.

emergency arrangements should be tested during the monitoring phase including first-aid provision.

It is crucial that the operator should be monitored to ascertain that all relevant procedures have been understood and followed. The operator may also be able to suggest improvements to the equipment or system of work. The supervisor is an important source of information during the monitoring process.

Where the organization is involved with shift work, it is essential that the risk controls are monitored on all shifts to ensure the uniformity of application.

The effectiveness and relevance of any training or instruction given should be monitored.

Periodically the risk control measures should be reviewed. Monitoring and other reports are crucial for the review to be useful. Reviews often take place at safety committee and/or at management meetings. A serious accident or incident should lead to an immediate review of the risk control measures in place.

6.4 Controlling health risks

6.4.1 Types of health risk

The principles of control for health risks are the same as those for safety. However, the nature of health risks can make the link between work activities and employee ill health less obvious than in the case of injury from an accident (Figure 6.16).

The COSHH Amendment Regulations 2004 set out the principles of good practice for the control of exposure (see the following box).

Figure 6.17 shows a route map for achieving adequate control.

Unlike safety risks, which can lead to immediate injury, the result of daily exposure to health risks may not manifest itself for months, years and, in some cases, decades. Irreversible health damage may occur before any symptoms are apparent. It is, therefore, essential to develop a preventive strategy to identify and control risks before anyone is exposed to them.

Risks to health from work activities include:

➤ skin contact with irritant substances, leading to dermatitis, etc.;
➤ inhalation of respiratory sensitizers, triggering immune responses such as asthma;
➤ badly designed workstations requiring awkward body postures or repetitive movements, resulting in upper limb disorders, repetitive strain injury and other musculoskeletal conditions;

Figure 6.16 Health risk – checking on the contents.

➤ noise levels which are too high, causing deafness and conditions such as tinnitus;
➤ too much vibration, for example from hand-held tools leading to hand–arm vibration syndrome and circulatory problems;
➤ exposure to ionizing and non-ionizing radiation including ultraviolet in the sun's rays, causing burns, sickness and skin cancer;
➤ infections ranging from minor sickness to life-threatening conditions, caused by inhaling or being contaminated with microbiological organisms;
➤ stress causing mental and physical disorders.

Some illnesses or conditions, such as asthma and back pain, have both occupational and non-occupational

Principles of good practice for the control of exposure to substances hazardous to health

(a) Design and operate processes and activities to minimize emission, release and spread of substances hazardous to health.
(b) Take into account all relevant routes of exposure – inhalation, skin absorption and ingestion – when developing control measures.
(c) Control exposure by measures that are proportionate to the health risk.
(d) Choose the most effective and reliable control options which minimize the escape and spread of substances hazardous to health.
(e) Where adequate control of exposure cannot be achieved by other means, provide, in

combination with other control measures, suitable PPE.
(f) Check and review regularly all elements of control measures for their continuing effectiveness.
(g) Inform and train all employees on the hazards and risks from the substances with which they work and the use of control measures developed to minimize the risks.
(h) Ensure that the introduction of control measures does not increase the overall risk to health and safety.

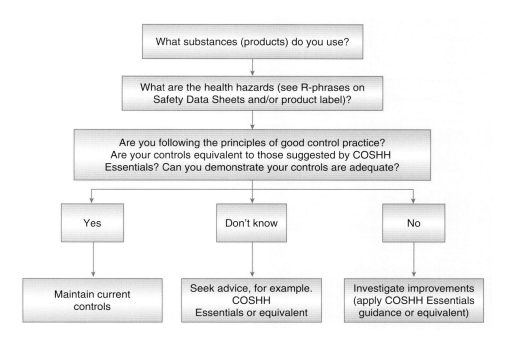

Figure 6.17 Route map for adequate control for SMEs/non-experts. *Source:* HSE.

causes and it may be difficult to establish a definite causal link with a person's work activity or their exposure to particular agents or substances. But, if there is evidence that shows the illness or condition is prevalent among the type of workers to which the person belongs or among workers exposed to similar agents or substances, it is likely that their work and exposure has contributed in some way.

6.4.2 Assessing exposure and health surveillance

Some aspects of health exposure will need input from specialist or professional advisers, such as occupational health hygienists, nurses and doctors. However, considerable progress can be made by taking straightforward measures such as:

➤ consulting the workforce on the design of workplaces;
➤ talking to manufacturers and suppliers of substances and work equipment about minimizing exposure;
➤ enclosing machinery to cut down dust, fumes and noise;
➤ researching the use of less hazardous substances;
➤ ensuring that employees are given appropriate information and are trained in the safe handling of all the substances and materials to which they may be exposed.

To assess health risks and to make sure that control measures are working properly, it may be necessary, for example, to measure the concentration of substances in air to make sure that exposures remain within the assigned WELs. Sometimes health surveillance of workers who may be exposed will be needed. This will enable data to be collected to check control measures and for early detection of any adverse changes to health. Health surveillance procedures available include biological monitoring for bodily uptake of substances, examination for symptoms and medical surveillance – which may entail clinical examinations and physiological or psychological measurements by occupationally qualified registered medical practitioners. The procedure chosen should be suitable for the case concerned. Sometimes a method of surveillance is specified for a particular substance, for example, in the COSHH ACOP. Whenever surveillance is undertaken, a health record has to be kept for the person concerned.

Health surveillance should be supervised by a registered medical practitioner or, where appropriate, it should be done by a suitably qualified person (e.g. an occupational nurse). In the case of inspections for easily detectable symptoms like chrome ulceration or early signs of dermatitis, health surveillance should be done by a suitably trained responsible person. If workers could be exposed to substances listed in Schedule 6 of the COSHH

Regulations, medical surveillance, under the supervision of an HSE employment medical adviser or a doctor appointed by HSE, is required.

6.5 Safe systems of work

6.5.1 What is a safe system of work?

A safe system of work has been defined as:

The integration of personnel, articles and substances in a laid out and considered method of working which takes proper account of the risks to employees and others who may be affected, such as visitors and contractors, and provides a formal framework to ensure that all of the steps necessary for safe working have been anticipated and implemented.

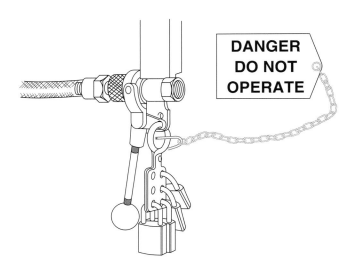

Figure 6.18 Multi-padlocked hasp for locking off an isolation valve.

In simple terms, a safe system of work is a defined method for doing a job in a safe way. It takes account of all foreseeable hazards to health and safety and seeks to eliminate or minimize these. Safe systems of work are normally formal and documented, for example in written operating procedures but, in some cases, they may be verbal.

The particular importance of safe systems of work stems from the recognition that most accidents are caused by a combination of factors (plant, substances, lack of training and/or supervision, etc.). Hence prevention must be based on an integral approach and not one which only deals with each factor in isolation. The adoption of a safe system of work provides this integral approach because an effective safe system:

➤ is based on looking at the job as a whole;
➤ starts from an analysis of all foreseeable hazards, for example physical, chemical, health;
➤ brings together all the necessary precautions, including design, physical precautions, training, monitoring, procedures and PPE.

It follows from this that the use of safe systems of work is in no way a replacement for other precautions, such as good equipment design, safe construction and the use of physical safeguards. However, there are many situations where these will not give adequate protection in themselves, and then a carefully thought-out and properly implemented safe system of work is especially important. The best example is maintenance and repair work, which will often involve, as a first-stage, dismantling the guard or breaking through the containment, which exists for the protection of the ordinary process operator. In some of these operations, a permit-to-work procedure will be the most appropriate type of safe system of work.

The operations covered may be simple or complex, routine or unusual.

Whether the system is verbal or written, and whether the operation it covers is simple or complex, routine or unusual, the essential features are forethought and planning – to ensure that all foreseeable hazards are identified and controlled. In particular, this will involve scrutiny of:

➤ the sequence of operations to be carried out;
➤ the equipment, plant, machinery and tools involved;
➤ chemicals and other substances to which people might be exposed in the course of the work;
➤ the people doing the work – their skill and experience;
➤ foreseeable hazards (health, safety, environment), whether to the people doing the work or to others who might be affected by it;

- practical precautions which, when adopted, will eliminate or minimize these hazards (Figure 6.18);
- the training needs of those who will manage and operate under the procedure;
- monitoring systems to ensure that the defined precautions are implemented effectively.

6.5.2 Legal requirements

The HSW Act section 2 requires employers to provide safe plant and systems of work. In addition, many Regulations made under the Act, such as the Provision and Use of Work Equipment Regulations 1998, require information and instruction to be provided to employees and others. In effect, this is also a more specific requirement to provide safe systems of work. Many of these safe systems, information and instructions will need to be in writing.

There is also a need for employers to provide a safe system of work to fulfil their common law duty of care.

6.5.3 Assessment of what safe systems of work are required

Requirement

It is the responsibility of the management in each organization to ensure that its operations are assessed to determine where safe systems of work need to be developed.

This assessment must, at the same time, decide the most appropriate form for the safe system; that is:

- Is a written procedure required?
- Should the operation only be carried out under permit to work?
- Is an informal system sufficient?

Factors to be considered

It is recognized that each organization must have the freedom to devise systems that match the risk potential of their operations and which are practicable in their situation. However, they should take account of the following factors in making their decision:

- types of risk involved in the operation;
- magnitude of the risk, including consideration of the worst foreseeable loss;
- complexity of the operation;
- past accident and loss experience;
- requirements and recommendations of the relevant health and safety authorities;
- the type of documentation needed;

- resources required to implement the safe system of work (including training and monitoring).

6.5.4 Development of safe systems

(a) Role of competent person

The competent person appointed under the MHSW Regulations and/or the CDM Coordinator appointed under the CDM 2007 Regulations should assist managers to draw up guidelines for safe systems of work. This will include, where necessary, particularly in construction work, method statements. The competent person should prepare suitable forms and should advise management on the adequacy of the safe systems produced.

(b) Role of managers

Primarily management is responsible for providing safe systems of work, as they will know the detailed way in which the task should be carried out.

Management is responsible to ensure that employees are adequately trained in a specific safe system of work and are competent to carry out the work safely. Managers need to provide sufficient supervision to ensure that the system of work is followed and the work is carried out safely. The level of supervision will depend on the experience of the particular employees concerned and the complexity and risks of the task.

When construction work is involved, principal contractors will need to monitor sub-contractors to check that they are providing suitable safe systems of work, have trained their employees and are carrying out the tasks in accordance with the safe systems.

(c) Role of employees/consultation

Many people operating a piece of machinery or a manufacturing process are in the best position to help with the preparation of safe systems of work. Consultation with those employees who will be exposed to the risks, either directly or through their representatives, is also a legal requirement. The importance of discussing the proposed system with those who will have to work under it, and those who will have to supervise its operation, cannot be emphasized enough.

Employees have a responsibility to follow the safe system of work.

(d) Analysis

The safe system of work should be based on a thorough analysis of the job or operation to be covered by the system. The way this analysis is done will depend on the nature of the job/operation.

If the operation being considered is a new one involving high loss potential, the use of formal hazard analysis techniques such as hazard and operability (HAZOP) study, fault tree analysis (FTA) or failure modes and effects analysis should be considered.

However, where the potential for loss is lower, a more simple approach, such as job safety analysis (JSA), will be sufficient. This will involve three key stages:

➤ identification of the key steps in the job/operation – what activities will the work involve?
➤ analysis and assessment of the risks associated with each stage – what could go wrong?
➤ definition of the precautions or controls to be taken – what steps need to be taken to ensure the operation proceeds without danger, either to the people doing the work, or to anyone else?

The results of this analysis are then used to draw up the safe operating procedure or method statement. (See Appendix 6.1 for a suitable form.)

(e) Introducing controls

There are a variety of controls that can be adopted in safe systems of work. They can be split into the following three basic categories:

(i) Technical – these are engineering or process type controls which engineer out or contain the hazard so that the risks are acceptable. For example exhaust ventilation, a machine guard, dust respirator.

(ii) Procedural – these are ways of doing things to ensure that the work is done according to the procedure, legislation or cultural requirements of the organization. For example a supervisor must be involved, the induction course must be taken before the work commences, a particular type of form or a person's signature must be obtained before proceeding, the names of the workforce must be recorded.

(iii) Behavioural – these are controls which require a certain standard of behaviour from individuals or groups of individuals. For example no smoking is permitted during the task, hard hats must be worn, all lifts are to be in tandem between two workers.

6.5.5 Preparation of safe systems

A checklist for use in the preparation of safe systems of work is set out as follows:

➤ What is the work to be done?
➤ What are the potential hazards?

➤ Is the work covered by any existing instructions or procedures? If so, to what extent (if any) do these need to be modified?
➤ Who is to do the work?
➤ What are their skills and abilities – is any special training needed?
➤ Under whose control and supervision will the work be done?
➤ Will any special tools, protective clothing or equipment (e.g. breathing apparatus) be needed? Are they ready and available for use?
➤ Are the people who are to do the work adequately trained to use the above?
➤ What isolations and locking-off will be needed for the work to be done safely?
➤ Is a permit to work required for any aspect of the work?
➤ Will the work interfere with other activities? Will other activities create a hazard to the people doing the work?
➤ Have other departments been informed about the work to be done, where appropriate?
➤ How will the people doing the work communicate with each other?
➤ Have possible emergencies and the action to be taken been considered?
➤ Should the emergency services be notified?
➤ What are the arrangements for handover of the plant/equipment at the end of the work? (For maintenance/project work, etc.)
➤ Do the planned precautions take account of all foreseeable hazards?
➤ Who needs to be informed about or receive copies of the safe system of work?
➤ What arrangements will there be to see that the agreed system is followed and that it works in practice?
➤ What mechanism is there to ensure that the safe system of work stays relevant and up-to-date?

6.5.6 Documentation

Safe systems of work should be properly documented.

Wherever possible, they should be incorporated into normal process operating procedures. This is so that:

➤ health and safety are seen as an integral part of, and not add-on to, normal production procedures;
➤ the need for operators and supervisors to refer to separate manuals is minimized.

Whatever method is used, all written systems of work should be signed by the relevant managers to indicate approval or authorization. Version numbers should be included so that it can quickly be verified that the most up-to-date version is in use. Records should be kept of copies of the documentation, so that all sets are amended when updates and other revisions are issued.

As far as possible, systems should be written in a non-technical style and should specifically be designed to be as intelligible and user-friendly as possible. It may be necessary to produce simple summary sheets which contain all the key points in an easy-to-read format.

6.5.7 Communication and training

People doing work or supervising work must be made fully aware of the laid-down safe systems that apply. The preparation of safe systems will often identify a training need that must be met before the system can be implemented effectively.

In addition, people should receive training in how the system is to operate. This applies not only to those directly involved in doing the work but also to supervisors/managers who are to oversee it.

In particular, the training might include:

➤ why a safe system is needed;
➤ what is involved in the work;
➤ the hazards which have been identified;
➤ the precautions which have been decided and, in particular
   ● the isolations and locking-off required, and how this is to be done
   ● details of the permit-to-work system, if applicable
   ● any monitoring (e.g. air testing) which is to be done during the work, or before it starts
   ● how to use any necessary PPE
   ● emergency procedures.

6.5.8 Monitoring safe systems

Safe systems of work should be monitored to ensure that they are effective in practice. This will involve:

➤ reviewing and revising the systems themselves, to ensure they stay up-to-date;
➤ inspecting to identify how fully they are being implemented.

In practice, these two things go together, as it is likely that a system that is out of date will not be fully implemented by the people who are intended to operate it.

All organizations are responsible for ensuring that their safe systems of work are reviewed and revised as appropriate. Monitoring of implementation is part of all line managers' normal operating responsibilities, and should also take place during health and safety audits.

6.6 Lone workers

People who work by themselves without close or direct supervision are found in many work situations. In some cases they are the sole occupant of small workshops or warehouses; they may work in remote sections of a large site; they may work out of normal hours, like cleaners or security personnel; they may be working away from their main base as installers, or maintenance people; they could be people giving a service, like domiciliary care workers, drivers and estate agents.

There is no general legal reason why people should not work alone but there may be special risks which require two or more people to be present, for example, during entry into a confined space in order to effect a rescue. It is important to ensure that a lone worker is not put at any higher risk than other workers. This is achieved by carrying out a specific risk assessment and introducing special protection arrangements for their safety. People particularly at risk, like young people or women, should also be considered. People's overall health and suitability to work alone should be taken into account.

Procedures may include:

➤ periodic visits from the supervisor to observe what is happening;
➤ regular voice contact between the lone worker and the supervisor;
➤ automatic warning devices to alert others if a specific signal is not received from the lone worker;
➤ other devices to raise the alarm, which are activated by the absence of some specific action;
➤ checks that the lone worker has returned safely home or to their base;
➤ special arrangements for first aid to deal with minor injuries – this may include mobile first-aid kits;
➤ arrangements for emergencies – these should be established and employees trained.

6.7 Permits to work

6.7.1 Introduction

Safe systems of work are crucial in work such as the maintenance of chemical plant where the potential risks are high and the careful coordination of activities and precautions is essential to safe working. In this situation and others of similar risk potential, the safe system of work is likely to take the form of a permit-to-work procedure.

The permit-to-work procedure is a specialized type of safe system of work for ensuring that potentially very dangerous work (e.g. entry into process plant and other confined spaces) is done safely.

Although this procedure has been developed and refined by the chemical industry, the principles of the permit-to-work procedure are equally applicable to the management of complex risks in other industries.

The fundamental principle is that certain defined operations are prohibited without the specific permission of a responsible manager, this permission being only granted once stringent checks have been made to ensure that all necessary precautions have been taken and that it is safe for work to go ahead.

The people doing the work take on responsibility for following and maintaining the safeguards set out in the permit, which will define the work to be done (no other work being permitted) and the timescale in which it must be carried out.

To be effective, the permit system requires the training needs of those involved to be identified and met, and that monitoring procedures ensure that the system is operating as intended.

6.7.2 Permit-to-work procedures

The permit-to-work procedure is a specialized type of safe system of work under which certain categories of high risk–potential work may only be done with the specific permission of an authorized manager. This permission (in the form of the permit to work) will be given only if the laid-down precautions are in force and have been checked.

The permit document will typically specify:

➤ what work is to be done;
➤ the plant/equipment involved, and how it is identified;
➤ who is authorized to do the work;
➤ the steps which have already been taken to make the plant safe;
➤ potential hazards which remain, or which may arise as the work proceeds;
➤ the precautions to be taken against these hazards;
➤ for how long the permit is valid;
➤ that the equipment is released to those who are to carry out the work.

In accepting the permit, the person in charge of doing the authorized work normally undertakes to take/maintain whatever precautions are outlined in the permit. The permit will also include spaces for:

➤ signature certifying that the work is complete;
➤ signature confirming re-acceptance of the plant/equipment. See Figure 6.19 and Appendix 6.3.

6.7.3 Principles

Permit systems must adhere to the following eight principles:

1. Wherever possible, and especially with routine jobs, hazards should be eliminated so that the work can be done safely without requiring a permit to work.
2. Although the Site Manager may delegate the responsibility for the operation of the permit system, the overall responsibility for ensuring safe operation rests with him/her.
3. The permit must be recognized as the master instruction which, until it is cancelled, overrides all other instructions.
4. The permit applies to everyone on site, including contractors.
5. Information given in a permit must be detailed and accurate. It must state:
   (a) which plant/equipment has been made safe and the steps by which this has been achieved;
   (b) what work may be done;
   (c) the time at which the permit comes into effect.
6. The permit remains in force until the work has been completed and the permit is cancelled by the person who issued it or by the person nominated by management to take over the responsibility (e.g. at the end of a shift or during absence).
7. No work other than that specified is authorized. If it is found that the planned work has to be changed, the existing permit should be cancelled and a new one issued.
8. Responsibility for the plant must be clearly defined at all stages.

6.7.4 Work requiring a permit

The nature of permit-to-work procedures will vary in their scope depending on the job and the risks involved. However, a permit-to-work system is unlikely to be needed where, for example:

(a) the assessed risks are low and can be controlled easily;
(b) the system of work is very simple;

Associated Permits (specify type, if none state N/A)	On the Job Copy	Best Practice Company Ltd
Type:	Prefix & Number:	
Type:	Prefix & Number:	
Type:	Prefix & Number:	**GENERAL WORK PERMIT** Number GWP 0000

Section 1 ISSUE PLEASE PRINT DETAILS

Permit Receiver/Competent Person in charge of work:	Location of work/Equipment to be worked on:
Names of persons detailed to carry out work:	Details of work to be done:
Risk Assessment attached? Yes ☐ No ☐ Reference:	Safety Method Statement or Safe System of Work attached? Yes ☐ No ☐ N/A ☐ Reference:

Section 2 ISOLATION of electrical or mechanical plant, liquid or gas pipeline or other energy source – give details

Item:	Lock location & reference:	Isolated by: Print name
Attached Isolation Sheet. Yes ☐ No ☐ Reference:	Location of Keys:	

Section 3 PREPARATIONS/PRECAUTIONS Tick box

	Yes	No	N/A	Section 3 Continued
				8. Other Precautions/Control Measures required (specify)
1. Has every source of energy been isolated?				
2. Have all isolations been tagged?				
3. Have all isolations been tested?				
4a. Does the standard of pipeline isolation meet minimum Corporate standards?				
4b. If not, specify additional precautions				

Section 4 PERSONAL PROTECTIVE EQUIPMENT (A.P. tick box)

				Site Standard		Other (specify):	
				Plus:			
5. Are vessels/pipes free of toxic/flammable, gas, dangerous sludge & depressurized?				Safety Goggles			
				Hearing protection			
6. Are asbestos containing materials present?				Respiratory protection			
7. Are the risk assessment control measures implemented?				Specify type:			

Section 5 TIME LIMITS

from _____ hrs on __/__/__ to _____ hrs on __/__/__

Operational Period Max 24 hours   Non-Operational Period Max 28 days   See overleaf for details of time extensions allowed.

Section 6 Authorization Permit Issuer/Authorized Person	**Section 7 RECEIPT** Permit Receiver/Competent Person in charge of work
I certify that it is safe to work in the area/on the equipment detailed in Section 1 above and that all safety measures detailed in Sections 2–4 have been carried out/complied with. **ALL OTHER PARTS ARE DANGEROUS**	I have read, understood and accept the requirements of this permit. I will ensure that everyone working under my supervision will strictly follow the requirements of this permit. I have checked the isolations.
Print Name: Date: Time:	Print Name: Date: Time:
Signed:	Signed:

Section 8 SUSPENSION OF GENERAL WORK PERMIT   This is an exception and must be signed on the front of the 'on the job' copy.

I certify that the task for which this Permit was issued has now been suspended. We have agreed and implemented a procedure which complies with the criteria noted in the checklist in Section 13 overleaf.

Signed Permit Receiver/Competent Person in charge of work	Signed Authorized Person
Date: Time:	Date: Time:

- -

The plant has been re-isolated and the original permit conditions apply

Signed Permit Receiver/Competent Person in charge of work	Signed Authorized Person
Date: Time:	Date: Time:

Section 9 CLEARANCE Permit Receiver/Competent Person in charge of work	**Section 10 CANCELLATION** Permit Issuer/Authorized Person
I certify that the work for which the permit was issued is now COMPLETED and that all persons at risk have been WITHDRAWN and WARNED that it is NO LONGER SAFE to work on the plant specified on this permit and that GEAR, TOOLS and EQUIPMENT are all CLEAR.	This permit to work is hereby CANCELLED, all plant is restored to safe operating conditions, including the replacement of guards.
Print Name: Date: Time:	Print Name: Date: Time:
Signed:	Signed:

GWPFrOTJCDraft7   20/12/2008

Figure 6.19 Permit to work.

(c) other work being done nearby cannot affect the work concerned in say a confined space entry, or a welding operation.

However, where there are high risks and the system of work is complex and other operations may interfere, a formal permit to work should be used.

The main types of permit and the work covered by each are identified below. Appendix 6.2 illustrates the essential elements of a permit-to-work form with supporting notes on its operation.

General permit

The general permit should be used for work such as:

➤ alterations to or overhaul of plant or machinery where mechanical, toxic or electrical hazards may arise;
➤ work on or near overhead crane tracks;
➤ work on pipelines with hazardous contents;
➤ repairs to railway tracks, tippers, conveyors;
➤ work with asbestos-based materials;
➤ work involving ionizing radiation;
➤ roof work;
➤ excavations to avoid underground services.

Confined space permit

Confined spaces include chambers, tanks (sealed and open-top), vessels, furnaces, ducts, sewers, manholes, pits, flues, excavations, boilers, reactors and ovens.

Many fatal accidents have occurred where inadequate precautions were taken before and during work involving entry into confined spaces (Figure 6.20). The two main hazards are the potential presence of toxic or other dangerous substances and the absence of adequate oxygen. In addition, there may be mechanical hazards (entanglement on agitators), ingress of fluids, risk of engulfment in a free flowing solid like grain or sugar, and raised temperatures. The work to be carried out may itself be especially hazardous when done in a confined space, for example cleaning using solvents, cutting/welding work. Should the person working in a confined space get into difficulties for whatever reason, getting help in and getting the individual out may prove difficult and dangerous.

Stringent preparation, isolation, air testing and other precautions are therefore essential and experience shows that the use of a confined space entry permit is essential to confirm that all the appropriate precautions have been taken.

The Confined Spaces Regulations 1997 are summarized in Chapter 17. They detail the specific controls that are necessary when people enter confined spaces.

Figure 6.20 Entering a confined space.

Work on high-voltage apparatus (including testing)

Work on high-voltage apparatus (over about 600V) is potentially high risk. Hazards include:

➤ possibly fatal electric shock/burns to the people doing the work;
➤ electrical fires/explosions;
➤ consequential danger from disruption of power supply to safety-critical plant and equipment.

In view of the risk, this work must only be done by suitably trained and competent people acting under the terms of a high-voltage permit.

Hot work

Hot work is potentially hazardous as:

➤ a source of ignition in any plant in which flammable materials are handled;
➤ a cause of fires in all processes, regardless of whether flammable materials are present.

Hot work includes cutting, welding, brazing, soldering and any process involving the application of a naked flame. Drilling and grinding should also be included where a flammable atmosphere is potentially present (Figure 6.21).

Hot work should therefore be done under the terms of a hot work permit, the only exception being where hot work is done in a designated maintenance area suitable for the purpose.

Figure 6.21 A Hot Work permit is usually essential except in designated areas.

6.7.5 Responsibilities

The effective operation of the permit system requires the involvement of many people. The following specific responsibilities can be identified:

(*Note*: all appointments, definitions of work requiring a permit, etc. must be in writing. All the categories of people identified below should receive training in the operation of the permit system as it affects them.)

Site manager

➤ has overall responsibility for the operation and management of the permit system;
➤ appoints a senior manager (normally the Chief Engineer) to act as a senior authorized person.

Senior authorized person

➤ is responsible to the Site Manager for the operation of the permit system;
➤ defines the work on the site which requires a permit;
➤ ensures that people responsible for this work are aware that it must only be done under the terms of a valid permit;

➤ appoints all necessary authorized persons;
➤ appoints a deputy to act in his/her absence.

Authorized persons

➤ issue permits to competent persons and retain copies;
➤ personally inspect the site to ensure that the conditions and proposed precautions are adequate and that it is safe for the work to proceed;
➤ accompany the competent person to the site to ensure that the plant/equipment is correctly identified and that the competent person understands the permit;
➤ cancel the permit on satisfactory completion of the work.

Competent persons

➤ receive permits from authorized persons;
➤ read the permit and make sure they fully understand the work to be done and the precautions to be taken;
➤ signify their acceptance of the permit by signing both copies;
➤ comply with the permit and make sure those under their supervision similarly understand and implement the required precautions;
➤ on completion of the work, return the permit to the authorized person who issued it;.

Operatives

➤ read the permit and comply with its requirements, under the supervision of the competent person.

Specialists

A number of permits require the advice/skills of specialists in order to operate effectively. Such specialists may include chemists, electrical engineers, health and safety advisers and fire officers. Their role may involve:

➤ isolations within his/her discipline – for example electrical work;
➤ using suitable techniques and equipment to monitor the working environment for toxic or flammable materials, or for lack of oxygen;
➤ giving advice to managers on safe methods of working.

Specialists must not assume responsibility for the permit system. This lies with the Site Manager and the senior authorized person.

Engineers (and others responsible for work covered by permits)

➤ ensure that permits are raised as required.

Contractors

The permit system should be applied to contractors in the same way as to direct employees.

The contractor must be given adequate information and training on the permit system, the restrictions it imposes and the precautions it requires.

6.8 Emergency planning procedures

6.8.1 Introduction

Most of this chapter is about the principles of control to prevent accidents and ill health. Emergency procedures, however, are about control procedures and equipment to limit the damage to people and property caused by an incident. Local fire and rescue authorities will often be involved and are normally prepared to give advice to employers.

Under Regulation 8 of the Management of Health and Safety at Work Regulations 1999, procedures must be established and set in motion when necessary to deal with serious and imminent danger to persons at work. Necessary links must be maintained with local authorities, particularly with regard to first aid, emergency medical care and rescue work (Figure 6.22).

Although fire is the most common emergency likely to be faced, there are many other possibilities, which should be considered including:

➤ gas explosion;
➤ electrical burn or electrocution;
➤ escape of toxic gases or fumes;

Figure 6.22 Emergency services at work.

➤ discovery of dangerous dusts like asbestos in the atmosphere;
➤ terrorist threat;
➤ large vehicle crashing into the premises;
➤ aircraft crash if near a flight path;
➤ spread of highly infectious disease;
➤ severe weather with high winds and flooding.

For fire emergencies, see Chapter 13.

6.8.2 Supervisory duties

A member of the site staff should be nominated to supervise and co-ordinate all emergency arrangements. This person should be in a senior position or at least have direct access to a senior manager. Senior members of the staff should be appointed as departmental fire/emergency procedure wardens, with deputies for every occasion of absence, however, brief. They should ensure that the following precautions are taken:

➤ Everyone on site can be alerted to an emergency.
➤ Everyone on site knows what signal will be given for an emergency and knows what to do.
➤ Someone who has been trained in what to do is on site and ready to co-ordinate activities.
➤ Emergency routes are kept clear, signed and adequately lit.
➤ There are arrangements for calling the fire and rescue services and to give them special information about high-hazard work, for example, in tunnels or confined spaces.
➤ There is adequate access to the site for the emergency services and this is always kept clear.
➤ Suitable arrangements for treating and recovering injured people are set up.
➤ Someone is posted to the site entrance to receive and direct the emergency services.

6.8.3 Assembly and roll call

Assembly points should be established for use in the event of evacuation. It should be at a position, preferably undercover, which is unlikely to be affected at the time of emergency. In some cases, it may be necessary to make mutual arrangements with the client or occupiers of nearby premises.

In the case of small sites, a complete list of the names of all staff should be maintained so that a roll call can be made if evacuation becomes necessary.

In those premises where the number of staff would make a single roll call difficult, each area warden should maintain a list of the names of employees and contractors in their area. Roll call lists must be updated regularly.

First aid at work

6.9.1 Introduction

People at work can suffer injuries or fall ill. It does not matter whether the injury or the illness is caused by the work they do. What is important is that they receive immediate attention and that an ambulance is called in serious cases. First aid at work (FAW) covers the arrangements employers must make to ensure this happens. It can save lives and prevent minor injuries becoming major ones.

The Health and Safety (First-Aid) Regulations 1981 require employers to provide adequate and appropriate equipment, facilities and personnel to enable first aid to be given to employees if they are injured or become ill at work.

What is adequate and appropriate will depend on the circumstances in a particular workplace.

The minimum first-aid provision on any work site is:

➤ a suitably stocked first-aid box;
➤ an appointed person to take charge of first-aid arrangements.

It is also important to remember that accidents can happen at any time. First-aid provision needs to be available at all times people are at work.

Many small firms will only need to make the minimum first-aid provision. However, there are factors which might make greater provision necessary. The following checklist covers the points that should be considered.

6.9.2 Aspects to consider

The risk assessments carried out under the MHSW and COSHH Regulations should show whether there are any specific risks in the workplace. The following should be considered:

➤ Are there hazardous substances, dangerous tools and equipment; dangerous manual handling tasks, electrical shock risks, dangers from neighbours or animals?
➤ Are there different levels of risk in parts of the premises or site?
➤ What is the accident and ill-health record, and type and location of incidents?
➤ What is the total number of persons likely to be on site?
➤ Are there young people, pregnant or nursing mothers on site, employees with disabilities or special health problems?
➤ Are the facilities widely dispersed with several buildings or compact in a multi-storey building?
➤ What is the pattern of working hours? Does it involve night work?
➤ Is the site remote from emergency medical services?
➤ Do employees travel a lot or work alone?
➤ Do any employees work at sites occupied by other employers?
➤ Are members of the public regularly on site?

6.9.3 Impact on first-aid provision if risks are significant

First-aiders may need to be appointed if risks are significant.

This will involve a number of factors which must be considered, including:

➤ training for first aiders;
➤ additional first-aid equipment and the contents of the first-aid box;
➤ siting of first-aid equipment to meet the various demands in the premises. For example provision of equipment in each building or on several floors; There needs to be first-aid provision at all times during working hours;
➤ informing local medical services of the site and its risks;
➤ any special arrangements that may be needed with the local emergency services.

Any first-aid room provided under these Regulations must be easily accessible to stretchers and to other equipment needed to convey patients to and from the room. They must be sign posted according to the Safety Signs and Signals Regulations (Figure 6.23).

If employees travel away from the site, the employer needs to consider:

➤ issuing personal first-aid kits and providing training;
➤ issuing mobile phones to employees;

Figure 6.23 (Left) First-aid and stretcher sign; (Right) First-aid sign.

➤ making arrangements with employers on other sites.

Although there are no legal responsibilities for non-employees, the HSE strongly recommends that they are included in any first-aid provision.

6.9.4 Contents of the first-aid box

There is no standard list of items to put in a first-aid box. It depends on what the employer assesses the needs to be. Where there is no special risk in the workplace, a minimum stock of first-aid items is listed (see Table 6.2).

Tablets or medicines should not be kept in the first-aid box. Table 6.2 shows a suggested contents list only; equivalent but different items will be considered acceptable.

Table 6.2 Contents of first-aid box – low risk	
Stock for up to 50 persons:	
A leaflet giving general guidance on first aid, for example HSE leaflet *Basic advice on first aid at work*.	
● Medical adhesive plasters	40
● Sterile eye pads	4
● Individually wrapped triangular bandages	6
● Safety pins	6
● Individually wrapped: medium sterile unmedicated wound dressings	8
● Individually wrapped: large sterile unmedicated wound dressings	4
● Individually wrapped wipes	10
● Paramedic shears	1
● Pairs of latex gloves	2
● Sterile eyewash if no clean running water	2

6.9.5 Appointed persons

An appointed person is someone who is appointed by management to:

➤ take charge when someone is injured or falls ill. This includes calling an ambulance if required;
➤ look after the first-aid equipment, for example keeping the first-aid box replenished;
➤ keeping records of treatment given.

Appointed persons should never attempt to give first aid for which they are not competent. Short emergency first-aid training courses are available. Remember that an appointed person should be available at all times when people are at work on site – this may mean appointing more than one. The training should be repeated every 3 years to keep up-to-date.

6.9.6 A first aider

A first aider is someone who has undergone an HSE-approved training course in administering FAW and holds a current FAW certificate. Lists of local training organizations are available from the local environmental officer or HSE offices. The training should be repeated every 3 years to maintain a valid certificate and keep the first aider up-to-date.

It is not possible to give hard and fast rules on when or how many first aiders or appointed persons might be needed. This will depend on the circumstances of each particular organization or work site. Table 6.3 offers suggestions on how many first aiders or appointed persons might be needed in relation to categories of risk and number of employees. The details in the table are suggestions only; they are not definitive, nor are they a legal requirement.

Employees must be informed of the first-aid arrangements. Putting up notices telling staff who and where the first aiders or appointed persons are and where the first-aid box is will usually be enough. Special arrangements will be needed for employees with reading or language difficulties.

To ensure cover at all times when people are at work and where there are special circumstances, such as remoteness from emergency medical services, shift work or sites with several separate buildings, there may need to be more first-aid personnel than set out in Table 6.3.

6.9.7 Implementation of changes to first-aid training and approval arrangements

HSE recently conducted a consultation exercise on draft guidance for employers and first-aid training providers to

Table 6.3 Number of first-aid personnel

Category of risk	Numbers employed at any location	Suggested number of first-aid personnel
Lower risk		
For example shops and offices, libraries	Fewer than 50	At least one appointed person
	50–100 More than 100	At least one first aider One additional first aider for every 100 employees
Medium risk		
For example light engineering and assembly work, food processing, warehousing	Fewer than 20	At least one appointed person
	20–100 More than 100	At least one first aider for every 50 employed (or part thereof) One additional first aider for every 100 employees
Higher risk		
For example most construction, slaughterhouses, chemical manufacture, extensive work with dangerous machinery or sharp instruments	Fewer than 5	At least one appointed person
	5–50 More than 50	At least one first aider One additional first aider for every 50 employees

Source: HSE.

support changes to first-aid training and approval arrangements. In June 2008, the HSE set a date for implementing the changes and identify the guidance that will be put in place to support employers and training providers.

HSE intends to introduce the changes to first-aid training courses from **1 October 2009** so employer duty holders will need to implement them from this date. Guidance on the changes to approval arrangements will be available for first-aid training providers at an earlier date to enable them to prepare for the new training course structure.

Detailed guidance for employers will be available as a revision of the current document produced by HSE: *First Aid at Work – The Health and Safety (First-Aid) Regulations 1981 – Approved Code of Practice and Guidance (L74)*. Within this, it is only the guidance that is being revised – the Regulations and Approved Code of Practice will remain the same. HSE also intends to revise its guidance for small- and medium-sized enterprises in *First Aid at Work – Your Questions Answered (INDG214)*. Both publications will be available for 1 October 2009.

Employers will not be required to retrain all their first aiders as soon as the implementation date is reached. First aiders with a valid FAW certificate will only enter the new arrangements when their certificate expires. This means that it will take 3 years post implementation before all first aiders in the workplace are captured within the new training structure.

Detailed guidance for first-aid training organizations is available on the HSE Web site: First-aid training and qualifications for the purposes of the Health and Safety (First-Aid) Regulations 1981, www.hse.gov.uk/pubns/web41.pdf.

The intention is that in future, employers will be able to send suitable employees on either a 6-hour (minimum) emergency first aid at work (EFAW) or an 18-hour (minimum) FAW course, based on the findings of their first-aid needs assessment (see Figure 6.24). After 3 years, first aiders will need to complete another course (either a 6-hour EFAW or 12-hour FAW requalification course, as appropriate) to obtain a new certificate. Within any 3-year certification period, first aiders should complete two annual refresher courses, covering basic life support/skills updates, that will each last for at least 3 hours.

6.10 Sources of reference

Successful Health and Safety Management HSG65, HSE Books 1997 ISBN 978 0 7176 1276 5.

The Management of Health and Safety at Work Regulations 1999 L21, HSE Books 2000 ISBN 978 0 7176 2488 1.

First Aid at Work, The Health and Safety (First-Aid) Regulations 1981 L74 HSE Books 1997 ISBN 978 0 7176 1050 1. (To be revised in October 2009.)

Safe work in confined spaces. Confined Spaces Regulations 1997 L101 HSE Books 1997 ISBN 978 0 7176 1405 9.

Safety Signs and Signals (L64) HSE Books 1996 ISBN 978 0 7176 0870 6.

Relevant statutory provisions

The Management of Health and Safety at Work Regulations 1999.

The Health and Safety at Work etc. Act 1974 – section 2.

The Personal Protective Equipment at Work Regulations 1992.

Health and Safety (First Aid) Regulations 1981.

Confined Spaces Regulations 1997.

The Health and Safety (Safety Signs and Signals) Regulations 1996.

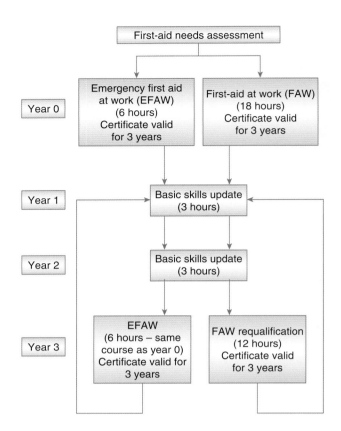

Figure 6.24 Flow chart showing courses to be completed over a 3-year certification period for EFAW and FAW. The dotted line indicates the route to be taken in subsequent years after completion of the relevant course at year 3.

6.11 Practice NEBOSH Questions for Chapter 6

1. With respect to the management of risk within the workplace:

   (i) Explain the meaning of the term 'hierarchy of control'.

   (ii) Outline, with examples, the standard hierarchy that should be applied with respect to controlling health and safety risks in the workplace.

2. **Outline** the possible effects on health and safety of poor standards of housekeeping in the workplace.

3. **Name** and **describe FOUR** classes of safety sign prescribed by the Health and Safety (Safety Signs and Signals) Regulations 1996.

4. **State** the shape and colour, **AND give** a relevant example, of **each** of the following types of safety sign:
   (i) prohibition
   (ii) warning
   (iii) mandatory
   (iv) emergency escape or first aid.

5. **Explain** why personal protective equipment (PPE) should be considered as a last resort in the control of occupational health hazards.

6. (i) **Explain** the meaning of the term 'safe system of work'.
   (ii) **Describe** the enforcement action that could be taken by an enforcing authority when a safe system of work has not been implemented.

7. **Outline** the factors that should be considered when developing a safe system of work.

8. **Identify EIGHT** sources of information that might usefully be consulted when developing a safe system of work.

9. In relation to a safe system of work, **outline** the factors that might influence the choice of risk control measures.

10. **Outline** the reasons why employees might fail to comply with safety procedures at work.

11. Due to an increase in knife related accidents among hotel kitchen staff who use the sharp tools in the preparation of food for the restaurant, a safe system of work is to be developed to minimize the risk of injury to this group of employees.
   (i) **Identify** the legal requirements under which the employer must provide a safe system of work.
   (ii) **Describe** the issues to be addressed when developing the safe system of work for the hotel kitchen staff who use the knives as part of their work.
   (iii) **Outline** the ways in which the employer could motivate the hotel kitchen staff to follow the safe system of work.

12. (i) **Define** the term 'permit-to-work system'.
   (ii) **Outline THREE** types of work situation that may require a permit-to-work system, giving reasons in **EACH** case for the requirement.
   (iii) **Outline** the specific details that should be included in a permit-to-work for entry into a confined space.

13. (i) **Identify TWO** specific work activities for which a permit-to-work system might be needed.
   (ii) **Outline** the key elements of a permit-to-work system.

14. **Outline** the issues to be addressed in a training session on the operation of a permit-to-work system.

15. With reference to the Confined Space Regulations 1997:
   (i) **Define** the meaning of 'confined space', giving **TWO** workplace examples.
   (ii) **Outline** specific hazards associated with working in confined spaces.
   (iii) **Identify FOUR** 'specified risks' that may arise from work in a confined space.

16. **Outline** the precautions that should be taken in order to ensure the safety of employees undertaking maintenance work in an underground storage vessel.

17. **Outline** the factors that should be considered when preparing a procedure to deal with a workplace emergency.

18. (i) **Identify FOUR** types of emergency procedure that an organization might need to have in place.
   (ii) **Explain** why visitors to a workplace should be informed of the emergency procedures.

19. (i) **Identify THREE** types of emergency in the workplace for which employees may need to be evacuated.
   (ii) **Explain** why it is important to develop workplace procedures to enable the safe evacuation of employees during an emergency.

20. A car maintenance workshop located adjacent to a river was flooded. Repairs were needed to structural steelwork, garage pits and basements, electrical equipment and flammable stores.
   (i) **Identify FOUR** emergencies that may occur during the repairs.
   (ii) **Outline** why emergency procedures are needed.
   (iii) **Identify FOUR** types of permit that may be required to undertake the repairs **AND outline** a relevant task that requires **EACH** of the specified permits.

21. (i) **Identify** the **TWO** main functions of first-aid treatment.
   (ii) **Outline** the factors to consider when making an assessment of first-aid provision in a workplace.

Appendix 6.1 Job safety analysis form

JOB SAFETY ANALYSIS					
Job			Date		
Department			Carried out by		
Description of job					
Legal requirements and guidance					
Task steps	Hazards	Consequence	Severity	Risk C × S	Controls
Safe system of work					
Job Instruction					
Training requirements					
Review date					

Appendix 6.2 Essential elements of a permit-to-work form

2. Should have any reference to other relevant permits or isolation certificates.	1 Permit Title	2 Permit Number
	3 Job Title	
	4 Plant identification	
5. Description of work to be done and its limitations.	5 Description of Work	
6. Hazard identification including residual hazards and hazards introduced by the work.	6 Hazard Identification	
7. Precautions necessary – person(s) who carry out precautions, e.g. isolations, should sign that precautions have been taken.	7 Precautions necessary	7 Signatures ....
8. Protective equipment needed for the task.	8 Protective Equipment	
9. Authorization signature confirming that isolations have been made and precautions taken, except where these can only be taken during the work. Date and time duration of permit.	9 Authorization	
10. Acceptance signature confirming understanding of work to be done, hazards involved and precautions required. Also confirming permit information has been explained to all workers involved.	10 Acceptance	
11. Extension/Shift handover signature confirming checks have been made that plant remains safe to be worked on, and new acceptance/workers made fully aware of hazards/precautions. New time expiry given.	11 Extension/Shift handover	
12. Hand back signed by acceptor certifying work completed. Signed by issuer certifying work completed and plant ready for testing and re-commissioning.	12 Hand back	
13. Cancellation certifying work tested and plant satisfactorily recommissioned.	13 Cancellation	

Source: HSE.

Appendix 6.3 Asbestos examples of safe systems of work

Standards required

All work involving asbestos in any form will be carried out in accordance with the current Control of Asbestos Regulations and Approved Code of Practice. Any asbestos removal must be done in accordance with the Asbestos Licensing Regulations. The use of new asbestos containing materials is prohibited.

Planning procedures

All work will be tendered for or negotiated in accordance with the approved standards.

The Contracts Manager will ascertain at an early stage whether asbestos in any form is likely to be present or used on the site. If details provided by the client are inconclusive, then an occupational hygiene specialist will be asked to take and analyse samples.

Method statements will be prepared by the Contracts Manager in conjunction with an occupational hygiene specialist, and, where necessary, a licensed asbestos removal contractor will be selected to carry out the work.

The Contracts Manager will ensure that any requirement to give notice of the work to the Health and Safety Executive is complied with.

Where any work involving asbestos materials not subject to the licensing requirements is to be carried out by employees, the working methods, precautions, safety equipment, protective clothing, special tools, etc. will be arranged by the Contracts Manager.

Supervision

Before work starts, all information on working methods and precautions agreed will be issued to site supervision by the Contracts Manager in conjunction with the Safety Adviser/Officer.

The Site Supervisor in conjunction with management will ensure that the licensed contractor contracted to carry out the removal work has set up operations in accordance with the agreed method statement and that the precautions required are fully maintained throughout the operation so that others not involved are not exposed to risk.

Where necessary, smoke testing of the enclosure and monitoring of airborne asbestos fibre concentrations outside the removal enclosure will be carried out by an occupational hygiene specialist.

The Site Supervisor will ensure that when removal operations have been completed, no unauthorized person enters the asbestos removal area until clearance samples have been taken by an occupational hygiene specialist and confirmation received that the results are satisfactory.

Where employees are required to use or handle materials containing asbestos not subject to the Licensing Regulations, the Site Supervisor will ensure that the appropriate safety equipment and protective clothing is provided and that the agreed safe working procedures are understood by employees and complied with.

All warning labels will be left in place on any asbestos materials used on site.

Examples

1. Painting undamaged asbestos insulating boards

    Description

    This task guidance sheet can be used where undamaged asbestos insulating boards need to be painted. This may be to protect them, or for aesthetic reasons.

    It is not appropriate where the material is damaged. Use a specialist contractor licensed by HSE.

    Carry out this work only if you are properly trained.

    PPE

    ➤ disposable overalls fitted with a hood;
    ➤ boots without laces (laced boots can be difficult to decontaminate);
    ➤ disposable particulate respirator (FF P3).

    Equipment

    ➤ 500 gauge polythene sheeting and duct tape;
    ➤ warning tape and notices;
    ➤ type H vacuum cleaner to BS5415 (if dust needs to be removed from the asbestos insulating board);
    ➤ paint conforming to the original specification, for example fire resistant. Select one low in hazardous constituents, for example solvents;
    ➤ low-pressure spray or roller/brush;
    ➤ bucket of water and rags;
    ➤ suitable asbestos waste container, for example a labelled polythene sack;
    ➤ appropriate lighting.

    Preparing the work area

    ➤ This work may be carried out at height; if so, the appropriate precautions MUST be taken.
    ➤ Carry out the work with the minimum number of people present.
    ➤ Restrict access, for example close the door and/or use warning tape and notices.

- Use polythene sheeting, secured with duct tape, to cover surfaces within the segregated area, which could become contaminated.
- Ensure adequate lighting.

Painting

- Never prepare surfaces by sanding.
- Before starting, check there is no damage.
- Repair any minor damage.
- If dust needs to be removed, use a Type H vacuum cleaner or rags.
- Preferably use the spray to apply the paint.
- Spray using a sweeping motion.
- Do not concentrate on one area as this could cause damage.
- Alternatively, apply the brush/roller lightly to avoid abrasion/damage.

Cleaning

- Use wet rags to clean the equipment.
- Use wet rags to clean the segregated area.
- Place debris, used rags, polythene sheeting and other waste in the waste container.

Personal decontamination

- Use suitable personal decontamination procedure.

Clearance procedure

- Visually inspect the area to make sure that it has been properly cleaned.
- Clearance air sampling is not normally required.

2. Removal of asbestos cement sheets, gutters, etc.

Description

This task guidance sheet can be used where asbestos cement sheets, gutters, drains and ridge caps, etc. need to be removed.

For the large-scale removal of asbestos cement, for example demolition, read *Working with Asbestos Cement* HSG189/2 HSE Books 1999 ISBN 978 0 7176 1667 1.

It is not appropriate for the removal of asbestos insulating board.

Carry out this work only if you are properly trained.

PPE

- Use disposable overalls fitted with a hood.
- Waterproof clothing may be required outside.
- Boots without faces (laced boots can be difficult to decontaminate).
- Use disposable particulate respirator (FIF P3).

Equipment

- 500 and 1000 gauge polythene sheeting and duct tape;
- warning tape and notices;
- bolt cutters;
- bucket of water, garden type spray and rags;
- suitable asbestos waste container, for example a labelled polythene sack;
- lockable skip for larger quantities of asbestos cement;
- asbestos warning stickers;
- appropriate lighting.

Preparing the work area

This work may be carried out at height; if so, the appropriate precautions to prevent the risk of fails MUST be taken.

- Carry out the work with the minimum number of people present.
- Restrict access, for example close the door and/or use warning tape and notices.
- Use 500-gauge polythene sheeting, secured with duct tape, to cover any surface within the segregated area, which could become contaminated.
- It is dangerous to seal over exhaust vents from heating units in use.
- Ensure adequate lighting.

Overlaying

- Instead of removing asbestos cement roofs, consider overlaying with a non-asbestos material.
- Attach sheets to existing purlings but avoid drilling through the asbestos cement.
- Note the presence of the asbestos cement so that it can be managed.

Removal

- Avoid breaking the asbestos cement products.
- If the sheets are held in place with fasteners, dampen and remove – take care not to create a risk of slips.
- If the sheets are bolted in place, use bolt cutters avoiding contact with the asbestos cement. Remove bolts carefully.
- Unbolt or use bolt cutters to release gutters, drain pipes and ridge caps, avoiding contact with the asbestos cement.
- Lower the asbestos cement to the ground. Do not use rubble chutes.
- Check for debris in fasteners or bolt holes. Clean with wet rags.

➤ Single asbestos cement products can be double wrapped in 1000 **gauge** polythene sheeting (or placed in waste containers if small enough), Attach asbestos warning stickers.

➤ Where there are several asbestos cement sheets and other large items, place in a lockable skip.

Cleaning

➤ Use wet rags to clean the equipment.

➤ Use wet rags to clean segregated area.

➤ Place debris, used rags, polythene sheeting and other waste in the waste container.

Personal decontamination

➤ Use a suitable personal decontamination system.

Clearance procedure

➤ Visually inspect the area to make sure that it has been properly cleaned.

➤ Clearance air sampling is not normally required.

3. Personal decontamination system
Description

This guidance sheet explains how you should decontaminate yourself after working with asbestos materials.

If you do not decontaminate yourself properly, you may take asbestos fibres home on your clothing. You or your family and friends could be exposed to them if they were disturbed and became airborne.

It is important that you follow the procedures given in the task guidance sheets and wear PPE such as overalls correctly; this will make cleaning easier.

Removing and decontaminating PPE

➤ Remove your respirator last.

➤ Clean your boots with wet rags.

➤ Where available, use a Type H vacuum cleaner to clean your overalls.

➤ Otherwise use a wet rag – use a 'patting' action – rubbing can disturb fibres.

➤ Where two or more workers are involved they can help each other by 'buddy' cleaning.

➤ Remove overalls by turning inside out – place in suitable asbestos waste container.

➤ Use wet rags to clean waterproof clothing.

➤ Disposable respirators can then be removed and placed in a suitable asbestos waste container.

Personal decontamination

➤ Site-washing facilities can be used but restrict access during asbestos work.

➤ Wash each time you leave the work area.

➤ Use wet rags to clean washing facilities at the end of the job.

➤ Clean facilities daily if the job lasts more than a day.

➤ Visually inspect the facilities once the job is finished.

➤ Clearance air sampling is not normally required.

4. Further Information

These examples are taken from Asbestos Essentials Task Manual HSG210 (now revised) HSE Books 2008 ISBN 978 0 7176 6263 0. Many more examples are contained in the publication including equipment and method guidance sheets. Obtainable from HSE Books.

Monitoring, review and audit

7

After reading this chapter you should be able to:

1. outline and differentiate between active (proactive) monitoring procedures, including inspections, sampling, tours and reactive monitoring procedures, explaining their role within a monitoring regime
2. carry out a workplace inspection, and communicate findings in the form of an effective and persuasive report
3. explain the purpose of regular reviews of health and safety performance, the means by which reviews might be undertaken and the criteria that will influence the frequency of such reviews
4. explain the meaning of the term 'health and safety audit' and describe the preparations that may be needed prior to an audit and the information that may be needed during an audit.

7.1 Introduction

This chapter concerns the monitoring of health and safety performance, including both positive measures like inspections and negative measures like injury statistics. It is about reviewing progress to see if something better can be done and auditing to ensure that what has been planned is being implemented.

Measurement is a key step in any management process and forms the basis of continuous improvement. If measurement is not carried out correctly, the effectiveness of the health and safety management system is undermined and there is no reliable information to show managers how well the health and safety risks are controlled.

Managers should ask key questions to ensure that arrangements for health and safety risk control are in place, comply with the law as a minimum, and operate effectively.

There are two basic types of monitoring:

➤ **Proactive or active monitoring**, by taking the initiative before things go wrong, involves routine inspections and checks to make sure that standards and policies are being implemented and that controls are working.
➤ **Reactive monitoring**, after things go wrong, involves looking at historical events to learn from mistakes and see what can be put right to prevent a recurrence.

The UK Health and Safety Executive's (HSE's) experience is that organizations find health and safety performance measurement a difficult subject. They struggle to develop health and safety performance measures which are not based solely on injury and ill-health statistics.

7.2 The traditional approach to measuring health and safety performance

Senior managers often measure company performance by using, for example, percentage profit, return on investment or market share. A common feature of the measures would be that they are generally positive in nature, which demonstrates achievement, rather than negative, which demonstrates failure.

Yet, if senior managers are asked how they measure their companies' health and safety performance, it is likely that the only measure would be accident or injury statistics. Although the general business performance of an organization is subject to a range of positive measures, for health and safety it too often comes down to one negative measure of failure.

Health and safety differs from many areas measured by managers because improvement in performance means fewer outcomes from the measure (injuries or ill health) rather than more. A low injury or ill-health rate trend over years is still no guarantee that risks are being controlled and that incidents will not happen in the future. This is particularly true in organizations where major hazards are present but there is a low probability of accidents.

There is no single reliable measure of health and safety performance. What is required is a 'basket' of measures, providing information on a range of health and safety issues.

There are some significant problems with the use of injury/ill-health statistics in isolation:

➤ There may be under-reporting – focusing on injury and ill-health rates as a measure, especially if a reward system is involved, can lead to non-reporting to keep up performance.
➤ It is often a matter of chance whether a particular incident causes an injury, and they may not show whether or not a hazard is under control. Luck, or a reduction in the number of people exposed, may produce a low injury/accident rate rather than good health and safety management.
➤ An injury is the particular consequence of an incident and often does not reflect the potential severity. For example an unguarded machine could result in a cut finger or an amputation.
➤ People can be absent from work for reasons which are not related to the severity of the incident.
➤ There is evidence to show that there is little relationship between 'occupational' injury statistics (e.g. slips, trip and falls) and the reasons for the lack of control of major accident hazards (e.g. loss of containment of flammable or toxic material).
➤ A small number of accidents may lead to complacency.
➤ Injury statistics demonstrate outcomes not causes.

Because of the potential shortcomings related to the use of accident/injury and ill-health data as a single measure

of performance, more proactive or 'upstream' measures are required. These require a systematic approach to deriving positive measures and how they link to the overall risk control process, rather than a quick-fix based on things that can easily be counted, such as the numbers of training courses or numbers of inspections, which has limited value. The resultant data provide no information on how the figure was arrived at, whether it is 'acceptable' (i.e. good/bad) or the quality and effectiveness of the activity. A more disciplined approach to health and safety performance measurement is required. This needs to develop as the health and safety management system develops.

7.3 Why measure performance?

7.3.1 Introduction

You can't manage what you can't measure

—Drucker

Measurement is an accepted part of the 'plan-do-check-act' management process. Measuring performance is as much part of a health and safety management system as financial, production or service delivery management. The HSG65 framework for managing health and safety, discussed at the end of Chapter 1 and illustrated in Figure 7.1, shows where measuring performance fits within the overall health and safety management system.

The main purpose of measuring health and safety performance is to provide information on the progress and current status of the strategies, processes and activities employed to control health and safety risks. Effective measurement not only provides information on what the levels are but also why they are at this level, so that corrective action can be taken.

7.3.2 Answering questions

Health and safety monitoring or performance measurement should seek to answer such questions as the following:

➤ Where is the position relative to the overall health and safety aims and objectives?
➤ Where is the position relative to the control of hazards and risks?

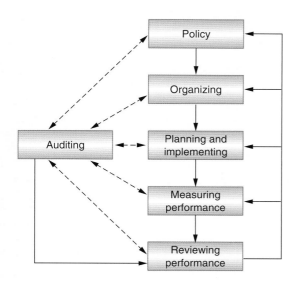

Figure 7.1 The health and safety management system (HSG65).

> How does the organization compare with others?
> What is the reason for the current position?
> Is the organization getting better or worse over time?
> Is the management of health and safety doing the right things?
> Is the management of health and safety doing things right consistently?
> Is the management of health and safety proportionate to the hazards and risks?
> Is the management of health and safety efficient?
> Is an effective health and safety management system in place across all parts of the organization?
> Is the culture supportive of health and safety, particularly in the face of competing demands?

These questions should be asked at all management levels throughout the organization. The aim of monitoring should be to provide a complete picture of an organization's health and safety performance.

7.3.3 Decision making

The measurement information helps in deciding:

> where the organization is in relation to where it wants to be;
> what progress is necessary and reasonable in the circumstances;
> how that progress might be achieved against particular restraints (e.g. resources or time);

> priorities – what should be done first and what is most important;
> effective use of resources.

7.3.4 Addressing different information needs

Information from the performance measurement is needed by a variety of people. These will include directors, senior managers, line managers, supervisors, health and safety professionals and employees/safety representatives. They each need information appropriate to their position and responsibilities within the health and safety management system.

For example what the Chief Executive Officer of a large organization needs to know from the performance measurement system will differ in detail and nature from the information needs of the manager of a particular location.

A coordinated approach is required so that individual measuring activities fit within the general performance measurement framework.

Although the primary focus for performance measurement is to meet the internal needs of an organization, there is an increasing need to demonstrate to external stakeholders (regulators, insurance companies, shareholders, suppliers, contractors, members of the public, etc.) that arrangements to control health and safety risks are in place, operating correctly and effectively.

7.4 What to measure

7.4.1 Introduction

In order to achieve an outcome of no injuries or work-related ill health, and to satisfy stakeholders, health and safety risks need to be controlled. Effective risk control is founded on an effective health and safety management system. This is illustrated in Figure 7.2.

7.4.2 Effective risk control

The health and safety management system comprises three levels of control (see Figure 7.2):

> Level 3 – effective workplace precautions provided and maintained to prevent harm to people who are exposed to the risks;
> Level 2 – risk control systems (RCSs): the basis for ensuring that adequate workplace precautions are provided and maintained;

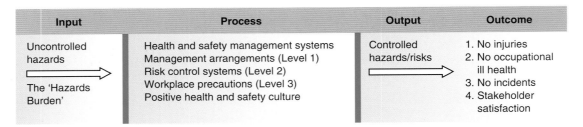

Figure 7.2 Health and safety management system.

➤ Level 1 – the key elements of the health and safety management system: the management arrangements (including plans and objectives) necessary to organize, plan, control and monitor the design and implementation of RCSs.

The health and safety culture must be positive to support each level.

Performance measurement should cover all elements of Figure 7.2 and be based on a balanced approach which combines:

➤ **Input: monitoring** the scale, nature and distribution of hazards created by the organization's activities – measures of the hazard burden.
➤ **Process: active monitoring** of the adequacy, development, implementation and deployment of the health and safety management system and the activities to promote a positive health and safety culture – measures of success.
➤ **Outcomes: reactive monitoring** of adverse outcomes resulting in injuries, ill health, loss and accidents with the potential to cause injuries, ill health or loss – measures of failure.

Proactive or active monitoring – how to measure performance

7.5.1 Introduction

The measurement process can gather information through:

➤ direct observation of conditions and of people's behaviour (sometimes referred to as unsafe acts and unsafe conditions monitoring);
➤ talking to people to elicit facts and their experiences as well as gauging their views and opinions';
➤ examining written reports, documents and records.

These information sources can be used independently or in combination. Direct observation includes inspection activities and the monitoring of the work environment (e.g. temperature, dust levels, solvent levels, noise levels) and people's behaviour. Each risk control system (RCS) should have a built-in monitoring element that will define the frequency of monitoring; these can be combined to form a common inspection system.

7.5.2 Inspections

General

This may be achieved by developing a checklist or inspection form that covers the key issues to be monitored in a particular department or area of the organization within a particular period. It might be useful to structure this checklist using the 'four Ps' (note that the examples are not a definitive list):

➤ **premises**, including:
  - work at height
  - access
  - working environment
  - welfare
  - services
  - fire precautions
➤ **plant and substances**, including:
  - work equipment
  - manual and mechanical handling
  - vehicles
  - dangerous/flammable substance
  - hazardous substance
➤ **procedures**, including:
  - risk assessments
  - safe systems of work
  - permits to work
  - personal protective equipment
  - contractors
  - notices, signs and posters
➤ **people**, including:
  - health surveillance
  - people's behaviour

- training and supervision
- appropriate authorized persons
- violence
- those especially at risk.

It is essential that people carrying out an inspection do not in any way put themselves or anyone else at risk. Particular care must be taken with regard to safe access. In carrying out these safety inspections, the safety of people's actions should be considered, in addition to the safety of the conditions they are working in – a ladder might be in perfect condition but it has to be used properly too.

Key points in becoming a good observer

To improve health and safety performance, managers and supervisors must eliminate unsafe acts by observing them, taking immediate corrective action and following up to prevent recurrence. To become a good observer, they must improve their observation skills and must learn how to observe effectively. Effective observation includes the following key points:

- be selective
- know what to look for
- practice
- keep an open mind
- guard against habit and familiarity
- do not be satisfied with general impressions
- record observations systematically.

Observation techniques

In addition, to become a good observer, a person must:

- stop for 10–30 seconds before entering a new area to ascertain where employees are working;
- be alert for unsafe practices that are corrected as soon as you enter an area;
- observe activity – do not avoid the action;
- remember ABBI – look above, below, behind, inside;
- develop a questioning attitude to determine what injuries might occur if the unexpected happened and how the job might be accomplished more safely. Ask 'why?' and 'what could happen if …?';
- use all senses: sight, hearing, smell, touch;
- maintain a balanced approach. Observe all phases of the job;
- be inquisitive;
- observe for ideas – not just to determine problems;
- recognize good performance.

Daily/weekly/monthly safety inspections

These will be aimed at checking conditions in a specified area against a fixed checklist drawn up by local management. It will cover specific items, such as the guards at particular

(a)

(b)

Figure 7.3 (a) Poor conditions – inspection needed; (b) inspection in progress.

machines, whether access/agreed routes are clear, whether fire extinguishers are in place, etc. The checks should be carried out by staff of the department who should sign off the checklist. They should not last more than half an hour, perhaps less. This is not a specific hazard-spotting operation, but there should be a space on the checklist for the inspectors to note down any particular problems encountered.

Reports from inspections

Some of the items arising from safety inspections will have been dealt with immediately; other items will require action by specified people. Where there is some doubt about the problem, and what exactly is required, advice should be sought from the Site Safety Adviser or external expert. A brief report of the inspection and any resulting action list should be submitted to the safety committee. While the committee may not have the time available to consider all reports in detail, it will want to be satisfied that appropriate action is taken to resolve all matters; it will be necessary for the committee to follow up the reports until all matters are resolved (Figure 7.3).

Essential elements of a report are:

➤ identification of the organization, workplace, inspector and date of inspection;
➤ list of observations;
➤ priority or risk level;
➤ actions to be taken;
➤ timescale for completion of the actions.

Appendix 7.1 gives examples of poor workplaces which can be used for practice exercises. Appendix 7.2 shows a specimen workplace inspection form which can be adapted for use at many workplaces. In Chapter 20 (Appendix 20.1) there is a completed version of this form used for the NEBOSH practical assessment.

Appendix 7.3 gives a workplace inspection checklist which can be used to assist in drawing up a specific checklist for any particular workplace or as an aid memoir for the workplace inspection report form.

Inspection standards

In order to get maximum value from inspection checklists, they should be designed so that they require objective rather than subjective judgements of conditions. For example asking the people undertaking a general inspection of the workplace to rate housekeeping as good or bad, begs questions as to what does good and bad mean, and what criteria should be used to judge this. If good housekeeping means there is no rubbish left on the floor, all waste bins are regularly emptied and not overflowing, floors are swept each day and cleaned once a week, decorations should be in good condition with no peeling paint, then this should be stated. Adequate expected standards should be provided in separate notes so that those inspecting know the standards that are required.

The checklist or inspection form should facilitate:

➤ the planning and initiation of remedial action, by requiring those doing the inspection to rank deficiencies in priority order (those actions which are most important rather than those which can easily be done quickly);
➤ identifying those responsible for taking remedial actions, with sensible timescales to track progress on implementation;
➤ periodic monitoring to identify common themes which might reveal underlying problems in the system;
➤ management information on the frequency or nature of the monitoring arrangements.

7.5.3 Safety sampling

Safety sampling is a helpful technique that helps organizations to concentrate on one particular area or subject at a time. A specific area is chosen which can be inspected in about 30 minutes. A checklist is drawn up to facilitate the inspection looking at specific issues. These may be different types of hazard: they may be unsafe acts or conditions noted; they may be proactive, good behaviour or practices noted.

The inspection team or person then carries out the sampling at the same time each day or week in the specified period. The results are recorded and analysed to see if the changes are good or bad over time. Of course, defects noted must be brought to the notice of the appropriate person for action on each occasion.

7.6 Measuring failure – reactive monitoring

So far, this chapter has concentrated on measuring activities designed to prevent the occurrence of injuries and work-related ill health (active monitoring). Failures in risk control also need to be measured (reactive monitoring), to provide opportunities to check performance, learn from failures and improve the health and safety management system.

Reactive monitoring arrangements include systems to identify and report:

➤ injuries and work-related ill health (details of the incident rate calculation is given in Chapter 4);
➤ other losses such as damage to property;
➤ incidents, including those with the potential to cause injury, ill health or loss (near misses);
➤ hazards and faults;
➤ weaknesses or omissions in performance standards and systems, including complaints from employees and enforcement action by the authorities.

Guidance on investigating and analysing these incidents is given in Chapter 8.

7.7 Who should monitor performance?

Performance should be measured at each management level from directors downwards. It is not sufficient to monitor by exception, where unless problems are raised, it is assumed to be satisfactory. Senior managers must satisfy themselves that the correct arrangements are in place and working properly. Responsibilities for both active and reactive monitoring must be laid down and managers need to be personally involved in making sure that plans and objectives are met and compliance with standards is achieved. Although systems may be set up with the guidance of safety professionals, managers should be personally involved and given sufficient training to be competent to make informed judgements about monitoring performance.

Other people, such as safety representatives, will also have the right to inspect the workplace. Each employee should be encouraged to inspect their own workplace frequently to check for obvious problems and rectify them if possible or report hazards to their supervisors.

Specific statutory (or thorough) examinations of, for example, lifting equipment or pressure vessels, have to be carried out at intervals laid down in written schemes by competent persons – usually specially trained and experienced inspection/insurance company personnel.

7.8 Frequency of monitoring and inspections

This will depend on the level of risk and any statutory inspection requirement. Directors may be expected to examine the premises formally at an annual audit, whereas departmental supervisors may be expected to carry out inspection every week. Senior managers should regularly monitor the health and safety plan to ensure that objectives are being met and to make any changes to the plan as necessary.

Data from reactive monitoring should be considered by senior managers at least once a month. In most organizations serious events would be closely monitored as they happen.

7.9 Report writing

There are three main aims to the writing of reports and they are all about communication. A report should aim to:

➤ get a message through to the reader;
➤ make the message and the arguments clear and easy to understand;
➤ make the arguments and conclusions persuasive.

Communication starts with trying to get into the mind of the reader, imagining what would most effectively catch the attention, what would be most likely to convince, and what will make this report stand out among others.

A vital part of this is presentation; so while a handwritten report is better than nothing if time is short, a well-organized, typed report is very much clearer. To the reader of the report, who may well be very busy with a great deal of written information to wade through, a clear, well-presented report will produce a positive attitude from the outset, with instant benefit to the writer.

Five factors which help to make reports effective are:

➤ structure;
➤ presentation of arguments;
➤ style;
➤ presentation of data;
➤ how the report itself is presented.

7.9.1 Structure

The structure of a report is the key to its professionalism. Good structuring will:

➤ help the reader to understand the information and follow the arguments contained in the report;
➤ increase the writer's credibility;
➤ ensure that the material contained in the report is organized to the best advantage.

The following list shows a frequently used method of producing a report, but always bear in mind that different organizations use different formats:

1. title page
2. summary
3. contents list
4. introduction
5. main body of the report
6. conclusions
7. recommendations
8. appendices
9. references.

It is important to check with the organization requesting the report in case their in-house format differs.

1. Title page

This will contain:

➤ a title and often a subtitle;
➤ the name of the person or organization to whom the report is addressed;
➤ the name of the writer(s) and their organization;
➤ the date on which the report was submitted.

As report writing is about communication, it is a good idea to choose a title that is eye-catching and memorable as well as being informative, if this is appropriate to the subject.

2. Summary

Limit the summary to between 150 and 500 words. Do not include any evidence or data. This should be kept for the main report. Include the main conclusions and principal recommendations and place the summary near the front of the report.

3. Contents list

Put the contents list near the beginning of the report. Short reports do not need a list but if there are several headings, it does help the reader to grasp the overall content of the report in a short time.

4. Introduction

The introduction should contain the following:

- information about who commissioned the report and when;
- the reason for the report;
- objectives of the report;
- terms of reference;
- preparation of the report (type of data, research undertaken, subjects interviewed, etc.);
- methodology used in any analysis;
- problems and the methods used to tackle them;
- details about consultation with clients, employees, etc.

There may be other items that are specific to the report.

5. The main body of the report

This part of the report should describe, in detail, what was discovered (the facts), and the significance of these discoveries (analysis) and their importance (evaluation). Graphs, tables and charts are often used at this stage in the report. These should have the function of summarizing information rather than giving large amounts of detail. The more detailed graphics should be made into appendices.

To make it more digestible, this part of the report should be divided into sections, using numbered headings and sub-headings. Very long and complex reports will need to be broken down into chapters.

6. Conclusions

The concluding part of the report should be a reasonably detailed 'summing up'. It should give the conclusions arrived at by the writer and explain why the writer has reached these conclusions.

7. Recommendations

The use of this section depends on the requirements of the person commissioning the report. If recommendations are required, provide as few as possible, to retain a clear focus. Report writers are often asked only to provide the facts.

8. Appendices

This part of the report should contain sections that may be useful to a reader who requires more detail. Examples would be the detailed charts, graphs and tables, any questionnaires used in constructing the evidence mentioned in item 5, forms, case studies and so on. The appendices are the background material of the report.

9. References

If any books, papers or journal articles have been used as source material, this should be acknowledged in a reference section. There are a number of accepted referencing methods used by academics.

Because the reader is likely to be a person with some degree of expertise in the subject, a report must be reliable, credible, relevant and thorough. It is therefore important to avoid emotional language, opinions presented as facts and arguments that have no supporting evidence. To make a report more persuasive, the writer needs to:

- present the information clearly;
- provide reliable evidence;
- present arguments logically;
- avoid falsifying, tampering with or concealing facts.

Expertise in an area of knowledge means that distortions, errors and omissions will quickly be spotted by the discerning reader and the presence of any of these will cast doubt on the credibility of the whole report.

Reports are usually used as part of a decision-making process. If this is the case, clear, unembellished facts are needed. Exceptions to this would be where the report is a proposal document or where a recommendation is specifically requested. Unless this is the case, it is better not to make recommendations.

A report should play a key role in organizing information for the use of decision makers. It should review a complex and/or extensive body of information and make a summary of all the important issues.

It is relatively straightforward to produce a report, as long as the writer keeps to a clear format. Using the format described here, it should be possible to tell the reader as clearly as possible:

- what happened and why
- who was involved

➤ what it cost if appropriate

➤ what the result was.

There may be a request for a special report and this is likely to be longer and more difficult to produce. Often it will relate to a 'critical incident' and the decision makers will be looking for information to help them:

➤ decide whether this is a problem or an opportunity;

➤ decide whether to take action;

➤ decide what action, if any, to take.

Finally, report writing should be kept simple. Nothing is gained in the use of long, complicated sentences, jargon and official-sounding language. When the report is finished, it is helpful to run through it with the express intention of simplifying the language and making sure that it says what was intended in a clear and straightforward way.

> KEEP IT SHORT AND SIMPLE

7.10   Review and audit

7.10.1 Audits – purpose

The final steps in the health and safety management control cycle are auditing and performance review. Organizations need to be able to reinforce, maintain and develop the ability to reduce risks. The 'feedback loop' produced by this final stage in the process enables them to do this and to ensure continuing effectiveness of the health and safety management system.

Audit is a business discipline which is frequently used, for example, in finance, environmental matters and quality. It can equally well be applied to health and safety.

The term is often used to mean inspection or other monitoring activity. Here, the following definition is used, which follows HSG65:

> *The structured process of collecting independent information on the efficiency, effectiveness and reliability of the total health and safety management system and drawing up plans for corrective action.*

Over time, it is inevitable that control systems will decay and may even become obsolete as things change. Auditing is a way of supporting monitoring by providing managers with information. It will show how effectively plans and the components of health and safety management systems are being implemented. In addition, it will provide a check on the adequacy and effectiveness of the management arrangements and risk control systems (RCS).

Auditing is critical to a health and safety management system, but it is not a substitute for other essential parts of the system. Companies need systems in place to manage cash flow and pay the bills – this cannot be managed through an annual audit. In the same way, health and safety needs to be managed on a day-to-day basis and for this, organizations need to have systems in place. A periodic audit will not achieve this.

The aims of auditing should be to establish that the three major components of a safety management system are in place and operating effectively. It should show that:

➤ appropriate management arrangements are in place;

➤ adequate RCSs exist, are implemented and are consistent with the hazard profile of the organization;

➤ appropriate workplace precautions are in place.

Where the organization is spread over a number of sites, the management arrangements linking the centre with the business units and sites should be covered by the audit.

There are a number of ways in which this can be achieved and some parts of the system do not need auditing as often as others. For example an audit to verify the implementation of RCSs would be made more frequently than a more overall audit of the capability of the organization or of the management arrangements for health and safety. Critical RCSs, which control the principal hazards of the business, would need to be audited more frequently. Where there are complex workplace precautions, it may be necessary to undertake technical audits. An example would be chemical process plant integrity and control systems.

A well-structured auditing programme will give a comprehensive picture of the effectiveness of the health and safety management system in controlling risks. Such a programme will indicate when and how each component part will be audited. Managers, safety representatives and employees, working as a team, will effectively widen involvement and cooperation needed to put together the programme and implement it.

The process of auditing involves:

➤ gathering information from all levels of an organization about the health and safety management system;

➤ making informed judgements about its adequacy and performance.

7.10.2 Gathering information

Decisions will need to be made about the level and detail of the audit before starting to gather information about the health and safety management of an organization. Auditing involves sampling; so initially it is necessary to decide how much sampling is needed for the assessment to be reliable. The type of audit and its complexity will relate to its objectives and scope, to the size and complexity of the organization and to the length of time that the existing health and safety management system has been in operation.

Information sources of interviewing people, looking at documents and checking physical conditions are usually approached in the following order:

Preparatory work

➤ meet with relevant managers and employee representatives to discuss and agree the objectives and scope of the audit;
➤ prepare and agree the audit procedure with managers;
➤ gather and consider documentation.

On site

➤ interviewing;
➤ review and assessment of additional documents;
➤ observation of physical conditions and work activities.

Conclusion

➤ assemble the evidence;
➤ evaluate the evidence;
➤ write an audit report.

7.10.3 Making judgements

It is essential to start with a relevant standard or benchmark against which the adequacy of a health and safety management system can be judged. If standards are not clear, assessment cannot be reliable. Audit judgements should be informed by legal standards, HSE guidance and applicable industry standards. HSG65 sets out benchmarks for management arrangements and for the design of RCSs. This book follows the same concepts.

Auditing should not be seen as a fault-finding activity. It should make a valuable contribution to the health and safety management system and to learning. It should recognize achievement as well as highlight areas where more needs to be done.

Scoring systems can be used in auditing along with judgements and recommendations. This can be seen as a useful way to compare sites or monitor progress over time. However, there is no evidence that quantified results produce a more effective response than the use of qualitative evidence. Indeed, the introduction of a scoring system can, the HSE believes, have a negative effect, as it may encourage managers to place more emphasis on high-scoring questions which may not be as relevant to the development of an effective health and safety management system.

To achieve the best results, auditors should be competent people who are independent of the area and of the activities being audited. External consultants or staff from other areas of the organization can be used. An organization can use its own auditing system or one of the proprietary systems on the market or, as it is unlikely that any ready-made system will provide a perfect fit, a combination of both. With any scheme, cost and benefits have to be taken into account. Common problems include:

➤ systems can be too general in their approach. These may need considerable work to make them fit the needs and risks of the organization;
➤ systems can be too cumbersome for the size and culture of the organization;
➤ scoring systems may conceal problems in underlying detail;
➤ organizations may design their management system to gain maximum points rather than using one which suits the needs and hazard profile of the business.

HSE encourages organizations to assess their health and safety management systems using in-house or proprietary schemes but without endorsing any particular one.

7.10.4 Performance review

When performance is reviewed, judgements are made about its adequacy and decisions are taken about how and when to rectify problems. The feedback loop is needed by organizations so that they can see whether the health and safety management system is working as intended. The information for review of performance comes from audits of RCSs and workplace precautions, and from the measurement of activities. There may be other influences, both internal and external, such as re-organization, new legislation or changes in current good practice. These may result in the necessity to redesign or change parts of the health and safety management system or to alter its direction or objectives.

Performance standards need to be established which will identify the systems requiring change, responsibilities, and completion dates. It is essential to feed back the information about success and failure so that employees are motivated to maintain and improve performance.

In a review, the following areas will need to be examined:

- the operation and maintenance of the existing system;
- how the safety management system is designed, developed and installed to accommodate changing circumstances.

Reviewing is a continuous process. It should be undertaken at different levels within the organization. Responses will be needed as follows:

- by first-line supervisors or other managers to remedy failures to implement workplace precautions which they observe in the course of routine activities;
- to remedy sub-standard performance identified by active and reactive monitoring;
- to the assessment of plans at individual, departmental, site, group or organizational level;
- to the results of audits.

The frequency of review at each level should be decided upon by the organization and reviewing activities should be devised which will suit the measuring and auditing activities. The review will identify specific remedial actions which establish who is responsible for implementation and set deadlines for completion.

7.11 Sources of reference

Successful Health and Safety Management (HSG65), HSE Books 1997 ISBN 978 0 7176 1276 5.

Relevant statutory provisions

The Management of Health and Safety at Work Regulations 1999.

7.12 Practice NEBOSH questions for Chapter 7

1. **Outline** ways in which an organization can monitor its health and safety performance.

2. **Identify EIGHT** measures that could be used by an organization in order to monitor its health and safety performance.

3. **Outline TWO** reactive measures **AND TWO** proactive (or active) measures that can be used in monitoring an organization's health and safety performance.

4. **Outline FOUR** proactive (or active) monitoring methods that can be used in assessing the health and safety performance of an organization.

5. **Identify FOUR** *active (or proactive)* **AND FOUR** *reactive* means by which an organization can monitor its health and safety performance.

6. An employer intends to implement a programme of regular workplace inspections following a workplace accident.
   (i) Outline the factors that should be considered when planning such inspections.
   (ii) Outline THREE additional proactive (active) methods that could be used in the monitoring of health and safety performance.
   (iii) Identify the possible costs to the organization as a result of the accident.

7. **Identify** the main topic areas that should be included in a planned health and safety inspection of a workplace.

8. A health and safety inspection has been carried out on one of a company's workshops. The inspection has found a number of unsafe conditions and practices and some positive issues.
   (i) In addition to the date and time the inspection was carried out, **state** other issues that should also be included in the report to enable management to make an informal decision on possible remedial action to be taken.
   (ii) Explain how the report should be structured and presented in order to make it more effective and to increase the likelihood of action being taken by management.

9. **Explain** how the following may be used to improve safety performance within an organization:
   (i) accident data
   (ii) safety inspections.

10. **Outline** factors that would determine the frequency with which health and safety inspections should be undertaken in the workplace.

11. **(i) Define** the term 'safety survey'.
    (ii) Outline the issues which should be considered when a safety survey of a workplace is to be undertaken.

12. **Outline** the *strengths* **AND** *weaknesses* of using a checklist to complete a health and safety inspection of a workplace.

13. **(i) Explain** the meaning of the term health and safety 'audit'.
    (ii) Outline the issues that need to be considered at the planning stage of the audit.
    (iii) State TWO methods of gathering information during an audit.

14. **(i) Outline** the differences between health and safety 'audits' and 'workplace inspections'.
    (ii) Identify issues to be considered when compiling an action plan on completion of a workplace inspection.

15. **Outline** the topics that should be included in a health and safety audit.

16. **Outline** the reasons why an organization should review and monitor its health and safety performance.

17. A health and safety audit of an organization has identified a general lack of compliance with procedures.
    (i) Describe the possible reasons for procedures not being followed.
    (ii) Outline the practical measures that could be taken to motivate employees to comply with health and safety procedures.

18. **List** the documents that are likely to be examined during a health and safety audit.

19. **Identify** the *advantages* **AND** *disadvantages* of carrying out a health and safety audit of an organization's activities by:
    (i) an internal auditor
    (ii) an external auditor.

Appendix 7.1 Workplace inspection exercises

Figures 7.4–7.7 shows workplaces with numerous inadequately controlled hazards. They can be used to practise workplace inspections and risk assessments.

Figure 7.4 Office.

Figure 7.5 Road repair.

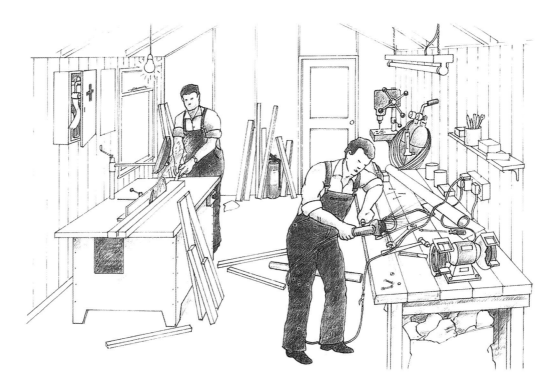

Figure 7.6 Workshop.

Figure 7.7 Roof repair/unloading.

Appendix 7.2 Specimen workplace inspection report form

See Appendix 7.3 for inspection check list and Appendix 5.1 for hazard checklist.

Workplace Inspection Form

Workplace Inspection Cover Sheet

Name of Company/ Organization	
Work area covered by this inspection	
Activity carried out in workplace	

Person carrying out inspection (PLEASE PRINT)		Date of Inspection	

| Observations

List hazards, unsafe practices and good practices | Priority/ risk

(H, M, L) | Actions to be taken (if any)

List all immediate and longer-term actions required | Timescale

Immediate 1 week, etc. |
|---|---|---|---|
| | | | |

Appendix 7.3 Workplace inspection checklist

PREMISES			
1	**Work at height**	Ladders/step ladders	Right equipment for the job? Level base? Correct angle? Secured at top and bottom? Equipment in good condition? Regularly inspected?
		Working platforms/ temporary scaffolds	Suitable for the task? Properly erected? Good access? Maintained and inspected?
		Use of mobile elevating work platforms	Suitable for task? Operators properly trained? Properly maintained?
2	**Access**	Access ways	Adequate for people, machinery and work in progress? Unobstructed? Properly marked? Stairs in good condition? Handrails provided?
3	**Working environment**	Housekeeping	Tidy, clean, well organized?
		Flooring	Even and in good condition? Non-slippery?
		Comfort /health	Crowded? Too hot/cold? Ventilation? Humidity? Dusty? Lighting?
		Cleaning	Slip risk controlled? Hygienic conditions?
		Noise	Normal conversation possible? Noise assessment needed/not needed? Noise areas designated?
		Ergonomics	Tasks require uncomfortable postures or actions? Frequent repetitive actions accompanying muscular strain?

		Visual display units	Workstation assessments needed/not needed? Chairs adjustable/comfortable/maintained properly? Cables properly controlled? Lighting OK? No glare?
4	**Welfare**	Toilets/Washing	Washing and toilet facilities satisfactory? Kept clean, with soap and towels? Adequate changing facilities?
		Eating facilities	Clean and adequate/Means of heating food?
		Rest room	For pregnant or nursing mothers? Kept clean?
		First aid	Suitably placed and provisioned? Appointed person? Trained first aider? Correct signs and notices? Eye wash bottles as necessary?
5	**Services**	Electrical equipment	Portable equipment tested? Leads tidy not damaged? Fixed installation inspected?
		Gas	Equipment serviced annually?
		Water	Hot and cold water provided? Drinking water provided?
6	**Fire precautions**	Fire extinguishers	In place? Full? Correct type? Maintenance contract?
		Fire instructions	Posted up? Not defaced or damaged?
		Fire alarms	Fitted and tested regularly?
		Means of escape/Fire exits	Adequate for the numbers involved? Unobstructed? Easily opened? Properly signed?
		Means of escape/Fire exits	Adequate for the numbers involved? Unobstructed? Easily opened? Properly signed?

		PLANT AND SUBSTANCES	
7	**Work equipment**	Lifting equipment	Thoroughly examined? Properly maintained? Slings, etc. properly maintained? Operators properly trained?
		Pressure systems	Written schemes for inspection? Safe working pressure marked? Properly maintained?
		Sharps	Safety knives used? Knives/needles/glass properly used/disposed of?
		Vibration	Any vibration problems with hand-held machinery or with whole body from vehicle seats, etc.?
		Tools and equipment	Right tool for the job? In good condition?
		Manual handling	Moving excessive weight? Assessments carried out? Using correct technique? Could it be eliminated or reduced?
8	**Manual & mechanical handling**	Mechanical handling	Fork-lifts and other trucks properly maintained? Drivers authorized and properly trained? Passengers only where specifically intended with suitable seat?
9	**Vehicles**	On site	Speeding limits? Following correct route? Properly serviced? Drivers authorized?
		Road risks	Suitable vehicles used? No use of mobile phones when driving? Properly serviced? Schedules managed properly?
10	**Dangerous substances**	Flammable liquids and gases	Stored properly? Used properly/minimum quantities in workplace? Sources of ignition? Correct signs used?

11	**Hazardous substances**	Chemicals	COSHH assessments OK? Exposures adequately controlled? Data sheet information available? Spillage procedure available? Properly stored and separated as necessary? Properly disposed of?
		Exhaust ventilation	Suitable and sufficient? Properly maintained? Inspected regularly?
		PROCEDURES	
12	**Risk assessments**		Carried out? General and fire? Suitable and sufficient?
13	**Safe systems of work**		Provided as necessary? Kept up to date/Followed?
14	**Permits to work**		Used for high risk maintenance? Procedure OK? Properly followed?
15	**Personal protective equipment**		Correct type? Worn correctly? Good condition?
16	**Contractors**		Is their competence checked thoroughly? Are there control rules and procedures? Are they followed?
17	**Notices, signs and posters**	Employers' liability insurance	Notice displayed? In date?
		Health and Safety law poster	Displayed?
		Safety Signs	Correct type of sign used/ signs in place and maintained?

PEOPLE			
18	**Health surveillance**		Specific surveillance required by law? Stress or fatigue?
19	**People's behaviour**		Are behaviour audits carried out? Is behaviour considered in the safety programme?
20	**Training and supervision**		Suitable and sufficient? Induction training? Refresher training?
21	**Appropriate authorized person**		Is there a system like permits to work for authorizing people for certain special tasks involving dangerous machinery, entry into confined spaces?
22	**Violence**		Any violence likely in workplace? Is it controlled? Are there policies in place?
23	**Especially at risk categories**	Young persons	Employed? Special risk assessments?
		New or expectant mothers	Employed? Special risk assessments?

Incident and accident investigation recording and reporting

8

8.1 Introduction

This chapter is concerned with the recording of incidents and accidents at work, their investigation, the legal reporting requirements and simple analysis of incidents to help managers benefit from the investigation and recording process.

Incidents and accidents rarely result from a single cause and many turn out to be complex. Most incidents involve multiple, interrelated causal factors. They can occur whenever significant deficiencies, oversights, errors, omissions or unexpected changes occur. Any one of these can be the precursor of an accident or incident. There is a value on collecting data on all incidents and potential losses as it helps to prevent more serious events. (See Chapter 5 for accident ratios and definitions.)

Incidents and accidents, whether they cause damage to property or more serious injury and/or ill health to people, should be properly and thoroughly investigated to allow an organization to take the appropriate action to prevent a recurrence (Figure 8.1). Good investigation is a key element to making improvements in health and safety performance.

Incident investigation is considered to be part of a **reactive monitoring system** because it is triggered after an event.

The range of adverse events includes:

1. **Accidents**: An event that results in injury or ill health, including sickness, absences.
2. **Incidents**:
   (a) Near miss – Events that have the potential to cause injury or ill health and may cause damage to property, personal effects, work in progress.

Figure 8.1 A dangerous occurrence – fire.

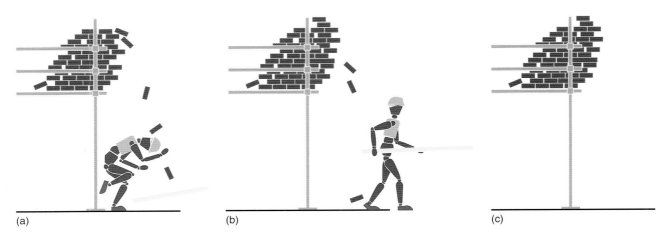

Figure 8.2 (a) Accident. (b) Near miss. (c) Undesired circumstances. *Source*: HSE.

(b) Undesired circumstance – A set of conditions or circumstances that have the potential to cause injury or ill health or damage.

Figure 8.2 demonstrates the difference between an accident, near miss and undesired circumstances.

Each type of event gives the opportunity to:

➤ check performance;
➤ identify underlying deficiencies in management systems and procedures;
➤ learn from mistakes and add to the corporate memory;
➤ reinforce key health and safety messages;
➤ identify trends and priorities for prevention;
➤ provide valuable information if there is a claim for compensation;
➤ help meet legal requirements for reporting certain incidents to the authorities.

8.2 Reasons for incident/accident investigation

8.2.1 Logic and understanding

Incident/accident investigation is based on the logic that:

➤ all incidents/accidents have causes – eliminate the cause and eliminate future incidents;
➤ the direct and indirect causes of an incident/accident can be discovered through investigation;
➤ corrective action indicated by the causation can be taken to eliminate future incidents/accidents.

Investigation is not intended to be a mechanism for apportioning blame. There are often strong emotions associated with injury or significant losses. It is all too easy to look for someone to blame without considering the reasons why a person behaved in a particular way. Often short cuts to working procedures that may have contributed to the accident give no personal advantage to the person injured. The short cut may have been taken out of loyalty to the organization or ignorance of a safer method.

Valuable information and understanding can be gained from carrying out accident/incident investigations. These include:

➤ an understanding of how and why problems arose which caused the accident/incident;
➤ an understanding of the ways people are exposed to substances or situations which can cause them harm;
➤ a snapshot of what really happens, for example why people take short cuts or ignore safety rules;
➤ identifying deficiencies in the control of risks in the organization.

8.2.2 Legal reasons

The legal reasons for conducting an investigation are:

➤ to ensure that the organization is operating in compliance with legal requirements;
➤ that it forms an essential part of the MHSW Regulation 5 requirements to plan, organize, control, monitor and review health and safety arrangements;
➤ to comply with the Woolf Report on civil action which changed the way cases are run. Full disclosure of the circumstances of an accident/incident has to be

made to the injured parties considering legal action. The fact that a thorough investigation was carried out and remedial action taken would demonstrate to a court that a company has a positive attitude to health and safety. The investigation will also provide essential information for insurers in the event of an employer's liability or other claim.

8.2.3 Benefits

There are many benefits from investigating accidents/incidents. These include:

➤ the prevention of similar events occurring again. Where the outcomes are serious injuries the enforcing authorities are likely to take a tough stance if previous warnings have been ignored;
➤ the prevention of business losses due to disruption immediately after the event, loss of production, loss of business through a lowering of reputation or inability to deliver, and the costs of criminal and legal actions;
➤ improvement in employee morale and general attitudes to health and safety particularly if they have been involved in the investigations;
➤ improving management skills to improve health and safety performance throughout the organization.

The case for investigating near misses and undesired circumstances may not be so obvious but it is just as useful and much easier as there are no injured people to deal with (Figure 8.3). There are no demoralized people at work or distressed families and seldom a legal action to answer. Witnesses will be more willing to speak the truth and help with the investigation.

8.2.4 What managers need to do

Managers need to:

➤ communicate the type of accident and incident that needs to be reported;
➤ provide a system for reporting and recording;
➤ check that proper reports are being made;
➤ make appropriate records of accidents and incidents;
➤ investigate all incidents and accidents reported;
➤ analyse the events routinely to check for trends in performance and the prevalence of types of incident or injury;
➤ monitor the system to make sure that it is working satisfactorily.

Figure 8.3 Near miss event likely – accident waiting to happen.

8.3 Which incidents/accidents should be investigated?

8.3.1 Injury accident

Should every accident be investigated or only those that lead to serious injury? In fact the main determinant is the potential of the accident to cause harm rather than the actual harm resulting. For example, a slip can result in an embarrassing flailing of arms or, just as easily, a broken leg. The frequency of occurrence of the accident type is also important – a stream of minor cuts from paper needs looking into.

As it is not possible to determine the potential for harm simply from the resulting injury, the only really sensible solution is to investigate all accidents. The amount of time and effort spent on the investigation should, however, vary depending on the level of risk (severity of potential harm, frequency of occurrence). The most effort should be focused on significant events involving serious injury, ill health or losses and events which have the potential for multiple or serious harm to people or substantial losses. These factors should become clear during the accident investigation and be used to guide how much time should be taken.

The following table has been developed by the Health and Safety Executive (HSE) to help to determine the level of investigation which is appropriate. The potential worst injury consequences in any particular situation should be considered when using the table. A particular incident like a scaffold collapse may not have caused an injury but had the potential to cause major or fatal injuries.

Likelihood of recurrence	Potential worst injury consequences of accident/incident			
	Minor	Serious	Major	Fatal
Certain				
Likely				
Possible				
Unlikely				
Rare				

Risk	Minimal	Low	Medium	High
Investigation level	Minimal level	Low level	Medium level	High level

➤ In a **minimal-level** investigation, the relevant supervisor will look into the circumstances of the accident/incident and try to learn any lessons which will prevent future incidents.

➤ A **low-level** investigation will involve a short investigation by the relevant supervisor or line manager into the circumstances and immediate underlying and root causes of the accident/incident, to try to prevent a recurrence and to learn any general lessons.

➤ A **medium-level** investigation will involve a more detailed investigation by the relevant supervisor or line manager, the health and safety adviser and

employee representatives and will look for the immediate, underlying and root causes.

➤ A **high-level** investigation will involve a team-based investigation, involving supervisors or line managers, health and safety advisers and employee representatives. It will be carried out under the supervision of senior management or directors and will look for the immediate, underlying and root causes.

8.4 Investigations and causes of accidents/incidents

8.4.1 Who should investigate?

Investigations should be led by supervisors, line managers or other people with sufficient status and knowledge to make recommendations that will be respected by the organization. The person to lead many investigations will be the Department Manager or Supervisor of the person/area involved because they:

➤ know about the situation;
➤ know most about the employees;
➤ have a personal interest in preventing further incidents/accidents affecting 'their' people, equipment, area, materials;
➤ can take immediate action to prevent a similar incident;
➤ can communicate most effectively with the other employees concerned;
➤ can demonstrate practical concern for employees and control over the immediate work situation.

8.4.2 When should the investigation be conducted?

The investigation should be carried out as soon as possible after the incident to allow the maximum amount of information to be obtained. There may be difficulties which should be considered in setting up the investigation quickly – if, for example, the victim is removed from the site of the accident, or if there is a lack of a particular expert. An immediate investigation is advantageous because:

➤ factors are fresh in the minds of witnesses;
➤ witnesses have had less time to talk (there is an almost automatic tendency for people to adjust their story of the events to bring it into line with a consensus view);

➤ physical conditions have had less time to change;
➤ more people are likely to be available, for example delivery drivers, contractors and visitors, who will quickly disperse following an incident, making contact very difficult;
➤ there will probably be the opportunity to take immediate action to prevent a recurrence and to demonstrate management commitment to improvement;
➤ immediate information from the person suffering the accident often proves to be most useful.

Consideration should be given to asking the person to return to site for the accident investigation if they are physically able, rather than wait for them to return to work. A second option, although not as valuable, would be to visit the injured person at home or even in hospital (with their permission) to discuss the accident.

8.4.3 Investigation method

There are four basic elements to a sound investigation:

1. Collect facts about what has occurred.
2. Assemble, and analyse the information obtained.
3. Compare the information with acceptable industry and company standards and legal requirements to draw conclusions.
4. Implement the findings and monitor progress.

Information should be gathered from all available sources, for example, witnesses, supervisors, physical conditions, hazard data sheets, written systems of work, training records. The amount of time spent should not, however, be disproportionate to the risk. The aim of the investigation should be to explore the situation for possible underlying factors, in addition to the immediately obvious causes of the accident. For example, in a machinery accident it would not be sufficient to conclude that an accident occurred because a machine was inadequately guarded. It is necessary to look into the possible underlying system failure that may have occurred.

Investigations have three facets, which are particularly valuable and can be used to check against each other:

➤ direct observation of the scene, premises, workplace, relationship of components, materials and substances being used, possible reconstruction of events and injuries or condition of the person concerned;
➤ documents including written instructions, training records, procedures, safe operating systems, risk assessments, policies, records of inspections or test and examinations carried out;

➤ interviews (including written statements) with persons injured, witnesses, people who have carried out similar functions or examinations and tests on the equipment involved and people with specialist knowledge.

Immediate causes

A detailed investigation should look at the following factors as they can provide useful information about **immediate causes** that have been manifested in the incident/accident.

➤ Personal factors:
  • behaviour of the people involved
  • suitability of people doing the work
  • training and competence.
➤ Task factors:
  • workplace conditions and precautions or controls
  • actual method of work adopted at the time
  • ergonomic factors
  • normal working practice either written or customary.

Underlying and root causes

A thorough investigation should also look at the following factors as they can provide useful information about **underlying and root causes** that have been manifested in the incident/accident:

Underlying causes are the less obvious system or organizational reasons for an accident or incident such as:

➤ pre-start-up machinery checks were not made by supervisors
➤ the hazard had not been considered in the risk assessment
➤ there was no suitable method statement
➤ pressures of production had been more important
➤ the employee was under a lot of personal pressure at the time
➤ have there been previous similar incidents?
➤ was there adequate supervision, control and coordination of the work involved?

Root causes involve an initiating event or failing from which all other causes or failings arise. Root causes are generally management, planning or organizational failings including:

➤ quality of the health and safety policy and procedures;
➤ quality of consultation and cooperation of employees;
➤ the adequacy and quality of communications and information;

➤ deficiencies in risk assessments, plans and control systems;

➤ deficiencies in monitoring and measurement of work activities;

➤ quality and frequency of reviews and audits.

8.4.4 Investigation interview techniques

It must be made clear at the outset and during the course of the interview that the aim is not to apportion blame but to discover the facts and use them to prevent similar accidents or incidents in the future.

A witness should be given the opportunity to explain what happened in their own way without too much interruption and suggestion. Questions should then be asked to elicit more information. These should be of the open type, which do not suggest the answer. Questions starting with the words in Figure 8.4 are useful.

'Why' should not be used at this stage. The facts should be gathered first, with notes being taken at the end of the explanation. The investigator should then read them or give a summary back to the witness, indicating clearly that they are prepared to alter the notes, if the witness is not content with them.

If possible, indication should be given to the witness about immediate actions that will be taken to prevent a similar occurrence and that there could be further improvements depending on the outcome of the investigation.

Seeing people injured can often be very upsetting for witnesses, which should be borne in mind. This does not mean they will not be prepared to talk about what has happened. They may in fact wish to help, but questions

Figure 8.4 Questions to be asked in an investigation.

should be sensitive; upsetting the witness further should be avoided.

8.4.5 Comparison with relevant standards

There are usually suitable and relevant standards which may come from the HSE, industry or the organization itself. These should be carefully considered to see if:

➤ suitable standards are available to cover legal standards and the controls required by the risk assessments;

➤ the standards are sufficient and available to the organization;

➤ the standards were implemented in practice;

➤ the standards were implemented, why there was a failure;

➤ changes should be made to the standards.

8.4.6 Recommendations

The investigation should have highlighted both immediate causes and underlying causes. Recommendations, both for immediate action and for longer term improvements, should come out of this, but it may be necessary to ensure that the report goes further up the management chain if the improvements recommended require authorization, which cannot be given by the investigating team.

8.4.7 Follow-up

It is essential that a follow-up is made to check on the implementation of the recommendations. It is also necessary to review the effect of the recommendations to check whether they have achieved the desired result and whether they have had unforeseen 'knock-on' effects, creating additional risks and problems.

8.4.8 Use of information

The accident or incident investigation should be used to generate recommendations but should also be used to generate safety awareness. The investigation report or a summary should therefore be circulated locally to relevant people and, when appropriate, summaries circulated throughout the organization. The accident or incident does not need to have resulted in a 3-day lost time injury for this system to be used.

8.4.9 Training

A number of people will potentially be involved in accident or incident investigation. For most of these people this will only be necessary on very few occasions. Training guidance and help will therefore be required. Training can be provided in accident/incident investigation in courses run on site and also in numerous off-site venues. Computer-based training courses are also available. These are intended to provide refresher training on an individual basis or complete training at office sites, for example, where it may not be feasible to provide practical training.

8.4.10 Investigation form

Headings which could be used to compile an accident/ incident investigation form are given below:

- ➤ date and location of accident/incident;
- ➤ circumstances of accident/incident;
- ➤ immediate cause of accident/incident;
- ➤ underlying cause of accident/incident;
- ➤ immediate action taken;
- ➤ recommendation for further improvement;
- ➤ report circulation list;
- ➤ date of investigation;
- ➤ signature of investigating team leader;
- ➤ names of investigating team.

Follow-up
- ➤ were the recommendations implemented?
- ➤ were the recommendations effective?

The following categories of immediate causes of accident are used in F2508:

1. contact with moving machinery or material being machined;
2. struck by moving, including flying or falling object;
3. struck by moving vehicle;
4. struck against something fixed or stationary;
5. injured whilst handling lifting or carrying;
6. slip, trip or fall on same level;
7. fall from height; indicate approximate distance of fall in metres;
8. trapped by something collapsing or overturning;
9. drowning or asphyxiation;
10. exposure to or contact with harmful substances;
11. exposure to fire;
12. exposure to an explosion;
13. contact with electricity or an electrical discharge;
14. injured by an animal;
15. violence;
16. other kind of accident.
    The investigation form shown in Appendix 8.1 uses these immediate causes which can be used for analysis purposes.

8.4.11 Key date for medium level of investigation

The HSE in HSG65 has suggested that the key data included in Box 8.1 should be covered in an investigation report. This level of data is more appropriate to a medium level of investigation (see Section 8.3.1).

8.5 Legal recording and reporting requirements

8.5.1 Accident book

Under the Social Security (Claims and Payments) Regulations 1979, Regulation 25, employers must keep a record of accidents at premises where more than 10 people are employed. Anyone injured at work is required to inform the employer and record information on the accident in an accident book, including a statement on how the accident happened.

The employer is required to investigate the cause and enter this in the accident book if they discover anything that differs from the entry made by the employee. The purpose of this record is to ensure that information is available if a claim is made for compensation.

The HSE produced a new Accident Book BI 510 (see Figure 8.5) in May 2003 with notes on these Regulations, and the Reporting of Injuries, Diseases and Dangerous Occurrences Regulations 1995 (RIDDOR) and now complies with the Data Protection Act 1998.

8.5.2 RIDDOR

RIDDOR requires employers, the self-employed and those in control of premises, to report certain more serious accidents and incidents to the HSE or other enforcing authority and to keep a record. There are no exemptions for small organizations. A full summary of the Regulations is given in Chapter 17. The reporting and recording requirements are as follows:

Death or major injury
If an accident occurs at work and:

- ➤ an employee or self-employed person working on the premise is killed or suffers a major injury (including the effects of physical violence) (see Chapter 17 under RIDDOR for definition of major injury);
- ➤ a member of the public is killed or taken to hospital.

The event

➤ Details of any injured person, including age, sex, experience, training, etc.

➤ A description of the circumstances, including the place, time of day and conditions.

➤ Details of the event, including:
- any actions which led directly to the event
- the direct causes of any injuries, ill health or other loss
- the immediate causes of the event
- the underlying causes – for example failures in workplace precautions, risk control systems or management arrangements.

➤ Details of the outcomes, including in particular:
- the nature of the outcome – for example injuries or ill health to employees or members of the public, damage to property, process disruption, emissions to the environment, creation of hazards
- the severity of the harm caused, including injuries, ill health and losses.

➤ The immediate management response to the situation and its adequacy:
- Was it dealt with promptly?

- Were continuing risks dealt with promptly and adequately?
- Was the first-aid response adequate?
- Were emergency procedures followed?

➤ Whether the event was preventable and if so how.

The potential consequences

➤ What was the worst that could have happened?

➤ What prevented the worst from happening?

➤ How often could such an event occur (the '**recurrence potential**')?

➤ What was the worst injury or damage which could have resulted (the '**severity potential**')?

➤ How many people could the event have affected (the '**population potential**')?

Recommendations

Prioritized actions with responsibilities and targets for completion.

Source: HSG65.

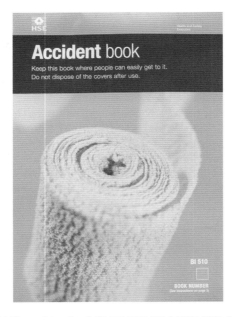

Figure 8.5 The accident book (BI 510 ISBN 978 0 7176 6249 4).

The responsible person must notify the enforcing authority without delay by the quickest practicable means, for example telephone. They will need to give brief details about the organization, the injured person(s) and the circumstances of the accident and, within 10 days, the responsible person must also send a completed accident report form, F2508.

Over 3-day lost time injury

If there is an accident connected with work (including physical violence), and an employee or self-employed person working on the premises suffers an injury and is away from work or not doing his/her normal duties for more than 3 days (including weekends, rest days or holidays but not counting the day of the accident), the responsible person must send a completed accident report form, F2508, to the enforcing authority within 10 days.

Disease

If a doctor notifies the responsible person that an employee suffers from a reportable work-related disease a completed disease report form, F2508A, must be sent to the enforcing authority. A summary is included in Chapter 17 and a full list is included with a pad of report forms. The HSE InfoLine or the Incident Contact Centre can be contacted to check whether a particular disease is reportable.

Dangerous occurrence

If an incident happens which does not result in a reportable injury, but obviously could have done, it could be a dangerous occurrence as defined by a list in the Regulations (see Chapter 17 for a summary of dangerous occurrences). All dangerous occurrences must be reported immediately by, for example, telephone, to the enforcing authorities. The HSE InfoLine or the Incident Contact Centre can be contacted to check whether a dangerous occurrence is reportable.

A completed Dangerous occurrence report form, F2508DO, must be sent to the enforcing authorities within 10 days. There are also special report forms for flammable gas incidents and incidents offshore.

8.5.3 Whom to report to

Until 2001 all reports had to be made to the local HSE office or Local Authority. This can still be done, but there is a centralized national system, called the Incident Contact Centre (ICC), which is a joint venture between the HSE and local authorities. All reports sent locally are now passed on to the ICC.

This means that employers no longer need be concerned about which authority to report to, as all incidents can be reported to the ICC directly. Reports can be sent by telephone, fax, the Internet or by post. If reporting by the Internet or telephone, a copy of the report is sent to the responsible person for correction, if necessary, and for their records.

➤ Postal reports should be sent to:
  - Incident Contact Centre
    Caerphilly Business Park
    Caerphilly
    CF83 3GG

➤ For Internet reports go to:
  - www.hse.gov.uk/riddor/online.htm

➤ or link through
  - www.hse.gov.uk

➤ By telephone (charged at local rates)
  - 0845 300 9923

➤ By fax (charged at local rates)
  - 0845 300 9924

➤ By e-mail: www.hse.gov.uk/forms/incident/index.htm. If the HSE is needed urgently to report a major incident

or fatality phone 0151 951 4000 to be directed to the appropriate HSE office.

Internal systems for collecting and analysing incident data

Introduction

Managers need effective internal systems to know whether the organization is getting better or worse, to know what is happening and why, and to assess whether objectives are being achieved. Chapter 16 deals with monitoring generally; but here, the basic requirements of a collection and analysis system for incidents are discussed.

The incident report form (discussed earlier) is the basic starting point for any internal system. Each organization needs to lay down what the system involves and who is responsible to do each part of the procedure. This will involve:

➤ what type of incidents should be reported;
➤ who completes the incident report form – normally the manager responsible for the investigation;
➤ how copies should be circulated in the organization;
➤ who is responsible to provide management measurement data;
➤ how the incident data should be analysed and at what intervals;
➤ the arrangements to ensure that action is taken on the data provided.

The data should seek to answer the following questions:

➤ are failure incidents occurring, including injuries, ill health and other loss incidents?
➤ where are they occurring?
➤ what is the nature of the failures?
➤ how serious are they?
➤ what are the potential consequences?
➤ what are the reasons for the failures?
➤ how much has it cost?
➤ what improvements in controls and the management system are required?
➤ how do these issues vary with time?
➤ is the organization getting better or worse?

Type of accident/incident

Most organizations will want to collect data on:

➤ all injury accidents;
➤ cases of ill health;
➤ sickness absence;

➤ damage to property, personal effects and work in progress;

➤ incidents with the potential to cause serious injury, ill health or damage (undesired circumstances).

Not all of these are required by law, but this should not deter the organization that wishes to control risks effectively.

Analysis

All the information, whether in accident books or report forms, will need to be analysed so that useful management data can be prepared. Many organizations look at the analysis every month and annually. However, where there are very few accidents/incidents, quarterly may be sufficient. The health and safety information should be used alongside other business measures and should receive equal status.

There are several ways in which data can be analysed and presented. The most common ways are:

➤ by causation using the classification used on the RIDDOR form F2508. This has been used on the example accident/incident report form (see Appendix 8.1);

➤ by the nature of the injury, such as cuts, abrasions, asphyxiation and amputations;

➤ by the part of the body affected, such as hands, arms, feet, lower leg, upper leg, head, eyes, back and so on. Sub-divisions of these categories could be useful if there were sufficient incidents;

➤ by age and experience at the job;

➤ by time of day;

➤ by occupation or location of the job;

➤ by type of equipment used.

There are a number of up-to-date computer recording programs which can be used to manipulate the data if significant numbers are involved. The trends can be shown against monthly, quarterly and annual past performance of, preferably, the same organization. If indices are calculated, such as Incident Rate, comparisons can be made nationally with HSE figures and with other similar organizations or businesses in the same industrial group. This is really of major value only to larger organizations with significant numbers of events. (See Chapters 4 and 5 for more information on definitions and calculation of Incident Rates.)

The HSE produces annual bulletins of national performance along with a detailed statistical report, which can be used for comparisons. There are difficulties in comparisons across Europe and say, with the USA, where the definitions of accidents or time lost vary.

Reports should be prepared with simple tables and graphs showing trends and comparisons. Line graphs, bar charts and pie charts are all used quite extensively with good effect. All analysis reports should be made available to employees as well as managers. This can often be done through the Health and Safety Committee and safety representatives, where they exist, or directly to all employees in small organizations. Other routine meetings, team briefings and notice boards can all be used to communicate the message.

It is particularly important to make sure that any actions recommended or highlighted by the reports are taken quickly and employees kept informed.

8.7 Compensation and insurance issues

Accidents/incidents arising out of the organization's activities resulting in injuries to people and incidents resulting in damage to property can lead to compensation claims. The second objective of an investigation should be to collect and record relevant information for the purposes of dealing with any claim. It must be remembered that, in the longer term, prevention is the best way to reduce claims and must be the first objective in the investigation. An overzealous approach to gathering information concentrating on the compensation aspect can, in fact, prompt a claim from the injured party where there was no particular intention to take this route before the investigation. Nevertheless, relevant information should be collected. Sticking to the collection of facts is usually the best approach.

As mentioned earlier the legal system in England and Wales changed dramatically with the introduction of the Woolf reforms in 1999. These reforms apply to injury claims. This date was feared by many because of the uncertainty and the fact that the pre-action protocols were very demanding.

The essence of the pre-legal action protocols is as follows:

1. 'letter of claim' to be acknowledged within 21 days;
2. ninety days from date of acknowledgement to either accept liability or deny. If liability is denied then full reasons must be given;
3. agreement to be reached on using a single expert.

The overriding message is that to comply with the protocols quick action is necessary. It is also vitally important that records are accurately kept and accessible.

Lord Woolf made it clear in his instructions to the judiciary that there should be very little leeway given to claimants and defendants who did not comply.

What has been the effect of these reforms on day to day activity?

At the present time there has been a dramatic drop in the number of cases moving to litigation. Whilst this was the primary objective of the reforms, the actual drop is far more significant than any one anticipated. In large part this may simply be due to general unfamiliarity with the new rules and we can expect an increase in litigation as time goes by.

However, some of the positive effects have been:

➤ the elimination of speculative actions due to the requirement to fully outline the claimant's case in the letter of claim;
➤ earlier and more comprehensive details of the claim allowing a more focused investigation and response;
➤ 'Part 36 offers' (payments into court) seeming to be having greater effect in deterring claimants from pursuing litigation;
➤ overall faster settlement.

The negative effects have principally arisen from failure to comply with the timescales, particularly relating to the gathering of evidence and records and having no time therefore to construct a proper defence.

Whilst it is still too early to make a full assessment of the impact of Woolf there are reasons to be cautiously optimistic.

Appendix 8.2 provides a checklist of headings, which may assist in the collection of information. It is not expected that all accidents and incidents will be investigated in depth and a dossier with full information prepared. Judgement has to be applied as to which incidents might give rise to a claim and when a full record of information is required. All accident/incident report forms should include the names of all witnesses as a minimum. Where the injury is likely to give rise to lost time, a photograph(s) of the situation should be taken.

8.8 Sources of reference

Guide to the Reporting of Injuries, Diseases and Dangerous Occurrences Regulations 1995 L73 HSE Books 2008 ISBN 978 0 7176 6290 6.
Investigating Incidents and Accidents at Work HSG245 HSE Books 2004 ISBN 978 0 7176 2827 8.
Incident at Work Misc769 HSE Web 2007.

Relevant statutory provisions
The Reporting of Injuries, Diseases and Dangerous Occurrences Regulations 1995.

8.9 Practice NEBOSH questions for Chapter 8

1. **(i)** **Give FOUR** reasons why an organization should have a system for the internal reporting of accidents.
   (ii) **Outline** factors that may discourage employees from reporting accidents at work.

2. An employee has been seriously injured after being struck by a reversing vehicle in a loading bay.
   (i) **Give FOUR** reasons why the accident should be investigated by the person's employer.
   (ii) **Outline** the information that should be included in the investigation report.
   (iii) Outline FOUR possible immediate causes **AND FOUR** possible underlying (root) causes of the accident.

3. **State** the issues that should be included in a typical workplace accident reporting procedure.

4. An organization has decided to conduct an internal investigation of an accident in which an employee was injured following the collapse of storage racking.
   (i) **Outline FOUR** benefits to the organization of investigating the accident.
   (ii) Giving reasons in **EACH** case, **identify FOUR** people who may be considered useful members of the investigation team.
   (iii) Having defined the team, **outline** the factors that should be considered when planning the investigation.

5. **Explain** the purpose and benefits of collecting 'near-miss' incident data.

6. **Explain** the differences between the *immediate* and the *root (underlying)* causes of an accident.

7. **Outline** the immediate and longer term actions that should be taken following a major injury accident at work.

8. An employee slipped on a patch of oil on a warehouse floor and was admitted to hospital where he remained for several days. The oil was found close to a stack of pallets that had been left abandoned on the designated pedestrian walkway.
   (i) **Outline** the legal requirements for reporting the accident to the enforcing authority.
   (ii) **Identify** the possible *immediate* **AND** *root* causes of the accident.
   (iii) **Outline** ways in which management could demonstrate their commitment to improve health and safety standards in the workplace following the accident.

9. An employee broke their arm when they fell down a flight of stairs at work while carrying a box of letters to the mail room:
   (i) **state** the requirements for reporting the accident to the enforcing authority **AND**
   (ii) **give** the information that should be contained in the accident report.

10. **Identify** the questions that might be included on a checklist to gather information following an accident involving slips, trips and falls.

11. **Outline** the key points that should be covered in a training session for employees on the reporting of accidents/incidents.

12. With reference to the Reporting of Injuries, Diseases and Dangerous Occurrences Regulations 1995:
    (i) **State** the legal requirements for reporting a fatality resulting from an accident at work to an enforcing authority.
    (ii) **Outline THREE** further categories of work-related injury (other than fatal injuries) that are reportable.
    (iii) **State** the requirements for reporting an 'over three-day' injury.

13. An employee has suffered a fatal injury at work.
    (i) **Outline** the procedure for reporting the accident to the enforcing authority.
    (ii) **State** who should be informed of the accident, other than the enforcing authority.

14. With reference to the Reporting of Injuries, Diseases and Dangerous Occurrences Regulations 1995:
    (i) **List FOUR** types of major injury.
    (ii) **Outline** the procedures for reporting a major injury to an enforcing authority.

15. With reference to the Reporting of Injuries, Diseases and Dangerous Occurrences Regulations (RIDDOR) 1995:
    (i) **Explain**, using **TWO** examples, the meaning of the term 'dangerous occurrence'.
    (ii) **State** the requirements for reporting a dangerous occurrence.

16. (i) **State THREE** occupational diseases reportable under the Reporting of Injuries, Diseases and Dangerous Occurrences Regulations (RIDDOR) 1995.
    (ii) **Outline** the legal requirements for reporting an occupational disease.

17. (i) **Identify TWO** ill-health conditions that are reportable under the Reporting of Injuries, Diseases and Dangerous Occurrences Regulations (RIDDOR) 1995.
    (ii) **Outline** reasons why employers should keep records of occupational ill health among employees.

18. An employee is claiming compensation for injuries received during an accident involving a fork-lift truck.
    (i) **Identify** the documented information that the employer might draw together when preparing a possible defence against the claim.

Appendix 8.1 Accident/incident report form

ACCIDENT/INCIDENT REPORT

INJURED PERSON: .... Date of Incident: / /20   Time: .... a.m./p.m.
POSITION:.. Place of Incident: ....
DEPARTMENT:.. Details of Injury: ....
Investigation carried out by: ....
Position:.. Estimated Absence:...

Brief details of Accident/Incident (A detailed report together with diagrams, photographs and any witness statements should be attached where necessary. Please complete all details requested overleaf.)

Immediate Causes	**Underlying Causes**

Conclusions (How can we prevent this kind of incident/accident occurring again?)

Action to be taken: **Completion Date:**   / /20

IMPORTANT

Please ensure that an accident/incident investigation and report is completed and forwarded to Human Resources **within 48 hours of the accident/incident occurring**.

Remember that accidents involving major injuries or dangerous occurrences have to be notified immediately by telephone to the HSE/Local authority.

Signature of manager making report:...   Copies: Personnel Manager
                                                                                                          Health and Safety Manager
                         Date: / /20   Payroll Controller

INJURED PERSON: Surname...   Forenames..

                                                                                                             Male/Female

Home address...Age.................................

Employee ☐   *Agency Temp* ☐   *Contractor* ☐   *Visitor* ☐   *Youth Trainee* ☐   (Tick one box)

(Continued)

Kind of Accident/incident Indicate what kind of accident/incident led to the injury or condition (tick one box)

Contact with moving machinery or material being machined 1	Injured while handling, lifting or carrying 5	Drowning or asphyxiation 9	Contact with electricity or an electrical discharge 13
Struck by moving, flying or falling object 2	Slip, trip or fall on same level 6	Exposure to or contact with harmful substance 10	Injured by an animal 14
Struck by moving vehicle 3	Fall from height; indicate approx distance of fall ...... m 7	Exposure to fire 11	Violence; physically assaulted by a person 15
Struck against something fixed or stationary 4	Trapped by something collapsing 8	Exposure to an explosion 12	Other kind of accident 16

Detail any machinery, chemicals, tools etc. involved

Incident first reported to: Name ....

Position ....   Dept ....

First Aid/medical attention by: First Aider Name ....   Dept ....

....   Doctor Name ....

     Medical centre ....   Hospital ....

WITNESSES

Name	Position & Dept	Statement obtained (yes/no) Attach all statements taken
....	....	yes/no
....	....	yes/no
....	....	yes/no
....	....	yes/no

For Office use only

If relevant: Date reported to HSE/LA   a) by telephone   .../ .../20

     b) on form F2508   .../ .../20

Date reported to organization's insurers   .../ .../20

WERE THE RECOMMENDATIONS EFFECTIVE? YES/NO
IF NO SAY WHAT ACTION SHOULD BE TAKEN.

Appendix 8.2 Information for insurance/compensation claims

Information for insurance/compensation purposes following accident or incident

Factual information needs to be collected where there is the likelihood of some form of claim either against the organization or by the organization (e.g. damage to equipment). This aspect should be considered as a second objective in accident/incident investigation, the first being to learn from the accident/incident to reduce the possibility of accidents/incidents occurring in the future.

Workplace claims

- accident book entry;
- first aider report;
- surgery record;
- foreman/supervisor accident report;
- safety representatives accident report;
- RIDDOR report to HSE;
- other communications between defendants and HSE;
- minutes of Health and Safety Committee meeting(s) where accident/incident considered;
- report to DSS;
- documents listed above relative to any previous Accident/incident identified by the claimant and relied upon as proof of negligence;
- earnings information where defendant is employer.

Documents produced to comply with requirements of the Management of Health and Safety at Work Regulations 1999:

- pre-accident/incident risk assessment;
- post-accident/incident re-assessment;
- Accident/incident Investigation Report prepared in implementing the requirements; Health Surveillance Records in appropriate cases; information provided to employees;
- documents relating to the employee's health and safety training.

Workplace claims – examples of disclosure where specific Regulations apply

Section A – Workplace (Health Safety and Welfare) Regulations 1992

- repair and maintenance records;
- housekeeping records;
- hazard warning signs or notices (traffic routes).

Section B – Provision and Use of Work Equipment Regulations 1998

- manufacturers' specifications and instructions in respect of relevant work equipment establishing its suitability to comply with Regulation 5;
- maintenance log/maintenance records required; documents providing information and instructions to employees; documents provided to the employee in respect of training for use;
- any notice, sign or document relied upon as a defence against alleged breaches dealing with controls and control systems.

Section C – Personal Protective Equipment at Work Regulations 1992

- documents relating to the assessment of the PPE;
- documents relating to the maintenance and replacement of PPE;
- record of maintenance procedures for PPE;
- records of tests and examinations of PPE;
- documents providing information, instruction and training in relation to the PPE;
- instructions for use of PPE to include the manufacturers' instructions.

Section D – Manual Handling Operations Regulations 1992

- manual handling risk assessment carried out;
- re-assessment carried out post-accident;
- documents showing the information provided to the employee to give general indications related to the load and precise indications on the weight of the load and the heaviest side of the load if the centre of gravity was not positioned centrally;
- documents relating to training in respect of manual handling operations and training records.

Section E – Health and Safety (Display Screen Equipment) Regulations 1992

- analysis of work stations to assess and reduce risks;
- re-assessment of analysis of work stations to assess and reduce risks following development of symptoms by the claimant;
- documents detailing the provision of training including training records;
- documents providing information to employees.

Section F – Control of Substances Hazardous to Health Regulations 2002

- risk assessments and any reviews;
- copy labels from containers used for storage handling and disposal of carcinogens;
- warning signs identifying designation of areas and installations which may be contaminated by carcinogens;
- documents relating to the assessment of the PPE;
- documents relating to the maintenance and replacement of PPE;
- record of maintenance procedures for PPE;
- records of tests and examinations of PPE;

- documents providing information, instruction and training in relation to the PPE;
- instructions for use of PPE to include the manufacturers' instructions;
- air monitoring records for substances assigned a workplace exposure limit;
- maintenance examination and test of control measures records;
- monitoring records;
- health surveillance records;
- documents detailing information, instruction and training including training records for employees;
- labels and health and safety data sheets supplied to the employers.

Movement of people and vehicles – hazards and control

9

After reading this chapter you should be able to:

1. identify the hazards that may cause injuries to pedestrians in the workplace and the control measures to reduce the risk of such injuries
2. identify the hazards presented by the movement of vehicles in the workplace and the control measures to reduce the risks they present.

9.1 Introduction

People are most often involved in accidents as they walk around the workplace or when they come into contact with vehicles in or around the workplace. It is therefore important to understand the various common accident causes and the control strategies that can be employed to reduce them. Slips, trips and falls account for the majority of accidents to pedestrians and the more serious accidents between pedestrians and vehicles can often be traced back to excessive speed or other unsafe vehicle practices, such as lack of driver training. Many of the risks associated with these hazards can be significantly reduced by an effective management system.

9.2 Hazards to pedestrians

The most common hazards to pedestrians at work are slips, trips and falls on the same level, falls from height,

Figure 9.1 Tripping hazards.

collisions with moving vehicles, being struck by moving, falling or flying objects and striking against fixed or stationary objects. Each of these will be considered in turn, including the conditions and environment in which the particular hazard may arise.

9.2.1 Slips, trips and falls on the same level

These are the most common of the hazards faced by pedestrians and accounted for 30% of all the major accidents every year and 20% of over 3-day injuries reported to the Health and Safety Executive (HSE), who have reported that every 25 minutes someone breaks or fractures a bone due to slipping, tripping or falling at work (Figure 9.1). It has been estimated that the annual cost of these accidents to the nation is £750 million and a direct cost to employers of £300 million. The highest reported injuries are reported in the food and related industries. Older workers, especially women, are the most severely injured group from falls resulting in fractures of the hips and/or femur. Civil compensation claims are becoming more common and costly to employers and such claims are now being made by members of the public who have tripped on uneven paving slabs on pavements or in shopping centres.

The HSE has been so concerned at the large number of such accidents that it has identified slips, trips and falls on the same level as a key risk area. The costs of slips, trips and falls on the same level are high to the injured employee (lost income and pain), the employer (direct and indirect costs including lost production) and to society as a whole in terms of health and social security costs.

Slip hazards are caused by:

➤ wet or dusty floors;
➤ the spillage of wet or dry substances – oil, water, flour dust and plastic pellets used in plastic manufacture (Figure 9.2);
➤ loose mats on slippery floors;
➤ wet and/or icy weather conditions;
➤ unsuitable footwear or floor coverings or sloping floors.

Trip hazards are caused by:

➤ loose floorboards or carpets;
➤ obstructions, low walls, low fixtures on the floor;
➤ cables or trailing leads across walkways or uneven surfaces; leads to portable electrical hand tools and other electrical appliances (vacuum cleaners and overhead projectors); raised telephone and electrical sockets – also a serious trip hazard (this can be a significant problem when the display screen workstations are re-orientated in an office);
➤ rugs and mats – particularly when worn or placed on a polished surface;
➤ poor housekeeping – obstacles left on walkways, rubbish not removed regularly;
➤ poor lighting levels – particularly near steps or other changes in level;

Figure 9.2 Cleaning must be done carefully to prevent slipping.

➤ sloping or uneven floors – particularly where there is poor lighting or no handrails;
➤ unsuitable footwear – shoes with a slippery sole or lack of ankle support.

The vast majority of major accidents involving slips, trips and falls on the same level result in dislocated or fractured bones.

9.2.2 Falls from work at height

These are the most common cause of serious injury or death in the construction industry and the topic is covered in Chapter 16. These accidents are often concerned with falls of greater than about 2 m and often result in fractured bones, serious head injuries, loss of consciousness and death. Twenty-five per cent of all deaths at work and 19% of all major accidents are due to falls from a height. Falls down staircases and stairways, through fragile surfaces, off landings and stepladders and from vehicles, all come into this category. Injury, sometimes serious, can also result from falls below 2 m, for example using swivel chairs for access to high shelves.

9.2.3 Collisions with moving vehicles

These can occur within the workplace premises or on the access roads around the building. It is a particular problem where there is no separation between pedestrians and vehicles or where vehicles are speeding. Poor lighting, blind corners, the lack of warning signs and barriers at road crossing points also increase the risk of this type of accident. Eighteen per cent of fatalities at work are caused by collisions between pedestrians and moving vehicles with the greatest number occurring in the service sector (primarily in retail and warehouse activities).

9.2.4 *Being struck by moving, falling or flying objects*

This causes 18% of fatalities at work and is the second highest cause of fatality in the construction industry. It also causes 15% of all major and 14% of over 3-day accidents. Moving objects include articles being moved, moving parts of machinery or conveyor belt systems, and flying objects are often generated by the disintegration of a moving part or a failure of a system under pressure. Falling objects are a major problem in construction (due to careless working at height) and in warehouse work (due to careless stacking of pallets on racking). The head is particularly vulnerable to these hazards. Items falling off high shelves and moving loads are also significant hazards in many sectors of industry.

9.2.5 *Striking against fixed or stationary objects*

This accounts for over 1000 major accidents every year. Injuries are caused to a person either by colliding with a fixed part of the building structure, work in progress, a machine member or a stationary vehicle or by falling against such objects. The head appears to be the most vulnerable part of the body to this particular hazard and this is invariably caused by the misjudgement of the height of an obstacle. Concussion in a mild form is the most common outcome and a medical check-up is normally recommended. It is a very common injury during maintenance operations when there is, perhaps, less familiarity with particular space restrictions around a machine. Effective solutions to all these hazards need not be expensive, time consuming or complicated. Employee awareness and common sense combined with a good housekeeping regime will solve many of the problems.

9.3 Control strategies for pedestrian hazards

9.3.1 *Slips, trips and falls on the same level*

These may be prevented or, at least, reduced by several control strategies. These and all the other pedestrian hazards discussed should be included in the workplace risk assessments required under the Management of Health and Safety at Work Regulations by identifying slip or trip hazards, such as poor or uneven floor/pavement surfaces, badly lit stairways and puddles from leaking roofs. There is also a legal requirement in the Workplace (Health, Safety and Welfare) Regulations for all floors to be suitable, in good condition and free from obstructions. Traffic routes must be so organized that people can move around the workplace safely.

The key elements of a health and safety management system are as relevant to these as to any other hazards:

➤ **Planning** – remove or minimize the risks by using appropriate control measures and defined working practices (e.g. covering all trailing leads).
➤ **Organization** – involve employees and supervisors in the planning process by defining responsibility for keeping given areas tidy and free from trip hazards.
➤ **Control** – record all cleaning and maintenance work. Ensure that anti-slip covers and cappings are placed on stairs, ladders, catwalks, kitchen floors and smooth walkways. Use warning signs when floor surfaces have recently been washed.
➤ **Monitoring and review** – carry out regular safety audits of cleaning and housekeeping procedures and include trip hazards in safety surveys. Check on accident records to see whether there has been an improvement or if an accident black spot can be identified.

Slip and trip accidents are a major problem for large retail stores for both customers and employees. The provision of non-slip flooring, a good standard of lighting and minimizing the need to block aisles during the re-stocking of merchandise are typical measures that many stores use to reduce such accidents. Other measures include the wearing of suitable footwear by employees, adequate handrails on stairways, the highlighting of any floor level changes and procedures to ensure a quick and effective response to any reports of floor damage or spillages. Good housekeeping procedures are essential. The design of the store layout and any associated warehouse can also ensure a reduction in all types of accident. Many of these measures are valid for a range of workplaces.

9.3.2 *Falls from work at height*

These may be controlled by the use of suitable guardrails and barriers and also by the application of the hierarchy of controls discussed in Chapters 6 and 16 which is:

➤ remove the possibility of falling a distance that could cause personal injury (e.g. by undertaking the work at ground level);

➤ protect against the hazard of falling a distance that could cause personal injury (e.g. by using handrails);

➤ stop the person from falling a distance that could cause personal injury (e.g. by the provision of safety harnesses);

➤ mitigate the consequences of falling a distance that could cause personal injury (e.g. by the use of air bags).

The principal means of preventing falls of people or materials includes the use of fencing, guardrails, toe boards, working platforms, access boards and ladder hoops (Figure 9.3). Safety nets and safety harnesses should only be used when all other possibilities are not practical. The use of banisters on open sides of stairways and handrails fitted on adjacent walls will also help to prevent people from falling. Holes in floors and pits should always be fenced or adequately covered. Precautions should be taken when working on fragile surfaces (see Chapter 16 for details).

Permanent staircases are also a source of accidents included within this category of falling from a height and the following design and safety features will help to reduce the risk of such accidents:

➤ adequate width of the stairway, depth of the tread and provision of landings and banisters or handrails and intermediate rails; the treads and risers should always be of uniform size throughout the staircase and designed to meet Building Regulations requirements for angle of incline (i.e. steepness of staircase);

➤ provision of non-slip surfaces and reflective edging;

➤ adequate lighting;

➤ adequate maintenance;

Figure 9.3 Falling from a height – tower scaffold.

➤ special or alternative provision for disabled people (e.g. personnel elevator at the side of the staircase).

Great care should be used when people are loading or unloading vehicles; as far as possible people should avoid climbing onto vehicles or their loads. For example sheeting of lorries should be carried out in designated places using properly designed access equipment.

9.3.3 Collisions with moving vehicles

These are best prevented by completely separating pedestrians and vehicles, providing well-marked, protected and laid-out pedestrian walkways. People should cross roads by designated and clearly marked pedestrian crossings. Suitable guardrails and barriers should be erected at entrances and exits from buildings and at 'blind' corners at the end of racking in warehouses. Particular care must be taken in areas where lorries are being loaded or unloaded. It is important that separate doorways are provided for pedestrians and vehicles and all such doorways should be provided with a vision panel and an indication of the safe clearance height, if used by vehicles. Finally, the enforcement of a sensible speed limit, coupled, where practicable, with speed governing devices, is another effective control measure.

9.3.4 Being struck by moving, falling or flying objects

These may be prevented by guarding or fencing the moving part (as discussed in Chapter 11) or by adopting the measures outlined for construction work (Chapter 16). Both construction workers and members of the public need to be protected from the hazards associated with falling objects. Both groups should be protected by the use of covered walkways or suitable netting to catch falling debris where this is a significant hazard. Waste material should be brought to ground level by the use of chutes or hoists. Waste should not be thrown from a height and only minimal quantities of building materials should be stored on working platforms. Appropriate personal protective equipment, such as hard hats or safety glasses, should be worn at all times when construction operations are taking place.

It is often possible to remove high-level storage in offices and provide driver protection on lift truck cabs in warehouses. Storage racking is particularly vulnerable and should be strong and stable enough for the loads it has to carry. Damage from vehicles in a warehouse can easily weaken the structure and cause collapse. Uprights need protection, particularly at corners.

The following action can be taken to keep racking serviceable:

➤ Inspect them regularly and encourage workers to report any problems.
➤ Post notices with maximum permissible loads and never exceed the loading.
➤ Use good pallets and safe stacking methods.
➤ Band, box or wrap articles to prevent items falling.
➤ Set limits on the height of stacks and regularly inspect to make sure that limits are being followed.
➤ Provide instruction and training for staff and special procedures for difficult objects.

9.3.5 Striking against fixed or stationary objects

This can only be effectively controlled by:

➤ having good standards of lighting and housekeeping;
➤ defining walkways and making sure they are used;
➤ the use of awareness measures, such as training and information in the form of signs or distinctive colouring;
➤ the use of appropriate personal protective equipment, such as head protection, as discussed previously.

9.3.6 General preventative measures for pedestrian hazards

Minimizing pedestrian hazards and promoting good work practices requires a mixture of sensible planning, good housekeeping and common sense. A few of the required measures are costly or difficult to introduce and, although they are mainly applicable to slips, trips and falls on the same level and collisions with moving vehicles, they can be adapted to all types of pedestrian hazard (Figure 9.4). Typical measures include the following:

➤ Develop a safe workplace as early as possible and ensure that suitable floor surfaces and lighting are selected and vehicle and pedestrian routes are carefully planned. Lighting should not dazzle approaching vehicles nor should pedestrians be obscured by stored products. Lighting is very important where there are changes of level or stairways. Any physical hazards, such as low beams, vehicular movements or pedestrian crossings, should be clearly marked. Staircases need particular attention to ensure that they are slip resistant and the edges of the stairs marked to indicate a trip hazard.

➤ Consider pedestrian safety when re-orientating the workplace layout (e.g. the need to reposition lighting and emergency lighting).
➤ Adopt and mark designated walkways (Figure 9.5).
➤ Apply good housekeeping principles by keeping all areas, particularly walkways, as tidy as possible and ensure that any spillages are quickly removed.
➤ Ensure that all workers are suitably trained in the correct use of any safety devices (such as machine guarding or personal protective equipment) or cleaning equipment provided by the employer.
➤ Only use cleaning materials and substances that are effective and compatible with the surfaces being cleaned, so that additional slip hazards are not created.
➤ Ensure that a suitable system of maintenance, cleaning, fault reporting and repair are in place and working effectively. Areas that are being cleaned must be fenced and warning signs erected. Care must also be taken with trailing electrical leads used with the cleaning equipment. Records of cleaning, repairs and maintenance should be kept.
➤ Ensure that all workers are wearing high visibility clothing and appropriate footwear with the correct type of slip-resistant soles suitable for the type of flooring.
➤ Consider whether there are significant pedestrian hazards present in the area when any workplace risk assessments are being undertaken.

It can be seen, therefore, that floors and traffic routes should be of sound construction. If there are frequent, possibly transient, slip hazards, the provision of slip-resistant coating and/or mats should be considered and warning notices posted. Any damaged areas must be cordoned off until repairs are completed. Risk assessments should review past accidents and near misses to enable relevant controls, such as suitable footwear, to be introduced. Employees can often indicate problem areas, so employee consultation is important.

9.4 Hazards in vehicle operations

Many different kinds of vehicle are used in the workplace including dumper trucks, heavy goods vehicles, all terrain vehicles and, perhaps the most common, the fork-lift truck. Approximately 70 persons are killed annually following vehicle accidents in the workplace. There are also

(a)

(b)

Figure 9.4 (a) Typical warehouse vehicle loading/unloading area with separate pedestrian access. (b) Barriers to prevent collision with tank surrounds/bunds.

over 1000 major accidents (involving serious fractures, head injuries and amputations) caused by:

➤ collisions between pedestrians and vehicles;
➤ people falling from vehicles;
➤ people being struck by objects falling from vehicles;
➤ people being struck by an overturning vehicle;
➤ communication problems between vehicle drivers and employees or members of the public.

A key cause of these accidents is the lack of competent and documented driver training. HSE investigations, for example, have shown that in over 30% of dumper truck accidents on construction sites, the drivers had little experience and no training. Common forms of these accidents

include driving into excavations, overturning while driving up steep inclines and runaway vehicles which have been left unattended with the engine running.

Risks of injuries to employees and members of the public involving vehicles could arise due to the following occurrences:

➤ collision with pedestrians;
➤ collision with other vehicles;
➤ overloading of vehicles;
➤ overturning of vehicles;
➤ general vehicle movements and parking;
➤ dangerous occurrences or other emergency incidents (including fire);
➤ access and egress from the buildings and the site.

(a) (b)

Figure 9.5 (a) Internal roadway with appropriate markings. (b) Unsafe stacks of heavy boxes.

There are several other more general hazardous situations involving pedestrians and vehicles. These include the following:

➤ reversing of vehicles, especially inside buildings;
➤ poor road surfaces and/or poorly drained road surfaces;
➤ roadways too narrow with insufficient safe parking areas;
➤ roadways poorly marked out and inappropriate or unfamiliar signs used;
➤ too few pedestrian crossing points;
➤ the non-separation of pedestrians and vehicles;
➤ lack of barriers along roadways;
➤ lack of directional and other signs;
➤ poor environmental factors, such as lighting, dust and noise;
➤ ill-defined speed limits and/or speed limits which are not enforced;
➤ poor or no regular maintenance checks;
➤ vehicles used by untrained and/or unauthorized personnel;
➤ poor training or lack of refresher training.

Vehicle operations need to be carefully planned so that the possibility of accidents is minimized.

9.5 Mobile work equipment

9.5.1 Hazards of mobile work equipment

Mobile work equipment is used extensively throughout industry – in factories, warehouses and construction sites. As mentioned in the previous section, the most common is the fork-lift truck (Figure 9.6).

Accidents, possibly causing injuries to people, often arise from one or more of the following events:

➤ poor maintenance with defective brakes, tyres and steering;
➤ poor visibility because of dirty mirrors and windows or loads which obstruct the driver's view;
➤ operating on rough ground or steep gradients which causes the mobile equipment to turn on its side 90° plus or rollover 180° or more;
➤ carrying of passengers without the proper accommodation for them;
➤ people being flung out as the vehicle overturns and being crushed by it;
➤ being crushed under wheels as the vehicle moves;
➤ being struck by a vehicle or an attachment;
➤ lack of driver training or experience;

Figure 9.7 Telescopic materials handler.

Figure 9.6 Industrial counter balanced lift truck.

➤ underlying causes of poor management procedures and controls, safe working practices, information, instruction, training and supervision;
➤ collision with other vehicles;
➤ overloading of vehicles;
➤ general vehicle movements and parking;
➤ dangerous occurrences or other emergency incidents (including fire);
➤ access and egress from the buildings and the site.

The machines most at risk of rollover according to the HSE are:

➤ compact dumpers frequently used in construction sites;
➤ agricultural tractors;
➤ variable reach rough terrain trucks (telehandlers) (Figure 9.7).

9.5.2 Mobile Work Equipment Legislation – PUWER 1998 Part III

(a) General

The main purpose of the mobile work equipment PUWER 1998 Part III, Regulations 25 to 30 is to require additional precautions relating to work equipment while it is travelling from one location to another or where it does work while moving. All appropriate sections of PUWER 98 will also apply to mobile equipment as it does to all work equipment; for example dangerous moving parts of the engine would be covered by Part II Regulations 10, 11 and 12 of PUWER 98. If the equipment is designed primarily for travel on public roads, the Road Vehicles (Construction and Use) Regulations will normally be sufficient to comply with PUWER 98.

Mobile equipment would normally move on wheels, tracks, rollers, skids, etc. Mobile equipment may be self-propelled, towed or remote controlled and may incorporate attachments. Pedestrian-controlled work equipment, such as lawn mowers, are not covered by Part III.

(b) Employees carried on mobile work equipment – Regulation 25

No employee may be carried on mobile work equipment:

➤ unless it is suitable for carrying persons;
➤ unless it incorporates features to reduce risks as low as is reasonably practicable, including risks from wheels and tracks.

Where there is significant risk of falling materials, falling-object protective structures (FOPS) must be fitted.

(c) Rolling over of mobile work equipment

Where there is a risk of overturning it must be minimized by:

➤ stabilizing the equipment;
➤ fitting a structure so that it only falls on its side;
➤ fitting a structure which gives sufficient clearance for anyone being carried if it turns over further – rollover protection structure (ROPS);
➤ a device giving comparable protection;
➤ fitting a suitable restraining system for people if there is a risk of being crushed by rolling over.

(d) Self-propelled work equipment

Where self-propelled work equipment may involve risks while in motion it shall have:

➤ facilities to prevent unauthorized starting;
➤ facilities to minimize the consequences of collision (with multiple rail-mounted equipment);
➤ a device for braking and stopping;
➤ emergency facilities for braking and stopping, in the event of failure of the main facility, which have readily accessible or automatic controls (where safety constraints so require);
➤ devices fitted to improve vision (where the driver's vision is inadequate);
➤ appropriate lighting fitted or otherwise it shall be made sufficiently safe for its use (if used at night or in dark places);
➤ if there is anything carried or towed that constitutes a fire hazard liable to endanger employees (particularly, if escape is difficult such as from a tower crane), appropriate fire-fighting equipment carried, unless it is sufficiently close by.

(e) Rollover and falling-object protection (ROPS and FOPS)

Rollover protective structures are now becoming much more affordable and available for most types of mobile equipment where there is a high risk of turning over. Their use is spreading across most developed countries and even some less well-developed areas. A ROPS is a cab or frame that provides a safe zone for the vehicle operator in the event of a rollover.

The ROPS frame must pass a series of static and dynamic crush tests. These tests examine the ability of the

Figure 9.8 Various construction plant with driver protection.

ROPS to withstand various loads to see if the protective zone around the operator remains intact in an overturn. A home-made bar attached to a tractor axle or simple shelter from the sun or rain cannot protect the operator if the equipment overturns.

The ROPS must meet International Standards such as ISO 3471:1994. All mobile equipment safeguards should comply with the essential health and safety requirements of the Supply of Machinery (Safety) Regulations 1992 (From December 2009, these become the 2008 Regulations) but need not carry a CE marking.

ROPS must also be correctly installed strictly following the manufacturers' instructions and using the correct strength bolts and fixings. They should never be modified by drilling, cutting, welding or other means as this may seriously weaken the structure.

FOPS are required where there is a significant risk of objects falling on the equipment operator or other authorized person using the mobile equipment. Canopies that protect against falling objects (FOPS) must be properly designed and certified for that purpose. Front loaders work in woods or construction sites near scaffolding or buildings under construction and high bay storage areas, these all being locations where there is a risk of falling objects. Purchasers of equipment should check that any canopies fitted are FOPS. ROPS should never be modified by the user to fit a canopy without consultation with the manufacturers.

ROPS provide some safety during overturning but operatives must be confined to the protective zone of the ROPS. So where ROPS are fitted, a suitable restraining system must be provided for all seats. The use of seat restraints could avoid accidents where drivers are thrown from machines, thrown through windows or doors or thrown around inside the cab. In agriculture and forestry, 50% of overturning accidents occur on slopes of less than 10° and 25% on slopes of 5° or less. This means that seat restraints should be used most of the time that the vehicle is being operated.

9.6 Safe driving

Drivers have an important role to play in the safe use of mobile equipment. They should include the following in their safe working practice checklist:

➤ Make sure they understand fully the operating procedures and controls on the equipment being used.
➤ Only operate equipment for which they are trained and authorized.
➤ Never drive if abilities are impaired by, for example, alcohol, poor vision or hearing, ill health or drugs whether prescribed or not.
➤ Use the seat restraints where provided.
➤ Know the site rules and signals.
➤ Know the safe operating limits relating to the terrain and loads being carried.
➤ Keep vehicles in a suitably clean and tidy condition with particular attention to mirrors and windows or loose items which could interfere with the controls.
➤ Drive at suitable speeds and following site rules and routes at all times.
➤ Allow passengers only when there are safe seats provided on the equipment.
➤ Park vehicles on suitable flat ground with the engine switched off and the parking brakes applied; use wheel chocks if necessary.
➤ Make use of visibility aids or a signaller when vision is restricted.
➤ Get off the vehicle during loading operations unless adequate protection is provided.
➤ Ensure that the load is safe to move.
➤ Do not get off vehicle until it is stationary, engine stopped and parking brake applied.
➤ Where practicable, remove the operating key when getting off the vehicle.
➤ Take the correct precautions such as not smoking and switching off the engine when refuelling.
➤ Report any defects immediately.

9.7 Control strategies for safe vehicle and mobile plant operations

Any control strategy involving vehicle operations will involve a risk assessment to ascertain where, on traffic routes, accidents are most likely to happen. It is important that the risk assessment examines both internal and external traffic routes, particularly when goods are loaded and unloaded from lorries. It should also assess whether designated traffic routes are suitable for the purpose and sufficient for the volume of traffic.

The following need to be addressed:

➤ Traffic routes, loading and storage areas need to be well designed with enforced speed limits, good visibility and the separation of vehicles and pedestrians whenever reasonably practicable.
➤ Environmental considerations, such as visibility, road surface conditions, road gradients and changes in road level, must also be taken into account.
➤ The use of one-way systems and separate site access gates for vehicles and pedestrians may be required.
➤ The safety of members of the public must be considered, particularly where vehicles cross public footpaths.
➤ All external roadways must be appropriately marked, particularly where there could be doubt on right of way, and suitable direction and speed limit signs erected along the roadways. While there may well be a difference between internal and external speed limits, it is important that all speed limits are observed.
➤ Induction training for all new employees must include the location and designation of pedestrian walkways and crossings and the location of areas in the factory where pedestrians and fork-lift trucks use the same roadways.
➤ The identification of recognized and prohibited parking areas around the site should also be given during these training sessions.

Many industries have vehicles designed and used for specific workplace activities. The safe system of work for those activities should include:

➤ details of the work area (e.g. vehicle routes, provision for pedestrians, signage);
➤ details of vehicles (e.g. type, safety features and checks, maintenance requirements);

➤ information and training for employees (e.g. driver training, traffic hazard briefing);

➤ type of vehicle activities (e.g. loading and unloading, refuelling or recharging, reversing, tipping).

9.8 The management of vehicle movements

The movement of vehicles should be properly managed, as should vehicle maintenance and driver training. The development of an agreed code of practice for drivers, to which all drivers should sign up, and the enforcement of site rules covering all vehicular movements are essential for effective vehicle management.

All vehicles should be subject to appropriate regular preventative maintenance programmes with appropriate records kept and all vehicle maintenance procedures properly documented. Many vehicles, such as mobile cranes, require regular inspection by a competent person and test certificates.

Certain vehicle movements, such as reversing, are more hazardous than others and particular safe systems should be set up. The reversing of lorries, for example, must be kept to a minimum (and then restricted to particular areas). Vehicles should be fitted with reversing warning systems as well as being able to give warning of approach. Refuges, where pedestrians can stand to avoid reversing vehicles, are a useful safety measure. Banksmen, who direct reversing vehicles, should also be alert to the possibility of pedestrians crossing in the path of the vehicle. Where there are many vehicle movements, consideration should be given to the provision of high visibility clothing. Pedestrians must keep to designated walkways and crossing points, observe safety signs and use doors that are separate to those used by vehicles. Visitors who are unfamiliar with the site and access points should be escorted through the workplace.

Fire is often a hazard which is associated with many vehicular activities, such as battery charging and the storage of warehouse pallets. All batteries should be recharged in a separate well-ventilated area.

As mentioned earlier, driver training, given by competent people, is essential. Only trained drivers should be allowed to drive vehicles and the training should be relevant to the particular vehicle (fork-lift truck, dumper truck, lorry, etc.). All drivers must receive specific training and instruction before they are permitted to drive

vehicles. They must also be given refresher training and medical examinations at regular intervals. This involves a management system for ensuring driver competence, which must include detailed records of all drivers with appropriate training dates and certification in the form of a driving licence or authorization. Competence and its definition are discussed in Chapter 4.

The HSE publications *Workplace Transport Safety. Guidance for Employers* HSG136, and *Managing Vehicle Safety at the Workplace* INDG199 (revised) provide useful checklists of relevant safety requirements that should be in place when vehicles are used in a workplace.

9.9 Managing occupational road safety

9.9.1 Introduction

This section has been added outside the NEBOSH Certificate because it is an important area of concern recently given more prominence by the UK's HSE. It has been estimated that up to a third of all road traffic accidents involve somebody who is at work at the time. This may account for over 20 fatalities and 250 serious injuries every week. It is estimated that between a quarter and a third of all road traffic crashes involve someone who was at work at the time. Based on 2006 statistics, this means around 800–1060 deaths a year on the road, compared with 241 fatal injuries to workers in the 'traditional workplace'. Some employers believe, incorrectly, that if they comply with certain road traffic law requirements, so that company vehicles have a valid MOT test certificate, and drivers hold a valid licence, this is enough to ensure the safety of their employees, and others, when they are on the road. However, health and safety law applies to on-the-road work activities as it does to all work activities, and the risks should be managed effectively within a health and safety management system.

These requirements are in addition to the duties employers have under road traffic law, for example the Road Traffic Act and Road Vehicle (Construction and Use) Regulations, which are administered by the police and other agencies such as the Vehicle and Operator Services Agency.

Health and safety law does not apply to commuting, unless the employee is travelling from their home to a location which is not their usual place of work.

9.9.2 Benefits of managing work-related road safety

The true costs of accidents to organizations are nearly always higher than just the costs of repairs and insurance claims. The benefits of managing work-related road safety can be considerable, no matter what the size of the organizations. There will be benefits in the area of:

➤ control: costs, such as wear and tear and fuel, insurance premiums and claims can be better controlled;
➤ driver training and vehicle purchase: better informed decisions can be made;
➤ lost time: fewer days will be lost due to injury, ill health and work rescheduling;
➤ vehicles: fewer will need to be off the road for repair;
➤ orders: fewer orders will be missed;
➤ key employees: there is likely to be a reduction in driving bans.

9.9.3 Managing occupational road risks

Where work-related road safety is integrated into the arrangements for managing health and safety at work, it can be managed effectively. The main areas to be addressed are policy, responsibility, organization, systems and monitoring. Employees should be encouraged to report all work-related road incidents and be assured that punitive action will not be taken against them (Figure 9.9).

The risk assessment should:

➤ consider the use, for example, of air or rail transport as a partial alternative to driving;
➤ attempt to avoid situations where employees feel under pressure;
➤ make sure that maintenance work is organized to reduce the risk of vehicle failure. This is particularly important when pool cars are used because pool car users often assume another user is checking on maintenance and the MOT. The safety critical systems that need to be properly maintained are the brakes, steering and tyres. Similarly, if the car is leased and serviced by the leasing company, a system should be in place to confirm that servicing is being done to a reasonable standard;
➤ insist that drivers and passengers are adequately protected in the event of an incident. Crash helmets and protective clothing for those who ride motorcycles and other two-wheeled vehicles should be of the appropriate colour and standard;
➤ ensure that company policy covers the important aspects of the Highway Code.

Figure 9.9 Occupational road risk.

9.9.4 Evaluating the risks

The following considerations can be used to check on work-related road safety management.

The driver
Competency

➤ Is the driver competent, experienced and capable of doing the work safely?
➤ Is his or her licence valid for the type of vehicle to be driven (Figure 9.10)?
➤ Is the vehicle suitable for the task or is it restricted by the driver's licence?
➤ Does recruitment procedure include appropriate pre-appointment checks?
➤ Is the driving licence checked for validity on recruitment and periodically thereafter?
➤ When the driver is at work, is he or she aware of company policy on work-related road safety?
➤ Are written instructions and guidance available?
➤ Has the company specified and monitored the standards of skill and expertise required for the circumstances for the job?

Training

Are drivers properly trained?

➤ Do drivers need additional training to carry out their duties safely?

Figure 9.10 Must have a licence valid for vehicle.

➤ Does the company provide induction training for drivers?

➤ Are those drivers whose work exposes them to the highest risk given priority in training?

➤ Do drivers need to know how to carry out routine safety checks such as those on lights, tyres and wheel fixings?

➤ Do drivers know how to adjust safety equipment correctly, for example seat belts and head restraints?

➤ Is the headrest 3.8 cm (1.5 in.) behind the driver's head?

➤ Is the front of the seat higher than the back and are the legs 45° to the floor?

➤ Is the steering wheel adjustable and set low to avoid shoulder stress?

➤ Are drivers able to use anti-lock brakes (ABS) properly?

➤ Do drivers have the expertise to ensure safe load distribution?

➤ If the vehicle breaks down, do drivers know what to do to ensure their own safety?

➤ Is there a handbook for drivers?

➤ Are drivers aware of the dangers of fatigue?

➤ Do drivers know the height of their vehicle, both laden and empty?

Fitness and health

➤ The driver's level of health and fitness should be sufficient for safe driving.

➤ Drivers of Heavy Goods Vehicles (HGVs) must have the appropriate medical certificate.

➤ Drivers who are most at risk should also undergo regular medicals. Staff should not drive, or undertake other duties, while taking a course of medicine that might impair their judgement.

➤ All drivers should have regular (every 2 years) eyesight tests. A recent survey has indicated that 25% of motorists have a level of eyesight below the legal standard for driving. Drivers should rest their eyes by taking a break of at least 15 minutes every 2 hours.

New offences under the Road Safety Act allow courts to imprison drivers who cause deaths by not paying due care to the road or to other road users. Avoidable distractions which courts will consider when sentencing motorists who have killed include:

➤ using a mobile phone (for either calling or texting);

➤ drinking and eating;

➤ applying make-up;

➤ anything else which takes their attention away from the road and which a court judges to have been an avoidable distraction.

Every year, over 87 000 motorists are disqualified for drink-driving or driving while under the influence of drugs and up to 20% of drink-drivers are caught the morning after drinking. The Department for Transport have calculated that 5% of drivers who failed a breath test after a crash were driving for work at the time.

The vehicle
Suitability

All vehicles should be fit for the purpose for which they are used. When purchasing new or replacement vehicles, the management should look for vehicles that are most suitable for driving and public health and safety. The fleet should be suitable for the job in hand. Where privately owned vehicles are used for work, they should be insured for business use and have an appropriate MOT certificate test (for vehicles over 3 years old, in the UK).

Condition and safety equipment

Are vehicles maintained in a safe and fit condition? There will need to be:

➤ maintenance arrangements to acceptable standards;

➤ basic safety checks for drivers;

➤ a method of ensuring that the vehicle does not exceed its maximum load weight;

➤ reliable methods to secure goods and equipment in transit;

➤ checks to make sure that safety equipment is in good working order;

➤ checks on seat belts and head restraints. (Are they fitted correctly and functioning properly?);

➤ drivers needing to know what action to take if they consider their vehicle is unsafe.

Ergonomic considerations

The health of the drivers, and possibly also their safety, may be put at risk from an inappropriate seating position or driving posture. Ergonomic considerations should therefore be considered before purchasing or leasing new vehicles. Information may need to be provided to drivers about good posture and, where appropriate, on how to set their seat correctly.

The load

For any lorry driving, most of the topics covered in this section are relevant. However, the load being carried is an additional issue. If the load is hazardous, emergency procedures (and possibly equipment) must be in place and the driver trained in those procedures. The load should be stacked safely in the lorry so that it cannot move during the journey. There must also be satisfactory arrangements for handling the load at either end of the journey.

The journey
Routes

Route planning is crucial. Safe routes should be chosen which are appropriate for the type of vehicle undertaking the journey wherever practicable. Motorways are the safest roads. Minor roads are suitable for cars, but they are less safe and could present difficulties for larger vehicles. Overhead restrictions, for example bridges, tunnels and other hazards such as level crossings, may present dangers for long and/or high vehicles, so route planning should take particular account of these.

Scheduling

There are danger periods during the day and night when people are most likely, on average, to feel sleepy. These are between 2 a.m. and 6 a.m. and between 2 p.m. and 4 p.m. Schedules need to take sufficient account of these periods. Where tachographs are carried, they should be checked regularly to make sure that drivers are not putting themselves and others at risk by driving for long periods without a break. Periods of peak traffic flow should be avoided if possible and new drivers should be given extra support while training.

Time

Has enough time been allowed to complete the driving job safely? A realistic schedule would take into account the type and condition of the road and allow the driver rest breaks. A non-vocational driver should not be expected to drive and work for longer than a professional driver. The recommendation of the Highway Code is for a 15-minute break every 2 hours.

➤ Are drivers put under pressure by the policy of the company? Are they encouraged to take unnecessary risks, for example exceeding safe speeds because of agreed arrival times?

➤ Is it possible for the driver to make an overnight stay? This may be preferable to having to complete a long road journey at the end of the working day.

➤ Are staff aware that working irregular hours can add to the dangers of driving? They need to be advised of the dangers of driving home from work when they are excessively tired. In such circumstances they may wish to consider an alternative, such as a taxi.

Distance

Managers need to satisfy themselves that drivers will not be put at risk from fatigue caused by driving excessive distances without appropriate breaks. Combining driving with other methods of transport may make it possible for long road journeys to be eliminated or reduced. Employees should not be asked to work an exceptionally long day.

Weather conditions

When planning journeys, sufficient consideration will need to be given to adverse weather conditions, such as snow, ice, heavy rain and high winds. Routes should be rescheduled and journey times adapted to take adverse weather conditions into consideration. Where poor weather conditions are likely to be encountered, vehicles should be properly equipped to operate, with, for example, ABS.

Where there are ways of reducing risk, for example when driving a high-sided vehicle in strong winds with a light load, drivers should have the expertise to deal with the situation. In addition, they should not feel pressurized to complete journeys where weather conditions are exceptionally difficult and this should be made clear by management.

9.9.5 Typical health and safety rules for drivers of cars on company business

The following example shows typical rules that have been prepared for use by car drivers.

At least 25% of all road accidents are work-related accidents involving people who are using the vehicle on company business. Drivers are expected to understand and comply with the relevant requirements of the current edition of the Highway Code. The following rules have been produced to reduce accidents at work. Any breach of these rules will be a disciplinary offence.

➤ All drivers must have a current and valid driving licence.
➤ All vehicles must carry comprehensive insurance for use at work.
➤ Plan the journey in advance to avoid, where possible, dangerous roads or traffic delays.
➤ Use headlights in poor weather conditions and fog lights in foggy conditions (visibility <100 m).
➤ Use hazard warning lights if an accident or severe traffic congestion is approached (particularly on motorways).

➤ All speed limits must be observed but speeds should always be safe for the conditions encountered.
➤ Drivers must not drive continuously for more than 2 hours without a break of at least 15 minutes.
➤ Mobile phones, including hands-free equipment, must not be used whilst driving. They must be turned off during the journey and only used during the rest periods or when the vehicle is safely parked and the handbrake on.
➤ No alcohol must be consumed during the day of the journey until the journey is completed. Only minimal amounts of alcohol should be consumed on the day before a journey is to be made.
➤ No recreational drugs should be taken on the day of a journey. Some prescribed and over the counter drugs and medicines can also affect driver awareness and speed of reaction. Always check with a doctor or pharmacist to ensure that it is safe to drive.

9.9.6 Further information on occupational road safety

The Highway Code The Stationery Office 2001 ISBN 0 11 552290 5 Can also be viewed on www.highwaycode.gov.uk

Managing Occupational Road Risk Royal Society for the Prevention of Accidents available from Edgbaston Park, 353 Bristol Road, Birmingham B5 7ST Tel: 0121 248 2000.

Driving at Work – Managing Work-Related Road Safety HSE 2003 INDG382 ISBN 9780 7176 2740 0.

9.10 Sources of reference

Essential of Health and Safety at Work, HSE Books 2006 ISBN 9780 0 7176 6179 4.

The Workplace (Health, Safety and Welfare) Regulations 1992 ACOP L24, HSE Books1992 ISBN 978 0 7176 0413 5.

Safe Use of Work Equipment ACOP L22, HSE Books 2008 ISBN 978 0 7176 6295 1.

Workplace Transport Safety – Guidance for Employers HSG136, HSE Books 2005 ISBN 978 0 7176 6154 1.

Lighting at Work HSG38, HSE Books 1998 ISBN 978 0 7176 1232 1.

The Health and Safety (Safety Signs and Signals) Regulations 1996. Guidance on Regulations L64, HSE Books 1996 ISBN 978 0 7176 0870 6.

Relevant statutory provisions

The Workplace (Health, Safety and Welfare) Regulations 1992.

The Provision and Use of Work Equipment Regulations 1998 – Part III in particular.

The Health and Safety (Safety Signs and Signals) Regulations 1996.

9.11 Practice NEBOSH questions for Chapter 9

1. **(i) State EIGHT** types of hazard that may cause slips or trips at work.
   (ii) Outline how slip and trip hazards in the workplace might be controlled.

2. A cleaner is required to polish floors using a rotary floor polisher.

   (i) Identify the hazards that might be associated with this operation.
   (ii) Outline suitable control measures that might be used to minimize the risk.

3. Staff are employed to clean a large science college in the mornings and evenings.

   Outline the specific hazards the cleaners could be exposed to.

4. **Outline** the precautionary measures that may be needed to prevent slips and trip hazards in an engineering factory.

5. **Outline** the measures that may be needed to reduce the risk of slip and trip accidents in a large supermarket.

6. **List EIGHT** design features and/or safe practices intended to reduce the risk of accidents on staircases used as internal pedestrian routes within work premises.

7. **(i) Give FOUR** reasons why accidents may occur on stairs.
   (ii) Outline ways in which accidents on stairs may be prevented.

8. **Outline** the factors that may increase the risk of injury to pedestrians who need to walk through a warehouse.

9. **Describe** the physical features of traffic routes within a workplace designed to ensure the safe movement of vehicles.

10. **Outline** the factors that should be taken into account when planning traffic routes for internal transport.

11. **Outline** measures to be taken to prevent accidents when pedestrians are required to work in vehicle manoevring areas.

12. In order to improve the safety of pedestrians in vehicle manoevring areas, identify the rules that should be followed by:
    (i) drivers
    (ii) pedestrians.

13. **Outline** the means by which the risk of accidents from reversing vehicles within a workplace can be reduced.

14. The warehouse of a ceramic tile manufacturer is to be developed to increase its storage capacity and to enable the use of internal transport to transfer the goods to and from the loading bays.
    (i) Outline the design features of the traffic routes that should be addressed in order to minimize the risk of fork-lift truck-related accidents.
    (ii) Describe *additional* measures that need to be taken to protect pedestrians from the risk of being struck by a fork-lift truck in the warehouse.

15. **Outline** the factors to consider when assessing the risks to a long distance delivery driver.

Manual and mechanical handling hazards and control

10

After reading this chapter you should be able to:

1. describe the hazards and the risk factors which should be considered when assessing risks from manual handling activities
2. suggest ways of minimizing manual handling risk
3. identify the hazards and explain the precautions and procedures to ensure safety in the use of lifting and moving equipment with specific reference to fork-lift trucks, manually operated load-moving equipment (sack trucks, pallet trucks), lifts, hoists, conveyors and cranes.

10.1 Introduction

Until a few years ago, accidents caused by the manual handling of loads were the largest single cause of over 3-day accidents reported to the Health and Safety Executive (HSE). The Manual Handling Operations Regulations recognized this fact and helped to reduce the number of these accidents. However, accidents due to poor manual handling technique still accounts for over 25% of all reported accidents and in some occupational sectors, such as the health service, the figure rises above 50%. An understanding of the factors causing some of these accidents is essential if they are to be further reduced. Mechanical handling methods should always be used whenever possible, but they are not without their hazards, many of which are outlined in Chapter 9 and later in Chapter 10. Much mechanical handling

Figure 10.1 Handling goods onto a truck in a typical docking bay.

involves the use of lifting equipment, such as cranes and lifts, which present specific hazards to both the users and bystanders (Figure 10.1). The risks from these hazards are reduced by thorough examinations and inspections as required by the Lifting Operations and Lifting Equipment Regulations (LOLER).

10.2 Manual handling hazards and injuries

The term 'manual handling' is defined as the movement of a load by human effort alone. This effort may be applied directly or indirectly using a rope or a lever. Manual handling may involve the transportation of the load or the direct support of the load including pushing, pulling, carrying, moving using bodily force and, of course, straightforward lifting. Back injury due to the lifting of heavy loads is very common and several million working days are lost each year as a result of such injuries.

Typical hazards of manual handling include:

➤ lifting a load which is too heavy or too cumbersome, resulting in back injury;
➤ poor posture during lifting or poor lifting technique, resulting in back injury;
➤ dropping a load, resulting in foot injury;
➤ lifting sharp-edged or hot loads resulting in hand injuries.

10.2.1 Injuries caused by manual handling

Manual handling operations can cause a wide range of acute and chronic injuries to workers (Figure 10.2). Acute injuries normally lead to sickness leave from work and a period of rest during which time the damage heals. Chronic injuries build up over a long period of time and are usually irreversible, producing illnesses such as arthritic and spinal disorders. There is considerable evidence to suggest that modern lifestyles, such as a lack of exercise and regular physical effort, have contributed to the long-term serious effects of these injuries.

The most common injuries associated with poor manual handling techniques are all musculoskeletal in nature and are:

➤ muscular sprains and strains – caused when a muscular tissue (or ligament or tendon) is stretched beyond its normal capability leading to a weakening, bruising and painful inflammation of the area affected. Such injuries normally occur in the back or in the arms and wrists;
➤ back injuries – include injuries to the discs situated between the spinal vertebrae (i.e. bones) and can lead to a very painful prolapsed disc lesion (commonly known as a slipped disc). This type of injury can lead to other conditions known as lumbago and sciatica (where pain travels down the leg);
➤ trapped nerve – usually occurring in the back as a result of another injury but aggravated by manual handling;
➤ hernia – this is a rupture of the body cavity wall in the lower abdomen, causing a protrusion of part of the intestine. This condition eventually requires surgery to repair the damage;
➤ cuts, bruising and abrasions – caused by handling loads with unprotected sharp corners or edges;
➤ fractures – normally of the feet due to the dropping of a load. Fractures of the hand also occur but are less common;

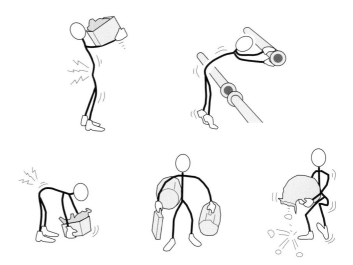

Figure 10.2 Manual handling – there are many potential hazards.

➤ work-related upper limb disorders (WRULDs) – cover a wide range of musculoskeletal disorders, which are discussed in detail in Chapter 15;
➤ rheumatism – this is a chronic disorder involving severe pain in the joints. It has many causes, one of which is believed to be the muscular strains induced by poor manual handling lifting technique.

The sites on the body of injuries caused by manual handling accidents are shown in Figure 10.3.

In general, pulling a load is much easier for the body than pushing one. If a load can only be pushed, then pushing backwards using the back is less stressful on body muscles. Lifting a load from a surface at waist level is easier than lifting from floor level and most injuries during lifting are caused by lifting and twisting at the same time. If a load has to be carried, it is easier to carry it at waist level and close to the body trunk. A firm grip is essential when moving any type of load.

10.3 Manual handling risk assessments

10.3.1 Hierarchy of measures for manual handling operations

With the introduction of the Manual Handling Operations Regulations, the emphasis during the assessment of lifting operations changed from a simple reliance on safe lifting

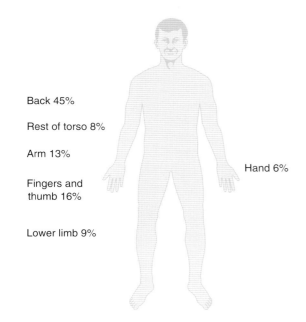

Back 45%

Rest of torso 8%

Arm 13%

Hand 6%

Fingers and
thumb 16%

Lower limb 9%

Figure 10.3 Main sites of injury caused by manual handling accidents.

techniques to an analysis, using risk assessment, of the need for manual handling. The Regulations established a clear hierarchy of measures to be taken when an employer is confronted with a manual handling operation:

➤ Avoid manual handling operations so far as is reasonably practicable by either redesigning the task to avoid moving the load or by automating or mechanizing the operations.
➤ If manual handling cannot be avoided, then a suitable and sufficient risk assessment should be made.
➤ Reduce the risk of injury from those operations so far as is reasonably practicable, either by the use of mechanical handling or making improvements to the task, the load and the working environment.

The guidance given to the Manual Handling Operations Regulations (available in the HSE Legal series – L23) is a very useful document. It gives very helpful advice on manual handling assessments and manual handling training. The advice is applicable to all occupational sectors. Appendix 10.1 gives an example of manual handling assessment forms.

10.3.2 Manual handling assessments

The Regulations specify four main factors which must be taken into account during the assessment. These are the task, the load, the working environment and the capability of the individual who is expected to do the lifting.

The **task** should be analysed in detail so that all aspects of manual handling are covered including the use of mechanical assistance. The number of people involved and the cost of the task should also be considered. Some or all of the following questions are relevant to most manual handling tasks:

➤ Is the load held or manipulated at a distance from the trunk? The further from the trunk, the more difficult it is to control the load and the stress imposed on the back is greater.
➤ Is a satisfactory body posture being adopted? Feet should be firmly on the ground and slightly apart and there should be no stooping or twisting of the trunk. It should not be necessary to reach upwards, as this will place additional stresses on the arms, back and shoulders. The effect of these risk factors is significantly increased if several are present while the task is being performed.
➤ Are there excessive distances to carry or lift the load? Over distances greater than 10 m, the physical demands of carrying the load will dominate the operation. The frequency of lifting and the vertical and horizontal distances the load needs to be carried (particularly if it has to be lifted from the ground and/or placed on a high shelf) are very important considerations.
➤ Is there excessive pulling and pushing of the load? The state of floor surfaces and the footwear of the individual should be noted so that slips and trips may be avoided.
➤ Is there a risk of a sudden movement of the load? The load may be restricted or jammed in some way.
➤ Is frequent or prolonged physical effort required? Frequent and prolonged tasks can lead to fatigue and a greater risk of injury.
➤ Are there sufficient rest or recovery periods? Breaks and/or the changing of tasks enables the body to recover more easily from strenuous activity.
➤ Is there an imposed rate of work on the task? This is a particular problem with some automated production lines and can be addressed by spells on other operations away from the line.
➤ Are the loads being handled while the individual is seated? In these cases, the legs are not used during the lifting processes and stress is placed on the arms and back.
➤ Does the handling involve two or more people? The handling capability of an individual reduces when he/she becomes a member of a team (e.g. for a three-person team, the capability is half the sum of the individual capabilities). Visibility, obstructions and the roughness of the ground must all be considered when team handling takes place.

The **load** must be carefully considered during the assessment and the following questions asked:

➤ Is the load too heavy? The maximum load that an individual can lift will depend on the capability of the individual and the position of the load relative to the body. There is therefore no safe load. Figure 10.4 is reproduced from the HSE guidance, which does give some advice on loading levels. It recommends that loads in excess of 25 kg should not be lifted or carried by a man (and this is only permissible when the load is at the level of and adjacent to the thighs). For women, the guideline figures should be reduced by about one-third.

➤ Is the load too bulky or unwieldy? In general, if any dimension of the load exceeds 0.75 m (c. 2 ft.), its handling is likely to pose a risk of injury. Visibility around the load is important. It may hit obstructions or become unstable in windy conditions. The position of the centre of gravity is very important for stable lifting – it should be as close to the body as possible.

➤ Is the load difficult to grasp? Grip difficulties will be caused by slippery surfaces, rounded corners or a lack of foot room.

➤ Are the contents of the load likely to shift? This is a particular problem when the load is a container full of smaller items, such as a sack full of nuts and bolts. The movements of people (in a nursing home) or animals (in a veterinary surgery) are loads which fall into this category.

➤ Is the load sharp, hot or cold? Personal protective equipment may be required.

The **working environment** in which the manual handling operation is to take place, must be considered during the assessment. The following areas will need to be assessed:

➤ any space constraints which might inhibit good posture. Such constraints include lack of headroom, narrow walkways and items of furniture;

➤ slippery, uneven or unstable floors;

➤ variations in levels of floors or work surfaces, possibly requiring the use of ladders;

➤ extremes of temperature and humidity. These effects are discussed in detail in Chapter 15;

➤ ventilation problems or gusts of wind;

➤ poor lighting conditions.

Finally, the capability of the **individual** to lift or carry the load must be assessed. The following questions will need to be asked:

➤ Does the task require unusual characteristics of the individual (e.g. strength or height)? It is important to remember that strength and general manual handling ability depends on age, gender, state of health and fitness.

➤ Are employees who might reasonably be considered to be pregnant or to have a health problem, put at risk by the task? Particular care should be taken to protect pregnant women or those who have recently given birth from handling loads. Allowance should also be given to any employee who has a health problem, which could be exacerbated by manual handling.

The assessment must be reviewed if there is reason to suspect that it is no longer valid or there has been a significant change to the manual handling operations to which it relates.

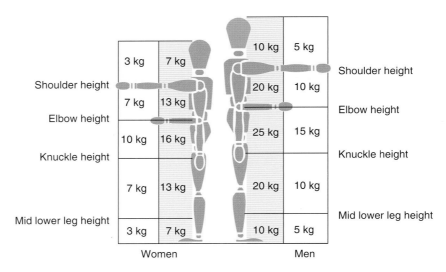

Figure 10.4 HSE guidance for manual lifting – recommended weights.

10.3.3 Reducing the risk of injury

This involves the introduction of control measures resulting from the manual handling risk assessment. The guidance to the Regulations (L23) and the HSE publication *Manual Handling – Solutions You Can Handle* (HSG115) contain many ideas to reduce the risk of injury from manual handling operations. An ergonomic approach, discussed in detail in Chapter 15, is generally required to design and develop the manual handling operation as a whole. The control measures can be grouped under five headings. However, the first consideration, when it is reasonably practicable, is mechanical assistance.

Mechanical assistance involves the use of mechanical aids to assist the manual handling operation such as wheelbarrows, hand-powered hydraulic hoists, specially adapted trolleys, hoists for lifting patients, roller conveyors and automated systems using robots (Figure 10.5).

The **task** can be improved by changing the layout of the workstation by, for example, storing frequently used loads at waist level. The removal of obstacles and the use of a better lifting technique that relies on the leg rather than back muscles should be encouraged. When pushing, the hands should be positioned correctly. The work routine should also be examined to see whether job rotation is being used as effectively as it could be. Special attention should be paid to seated manual handlers to ensure that loads are not lifted from the floor while they are seated. Employees should be encouraged to seek help if a difficult load is to be moved so that a team of people can move the load. Adequate and suitable personal protective equipment should be provided where there is a risk of loss of grip or injury. Care must be taken to ensure that the clothing does not become a hazard in itself (e.g. the snagging of fasteners and pockets).

The **load** should be examined to see whether it could be made lighter, smaller or easier to grasp or manage. This could be achieved by splitting the load, the positioning of handholds or a sling, or ensuring that the centre of gravity is brought closer to the handler's body. Attempts should be made to make the load more stable and any surface hazards, such as slippery deposits or sharp edges, should be removed. It is very important to ensure that any improvements do not, inadvertently, lead to the creation of additional hazards (Figure 10.6).

The **working environment** can be improved in many ways. Space constraints should be removed or reduced. Floors should be regularly cleaned and repaired when damaged. Adequate lighting is essential and working at more than one level should be minimized so that hazardous ladder work is avoided. Attention should be given to the need for suitable temperatures and ventilation in the working area.

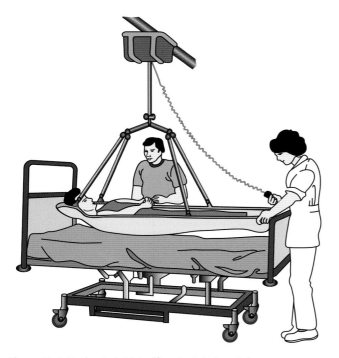

Figure 10.5 Mechanical aids to lift patients in hospital.

Figure 10.6 A pallet truck.

The **capability of the individual** is the fifth area where control measures can be applied to reduce the risk of injury. The state of health of the employee and his/her medical record will provide the first indication as to whether the individual is capable of undertaking the task. A period of sick leave or a change of job can make an individual vulnerable to manual handling injury. The Regulations require that the employee be given information and training. The information includes the provision, where it is reasonably practicable to do so, of precise information on the weight of each load and the heaviest side of any load whose centre of gravity is not centrally positioned. In a more detailed risk assessment, other factors will need to be considered such as the effect of personal protective equipment and psychosocial factors in the work organization. The following points may need to be assessed:

➤ Does protective clothing hinder movement or posture?
➤ Is the correct personal protective equipment being worn?
➤ Is proper consideration given to the planning and scheduling of rest breaks?
  • Is there good communication between managers and employees during risk assessment or workstation design?
➤ Is there a mechanism in place to deal with sudden changes in the volume of workload?
➤ Have employees been given sufficient training and information?
➤ Does the worker have any learning disabilities and, if so, has this been taken into account in the assessment?

Recent amendments to the Manual Handling Operations Regulations have emphasized that a worker may be at risk if he/she:

➤ is physically unsuited to carry out the tasks in question;
➤ is wearing unsuitable clothing, footwear or other personal effects;
➤ does not have adequate or appropriate knowledge or training.

HSE has developed a Manual Handling Assessment Chart (MAC) to help with the assessment of common risks associated with lifting, carrying and handling. The MAC is available on the HSE website.

The training requirements are given in the following section.

10.3.4 Manual handling training

Training alone will not reduce manual handling injuries – there still need to be safe systems of work in place and the full implementation of the control measures highlighted in the manual handling assessment. The following topics should be addressed in a manual handling training session:

➤ types of injuries associated with manual handling activities;
➤ the findings of the manual handling assessment;
➤ the recognition of potentially hazardous manual handling operations;
➤ the correct use of mechanical handling aids;
➤ the correct use of personal protective equipment;
➤ features of the working environment which aid safety in manual handling operations;
➤ good housekeeping issues;
➤ factors which affect the capability of the individual;
➤ good lifting or manual handling technique as shown in Figure 10.7.

Finally, it needs to be stressed that if injuries involving manual handling operations are to be avoided, planning, control and effective supervision are essential.

10.4 Safety in the use of lifting and moving equipment

An amendment to LOLER included the positioning and installation of lifting equipment and the organization of lifting operations.

10.4.1 Positioning and installation of lifting equipment

Lifting equipment must be positioned and installed so as to reduce the risks, so far as is reasonably practicable, from:

➤ equipment or a load striking a person;
➤ a load drifting, falling freely or being released unintentionally.

Lifting equipment should be positioned and installed to minimize the need to lift loads over people and to prevent crushing in extreme positions. It should be designed to stop safely in the event of a power failure and not release its load. Lifting equipment, which follows a fixed path, should be enclosed with suitable and substantial interlocked gates and any necessary protection in the event of power failure.

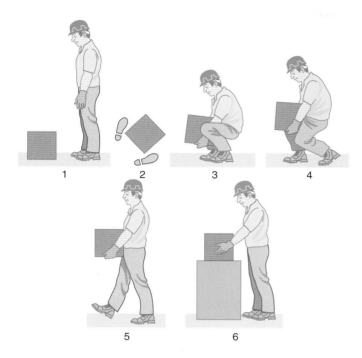

1. Check suitable clothing and assess load. Heaviest side to body.

2. Place feet apart – bend knees.

3. Firm grip – close to body Slight bending of back, hips and knees at start.

4. Lift smoothly to knee level and then waist level. No further bending of back.

5. With clear visibility move forward without twisting. Keep load close to the waist. Turn by moving feet. Keep head up. Do not look at load.

6. Set load down at waist level or to knee level and then on the floor.

Figure 10.7 The main elements of a good lifting technique.

10.4.2 The organization of lifting operations

Every lifting operation, that is lifting or lowering of a load, shall be:

➤ properly planned by a competent person;
➤ appropriately supervised;
➤ carried out in a safe manner.

The person planning the operation should have adequate practical and theoretical knowledge and experience of planning lifting operations. The plan needs to address the risks identified by the risk assessment and identify the resources, the procedures and the responsibilities required so that any lifting operation is carried out safely. For routine simple lifts, a plan will normally be the responsibility of the people using the lifting equipment. For complex lifting operations, for example, where two cranes are used to lift one load, a written plan may need to be produced each time.

The planning should include: the need to avoid suspending loads over occupied areas, visibility, the attaching/detaching and securing of loads, the environment, the location, the possibility of overturning, the proximity to other objects, any lifting of people and the pre-use checks required for the equipment.

10.4.3 Summary of the requirements for lifting operations

There are four general requirements for all lifting operations:

➤ use strong, stable and suitable lifting equipment;
➤ the equipment should be positioned and installed correctly;
➤ the equipment should be visibly marked with the safe working load (SWL);
➤ lifting operations must be planned, supervised and performed in a safe manner by competent people.

10.5 Types of mechanical handling and lifting equipment

There are four elements to mechanical handling, each of which can present hazards. These are handling equipment, the load, the workplace and the employees involved.

The **mechanical handling equipment** must be capable of lifting and/or moving the load. It must be fault-free, well maintained and inspected on a regular basis. The hazards related to such equipment include collisions between people and the equipment and personal injury from being trapped in moving parts of the equipment (such as belt and screw conveyors).

The **load** should be prepared for transportation in such a way as to minimize the possibility of accidents. The hazards will be related to the nature of the load (e.g. substances which are flammable or hazardous to health) or the security and stability of the load (e.g. collapse of bales or incorrectly stacked pallets).

The **workplace** should be designed so that, whenever possible, workers and the load are kept apart. If, for example,

an overhead crane is to be used, then people should be segregated away or barred from the path of the load.

The **employees** and others and any other people who are to use the equipment must be properly trained and competent in its safe use.

10.5.1 Conveyors and elevators

Conveyors transport loads along a given level which may not be completely horizontal, whereas elevators move loads from one level or floor to another.

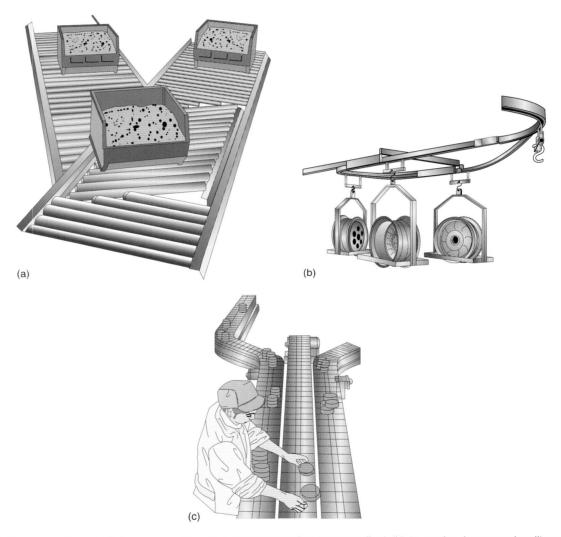

(a)

(b)

(c)

Figure 10.8 Conveyor systems. (a) Roller conveyor (may have powered and free running rollers). (b) An overhead conveyor handling wheels. Other designs of overhead conveyor are useful for transferring components and garments between workstations in, for example, manufacture of machines or clothing. (c) A slat conveyor in use in the food industry.

There are three common forms of **conveyor** – belt, roller and screw conveyors. The most common hazards and preventative measures are:

- the in-running nip, where a hand is trapped between the rotating rollers and the belt. Protection from this hazard can be provided by nip guards and trip devices;
- entanglement with the power drive requiring the fitting of fixed guards and the restriction of loose clothing which could become caught in the drive;
- loads falling from the conveyor. This can be avoided by edge guards and barriers;
- impact against overhead systems. Protection against this hazard may be given by the use of bump caps, warning signs and restricted access;
- contact hazards prevented by the removal of sharp edges, conveyor edge protection and restricted access;
- manual handling hazards;
- noise and vibration hazards.

Screw conveyors, often used to move very viscous substances, must be provided with either fixed guards or covers to prevent accidental access. People should be prohibited from riding on belt conveyors, and emergency trip wires or stop buttons must be fitted and be operational at all times.

Elevators are used to transport goods between floors, such as the transportation of building bricks to upper storeys during the construction of a building or the transportation of grain sacks into the loft of a barn (Figure 10.9). Guards should be fitted at either end of the elevator and around the power drive. The most common hazard is injury due to loads falling from elevators. There are also potential manual handling problems at both the feed and discharge ends of the elevator.

10.5.2 Fork-lift trucks

The most common form of mobile handling equipment is the fork-lift truck [shown in Figure 10.10 (see also Figure 9.6)]. It comes from the group of vehicles, known as lift trucks, and can be used in factories, on construction sites and on farms. The term fork-lift truck is normally applied to the counterbalanced lift truck, where the load on the forks is counterbalanced by the weight of the vehicle over the rear wheels. The reach truck is designed to operate in narrower aisles in warehouses and enables the load to be retracted within the wheelbase. The very narrow aisle (VNA) truck does not turn within the aisle to deposit or retrieve a load. It is often guided by guides or rails on the floor. Other forms of lift truck include the pallet truck and the pallet stacker truck, both of which may be pedestrian or rider controlled.

There are many hazards associated with the use of fork-lift trucks. These include:

- overturning – maneuvering at too high a speed (particularly cornering); wheels hitting an obstruction such as a kerb; sudden braking; poor tyre condition leading to skidding; driving forwards down a ramp; movement of the load; insecure, excessive or uneven loading; incorrect tilt or driving along a ramp;
- overloading – exceeding the rated capacity of the machine;
- collisions – particularly with warehouse racking which can lead to a collapse of the whole racking system;
- silent operation of the electrically powered fork-lift truck – can make pedestrians unaware of its presence;
- uneven road surface – can cause the vehicle to overturn and/or cause musculoskeletal problems for the driver;
- overhead obstructions – a particular problem for inexperienced drivers;
- loss of load – shrink wrapping or sheeting will reduce this hazard;

Figure 10.9 A brick elevator.

Figure 10.10 Reach truck designed so that the load retracts within the wheelbase to save space.

➤ inadequate maintenance leading to mechanical failure;
➤ use as a work platform;
➤ speeding – strict enforcement of speed limits is essential;
➤ poor vision around the load;
➤ pedestrians – particularly when pedestrians and vehicles use the same roadways. Warning signs, indicating the presence of fork-lift trucks, should be posted at regular intervals;
➤ dangerous stacking or de-stacking technique – this can destabilize a complete racking column;
➤ carrying passengers – this should be a disciplinary offence;
➤ battery charging – presents an explosion and fire risk;
➤ fire – often caused by poor maintenance resulting in fuel leakages or engine/motor burnout, or through using an unsuitable fork-lift truck in areas where flammable liquids or gases are used and stored;
➤ lack of driver training.

If fork-lift trucks are to be used outside, visibility and lighting, weather conditions and the movement of other vehicles become additional hazards.

There are also the following physical hazards:

➤ noise – caused by poor silencing of the power unit;
➤ exhaust fumes – should only be a problem when the maintenance regime is poor;
➤ vibrations – often caused by a rough road surface or wide expansion joints. Badly inflated tyres will exacerbate this problem;
➤ manual handling – resulting from manoeuvring the load by hand or lifting batteries or gas cylinders;
➤ ergonomic – musculoskeletal injuries caused by soft tyres and/or undulating road surface or holes or cracks in the road surface (e.g. expansion joints).

Regular and documented maintenance by competent mechanics is essential. However, the driver should undertake the following checks at the beginning of each shift:

➤ condition of tyres and correct tyre pressures;
➤ effectiveness of all brakes;
➤ audible reversing horn and light working properly;
➤ lights, if fitted, working correctly;
➤ mirrors, if fitted, in good working order and properly set;
➤ secure and properly adjusted seat;
➤ correct fluid levels, when appropriate;
➤ fully charged batteries, when appropriate;
➤ correct working of all lifting and tilting systems.

A more detailed inspection should be undertaken by a competent person within the organization on a weekly basis to include the mast and the steering gear. Driver training is essential and should be given by a competent trainer. The training session must include the site rules covering items such as the fork-lift truck driver code of practice for the organization, speed limits, stacking procedures and reversing rules. Refresher training should be provided at regular intervals and a detailed record kept of all training received. Table 10.1 illustrates some key requirements of fork-lift truck drivers and the points listed should be included in most codes of practice.

Finally, care must be taken with the selection of drivers, including relevant health checks and previous experience. Drivers should be at least 18 years of age and their fitness to drive should be reassessed regularly (every five years after the age of 40 and every year after 65 (HSG6)).

10.5.3 Other forms of lifting equipment

The other types of lifting equipment to be considered are cranes (mobile overhead and jib), lifts and hoists and lifting tackle. A sample risk assessment for the use of lifting equipment is given in Appendix 10.2.

Table 10.1 Safe driving of lift trucks

Drivers must:

- drive at a suitable speed to suit road conditions and visibility
- use the horn when necessary (at blind corners and doorways)
- always be aware of pedestrians and other vehicles
- take special care when reversing (do not rely on mirrors)
- take special care when handling loads which restrict visibility
- travel with the forks (or other equipment fitted to the mast) lowered
- use the prescribed lanes
- obey the speed limits
- take special care on wet and uneven surfaces
- use the handbrake, tilt and other controls correctly
- take special care on ramps
- always leave the truck in a state which is safe and discourages unauthorized use (brake on, motor off, forks down, key out).

Drivers must not:

- operate in conditions in which it is not possible to drive and handle loads safely (e.g. partially blocked aisles)
- travel with the forks raised
- use the forks to raise or lower persons unless a purpose-built working cage is used
- carry passengers
- park in an unsafe place (e.g. obstructing emergency exits)
- turn round on ramps
- drive into areas where the truck would cause a hazard (flammable substance store)
- allow unauthorized use.

The lifting operation should be properly prepared and planned. This involves the selection of a suitable crane having up-to-date test certificates and examination reports that have been checked. A risk assessment of the task will be needed which would ascertain the weight, size and shape of the load and its final resting place. A written plan for completing the lift should be drafted and a competent person appointed to supervise the operation.

Cranes may be either a jib crane or an overhead gantry travelling crane. The safety requirements are similar for each type. All cranes need to be properly designed, constructed, installed and maintained. They must also be operated in accordance with a safe system of work. They should only be driven by authorized persons who are fit and trained. Each crane is issued with a certificate by its manufacturer giving details of the safe working load (SWL). The SWL must **never** be exceeded and should be marked on the crane structure. If the SWL is variable, as with a jib crane (the SWL decreases as the operating radius increases), an SWL indicator should be fitted. Care should be taken to avoid sudden shock loading, as this will impose very high stresses on the crane structure. It is also very important that the load is properly shackled and all eyebolts tightened. Safe slinging should be included in any training programme. All controls should be clearly marked and be of the 'hold-to-run' type.

Large cranes, which incorporate a driving cab, often work in conjunction with a banksman, who will direct the lifting operation from the ground. It is important that banksmen are trained so that they understand recognized crane signals. The lifting operation should be properly prepared and planned. This involved the selection of a suitable crane having up-to-date test certificates and examination reports that have been checked. A risk assessment of the task will be needed which would ascertain the weight, size and shape of the load and its final resting place. A written plan for completing the lift should be drafted and a competent person appointed to supervise the operation.

The basic principles for the safe operation of cranes are as follows. **For all cranes**, the driver must:

- undertake a brief inspection of the crane and associated lifting tackle each time before it is used;
- check that all lifting accessory statutory inspections are in place and up-to-date;
- check that tyre pressures, where appropriate, are correct;
- ensure that loads are not left suspended when the crane is not in use;
- before a lift is made, ensure that nobody can be struck by the crane or the load;
- ensure that loads are never carried over people;
- ensure good visibility and communications;
- lift loads vertically – cranes must not be used to drag a load;
- travel with the load as close to the ground as possible;
- switch off power to the crane when it is left unattended.

Figure 10.11 Manoeuvring a yacht using a large overhead travelling gantry and slings in a marina.

For mobile jib cranes, the following points should be considered:

➤ Each lift must be properly planned, with the maximum load and radius of operation known.
➤ Overhead obstructions or hazards must be identified; it may be necessary to protect the crane from overhead power lines by using goal posts and bunting to mark the safe headroom.
➤ The ground on which the crane is to stand should be assessed for its load-bearing capacity.
➤ If fitted, outriggers should be used.

The principal reasons for crane failure, including loss of load, are:

➤ overloading;
➤ poor slinging of load;
➤ insecure or unbalanced load;
➤ loss of load;
➤ overturning;
➤ collision with another structure or overhead power lines;
➤ foundation failure;
➤ structural failure of the crane;
➤ operator error;
➤ lack of maintenance and/or regular inspections;
➤ no signaller used when driver's view is obscured;
➤ incorrect signals given.

Typical causes of recent serious incidents with tower cranes include:

➤ mechanical failure of the brake or lifting ram;
➤ overturn of the crane;

➤ jib collapse;
➤ a load or dropped load striking a worker;
➤ sling failure.

The reasons for some of these incidents were:

➤ poor site induction training – not dealing with site-specific risks and lasting too long (20–30 minutes maximum is sufficient time);
➤ problems with crane maintenance and thorough examinations;
➤ operators working long hours without a break;
➤ poor operator cabin design and too high a climbing distance;
➤ operator health problems;
➤ problems in communicating health and safety issues by crane operators on site.

During lifting operations using cranes, it must be ensured that:

➤ the driver has good visibility;
➤ there are no pedestrians below the load by using barriers, if necessary;
➤ an audible warning is given prior to the lifting operation.

If lifting takes place in windy conditions, tag lines may need to be attached to the load to control its movement.

A **lift or hoist** incorporates a platform or cage and is restricted in its movement by guides (Figure 10.12). Hoists are generally used in industrial settings (e.g. construction sites and garages), whereas lifts are normally used inside buildings. Lifts and hoists may be designed to carry passengers and/or goods alone. They should be of sound mechanical construction and have interlocking doors or gates, which must be completely closed before the lift or hoist moves. The hoistway should be properly enclosed so that the moving parts of the hoist are guarded. The loads should be secured on the hoist platform so that they cannot fall and the operator should have a clear view of the landing levels. There should be no unauthorized use of the hoist by untrained personnel. Passenger carrying lifts must be fitted with an automatic braking system to prevent overrunning, at least two suspension ropes, each capable alone of supporting the maximum working load, and a safety device which could support the lift in the event of suspension rope failure. Maintenance procedures must be rigorous, recorded and only undertaken by competent persons. It is very important that a safe system of work is employed during maintenance operations to protect others, such as members of the public, from falling down the lift shaft and other hazards.

Figure 10.12 Hoist for lifting a car.

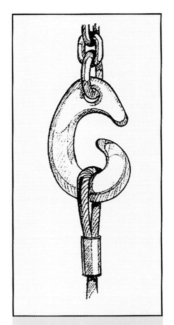

A hook designed to prevent displacement of the load.

A hook with a safety catch.

Figure 10.13 Specially designed safety hooks.

Other **items of lifting tackle**, usually used with cranes, include chain slings and hooks, wire and fibre rope slings, eyebolts and shackles (Figure 10.13). Special care should be taken, when slings are used, to ensure that the load is properly secured and balanced. Lifting hooks should be checked for signs of wear and any distortion of the hook. Shackles and eyebolts must be correctly tightened. Slings should always be checked for any damage before they are used and only competent people should use them. Training and instruction in the use of lifting tackle is essential and should include regular inspections of the tackle, in addition to the mandatory thorough examinations. Finally, care should be taken when these items are being stored between use.

10.6 Requirements for the statutory examination of lifting equipment

The LOLER specify examinations required for lifting equipment by using two terms – an inspection and a thorough examination. Both terms are defined by the HSE in guidance accompanying various Regulations.

Lifting equipment includes any equipment used at work to lift or lower loads including any anchoring, fixing or supporting attachments.

An **inspection** is used to identify whether the equipment can be operated, adjusted and maintained safely so that any defect, damage or wear can be detected before it results in unacceptable risks. It is normally performed by a competent person appointed by the employer (often an employee).

A **thorough examination** is a detailed examination, which may involve a visual check, a disassembly and testing of components and/or an equipment test under operating conditions. Such an examination must normally be carried out by a competent person who is independent of the employer. The examination is usually carried out according to a written scheme and a written report is submitted to the employer.

The Code of Practice for the Safe Use of Lifting Equipment defines a thorough examination as a visual examination carried out by a competent person carefully and critically and, where appropriate, supplemented by other means, such as measurement and testing, in order to check whether the equipment is safe to use.

A detailed summary of the thorough examination and inspection requirements of the LOLER is given in Chapter 17 (Section 17.21).

A thorough examination of lifting equipment should be undertaken at the following times:

➤ before the equipment is used for the first time;
➤ after it has been assembled at a new location;
➤ at least every 6 months for equipment used for lifting persons or a lifting accessory;
➤ at least every 12 months for all other lifting equipment including the lifting of loads over people;
➤ in accordance with a particular examination scheme drawn up by an independent competent person;
➤ each time that exceptional circumstances, which are likely to jeopardize the safety of the lifting equipment, have occurred (such as severe weather).

The person making the thorough examination of lifting equipment must:

➤ notify the employer forthwith of any defect which, in their opinion, is or could become dangerous;
➤ as soon as practicable (within 28 days) write an authenticated report to the employer and any person who leased or hired the equipment.

The Regulations specify the information that should be included in the report.

The initial report should be kept for as long as the lifting equipment is used (except for a lifting accessory which need only be kept for 2 years). For all other examinations, a copy of the report should be kept until the next thorough examination is made or for 2 years (whichever is the longer). If the report shows that a defect exists, which could lead to an existing or imminent risk of serious personal injury, a copy of the report must be sent, by the person making the thorough examination, to the appropriate enforcing authority.

The equipment should be inspected at suitable intervals between thorough examinations. The frequency and the extent of the inspection are determined by the level of risk presented by the lifting equipment. A report or record should be made of the inspection which should be kept until the next inspection. Unless stated otherwise, lifts and hoists should be inspected every week.

It is important to stress that thorough examinations must be accompanied by meticulous in-service inspection that will detect any damage every time the equipment is used. For most workplaces, it is better to store every item of portable lifting equipment in a central store where these inspections can take place and records of the equipment be kept. Users must also be encouraged to report any defects in equipment that they have used.

10.7 Sources of reference

Safe Use of Work Equipment ACOP L22, HSE Books 2008 ISBN 978 0 7176 6295 1.
Manual Handling (Guidance) L23, HSE Books 2004 ISBN 978 0 7176 2823 0.
Safety in Working with Lift Trucks HSG6, HSE Books 2000 ISBN 978 0 7176 1781 4.
Rider-operated Lift Trucks – Operator Training (ACOP and Guidance) L117, HSE Books 1999 ISBN 978 0 7176 2455 3.
Safe Use of Lifting Equipment L113, HSE Books 1998 ISBN 978 0 7176 1628 2.

Relevant statutory provisions
The Manual Handling Operations Regulations 1992.
The Provision and Use of Work Equipment Regulations 1998.
The Lifting Operations and Lifting Equipment Regulations 1998.

10.8 Practice NEBOSH questions for Chapter 10

1. (i) **List TWO** types of injury that may be caused by the incorrect manual handling of loads.
   (ii) **Outline** a good lifting technique that could be adopted by a person required to lift a load from the ground.
   (iii) **Give TWO** examples of how a manual handling task might be avoided.

2. (i) **List FOUR** specific types of injury that may be caused by the incorrect manual handling of loads.
   (ii) **Outline** the factors in relation to the load that will affect the risk of injury.

3. A storeman is required to place boxes of metal components by hand on to shelved racking.
   (i) **List FOUR** types of injury to which the storeman may be at risk while carrying out this operation.
   (ii) **Outline** the factors in relation to the task that may affect the risk of injury.

4. Employees working for a charity are required to collect plastic bags of clothes, books and other donated goods from outside householders' premises and carry them to a waiting company van.

Giving reasons in **EACH** case, **outline** the types of injury the employees may sustain from the activity.

5. **(i) Outline** the main requirements of the Manual Handling Operations Regulations 1992.
(ii) List the possible indications of a manual handling problem in a workplace.

6. An assessment has concluded that the person carrying out a particular manual handling task is fit and capable of lifting the loads involved.
Outline the factors associated with the task and work environment that would need to be considered in order to complete the assessment.

7. **Outline** the factors that may affect the risk from manual handling activities in relation to:
(i) the load
(ii) the individual.

8. An automated piece of equipment designed to lift boxes weighing 20 kg from a conveyor and place them on to pallets has failed and is likely to be out of action for several weeks. During this time, the task will be carried out manually.
(i) Outline the factors that should be considered when undertaking a manual handling assessment of the task.
(ii) Outline the measures that may be needed in order to reduce the risk of injury to employees carrying out the manual handling task.

9. **Outline** the issues to be considered when undertaking a manual handling assessment of a task that involves lifting buckets of water out of a sink.

10. Newspapers are printed, bundled and placed onto roller conveyors at a publisher's where they are transported to an area for dispatch to the customer.
Identify EIGHT hazards to which employees involved with the transport of the newspapers from the printing room to the dispatch room may be exposed.

11. A telecommunications worker needs to lift a 20 kg manhole cover in order to repair fibre optic cables. The working area is 1.5 m below ground and 1 m from the kerb of a busy road.
(i) Identify FOUR factors associated with the *worker* that may affect the risk of injury when removing the manhole cover.

(ii) Outline SIX types of hazard associated with these activities.
(iii) Outline possible control measures to minimize the risk to the *worker* and *pedestrians* whilst carrying out these activities.

12. **Outline** factors to be considered when undertaking a manual handling assessment of the work undertaken by baggage handlers at a large, busy airport.

13. **Outline FOUR** hazards and the corresponding precautions to be taken when using conveyor systems for moving materials within a workplace.

14. Drivers of internal transport are required to be competent before they are permitted to drive vehicles within the workplace.
(i) Explain the meaning of the term 'competence'.
(ii) Identify measures an employer can take in order to ensure that their drivers of internal transport are, and remain, competent.

15. **Outline** the health and safety topics that should be covered in a training course for fork-lift truck operators.

16. **Outline** the health and safety considerations when a fork-lift truck is to be used to unload palletized goods from a vehicle parked in a factory car park.

17. **Outline EIGHT** types of hazard associated with the operation of fork-lift trucks.

18. **List EIGHT** items to be included on a checklist for the routine inspection of a fork-lift truck at the beginning of a shift.

19. Battery operated fork-lift trucks are used to move materials within a warehouse.
(i) Describe FOUR hazards associated specifically with battery powered fork-lift trucks.
(ii) Outline the precautions that may be needed to ensure the safety of pedestrians in areas where the fork-lift trucks are operating.
(iii) Outline EIGHT rules to follow when a fork-lift truck is left unattended during a driver's work break.

20. **Identify EIGHT** ways in which a fork-lift truck may become unstable whilst in operation.

21. A driver of a fork-lift truck has been seriously injured after the vehicle he was driving overturned. **Outline**

the possible *immediate* causes of the accident associated with:
(i) the way in which the vehicle was driven
(ii) the workplace
(iii) the vehicle.

22. An organization is about to purchase a fork-lift truck. With reference to its possible intended use and working environment, **outline** the particular features of the vehicle that should be taken into account when determining its suitability for the job.

23. **(i)** **List THREE** types of crane used for lifting operations.
(ii) **Outline** a procedure for the safe lifting of a load by a crane, having ensured that the crane has been correctly selected and positioned for the job.

24. A modular, portable office building has been delivered to a company's premises. It is to be lifted from the delivery vehicle, to its designated resting place, by a mobile crane. To assist with the health and safety of persons who may be on the premises during the lifting operation:
(i) **Identify** the issues that should be considered when planning and preparing for the lifting operation to be undertaken.
(ii) **Outline** a procedure for the safe lifting and lowering of the portable office.
(iii) **Outline** the statutory requirements for the inspection and examination of lifting equipment.

25. **Outline** the precautions to be taken when employees are working at ground level in a workshop where loads are lifted and transported by means of an overhead gantry crane.

26. An engineering workshop uses an overhead gantry crane to transport materials.
(i) **Identify TWO** reasons why loads may fall from this crane.
(ii) **Outline** precautions to prevent accidents to employees working at ground level when overhead cranes are in use.

27. **Outline** the precautions that should be taken to reduce the risk of injury during lifting operations with a mobile jib crane.

28. A mechanical hoist is to be used to remove an engine from a vehicle in a motor repair shop.
(i) **State** the requirements for the statutory examination and inspection of the hoist.
(ii) **Outline** the precautions to be taken to reduce the risk of injury to employees and others during the lifting operation.

29. **State** the hazards associated with the use of a materials hoist on a construction site.

30. **(i)** **Identify TWO** occasions when a thorough examination of a fork-lift truck is required under the Lifting Operations and Lifting Equipment Regulations 1998.
(ii) **Outline** a range of circumstances that may cause a fork-lift truck to become unstable.
(iii) Other than those associated with instability, **identify FIVE** hazards presented by a diesel-powered fork-lift truck **AND describe** the precautions that might be necessary in **EACH** case.

31. **Describe** the occasions when a thorough examination of workplace lifting equipment should be undertaken.

Appendix 10.1 Manual handling of load assessment checklist

Section A – Preliminary assessment	
Task description: Factors beyond the limits of the guidelines?	Is an assessment needed? (i.e. is there a potential risk of injury, and are the factors beyond the limits of the guidelines?) Yes/No
If 'YES' continue. If 'NO' no further assessment required.	
Tasks covered by this assessment: Location: People involved: Date of assessment:	Diagram and other information
Section B – See detailed analysis form	
Section C – Overall assessment of the risk of injury? Low/Med/High	
Section D – Remedial action needed: Remedial action, in priority order: a b c d e f g h	
Date action should be completed:	Date for review:
Assessor's name:	Signature:

Manual Handling Risk Assessment	
Task description...	Employee.. I.D. Number..

Risk Factors					
A. Task Characteristics	**Yes/No**	**Risk Level**			**Current Controls**
		H	**M**	**L**	**See guidance**
1. Loads held away from trunk?					
2. Twisting?					

(Continued)

A. Task Characteristics	Yes/No	Risk Level			Current Controls
		H	M	L	See guidance
3. Stooping?					
4. Reaching upwards?					
5. Extensive vertical movements?					
6. Long carrying distances?					
7. Strenuous pushing or pulling?					
8. Unpredictable movements of loads?					
9. Repetitive handling operations?					
10. Insufficient periods of rest/ recovery?					
11. High work rate imposed?					
B. Load characteristics					
1. Heavy?					
2. Bulky?					
3. Difficult to grasp?					
4. Unstable/unpredictable?					
5. Harmful (sharp/hot)?					
C. Work environment characteristics					
1. Postural constraints?					
2. Floor suitability?					
3. Even surface?					
4. Thermal/humidity suitability?					
5. Lighting suitability?					
D. Individual characteristics					
1. Unusual capability required?					
2. Hazard to those with health problems?					
3. Hazard to pregnant workers?					
4. Special information/training required?					

(Continued)

	Yes/No	Risk Level			Current Controls
		H	M	L	See guidance
E. Other factors to consider					
1. Movement or posture hindered by protective clothing?					
2. Absence of correct/suitable PPE?					
3. Lack of planning and scheduling of basics/rest basics?					
4. Poor communication between managers and employees?					
5. Sudden changes in workload or seasonal changes in volume without mechanisms to deal with change?					
6. Lack of training and/or information?					
7. Any learning disabilities?					
Any further action needed?	**Yes/No**				
Details:					

Appendix 10.2 A typical risk assessment for the use of lifting equipment

INITIAL RISK ASSESSMENT	Use of Lifting Equipment			
SIGNIFICANT HAZARDS		**Low**	**Medium**	**High**
1. Unintentional release of load				✓
2. Unplanned movement of load		✓		
3. Damage to equipment		✓		
4. Crush injuries to personnel			✓	
5.				

ACTION ALREADY TAKEN TO REDUCE THE RISKS:

Compliance with:

Lifting Operations & Lifting Equipment Regulations (LOLER)
Safety Signs and Signals Regulations (SSSR). Provision and Use of Work Equipment Regulations (PUWER)
British Standard – Specification for flat woven webbing slings
BS – Guide to selection & use of lifting slings for multi purposes

Planning:

Copies of statutory thorough examinations of lifting equipment will be kept on site. Before selection of lifting equipment, the above standards will be considered as well as the weight, size, shape and centre of gravity of the load. Lifting equipment is subject to the planned maintenance programme.

Physical:

All items of lifting equipment will be identified individually and stored so as to prevent physical damage or deterioration. Safe working loads of lifting equipment will be established before use. Packing will be used to protect slings from sharp edges on the load. All items of lifting equipment will be visually examined for signs of damage before use. Ensuring the eyes of strops are directly below the appliance hook and that tail ropes are fitted to larger loads will check swinging of the load. Banksmen will be used where the lifting equipment operator's vision is obstructed. Approved hand signals will be used.

Managerial/Supervisory:

Only lifting equipment that is in date for statutory examination will be used. Manufacturer's instructions will be checked to ensure that methods of sling attachment and slinging arrangements generally are correct.

Training:

Personnel involved in the slinging of loads and use of lifting equipment will be required to be trained to CITB or equivalent standard. Supervisors will be trained in the supervision of lifting operations.

Date of Assessment........................	Assessment made by.................................
Risk Re-Assessment Date	Site Manager's Comments:

Work equipment hazards and control

11

11.1 Introduction

This chapter covers the scope and main requirements for work equipment as covered by Parts II and III of the Provision and Use of Work Equipment Regulations (PUWER). The requirements for the supply of new machinery are also included. Summaries of PUWER and The Supply of Machinery (Safety) Regulations are given in Chapter 17. The safe use of hand tools, hand-held power tools and the proper safeguarding of a small range of machinery used in industry and commerce are included.

Any equipment used by an employee at work is generally covered by the term 'work equipment'. The scope is

Figure 11.1 Typical machinery safety notice.

extremely wide and includes hand tools, power tools, ladders, photocopiers, laboratory apparatus, lifting equipment, fork-lift trucks, and motor vehicles (which are not privately owned). Virtually anything used to do a job of work, including employees' own equipment, is covered. The uses covered include starting or stopping the equipment, repairing, modifying, maintaining, servicing, cleaning and transporting.

Employers and the self-employed must ensure that work equipment is suitable, maintained, inspected if necessary, provided with adequate information and instruction and only used by people who have received sufficient training.

11.2 Suitability of work equipment and CE marking

11.2.1 Standards and requirements

When work equipment is provided it has to conform to standards which cover its supply as a new or second-hand piece of equipment and its use in the workplace. This involves:

➤ its initial integrity;
➤ the place where it will be used;
➤ the purpose for which it will be used.

There are two groups of law that deal with the provision of work equipment:

➤ One deals with what manufacturers and suppliers have to do. This can be called the 'supply' law. One of the most common is the Supply of Machinery (Safety) Regulations 1992 (in December 2009 the Supply of Machinery (Safety) Regulations 2008 come into force), which requires manufacturers and suppliers to ensure that machinery is safe when supplied and has CE marking. Its **primary** purpose is to prevent barriers to trade across the EU, and not to protect people at work

➤ The other deals with what the users of machinery and other work equipment have to do. This can be called the 'user' law, PUWER 98, and applies to most pieces of work equipment. Its **primary** purpose is to protect people at work.

Under 'user' law employers have to provide safe equipment of the correct type, ensure that it is correctly used and maintain it in a safe condition. When buying new equipment, the 'user' has to check that the equipment complies with all the 'supply' law that is relevant. **The user must check that the machine is safe before it is used**.

Most new equipment, including machinery in particular, should have 'CE' marking when purchased (Figure 11.2). 'CE' marking is only a claim by the manufacturer that the equipment is safe and that they have met relevant supply law. If this is done properly manufacturers will have to do the following:

➤ find out about the health and safety hazards (trapping, noise, crushing, electrical shock, dust, vibration, etc.) that are likely to be present when the machine is used;
➤ assess the likely risks;
➤ design out the hazards that result in risks; or, if that is not possible;
➤ provide safeguards (e.g. guarding dangerous parts of the machine, providing noise enclosures for noisy parts); or, if that is not possible;
➤ use warning signs on the machine to warn of hazards that cannot be designed out or safeguarded (e.g. 'noisy machine' signs).

Manufacturers also have to:

➤ keep information, explaining what they have done and why, in a technical file.
➤ fix CE marking to the machine where necessary, to show that they have complied with all the relevant supply laws;

Figure 11.2 CE mark.

➤ issue a 'Declaration of Conformity' for the machine (see Figure 11.3). This is a statement that the machine complies with the relevant essential health and safety requirements or with the example that underwent type-examination. A declaration of conformity must

- state the name and address of the manufacturer or importer into the EU
- contain a description of the machine, and its make, type and serial number
- indicate all relevant European Directives with which the machinery complies
- state details of any notified body that has been involved
- specify which standards have been used in the manufacture (if any)
- be signed by a person with authority to do so

EC Declaration of Conformity

We declare that units:
BD710,BD711,BD713,BD725,BD735E, BD750,SR600

conform to:
89/392/EEC,EN50144

A weighted sound pressure   103dB (A)
A weighted sound power   116dB (A)
Hand/arm weighted vibration   <2.5m/s²

Brian Cooke

Director of Engineering
Spennymoor, County Durham DL16 6JG
United Kingdom

Figure 11.3 Typical certificate of conformity.

Before buying new equipment the buyer will need to think about:

➤ where and how it will be used
➤ what it will be used for
➤ who will use it (skilled employees, trainees)
➤ what risks to health and safety might result
➤ how well health and safety risks are controlled by different manufacturers.

- provide the buyer with instructions to explain how to install, use and maintain the machinery safely.

This can help in deciding which equipment may be suitable, particularly if buying a standard piece of equipment 'off the shelf'.

If buying a more complex or custom-built machine the buyer should discuss their requirements with potential suppliers. For a custom-built piece of equipment, there is the opportunity to work with the supplier to design out the causes of injury and ill health. Time spent now on agreeing the necessary safeguards, to control health and safety risks, could save time and money later.

Note: Sometimes equipment is supplied via another organization, for example, an importer, rather than direct from the manufacturer, so this other organization is referred to as the supplier. It is important to realize that the supplier may not be the manufacturer.

When the equipment has been supplied the buyer should look for CE marking, check for a copy of the Declaration of Conformity and that there is a set of instructions in English on how the machine should be used and most important of all, **check to see if they think that it is safe**.

11.2.2 Limitations of CE marking

CE marking is not a guarantee that the machine is safe. It is a claim by the manufacturer that the machinery complies with the law. CE marking has many advantages if done properly, for example:

- it allows a common standard across Europe;
- it provides a means of selling to all European Community member states without barriers to trade;
- it ensures that instructions and safety information is supplied in a fairly standard way in most languages in the EC;
- it has encouraged the use of diagrams and pictorials which are common to all languages;
- it allows for independent type-examination for some machinery like woodworking machinery which has not been made to an EC-harmonized standard – identified by an EN marking before the standard number (e.g. BS EN ....).

Clearly there are disadvantages as well, for example:

- instruction manuals have become very long. Sometimes two volumes are provided, because of the number of languages required;
- translations can be very poor and disguise the proper meaning of the instruction;

- manufacturers can fraudulently put on the CE marking;
- manufacturers might make mistakes in claiming conformity with safety laws.

11.3 Use and maintenance of equipment with specific risks

Some pieces of work equipment involve specific risks to health and safety where it is not possible to control adequately the hazards by physical measures alone, for example the use of a bench-mounted circular saw or an abrasive wheel (Figure 11.4). In all cases the hierarchy of controls should be adopted to reduce the risks by:

- eliminating the risks; or, if this is not possible;
- taking physical measures to control the risks such as guards, but if the risks cannot be adequately controlled;
- taking appropriate software measures, such as a safe system of work.

PUWER 98 Regulation 7 restricts the use of such equipment to the persons designated to use it. These people need to have received sufficient information, instruction and training so that they can carry out the work using the equipment safely.

Repairs, modifications, maintenance or servicing is also restricted to designated persons. A designated person may be the operator if he/she has the necessary skills and has received specific instruction and training. Another person specifically trained to carry out a particular maintenance task, for example dressing an abrasive wheel, may not be the operator but may be designated to do this type of servicing task on a range of machines.

Figure 11.4 Using a bench-mounted abrasive wheel.

11.4 Information, instruction and training

People using and maintaining work equipment, where there are residual risks that cannot be sufficiently reduced by physical means, require enough information, instruction and training to operate safely.

The information and instructions are likely to come from the manufacturer in the form of operating and maintenance manuals. It is up to the employer to ensure that what is provided is easily understood, and set out logically with illustrations and standard symbols where appropriate. The information should normally be in good plain English but other languages may be necessary in some cases.

The extent of the information and instructions will depend on the complexity of the equipment and the specific risks associated with its use. They should cover:

➤ all safety and health aspects;
➤ any limitations on the use of the equipment;
➤ any foreseeable problems that could occur;
➤ safe methods to deal with the problems;
➤ any relevant experience with the equipment that would reduce the risks or help others to work more safely, should be recorded and circulated to everyone concerned.

Everyone who uses and maintains work equipment needs to be adequately trained. The amount of training required will depend on:

➤ the complexity and level of risk involved in using or maintaining the equipment;
➤ the experience and skills of the person doing the work, whether it is normal use or maintenance.

Training needs will be greatest when a person is first recruited but will also need to be considered:

➤ when working tasks are changed, particularly if the level of risk changes;
➤ if new technology or new equipment is introduced;
➤ where a system of work changes;
➤ when legal requirements change;
➤ periodically to update and refresh people's knowledge and skills;
➤ it may also be necessary following an accident.

Supervisors and managers also require adequate training to carry out their function, particularly if they only supervise a particular task occasionally. The training and supervision of young persons is particularly important because of their relative immaturity, unfamiliarity with a working environment and lack of awareness of existing or potential risks. Some Approved Codes of Practice, for example on the *Safe Use of Woodworking Machinery,* restrict the use of high-risk machinery, so that only young persons with sufficient maturity and competence who have finished their training may use the equipment unsupervised.

11.5 Maintenance and inspection

11.5.1 Maintenance

Work equipment needs to be properly maintained so that it continues to operate safely and in the way it was designed to perform. The amount of maintenance will be stipulated in the manufacturer's instructions and will depend on the amount of use, the working environment and the type of equipment. High-speed, high-risk machines, which are heavily used in an adverse environment like salt water, may require very frequent maintenance, whereas a simple hand tool, like a shovel, may require very little (Figure 11.5).

Maintenance management schemes can be based around a number of techniques designed to focus on those parts which deteriorate and need to be maintained to prevent health and safety risks. These techniques include the following:

➤ **Preventative planned maintenance** – this involves replacing parts and consumables or making necessary adjustments at preset intervals normally set by the manufacturer, so that there are no hazards created by component deterioration or failure. Vehicles are normally maintained on this basis.
➤ **Condition based maintenance** – this involves monitoring the condition of critical parts and carrying out maintenance whenever necessary to avoid hazards which could otherwise occur.

Figure 11.5 Typical maintenance notice.

➤ **Breakdown based maintenance** – here maintenance is only carried out when faults or failures have occurred. This is only acceptable if the failure does not present an immediate hazard and can be corrected before the risk is increased. If, for example, a bearing overheating can be detected by a monitoring device, it is acceptable to wait for the overheating to occur as long as the equipment can be stopped and repairs carried out before the fault becomes dangerous to persons employed.

In the context of health and safety, maintenance is not concerned with operational efficiency but only with avoiding risks to people. It is essential to ensure that maintenance work can be carried out safely. This will involve the following:

➤ competent well-trained maintenance people;
➤ the equipment being made safe for the maintenance work to be carried out. In many cases, the normal safeguards for operating the equipment may not be sufficient as maintenance sometimes involves going inside guards to observe and subsequently adjust, lubricate or repair the equipment. Careful design allowing adjustments, lubrication and observation from outside the guards, for example, can often eliminate the hazard. Making equipment safe will usually involve disconnecting the power supply and then preventing anything moving, falling or starting during the work. It may also involve waiting for equipment to cool or warm up to room temperature;
➤ a safe system of work being used to carry out the necessary procedures to make and keep the equipment safe and perform the maintenance tasks. This can often involve a formal 'permit-to-work' scheme to ensure that the correct sequence of safety critical tasks has been performed and all necessary precautions taken;
➤ correct tools and safety equipment being available to perform the maintenance work without risks to people. For example special lighting or ventilation may be required.

11.5.2 Inspection under PUWER

Complex equipment and/or high-risk equipment will probably need a maintenance log and may require a more rigid inspection regime to ensure continued safe operation. This is covered by PUWER 98, Regulation 6.

PUWER requires, where safety is dependent on the installation conditions and/or the work equipment is exposed to conditions causing deterioration, which may result in a significant risk and a dangerous situation developing, that the equipment is inspected by a competent person. In this case the competent person would normally be an employee, but there might be circumstances where an outside competent person would be used.

The inspection must be done:

➤ after installation for the first time;
➤ after assembly at a new site or in a new location and thereafter;
➤ at suitable intervals;
➤ each time exceptional circumstances occur which could affect safety.

The inspection under PUWER will vary from a simple visual inspection to a detailed comprehensive inspection, which may include some dismantling and/or testing. The level of inspection required would normally be less rigorous and intrusive than thorough examinations under the Lifting Operations and Lifting Equipment Regulations 1998 (LOLER) for certain lifting equipment. In the case of boilers and air receivers the inspection is covered by the Pressure Systems Safety Regulations 2000, which involves a thorough examination (see the summary in Chapter 17). The inspection under PUWER would only be needed in these cases if the examinations did not cover all the significant health and safety risks that are likely to arise.

11.5.3 Examination of boilers and air receivers

Under the Pressure Systems Safety Regulations 2000, a wide range of pressure vessels and systems require thorough examination by a competent person to an agreed specifically written scheme. This includes steam boilers, pressurized hot water plants and air receivers which are contained in the Certificate syllabus (Figure 11.6).

The Regulations place duties on designers and manufacturers but this section is only concerned with the duty on users to have the vessels examined. An employer who operates a steam boiler and/or a pressurized hot water plant and/or an air receiver must ensure:

➤ that it is supplied with the correct written information and markings;
➤ that the equipment is properly installed;
➤ that it is used within its operating limits;
➤ there is a written scheme for periodic examination of the equipment certified by a competent person (in the case of standard steam boilers and air receivers the scheme is likely to be provided by the manufacturer);

Figure 11.6 Typical electrically powered compressor with air receiver.

➤ that the equipment is examined in accordance with the written scheme by a competent person within the specified period;
➤ that a report of the periodic examination is held on file giving the required particulars;
➤ that the actions required by the report are carried out;
➤ that any other safety critical maintenance work is carried out, whether or not covered by the report.

In these cases the competent person is usually a specialist inspector from an external inspection organization. Many of these organizations are linked to the insurance companies who cover the financial risks of use of the pressure vessel.

11.6 Operation and working environment

To operate work equipment safely it must be fitted with easily reached and operated controls, kept stable, properly lit, kept clear and provided with adequate markings and warning signs. These are covered by PUWER 98, which applies to all types of work equipment.

11.6.1 Controls

Equipment should be provided with efficient means of:

➤ starting or making a significant change in operating conditions;

Equipment controls should:

➤ be easily reached from the operating positions
➤ not permit accidental starting of equipment
➤ move in the same direction as the motion being controlled
➤ vary in mode, shape and direction of movement to prevent inadvertent operation of the wrong control
➤ incorporate adequate red emergency stop buttons
➤ incorporate adequate red emergency stop buttons of the mushroom-headed type with lock-off
➤ have shrouded or sunken green start buttons to prevent accidental starting of the equipment
➤ be clearly marked to show what they do.

Figure 11.7 Equipment controls – design features.

➤ stopping in normal circumstances;
➤ emergency stopping as necessary to prevent danger.

All controls should be well positioned, clearly visible and identifiable, so that it is easy for the operator to know what each control does. Markings should be clearly visible and remain so under the conditions met at the workplace. See Figure 11.7 for more information on controls.

(a) Start controls

It should only be possible to start the work equipment by using the designed start control. Equipment may well have a start sequence which is electronically controlled to meet certain conditions before starting can be achieved, for example preheating a diesel engine, or purge cycle for gas-fed equipment. Restarting after a stoppage will require the same sequence to be performed.

Stoppage may have been deliberate or as a result of opening an interlocked guard or tripping a switch accidentally. In most cases it should not be possible to restart the equipment simply by shutting the guard or resetting the trip. Operation of the start control should be required.

Any other change to the operating conditions such as speed, pressure or temperature should only be done by using a control designed for the purpose.

(b) Stop controls

The action of normal stopping controls should bring the equipment to a safe condition in a safe manner. In some cases immediate stopping may cause other risks to occur. The stop controls do not have to be instantaneous and can bring the equipment to rest in a safe sequence or at the end of an operating cycle. It is only the parts necessary for safety, that is, accessible dangerous parts, that have to be stopped. So, for example, suitably guarded cooling fans may need to run continuously and be left on.

In some cases where there is, for example, stored energy in hydraulic systems, it may be necessary to insert physical scotches to prevent movement and/or to exhaust residual hydraulic pressure. These should be incorporated into the stopping cycle, which should be designed to dissipate or isolate all stored energy to prevent danger.

It should not be possible to reach dangerous parts of the equipment until it has come to a safe condition, e.g. stopped, cooled, electrically safe.

(c) Emergency stop controls

Emergency stop must be provided where the other safeguards in place are not sufficient to prevent danger to operatives and any other persons who may be affected. Where appropriate, there should be an emergency stop at each control point and at other locations around the equipment so that action can be taken quickly. Emergency stops should bring the equipment to a halt rapidly but this should be controlled where necessary so as not to create any additional hazards. Crash shutdowns of complex systems have to be carefully designed to optimize safety without causing additional risks.

Emergency stops are not a substitute for effective guarding of dangerous parts of equipment and should not be used for normal stopping of the equipment.

Emergency stop buttons should be easily identified, reached and operated. Common types are mushroom-headed buttons, bars, levers, kick-plates or pressure-sensitive cables. They are normally red and should need to be reset after use. With stop buttons this is either by twisting or a security key. (See Figure 11.8.)

Mobile work equipment is normally provided with effective means of stopping the engine or power source. In some cases large equipment may need emergency stop controls away from the operator's position.

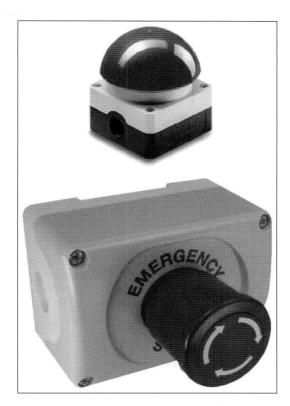

Figure 11.8 Emergency stop buttons.

(d) Isolation of equipment

Equipment should be provided with efficient means of isolating it from all sources of energy. The purpose is to make the equipment safe under particular conditions, for example when maintenance is to be carried out or where adverse weather conditions may make it unsafe to use. On static equipment isolation will usually be for mains electrical energy; however, in some cases there may be additional or alternative sources of energy. The isolation should cover all sources of energy such as diesel and petrol engines, LPG, steam, compressed air, hydraulics, batteries, heat. In some cases special consideration is necessary where, for example, hydraulic pumps are switched off, so as not to allow heavy pieces of equipment to fall under gravity. An example would be the loading shovel on an excavator.

11.6.2 Stability

Stability is important and is normally achieved by bolting equipment in place or, if this is not possible, by using clamps. Some equipment can be tied down, counterbalanced or weighted, so that it remains stable under all operating conditions. If portable equipment is weighted or counterbalanced, it should be reappraised when the equipment is moved to another position. If outriggers are needed for stability in certain conditions, for example to stabilize mobile access towers, they should be employed whenever conditions warrant the additional support. In severe weather conditions it may be necessary to stop using the equipment or reappraise the situation to ensure stability is maintained.

The quality of general and local lighting will need to be considered to ensure the safe operation of the equipment. The level of lighting and its position relative to the working area are often critical to the safe use of work equipment. Poor levels of lighting, glare and shadows can be dangerous when operating equipment. Some types of lighting, for example sodium lights, can change the colour of equipment, which may increase the level of risk. This is particularly important if the colour coding of pipe work or cables is essential for safety.

11.6.3 Markings

Markings on equipment must be clearly visible and durable. They should follow international conventions for some hazards like radiation and lasers and, as far as possible, conform to the Health and Safety (Safety Signs and Signals) Regulations (see Chapter 17 for a summary). The contents, or the hazards of the contents, as well as controls, will need to be marked on some equipment. Warnings or warning devices are required in some cases to alert operators or people nearby to any dangers, for example 'wear hardhats', a flashing light on an airport vehicle or a reversing horn on a truck.

11.7 User responsibilities

The responsibilities for users of work equipment are covered in section 7 of the Health and Safety at Work (HSW) Act and Regulation 14 of the Management Regulations. Section 7 requires employees to take reasonable care for themselves and others who may be affected and to cooperate with the employer. Regulation 14 requires employees to use equipment properly in accordance with instructions and training. They must also inform employers of dangerous situations and shortcomings in protection arrangements. The self-employed who are users have similar responsibilities to employers and employees combined.

11.8 Hand-held tools

11.8.1 Introduction

Work equipment includes hand tools and hand-held power tools. This section deals with hand tools. These tools need to be correct for the task, well maintained and properly used by trained people.

Five basic safety rules can help prevent hazards associated with the use of hand-held tools:

➤ keep all tools in good condition with regular maintenance;
➤ use the right tool for the job;
➤ examine each tool for damage before use and do not use damaged tools;
➤ use tools according to the manufacturer's instructions;
➤ provide and use properly the right personal protective equipment (PPE).

11.8.2 Hazards of hand tools

Hazards from the misuse or poor maintenance of hand tools (Figure 11.9) include:

➤ broken handles on files/chisels/screwdrivers/hammers which can cause cut hands or hammer heads to fly off;
➤ incorrect use of knives, saws and chisels with hands getting injured in the path of the cutting edge;
➤ tools that slip causing stab wounds;
➤ poor-quality uncomfortable handles that damage hands;
➤ splayed spanners that slip and damage hands or faces;
➤ chipped or loose hammer heads that fly off or slip;
➤ incorrectly sharpened or blunt chisels or scissors that slip and cut hands. Dull tools can cause more hazards than sharp ones. Cracked saw blades must be removed from service;
➤ flying particles that damage eyes from breaking up stone or concrete;
➤ electrocution or burns by using incorrect or damaged tools for electrical work;
➤ use of poorly insulated tools for hot work in the catering or food industry;

correct tool for the job is the first step in safe hand tool use. Tools are designed for specific needs. That is why screwdrivers have various lengths and tip styles and pliers have different head shapes. Using any tool inappropriately is a step in the wrong direction. To avoid personal injury and tool damage, select the proper tool to do the job well and safely.

High-quality professional hand tools will last many years if they are taken care of and treated with respect. Manufacturers design tools for specific applications. Use tools only for their intended purpose.

Suitability will include:

➤ specially protected and insulated tools for electricians;
➤ non-sparking tools for flammable atmospheres and the use of non-percussion tools and cold cutting methods;
➤ tools made of suitable quality materials which will not chip or splay in normal use;
➤ the correct tools for the job, for example using the right-sized spanner and the use of mallets not hammers on chisel heads. The wooden handles of tools must not be splintered;
➤ safety knives with enclosed blades for regular cutting operations;
➤ impact tools such as drift pins, wedges and cold-chisels being kept free of mushroomed heads;
➤ spanners not being used when jaws are sprung to the point that slippage occurs.

Inspection – All tools should be maintained in a safe and proper condition. This can be achieved through:

➤ the regular inspection of hand tools;
➤ discarding or prompt repair of defective tools;
➤ taking time to keep tools in the proper condition and ready for use;
➤ proper storage to prevent damage and corrosion;
➤ locking tools away when not in use to prevent them being used by unauthorized people.

Training – All users of hand tools should be properly trained in their use. This may well have been done through apprenticeships and similar training. This will be particularly important with specialist working conditions or work involving young people.

Always wear approved eye protection when using hand tools, particularly when percussion tools are being used. Metal and wood particies will fly out when the material is cut or planed, so other workers in the vicinity should wear eye protection, as well.

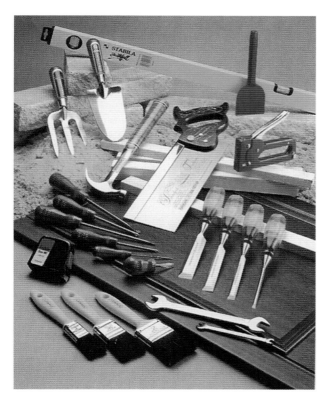

Figure 11.9 Typical range of hand tools.

➤ use of pipes or similar equipment as extension handles for a spanner which is likely to slip causing hand or face injury;
➤ mushroomed headed chisels or drifts which can damage hands or cause hammers (not suitable for chisels) and mallets to slip;
➤ use of spark-producing or percussion tools in flammable atmospheres;
➤ painful wrists and arms (upper limb disorders) from the frequent twisting from using screwdrivers;
➤ when using saw blades, knives or other tools, they should be directed away from aisle areas and away from other people working in close proximity.

11.8.3 Hand tools safety considerations

Use of hand tools should be properly controlled including those tools owned by employees. The following controls are important:

Suitability – All tools should be suitable for the purpose and location in which they are to be used. Using the

Use well-designed, high-quality tools – Finally, investing in high-quality tools makes the professional's job safer and easier:

➤ if extra leverage is needed, use high-leverage pliers, which give more cutting and gripping power than standard pliers. This helps, in particular, when making repetitive cuts or twisting numerous wire pairs;

➤ serrated jaws provide sure gripping action when pulling or twisting wires;

➤ some side-cutting and diagonal-cutting pliers are designed for heavy-duty cutting. When cutting screws, nails and hardened wire, only use pliers that are recommended for that use;

➤ pliers with hot riveting at the joint ensure smooth movement across the full action range of the pliers, which reduces handle wobble, resulting in a positive cut. The knives align perfectly every time;

➤ induction hardening on the cutting knives adds to long life, so the pliers cut cleanly day after day;

➤ sharp cutting knives and tempered handles also contribute to cutting ease;

➤ some pliers are designed to perform special functions. For example some high-leverage pliers have features that allow crimping connectors and pulling fish tapes;

➤ tool handles with dual-moulded material allow for a softer, more comfortable grip on the outer surface and a harder, more durable grip on the inner surface and handle ends;

➤ well-designed tools often include a contoured thumb area for a firmer grip or colour-coded handles for easy tool identification;

➤ insulated tools reduce the chance of injury where the tool may make contact with an energized source.

Well-designed tools are a pleasure to use. They save time, give professional results and help to do the job more safely.

11.9 Hand-held power tools

11.9.1 Introduction

The electrical hazards of portable hand-held tools and portable appliance testing (PAT) are covered in more detail in Chapter 12. This section deals mainly with other physical hazards and safeguards relating to hand-held power tools (Figure 11.10).

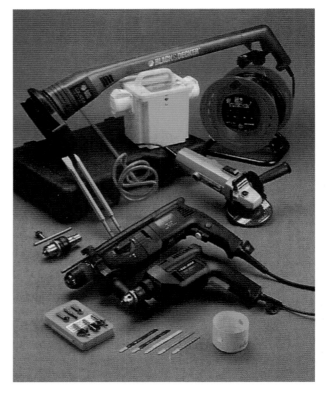

Figure 11.10 Typical range of hand-held power tools.

The section covers, in particular, electric drills, sanders, and chainsaws which are commonly used in the workplace.

11.9.2 General hazards of hand-held power tools

The general hazards involve:

➤ mechanical entanglement in rotating spindles or sanding discs;

➤ waste material flying out of the cutting area;

➤ coming into contact with the cutting blades or drill bits;

➤ risk of hitting electrical, gas or water services when drilling into building surfaces;

➤ electrocution/electric shock from poorly maintained equipment and cables or cutting the electrical cable;

➤ manual handling problem with a risk of injury if the tool is heavy or very powerful;

➤ hand–arm vibration, especially with pneumatic drill and chainsaws, disc cutters and petrol-driven units;

➤ tripping hazard from trailing cables, hoses or power supplies;

➤ eye hazard from flying particles;

➤ injury from poorly secured or clamped workpieces;
➤ fire and explosion hazard with petrol-driven tools or when used near flammable liquids, explosive dusts or gases;
➤ high noise levels with pneumatic chisels, planes and saws in particular (see Chapter 17);
➤ dust and fumes given off during the use of the tools levels (but see Chapter 16).

11.9.3 Typical safety controls and instructions

(a) Guarding

The exposed moving parts of power tools need to be safe-guarded. Belts, gears, shafts, pulleys, sprockets, spindles, drums, flywheels, chains or other reciprocating, rotating or moving parts of equipment must be guarded. Machine guards, as appropriate, must be provided to protect the operator and others from the following:

➤ point of operation;
➤ in-running nip points;
➤ rotating parts;
➤ flying chips and sparks.

Safety guards must never be removed when a tool is being used. Portable circular saws must be equipped at all times with guards. An upper guard must cover the entire blade of the saw. A retractable lower guard must cover the teeth of the saw, except where it makes contact with the work material. The lower guard must automatically return to the covering position when the tool is withdrawn from the work material.

(b) Operating controls and switches

Most hand-held power tools should be equipped with a constant-pressure switch or control that shuts off the power when pressure is released. On/off switches should be easily accessible without removing hands from the equipment.

Handles should be designed to protect operators from excessive vibration and keep their hands away from danger areas. In some cases handles are also designed to activate a brake of the cutting chain or blade, for example in a chainsaw. Equipment should be designed to reduce lifting and manual handling problems, with special harnesses being used as necessary, for example when using large strimmers.

Means of starting engines and holding equipment should be designed to minimize any musculoskeletal problems.

(c) Safe operations/instructions

When using power tools, the following basic safety measures should be observed to protect against electrical shock, personal injury, ill health and risk of fire. See also more detailed electrical precautions in Chapter 14. Operators should read these instructions before using the equipment and ensure that they are followed:

➤ maintain a clean and tidy working area that is well lit and clear of obstructions;
➤ never expose power tools to rain. Do not use power tools in damp or wet surroundings;
➤ do not use power tools in the vicinity of combustible fluids, dusts or gases unless they are specially protected and certified for use in these areas;
➤ protect against electric shock (if tools are electrically powered) by avoiding body contact with grounded objects such as pipes, scaffolds and metal ladders (see Chapter 14 for more electrical safety precautions);
➤ keep children away;
➤ do not let other persons handle the tool or the cable. Keep them away from the working area;
➤ store tools in a safe place when not in use where they are in a dry, locked area which is inaccessible to children;
➤ tools should not be overloaded as they operate better and safer in the performance range for which they were intended;
➤ use the right tool. Do not use small tools or attachments for heavy work. Do not use tools for purposes and tasks for which they were not intended; for example, do not use a hand-held circular saw to cut down trees or cut off branches;
➤ wear suitable work clothes. Do not wear loose-fitting clothing or jewellery. They can get entangled in moving parts. For outdoor work, rubber gloves and non-skid footwear are recommended. Long hair should be protected with a hair net;
➤ use safety glasses;
➤ also use filtering respirator mask for work that generates dust;
➤ do not abuse the power cable;
➤ do not carry the tool by the power cable and do not use the cable to pull the plug out of the power socket. Protect the cable from heat, oil and sharp edges;
➤ secure the workpiece. Use clamps or a vice to hold the workpiece. It is safer than using hands and it frees both hands for operating the tool;

➤ do not overreach the work area. Avoid abnormal body postures. Maintain a safe stance and maintain a proper balance at all times;

➤ maintain tools with care. Keep your tools clean and sharp for efficient and safe work. Follow the maintenance Regulations and instructions for the changing of tools. Check the plug and cable regularly and in the case of damage, have them repaired by a qualified service engineer. Also inspect extension cables regularly and replace if damaged;

➤ keep the handle dry and free of oil or grease;

➤ disconnect the power plug when not in use, before servicing and when changing the tool, that is blade, bits, cutter, sanding disc, etc;

➤ do not forget to remove key. Check before switching on that the key and any tools for adjustment are removed;

➤ avoid unintentional switch-on. Do not carry tools that are connected to power with your finger on the power switch. Check that the switch is turned off before connecting the power cable;

➤ outdoors, use extension cables. When working outdoors, use only extension cables which are intended for such use and marked accordingly;

➤ stay alert, keep eyes on the work. Use common sense. Do not operate tools when there are significant distractions;

➤ check the equipment for damage. Before further use of a tool, check carefully the protection devices or lightly damaged parts for proper operation and performance of their intended functions. Check movable parts for proper function, for whether there is binding or for damaged parts. All parts must be correctly mounted and meet all conditions necessary to ensure proper operation of the equipment;

➤ damaged protection devices and parts should be repaired or replaced by a competent service centre unless otherwise stated in the operating instructions. Damaged switches must be replaced by a competent service centre. Do not use any tool which cannot be turned on and off with the switch;

➤ only use accessories and attachments that are described in the operating instructions or are provided or recommended by the tool manufacturer. The use of tools other than those described in the operating instructions or in the catalogue of recommended tool inserts or accessories can result in a risk of personal injury;

➤ use engine-driven power tools in well-ventilated areas. Store petrol in a safe place in approved storage cans. Stop and let engines cool before refuelling.

11.9.4 Specific hazards and control measures for specified hand-held power tools

The following hand-held power tools have been put in the General Certificate syllabus: electric drill, sanders and chainsaws, all of which are commonly used. The hazards and safety control measures in addition to the general ones covered earlier are set out for each type of equipment.

(a) Electric drills

Hazards are (Figure 11.11):

➤ entanglement, particularly of loose clothing or long hair in rotating drill bits;

➤ high noise levels from the drill or attachment;

➤ eye injury from flying particles and chips, particularly from chisels;

➤ injury from poorly secured workpieces;

➤ electrocution/electric shock from poorly maintained equipment;

➤ electric shock from drilling into a live hidden cable;

➤ hand–arm vibration hazard in hammer mode;

➤ dust given off from material being worked on;

➤ tripping hazard from trailing cables;

➤ upper limb disorder from powerful machines with a strong torque, particularly if they jam and kick back;

➤ foot injury hazards from dropping heavy units onto unprotected feet;

➤ manual handling hazards. particularly with heavy machines and intensive use or using at awkward heights and/or reaches;

➤ fire and explosion hazard when used near flammable liquids, explosive dusts or gases;

➤ using equipment in poor weather conditions with wet, slippery surfaces, poor visibility and cold conditions.

Specific control measures include the following:

➤ use double-insulated tools or earthed reduced voltage tools with a residual current device (see Chapter 12 for electrical safeguards in detail);

➤ use a pilot hole or punch to start holes whenever possible;

➤ select the correct drill bit for the material being drilled;

➤ secure small pieces to be drilled to prevent spinning;

➤ protect against damage or injury on the far side if the bit is long enough to pass through the material or there are buried services in, say, plaster walls;

(a)

(b)

Figure 11.11 Electric drills.

➤ take care to prevent loose sleeves or long hair from being wound around the drill bit; for example wear short or close-fitting sleeves;
➤ wear suitable eye protection.

Figure 11.12 Disc sander.

(b) Sanders

There is a large range of hand-held sanders on the market, from rotating discs (Figure 11.12), random orbital (Figure 11.13b), rectangular orbital, belt sanders and heavy-duty floor sanders of both the rotating drum (Figure 11.13a) and, recently, the orbital floor sander (Figure 11.13). The high-speed rotating discs and drum types are the most hazard-ous but they all need using with care.

Hazards include:

➤ high noise levels from the sander in operation;
➤ injury from poorly secured workpieces;
➤ electrocution/electric shock from poorly maintained electrical equipment;
➤ potential of entanglement with rotating disc and drum sanders;
➤ sanding attachments becoming loose in the chuck and can fling off;
➤ injury from contact with abrasive surfaces, particularly with course abrasives and high-speed rotating sanding discs and drums;
➤ hand–arm vibration hazard, particularly from reciprocating equipment;
➤ health hazards from extensive dust given off from material being worked on;
➤ fire and health hazards from overheating of abraded surfaces, particularly if plastics are being sanded;
➤ tripping hazard from trailing cables;
➤ large powerful sanders suddenly gripping the surface and pulling the operative off their feet;
➤ foot injury hazards from dropping heavy units onto unprotected feet;

(a)

(b)

Figure 11.13 (a) Rotary drum floor sander. (b) Orbital finishing sander.

➤ manual handling hazards, particularly with heavy machines and intensive use or using at awkward heights and/or reaches;

➤ fire and explosion hazard when used near flammable liquids, explosive dusts or gases;

➤ using equipment in poor weather conditions with wet, slippery surfaces, poor visibility and cold conditions.

Specific control measures include the following:

➤ the workpieces must be securely clamped or held in position during sanding. In some cases a jig will be necessary. The direction of spin of disc sanders (normally anti-clockwise) is important to ensure that a small workpiece is pushed towards a stop or fence which is normally at the left side of the workpiece, particularly when clamping is impossible;

➤ abrasive sanding belts, discs and sheets should be properly and firmly attached to the machine without any torn parts or debris underneath. The manufacturer's instructions and fixing accessories should be used to ensure correct attachment of the abrasive. Operators should be trained, competent and registered to fit abrasive discs;

➤ old nails and fixings should be sunk below the surface or removed to prevent the sander snagging;

➤ always hold the equipment by the proper handles and particularly on large disc and floor sanders always use both hands. Excessive pressure should not be used as the surface will be rutted and the machine may malfunction (Figure 11.12);

➤ ensure that the dust extraction is working properly and has been emptied when about one-third full. Some extraction systems draw dust and air through the sanding sheet, which must have correctly pre-punched holes to allow the passage of air;

➤ operators should wear suitable dust respirators, eye protection and, where necessary, hearing protection;

➤ operators should wear suitable clothing avoiding loose garments, long hair and jewellery, which could catch in the equipment. Protective gloves and footwear are recommended.

11.10 Mechanical machinery hazards

Most machinery has the potential to cause injury to people, and machinery accidents figure prominently in official accident statistics. These injuries may range in severity from a minor cut or bruise, through various degrees of wounding and disabling mutilation, to crushing, decapitation or other fatal injury. It is not solely powered machinery that is hazardous, for many manually operated machines (e.g. hand-operated guillotines and fly presses) can still cause injury if not properly safeguarded.

Machinery movement basically consists of rotary, sliding or reciprocating action, or a combination of these. These movements may cause injury by entanglement, friction or abrasion, cutting, shearing, stabbing or puncture, impact, crushing, or by drawing a person into a position where one or more of these types of injury can occur. The hazards of machinery are set out in BS EN ISO 12100 – Part 2: 2003, which covers the classification of machinery hazards and how harm may occur. The following machinery hazards follow this standard (Figure 11.14).

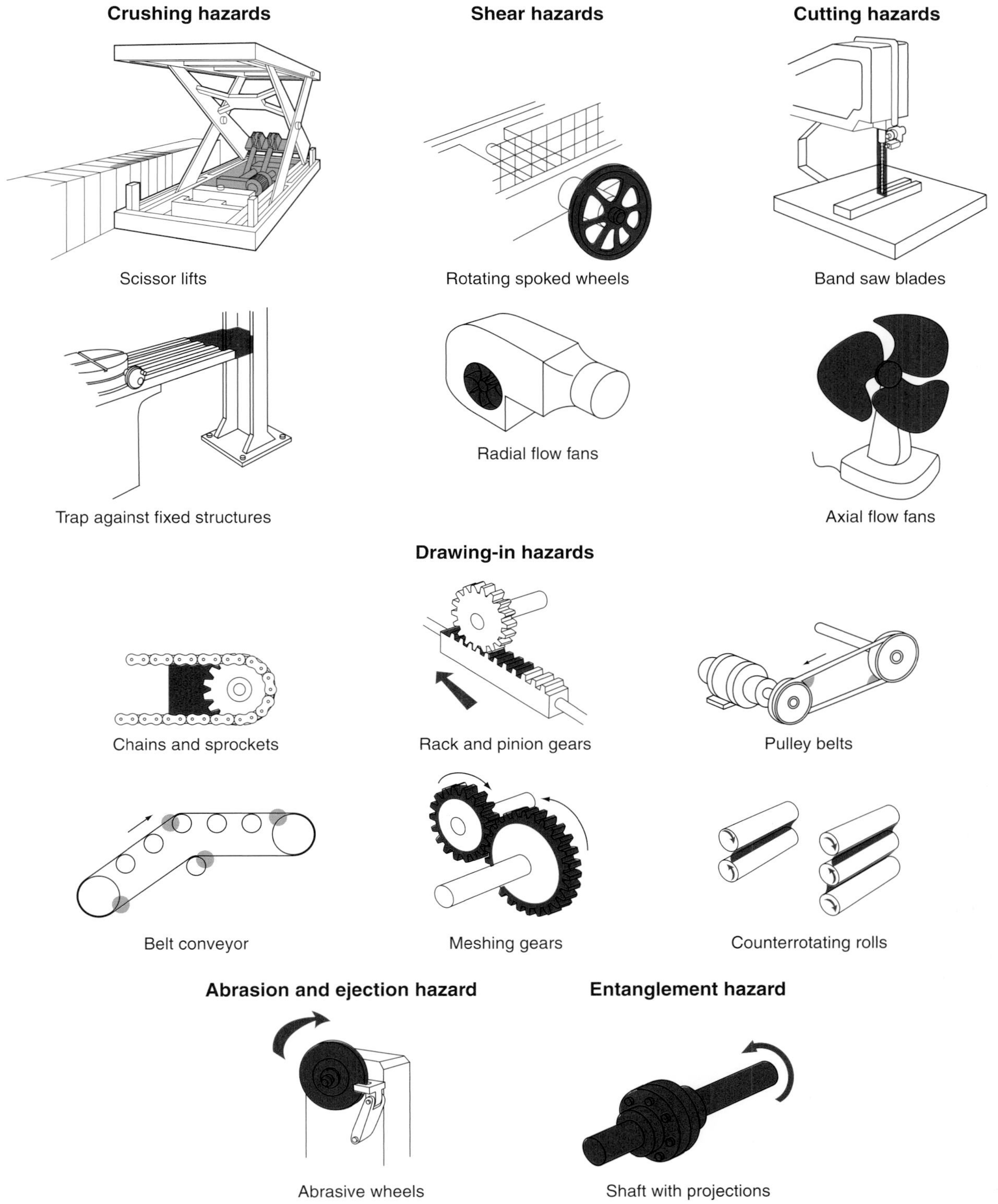

Crushing hazards

Scissor lifts

Trap against fixed structures

Shear hazards

Rotating spoked wheels

Radial flow fans

Cutting hazards

Band saw blades

Axial flow fans

Drawing-in hazards

Chains and sprockets

Rack and pinion gears

Pulley belts

Belt conveyor

Meshing gears

Counterrotating rolls

Abrasion and ejection hazard

Abrasive wheels

Entanglement hazard

Shaft with projections

Figure 11.14 Range of mechanical hazards.

A person may be injured at machinery as a result of:

➤ a **crushing hazard** through being trapped between a moving part of a machine and a fixed structure, such as a wall or any material in a machine;

➤ a **shearing hazard** which traps part of the body, typically a hand or fingers, between moving and fixed parts of the machine;

➤ a **cutting or severing hazard** through contact with a cutting edge, such as a band saw or rotating cutting disc;

➤ an **entanglement hazard** with the machinery which grips loose clothing, hair or working material, such as emery paper, around revolving exposed parts of the machinery. The smaller the diameter of the revolving part the easier it is to get a wrap or entanglement;

➤ a **drawing-in or trapping hazard** such as between in-running gear wheels or rollers or between belts and pulley drives;

➤ an **impact hazard** when a moving part directly strikes a person, such as with the accidental movement of a robot's working arm when maintenance is taking place;

➤ a **stabbing or puncture hazard** through ejection of particles from a machine or a sharp operating component like a needle on a sewing machine;

➤ contact with a **friction or abrasion hazard**, for example, on grinding wheels or sanding machines;

➤ a **high-pressure fluid injection (ejection hazard)**, for example, from a hydraulic system leak.

In practice, injury may involve several of these at once, for example, contact, followed by entanglement of clothing, followed by trapping.

11.11 Non-mechanical machinery hazards

Non-mechanical hazards include:

➤ access: slips, trips and falls; falling and moving objects; obstructions and projections;
➤ lifting and handling;
➤ electricity (including static electricity): shock, burns;
➤ burns and other injuries from fire and explosion;
➤ noise and vibration;
➤ pressure and vacuum;
➤ high/low temperature;
➤ inhalation of dust/fume/mist;
➤ suffocation;

➤ radiation: ionizing and non-ionizing;
➤ biological: viral or bacterial;
➤ physiological effects (e.g. musculoskeletal disorders);
➤ psycho-physiological effects (e.g. mental overload or underload);
➤ human errors;
➤ hazards from the environment where the machine is used (e.g. temperature, wind, snow, lightning).

In many cases it will be practicable to install safeguards which protect the operator from both mechanical and non-mechanical hazards.

For example a guard may prevent access to hot or electrically live parts as well as to moving ones. The use of guards which reduce noise levels at the same time is also common.

As a matter of policy, machinery hazards should be dealt with in this integrated way instead of dealing with each hazard in isolation.

11.12 Examples of machinery hazards

The following examples are given to demonstrate a small range of machines found in industry and commerce, which are included in the Certificate syllabus.

Office – photocopier
The hazards are:

➤ contact with moving parts when clearing a jam;
➤ electrical – when clearing a jam, maintaining the machine or through poorly maintained plug and wiring;
➤ heat through contact with hot parts when clearing a jam;
➤ health hazard from ozone or lack of ventilation in the area.

Office – document shredder
The hazards are:

➤ drawing in between the rotating cutters when feeding paper into the shredder;
➤ contact with the rotating cutters when emptying the waste container or clearing a jam;
➤ electrical through faulty plug and wiring or during maintenance;
➤ possible noise from the cutting action of the machine;
➤ possible dust from the cutting action.

Manufacturing and maintenance – bench-top grinding machine
The hazards are:

➤ contact with the rotating wheel causing abrasion;
➤ drawing in between the rotating wheel and a badly adjusted tool rest;
➤ bursting of the wheel, ejecting fragments which puncture the operator;
➤ electrical through faulty wiring and/or earthing or during maintenance;
➤ fragments given off during the grinding process causing eye injury;
➤ hot fragments given off which could cause a fire or burns;
➤ noise produced during the grinding process;
➤ possible health hazard from dust/particles/fumes given off during grinding.

Manufacturing and maintenance – pedestal drill
The hazards are:

➤ entanglement around the rotating spindle and chuck;
➤ contact with the cutting drill or workpiece;
➤ being struck by the workpiece if it rotates;
➤ being cut or punctured by fragments ejected from the rotating spindle and cutting device;
➤ drawing in to the rotating drive belt and pulley;
➤ contact or entanglement with the rotating motor;
➤ electrical from faulty wiring and/or earthing or during maintenance;
➤ possible health hazard from cutting fluids or dust given off during the process.

Agriculture/horticultural – cylinder mower
The hazards are:

➤ trapping, typically hands or fingers, in the shear caused by the rotating cutters;
➤ contact and entanglement with moving parts of the drive motor;
➤ drawing in between chain and sprocket drives;
➤ impact and cutting injuries from the machine starting accidentally;
➤ burns from hot parts of the engine;
➤ fire from the use of highly flammable petrol as a fuel;
➤ possible noise hazard from the drive motor;
➤ electrical if electrically powered, but this is unlikely;
➤ possible sensitization health hazard from cutting grass, for example, hay fever;
➤ possible health hazard from exhaust fumes.

Agriculture/horticultural – brush cutter/strimmer
The hazards are:

➤ entanglement with rotating parts of motor and shaft;
➤ cutting from contact with cutting head/line;
➤ electric shock, if electrically powered but this is unlikely;
➤ burns from hot parts of the engine;
➤ fire from the use of highly flammable petrol as a fuel;
➤ possible noise hazard from the drive motor and cutting action;
➤ eye and face puncture wounds from ejected particles;
➤ health hazard from hand–arm vibration causing white finger and other problems;
➤ back strain from carrying the machine while operating;
➤ health hazards from animal faeces.

Chainsaw
The hazards are:

➤ very serious cutting by contact with the high-speed cutting chain;
➤ kick back due to being caught on the wood being cut or contact with the top front corner of the chain in motion with the saw chain being kicked upwards towards the face in particular;
➤ pull in when the chain is caught and the saw is pulled forward;
➤ push back when the chain on the top of the saw bar is suddenly pinched and the saw is driven straight back towards the operator;
➤ burns from hot parts of the engine;
➤ high noise levels;
➤ hand–arm vibration causing white finger and other problems;
➤ fire from the use of highly flammable petrol as a fuel;
➤ eye and face puncture wounds from ejected particles;
➤ back strain while supporting the weight of the chainsaw while operating;
➤ electric shock if electrically powered;
➤ falls from height if using the chainsaw in trees and the like;
➤ lone working and the risk of serious injury;
➤ contact with overhead power lines if felling trees;
➤ being hit by falling branches or whole trees while felling;
➤ possible health hazards from cutting due to wood dust, particularly if wood has been seasoned;
➤ using chainsaw in poor weather conditions with slippery surfaces, poor visibility and cold conditions;
➤ health hazards from the engine fumes (carbon dioxide and carbon monoxide), particularly if used inside a shed or other building.

Retail – compactor

The hazards are:

➤ crushing hazard between the ram and the machine sides;
➤ trapping between the ram and machine frame (shear action);
➤ crushing when the waste unit is being changed if removed by truck;
➤ entanglement with moving parts of pump motor;
➤ electrical from faulty wiring and/or earthing or during maintenance;
➤ failure of hydraulic hoses with liquid released under pressure causing puncture to eyes or other parts of body;
➤ falling of vertical ram under gravity if the hydraulic system fails;
➤ handling hazards during loading and unloading.

Retail – checkout conveyor system

The hazards are:

➤ entanglement with belt fasteners if fitted;
➤ drawing in between belt and rollers if under tension;
➤ drawing in between drive belt and pulley;
➤ contact or entanglement with motor drive;
➤ electrical from faulty wiring and/or earthing or during maintenance.

Construction – cement/concrete mixer

The hazards are:

➤ contact and entanglement with moving parts of the drive motor;
➤ crushing between loading hopper (if fitted) and drum;
➤ drawing in between chain and sprocket drives;
➤ electrical, if electrically powered;
➤ burns from hot parts of engine;
➤ fire if highly flammable liquids used as fuel;
➤ possible noise hazards from the motor and dry mixing of aggregates;
➤ eye injury from splashing cement slurry;
➤ possible health hazard from cement dust and cement slurry while handling.

Construction – bench-mounted circular saw

The hazards are:

➤ contact with the cutting blade above and below the bench;
➤ ejection of the workpiece or timber as it closes after passing the cutting blade;

➤ drawing in between chain and sprocket or V belt drives;
➤ contact and entanglement with moving parts of the drive motor;
➤ likely noise hazards from the cutting action and motor;
➤ health hazards from wood dust given off during cutting;
➤ electric shock from faulty wiring and/or earthing or during maintenance;
➤ climatic conditions such as wet/extreme heat or cold;
➤ possible fire and/or explosion from wood dust caused by overheated blades from excessive friction or an electrical fault.

11.13 Practical safeguards

PUWER requires that access to dangerous parts of machinery should be prevented in a preferred order or hierarchy of control methods. The standard required is a 'practicable' one, so that the only acceptable reason for non-compliance is that there is no technical solution. Cost is not a factor. (See Chapter 1 for more details on standards of compliance.)

The levels of protection required are, in order of implementation:

➤ fixed enclosing guarding;
➤ other guards or protection devices, such as interlocked guards and pressure-sensitive mats;
➤ protection appliances, such as jigs, holders and push sticks;
➤ the provision of information, instruction, training and supervision.

As the mechanical hazard of machinery arises principally from someone coming into contact or entanglement with dangerous components, risk reduction is based on preventing this contact occurring.

This may be by means of:

➤ a physical barrier between the individual and the component (e.g. a fixed enclosing guard);
➤ a device which allows access only when the component is in a safe state (e.g. an interlocked guard which prevents the machine starting unless a guard is closed and acts to stop the machine if the guard is opened);
➤ a device which detects that the individual is entering a risk area and then stops the machine (e.g. certain photoelectric guards and pressure-sensitive mats).

The best method should, ideally, be chosen by the designer as early in the life of the machine as possible. It is often found that safeguards which are 'bolted on' instead of 'built in' are not only less effective in reducing risk, but are also more likely to inhibit the normal operation of the machine. In addition, they may in themselves create hazards and are likely to be difficult and hence expensive to maintain.

11.13.1 Fixed guards

Fixed guards have the advantage of being simple, always in position, difficult to remove and almost maintenance-free. Their disadvantage is that they do not always properly prevent access, they are often left off by maintenance staff and they can create difficulties for the operation of the machine.

A fixed guard has no moving parts and should, by its design, prevent access to the dangerous parts of the machinery. It must be of robust construction and sufficient to withstand the stresses of the process and environmental conditions. If visibility or free air flow (e.g. for cooling) is necessary, this must be allowed for in the design and construction of the guard. If the guard can be opened or removed, this must only be possible with the aid of a tool.

An alternative fixed guard is the distance fixed guard, which does not completely enclose a hazard, but which reduces access by virtue of its dimensions and its distance from the hazard. Where perimeter fence guards are used, the guard must follow the contours of the machinery as far as possible, thus minimizing space between the guard and the machinery. With this type of guard, it is important that the safety devices and operating systems prevent the machinery being operated with the guards closed and someone inside the guard, that is in the danger area. Figure 11.15 shows a range of fixed guards for some of the examples shown in Figure 11.14.

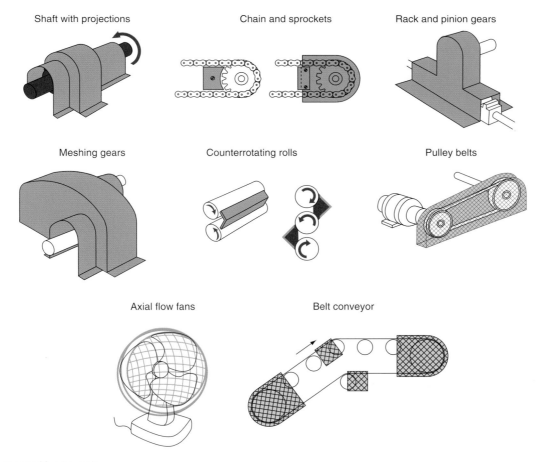

Figure 11.15 Range of fixed guards.

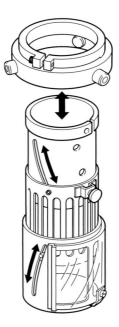

Figure 11.16 Adjustable guard for a rotating shaft, such as a pedestal drill.

11.13.2 Adjustable guards

User-adjusted guard
These are fixed or movable guards, which are adjustable for a particular operation during which they remain fixed. They are particularly used with machine tools where some access to the dangerous part is required (e.g. drills, circular saws, milling machines) and where the clearance required will vary (e.g. with the size of the cutter in use on a horizontal milling machine or with the size of the timber being sawn on a circular saw bench) (Figure 11.16).

Adjustable guards may be the only option with cutting tools, which are otherwise very difficult to guard, but they have the disadvantage of requiring frequent readjustment. By the nature of the machines on which they are most frequently used, there will still be some access to the dangerous parts, so these machines must only be used by suitably trained operators. Jigs, push sticks and false tables must be used wherever possible to minimize hazards during the feeding of the workpiece. The working area should be well lit and kept free of anything which might cause the operator to slip or trip.

Self-adjusting guard
A self-adjusting guard is one which adjusts itself to accommodate, for example, the passage of material.

A good example is the spring-loaded guard fitted to many portable circular saws.

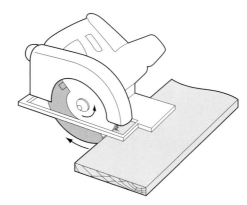

Figure 11.17 Self-adjusting guard on a wood saw.

As with adjustable guards (see Figure 11.16) they only provide a partial solution in that they may well still allow access to the dangerous part of the machinery. They require careful maintenance to ensure they work to the best advantage (Figure 11.17).

11.13.3 Interlocking guard

The advantages of interlocked guards are that they allow safe access to operate and maintain the machine without dismantling the safety devices. Their disadvantage stems from the constant need to ensure that they are operating correctly and designed to be fail-safe. Maintenance and inspection procedures must be very strict.

This is a guard which is movable (or which has a movable part) whose movement is connected with the power or control system of the machine.

An interlocking guard must be connected to the machine controls such that:

➤ until the guard is closed the interlock prevents the machinery from operating by interrupting the power medium;
➤ either the guard remains locked until the risk of injury from the hazard has passed or opening the guard causes the hazard to be eliminated before access is possible.

A passenger lift or hoist is a good illustration of these principles: the lift will not move unless the doors are closed, and the doors remain closed and locked until the lift is stationary and in such a position that it is safe for the doors to open.

Special care is needed with systems which have stored energy. This might be the momentum of a heavy moving part, stored pressure in a hydraulic or pneumatic system, or even the simple fact of a part being able to move under gravity even though the power is disconnected. In these

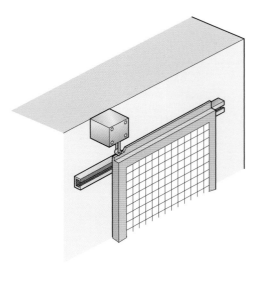
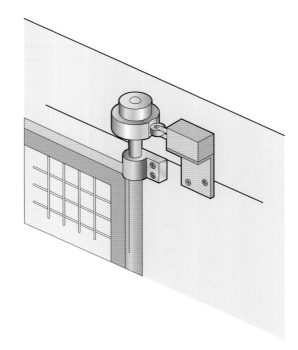

Figure 11.18 Typical sliding and hinged interlocking guards.

situations, dangerous movement may continue or be possible with the guard open, and these factors need to be considered in the overall design. Braking devices (to arrest movement when the guard is opened) or delay devices (to prevent the guard opening until the machinery is safe) may be needed. All interlocking systems must be designed to minimize the risk of failure-to-danger and should not be easy to defeat (Figure 11.18).

11.14 Other safety devices

11.14.1 Trip devices

A trip device does not physically keep people away but detects when a person approaches close to a danger point. It should be designed to stop the machine before injury occurs. A trip device depends on the ability of the machine to stop quickly and in some cases a brake may need to be fitted. Trip devices can be:

➤ mechanical in the form of a bar or barrier;
➤ electrical in the form of a trip switch on an actuator rod, wire or other mechanism;
➤ photoelectric or other type of presence-sensing device;
➤ pressure-sensitive mat.

They should be designed to be self-resetting so that the machine must be restarted using the normal procedure (Figure 11.19).

11.14.2 Two-handed control devices

These are devices which require the operator to have both hands in a safe place (the location of the controls) before the machine can be operated. They are an option on machinery that is otherwise very difficult to guard but they have the drawback that they only protect the operator's hands. It is therefore essential that the design does not allow any other part of the operator's body to enter the danger zone during operation. More significantly, they give no protection to anyone other than the operator.

Where two-handed controls are used, the following principles must be followed:

➤ the controls should be so placed, separated and protected as to prevent spanning with one hand only, being operated with one hand and another part of the body, or being readily bridged;
➤ it should not be possible to set the dangerous parts in motion unless the controls are operated within approximately 0.5 seconds of each other. Having set the dangerous parts in motion, it should not be

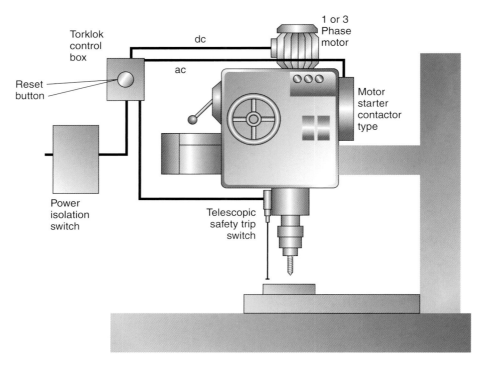

Figure 11.19 Schematic diagram of a telescopic trip device fitted to a radial drill.

Figure 11.20 Pedestal-mounted free-standing two-hand control device.

possible to do so again until both controls have been returned to their off position;

➤ movement of the dangerous parts should be arrested immediately or, where appropriate, arrested and reversed if one or both controls are released while there is still danger from movement of the parts;

➤ the hand controls should be situated at such a distance from the danger point that, on releasing the controls, it is not possible for the operator to reach the danger point before the motion of the dangerous parts has been arrested or, where appropriate, arrested and reversed (Figure 11.20).

11.14.3 Hold-to-run control

This is a control which allows movement of the machinery only as long as the control is held in a set position. The control must return automatically to the stop position when released. Where the machinery runs at crawl speed, this speed should be kept as low as practicable.

Hold-to-run controls give even less protection to the operator than two-handed controls and have the same main drawback in that they give no protection to anyone other than the operator.

However, along with limited movement devices (systems which permit only a limited amount of machine movement on each occasion that the control is operated and are often called 'inching devices'), they are extremely relevant to operations such as setting, where access may well be necessary and safeguarding by any other means is difficult to achieve.

Figure 11.21 Typical photocopier.

11.15 Application of safeguards to the range of machines

The application of the safeguards to the Certificate range of machines is as follows.

Office – photocopier
Application of safeguards (Figure 11.21)

➤ the machines are provided with an all-enclosing case which prevents access to the internal moving, hot or electrical parts;
➤ the access doors are interlocked so that the machine is automatically switched off when gaining access to clear jams or maintain the machine. It is good practice to switch off when opening the machine;
➤ internal electrics are insulated and protected to prevent contact;
➤ regular inspection and maintenance should be carried out;
➤ the machine should be on the PAT schedule;
➤ good ventilation in the machine room should be maintained.

Office – document shredder
Application of safeguards (Figure 11.22)

➤ enclosed fixed guards surround the cutters with restricted access for paper only, which prevents fingers reaching the dangerous parts;
➤ interlocks are fitted to the cutter head so that the machine is switched off when the waste bin is emptied;
➤ a trip device is used to start the machine automatically when paper is fed in;

Figure 11.22 Typical office paper shredder.

➤ machine should be on PAT schedule and regularly checked;
➤ general ventilation will cover most dust problems except for very large machines where dust extraction may be necessary;
➤ noise levels should be checked and the equipment perhaps placed on a rubber mat if standing on a hard reflective floor.

Manufacturing and maintenance – bench-top grinder
Application of safeguards (Figure 11.23)

➤ wheel should be enclosed as much as possible in a strong casing capable of containing a burst wheel;
➤ grinder should be bolted down to prevent movement;
➤ an adjustable tool rest should be adjusted as close as possible to the wheel;
➤ adjustable screen should be fitted over the wheel to protect the eyes of the operator. Goggles should also be worn;
➤ only properly trained competent and registered people should mount an abrasive wheel;
➤ the maximum speed should be marked on the machine so that the abrasive wheel can be matched to the machine speed to ensure that the wheel permitted speed exceeds or equals the machine max speed.

➤ noise levels should be checked and attenuating screens used if necessary;
➤ the machine should be on the PAT schedule and regularly checked;
➤ if necessary extract ventilation should be fitted to the wheel encasing to remove dust at source.

Figure 11.23 Typical bench-mounted grinder.

Manufacturing and maintenance – pedestal drill
Application of safeguards (Figure 11.24)

➤ motor and drive should be fitted with fixed guard;
➤ machine should be bolted down to prevent movement;
➤ the spindle should be guarded by an adjustable guard, which is fixed in position during the work;
➤ a clamp should be available on the pedestal base to secure workpieces;
➤ the machine should be on the PAT schedule and regularly checked;
➤ cutting fluid, if used, should be contained and not allowed to get onto clothing or skin. A splash guard may be required but is unlikely;
➤ goggles should be worn by the operator.

Agricultural/horticultural – cylinder mower
Application of safeguards (Figure 11.25)

➤ machine should be designed to operate with the grass collection box in position to restrict access to the bottom blade trap. A warning sign should be fitted on the machine;

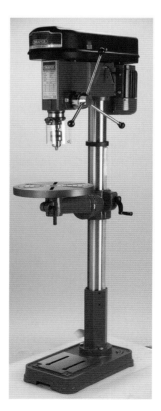

Figure 11.24 Typical pedestal drill.

Figure 11.25 Typical large cylinder mower.

➤ on pedestrian-controlled machines the control handle should automatically stop the blade rotation when the operator's hands are removed. It should take two separate actions to restart;

➤ ride-on machines should be fitted with a device to automatically stop the blades when the operator leaves the operating position. This is normally a switch under the seat and it should be tested to ensure that it is functioning correctly and is not been defective;

➤ drives and motor should be completely encased with a fixed guard;

➤ the machine should only be refuelled in the open air with a cool engine, using a proper highly flammable fuel container with pourer to restrict spillage. No smoking should be allowed;

➤ hot surfaces like the exhaust should be covered;

➤ engine must only be run in the open air to prevent a build up of fumes;

➤ noise levels should be checked and if necessary an improved silencer fitted to the engine and where required hearing protection used;

➤ hay-fever-like problems from grass cutting are difficult to control. A suitable dust mask may be required to protect the user.

Agricultural/horticultural – brush cutter/strimmer
Application of safeguards

➤ moving engine parts should be enclosed;

➤ rotating shafts should be encased in a fixed drive shaft cover;

➤ rotating cutting head should have a fixed top guard, which extends out on the user side of the machine;

➤ line changes must only be done either automatically or with the engine switched off;

➤ engine should only be run in the open air;

➤ refuelling should only be done in the open air using a proper highly flammable liquid container with pouring spout;

➤ boots with steel toe cap and good grip, stout trousers and non-snagging upper garments should be worn in addition to hard hat fitted with full face screen and safety glasses;

➤ if the noise levels are sufficiently high (normally they are with petrol-driven units) suitable hearing protection should be worn;

➤ low-vibration characteristics should be balanced with engine power and speed of work to achieve the minimum overall vibration exposure. Handles should be of an anti-vibration type. Engines should be mounted on flexible mountings. Work periods should be limited to allow recovery;

➤ washing arrangements and warm impervious gloves should be provided to guard against health risks;

➤ properly constructed harness should be worn which comfortably balances the weight of the machine.

Agriculture/horticulture – chainsaw
Application of safeguards (Figure 11.26)

➤ may only be operated by fully trained, fit and competent people. Using chainsaws in tree work requires a relevant certificate of competence or national competence award, unless the users are undergoing such training and are adequately supervised. However, in the agricultural sector, this requirement applies only to first-time users of a chainsaw.

This means everyone working with chainsaws on or in trees should hold such a certificate or award **unless**:

➤ it is being done as part of agricultural operations (e.g. hedging, clearing fallen branches, pruning trees to maintain clearance for machines);

➤ the work is being done by the occupier or their employees;

➤ they have used a chainsaw before 5 December 1998.

In any case, operators using chainsaws for any task in agriculture or any other industry must be competent under PUWER 98.

Competence assessment
Lantra Awards, Stoneleigh Park, Kenilworth, Warwickshire CV8 2LG, Tel: 02476 419703, Fax: 02476 411655.

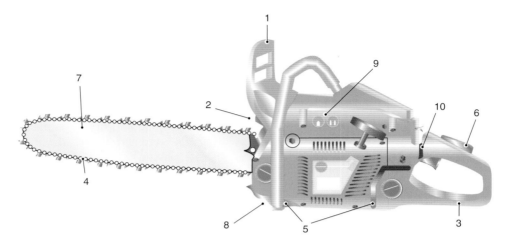

Figure 11.26 Typical chainsaw with rear handle. The rear handle project from the back of the saw. It is designed to always be gripped with both hands, with the right hand on the rear handle. It may be necessary to have a range of saws with different guide bar lengths available. As a general rule, choose a chainsaw with the shortest guide bar suitable for the work. 1 – hand guard with integral chain brake; 2 – exhaust outlet directed to the RHS away from the operator; 3 – chain breakage guard at bottom of rear handle; 4 – chain designed to have low-kickback tendency; 5 – rubber anti-vibration mountings; 6 – lockout for the throttle trigger; 7 – guide bar (should be protected when transporting chainsaw); 8 – bottom chain catcher; 9 – PPE hand/eye/ear defender signs; 10 – on/off switch.

NPTC, Stoneleigh Park, Kenilworth, Warwickshire CV8 2LG.

➤ avoid working alone with a chainsaw. Where this is not possible, establish procedures to raise the alarm if something goes wrong. These may include:
  • regular contact with others using either a radio or telephone
  • someone regularly visiting the work site
  • carrying a whistle to raise the alarm
  • an automatic signalling device which sends a signal at a preset time unless prevented from doing so
  • checks to ensure operators return to base or home at an agreed time.
➤ moving engine parts should be enclosed;
➤ electrical units should be double-insulated and cables fitted with residual current devices;
➤ the saw must be fitted with a top handle and effective brake mechanism;
➤ chainsaws expose operators to high levels of noise and hand-arm vibration which can lead to hearing loss and conditions such as vibration white finger. These risks can be controlled by good management practice including:
  • purchasing policies for low-noise/low-vibration chainsaws (e.g. with anti-vibration mounts and heated handles)
  • providing suitable hearing protection
  • proper maintenance schedules for chainsaws and protective equipment

  • giving information and training to operators on the health risks associated with chainsaws and use of PPE, etc.
➤ proper maintenance is essential for safe use and protection against ill health from excessive noise and vibration. The saw must be maintained in its manufactured condition with all the safety devices in efficient working order and all guards in place. It should be regularly serviced by someone who is competent to do the job.

Operators need to be trained in the correct chain-sharpening techniques and chain and guide bar maintenance to keep the saw in safe working condition. Operators should report any damage or excessive wear from daily checks on the following:

  • on/off switch
  • chain brake
  • chain catcher
  • silencer
  • guide bar, drive sprocket and chain links
  • side plate, front and rear hand guards
  • anti-vibration mounts
  • starting cord for correct tension.
➤ make sure petrol containers are in good condition and clearly labelled, with securely fitting caps. Use containers which are specially designed for chainsaw fuelling and lubrication. Fit an auto-filler spout to the

Figure 11.27 Kevlar gloves, overtrousers and overshoes providing protection against chainsaw cuts. Helmet, ear and face shields protect the head. Apprentice under instruction – first felling.

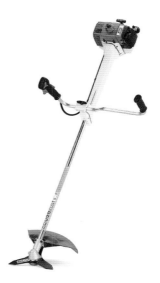

Figure 11.28 STIHL petrol-driven brush cutter for professional users.

outlet of a petrol container to reduce the risk of spillage from over-filling. Operators should:

● avoid getting dirt in the fuel system (this may cause the chainsaw to be unreliable)
● securely replace all filler caps immediately after fuelling/oiling
● wipe up any spilt petrol/oil
● during starting and use, keep fuel containers well away from fires and other sources of ignition, including the saw itself (at least 4 m is recommended).

Do not allow operators to use discarded engine oil as a chain lubricant – it is a very poor lubricant and may cause cancer if it is in regular contact with an operator's skin.

➤ when starting the saw, operators should maintain a safe working distance from other people and ensure the saw chain is clear of obstructions;
➤ kickback is the sudden uncontrolled upward and backward movement of the chain and guide bar towards the operator. This can happen when the saw chain at the nose of the guide bar hits an object. Kickback is responsible for a significant proportion of chainsaw accidents, many of which are to the face and parts of the upper body where it is difficult to provide protection. A properly maintained chain brake and use of low-kickback chains (safety chains) reduce the effect, but cannot entirely prevent it.

Make sure operators use the saw in a way which avoids kickback by:

● not allowing the nose of the guide bar to accidentally come into contact with any obstruction, for example branches, logs, stumps
● not overreaching
● keeping the saw below chest height
● keeping the thumb of the left hand around the back of the front handle
● using the appropriate chain speed for the material being cut.

➤ to avoid pull-in, always hold the spiked bumper securely against the tree or limb;
➤ to avoid push-back, be alert to conditions that may cause the top of the guide bar to be pinched and do not twist the guide bar in the cut;
➤ training in good manual handling techniques and using handling aids/tools should reduce the risk of back injuries;
➤ to avoid overhead and other service hazards before felling starts on the work site:

● contact the owners of any overhead power lines within a distance equal to twice the height of any tree to be felled to discuss whether the lines need to be lowered or made dead
● do not start work until agreement has been reached on the precautions to be taken
● check whether there are underground services such as power cables or gas pipes which could be damaged when the tree strikes the ground

- if there are roads or public rights or way within a distance equal to twice the height of the tree to be felled, ensure that road users and members of the public do not enter the danger zone. You may need to arrange warning notices, diversions or traffic control.

Safe working practices for felling are given in Health and Safety Executive's (HSE's) AFAG leaflets 300 series.

➤ Suitable PPE should always be worn, no matter how small the job. European standards for chainsaw PPE are published as part of EN 381 *Protective Clothing for users of Hand-Held Chainsaws* (Figure 11.27).

Protective clothing complying with this standard should provide a consistent level of resistance to chainsaw cut-through. Other clothing worn with the PPE should be close-fitting and non-snagging.

> **NB No protective equipment with this standard should provide a consistent level of resistance to chainsaw cut-through cutting by a hand-held chainsaw.**

Safety helmet – to EN 397 (arborists working from a rope and harness may use a suitable adapted rock-climbing helmet).
Hearing protection – to EN 352-1.
Eye protection – mesh visors to EN 1731 or safety glasses to EN 166.
Upper body protection[1] – chainsaw jackets to BS EN 381-11.
Gloves – to EN 381-7. The use of appropriate gloves is recommended under most circumstances. The type of glove will depend on a risk assessment of the task and the machine. Consider the need for protection from cuts from the chainsaw, thorny material and cold/wet conditions.
Leg protection – to EN 381-5 (all-round protection is recommended for arborists working in trees, and for occasional users such as those working in agriculture).
Chainsaw boots – to BS EN ISO 20345:2004 and bearing a shield depicting a chainsaw to show compliance with EN 381-3 (for occasional users working on even ground where there is little risk of tripping or snagging on

[1]Chainsaw jackets can provide additional protection where operators are at increased risk (e.g. trainees, unavoidable use of a chainsaw above chest height). However, this needs to be weighed against increased heat stress generated by physical exertion (e.g. working from a rope and harness).

undergrowth or brash, protective gaiters conforming to EN 381-9 worn in combination with steel-toe-capped safety boots).

➤ If conditions are dusty, suitable filtering face masks should be worn.

Retail – compactor
Application of safeguards (Figure 11.29)

➤ access doors to the loading area, which gives access to the ram, should be positively interlocked with electrical and/or hydraulic mechanisms;
➤ the ram pressure should be dumped if it is hydraulic;
➤ drives of motors should be properly guarded;
➤ if the waste unit is removed by truck, the ram mechanism should be interlocked with the unit so that it cannot operate when the unit is changed for an empty one;
➤ the machine should be regularly inspected and tested by a competent person;
➤ if the hydraulic ram can fall under gravity, mechanical restraints should automatically move into position when the doors are opened;
➤ emergency stop buttons should be fitted on each side.

Retail – checkout conveyor
Application of safeguards

➤ all traps between belt and rollers are provided with either fixed guards or interlocked guards;
➤ motor and drive unit should be provided with a fixed guard and access to the underside of the conveyor is prevented by enclosure;
➤ adequate emergency stop buttons must be provided by the checkout operator;

Figure 11.29 Typical retail compactor.

➤ machine should be on the PAT electrical inspection schedule and regularly checked;

➤ auto stop system should be fitted so that conveyor does not push products into the operator's working zone.

Construction – cement mixer

Application of safeguards (Figure 11.30)

➤ operating position for the hopper hoist should be designed so that anyone in the trapping area is visible to the operator. The use of the machine should be restricted to designated operators only. As far as possible the trapping point should be designed out. The hoist operating location should be fenced off just allowing access for barrows, etc., to the unloading area;

➤ drives and rotating parts of engine should be enclosed;

➤ the drum gearing should be enclosed and persons kept away from the rotating drum, which is normally fairly high on large machines;

➤ no one should be allowed to stand on the machine while it is in motion;

➤ goggles should be worn to prevent cement splashes;

➤ if petrol-driven, care is required with flammable liquids and refuelling;

➤ engines must only be run in the open air;

➤ electric machines should be regularly checked and be on the PAT schedule;

➤ noise levels should be checked and noise attenuation, for example silencers and damping, fitted if necessary.

Figure 11.30 Typical cement/concrete mixer.

Construction – bench-mounted circular saw

Application of safeguards (Figure 11.31)

➤ a fixed guard should be fitted to the blade below the bench;

➤ fixed guards should be fitted to the motor and drives;

➤ an adjustable top guard should be fitted to the blade above the bench which encloses as much of the blade as possible. An adjustable front section should also be fitted;

➤ a riving knife should be fitted behind the blade to keep the cut timber apart and prevent ejection;

➤ a push stick should be used on short workpieces (under 300 mm) or for the last 300 mm of longer cuts;

➤ blades should be kept properly sharpened and set with the diameter of the smallest blade marked on the machine;

➤ noise attenuation should be applied to the machine, for example damping, special quiet saw blades, and, if necessary, fitting in an enclosure. Hearing protection may have to be used;

➤ protection against wet weather should be provided;

➤ the electrical parts should be regularly checked in addition to all the mechanical guards;

➤ extraction ventilation will be required for the wood dust and shavings;

➤ suitable dust masks should be worn;

➤ suitable warm or cool clothing will be needed when used in hot or cold locations;

➤ space around machine should be kept clear.

11.16 Guard construction

The design and construction of guards must be appropriate to the risks identified and the mode of operation of the machinery in question.

The following factors should be considered:

➤ strength – adequate guards for the purpose, able to resist the forces and vibration involved and able to withstand impact (where applicable);

➤ weight and size – in relation to the need to remove and replace the guard during maintenance;

➤ compatibility with materials being processed and lubricants, etc.;

➤ hygiene and the need to comply with food safety Regulations;

➤ visibility – may be necessary to see through the guard for both operational and safety reasons;

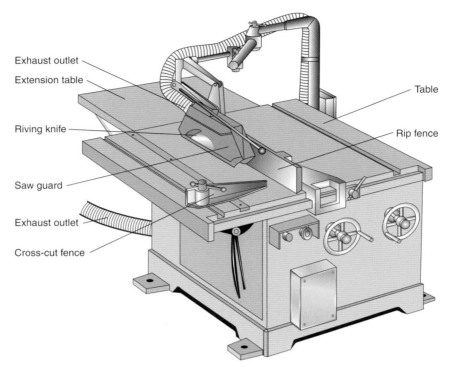

Figure 11.31 Bench-mounted circular saw.

➤ noise attenuation – guards often utilsed to reduce the noise levels produced by a machine. Conversely, the resonance of large undamped panels may exacerbate the noise problem;

➤ enabling a free flow of air – where necessary (e.g. for ventilation);

➤ avoidance of additional hazards – for example free of sharp edges;

➤ ease of maintenance and cleanliness;

➤ openings – the size of openings and their distance from the dangerous parts should not allow anyone to be able to reach into a danger zone. These values can be determined by experiment or by reference to standard tables. If doing so by experiment, it is essential that the machine is first stopped and made safe (e.g. by isolation). The detailed information on openings is contained in EN 294:1992, EN 349:1993 and EN 811:1997.

11.17 Sources of reference

Safe Use of Work Equipment (ACOP) L22 (3rd edition) HSE Books 2008 ISBN 978 0 7176 6295 1.

BS EN ISO 12100 (formerly BS EN 292) and also PD 5304; 2005 Guidance on Safe Use of Machinery BSI2005 ISBN 0 580 46818 6.

Relevant statutory provisions

The Supply of Machinery (Safety) Regulations 1992 (From December 2009 – the 2008 Regulations come into force) Scope and application, and relationship to CE marking.

The Provision and Use of Work Equipment Regulations 1998 – Part II.

The Personal Protective Equipment at Work Regulations 1992.

11.18 Practice NEBOSH questions for Chapter 11

1. **List** the main requirements of the Provision and Use of Work Equipment Regulations 1998.

2. **(i) Outline** a hierarchy of control measures that may be used to prevent contact with dangerous parts of machinery.

(ii) Identify FOUR non-mechanical hazards that may be encountered on woodworking machines and **outline** the possible health and safety effects from exposure in **EACH** case.

3. In order to meet a production deadline, a supervisor instructed an employee to operate a machine which they both knew to be defective.

   Giving reasons in **EACH** case, **identify** possible breaches of the Health and Safety at Work etc. Act 1974 in relation to this scenario.

4. An employee was seriously injured while operating a machine, having disabled an interlocking device. The accident was properly reported to the enforcing authority and an HSE inspector has arrived to investigate the accident.

   (i) Outline the powers the inspector may use in order to gather information for the investigation.

   (ii) Giving reasons in **EACH** case, **identify** the specific health and safety legislation that the machine operator may have breached.

   (iii) Outline the action management may take to prevent similar accidents.

5. **Identify** the factors to consider when assessing the suitability of controls (including emergency controls) of an item of work equipment.

6. **Outline EIGHT** factors that may be important in determining the maintenance requirements for an item of work equipment.

7. **Identify** the factors that would help to determine the maintenance requirements of an item of work equipment.

8. **Outline** the measures to be taken to reduce the risk of accidents associated with the routine maintenance of machinery.

9. **Outline** the practical precautions that might be needed prior to the repair of a large item of process machinery.

10. A decorator uses a hand-held electric sander for the preparation of wood prior to painting.

    (i) Outline the checks that should be made to ensure the electrical safety of the sander.

    (ii) Other than electricity, **identify FOUR** hazards associated with the use of the sander.

11. A carpenter is using a hammer and chisel to cut out a recess in a wooden door.

    (i) Identify FOUR unsafe conditions, associated with the tools, which could affect the safety of the carpenter.

    (ii) Outline suitable control measures for minimizing the risk to the carpenter when using the tools.

12. **Provide sketches** to show clearly the nature of the following mechanical hazards from moving parts of machinery:

    (i) entanglement

    (ii) crushing

    (iii) drawing-in

    (iv) shear.

13. **Identify:**

    (i) TWO mechanical hazards associated with moving parts of machinery.

    (ii) TWO non-mechanical hazards to which a machine operator may be exposed.

14. **List EIGHT** types of mechanical hazard associated with moving parts of machinery.

15. **(i) Identify THREE** mechanical hazards associated with the use of a bench-top grinder.

    (ii) Outline the precautions to be taken to minimize the risk of injury to operators of bench-top grinders.

16. **Identify FOUR** mechanical hazards presented by pedestal drills and **outline** in each case how injury might occur.

17. A new pedestal (pillar) drill has been installed in an engineering workshop. **Identify** the factors that should be considered before it is first used, to reduce the risk of injury to the operators.

18. With reference to an accident involving an operator who comes into contact with a dangerous part of a machine, **describe:**

    (i) the possible immediate causes

    (ii) the possible root (underlying) causes.

19. In relation to cutting timber using a bench-mounted circular saw:

    (i) outline the mechanical hazards to which an operator might be exposed

(ii) identify the guards and protective devices designed to prevent contact with the saw blade, and in **EACH** case, **explain** how the operator is protected

(iii) outline FOUR non-mechanical hazards presented by the operation, identifying the possible health and safety effects in **EACH** case.

20. Bench-mounted circular saws are operated in a workshop to cut wood to size in the manufacture of wooden pallets.

    (i) In relation to the use of the circular saw, **identify FOUR** risks to the health **AND FOUR** risks to the safety of the saw operators

    (ii) Outline the measures that can be taken to minimize the health and safety risks to the circular saw operators.

21. A chainsaw is to be used to fell a tree from ground level. In relation to this task:

    (i) identify FOUR hazards associated with the use of the chainsaw

    (ii) list the items of personal protective equipment that should be used by the chainsaw operative.

22. An employee is to use a petrol-driven chainsaw to fell a tree from ground level. Outline the hazards faced by the employee in carrying out this task.

23. **Identify FOUR** hazards when cutting grass on roadside verges with a rider-operated motor-mower and **outline** the precautions to be taken in **EACH** one.

24. **Outline** the **FOUR** main categories of guards and safeguarding devices that can be used to minimize the risk of contact with dangerous parts of machinery.

25. **(i) Describe** when a fixed guard would be an appropriate means of providing protection against mechanical hazards.

    (ii) Outline the features of fixed guards designed to minimize the risk of injury *or* ill health from dangerous parts of machinery.

26. **(i)** In relation to machine safety, **outline** the principles of operation of:

    (a) interlocked guards

    (b) trip devices.

    (ii) Other than contact with dangerous parts, **identify FOUR** types of hazard against which fixed guards on machines may provide protection.

27. **(i) Outline** the principles of the following types of machine guard:

    (a) fixed guard

    (b) interlocked guard.

    (ii) Identify TWO advantages **and TWO** possible disadvantages of a fixed machine guard.

Electrical hazards and control

12

After reading this chapter you should be able to:

1. identify the hazards and evaluate the consequential risks from the use of electricity in the workplace
2. describe the control measures that should be taken when working with electrical systems or using electrical equipment.

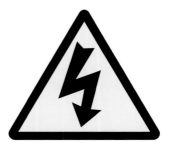

Figure 12.1 Beware of electricity – typical sign.

12.1 Introduction

Electricity is a widely used, efficient and convenient, but potentially hazardous method of transmitting and using energy. It is in use in every factory, workshop, laboratory and office in the country. Any use of electricity has the potential to be very hazardous with possible fatal results. Legislation has been in place for many years to control and regulate the use of electrical energy and the activities associated with its use. Such legislation provides a framework for the standards required in the design, installation, maintenance and use of electrical equipment and systems and the supervision of these activities to minimize the risk of injury. Electrical work from the largest to the smallest installation must be carried out by people known to be competent to undertake such work. New installations always require expert advice at all appropriate levels to cover both design aspects of the system and its associated equipment. Electrical systems and equipment must be properly selected, installed, used and maintained.

Approximately 8% of all fatalities at work are caused by electric shock. Over the last few years, there have been 1000 electrical accidents each year and 25 people die of their injuries. The majority of the fatalities occur in the agriculture, extractive and utility supply and service industries, whereas the majority of the major accidents happen in the manufacturing, construction and service industries.

Only voltages up to and including **mains voltage** (220/240V) and the three principal electrical hazards – electric shock, electric burns and electrical fires and explosions – are considered in detail in this chapter.

12.2 Principles of electricity and some definitions

12.2.1 Basic principles and measurement of electricity

In simple terms, electricity is the flow or movement of electrons through a substance which allows the transfer of electrical energy from one position to another. The substance through which the electricity flows is called a **conductor**. This flow or movement of electrons is known as the **electric current**. There are two forms of electric current – direct and alternating. **Direct current (dc)** involves the flow of electrons along a conductor from one end to the other. This type of current is mainly restricted to batteries and similar devices. **Alternating current (ac)** is produced by a rotating alternator and causes an oscillation of the electrons rather than a flow of electrons so that energy is passed from one electron to the adjacent one and so on through the length of the conductor.

It is sometimes easier to understand the basic principles of electricity by comparing its movement with that of water in a pipe flowing downhill. The flow of water through the pipe (measured in litres per second) is similar to the current flowing through the conductor which is measured in amperes, normally abbreviated to **amps (A)**. Sometimes very small currents are used and these are measured in milliamps (mA).

The higher the pressure drop is along the pipeline, the greater will be the flow rate of water and, in a similar way, the higher the electrical 'pressure difference' along the conductor, the higher the current will be. This electrical 'pressure difference' or potential difference is measured in **volts**.

The flow rate through the pipe will also vary for a fixed pressure drop as the roughness on the inside surface of the pipe varies – the rougher the surface, the slower the flow and the higher the resistance to flow becomes. Similarly, for electricity, the poorer the conductor, the higher the resistance is to electrical current and the lower the current becomes. Electrical resistance is measured in **ohms**.

The voltage (V), the current (I) and the resistance (R) are related by the following formula, known as Ohm's law:

$$V = I \times R \text{ (volts)}$$

and, electrical power (P) is given by:

$$P = V \times I \text{ (watts)}$$

These basic formulae enable simple calculations to be made so that, for example, the correct size of fuse may be ascertained for a particular piece of electrical equipment.

Conductors and insulators

Conductors are nearly always metals, copper being a particularly good conductor, and are usually in wire form but they can be gases or liquids, water being a particularly good conductor of electricity. Superconductors is a term given to certain metals which have a very low resistance to electricity at low temperatures.

Very poor conductors are known as **insulators** and include materials such as rubber, timber and plastics. Insulating material is used to protect people from some of the hazards associated with electricity.

Short circuit

Electrical equipment components and an electrical power supply (normally the mains or a battery) are joined together by a conductor to form a **circuit**. If the circuit is broken in some way so that the current flows directly to earth rather than to a piece of equipment, a **short circuit** is made. As the resistance is greatly reduced but the voltage remains the same, a rapid increase in current occurs which could cause significant problems if suitable protection were not available.

Earthing

The electricity supply company has one of its conductors solidly connected to the earth and every circuit supplied by the company must have one of its conductors connected to earth. This means that if there is a fault, such as a break in the circuit, the current, known as the earth fault current, will return directly to earth, which forms the circuit of least resistance, thus maintaining the supply circuit. This process is known as **earthing**. Other devices, such as fuses and residual current devices (RCDs), which will be described later, will also be needed within the circuit to interrupt the current flow to earth so as to protect people from electric shock and equipment from overheating. Good and effective earthing is absolutely essential and must be connected and checked by a competent person. Where a direct contact with earth is not possible, for example in a motor car, a common voltage reference point is used, such as the vehicle chassis.

Where other potential metallic conductors exist near to electrical conductors in a building, they must be connected to the main earth terminal to ensure **equipotential bonding** of all conductors to earth. This applies to gas, water and central heating pipes and other devices such as lightning protection systems. **Supplementary bonding** is required in bathrooms and kitchens where, for example, metal sinks and other metallic equipment surfaces are present. This involves the connection of a conductor from the sink to a water supply pipe which has been earthed by equipotential bonding. There have been several fatalities due to electric shocks from 'live' service pipes or kitchen sinks.

12.2.2 Some definitions

Certain terms are frequently used with reference to electricity and the more common ones are defined here.

Low voltage – A voltage normally not exceeding 600V ac between conductors and earth or 1000V ac between phases. Mains voltage falls into this category.

High voltage – A voltage normally exceeding 600V ac between conductors and earth or 1000V ac between phases.

Mains voltage – The common voltage available in domestic premises and many workplaces and normally taken from three pin socket points. In the UK, it is distributed by the national grid and is usually supplied between 220 and 240V ac and at 50 cycles/s.

Maintenance – A combination of any actions carried out to retain an item of electrical equipment in, or restore it to, an acceptable and safe condition.

Testing – A measurement carried out to monitor the conditions of an item of electrical equipment without physically altering the construction of the item or the electrical system to which it is connected.

Inspection – A maintenance action involving the careful scrutiny of an item of electrical equipment, using, if necessary, all the senses to detect any failure to meet an acceptable and safe condition. An inspection does not include any dismantling of the item of equipment.

Examination – An inspection together with the possible partial dismantling of an item of electrical equipment, including measurement and non-destructive testing as required, in order to arrive at a reliable conclusion as to its condition and safety.

Isolation – Involves cutting off the electrical supply from all or a discrete section of the installation by separating the installation or section from every source of electrical energy. This is the normal practice so as to ensure the safety of persons working on or in the vicinity of electrical components which are normally live and where there is a risk of direct contact with live electricity.

Competent electrical person – A person possessing sufficient electrical knowledge and experience to avoid the risks to health and safety associated with electrical equipment and electricity in general.

12.3 Electrical hazards and injuries

Electricity is a safe, clean and quiet method of transmitting energy. However, this apparently benign source of energy when accidentally brought into contact with conducting material, such as people, animals or metals, permits releases of energy which may result in serious damage or loss of life. Constant awareness is necessary to avoid and prevent danger from accidental releases of electrical energy.

The principal hazards associated with electricity are:

➤ electric shock
➤ electric burns
➤ electrical fires and explosions
➤ arcing
➤ secondary hazards.

The use of portable electrical equipment can lead to a higher likelihood of these hazards ocurring.

12.3.1 Electric shock and burns

Electric shock is the convulsive reaction by the human body to the flow of electric current through it. This sense of shock is accompanied by pain and, in more severe cases, by burning. The shock can be produced by low voltages, high voltages or lightning. Most incidents of electric shock occur when the person becomes the route to earth for a live conductor. The effect of electric shock and the resultant severity of injury depend upon the size of the electric current passing through the body which, in turn, depends on the voltage and the electrical resistance of the skin. If the skin is wet, a shock from mains voltage (220/240V) could well be fatal. The effect of shock is very dependent on conditions at the time but it is always dangerous and must be avoided. Electric burns are usually more severe than those caused by heat, since they can penetrate deep into the tissues of the body.

The effect of electric current on the human body depends on its pathway through the body (e.g. hand to hand or hand to foot), the frequency of the current, the length of time of the shock and the size of the current. Current size is dependent on the duration of contact and the electrical resistance of body tissue. The electrical resistance of the body is greatest in the skin and is approximately 100 000 ohm; however, this may be reduced by a factor of 100 when the skin is wet. The body beneath the skin offers very little resistance to electricity due to its very high water content and, while the overall body resistance varies considerably between people and during the lifetime of each person, it averages at 1000 ohm. Skin that is wounded, bruised or damaged will considerably reduce human electrical resistance and work should not be undertaken on electrical equipment if damaged skin is unprotected.

An electric current of 1 mA is detectable by touch and one of 10 mA will cause muscle contraction which may prevent the person from being able to release the conductor, and if the chest is in the current path, respiratory movement may be prevented, causing asphyxia. Current passing through the chest may also cause fibrillation of the heart (vibration of the heart muscle) and disrupt the normal rhythm of the heart, though this is likely only within a particular range of currents. The shock can also cause the heart to stop completely (cardiac arrest) and this will lead to the cessation of breathing. Current passing through the respiratory centre of the brain may cause respiratory arrest that does not quickly respond to the breaking of the electrical contact. These effects on the heart and respiratory system can be caused by currents as low as 25 mA. It is not possible to be precise on the threshold current because it is dependent on the environmental conditions at the time, as well as the age, sex, body weight and health of the person.

Burns of the skin occur at the point of electrical contact due to the high resistance of skin. These burns may be deep, slow to heal and often leave permanent scars. Burns may also occur inside the body along the path of the electric current, causing damage to muscle tissue and blood cells. Burns associated with radiation and microwaves are dealt with in Chapter 15.

12.3.2 Treatment of electric shock and burns

There are many excellent posters available which illustrate a first-aid procedure for treating electric shock and such posters should be positioned close to electrical junction boxes or isolation switches (Figure 12.2). The recommended procedure for treating an unconscious person who has received a **low-voltage** electric shock is as follows:

1. On finding a person suffering from electric shock, raise the alarm by calling for help from colleagues (including a trained first aider).
2. Switch off the power if it is possible and/or the position of the emergency isolation switch is known.
3. Call for an ambulance.
4. If it is not possible to switch off the power, then push or pull the person away from the conductor using an object made from a good insulator, such as a wooden chair or broom. Remember to stand on dry insulating material, for example, a wooden pallet, rubber mat or wooden box. If these precautions are not taken, then the rescuer will also be electrocuted.
5. If the person is breathing, place him/her in the recovery position so that an open airway is maintained and the mouth can drain if necessary.
6. If the person is not breathing, apply mouth-to-mouth resuscitation and, in the absence of a pulse, chest compressions. When the person is breathing normally place them in the recovery position.

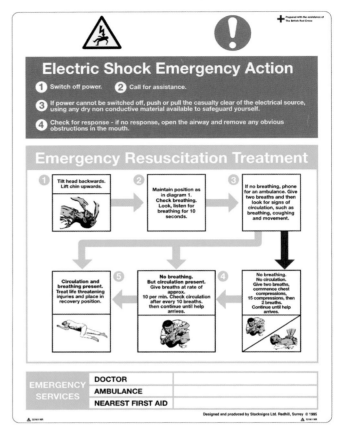

Figure 12.2 Typical electric shock treatment poster.

7. Treat any burns by placing a sterile dressing over the burn and secure with a bandage. Any loose skin or blisters should not be touched nor any lotions or ointments applied to the burn wound.
8. If the person regains consciousness, treat for normal shock.
9. Remain with the person until they are taken to a hospital or local surgery.

It is important to note that electrocution by high-voltage electricity is normally instantly fatal. On discovering a person who has been electrocuted by high-voltage electricity, the police and electricity supply company should be informed. If the person remains in contact with or within 18 m of the supply, then he/she should not be approached to within 18 m by others until the supply has been switched off and clearance has been given by the emergency services. High-voltage electricity can 'arc' over distances less than 18 m, thus electrocuting the would-be rescuer (Figure 12.3).

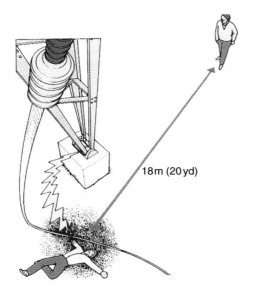

Figure 12.3 Keep 18 m clear on high-voltage lines.

Figure 12.4 Over 25% of fires are caused by electrical malfunction.

12.3.3 Electrical fires and explosions

Over 25% of all fires have a cause linked to a malfunction of either a piece of electrical equipment or wiring or both. Electrical fires are often caused by a lack of reasonable care in the maintenance and use of electrical installations and equipment. The electricity that provides heat and light and drives electric motors is capable of igniting insulating or other combustible material if the equipment is misused, is not adequate to carry the electrical load or is not properly installed and maintained. The most common causes of fire in electrical installations are short circuits, overheating of cables and equipment, the ignition of flammable gases and vapours and the ignition of combustible substances by static electrical discharges.

Short circuits happen, as mentioned earlier, if insulation becomes faulty, and an unintended flow of current between two conductors or between one conductor and earth occurs. The amount of the current depends, among other things, upon the voltage, the condition of the insulating material and the distance between the conductors. At first the current flow will be low, but as the fault develops the current will increase and the area surrounding the fault will heat up. In time, if the fault persists, a total breakdown of insulation will result and excessive current will flow through the fault. If the fuse fails to operate or is in excess of the recommended fuse rating, overheating will occur and a fire will result. A fire can also be caused if combustible material is in close proximity to the heated wire or hot sparks are

ejected. Short circuits are most likely to occur where electrical equipment or cables are susceptible to damage by water leaks or mechanical damage. Twisted or bent cables can also cause breakdowns in insulation materials.

Inspection covers and cable boxes are particular problem areas. Effective steps should be taken to prevent the entry of moisture as this will reduce or eliminate the risk. Covers can themselves be a problem especially in dusty areas where the dust can accumulate on flat insulating surfaces resulting in tracking between conductors at different voltages and a subsequent insulation failure. The interior of inspection panels should be kept clean and dust-free by using a suitable vacuum cleaner.

Overheating of cables and equipment will occur if they become overloaded. Electrical equipment and circuits are normally rated to carry a given safe current which will keep the temperature rise of the conductors in the circuit or appliance within permissible limits and avoid the possibility of fire. These safe currents define the maximum size of the fuse (the fuse rating) required for the appliance. A common cause of circuit overloading is the use of equipment and cables which are too small for the imposed electrical load. This is often caused by the addition of more and more equipment to the circuit, thus taking it beyond its original design specification. In offices, the overuse of multisocket unfused outlet adaptors can create overload problems (sometimes known as the Christmas tree effect). The more modern multiplugs are much safer as they lead to one fused plug and cannot be easily overloaded (see Figure 12.5). Another cause of overloading is mechanical breakdown or wear of an electric motor and the driven machinery. Motors must be maintained in good condition with particular attention paid to bearing surfaces. Fuses do not always provide total protection against the overloading of motors and, in some cases, severe heating may occur without the fuses being activated.

Loose cable connections are one of the most common causes of overheating and may be readily detected (as well as overloaded cables) by a thermal imaging survey (a technique which indicates the presence of hot spots). The bunching of cables together can also cause excessive heat to be developed within the inner cable leading to a fire risk. This can happen with cable extension reels, which have only been partially unwound, used for high-energy appliances like an electric heater.

Ventilation is necessary to maintain safe temperatures in most electrical equipment and overheating is liable to occur if ventilation is in any way obstructed or reduced. All electrical equipment must be kept free of any obstructions that restrict the free supply of air to the equipment and, in particular, to the ventilation apertures.

Most electrical equipment either sparks in normal operation or is liable to spark under fault conditions. Some electrical appliances such as electric heaters, are specifically designed to produce high temperatures. These circumstances create fire and explosion hazards, which demand very careful assessment in locations where processes capable of producing flammable concentrations of gas or vapour are used, or where flammable liquids are stored.

It is likely that many fires are caused by static electrical discharges. Static electricity can, in general, be eliminated by the careful design and selection of materials used in equipment and plant, and the materials used in products being manufactured. When it is impractical to avoid the generation of static electricity, a means of control must be devised. Where flammable materials are present, especially

if they are gases or dusts, then there is a great danger of fire and explosion, even if there is only a small discharge of static electricity. The control and prevention of static electricity is considered in more detail later.

The use of electrical equipment in potentially flammable atmospheres should be avoided as far as possible. However, there will be many cases where electrical equipment must be used and, in these cases, the standards for the construction of the equipment should comply with the Equipment and Protective Systems Intended for Use in Potentially Explosive Atmospheres Regulations, known as ATEX. Details on the classification or zoning of areas is contained in the Dangerous Substances and Explosive Atmospheres Regulations and ACOPs.

Before electrical equipment is installed in any location where flammable vapours or gases may be present, the area must be zoned in accordance with the Dangerous Substances and Explosive Atmosphere Regulations, and records of the zoned areas must be marked on building drawings and revised when any zoned area is changed. The installation and maintenance of electrical equipment in potentially flammable atmospheres is a specialized task. It must only be undertaken by electricians or instrument mechanics who are trained to ATEX standards.

In the case of a fire involving electrical equipment, the first action must be the isolation of the power supply so that the circuit is no longer live. This is achieved by switching off the power supply at the mains isolation switch or at another appropriate point in the system. Where it is not possible to switch off the current, the fire must be attacked in a way which will not cause additional danger. The use of a non-conducting extinguishing medium, such as carbon dioxide or powder, is necessary. After extinguishing such a fire, careful watch should be kept for renewed outbreaks until the fault has been rectified. Re-ignition is a particular problem when carbon dioxide extinguishers are used, although less equipment may be damaged than is the case when powder is used.

Finally, the chances of electrical fires occurring are considerably reduced if the original installation was undertaken by competent electricians working to recognized standards, such as the Institution of Electrical Engineers' Code of Practice. It is also important to have a system of regular testing and inspection in place so that any remedial maintenance can take place.

12.3.4 Electric arcing

A person who is standing on earth too close to a high-voltage conductor may suffer flash burns as a result of arc

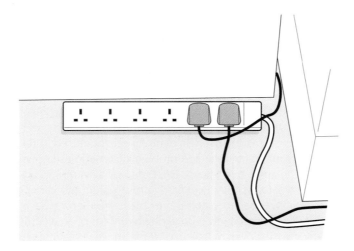

Figure 12.5 Modern multiplug.

formation. Such burns may be extensive and lower the resistance of the skin so that electric shock may add to the ill effects. Electric arc faults can cause temporary blindness by burning the retina of the eye and this may lead to additional secondary hazards. The quantity of electrical energy is as important as the size of the voltage since the voltage will determine the distance over which the arc will travel. The risk of arcing can be reduced by the insulation of live conductors.

Strong electromagnetic fields induce surface charges on people. If these charges accumulate, skin sensation is affected and spark discharges to earth may cause localized pain or bruising. Whether prolonged exposure to strong fields has any other significant effects on health has not been proved. However, the action of an implanted cardiac pacemaker may be disturbed by the close proximity of its wearer to a powerful electromagnetic field. The health effects of arcing and other non-ionizing radiation are covered in Chapter 15.

12.3.5 Static electricity

Static electricity is produced by the build-up of electrons on weak electrical conductors or insulating materials. These materials may be gaseous, liquid or solid and may include flammable liquids, powders, plastic films and granules. Plastics have a high resistance that enables them to retain static charges for long periods of time. The generation of static may be caused by the rapid separation of highly insulated materials by friction or by transfer from one highly charged material to another in an electric field by induction (see Figure 12.6).

Figure 12.6 Prevention of static discharge – container connected to earthed drum.

A static electric shock, perhaps caused by closing a door with a metallic handle, can produce a voltage greater than 10 000 V. Since the current flows for a very short period of time, there is seldom any serious harm to an individual. However, discharges of static electricity may be sufficient to cause serious electric shock and are always a potential source of ignition when flammable liquid, dusts or powders are present. This is a particular problem in the parts of the printing industry where solvent-based inks are used on high-speed web presses. Flour dust in a mill has also been ignited by static electricity.

Static electricity may build up on both materials and people. When a charged person approaches flammable gases or vapours and a spark ignites the substance, the resulting explosion or fire often causes serious injury. In these situations, effective static control systems must be used.

Lightning strikes are a natural form of static electricity and result in large amounts of electrical energy being dissipated in a short time in a limited space with a varying degree of damage. The current produced in the vast majority of strikes exceeds 3000 A over a short period of time. Before a strike, the electrical potential between the cloud and earth might be about 100 million volts and the energy released at its peak might be about 100 million watts per metre of strike.

The need to provide lightning protection depends on a number of factors, which include:

- the risk of a strike occurring;
- the number of people likely to be affected;
- the location of the structure and the nearness of other tall structures in the vicinity;
- the type of construction, including the materials used;
- the contents of the structure or building (including any flammable substances);
- the value of the building and its contents.

Expert advice will be required from a specialist company in lightning protection, especially when flammable substances are involved. Lightning strikes can also cause complete destruction and/or significant disruption of electronic equipment.

12.3.6 Portable electrical equipment

Portable and transportable electrical equipment is defined by the Health and Safety Executive as 'not part of a fixed installation but may be connected to a fixed installation by means of a flexible cable and either a socket and plug or a spur box or similar means'. It may be hand-held or hand-operated while connected to the supply, or is intended or likely to be moved while connected to the supply. The auxiliary equipment, such as extension leads, plugs and sockets, used with portable tools, is also classified as portable equipment. The term 'portable' means both portable and transportable (Figure 12.7).

Almost 25% of all reportable electrical accidents involve portable electrical equipment (known as portable appliances). While most of these accidents were caused by electric shock, over 2000 fires each year are started by faulty cables used by portable appliances, caused by a lack of effective maintenance. Portable electrical tools often present a high risk of injury, which is frequently caused by the conditions under which they are used. These conditions include the use of defective or unsuitable equipment and, indeed, the misuse of equipment. There must be a system to record the inspection, maintenance and repair of these tools.

Where plugs and sockets are used for portable tools, sufficient sockets must be provided for all the equipment and adaptors should not be used. Many accidents are caused by faulty flexible cables, extension leads, plugs and sockets, particularly when these items become damp or worn. Accidents often occur when contact is made with some part of the tool which has become live (probably at mains voltage), while the user is standing on, or in contact with, an earthed conducting surface. If the electricity supply

is at more than 50V ac, then the electric shock that a person may receive from such defective equipment is potentially lethal. In adverse environmental conditions, such as humid or damp atmospheres, even lower voltages can be dangerous. Portable electrical equipment should not be used in flammable atmospheres if it can be avoided and it must also comply with any standard relevant to the particular environment. Air-operated equipment should also be used as an alternative whenever it is practiable.

Some portable equipment requires substantial power to operate and may require voltages higher than those usually used for portable tools, so that the current is kept down to reasonable levels. In these cases, power leads with a separate earth conductor and earth screen must be used. Earth leakage relays and earth monitoring equipment must also be used, together with substantial plugs and sockets designed for this type of system.

Electrical equipment is safe when properly selected, used and maintained. It is important, however, that the environmental conditions are always carefully considered. The hazards associated with portable appliances increase with the frequency of use and the harshness of the environment (construction sites are often particularly hazardous in this

(a)

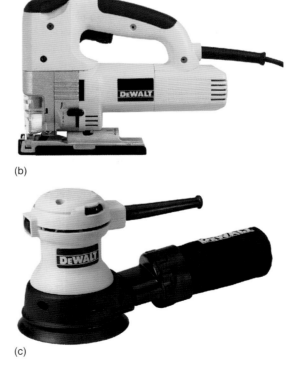

(b)

(c)

Figure 12.7 Portable hand-held electric power tools.

respect). These factors must be considered when inspection, testing and maintenance procedures are being developed.

12.3.7 Secondary hazards

It is important to note that there are other hazards associated with portable electrical appliances, such as abrasion and impact, noise and vibration. Trailing leads used for portable equipment and raised socket points offer serious trip hazards and both should be used with great care near pedestrian walkways. Power drives from electric motors should always be guarded against entanglement hazards.

Secondary hazards are those additional hazards which present themselves as a result of an electrical hazard. It is very important that these hazards are considered during a risk assessment. An electric shock could lead to a fall from height if the shock occurred on a scaffold or it could lead to a collision with a vehicle if the victim collapsed on to a roadway.

Similarly, an electrical fire could lead to all the associated fire hazards outlined in Chapter 13 (e.g. suffocation, burns and structural collapse) and electrical burns can easily lead to infections.

12.4 General control measures for electrical hazards

The principal control measures for electrical hazards are contained in the statutory precautionary requirements covered by the Electricity at Work Regulations, the main provisions of which are outlined in Chapter 17. They are applicable to all electrical equipment and systems found at the workplace and impose duties on employers, employees and the self-employed.

The Regulations cover the following topics:

- the design, construction and maintenance of electrical systems, work activities and protective equipment;
- the strength and capability of electrical equipment;
- the protection of equipment against adverse and hazardous environments;
- the insulation, protection and placing of electrical conductors;
- the earthing of conductors and other suitable precautions;
- the integrity of referenced conductors;
- the suitability of joints and connections used in electrical systems;
- means for protection from excess current;
- means for cutting off the supply and for isolation;

- the precautions to be taken for work on equipment made dead;
- working on or near live conductors;
- adequate working space, access and lighting;
- the competence requirements for persons working on electrical equipment to prevent danger and injury.

Detailed safety standards for designers and installers of electrical systems and equipment are given a code of practice, published by the Institute of Electrical Engineers, known as the IEE Wiring Regulations. While these Regulations are not legally binding, they are recognized as a code of good practice and widely used as an industry standard.

The risk of injury and damage inherent in the use of electricity can only be controlled effectively by the introduction of employee training, safe operating procedures (safe systems of work) and guidance to cover specific tasks.

Training is required at all levels of the organization ranging from simple on-the-job instruction to apprenticeship for electrical technicians and supervisory courses for experienced electrical engineers. First-aid training related to the need for cardiovascular resuscitation and treatment of electric burns should be available to all people working on electrical equipment and their supervisors.

A **management system** should be in place to ensure that the electrical systems are installed, operated and maintained in a safe manner. All managers should be responsible for the provision of adequate resources of people, material and advice to ensure that the safety of electrical systems under their control is satisfactory and that **safe systems of work** are in place for all electrical equipment. (Chapter 6 gives more information on both safe systems of work and permits to work.)

For small factories and office or shop premises where the system voltages are normally at mains voltage, it may be necessary for an external competent person to be available to offer the necessary advice. Managers must set up a high-voltage permit-to-work system for all work at and above 600 V. The system should be appropriate to the extent of the electrical system involved. Consideration should also be given to the introduction of a permit system for voltages under 600 V when appropriate and for all work on live conductors.

The additional control measures that should be taken when working with electricity or using electrical equipment are summarized by the following topics:

- the selection of suitable equipment;
- the use of protective systems;
- inspection and maintenance strategies.

These three groups of measures will be discussed in detail.

12.5 The selection and suitability of equipment

Many factors which affect the selection of suitable electrical equipment, such as flammable, explosive and damp atmospheres and adverse weather conditions, have already been considered. Other issues include high or low temperatures, dirty or corrosive processes or problems associated with vegetation or animals (e.g. tree roots touching and displacing underground power cables, farm animals urinating near power supply lines and rats gnawing through cables). Temperature extremes will affect, for example, the lubrication of motor bearings and corrosive atmospheres can lead to the breakdown of insulating materials. The equipment selected must be suitable for the task demanded or either it will become overloaded or running costs will be too high.

The equipment should be installed to a recognized standard and capable of being isolated in the event of an emergency. It is also important that the equipment is effectively and safely earthed. Electric supply failures may affect process plant and equipment. These are certain to happen at some time and the design of the installation should be such that a safe shutdown can be achieved in the event of a total mains failure. This may require the use of a battery-backed shutdown system or emergency standby electric generators (assuming that this is cost-effective).

Finally, it is important to stress that electrical equipment must only be used within the rating performance given by the manufacturer and any accompanying instructions from the manufacturer or supplier must be carefully followed.

12.5.1 The advantages and limitations of protective systems

There are several different types of protective system and technique that may be used to protect people, plant and premises from electrical hazards, some of which, for example earthing, have already been considered earlier in this chapter. However, only the more common types of protection will be considered here.

Fuse

A fuse will provide protection against faults and continuous marginal current overloads. It is basically a thin strip of conducting wire which will melt when an excess of the rated current passes through it, thus breaking the circuit. A fuse rated at 13 A will melt when a current in excess of 13 A passes through the fuse thus stopping the flow of current. A **circuit breaker** throws a switch off when excess current passes and is similar in action to a fuse (Figure 12.8). Protection against overload is provided by fuses which detect a continuous marginal excess flow of current and energy. This overcurrent protection is arranged to operate before damage occurs, either to the supply system or to the load which may be a motor or heater. When providing protection against overload, consideration needs to be made as to whether tripping the circuit could give rise to an even more dangerous situation, such as with fire-fighting equipment.

The prime objective of a fuse is to protect equipment or an installation from overheating and becoming a fire hazard. It is not an effective protection against electric shock due to the time that it takes to cut the current flow.

(a)

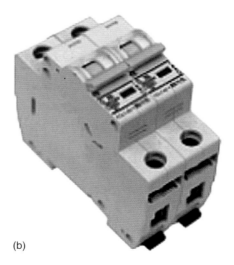

(b)

Figure 12.8 (a) Typical volt fuses. (b) Mini circuit breaker.

The examination of fuses is a vital part of an inspection programme to ensure that the correct size or rating is fitted at all times.

Insulation

Insulation is used to protect people from electric shock, the short circuiting of live conductors and the dangers associated with fire and explosions. Insulation is achieved by covering the conductor with an insulating material. Insulation is often accompanied by the enclosure of the live conductors so that they are out of reach to people. A breakdown in insulation can cause electric shock, fire, explosion or instrument damage.

Isolation

The isolation of an electrical circuit involves more than 'switching off' the current in that the circuit is made dead and cannot be accidentally re-energized. It, therefore, creates a barrier between the equipment and the electrical supply which only an authorized person should be able to remove. When it is intended to carry out work, such as mechanical maintenance or a cleaning operation on plant or machinery, isolation of electrical equipment will be required to ensure safety during the work process. Isolators should always be locked off when work is to be done on electrical equipment.

Before earthing or working on an isolated circuit, checks must be made to ensure that the circuit is dead and that the isolation switch is 'locked off' and clearly labelled [Figure 12.9(c) and 12.9(d)].

Reduced low-voltage systems

When the working conditions are relatively severe, either due to wet conditions or heavy and frequent usage of equipment, reduced voltage systems should be used.

All portable tools used on construction sites, in vehicle washing stations or near swimming pools should operate on 110V or less, preferably with a centre tapped to earth at 55V. This means that while the full 110V are available to power the tool, only 55V are available should the worker suffer an electric shock. At this level of voltage, the effect of any electric shock should not be severe. For lighting, even lower voltages can be used and are even safer. Another way to reduce the voltage is to use **battery (cordless) operated hand tools**.

Residual current devices

If electrical equipment must operate at mains voltage, the best form of protection against electric shock is the Residual Current Device (RCD). RCDs, also known as earth leakage circuit breakers, monitor and compare the current flowing in the live and neutral conductors supplying the protected equipment. Such devices are very sensitive to differences of current between the live and neutral power lines and will cut the supply to the equipment in a very short period of time when a difference of only a few milliamperes occurs. It is the speed of the reaction which offers the protection against electric shock.

RCDs can be used to protect installations against fire, as they will interrupt the electrical supply before sufficient energy to start a fire has accumulated. For protection against electric shock, the RCD must have a rated residual current of 30mA or less and an operating time of 40 milliseconds or less at a residual current of 250mA. The protected equipment must be properly protected by insulation and enclosure in addition to the RCD. The RCD will not prevent shock or limit the current resulting from an accidental contact, but it will ensure that the duration of the shock is limited to the time taken for the RCD to operate. The RCD has a test button which should be tested frequently to ensure that it is working properly [Figure 12.9(b)].

Double insulation

To remove the need for earthing on some portable power tools, double insulation is used. Double insulation employs two independent layers of insulation over the live conductors, each layer alone being adequate to insulate the electrical equipment safely. As such tools are not protected by an earth, they must be inspected and maintained regularly and must be discarded if damaged.

The below figure shows the symbol which is marked on double-insulated portable power tools.

Double insulation sign.

12.6 Inspection and maintenance strategies

12.6.1 Maintenance strategies

Regulation 4(2) of the Electricity at Work Regulations requires that 'as may be necessary to prevent danger, all systems shall be maintained so as to prevent so far as is

(a) 110 volt transportable transformer from 240 volt supply.

(b) 240 volt socket with RCD device built in.

(c) Lockable electrical cabinets.

(d) Electrical cabinet with multihasp and lock off padlocks.

Figure 12.9 (a) Transformer, (b) RCD and (c) & (d) isolators.

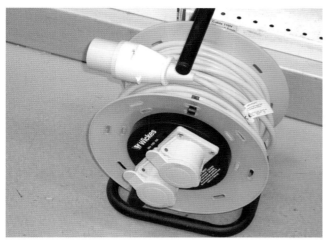

(a) Reduced voltage normally 110 volts.

(b) 240 volts mains with cut out.

Figure 12.10 (a) & (b) Multiplugs, extension lead and special plugs and sockets.

reasonably practicable, such danger'. Regular maintenance is, therefore, required to ensure that a serious risk of injury or fire does not result from installed electrical equipment. Maintenance standards should be set as high as possible so that a more reliable and safe electrical system will result. Inspection and maintenance periods should be determined by reference to the recommendations of the manufacturer, consideration of the operating conditions and the environment in which equipment is located. The importance of equipment within the plant, from the plant safety and operational viewpoint, will also have a bearing on inspection and maintenance periods. The mechanical safety of driven machinery is vital and the electrical maintenance and isolation of the electrically powered drives is an essential part of that safety.

The particular areas of interest for inspection and maintenance are:

➤ the cleanliness of insulator and conductor surfaces;
➤ the mechanical and electrical integrity of all joints and connections;
➤ the integrity of mechanical mechanisms, such as switches and relays;
➤ the calibration, condition and operation of all protection equipment, such as circuit breakers, RCDs and switches.

Safe operating procedures for the isolation of plant and machinery during both electrical and mechanical maintenance must be prepared and followed. All electrical isolators must, wherever possible, be fitted with mechanisms which can be locked in the 'open/off' position and there must be a procedure to allow fuse withdrawal wherever isolators are not fitted.

Working on live equipment with voltages in excess of 110 V must not be permitted except where fault-finding or testing measurements cannot be done in any other way. Reasons such as the inconvenience of halting production are not acceptable.

Part of the maintenance process should include an appropriate system of visual inspection. By concentrating on a simple, inexpensive system of looking for visible signs of damage or faults, many of the electrical risks can be controlled, although more systematic testing may be necessary at a later stage.

All fixed electrical installations should be inspected and tested periodically by a competent person, such as a member of the National Inspection Council for Electrical Installation Contracting (NICEIC).

12.6.2 Inspection strategies

Regular inspection of electrical equipment is an essential component of any preventative maintenance programme and, therefore, regular inspection is required under Regulation 4(2) of the Electricity at Work Regulations, which has been quoted previously. Any strategy for the inspection of electrical equipment, particularly portable appliances, should involve the following considerations:

➤ a means of identifying the equipment to be tested;
➤ the number and type of appliances to be tested;

- the competence of those who will undertake the testing (whether in-house or brought in);
- the legal requirements for portable appliance testing (PAT) and other electrical equipment testing and the guidance available;
- organizational duties of those with responsibilities for PAT and other electrical equipment testing;
- test equipment selection and re-calibration;
- the development of a recording, monitoring and review system;
- the development of any training requirements resulting from the test programme.

12.7 Portable electrical appliances testing

Portable appliances should be subject to three levels of inspection – a user check, a formal visual inspection and a combined inspection and test.

12.7.1 User checks

When any portable electrical hand tool, appliance, extension lead or similar item of equipment is taken into use, at least once each week or, in the case of heavy work, before each shift, the following visual check and associated questions should be asked:

- Is there a recent PAT label attached to the equipment?
- Are any bare wires visible?
- Is the cable covering undamaged and free from cuts and abrasions (apart from light scuffing)?
- Is the cable too long or too short? (Does it present a trip hazard?)
- Is the plug in good condition (for example the casing is not cracked and the pins are not bent) (see Figure 12.11)?
- Are there no taped or other non-standard joints in the cable?
- Is the outer covering (sheath) of the cable gripped where it enters the plug or the equipment? (The coloured insulation of the internal wires should not be visible.)
- Is the outer case of the equipment undamaged or loose and are all screws in place?
- Are there any overheating or burn marks on the plug, cable, sockets or the equipment?
- Are the trip devices (RCDs) working effectively (by pressing the 'test' button)?

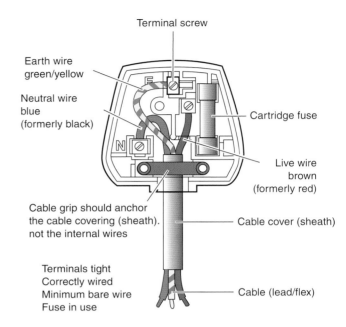

Figure 12.11 UK standard 3-pin plug wiring.

12.7.2 Formal visual inspections and tests

There should be a **formal visual inspection** routinely carried out on all portable electrical appliances. Faulty equipment should be taken out of service as soon as the damage is noticed. At this inspection the plug cover (if not moulded) should be removed to check that the correct fuse is included, but the equipment itself should not be taken apart. This work can normally be carried out by a trained person who has sufficient information and knowledge.

Some faults, such as the loss of earth continuity due to wires breaking or loosening within the equipment, the breakdown of insulation and internal contamination (e.g. dust containing metal particles may cause short circuiting if it gets inside the tool) will not be spotted by visual inspections. To identify these problems, a programme of testing and inspection will be necessary.

This formal **combined testing and inspection** should be carried out by a competent person when there is reason to suspect the equipment may be faulty, damaged or contaminated, but this cannot be confirmed by visual inspection or after any repair, modification or similar work to the equipment, which could have affected its electrical safety. The competent person could be a person who has been specifically trained to carry out the testing of portable appliances using a simple 'pass/fail' type of tester. When more sophisticated tests are required, a competent person with the necessary technical electrical knowledge and experience would be needed. The inspection and testing should normally include the following checks:

- that the polarity is correct;
- that the correct fuses are being used;
- that all cables and cores are effectively terminated;
- that the equipment is suitable for its environment.

Testing need not be expensive in many low-risk premises like shops and offices, if an employee is trained to perform the tests and appropriate equipment is purchased.

12.7.3 Frequency of inspection and testing

The frequency of inspection and testing should be based on a risk assessment which is related to the usage, type and operational environment of the equipment. The harsher the working environment is, the more frequent the period of inspection. Thus tools used on a construction site should be tested much more frequently than a visual display unit which is never moved from a desk. Manufacturers or suppliers may recommend a suitable testing period. Table 12.1 lists the suggested intervals for inspection and testing derived from HSE publications *Maintaining Portable and Transportable Electrical Equipment* (HSG107 and INDG236 and 237).

It is very important to stress that there is no 'correct' interval for testing – it depends on the frequency of usage, type of equipment, and how and where it is used. A few years ago, a young trainee was badly scalded by a boiling kettle of water which exploded while in use. On investigation, an inspection report indicated that the kettle had been checked by a competent person and passed just a few weeks before the accident. Further investigation showed that this kettle was the only method of boiling water on the premises and was in use continuously for 24 hours each day. It was therefore unsuitable for the purpose and a plumbed-in continuous-use hot water heater would have been far more suitable.

12.7.4 Records of inspection and testing

Schedules which give details of the inspection and maintenance periods and the respective programmes must be kept together with records of the inspection findings and the work done during maintenance. Records must include both individual items of equipment and a description of the complete system or section of the system. They should always be kept up-to-date and with an audit procedure

in place to monitor the records and any required actions. The records do not have to be paper based but could be stored electronically on a computer. It is good practice to label the piece of equipment with the date of the last combined test and inspection.

The effectiveness of the equipment maintenance programme may be monitored and reviewed if a record of tests is kept. It can also be used as an inventory of portable appliances and help to regulate the use of unauthorized appliances. The record will enable any adverse trends to be monitored and to check that suitable equipment has been selected. It may also give an indication as to whether the equipment is being used correctly.

12.7.5 Advantages and limitations of PAT

The advantages of PAT include:

- an earlier recognition of potentially serious equipment faults, such as poor earthing, frayed and damaged cables and cracked plugs;
- discovery of incorrect or inappropriate electrical supply and/or equipment;
- discovery of incorrect fuses being used;
- a reduction in the number of electrical accidents;
- monitoring the misuse of portable appliances;
- equipment selection procedures checkable;
- an increased awareness of the hazards associated with electricity;
- a more regular maintenance regime should result.

The limitations of PAT include:

- some fixed equipment is tested too often leading to excessive costs;
- some unauthorized portable equipment, such as personal kettles, are never tested as there is no record of them;
- equipment may be misused or overused between tests due to a lack of understanding of the meaning of the test results;
- all faults, including trivial ones, are included on the action list, so the list becomes very long and the more significant faults are forgotten or overlooked;
- the level of competence of the tester can be too low;
- the testing equipment has not been properly calibrated and/or checked before testing takes place.

Most of the limitations may be addressed and the reduction in electrical accidents and injuries enables the advantages of PAT to greatly outweigh the limitations.

Table 12.1 Suggested intervals for portable appliance inspection and testing			
Type of business/equipment	User checks	Formal visual inspection	Combined inspection and electrical tests
Equipment hire	Yes	Before issue and after return	Before issue
Construction	Yes	Before initial use and then every month	3 months
Industrial	Yes	Before initial use and then every 3 months	6–12 months
Hotels and offices, low-risk environments			
Battery operated (less than 20V)	No	No	No
Extra low voltage (less than 50V ac), for example telephone equipment, low-voltage desk lights	No	No	No
Computers/photocopiers/fax machines	No	Yes 2–4 years	No if double insulated, otherwise up to 5 years
Double-insulated equipment: not hand-held. Moved occasionally, for example fans, table lamps, slide projectors	No	Yes 2–4 years	No
Double-insulated equipment: hand-held, for example some floor cleaners, some kitchen equipment and irons	Yes	Yes 6 months – 1 year	Yes 1–2 years
Earthed equipment (class 1): for example electric kettles, some floor cleaners, portable electric heaters	Yes	Yes 6 months – 1 year	
Cables (leads) and plug connected to above. Extension leads (mains voltage)	Yes	Yes 6 months – 4 years depending on the type of equipment it is connected to	Yes 1–5 years depending on the type of equipment it is connected to

Source: Derived from HSE. Note: Operational experience may demonstrate that the above intervals can be reviewed.

12.8 Sources of reference

Electricity at Work – Safe Working Practices HSG85, HSE Books 2003 ISBN 978 0 7176 2164 4.

Maintaining Portable and Transportable Electrical Equipment HSG107, HSE Books 2004 ISBN 978 0 7176 2805 6.

Relevant statutory provisions

The Electricity at Work Regulations 1989.
The Health and Safety (First-Aid) Regulations 1981.

12.9 Practice NEBOSH questions for Chapter 12

1. In relation to electrical safety, **explain** the meaning of the following terms:
   (i) 'isolation'
   (ii) 'earthing'
   (iii) 'reduced low voltage'
   (iv) 'overcurrent protection'.

2. (i) With reference to an electric circuit, state the relationship between voltage, current and resistance.
   (ii) **Outline** the factors that could affect the severity of injury from an electric shock received from a defective hand-held tool.

3. (i) **Outline** the dangers associated with electricity.
   (ii) **Outline** the emergency action to take if a person suffers a severe electric shock.

4. (i) **Outline** the effects on the human body from a severe electric shock.
   (ii) **Describe** how earthing can reduce the risk of receiving an electric shock.
   (iii) **Outline FOUR** factors that may affect the severity of injury from contact with electricity.

5. In relation to the use of electrical cables and plugs in the workplace:
   (i) **identify FOUR** examples of faults and bad practices that could contribute to electrical accidents
   (ii) **outline** the corresponding precautions that should be taken for **EACH** of the examples identified in (i).

6. **Outline** a range of checks that should be made to ensure electrical safety in an office environment.

7. **Outline** practical measures to reduce the risk of injury from electricity when using portable electric tools.

8. When using a portable electrical grinder to cut a support beam at height, the operator accidentally makes direct contact with the supply cable.
   (i) **Identify FOUR** possible outcomes associated with this action.
   (ii) **Identify FOUR** protective devices/systems that could minimize the risk to the operator in these circumstances.

9. A joiner has received an electric shock from a hand-held 230V drill while fitting floorboards to an upstairs room of a new property. The drill is 5 years old but has not been tested during this time. The injury to the joiner was fortunately not serious.
   (i) **Identify** the factors that may have limited the severity of injury on this occasion.
   (ii) **Outline** the physical effects on the body that such contact with electricity could have caused under different circumstances.
   (iii) **Outline** ways of minimizing the risk of serious injury from electricity when using a portable electric drill.
   (iv) **Describe** the types of inspection and/or test to which the drill should have been subjected, identifying the particular features that should be checked for **EACH** type and the factors that might affect the frequencies required.

10. **Outline** measures that should be taken to minimize the risk of fire from electrical equipment.

11. **List** the items that should be included on a checklist for the routine visual inspection of portable electrical appliances.

12. Giving a suitable example in **EACH** case, **identify** particular conditions of a working environment that may increase the risks from the use of portable electrical equipment.

13. With respect to the use of portable electrical appliances in the workplace, **identify EIGHT** examples of faults and bad practices that could contribute to electrical accidents.

14. (i) **Outline** the **THREE** levels of inspection that should be included in a maintenance and inspection strategy for portable electrical appliances.
   (ii) Identify the reasons for keeping centralized records of the results of portable appliance testing within an organization.

Fire hazards and control

13

Figure 13.1 Fire is still a major risk in many workplaces.

13.1 Introduction

This chapter covers fire prevention in the workplace and how to ensure that people are properly protected if fire does occur. Each year UK fire and rescue services attend over 35 000 fires at work in which about 30 people are killed and over 2500 are injured. Fire and explosions at work account for about 2% of the major injuries reported under Reporting of Injuries, Diseases and Dangerous Occurrences Regulations (RIDDOR 1995) (Figure 13.1).

The financial costs associated with serious fires are very high including, in many cases (believed to be over 40%), the failure to start up business again. Never underestimate the potential of any fire. What may appear to be a small fire in a waste bin, if not dealt with, can quickly spread through a building or a structure. The Bradford City Football ground in 1985 or King's Cross Underground station in 1987 are examples of where small fires quickly became raging infernos, resulting in many deaths and serious injuries. The Buncefield fuel storage depot fire in December 2005 is a further recent example of the destruction which a large fire and explosion can cause.

Since the introduction of the Fire Services Act of 1947, the fire authorities have had the responsibility for fighting fires in all types of premises. In 1971, the Fire Precautions Act gave the fire authorities control over certain fire procedures, means of escape and basic fire protection equipment through the drawing up and issuing of Fire Certificates in certain categories of building. The Fire Certification was mainly introduced to combat a number of serious industrial fires that had occurred, with a needless loss of life, where simple well-planned protection would have allowed people to escape unhurt.

Following a Government review in the 1990s, the Fire Precautions (Workplace) Regulations 1999 came into force in December of that year. These Regulations were made under the Health and Safety at Work (HSW) Act. They were amended in 1999 so as to apply to a wider range of premises including those already subject to the Fire Precautions Act 1971. In many ways these Regulations established the principles of fire risk assessment which would underpin a reformed legislative framework for fire safety. There remained a difference of opinion within Government, as to the right home for fire legislation. The Home Office believed that process fire issues should remain with the HSW Act, while general fire safety should have a different legislative vehicle.

In 2000 the Fire Safety Advisory Board was established to reform the fire legislation to simplify, rationalize and consolidate existing legislation. It would provide for a risk-based approach to fire safety allowing more efficient, effective enforcement by the fire and rescue service and other enforcing authorities.

The Regulatory Reform (Fire Safety) Order 2005 (RRFSO) SI No 2005 1541 was made on 7 June 2005. The Order came into force on the delayed date of 1 October 2006. A summary of the Order has been included in Chapter 17. The Fire Precautions Act 1971 was repealed and the Fire Precautions (Workplace) Regulations 1997 were revoked by the Order. Since the CDM 2007 Regulations were approved, the RRFSO has been amended, as the fire sections within the Construction (Health, Safety and Welfare) Regulations 1997 (now revoked) have been transferred to the CDM 2007.

The RRFSO reforms the law relating to fire safety in non-domestic premises. The main emphasis of the changes is to move towards fire prevention. Fire Certificates under the Fire Precautions Act 1971 are abolished by the Order and cease to have legal status. The RRFSO replaces fire certification with a general duty to ensure, so far as is reasonably practicable, the safety of employees.

There is a duty to carry out fire risk assessments that should consider the safety of both employees and non-employees.

The Order imposes a number of specific duties in relation to the fire precautions to be taken. The Order provides for the enforcement of the Order, appeals, offences and connected matters. The Order also gives effect in England and Wales to parts of a number of EC Directives including the Framework, Workplace, Chemical Agents and Explosive Atmospheres Directives.

13.2 The Regulatory Reform (Fire Safety) Order (RRFSO) – requirements

13.2.1 Outline of RRFSO

Part 1 General – The RRFSO applies to all non-domestic premises other than those listed in Article 6. The main duty-holder is the 'responsible person' in relation to the premises, defined in Article 3. The duties on the responsible person are extended to any person who has, to any extent, control of the premises to the extent of their control (Article 5).

Part 2 imposes duties on the responsible person in relation to fire safety in premises.

Part 3 provides for enforcement.

Part 4 provides for offences and appeals.

Part 5 provides for miscellaneous matters including fire-fighters' switches for luminous tube signs, maintenance measures provided to ensure the safety of fire-fighters, civil liability for breach of statutory duty by an employer, special requirements for licensed premises and consultation by other authorities.

Schedule 1 sets out the matters to be taken into account in carrying out a risk assessment (Parts 1 and 2), the general principles to be applied in implementing fire safety measures (Part 3) and the special measures to be taken in relation to dangerous substances (Part 4).

The Order is mainly enforced by the fire and rescue authorities.

13.2.2 PART 1 General

Meanings

The Order defines a responsible person as the person who is in control of the premises – this may be the owner or somebody else.

The meaning of general fire precautions is set out in the Order, which covers:

➤ reduction of fire risks and fire spread;
➤ means of escape;
➤ keeping means of escape available for use;
➤ fire-fighting;
➤ fire detection and fire warning;
➤ action to be taken in the event of fire;
➤ instruction and training of employees.

But they do not cover **process-related fire precautions**. These include the use of plant or machinery or the use or storage of any dangerous substances. Process-related fire precautions still come under either the Health and Safety Executive (HSE) or local authority, and are covered by the general duties imposed by the HSW Act 1974 or other specific Regulations like the Dangerous Substances and Explosive Atmosphere Regulations (DSEAR).

Duties

Duties are placed on a 'responsible person' who is:

➤ the employer in a workplace, to the extent they have control;
➤ any other person who has control of the premises; or
➤ the owner of the premises.

The obligations of a particular responsible person relate to matters within their control. It is therefore advisable that arrangements between responsible persons, where there may be more than one (such as contracts or tenancy agreements) should clarify the division of responsibilities.

Premises covered

The RRFSO does not cover domestic premises, offshore installations, a ship (normal shipboard activities under a master), remote fields, woods or other land forming part of an agricultural or forestry operation (it does cover the buildings), means of transport, a mine (it does cover the buildings at the surface) and a borehole site.

Other alternative provisions cover premises such as sports grounds.

13.2.3 PART 2 Fire Safety Duties

The responsible person must take appropriate general fire precautions to protect both employees and persons who are not employees to ensure that the premises are safe. Details are given in Chapter 17.

Risk assessment and arrangements

The responsible person must:

➤ make a 'suitable and sufficient' risk assessment to identify the general fire precautions required (usually known as a fire risk assessment)
   • if a dangerous substance is or is liable to be present, this assessment must include the special provisions of Part 1 of Schedule 1 to the Order;
➤ review the risk assessment regularly or when there have been significant changes;
➤ with regard to young persons, take into account Part 2 of Schedule 1;

➤ record the significant findings where five or more people are employed or where there is a license or where Alterations Notice in place for the premises;
➤ apply the principles of prevention Part 3 of Schedule 1;
➤ set out appropriate fire safety arrangements for planning, organization, control, monitoring and review;
➤ eliminate or reduce risks from dangerous substances in accordance with Part 4 of Schedule 1.

Fire-fighting and fire detection

The responsible person must ensure that the premises are provided with appropriate:

➤ fire-fighting equipment (FFE);
➤ fire detectors and alarms;
➤ measures for fire-fighting which are adapted to the size and type of undertaking;
➤ trained and equipped competent persons to implement fire-fighting measures;
➤ contacts with external emergency services, particularly as regards fire-fighting, rescue work, first-aid and emergency medical care.

Emergency routes, exits and emergency procedures

The responsible person must ensure that routes to emergency exits and the exits themselves are kept clear and ready for use. See Chapter 17 for specific requirements for means of escape.

The responsible person must establish suitable and appropriate emergency procedures and appoint a sufficient number of competent persons to implement the procedures. This includes arrangements for undertaking fire drills and the provision of information and action regarding exposure to serious, imminent and unavoidable dangers. Where dangerous substances are used and/or stored, additional emergency measures covering the hazards must be set up. There should be appropriate visual or audible warnings and other communications systems to effect a prompt and safe exit from the endangered area.

Where necessary to protect persons' safety, all premises and facilities must be properly maintained and subject to a suitable system of maintenance.

Safety assistance

The responsible person must appoint (except in the case of the self-employed or partnerships where a person has the competence themselves) one or more people to assist in undertaking the preventative and protective measures. Competent persons must be given the time and means to fulfil their responsibilities. Competent persons must be kept informed of anything relevant to their role and have access to information on any dangerous substances present on the premises.

Provision of information

The responsible person must provide:

➤ their own employees; and
➤ the employer of any employees of an outside undertaking

with comprehensible and relevant information on the risks, precautions taken, persons appointed for fire-fighting and fire drills, and appointment of competent persons. Additional information is also required for dangerous substances used and/or stored on the premises.

Before a child (16–18 years old) is employed, information on special risks to children must be given to the child's parent or guardian.

Capabilities, training and cooperation

The responsible person must ensure that adequate training is provided when people are first employed or exposed to new or increased risks. This may occur with new equipment, changes of responsibilities, new technologies, new systems of work and new substances used. The training has to be done in working hours and repeated periodically as appropriate.

Where two or more responsible persons share duties or premises, they are required to coordinate their activity and co-operate with each other, including keeping each other informed of risks.

Duties of employees

These are covered in Chapter 17 and are similar to requirements under the HSW Act. Employees must take care of themselves and other relevant persons. They must co-operate with the employer and inform them of any situation which they would reasonably consider to present a serious and immediate danger, or a shortcoming in the protection arrangements.

13.2.4 PART 3 Enforcement

Enforcing authorities

The fire and rescue authority in the area local to the premises is normally the enforcing authority. However, in stand-alone construction sites, premises which require a Nuclear license, and where a ship is under repair, the HSE is the enforcing authority, except for the general means of escape and fire emergency procedures under CDM 2007.

Fire and Rescue Authorities officers appointed under the RRFSO have similar powers to those under the HSW Act.

These include the power to enter premises, make enquiries, require information, require facilities and assistance, take samples, require an article or substance to be dismantled or subjected to a process or test and the power to issue Enforcement Notices.

In addition to fire safety legislation, health and safety at work legislation also covers the elimination or minimization of fire risks. As well as the particular and main general duties under the HSW Act, fire risks are covered by specific rules, such as for dangerous substances and explosive atmospheres, work equipment, electricity and other hazards. Thus, environmental health officers or HSE inspectors may enforce health and safety standards for the assessment and removal or control of process-related fire risks, where it is necessary, for the protection of workers and others.

Alterations and Enforcement Notices

Enforcing authorities can issue (see Chapter 17 for details):

Alterations Notices – where the premises constitute a serious risk to people either due to the features of the premises, or hazards present on the premises. The Notices require the responsible person to notify changes and may require them to record the significant findings of the risk assessment and safety arrangements and to supply a copy of the risk assessment before making changes.

Enforcement Notices – where there is a failure to comply with the requirements of the RRFSO. The Notice must specify the provisions concerned and MAY include directions on remedial action.

Prohibition Notices – where the risks are so serious that the use of the premises should be prohibited or restricted. The Notice MAY include directions on remedial action.

13.2.5 PART 4 Offences

Cases can be tried on summary conviction in a magistrates court or on indictment in the Crown Court. The responsible person can be liable:

➤ on summary conviction to a fine not exceeding the statutory maximum;
➤ on indictment to an unlimited fine and/or imprisonment for up to 2 years for failure to comply with fire safety duties where there is a risk of death or serious injury and for failure to comply with an Alterations, Enforcement or Prohibition Notice.

Any person who fails to comply with their duties under the Order as regards fire risks can be prosecuted alongside or instead of a responsible person. Fines are limited to the statutory maximum (or levels 3 or 5 on the standard scale) on summary conviction but on indictment, the fine is unlimited.

13.3 Construction Design and Management Regulations 2007

Details of CDM 2007 are given in Chapter 17. Regulation 46 makes the Fire and Rescue authority responsible for the non-process fire-related sections of CDM 2007 on construction sites where other people are working within or as part of the site – for example, the work is being done in an occupied factory or office building.

The Regulations concerned are as follows.

(a) Regulation 39: Emergency procedures
In so far as they concern fire emergency procedures and involve suitable procedures for evacuation, familiarization of workers and testing the arrangements.

(b) Regulation 40: Emergency routes and exits
In so far as they concern fire emergency routes and exits and involve the following:

➤ a sufficient number of emergency routes and exits;
➤ must lead directly as possible to an identified safe place;
➤ must be kept clear and free of obstruction;
➤ provided with emergency lights as necessary;
➤ suitably signed..

(c) Regulation 41: Fire detection and fire-fighting
This involves the following:

➤ suitably located fire fighting equipment (FFE);
➤ suitably located fire detection and alarm systems;
➤ proper maintenance, examination and testing;
➤ ease of access unless automatically activated;
➤ training for every person on the construction site in the operation of fire equipment that they are likely to use;
➤ where there is a particular risk of fire, people being instructed before starting the work;
➤ equipment suitably signed.

See Table 13.1 for the various enforcement authorities and their responsibilities on construction sites.

13.4 Basic principles of fire

13.4.1 Fire triangle

Fire cannot take place unless three things are present. These are shown in Figure 13.2.

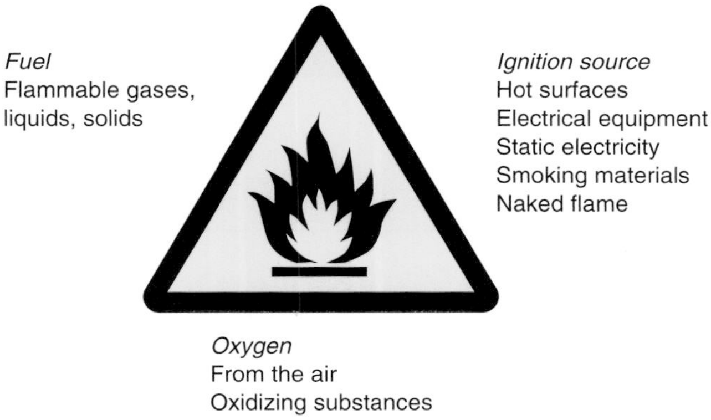

Fuel
Flammable gases, liquids, solids

Ignition source
Hot surfaces
Electrical equipment
Static electricity
Smoking materials
Naked flame

Oxygen
From the air
Oxidizing substances

Figure 13.2 Fire triangle.

The absence of any one of these elements will prevent a fire from starting. Prevention depends on avoiding these three coming together. Fire extinguishing depends on removing one of the elements from an existing fire, and is particularly difficult if an oxidizing substance is present.

Once a fire starts, it can spread very quickly from fuel to fuel as the heat increases.

13.4.2 Sources of ignition

Workplaces have numerous sources of ignition, some of which are obvious but others may be hidden inside machinery. Most of the sources may cause an accidental fire from sources inside but, in the case of arson (about 13% of industrial fires), the source of ignition may be brought from outside the workplace and will be deliberately used. The following are potential sources of ignition in the typical workplace:

➤ **Naked flames** – from smoking materials, cooking appliances, heating appliances and process equipment.
➤ **External sparks** – from grinding metals, welding, impact tools, electrical switch gear.
➤ **Internal sparking** – from electrical equipment (faulty and normal), machinery, lighting.
➤ **Hot surfaces** – from lighting, cooking, heating appliances, process equipment, poorly ventilated equipment, faulty and/or badly lubricated equipment, hot bearings and drive belts.
➤ **Static electricity** – causing significant high-voltage sparks from the separation of materials such as unwinding plastic, pouring highly flammable liquids, walking across insulated floors or removing synthetic overalls.

Table 13.1 Enforcement in respect of fire on construction sites

Type of premises fire issue		Fire and Rescue Authorities Under RRFSO and CDM 2007	HSE under HSW Act, CDM 2007 and RRFSO	Local Authorities, Inspectors under HSW Act, and RRFSO	Defence fire service	A fire inspector or person appointed by the Secretary of State
Fire emergency procedures; escape routes and exits; fire detection and fire fighting; and training issues on:	Stand-alone construction site		✓			
	Construction site where there are other activities (e.g. an occupied factory or office)	✓				
Process fire and explosion risks, where there are, for example, tar boilers/dangerous substances on:	Stand-alone sites		✓	✓		
	Sites where there are other activities (e.g. an occupied factory or office)					
Licensed nuclear sites and ships under construction or repair			✓			
Defence and Crown armed forces premises other than a ship					✓	✓
Sports grounds and Sports stands				✓		
Crown premises except nuclear; UK Atomic Energy premises except nuclear						✓

13.4.3 Sources of fuel

If it will burn it can be fuel for a fire. The things which will burn easily are the most likely to be the initial fuel, which then burns quickly and spreads the fire to other fuels. The most common things that will burn in a typical workplace are:

Solids – these include, wood, paper, cardboard, wrapping materials, plastics, rubber, foam (e.g. polystyrene tiles and furniture upholstery), textiles (e.g. furnishings and clothing), wall paper, hardboard and chipboard used as building materials, waste materials (e.g. wood shavings, dust, paper), hair.

Liquids – these include, paint, varnish, thinners, adhesives, petrol, white spirit, methylated spirits, paraffin, toluene, acetone and other chemicals. Most flammable liquids give off vapours which are heavier than air so they will fall to the lowest levels. A flash flame or an explosion can occur if the vapour catches fire in the correct concentrations of vapour and air.

Gases – flammable gases include LPG (liquefied petroleum gas in cylinders, usually butane or propane), acetylene (used for welding) and hydrogen. An explosion can occur if the air/gas mixture is within the explosive range.

13.4.4 Oxygen

Oxygen is of course provided by the air all around but this can be enhanced by wind, or by natural or powered ventilation systems which will provide additional oxygen to continue burning.

Cylinders providing oxygen for medical purposes or welding can also provide an additional very rich source of oxygen.

In addition, some chemicals such as nitrates, chlorates, chromates and peroxides can release oxygen as they burn and therefore need no external source of air.

13.5 Methods of extinction

There are four main methods of extinguishing fires, which are explained as follows:

➤ **Cooling** – reducing the ignition temperature by taking the heat out of the fire – using water to limit or reduce the temperature.

➤ **Smothering** – limiting the oxygen available by smothering and preventing the mixture of oxygen and flammable vapour – by the use of foam or a fire blanket.
➤ **Starving** – limiting the fuel supply – by removing the source of fuel by switching off electrical power, isolating the flow of flammable liquids or removing wood and textiles, etc.
➤ **Chemical reaction** – by interrupting the chain of combustion and combining the hydrogen atoms with chlorine atoms in the hydrocarbon chain, for example with Halon extinguishers. (Halons have generally been withdrawn because of their detrimental effect on the environment, as ozone depleting agents.)

13.6 Classification of fire

Fires are classified in accordance with British Standard *EN 2:1992 Classification of Fires*. However for all practical purposes there are FIVE main classes of fire – A, B, C, D and F, plus fires involving electrical equipment. BS 7937:2000 *The Specification of Portable Fire Extinguishers for Use on Cooking Oil Fires* introduced the new class F. The categories based on fuel and the means of extinguishing are as follows:

Class A – Fires which involve solid materials such as wood, paper, cardboard, textiles, furniture and plastics where there are normally glowing embers during combustion. Such fires are extinguished by cooling, which is achieved using water.
Class B – Fires which involve liquids or liquefied solids such as paints, oils or fats. These can be further subdivided into:
   Class B1 – fires which involve liquids that are soluble in water such as methanol. They can be extinguished by carbon dioxide, dry powder, water spray, light water and vaporizing liquids
   Class B2 – fires which involve liquids not soluble in water, such as petrol and oil. They can be extinguished by using foam, carbon dioxide, dry powder, light water and vaporizing liquid.
Class C – Fires which involve gases such as natural gas, or liquefied gases such as butane or propane. They can be extinguished using foam or dry powder in conjunction with water to cool any containers involved or nearby.
Class D – Fires which involve metals such as aluminium or magnesium. Special dry powder extinguishers are required to extinguish these fires, which may contain powdered graphite or talc.
Class F – Fires which involve high-temperature cooking oils or fats in large catering establishments or restaurants.

Electrical fires – Fires involving electrical equipment or circuitry do not constitute a fire class on their own, as electricity is a source of ignition that will feed a fire until switched off or isolated. But there are some pieces of equipment that can store, within capacitors, lethal voltages even when isolated. Extinguishers specifically designed for electrical use like carbon dioxide or dry powder units should always be used for this type of fire hazard.

Fire extinguishers are usually designed to tackle one or more classes of fire. This is discussed in 13.14.

13.7 Principles of heat transmission and fire spread

Fire transmits heat in several ways, which needs to be understood in order to prevent, plan escape from, and fight fires. Heat can be transmitted by convection, conduction, radiation and direct burning (Figure 13.3).

13.7.1 Convection

Hot air becomes less dense and rises drawing in cold new air to fuel the fire with more oxygen. The heat is transmitted upwards at sufficient intensity to ignite combustible materials in the path of the very hot products of combustion and flames. This is particularly important inside buildings or other structures where the shape may effectively form a chimney for the fire.

13.7.2 Conduction

This is the transmission of heat through a material with sufficient intensity to melt or destroy the material and ignite combustible materials which come into contact or close to a hot section. Metals like copper, steel and aluminium are very effective or good conductors of heat. Other materials like concrete, brickwork and insulation materials are very ineffective or poor conductors of heat.

Poor conductors or good insulators are used in fire protection arrangements. When a poor conductor is also incombustible, it is ideal for fire protection. Care is necessary to ensure that there are no other issues, such as health risks, with these materials. Asbestos is a very poor conductor of heat and is incombustible. However, very severe health effects, which now outweighs its value as a fire protection material and it is banned in the United Kingdom. Although still found in many buildings where it was used extensively

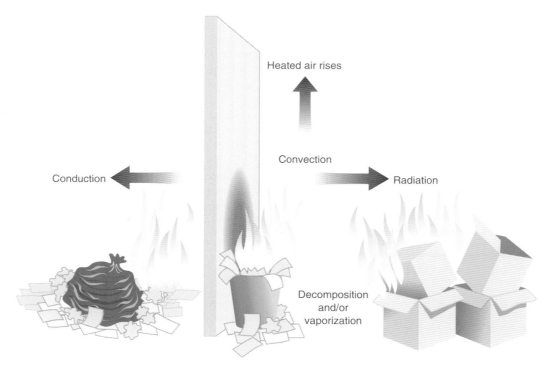

Figure 13.3 Principles of heat transmission.

for fire protection, it now has to be managed under the Control of Asbestos Regulations 2006.

13.7.3 Radiation

Often in a fire, the direct transmission of heat through the emission of heat waves from a surface can be so intense that adjacent materials are heated sufficiently to ignite. A metal surface glowing red-hot would be typical of a severe radiation hazard in a fire.

13.7.4 Direct burning

This is the effect of combustible materials catching fire through direct contact with flames which causes fire to spread, in the same way that lighting an open fire, with a range of readily combustible fuels, results in its spread within a grate.

13.7.5 Fire and Smoke spread in buildings

Where fire is not contained and people can move away to a safe location, there is little immediate risk to those people. However, where fire is confined inside buildings the fire behaves differently (Figure 13.4).

The smoke rising from the fire gets trapped inside the space by the immediate ceiling, then spreads horizontally across the space deepening all the time until the entire space is filled. The smoke will also pass through any holes or gaps in the walls, ceiling or floor and get into other parts of the building. It moves rapidly up staircases or lift wells and into any areas that are left open, or rooms which have open doors connecting to the staircase corridors. The heat from the building gets trapped inside, raising the temperature very rapidly. The toxic smoke and gases are an added danger to people inside the building, who must be able to escape quickly to a safe location.

13.8 Common causes of fire and consequences

13.8.1 Causes

The Home Office statistics show that the causes of fires in buildings, excluding dwellings, in 2005 was as shown in Figure 13.5. The total was 35 300, 6% less than the previous year. This follows a 10% fall between 2003 and 2004 and follows the general trend downwards since 1995.

Figure 13.4 Fire and smoke spread in buildings.

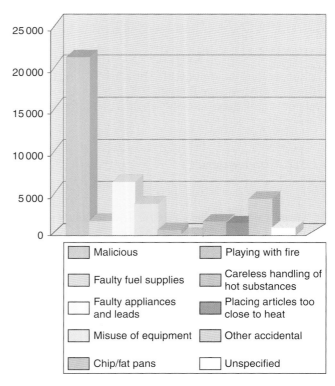

Figure 13.5 Causes of fire, 2005.

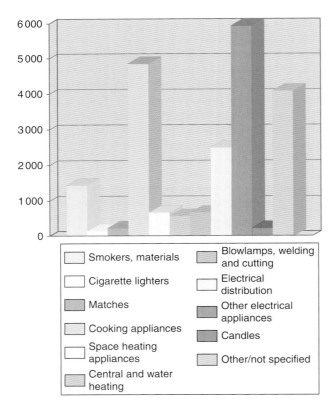

Figure 13.6 Accidental fires – sources of ignition, 2005.

The sources of ignition are shown in Figure 13.6. Out of the 21 200 accidental fires in 2005, it shows that cooking appliances and electrical equipment account for over 60% of the total.

13.8.2 Consequences

The main consequence of fire is:

➤ Death – although this is a very real risk, relatively few people die in building fires that are not in dwellings. In 2005, 21 (4%) people died out of a total of 485 in all fires. The main causes of all deaths were:
  • overcome by gas or smoke – 46%
  • burns – 27%
  • burns and overcome by gas or smoke – 20%
  • other – 7%.

Clearly gas and smoke are the main risks:

➤ Personal injury – some 1395 people were injured (12%) of total injuries in all fires.

➤ Building damage – can be very significant, particularly if the building materials have poor resistance to fire and there is little or no built-in fire protection.
➤ Flora and fauna damage – can be significant, particularly in a hot draught or forest fire.
➤ Loss of business and jobs – it is estimated that about 40% of businesses do not start up again after a significant fire. Many are under- or not insured and small companies often cannot afford the time and expense of setting up again when they probably still have old debts to service.
➤ Transport disruption – rail routes, roads and even airports are sometimes closed because of a serious fire. The worst case was of course 11 September 2001 in the USA when airports around the world were disrupted.
➤ Environmental damage from the fire and/or fighting the fire – fire-fighting water, the products of combustion and exploding building materials, such as asbestos cement roofs, can contaminate significant areas around the fire site.

13.9 Fire risk assessment

13.9.1 General

Fire risk assessments are now under the RRFSO and not under that Management of Health and Safety at Work Regulations 1999.

A fire risk assessment will indicate what fire precautions are needed. There are numerous ways of carrying out a fire risk assessment: the one described below is based on the method contained within Fire safety Guides published by the Department of Communities and Local Government (see Appendix 13.1). A systematic approach, considered in five simple stages, is generally the best practical method. The RRFSO guides are downloadable on www.firesafetyguides.communities.gov.uk.

13.9.2 Stage 1 – identify fire hazards

There are five main hazards produced by fire that should be considered when assessing the level of risk:

➤ oxygen depletion;
➤ flames and heat;
➤ smoke;
➤ gaseous combustion products;
➤ structural failure of buildings.

Of these, smoke and other gaseous combustion products are the most common cause of death in fires.

For a fire to occur, it needs sources of heat and fuel. If these hazards can be kept apart, removed or reduced, then the risks to people and businesses are minimized. **Identifying fire hazards** in the workplace is the first stage as follows:

Identify any combustibles – Most workplaces contain combustible materials. Usually, the presence of normal stock in trade should not cause concern, provided the materials are used safely and stored away from sources of ignition. Good standards of housekeeping are essential to minimize the risk of a fire starting or spreading quickly.

The amount of combustible material in a workplace should be kept as low as is reasonably practicable. Materials should not be stored in gangways, corridors or stairways or where they may obstruct exit doors and routes. Fires often start and are assisted to spread by combustible waste in the workplace. Such waste should be collected frequently and removed from the workplace, particularly where processes create large quantities of it.

Some combustible materials, such as flammable liquids, gases or plastic foams, ignite more readily than others and quickly produce large quantities of heat and/or dense toxic smoke. Ideally, such materials should be stored away from the workplace or in fire-resisting stores. The quantity of these materials kept or used in the workplace should be as small as possible, normally no more than half a day's supply.

Identify any sources of heat – All workplaces will contain heat/ignition sources; some will be obvious such as cooking sources, heaters, boilers, engines, smoking materials or heat from processes, whether in normal use or through carelessness or accidental failure. Others may be less obvious such as heat from chemical processes or electrical circuits and equipment.

Where possible, sources of ignition should be removed from the workplace or replaced with safer forms. Where this cannot be done, the ignition source should be kept well away from combustible materials or made the subject of management controls.

Particular care should be taken in areas where portable heaters are used or where smoking is permitted (now banned inside premises). Where heat is used as part of a process, it should be used carefully to reduce the chance of a fire as much as possible. Good security both inside and outside the workplace will help to combat the risk of arson.

Under smoke-free legislation, smoking is not permitted in enclosed or significantly enclosed areas. Outside designated safe areas should be provided for those who still require to smoke. The smoking rules should be rigorously enforced (Figure 13.7).

Demolition work can involve a high risk of fire and explosion. In particular:

➤ Dismantling tank structures can cause the ignition of flammable residues. This is especially dangerous if hot methods are used to dismantle tanks before residues are thoroughly cleaned out. The work should only be done by specialists.

➤ Disruption and ignition of buried gas and electrical services is a common problem. It should always be assumed that buried services are present unless it is positively confirmed that the area is clear. A survey using service detection equipment must be carried out by a competent person to identify any services. The services should then be marked, competently purged or made dead, before any further work is done. A permit to excavate or dig is the normal formal procedure to cover buried services.

Identify any unsafe acts – Persons undertaking unsafe acts such as smoking next to combustible materials, etc.

Identify any unsafe conditions – These are hazards that may assist a fire to spread in the workplace, for example if there are large areas of hardboard or polystyrene tiles etc., or open stairs that can enable a fire to spread quickly, trapping people and engulfing the whole building.

An ideal method of identifying and recording these hazards is by means of a simple single-line plan, an example of which is illustrated in Figure 13.8. Checklists may also be used. See Appendix 13.1.

13.9.3 Stage 2 – identify persons who are at significant risk

Consider the risk to any people who may be present. In many instances and particularly for most small workplaces the risk(s) identified will not be significant, and specific measures for persons in this category will not be required. There will, however, be some occasions when certain people may be especially at risk from the fire, because of their specific role, disability, sleeping, location or the workplace activity (see Section 13.17 for more information). Special consideration is needed if:

➤ sleeping accommodation is provided;
➤ persons are physically, visually or mentally challenged;
➤ people are unable to react quickly;
➤ persons are isolated.

People, such as visitors, the public or other workers, may come into the workplace from outside. The assessor

Figure 13.7 Smoke-free sign.

must decide whether the current arrangements are satisfactory or if changes are needed.

Because fire is a dynamic event, which, if unchecked, will spread throughout the workplace, all people present will eventually be at risk if fire occurs. Where people are at risk, adequate means of escape from fire should be provided together with arrangements for detecting and giving warning of fire. Fire fighting equipment suitable for the hazards in the workplace should be provided.

Some people may be at significant risk because they work in areas where fire is more likely or where rapid fire growth can be anticipated. Where possible, the hazards creating the high level of risk should be reduced. Specific steps should be taken to ensure that people affected are made aware of the danger and the action they should take to ensure their safety and the safety of others.

13.9.4 Stage 3 – evaluate and reduce the risks

If the building has been built and maintained in accordance with Building Regulations and is being put to its designed use, it is likely that the means of escape provisions will either be adequate, or it will be easy to decide what is required in relation to the risk. Having identified the hazards and the persons at risk, the next stage is to reduce the chance of a fire occurring and spreading, thereby minimizing the chance of harm to persons in the workplace. The principles of prevention laid down in the RRFSO should be followed at this stage (see Section 17.16.5). These are based on EC Directive requirements and are therefore the same as those used in the Management Regulations. (See Chapter 6.)

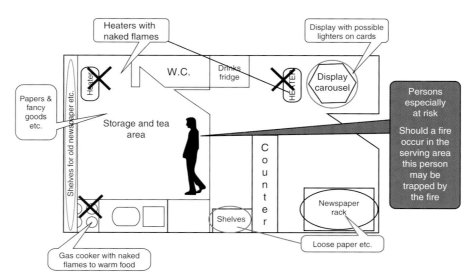

Figure 13.8 Simple line diagram to identify fire hazards.

Evaluate the risks

Attempt to classify each area as 'high', 'normal' or 'low risk'. If 'high risk', it may be necessary to reconsider the principles of prevention, otherwise additional compensatory measures will be required.

Low risk – Areas where there is minimal risk to persons' lives; where the risk of fire occurring is low; or the potential for fire, heat and smoke spreading is negligible and people would have plenty of time to react to an alert of fire.

Normal risk – Such areas will account for nearly all parts of most workplaces; where an outbreak of fire is likely to remain confined or spread slowly, with an effective fire warning allowing persons to escape to a place of safety.

High risk – Areas where the available time needed to evacuate the area is reduced by the speed of development of a fire, for example highly flammable or explosive materials stored or used (other than small quantities under controlled conditions); also where the reaction time to the fire alarm is slower because of the type of person present or the activity in the workplace, for example the infirm and elderly or persons sleeping on the premises.

Determine if the existing arrangements are adequate, or need improvement.

Matters that will have to be considered are:

➤ Means for detecting and giving warning in case of fire – can it be heard by all occupants? (See Section 13.11).
➤ Means of escape – are they adequate in size, number, location, well lit, unobstructed, safe to use, etc.? (See Section 13.12).
➤ Signs – for exits, fire routines, etc. (See Sections 13.12 and 6.3).
➤ FFE – wall-mounted or in a cradle on fire exit routes, suitable types for hazards present and sufficient in number? (See Section 13.14.)

13.9.5 Stage 4 – the findings (always recommended, see Stage 5 – review)

The findings of the assessment and the actions (including maintenance) arising from it should be recorded. If five or more people are employed, or an Alterations Notice is required, a formal record of the significant findings and any measures proposed to deal with them must be recorded. (See Appendix 13.2 and summary of the Order in Section 17.16.) The record should indicate:

➤ the date the assessment was made;
➤ the hazards identified;
➤ any staff and other people especially at risk;

➤ what action needs to be taken, and by when (**action plan**);
➤ the conclusions arising.

The above guidelines are to be used with caution. Each part of the workplace must be looked at and a decision made on how quickly persons would react to an alert of fire in each area. Adequate safety measures will be required if persons are identified as being at risk. Where maximum travel distances (see 13.12.3 and Table 13.2) cannot be achieved, extra fire safety precautions will be needed.

Where persons are at risk or an unacceptable hazard still exists, additional fire safety precautions will be required to compensate for this, or alternatively repeat previous stages to manage risk to an acceptable level.

13.9.6 Stage 5 – monitor and review on a regular basis

The fire risk assessment is not a one-off procedure. It should be continually monitored to ensure that the existing fire safety arrangements and fire risk assessment remain realistic. The assessment should be reviewed if there is a significant change in the occupancy, work activity, the materials used or stored when building works are proposed, or when it is no longer thought to be valid.

13.9.7 Structural features

The workplace may contain features that could promote the rapid spread of fire, heat or smoke and affect escape routes. These features may include ducts or flues, openings in floors or walls, or combustible wall or ceiling linings. Where people are put at risk from these features, appropriate steps should be taken to reduce the potential for rapid fire spread by, for example, non-combustible automatic dampers fitted in ducts or to provide an early warning of fire so that people can leave the workplace before their escape routes become unusable.

Combustible wall or ceiling linings should not be used on escape routes and large areas should be removed wherever they are found. Other holes in fire-resisting floors, walls or ceilings should be filled in with fire-resisting material to prevent the passage of smoke, heat and flames.

13.9.8 Maintenance and refurbishment

Sources of heat or combustible materials may be introduced into the workplace during periods of maintenance or refurbishment. Where the work involves the introduction of heat, such as welding, this should be carefully controlled by a safe system of work, for example Hot Work Permit (see Chapter 6 for details). All materials brought into the workplace in connection with the work being carried out should be stored away from sources of heat and not obstruct exit routes.

13.9.9 Fire plans

Fire plans should be produced and attached to the fire risk assessment. A copy should be posted in the workplace. A single-line plan of the area or floor should be produced or an existing plan should be used which needs to show:

➤ escape routes, numbers of exits, number of stairs, fire-resisting doors, fire-resisting walls and partitions, places of safety, and the like;
➤ fire safety signs and notices including pictorial fire exit signs and fire action notices;
➤ the location of fire warning call points and sounders or rotary gongs;
➤ the location of emergency lights;
➤ the location and type of FFE.

13.10 Dangerous substances

13.10.1 Introduction

The DSEAR apply at most workplaces where a dangerous substance is present or could be present.

The employer must:

➤ carry out a risk assessment of any work activities involving dangerous substances;
➤ provide a way of eliminating or reducing risks as far as is reasonably practicable;
➤ provide procedures and equipment to deal with accidents and emergencies;
➤ provide training and information for employees;
➤ classify places where explosive atmospheres may occur into zones and mark the zones where necessary. Phased in since 2006.

The Regulations give a detailed definition of 'dangerous substance', which should be referred to for more information. They include any substance or preparation which, because of its properties or the way it is used, could be harmful because of fires and explosions. The list includes petrol, LPG, paints, varnishes, solvents and some dusts. These are dusts which, when mixed with air, can cause an explosive atmosphere. Dusts from milling and sanding operations are examples of this. Most workplaces contain a certain amount of dangerous substances.

An explosive atmosphere is an accumulation of gas, mist, dust or vapour, mixed with air, which has the potential to catch fire or explode. Although an explosive atmosphere does not always result in an explosion (detonation), if it catches fire, flames can quickly travel through the workplace. In a confined space (e.g. in plant or equipment) the rapid spread of the flame front or rise in pressure can itself cause an explosion and rupture of the plant and/or building. For a detailed summary of the Regulations, see Chapter 17.

13.10.2 Risk assessment

This is the process of identifying and carefully examining the dangerous substances present or likely to be present in the workplace, the work activities involving them and how they might fail and cause fire, explosion and similar events that could harm employees and the public. The purpose of a risk assessment is to enable the employer to decide what needs to be done to eliminate or reduce the safety risks from dangerous substances as far as is reasonably practicable. It should take account of the following:

➤ what hazardous properties the substances have;
➤ the way they are used and stored;
➤ the possibility of hazardous explosive atmospheres occurring;
➤ any potential ignition sources.

Regardless of the quantity of dangerous substance present, the employer must carry out a risk assessment. This will enable them to decide whether existing measures are sufficient or whether they need to make any additional controls or precautions. Non-routine activities need to be assessed as well as the normal activities within the workplace. For example in maintenance work, there is often a higher potential for fire and explosion incidents to occur.

Unless the employer has already carried out a detailed assessment under the Management Regulations, there must be an assessment of the risks from fire, explosion and other events arising from dangerous substances, including addressing requirements specified by DSEAR. This is separate from the Fire Risk assessment made under the RRFSO as Dangerous Substances are considered to be process risks covered by the HSW Act.

Employers are required to ensure that the safety risks from dangerous substances are eliminated or, when this is not reasonably practicable, to take measures to control risks **and** to reduce the harmful effects of any fire, explosion or similar events, so far as is reasonably practicable.

13.10.3 Substitution

Substitution is the best solution. It is much better to replace a dangerous substance with a substance or process that totally eliminates the risk. In practice this is difficult to achieve; so it is more likely that the dangerous substance will be replaced with one that is less hazardous (e.g. by replacing a low-flashpoint solvent with a high-flashpoint one).

Designing the process so that it is less dangerous is an alternative solution. For example a change could be made from a batch production to a continuous production process, or the manner or sequence in which the dangerous substance is added could be altered. However, care must be taken when carrying out these steps to make sure that no other new safety or health risks are created or increased, as this would outweigh the improvements implemented as a result of DSEAR.

The fact is that where a dangerous substance is handled or stored for use as a fuel, there is often no scope to eliminate it and very little chance to reduce the quantities handled.

Where risk cannot be entirely eliminated, control and mitigation measures should be applied. This should reduce risk as follows.

13.10.4 Control measures

Control measures should be applied in the following order of priority:

➤ reduce the amount of dangerous substances to a minimum;
➤ avoid or minimize releases;
➤ control releases at source;
➤ prevent the formation of an explosive atmosphere;
➤ use a method such as ventilation to collect, contain and remove any releases to a safe place;
➤ avoid ignition sources;
➤ avoid adverse conditions (e.g. exceeding the limits of temperature or other control settings) that could lead to danger;
➤ keep incompatible substances apart.

13.10.5 Mitigation measures

Choose mitigation measures which are consistent with the risk assessment and appropriate to the nature of the activity or operation. These can include:

➤ preventing fires and explosions from spreading to other plant and equipment or to other parts of the workplace;
➤ making sure that a minimum number of employees is exposed;
➤ in the case of a process plant, providing plant and equipment that can safely contain or suppress an explosion, or vent it to a safe place.

In workplaces where explosive atmospheres may occur, you should ensure that:

➤ areas where hazardous explosive atmospheres may occur are classified into zones based on their likelihood and persistence;
➤ areas classified into zones are protected from sources of ignition by selecting suitable special equipment and protective systems;
➤ where necessary, areas classified into zones are marked with a specified 'EX' sign at their points of entry;
➤ employees working in zoned areas must be provided with appropriate clothing that does not create a risk of an electrostatic discharge igniting the explosive atmosphere;
➤ before coming into operation for the first time, areas where hazardous explosive atmosphere may be present are confirmed as being safe (verified) by a person (or organization) competent in the field of explosion protection. The person carrying out the verification must be competent to consider the particular risks at the workplace and the adequacy of control and other measures put in place.

13.10.6 Storage

Dangerous substances should be kept in a safe place in a separate building or the open air. Only small quantities of dangerous substances should be kept in a workroom or area as follows:

➤ For flammable liquids that have a flashpoint above the maximum ambient temperature (normally taken as 32°C), the small quantity that may be stored in the workroom is considered to be an amount up to 250 litres.
➤ For extremely and highly flammable liquids and those flammable liquids with a flashpoint below the maximum ambient temperature, the small quantity is considered to be up to 50 litres and this should be held in a special metal cupboard or container (Figure 13.9).

Any larger amounts should be kept in special fire-resisting store, which should be:

➤ properly ventilated;
➤ provided with spillage retaining arrangements such as sills;
➤ free of sources of ignition, such as unprotected electrical equipment, sources of static electrical sparks, naked flames or smoking materials;
➤ arranged so that incompatible chemicals do not become mixed together either in normal use or in a fire situation;

Figure 13.9 Various storage arrangements for highly flammable liquids.

➤ of fire-resisting construction;
➤ used for empty as well as full containers – all containers must be kept closed;
➤ kept clear of combustible materials such as cardboard or foam plastic packaging materials.

13.10.7 Flammable gases

Flammable gas cylinders also need to be stored and used safely. The following guidance should be adopted:

➤ Both full and empty cylinders should be stored outside. They should be kept in a separate secure compound at ground level with sufficient ventilation. Open mesh is preferable.

➤ Valves should be uppermost during storage to retain them in the vapour phase of the LPG.
➤ Cylinders must be protected from mechanical damage. Unstable cylinders should be together, for example, and cylinders must be protected from the heat of the summer sun.
➤ The correct fittings must be used. These include hoses couplers, clamps and regulators.
➤ Gas valves must be turned off after use at the end of the shift.
➤ Precautions must be taken to avoid welding flame 'flash back' into the hoses or cylinders. People need training in the proper lighting up and safe systems of work procedures; non-return valves and flame arrestors also need to be fitted.
➤ Cylinders must be changed in a well-ventilated area remote from any sources of ignition.
➤ Joints should be tested for gas leaks using soapy/detergent water – never use a flame.
➤ Flammable material must be removed or protected before welding or similar work.
➤ Cylinders should be positioned outside buildings with gas piped through in fixed metal piping.
➤ Both high and low ventilation must be maintained where LPG applications are being used.
➤ Flame failure devices are necessary to shut off the gas supply in the event of flame failure.

13.11   Fire detection and warning

In the event of fire, it is vital that everyone in the workplace is alerted as soon as possible. The earlier the fire is discovered, the more likely it is that people will be able to escape before the fire takes hold and before it blocks escape routes or makes escape difficult (Figure 13.10).

Every workplace should have detection and warning arrangements. Usually the people who work there will detect the fire and in many workplaces nothing further will be needed.

It is important to consider how long a fire is likely to burn before it is discovered. Fires are likely to be discovered quickly if they occur in places that are frequently visited by employees, or in occupied areas of a building. For example, employees are likely to smell burning or see smoke if a fire breaks out in an office.

Where there is concern that fire may break out in an unoccupied part of the premises, for example in a basement, some form of automatic fire detection should be

Figure 13.10 Typical fire point in offices with extinguishers, fire notice and alarm break-glass call point.

fitted. Commercially available heat or smoke detection systems can be used. In small premises, a series of inter-linked domestic smoke alarms that can be heard by everyone present will be sufficient. In most cases, staff can be relied upon to detect a fire.

In small workplaces where occupancy is low, a shouted warning should be all that is needed, so long as the warning can be heard and understood everywhere on the premises.

If the size or occupancy of a workplace means that a shouted warning is insufficient, hand-operated devices such as bells, gongs or sirens can be used. They should be installed on exit routes and should be clearly audible throughout the workplace. In all other places an electrically operated fire alarm system should be fitted. This will have call points adjacent to exit doors and enough bells or sounders to be clearly audible throughout the premises.

If it is thought that there might be some delay in fire being detected, automatic fire detection should be considered, linked into an electrical fire alarm system. Where a workplace provides sleeping accommodation or where fires may develop undetected, automatic detection must be provided. If a workplace provides sleeping accommodation for fewer than six people, interlinked domestic smoke alarms (wired to the mains electricity supply) can be used provided that they are audible throughout the workplace while people are present.

13.12 Means of escape in case of fire

13.12.1 General

It is essential to ensure that people can escape quickly from a workplace if there is a fire. Normally the entrances and exits to the workplace will provide escape routes, particularly if staff have been trained in what to do in case of fire and if it is certain that an early warning will be given.

In modern buildings which have had Building Regulation approval, and where there have not been significant changes to the building or where the workplace has recently been inspected by the fire authorities and found to be satisfactory, it is likely that the means of escape will be adequate. It may occasionally be necessary to improve the fire protection on existing escape routes, or to provide additional exits. In making a decision about the adequacy of means of escape, the following points should be considered:

➤ People need to be able to turn away from a fire as they escape or be able to pass a fire when it is very small.

➤ If a single-direction escape route is in a corridor, the corridor may need to be protected from fire by fire-resisting partitions and self-closing fire doors.

➤ Stair openings can act as natural chimneys in fires. This makes escape from the upper parts of some workplaces difficult. Most stairways, therefore, need to be separated from the workplace by fire-resisting partitions and self-closing fire doors. Where stairways serve no more than two open areas, in shops for example, which people may need to use as escape routes, there may be no need to use this type of protection.

13.12.2 Doors

Some doors may need to open in the direction of travel, such as:

➤ doors from a high-risk area, such as a paint spraying room or large kitchen;

➤ doors that may be used by more than 50 persons;

➤ doors at the foot of stairways where there may be a danger of people being crushed;

➤ Some sliding doors may be suitable for escape purposes provided that they do not put people using them at additional risk, slide easily and are marked with the direction of opening.

➤ Doors which only revolve and do not have hinged segments are not suitable as escape doors.

13.12.3 Escape routes

Escape routes should meet the following criteria:

➤ Where two or more escape routes are needed they should lead in different directions to places of safety.

➤ Escape routes need to be short and to lead people directly to a place of safety, such as the open air or an area of the workplace where there is no immediate danger.

➤ It should be possible for people to reach the open air without returning to the area of the fire. They should then be able to move well away from the building.

➤ Escape routes should be wide enough for the volume of people using them. A 750-mm door will allow up to 40 people to escape in 1 minute, so most doors and corridors will be wide enough. If the routes are likely to be used by people in wheelchairs, the minimum width will need to be 800 mm.

While the workplace is in use, it must be possible to open all doors easily and immediately from the inside, without using a key or similar device. Doors must be readily opened in the direction of escape. Fire doors should be self-closing (fire doors to cupboards or lockers can be simply latched or locked).

Make sure that there are no obstructions on escape routes, especially on corridors and stairways where people who are escaping could dislodge stored items or be caused to trip. Any fire hazards must be removed from exit routes as a fire on an exit route could have very serious consequences.

Escape routes need regular checks to make sure that they are not obstructed and that exit doors are not locked. Self-closing fire-resisting doors should be checked to ensure doors close fully, including those fitted with automatic release mechanisms.

The maximum advisable travel distances from any area in a workplace to a fire exit door leading out to a relative place of safety should be in accordance with Table 13.2.

13.12.4 Lighting

Escape routes must be well lit. If the route has only artificial lighting or if it is used during the hours of darkness, alternative sources of lighting should be considered in case the power fails during a fire. Check the routes when it is dark as, for example, there may be street lighting outside that provides sufficient illumination. In small workplaces it may be enough to provide the staff with torches that they can use if the power fails. However, it may be necessary to provide battery-operated emergency lights so that if the mains lighting fails the lights will operate automatically. Candles, matches and cigarette lighters are not adequate forms of emergency lighting.

13.12.5 Signs

Exit signs on doors or indicating exit routes should be provided where they will help people to find a safe escape route. Signs on exit routes should have directional arrows, 'up' for straight on and 'left', 'right' or 'down' according to the route to be taken. Advice on the use of all signs including exit signs can be found in Chapters 6 and 17.

Table 13.2					
Maximum Travel Distances* (Measured to a relative place of safety)	**Low Fire Risk**	**Normal Risk – Production Areas (Factory Only)**	**Normal Fire Risk**	**Normal Fire Risk Sleeping**	**High Fire Risk**
More than one route is provided	60 m	45 m	45 m	32 m	25 m
Only a single escape route is provided	45 m	25 m	18 m	16 m	12 m

*To an exit (open air where persons can disperse safely) storey exit (staircase separated from the remainder of the premises by fire resisting walls and self closing fire doors, etc.) or a compartment wall (fire resisting wall and self closing fire doors).

13.12.6 Escape times

Everyone in the building should be able to get to the nearest place of safety in between 2 and 3 minutes. This means that escape routes should be kept short. Where there is only one means of escape, or where the risk of fire is high, people should be able to reach a place of safety, or a place where there is more than one route available, in 1 minute.

The way to check this is to pace out the routes, walking slowly and noting the time. Start from where people work and walk to the nearest place of safety. Remember that the more people there are using the route, the longer they will take. People take longer to negotiate stairs and they are also likely to take longer if they have a disability.

Where fire drills are held, check how long it takes to evacuate each floor in the workplace. This can be used as a basis for assessment. If escape times are too long, it may be worth re-arranging the workplace so that people are closer to the nearest place of safety, rather than undertake expensive alterations to provide additional escape routes.

Reaction time needs to be considered. This is the amount of time people will need for preparation before they escape. It may involve, for example, closing down machinery, issues of security or helping visitors or members of the public out of the premises. Reaction time needs to be as short as possible to reduce risk to staff. Assessment of escape routes should include this. If reaction times are too long, additional routes may need to be provided. It is important that people know what to do in case of fire as this can lessen the time needed to evacuate the premises.

13.13 Principles of fire protection in buildings

13.13.1 General

The design of all new buildings and the design of extensions or modifications to existing buildings must be approved by the local planning authority.

Design data for new and modified buildings must be retained throughout the life of the structure.

Building legal standards are concerned mainly with safety of life. Therefore, it is necessary to consider the early stage of fire and how it affects the means of escape, and, also, aim to prevent eventual spread to other buildings.

Asset protection requires extra precautions that will have an effect at both early and later stages of the fire growth by controlling fire spread through and between buildings and preventing structural collapse. However,

this extra fire protection will also improve life safety not only for those escaping at the early stages of the fire but also for fire-fighters who will subsequently enter.

If a building is carefully designed and suitable materials are used to build it and maintain it, then the risk of injury or damage from fire can be substantially reduced. Three objectives must be met:

➤ it must be possible for everyone to leave the building quickly and safely;
➤ the building must remain standing for as long as possible;
➤ the spread of fire and smoke must be reduced.

These objectives can be met through the selection of materials and design of buildings.

13.13.2 Fire loading

The fire load of a building is used to classify types of building use. It may be calculated simply by multiplying the weight of all combustible materials by their energy values and dividing by the floor area under consideration. The higher the fire load, the more the effort needed to offset this by building to higher standards of fire resistance.

13.13.3 Surface spread of fire

Combustible materials, when present in a building as large continuous areas, such as for lining walls and ceilings, readily ignite and contribute to spread of fire over their surfaces. This can represent a risk to life in buildings, particularly where walls of fire-escape routes and stairways are lined with materials of this nature.

Materials are tested by insurance bodies and fire research establishments. The purpose of the test is to classify materials according to the tendency for flame to spread over their surfaces. As with all standardized test methods, care must be taken when applying test results to real applications.

In the UK, a material is classified as having a surface in one of the following categories:

➤ Class 1 – Surface of very low flame spread;
➤ Class 2 – Surface of low flame spread;
➤ Class 3 – Surface of medium flame spread;
➤ Class 4 – Surface of rapid flame spread.

The test shows how a material would behave in the initial stages of a fire.

As all materials tested are combustible, in a serious fire they would burn or be consumed. Therefore, there is an additional Class 0 of materials which are non-combustible throughout or, under specified conditions, non-combustible on one face and combustible on the other. The spread of flame rating of the combined Class 0 product must not be worse than that for Class 1.

Internal partitions of walls and ceilings should be Class 0 materials wherever possible and must not exceed Class 1.

13.13.4 Fire resistance of structural elements

If structural elements such as walls, floors, beams, columns and doors are to provide effective barriers to fire spread and to contribute to the stability of a building, they should be of a required standard of fire resistance.

In the UK, tests for fire resistance are made on elements of structure, full size if possible, or on a representative portion having minimum dimensions of 3 m long for columns and beams and 1 m² for walls and floors. All elements are exposed to the same standard fire provided by furnaces in which the temperature increases with time at a set rate. The conditions of exposure are appropriate to the element tested. Freestanding columns are subjected to heat all round, and walls and floors are exposed to heat on one side only. Elements of structure are graded by the length of time they continue to meet three criteria:

➤ the element must not collapse;
➤ the element must not develop cracks through which flames or hot gases can pass;
➤ the element must have enough resistance to the passage of heat so the temperature of the unexposed face does not rise by more than a prescribed amount.

The term fire resistance has a precise meaning. It should not be applied to such properties of materials as resistance to ignition or resistance to flame propagation. For example steel has a high resistance to ignition and flame propagation but will distort quickly in a fire and allow the structure to collapse – it therefore has poor 'fire resistance' (Figure 13.11). It must be insulated to provide good fire protection. This is normally done by encasing steel frames in concrete.

In the past, asbestos has been made into a paste and plastered onto steel frames, giving excellent fire protection, but it has caused major health problems and its use in new work is banned.

Building materials with high fire resistance are, for example, brick, stone, concrete, very heavy timbers (the outside chars and insulates the inside of the timber), and some specially made composite materials used for fire doors.

13.13.5 Insulating materials

Building materials used for thermal or sound insulation could contribute to the spread of fire. Only approved fire-resisting materials should be used.

Figure 13.11 Steel structures can collapse in the heat of a fire.

13.13.6 Fire compartmentation

A compartment is a part of a building that is separated from all other parts by walls and floors, and is designed to contain a fire for a specified time.

The principal object is to limit the effect of both direct fire damage and consequential business interruption caused by not only fire spread but also smoke and water damage in the same floor and other storeys.

Buildings are classified into purpose groups, according to their size. To control the spread of fire, any building whose size exceeds that specified for its purpose group must be divided into compartments that do not exceed the prescribed limits of volume and floor area. Otherwise, they must be provided with special fire protection. In the UK, the normal limit for the size of a compartment is 7000 m³. Compartments must be separated by walls and floors of sufficient fire resistance. Any openings needed in these walls or floors must be protected by fire-resisting doors to ensure proper fire-tight separation.

Ventilation and heating ducts must be fitted with fire dampers where they pass through compartment walls and floors. Firebreak walls must extend completely across a building from inside wall to outside wall. They must be stable; they must be able to stand even when the part of the building on one side or the other is destroyed.

No portion of the wall should be supported on unprotected steelwork nor should it have the ends of unprotected steel members embedded in it. The wall must extend up to the underside of a non-combustible roof surface, and sometimes above it. Any openings must be protected to the required minimum grade of fire resistance. If an external wall joins a firebreak wall and has an opening near the join, the firebreak wall may need to extend beyond the external wall.

An important function of external walls is to contain a fire within a building, or to prevent fire spreading from outside.

The fire resistance of external walls should be related to the:

➤ purpose for which the building is used;
➤ height, floor area and volume of the building;
➤ distance of the building from relevant boundaries and other buildings;
➤ extent of doors, windows and other openings in the wall.

A wall which separates properties from each other should have no doors or other openings in it.

13.14 Provision of fire fighting equipment

13.14.1 General

Since 1 April 2006, employers or those who have control of non-domestic premises have had a statutory duty under fire safety and health and safety legislation to ensure that there are appropriate means of fighting fires. The employer or controller of non-domestic premises is known as the 'responsible person'. The responsible person also has to provide suitably trained people to operate non-automatic FFE.

As Fire Certificates issued by fire authorities under the Fire Precautions Act 1971 have been abolished, instead it is up to the responsible person to identify the fire-fighting requirement for their premises.

If fire breaks out in the workplace and trained staff can safely extinguish it using suitable FFE, the risk to others will be removed. Therefore, all workplaces where people are at risk from fire should be provided with suitable FFE.

The most useful form of FFE for general fire risks is the water-type extinguisher or suitable alternative. One such extinguisher should be provided for around each 200 m^2 of floor space with a minimum of one per floor. If each floor has a hose reel, which is known to be in working order and of sufficient length for the floor it serves, there may be no need for water-type extinguishers to be provided.

Areas of special risks involving the use of oil, fats or electrical equipment may need carbon dioxide, dry powder or other types of extinguisher (Figure 13.12).

Fire extinguishers should be sited on exit routes, preferably near to exit doors or where they are provided for specific risks, near to the hazards they protect. Notices indicating the location of FFE should be displayed where the location of the equipment is not obvious or in areas of high fire risk where the notice will assist in reducing the risk to people in the workplace.

All Halon fire extinguishers should have been decommissioned as from December 2003 and disposed of safely.

Those carrying out hot work should have appropriate fire extinguishers with them and know how to use them.

The primary purpose of fire extinguishers is to tackle fires at a very early stage to enable people to make their escape. Putting out larger fires is the role of the fire and rescue services.

Extinguishers should conform to a recognized standard BS EN 3:7 Portable Fire Extinguishers. Characteristics, performance requirements and test methods.

13.14.2 Advantages and limitations of the main extinguishing media

Fire extinguishers are all red with 5% of the cylindrical area taken up with the colour code. The colour code (band) denotes on which class of fire the extinguisher can be used.

Water extinguishers (red band)
This type of extinguisher can only be used on Class A fires. They allow the user to direct water onto a fire from a considerable distance.

A 9 litre water extinguisher can be quite heavy and some water extinguishers with additives can achieve the same rating, although they are smaller and therefore considerably lighter. This type of extinguisher is not suitable for use on live electrical equipment, liquid or metal fires.

Water extinguishers with additives (red band)
This type of extinguisher is suitable for Class A fires. They can also be suitable for use on Class B fires and, where appropriate, this will be indicated on the extinguisher. They are generally more efficient than conventional water extinguishers.

Foam extinguishers (cream band)
This type of extinguisher can be used on Class A or B fires and is particularly suited to extinguishing liquid fires such as petrol and diesel. They should not be used on free-flowing liquid fires unless the operator has been specially trained, as these have the potential to rapidly spread the fire to adjacent material. This type of extinguisher is not suitable for deep-fat fryers or chip pans. They should not be used on electrical or metal fires.

	 Old colour BS 5406 / New colour BS EN3	 Class A Paper or wood etc.	 Class B Flammable liquids	 Class C Flammable gas fires	 Class D Metal fires	 Electrical fires
Red	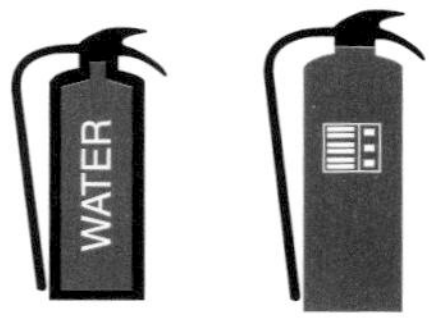 WATER	✓	Do not use ✗			Do not use ✗
Red	 Fire hose reel	✓	Do not use ✗			Do not use ✗
Cream	 FOAM	Note: Multi-purpose foams may be used ✓	Note: Specialist foams required for industrial alcohol ✓			Do not use ✗
Black	 CO₂ GAS		Secondary ✓			Primary ✓
Blue	 POWDER	✓	Note: Specialist DP required for solvents and esters ✓	✓	Note: Specialist dry powders may be required ✓	✓
Red	 Fire blanket		Primary ✓	*General note* – May be used in conjunction with other extinguishing agents or fire extinguishing techniques		
Canary yellow		 F	**Specialist hot cooking oil fires only** Specifically for dealing with high-temperature (360°C+) cooking oils used in large industrial size catering kitchens, restaurants and takeaway establishments with deep-fat frying facilities			

Figure 13.12 Types of fire extinguishers and labels. (*Note*: Main colour of all extinguishers is red with 5% for label.)

Powder extinguishers (blue band)

This type of extinguisher can be used on most classes of fire and achieve a good 'knock down' of the fire. They can be used on fires involving electrical equipment but will almost certainly render that equipment useless.

Because they do not cool the fire appreciably, it can re-ignite. Powder extinguishers can create a loss of visibility and may affect people who have breathing problems and are not generally suitable for confined spaces. They should not be used on metal fires.

Carbon dioxide extinguishers (black band)

This type of extinguisher is particularly suitable for fires involving electrical equipment as they will extinguish a fire without causing any further damage (except in the case of some electronic equipment, for example computers). As with all fires involving electrical equipment, the power should be disconnected if possible. These extinguishers should not be used on metal fires.

Wet chemical – class 'F' extinguishers

This type of extinguisher is particularly suitable for commercial catering establishments with deep-fat fryers. The intense heat in the fluid generated by fat fires means that when standard foam or carbon dioxide extinguishers stop discharging, re-ignition tends to occur.

Wet chemical extinguishers starve the fire of oxygen by sealing the burning fluid, which prevents flammable vapour reaching the atmosphere (Figures 13.12 and 13.13).

13.14.3 Fixed fire fighting equipment – sprinkler installations

Sprinklers should be considered as merely one component part of a total fire safety strategy, which is tailored to the existing and projected needs of a building. They have significant benefits to offer in suppressing fires until those best trained to deal with major incidents are on the scene to extinguish them.

Sprinklers, however, are an emotive topic. In some buildings, they have been used for a long time as the most significant element of a fire safety system. This situation is probably most prolific in warehouses and retail premises. They have unfortunately been resisted in most other buildings because of the initial capital cost and the perceived inherent risk of accidental water discharge.

Sprinkler systems can be very effective in controlling fires. They can be designed to protect life and/or property and may be regarded as a cost-effective solution for reducing the risks created by fire. If a building has a sprinkler installation,

Figure 13.13 Various fire fighting equipment.

it may have been installed as a result of a business decision, for example for the protection of business assets, or it may have been installed as a requirement, for example imposed under a local Act, or an integral part of the building design.

Sprinkler systems should normally extend to the entire building. In a well-designed system, only those heads in the immediate vicinity of the fire will actually operate. Sprinkler installations typically comprise a water supply (preferably a stored water supply incorporating tanks), pumps, pipe work and sprinkler heads. There are different types of sprinkler design; sprinklers can be operated to discharge water at roof or ceiling level or within storage racks. Other design types such as ESFR (early suppression fast response) and dry pipe may also be appropriate. In all cases, a competent person/contractor should be used to provide guidance.

The installation should be designed for the fire hazard; taking into account the building occupancy, the fire load and its burning characteristics and the sprinkler control characteristics. For each hazard the sprinkler installation design should take account of specific matters such as storage height, storage layout, ceiling clearance and sprinkler type (e.g. sprinkler orifice, sprinkler sensitivity).

There are some hazards where sprinklers should not be fitted, such as over salt baths and metal melt pans, because water will possibly cause an explosive reaction.

If any significant changes are being made to the premises, for example changing storage arrangements or material stored, the sprinkler installation should be checked to see that it is still appropriate and expert advice sought as necessary.

Sprinkler protection could give additional benefits, such as a reduction in the amount of portable fire fighting equipment necessary, and the relaxation of restrictions in the design of buildings.

Table 13.3 Maintenance and testing of fire equipment

Equipment	Period	Action
Fire-detection and fire-warning systems including self-contained smoke alarms and manually operated devices	Weekly	• Check all systems for state of repair and operation • Repair or replace defective units • Test operation of systems, self-contained alarms and manually operated devices
	Annually	• Full check and test of system by competent service engineer • Clean self-contained smoke alarms and change batteries
Emergency lighting including self-contained units and torches	Weekly	• Operate torches and replace batteries as required • Repair or replace any defective unit
	Monthly	• Check all systems, units and torches for state of repair and apparent function
	Annually	• Full check and test of systems and units by competent service engineer • Replace batteries in torches
Fire-fighting equipment installation and including hose-reels	Weekly	• Check all extinguishers including hose-reels for correct apparent function
	Annually	• Full check and test by competent service engineer

Guidance on the design and installation of new sprinkler systems and the maintenance of all systems is given in the Loss Prevention Council (LPC) Rules, BS EN 12845 or BS 5306 and should be carried out only by a competent person.

Routine maintenance by on-site personnel may include checking of pressure gauges, alarm systems, water supplies, any anti-freezing devices and automatic booster pump(s). For example diesel fire pumps should be given a test run for 30 minutes each week.

13.15 Maintenance and testing of fire equipment

It is important that equipment is fit for its purpose and is properly maintained and tested. One way in which this can be achieved is through companies that specialize in the test and maintenance of FFE.

All equipment provided to assist escape from the premises, such as fire detection and warning systems and emergency lighting, and all equipment provided to assist with fighting fire, should be regularly checked and maintained by a suitably competent person in accordance with the manufacturer's recommendations. Table 13.3 gives guidance on the frequency of test and maintenance and provides a simple guide to good practice.

13.16 Fire emergency plans

13.16.1 Introduction

Each workplace should have an emergency plan. The plan should include the action to be taken by staff in the event of fire, the evacuation procedure and the arrangements for calling the Fire and Rescue Authority.

For small workplaces, this could take the form of a simple fire action notice posted in positions where staff can read it and become familiar with it.

High-fire-risk or larger workplaces will need more detailed plans, which take account of the findings of the risk assessment, for example the staff significantly at risk and their location. For large workplaces, notices giving clear and concise instructions of the routine to be followed in case of fire should be prominently displayed. The notice should include the method of raising an alarm in the case of fire and the location of an assembly point to which staff escaping from the workplace should report.

13.16.2 Fire routines and fire notices

Site managers must make sure that all employees are familiar with the means of escape in case of fire and their use, and with the routine to be followed in the event of fire.

To achieve this, routine procedures must be set up and made known to all employees, generally outlining the action to be taken in case of fire and specifically laying down the duties of certain nominated persons. Notices should be posted throughout the premises.

While the need in individual premises may vary, there are a number of basic components which should be considered when designing any fire routine procedures:

- the action to be taken on discovering a fire;
- the method of operating the fire alarm;
- the arrangements for calling the fire brigade;
- the stopping of machinery and plant;
- first-stage fire-fighting by employees;
- evacuation of the premises;
- assembly of staff, customers and visitors, and carrying out a roll call to account for everyone on the premises.

The procedures must take account of those people who may have difficulty in escaping quickly from a building because of their location or a disability. Insurance companies and other responsible people may need to be consulted where special procedures are necessary to protect buildings and plant during or after people have been evacuated. For example there may be some special procedures necessary to ensure that a sprinkler system is operating in the event of fire.

13.16.3 Supervisory duties

A member of the staff should be nominated to supervise all fire and emergency arrangements. This person should be in a senior position or at least have direct access to a senior manager. Senior members of the staff should be appointed as departmental fire wardens, with deputies for every occasion of absence, however brief. In the event of fire or other emergency, their duties would be, while it remains safe to do so, to ensure that:

- the alarm has been raised;
- the whole department, including toilets and small rooms, has been evacuated;
- the fire and rescue service has been called;
- fire doors are closed to prevent fire spread to adjoining compartments and to protect escape routes;
- plant and machinery are shut down wherever possible and any other actions required to safeguard the premises are taken where they do not expose people to undue risks;
- a roll call is carried out at the assembly point and the result reported to whoever is in control of the evacuation.

Under normal conditions, fire wardens should check that good standards of housekeeping and preventive maintenance exist in their department, that exits and escape routes are kept free from obstruction, that all fire-fighting appliances are available for use and fire points are not obstructed, that smoking is rigidly controlled, and that all members of staff under their control are familiar with the emergency procedure and know how to use the fire alarm and FFE.

13.16.4 Assembly and roll call

Assembly points should be established for use in the event of evacuation. They should be in positions, preferably under cover, which are unlikely to be affected at the time of fire. In some cases, it may be necessary to make mutual arrangements with the occupiers of nearby premises.

In the case of small premises, a complete list of the names of all staff should be maintained so that a roll call can be made if evacuation becomes necessary.

In those premises where the number of staff would make a single roll call difficult, each departmental fire warden should maintain a list of the names of staff in their area. Roll call lists must be updated regularly.

13.16.5 Fire notices

Printed instructions for the action to be taken in the event of fire should be displayed throughout the premises. The information contained in the instructions should be stated briefly and clearly. The staff and their deputies to whom specific duties are allocated should be identified.

Instruction for the immediate calling of the fire brigade in case of fire should be displayed at telephone switchboards, exchange telephone instruments and security lodges.

A typical fire notice is given in Appendix 15.3.

13.16.6 Testing

The alarm system should be tested every week, while the premises are in normal use. The test should be carried out by activating a different call point each week, at a fixed time.

13.16.7 Fire drills

Once a fire routine has been established, it must be tested at regular intervals in order to ensure that all staff are familiar with the action to be taken in an emergency.

The most effective way of achieving this is by carrying out fire drills at prescribed intervals. Drills should be held at least twice a year other than in areas dealing with hazardous processes, where they should be more frequent. A programme of fire drills should be planned to ensure that all employees, including shift workers and part-time employees, are covered.

13.17 People with special needs

13.17.1 General

Of all the people who may be especially at risk, employers will need to pay particular attention to people who have special needs, including those with a disability.

The Disability Rights Commission estimates that 11 million people in this country have some form of disability, which may mean that they find it more difficult to leave a building if there is a fire. Under the Disability Discrimination Act (DDA), if disabled people could realistically expect to use premises, then employers or those in charge of premises must anticipate any reasonable adjustments that would make it easier for that right to be exercised.

The DDA includes the concept of 'reasonable adjustments' and this can be carried over into fire safety law. It can mean different things in different circumstances. For a small business, it may be considered reasonable to provide contrasting colours on a handrail to help people with vision impairment to follow an escape route more easily. However, it might be unreasonable to expect that same business to install an expensive voice alarm system. Appropriate 'reasonable adjustments' for a large business or organization may be much more significant.

Where people with special needs use or work on the premises, their needs should, so far as is practicable, be discussed with them. These will often be modest and may require only changes or modifications to existing procedures. You may need to develop individual 'personal emergency evacuation plans' (PEEPs) for disabled persons who frequently use a building. They will need to be confident of any plan/PEEP that is put in place after consultation with them. As part of your consultation exercise you will need to consider the matter of personal dignity.

If members of the public use the building, then those in control may need to develop a range of standard PEEPs which can be provided on request to a disabled person or others with special needs.

Guidance on removing barriers to the everyday needs of disabled people is in BS 8300. Much of this advice will also help disabled people during an evacuation.

Advice on the needs of people with a disability, including sensory impairment, is available from the organizations which represent various groups. (Names and addresses can be found in the telephone directory.)

Many businesses have discovered the problem of reconciling their duties under the DDA and RRFSO.

Employers need to consider the needs of all their staff and users of their building when fire evacuation strategies are being considered. This can include people with a surprisingly diverse range of access needs, not just those with perhaps more obvious disabilities such as wheelchair users. But access and egress for each of these groups can sometimes seem hard to reconcile with tightly controlled fire safety. For example Building Regulations require internal doors held open by electromagnetic devices to self-close when activated by smoke detectors, but a disabled person may not be able to open a heavy fire door once it has closed.

So what is more important, fulfilling fire safety requirements or fulfilling the requirements of the DDA?

Essentially, fire safety concerns life-threatening incidents, the DDA is about dignity and equal treatment. Basically, health and safety (including fire) overrides DDA if there is a conflict. However, there really should not be a problem achieving both. In respect of fire, it is never acceptable to refuse someone entry to an upper level on the basis of there not being an evacuation lift available. However, it is acceptable to pre-plan and decide to hold public meetings on the ground floor for this reason.

In other words, it should be possible to comply with fire safety legislation and comply with the DDA, as long as there are management procedures in place to make sure both are adhered to successfully. In many cases, the answer lies in planning ahead, and implementing procedures which may involve other staff in Personal Evacuation plans, for example.

13.17.2 Special needs, fire emergencies and precautions

If disabled people are going to be in your premises, then you must also provide a safe means for them to leave if there is a fire. You and your staff should be aware that disabled people may not react, or can react differently, to a fire warning or a fire. You should give similar consideration to others with special needs such as parents with young children or the elderly.

In premises with a simple layout, a common-sense approach, such as offering to help lead a blind person or helping an elderly person down steps may be enough. In more complex premises, more elaborate plans and procedures will be needed, with trained staff assigned to specified duties.

Consider the needs of people with mental disabilities or spatial recognition problems. The range of disabilities encountered can be considerable, extending from mild epilepsy to complete disorientation in an emergency situation. Many of these can be addressed by properly trained staff, discreet and empathetic use of the 'buddy system' or by careful planning of colour and texture to identify escape routes.

People with special needs (including members of the public) need special consideration when planning for emergencies. But the problems this raises are seldom great. Employers should:

➤ identify everyone who may need special help to get out;
➤ allocate responsibility to specific staff to help people with a disability in emergency situations;
➤ consider possible escape routes;
➤ enable the safe use of lifts;
➤ enable people with a disability to summon help in emergencies;
➤ train staff to be able to help their colleagues;
➤ consider safe havens.

People with impaired vision must be encouraged to familiarize themselves with escape routes, particularly those not in regular use. A 'buddy' system would be helpful. But, to take account of absences, more than one employee working near anyone with impaired vision should be taught how to help them.

Where people have hearing difficulties, particularly those who are profoundly deaf, then simply hearing the fire warning is likely to be the major difficulty. If these persons are never alone while on the premises then this may not be a serious problem, as it would be reasonable for other occupants to let them know that the building should be evacuated. If a person with hearing difficulties is likely to be alone, then consider other means of raising the alarm. Among the most popular are visual beacons and vibrating devices or pagers that are linked to the existing fire alarm.

People with impaired hearing may not hear alarms in the same way as those with normal hearing but may still be able to recognize the sound. This may be tested during the weekly alarm audibility test. There are alternative means of signalling, such as lights or other visual signs, vibrating devices or specially selected sound signals.

The Royal National Institute for Deaf People (19–23 Featherstone St, London EC1) can advise. Ask the fire brigade before installing alternative signals.

Wheelchair users or others with impaired mobility may need help to negotiate stairs, etc. Anyone selected to provide this help should be trained in the correct methods. Advice on the lifting and carrying of people can be obtained from the Fire and Rescue Service, Ambulance Service, British Red Cross Society, St John Ambulance Brigade or certain disability organizations.

Lifts should not be used as a means of escape in the event of a fire. If the power fails, the lift could stop between floors, trapping occupants in what may become a chimney of fire and smoke. But BS 5588: Part 8 provides advice on specially designed lifts for use by people with a disability in the event of a fire.

Employees with learning difficulties may also require special provision. Management should ensure that the colleagues of any employee with a learning difficulty know how to reassure them and lead them to safety.

Voice alarms

Research has shown that some people and, in particular, members of the public, do not always react quickly to a conventional fire alarm. Voice alarms are therefore becoming increasingly popular and can also incorporate a public address facility. The message or messages sent must be carefully considered. It is therefore essential to ensure that voice alarm systems are designed and installed by a person with specialist knowledge of these systems.

13.18 Sources of reference

Storage of dangerous substances, L135 Dangerous Substances and Explosive Atmospheres Regulations

2002. Approved Code of Practice and guidance, 2003, HSE Books ISBN 978 0 7176 2200 9. Dangerous substances and explosive atmospheres, L138. Dangerous Substances and Explosive Atmospheres Regulations 2002. Approved Code of Practice, 2003, HSE Books ISBN 978 0 7176 2203 2.

Fire Risk Assessment – Offices and Shops DCLG 2006 ISBN 978 1 85112 815 0.

Fire Risk Assessment – Factories and Warehouses DCLG 2006 ISBN 978 1 85112 816 7.

Relevant statutory provisions

The Management of Health and Safety at Work Regulations 1999.

The Health and Safety (Safety Signs and Signals) Regulations 1996.

Regulatory Reform (Fire Safety) Order 2005.

Fire (Scotland) Act 2005.

The Dangerous Substances and Explosive Atmospheres Regulations 2002 SI 2002 No 2776 ISBN 9780 11 042957 5.

13.19 Practice NEBOSH questions for Chapter 13

1. **Explain** the differences between the three types of Enforcement Notice which may be issued under the Regulatory Reform (Fire Safety) Order.

2. **(i)** With reference to the fire triangle, **outline TWO** methods of extinguishing fires.
   (ii) State the ways in which persons could be harmed by a fire in work premises.

3. **(i) Explain**, using a suitable sketch, the significance of the 'fire triangle'.
   (ii) Identify FOUR different types of ignition source that may cause a fire to occur, **AND** give a typical workplace example of **EACH** type.
   (iii) For each type of ignition source identified in (b), **outline** the precautions that could be taken to prevent a fire starting.

4. **(i) Identify FOUR** classes of fire **AND** the associated fuel sources.
   (ii) Identify FOUR types of portable fire extinguisher **AND** in **EACH** case state the class of fire on which they should be used.

5. **Identify** the **FOUR** methods of heat transfer and **explain** how **EACH** can cause the spread of fire.

6. Electricity is one of the causes of workplace fires.
   (i) Outline how fires could be caused by electricity.
   (ii) List TWO types of extinguisher that can be used safely on 'electrical' fires.
   (iii) Outline measures that should be taken to minimize the risk of fire from electrical equipment.
   (iv) Explain why water should not be used on fires involving electrical equipment.

7. **List EIGHT** ways of reducing the risk of a fire starting in a workplace.

8. **Identify** the **FIVE** stages of a fire risk assessment in a low-risk workplace.

9. **Outline** the factors to consider when carrying out a fire risk assessment of a workplace.

10. In relation to a workplace fire risk assessment, **outline** the issues that should be taken into account when assessing the means of escape.

11. **Outline** the precautions necessary for the safe storage and handling of small containers containing flammable solvents.

12. With respect to the handling of flammable solvents in a workshop, **outline** types of inadequate working practices that could increase the risk of a fire or explosion.

13. **Outline** the main requirements for a safe means of escape from a building in the event of a fire.

14. **List EIGHT** features of a safe means of escape from a building in the event of fire.

15. **Give** reasons that may delay the safe evacuation of employees from a workplace during a fire.

16. **Outline** specific measures that may be needed to ensure that persons with sensory impairment and/or physical disabilities may be evacuated safely in the event of a fire.

17. **(i) Outline** the main factors to be considered in the siting of fire extinguishers.
    (ii) Outline suitable arrangements for the inspection and maintenance of fire extinguishers in the workplace.

18. (i) **Identify TWO** ways in which an alarm can be raised in the event of a fire in a workplace.
    (ii) **Outline** the issues to consider in the selection and siting of portable fire extinguishers.

19. (i) **Outline TWO** advantages **AND TWO** disadvantages of using hose reels as a means of extinguishing fires.
    (ii) **Outline** the main factors to consider in the siting of hose reels.

20. With respect to the design features of a building:
    (i) **identify TWO** types of emergency warning systems that can be installed in the building to ensure that all employees can be made aware of the need to evacuate the building
    (ii) **outline SIX** structural measures that can help to prevent the spread of fire and smoke.

21. **Outline:**
    (i) the various ways in which persons might be harmed by a fire in work premises
    (ii) the additional measures that may be required to ensure the safe evacuation, in the event of fire, of employees with a range of physical impairments.

22. **Outline** reasons and benefits for undertaking regular fire drills in the workplace.

23. **Outline** the issues that should be included in a training programme for employees on the emergency action to take in the event of fire.

Appendix 13.1 Fire risk assessment as recommended in Fire Safety Guides published by the Department for Communities and Local Government in 2006

Checklist
Follow the five key steps – Fill in the checklist – Assess your fire risk and plan fire safety

1. Fire safety risk assessment
Fire starts when heat (source of ignition) comes into contact with fuel (anything that burns) and oxygen (air).

You need to keep sources of ignition and fuel apart.

How could a fire start?
Think about heaters, lighting, naked flames, electrical equipment, hot processes such as welding or grinding, cigarettes, matches and anything else that gets very hot or causes sparks.

What could burn?
Packaging, rubbish and furniture could all burn just like the more obvious fuels such as petrol, paint, varnish and white spirit. Also think about wood, paper, plastic, rubber and foam. Do the walls or ceilings have hardboard, chipboard or polystyrene? Check outside, too.

Check:
➤ Have you found anything that could start a fire? Make a note of it.

➤ Have you found anything that could burn? Make a note of it.

2. People at risk

People at risk
Everyone is at risk if there is a fire. Think whether the risk is greater for some because of when or where they work, such as night staff, or because they're not familiar with the premises, such as visitors or customers.

Children, the elderly or disabled people are especially vulnerable.

Check:
Have you identified:

➤ who could be at risk?
➤ who could be especially at risk?

Make a note of what you have found.

3. Evaluate and act

Evaluate
First, think about what you have found in Steps 1 and 2: what are the risks of a fire starting, and what are the risks to people in the building and nearby?

Remove and reduce risk
How can you avoid accidental fires? Could a source of heat or sparks fall, be knocked or pushed into something that would burn? Could that happen the other way round?

Protect
Take action to protect your premises and people from fire.

Check:
➤ Have you assessed the risks of fire in your workplace?
➤ Have you assessed the risk to staff and visitors?
➤ Have you kept any source of fuel and heat/sparks apart?
➤ If someone wanted to start a fire deliberately, is there anything around they could use?
➤ Have you removed or secured any fuel an arsonist could use?
➤ Have you protected your premises from accidental fire or arson?
➤ **How can you make sure everyone is safe in case of fire?**
  • Will you know there is a fire?
  • Do you have a plan to warn others?
  • Who will make sure everyone gets out?
  • Who will call the fire service?
  • Could you put out a small fire quickly and stop it spreading?
➤ **How will everyone escape?**
  • Have you planned escape routes?
  • Have you made sure people will be able to safely find their way out, even at night if necessary?
  • Does all your safety equipment work?
  • Will people know what to do and how to use equipment?

Make a note of what you have found.

4. Record, plan and train

Record

Keep a record of any fire hazards and what you have done to reduce or remove them. If your premises are small, a record is a good idea. If you have five or more staff or have a licence, then you must keep a record of what you have found and what you have done.

Plan

You must have a clear plan of how to prevent fire and how you will keep people safe in case of fire. If you share a building with others, you need to coordinate your plan with them.

Train

You need to make sure your staff know what to do in case of fire, and if necessary, are trained for their roles.

Check:

- ➤ Have you made a record of what you have found, and action you have taken?
- ➤ Have you planned what everyone will do if there is a fire?
- ➤ Have you discussed the plan with all staff?
- ➤ **Have you:**
  - informed and trained people (practised a fire drill and recorded how it went)?

- nominated staff to put in place your fire prevention measures, and trained them?
- made sure everyone can fulfil their role?
- informed temporary staff?
- consulted others who share a building with you, and included them in your plan?

5. Review

Keep your risk assessment under regular review. Over time, the risks may change.

If you identify significant changes in risk or make any significant changes to your plan, you must tell others who share the premises and, where appropriate, retrain staff.

Have you:

- ➤ made any changes to the building inside or out?
- ➤ had a fire or near miss?
- ➤ changed work practices?
- ➤ begun to store chemicals or dangerous substances?
- ➤ significantly changed your stock or stock levels? Have you planned your next fire drill?

The checklist above can help you with the fire risk assessment but you may need additional information especially if you have large or complex premises.

Source: Department for Community and Local government.

Appendix 13.2 Example form for recording significant findings as published in 2006 by the Department for Communities and Local Government in their Fire Safety Guides

Risk Assessment – Record of significant findings		
Rank assessment for	Assessment undertaken by	
Company	Date	
Address	Completed by	
	Signature	
Sheet number	Floor/area	Use
Step 1 – Identify fire hazards		
Sources of ignition	Sources of fuel	Sources of oxygen
Step 2 – People at risk		
Step 3 – Evaluate, remove, reduce and protect from risk		
(3.1) Evaluate the risk of the fire occurring		
(3.2) Evaluate the risk to people from a fire starting on the premises		
(3.3) Remove, reduce the hazards that may cause a fire		
(3.4) Remove, reduce the risks to people from a fire		
Assessment review		
Assessment review date	Completed by	Signature
Review outcome (where substantial changes have occurred a new record sheet should be used)		

Notes: 1. The risk assessment record findings should refer to plans, records or other documentary as necessary. 2. The information in this record should assist you to develop an emergency plan, to coordinate measures with other 'responsible persons' in the building, and to inform and train staff and inform other relevant persons.

Appendix 13.3 Typical fire notice

From Stocksigns.

Chemical and biological health hazards and control

14

After reading this chapter you should be able to:

1. recognize the forms of, and classification of, substances hazardous to health

2. explain the factors to be considered when undertaking a preliminary assessment of the health risks from substances commonly encountered in the workplace

3. describe the use and limitations of workplace exposure limits including the purpose of long-term and short-term exposure limits

4. distinguish between acute and chronic health effects

5. outline control measures that should be used to reduce the risk of ill-health from exposure to hazardous substances

6. outline the basic requirements related to disposal of waste and effluent (and the control of atmospheric pollution).

14.1 Introduction

Occupational health is as important as occupational safety but generally receives less attention from managers. Every year twice as many people suffer ill-health caused or exacerbated by the workplace than suffer workplace injury. Although these illnesses do not usually kill people, they can lead to many years of discomfort and pain. Such illnesses include respiratory disease, hearing problems, asthmatic conditions and back pain. Furthermore, it has been estimated in the UK that 30% of all cancers probably have an occupational link – that linkage is known for certain in 8% of cancer cases.

Work in the field of occupational health has been taking place for the last four centuries and possibly longer. The main reason for the relatively low profile for occupational health over the years has been the difficulty in linking the ill-health effect with the workplace cause. Many illnesses, such as asthma or back pain, can have a workplace cause but can also have other causes. Many of the advances in occupational health have been as a result of statistical and epidemiological studies (one well-known such study linked the incidence of lung cancer to cigarette smoking). While such studies are invaluable in the assessment of health risk, there is always an element of doubt when trying to link cause and effect. The measurement of gas and dust concentrations is also subject to doubt when a correlation is made between a measured sample and the workplace environment from which it was taken. Occupational health, unlike occupational safety, is generally more concerned with probabilities than certainties.

In this chapter, chemical and biological health hazards will be considered – other forms of health hazard will be covered in Chapter 15.

The chemical and biological health hazards described in this chapter are covered by the following health and safety Regulations:

➤ control of Substances Hazardous to Health Regulations;
➤ control of Lead at Work Regulations;
➤ control of Asbestos at Work Regulations.

14.2 Forms of chemical agent

Chemicals can be transported by a variety of agents and in a variety of forms. They are normally defined in the following ways.

Dusts are solid particles slightly heavier than air but often suspended in it for a period of time. The size of the particles ranges from about 0.4 µm (fine) to 10 µm (coarse). Dusts are created either by mechanical processes (e.g. grinding or pulverizing) or construction processes (e.g. concrete laying, demolition or sanding), or by specific tasks (e.g. furnace ash removal). The fine dust is much more hazardous because it penetrates deep into the lungs and remains there – known as **respirable dust**. In rare cases, respirable dust enters the bloodstream directly causing damage to other organs. Examples of such fine dust are cement, granulated plastic materials and silica dust produced from stone or concrete dust. Repeated exposure may lead to permanent lung disease. Any dusts which are capable of entering the nose and mouth during breathing, are known as **inhalable dusts**.

Gases are any substances at a temperature above their boiling point. Steam is the gaseous form of water. Common gases include carbon monoxide, carbon dioxide, nitrogen and oxygen. Gases are absorbed into the bloodstream where they may be beneficial (oxygen) or harmful (carbon monoxide).

Vapours are substances which are at or very close to their boiling temperatures. They are gaseous in form. Many solvents, such as cleaning fluids, fall into this category. The vapours, if inhaled, enter the bloodstream and some can cause short-term effects (dizziness) and long-term effects (brain damage).

Liquids are substances which normally exist at a temperature between freezing (solid) and boiling (vapours and gases). They are sometimes referred to as fluids in health and safety Regulations.

Mists are similar to vapours in that they exist at or near their boiling temperature but are closer to the liquid phase. This means that there are suspended very small liquid droplets present in the vapour. A mist is produced during a spraying process (such as paint spraying). Many industrially produced mists can be very damaging if inhaled, producing similar effects to vapours. It is possible for some mists to enter the body through the skin or by ingestion with food.

Fume is a collection of very small metallic particles (less than 1 µm) which have condensed from the gaseous state. They are most commonly generated by the welding process. The particles tend to be within the respirable range (approximately 0.4–1.0 µm) and can lead to long-term permanent lung damage. The exact nature of any harm depends on the metals used in the welding process and the duration of the exposure.

14.3 Forms of biological agent

As with chemicals, biological hazards may be transported by any of the following forms of agent.

Fungi are very small organisms, sometimes consisting of a single cell, and can appear as plants (e.g. mushrooms and yeast). Unlike other plants, they cannot produce their own food but either live on dead organic matter or on living animals or plants as parasites. Fungi reproduce by producing spores, which can cause allergic reactions when inhaled. The infections produced by fungi in humans may be mild, such as athlete's foot, or severe, such as ringworm. Many fungal infections can be treated with antibiotics.

Moulds are a particular group of very small fungi which, under damp conditions, will grow on surfaces such as walls, bread, cheese, leather and canvas. They can be beneficial (penicillin) or cause allergic reactions (asthma). Asthma attacks, athlete's foot and farmer's lung are all examples of fungal infections.

Bacteria are very small single-celled organisms which are much smaller than cells within the human body. They can live outside the body and be controlled and destroyed by antibiotic drugs. There is evidence that bacteria are becoming resistant to most antibiotics. This has been caused by the widespread misuse of antibiotics. It is important to note that not all bacteria are harmful to humans. Bacteria aid the digestion of food, and babies would not survive without their aid to break down the milk in their digestive system. Legionellosis, tuberculosis and tetanus are all bacterial diseases.

Viruses are minute non-cellular organisms which can only reproduce within a host cell. They are very much smaller than bacteria and cannot be controlled by antibiotics. They appear in various shapes and are continually developing new strains. They are usually only defeated by the defence and healing mechanisms of the body. Drugs can be used to relieve the symptoms of a viral attack but cannot cure it. The common cold is a viral infection as are hepatitis, AIDS (HIV) and influenza.

14.4 Classification of hazardous substances and their associated health risks

A hazardous substance is one which can cause ill-health to people at work. Such substances may include those used directly in the work processes (glues and paints), those produced by work activities (welding fumes) or those which occur naturally (dust). Hazardous substances are classified according to the severity and type of hazard which they may present to people who may come into contact with them. The contact may occur while working or transporting the substances or might occur during a fire or accidental spillage. There are several classifications but here only the five most common will be described.

Irritant is a non-corrosive substance which can cause skin (dermatitic) or lung (bronchial) inflammation after repeated contact. People who react in this way to a particular substance are **sensitized** or **allergic** to that substance. In most cases, it is likely that the concentration of the irritant may be more significant than the exposure time. Many household substances, such as wood preservatives, bleaches and glues are irritants. Many chemicals used as solvents are also irritants (white spirit, toluene and acetone). Formaldehyde and ozone are other examples of irritants.

Corrosive substances are ones which will attack, normally by burning, living tissue. Usually strong acids or alkalis, examples include sulphuric acid and caustic soda. Many tough cleaning substances, such as kitchen oven cleaners, are corrosive as are many dishwasher crystals.

Harmful is the most commonly used classification and describes a substance which, if swallowed, inhaled or penetrates the skin, **may** pose limited health risks. These risks can usually be minimized or removed by following the instruction provided with the substance (e.g. by using personal protective equipment). There are many household substances which fall into this category including bitumen-based paints and paint brush restorers. Many chemical cleansers are categorized as harmful. It is very common for substances labelled harmful also to be categorized as irritant.

Toxic substances are ones which impede or prevent the function of one or more organs within the body, such as the kidney, liver and heart. A toxic substance is, therefore, a poisonous one. Lead, mercury, pesticides and the gas carbon monoxide are toxic substances. The effect on the health of a person exposed to a toxic substance depends on the concentration and toxicity of the substance, the frequency of the exposure and the effectiveness of the control measures in place. The state of health and age of the person and the route of entry into the body have influence on the effect of the toxic substance.

Carcinogenic substances are ones which are known or suspected of promoting abnormal development of body cells to become cancers. Asbestos, hardwood dust, creosote and some mineral oils are carcinogenic. It is very important that the health and safety rules accompanying the substance are strictly followed.

Mutagenic substances are those which damage genetic material within cells, causing abnormal changes that can be passed from one generation to another.

Each of the classifications may be identified by a symbol and a symbolic letter – the most common of these are shown in Figure 14.1.

The effects on health of hazardous substances may be either acute or chronic.

Acute effects are of short duration and appear fairly rapidly, usually during or after a single or short-term exposure to a hazardous substance. Such effects may be severe and require hospital treatment but are usually reversible. Examples include asthma-type attacks, nausea and fainting.

Chronic effects develop over a period of time which may extend to many years. The word 'chronic' means 'with time' and should not be confused with 'severe' as its use in everyday speech often implies. Chronic health effects

| Harmful (Xn) | Irritant (Xi) | Corrosive (C) | Toxic (T) | Carcinogenic (T) |

Figure 14.1 Classification symbols.

are produced from prolonged or repeated exposures to hazardous substances resulting in a gradual, latent and often irreversible illness, which may remain undiagnosed for many years. Many cancers and mental diseases fall into the chronic category. During the development stage of a chronic disease, the individual may experience no symptoms.

14.4.1 The role of COSHH

The Control of Substances Hazardous to Health Regulations (COSHH) 1988 were the most comprehensive and significant piece of health and safety legislation to be introduced since the Health and Safety at Work Act 1974. They were enlarged to cover biological agents in 1994 and further amended in 1997, 1999, 2002 and 2005. A detailed summary of these Regulations appears in Chapter 17 (Section 17.11). The Regulations impose duties on employers to protect employees and others who may be exposed to substances hazardous to their health and requires employers to control exposure to such substances.

The COSHH Regulations offer a framework for employers to build a management system to assess health risks and to implement and monitor effective controls. Adherence to these Regulations will provide the following benefits to the employer and employee:

➤ improved productivity due to lower levels of ill-health and more effective use of materials;
➤ improved employee morale;
➤ lower numbers of civil court claims;
➤ better understanding of health and safety legal requirements.

Organizations which ignore COSHH requirements will be liable for enforcement action, including prosecution, under the Regulations.

14.5 Routes of entry to the human body

There are three principal routes of entry of hazardous substances into the human body:

➤ **inhalation** – breathing in the substance with normal air intake. This is the main route of contaminants into the body. These contaminants may be chemical (e.g. solvents or welding fume) or biological (e.g. bacteria or fungi) and become airborne by a variety of modes, such as sweeping, spraying, grinding and bagging. They

enter the lungs where they have access to the bloodstream and many other organs;
➤ **absorption through the skin** – the substance comes into contact with the skin and enters through either the pores or a wound. Tetanus can enter in this way as can toluene, benzene and various phenols;
➤ **ingestion** – through the mouth and swallowed into the stomach and the digestive system. This is not a significant route of entry to the body. The most common occurrences are due to airborne dust or poor personal hygiene (not washing hands before eating food) (see Figure 14.2).

Another very rare entry route is by **injection**. The abuse of compressed air lines by shooting high pressure air at the skin can lead to air bubbles entering the bloodstream. Accidents involving hypodermic syringes in a health or veterinary service setting are rare but illustrate this form of entry route.

The most effective control measures which can reduce the risk of infection from biological organisms are disinfection, proper disposal of clinical waste (including syringes), good personal hygiene and, where appropriate, personal protective equipment. Other measures include vermin control, water treatment and immunization.

There are five major functional systems within the human body – respiratory, nervous, cardiovascular (blood), urinary and integumentary (the skin).

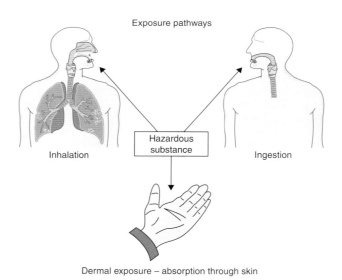

Figure 14.2 Hazardous substances – principal routes of entry into the human body.

14.5.1 The respiratory system

This comprises the lungs and associated organs (e.g. the nose). Air is breathed in through the nose, and passes through the trachea (windpipe) and the bronchi into the two lungs. Within the lungs, the air enters many smaller passageways (bronchioles) and thence to one of 300 000 terminal sacs called alveoli. The alveoli are approximately 0.1 mm across, although the entrance is much smaller. On arrival in the alveoli, there is a diffusion of oxygen into the bloodstream through blood capillaries and an effusion of carbon dioxide from the bloodstream. While soluble dust which enters the alveoli will be absorbed into the bloodstream, insoluble dust (respirable dust) will remain permanently, leading to possible chronic illness (Figure 14.3).

The whole of the bronchial system is lined with hairs, known as cilia. The cilia offer some protection against insoluble dusts. These hairs will arrest all non-respirable dust (above 5 μm) and, with the aid of mucus, pass the dust from one hair to a higher one and thus bring the dust back to the throat (this is known as the ciliary escalator). It has been shown that smoking damages this action. The nose will normally trap large particles (greater than 20 mm) before they enter the trachea. There are over 40 conditions that can affect the lungs and/or airways and impinge on the ability of a person to breathe normally.

Respirable dust tends to be long thin particles with sharp edges which puncture the alveoli walls. The puncture heals producing scar tissues which are less flexible than the original walls – this can lead to fibrosis. Such dusts include asbestos, coal, silica, some plastics and talc. The possible indicators of a dust problem in the workplace are fine deposits on surfaces, people and products or blocked filters on extraction equipment. Ill-health reports or complaints from the workforce could also indicate a dust problem.

Acute effects on the respiratory system include bronchitis and asthma and chronic effects include fibrosis and cancer. Hardwood dust, for example, can produce asthma attacks and nasal cancer.

Finally, asphyxiation, due to a lack of oxygen, is a problem in confined spaces particularly when MIG (metal inert gas) welding is taking place.

14.5.2 The nervous system

The nervous system consists primarily of the brain, the spinal cord and nerves extending throughout the body (Figure 14.4). Any muscle movement or sensation (e.g. hot and cold)

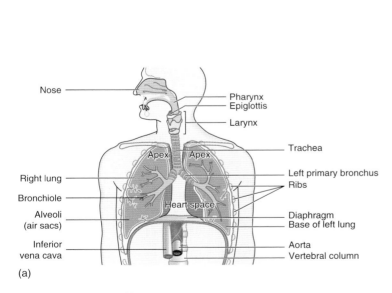

(a)

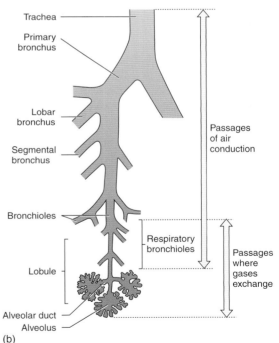

(b)

Figure 14.3 The upper and lower respiratory system.

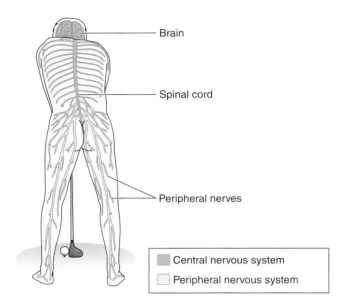

Figure 14.4 The nervous system.

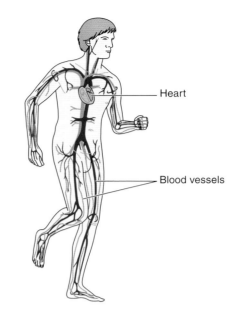

Figure 14.5 The cardiovascular system.

is controlled or sensed by the brain through small electrical impulses transmitted through the spinal cord and nervous system. The effectiveness of the nervous system can be reduced by **neurotoxins** and lead to changes in mental ability (loss of memory and anxiety), epilepsy and narcosis (dizziness and loss of consciousness). Organic solvents (trichloroethylene) and heavy metals (mercury) are well-known neurotoxins. The expression 'mad hatters' originated from the mental deterioration of top hat polishers in the 19th century who used mercury to produce a shiny finish on the top hats.

14.5.3 The cardiovascular system

The blood system uses the heart to pump blood around the body through arteries, veins and capillaries (Figure 14.5). Blood is produced in the bone marrow and consists of a plasma within which are red cells, white cells and platelets. The system has three basic objectives:

➤ to transport oxygen to vital organs, tissues and the brain and carbon dioxide back to the lungs (red cell function);
➤ to attack foreign organisms and build up a defence system (white cell function);
➤ to aid the healing of damaged tissue and prevent excessive bleeding by clotting (platelets).

There are several ways in which hazardous substances can interfere with the cardiovascular system. Benzene can affect the bone marrow by reducing the number of blood cells produced. Carbon monoxide prevents the red cells from absorbing sufficient oxygen and the effects depend on its concentration. Symptoms begin with headaches and end with unconsciousness and possibly death.

14.5.4 The urinary system

The urinary system extracts waste and other products from the blood. The two most important organs are the liver (normally considered part of the digestive system) and kidney, both of which can be affected by hazardous substances within the bloodstream (Figure 14.6).

The liver removes toxins from the blood, maintains the levels of blood sugars and produces protein for the blood plasma. Hazardous substances can cause the liver to be too active or inactive (e.g. xylene), lead to liver enlargement (e.g. cirrhosis caused by alcohol) or liver cancer (e.g. vinyl chloride).

The kidneys filter waste products from the blood as urine, regulate blood pressure and liquid volume in the body and produce hormones for making red blood cells. Heavy metals (e.g. cadmium and lead) and organic solvents (e.g. glycol ethers used in screen printing) can restrict the operation of the kidneys possibly leading to failure.

14.5.5 The skin

The skin holds the body together and is the first line of defence against infection. It regulates body temperature, is a sensing mechanism, provides an emergency food store (in the form of fat) and helps to conserve water. There are two layers – an outer layer called the epidermis (0.2 mm) and an inner layer called the dermis (4 mm). The epidermis is a tough protective layer and the dermis contains the sweat glands, nerve endings and hairs (Figure 14.7).

The most common industrial disease of the skin is **dermatitis** (non-infective dermatitis). It begins with a mild irritation on the skin and develops into blisters which can peel and weep becoming septic. It can be caused by various chemicals, mineral oils and solvents. There are two types:

➤ **irritant contact dermatitis** – occurs soon after contact with the substance and the condition reverses after contact ceases (detergents and weak acids);
➤ **allergic contact dermatitis** – caused by a sensitizer such as turpentine, epoxy resin, solder flux and formaldehyde.

Dermatitis is on the rise and is costing business more than £20 m a year, even though the cost of control measures to prevent the disease is minimal. Workers in the hotel and catering industry are particularly vulnerable to this debilitating disease.

For many years, dermatitis was seen as a 'nervous' disease which was psychological in nature. Nowadays, it

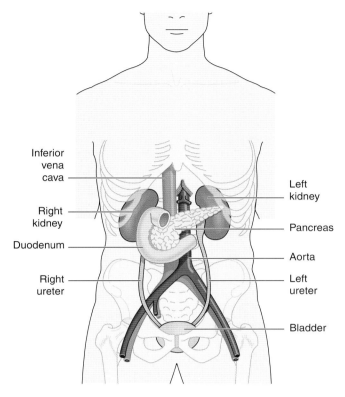

Figure 14.6 Parts of the urinary system.

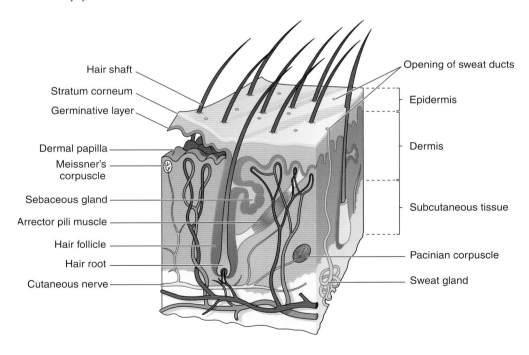

Figure 14.7 The skin – main structures in the dermis.

is recognized as an industrial disease which can be controlled by good personal hygiene, personal protective equipment, use of barrier creams and health screening of employees. Dermatitis can appear anywhere on the body but it is normally found on the hands. Therefore, gloves should always be worn when there is a risk of dermatitis.

The risks of dermatitis occurring increases with the presence of skin cuts or abrasions, which allow chemicals to be more easily absorbed, and also depend on the type, sensitivity and existing condition of the skin.

14.6 Health hazards of specific agents

The health hazards associated with hazardous substances can vary from very mild (momentary dizziness or a skin irritation) to very serious, such as a cancer.

Cancer is a serious body cell disorder in which the cells develop into tumours. There are two types of tumour – benign and malignant. Benign tumours do not spread but remain localized within the body and grow slowly. Malignant tumours are called cancers and often grow rapidly, spreading to other organs using the bloodstream and lymphatic system. Survival rates have improved dramatically in recent years as detection methods have improved and the tumours can be found in their early stages of development. A minority of cancers are believed to be occupational in origin.

Occupational asthma has approximately 4 million sufferers in the UK and it is estimated that 13 million working days are lost each year as a result of it. It is mainly caused by breathing in respiratory sensitizers, such as wood dusts, organic solvents, solder flux fumes or animal hair. The symptoms are coughing, wheezing, tightness of the chest and breathlessness due to a constriction of the airways. It can be a mild attack or a serious one that requires hospitalization. There is some evidence that stress can trigger an attack.

The following common agents of health hazards will be described together with the circumstances in which they may be found:

Ammonia is a colourless gas with a distinctive odour which, even in small concentrations, causes the eyes to smart and run and a tightening of the chest. It is a corrosive substance which can burn the skin, burn and seriously damage the eye, cause soreness and ulceration of the throat and severe bronchitis and oedema (excess of fluid) of the lungs. Good eye and respiratory protective equipment (RPE) is essential when maintaining equipment containing ammonia. Any such equipment should be tested regularly for leaks

and repaired promptly if required. Ammonia is also used in the production of fertilizers and synthetic fibres. Most work on ammonia plant should require a permit-to-work procedure.

Chlorine is a greenish, toxic gas with a pungent smell which is highly irritant to the respiratory system, producing severe bronchitis and oedema of the lungs and may also cause abdominal pain, nausea and vomiting. It is used as a disinfectant for drinking water and swimming pool water and in the manufacture of chemicals.

Organic solvents are used widely in industry as cleansing and degreasing agents. There are two main groups – the hydrocarbons (includes the aromatic and aliphatic hydrocarbons, such as toluene and white spirit) and the non-hydrocarbons (such as trichloroethylene and carbon tetrachloride). All organic solvents are heavier than air and most are sensitizers and irritants. Some are narcotics, while others can cause dermatitis and after long exposure periods liver and kidney failure. It is very important that the hazard data sheet accompanying the particular solvent is read and the recommended personal protective equipment is worn at all times.

Solvents are used extensively in a wide variety of industries as varnishes, paints, adhesives, glue strippers, printing inks and thinners. They are highest risk when used as sprays. One of the most hazardous is dichloromethane (DCM), also known as methylene chloride. It is used as a paint stripper, normally as a gel. It can produce narcotic effects and has been classified as a Category 3 carcinogen in the European Community. The minimum personal protective equipment requirements are impermeable overalls, apron, footwear, long gloves and gauntlet and chemically resistant goggles or visor. Respiratory protective equipment is also required if it cannot be demonstrated that exposure is below the workplace exposure limit (WEL).

Carbon dioxide is a colourless and odourless gas which is heavier than air. It represses the respiratory system, eventually causing death by asphyxiation. At low concentrations it will cause headaches and sweating followed by a loss of consciousness. The greatest hazard occurs in confined spaces, particularly where the gas is produced as a by-product.

Carbon monoxide is a colourless, tasteless and odourless gas which makes it impossible to detect without special measuring equipment. As explained earlier, carbon monoxide enters the blood (red cells) more readily than oxygen and restricts the supply of oxygen to vital organs. At low concentrations in the blood (less than 5 %), headaches and breathlessness will occur, while at higher concentrations unconsciousness and death will result. The most common occurrence of carbon monoxide is as an exhaust gas either from a vehicle or a heating system. In either case, it results from inefficient combustion and, possibly, poor maintenance.

Isocyanates are volatile organic compounds widely used in industry for products such as printing inks, adhesives, and two-pack paints (particularly in vehicle body shops) and in the manufacture of plastics (polyurethane products). They are irritants and sensitizers. Inflammation of the nasal passages, the throat and bronchitis are typical reactions to many isocyanates. When a person becomes sensitized to an isocyanate, very small amounts of the substance often provoke a serious reaction similar to an extreme asthma attack. Isocyanates also present a health hazard to fire fighters. They are subject to a Workplace Exposure Limit (WEL) and Respiratory Protective Equipment (RPE) should be worn.

Asbestos appears in three main forms – crocidolite (blue), amosite (brown) and chrysotile (white). The blue and brown asbestos are considered to be the most dangerous and may be found in older buildings where they were used as heat insulators around boilers and hot water pipes and as fire protection of structure. White asbestos has been used in asbestos cement products and brake linings. It is difficult to identify an asbestos product by its colour alone – laboratory identification is usually required. Many asbestos-containing materials (ACMs) are difficult to distinguish from other materials. It is easy to drill or cut ACMs unwittingly and release large quantities of airborne fibres that could cause long-term health problems to the operator. Asbestos produces a fine fibrous dust of respirable dust size which can become lodged in the lungs. The fibres can be very sharp and hard causing damage to the lining of the lungs over a period of many years. This can lead to one of the following diseases:

➤ asbestosis or fibrosis (scarring) of the lungs;
➤ lung cancer;
➤ mesothelioma – cancer of the lining of the lung or, in rarer cases, the abdominal cavity.

If asbestos is discovered during the performance of a contract, work should cease immediately and the employer informed. Typical sites of asbestos include ceiling tiles, asbestos cement roof and wall sheets, sprayed asbestos coatings on structural members, loft insulation and asbestos gaskets. Asbestos has its own Regulations (Control of Asbestos Regulations) and a summary of these is given in Chapter 17. These cover the need for a risk assessment, a method statement covering removal and disposal, air-monitoring procedures and the control measures (including personal protective equipment and training) to be used.

Training is required for the majority of workers involved in maintenance, refurbishment and demolition. The Health and Safety Executive (HSE) has estimated that approximately 50% of buildings still contain some form of asbestos and about 1.5 million workers require asbestos training.

The most recent Asbestos Regulations have removed textured coatings (decorative products such as Artex, Wondertex and Pebblecoat) from the asbestos licensing regime. Until 1992, these products contained white asbestos. This amendment by the HSE followed research work by the Health and Safety Laboratory. The new Regulations introduce an additional training requirement for asbestos awareness training. Such training should include:

➤ the health risks caused by exposure to asbestos;
➤ the materials that are likely to contain asbestos and where they are likely to be found;
➤ the methods to reduce asbestos risks during work;
➤ the action to take in an emergency, such as an uncontrolled release of asbestos dust.

Asbestos is the single biggest workplace killer. According to HSE statistics, there are 15 times as many deaths from asbestos as there are deaths from workplace accidents. Asbestos is responsible for at least 4000 deaths in the UK each year, and the HSE felt that there was a need to increase awareness amongst the workforce of the risks associated with this material.

Lead is a heavy, soft and easily worked metal. It is used in many industries but is most commonly associated with plumbing and roofing work. Lead enters the body normally by inhalation but can also enter by ingestion and skin contact. The main targets for lead are the spinal cord and the brain, the blood and blood production. The effects are normally chronic and develop as the quantity of lead builds up. Headaches and nausea are the early symptoms followed by anaemia, muscle weakening and (eventually) coma. Regular blood tests are a legal and sensible requirement as are good ventilation and the use of appropriate personal protective equipment. High personal hygiene standards and adequate welfare (washing) facilities are essential and must be used before smoking or food is consumed. The reduction in the use of leaded petrol was an acknowledgement of the health hazard represented by lead in the air. Lead is covered by its own set of regulations – the Control of Lead at Work Regulations (summarized in Chapter 17). These Regulations require risk assessments to be undertaken and engineering controls to be in place. They also recognize that lead can be transferred to an unborn child through the placenta and, therefore, offer additional protection to women of reproductive capacity. Medical surveillance in the form of a blood test of all employees who come into contact with lead operations, is required by the Regulations. Such tests should take place at least once a year.

Silica is the main component of most rocks and is a crystalline substance made of silicon and oxygen. It occurs in quartz (found in granite), sand and flint. Harm is caused by the inhalation of silica dust, which can lead to silicosis (acute and chronic), fibrosis and pneumoconiosis. The dust which causes the most harm is respirable dust which becomes trapped in the alveoli. This type of dust is sharp and very hard and, probably, causes wounding and scarring of lung tissue. As silicosis develops, breathing becomes more and more difficult and eventually as it reaches its advanced stage, lung and heart failure occur. It has also been noted that silicosis can result in the development of tuberculosis as a further complication. Hard rock miners, quarrymen, stone and pottery workers are most at risk. Health surveillance is recommended for workers in these occupations at initial employment and at subsequent regular intervals. Prevention is best achieved by the use of good dust extraction systems and respiratory protective equipment.

Leptospirosis and Weil's disease is caused by a bacterium found in the urine of rats. In humans, the kidneys and liver are attacked causing high temperatures and headaches followed by jaundice and, in up to 20% of cases, it can be fatal. It enters the body either through the skin or by ingestion. The most common source is contaminated water in a river, sewer or ditch and workers, such as canal or sewer workers, are most at risk. Leptospirosis is always a risk where rats are present, particularly if the associated environment is damp. Good, impervious protective clothing, particularly wellington boots, is essential in these situations and the covering of any skin wounds. For workers who are frequently in high-risk environments (sewer workers), immunization with a vaccine may be the best protection. Weil's disease is, strictly, a severe form of leptospirosis. The symptoms of leptospirosis are similar to influenza but those for Weil's disease are anaemia, nose bleeds and jaundice. While the most common source of infection is from the urine of rats, Weil's disease has been found in other animals, such as cattle; therefore, farm and veterinary workers may also be at risk.

Legionella is an airborne bacterium and is found in a variety of water sources. It produces a form of pneumonia caused by the bacteria penetrating to the alveoli in the lungs. This disease is known as Legionnaires' disease, named after the first documented outbreak at a State Convention of the American Legion held at Pennsylvania in 1976. During this outbreak, 200 men were affected, of whom 29 died. That outbreak and many subsequent ones were attributed to air-conditioning systems. The legionella bacterium cannot survive at temperatures above 60°C but grows between 20°C and 45°C, being most virulent at 37°C. It also requires food in the form of algae and other bacteria. Control of the bacteria involves the avoidance of water temperatures between 20°C and 45°C, avoidance of water stagnation and the build-up of algae and sediments and the use of suitable water-treatment chemicals. This work is often done by a specialist contractor.

The most common systems at risk from the bacterium are:

➤ water systems incorporating a cooling tower;
➤ water systems incorporating an evaporative condenser;
➤ hot and cold water systems and other plant where the water temperature may exceed 20°C.

An Approved Code of Practice (*Legionnaires' disease – The control of legionella bacteria in water systems* – L8) was produced by the HSE in 2000. Where plant at risk of the development of legionella exists, the following is required:

➤ a written 'suitable and sufficient' risk assessment;
➤ the preparation and implementation of a written control scheme involving the treatment, cleaning and maintenance of the system;
➤ appointment of a named person with responsibility for the management of the control scheme;
➤ the monitoring of the system by a competent person;
➤ record keeping and the review of procedures developed within the control scheme.

The code of practice also covers the design and construction of hot and cold water systems and cleaning and disinfection guidance. There have been several cases of members of the public becoming infected from a contaminated cooling tower situated on the roof of a building. It is required that all cooling towers are registered with the local authority. People are more susceptible to the disease if they are older or weakened by some other illness. It is, therefore, important that residential and nursing homes and hospitals are particularly vigilant.

Hepatitis is a disease of the liver and can cause high temperatures, nausea and jaundice. It can be caused by hazardous substances (some organic solvents) or by a virus. The virus can be transmitted from infected faeces (hepatitis A) or by infected blood (hepatitis B and C). The normal precautions include good personal hygiene, particularly when handling food and in the use of blood products. Hospital workers who come into contact with blood products are at risk of hepatitis as are drug addicts who share needles. It is also important that workers at risk regularly wash their hands and wear protective disposable gloves.

Requirements of the COSHH Regulations

When hazardous substances are considered for use (or in use) at a place of work, the COSHH Regulations impose certain duties on employers and require employees to cooperate with the employer by following any measures taken to fulfil those duties. The principal requirements are as follows.

1. Employers must undertake a suitable and sufficient assessment of the health risks created by work which is liable to expose their employees to substances hazardous to health and of the steps that need to be taken by employers to meet the requirements of these Regulations (Regulation 6).
2. Employers must prevent, or where this is not reasonably practicable, adequately control the exposure of their employees to substances hazardous to health. WELs, which should not be exceeded, are specified by the HSE for certain substances. As far as inhalation is concerned, control should be achieved by means other than personal protective equipment. If, however, respiratory protective equipment, for example, is used, then the equipment must conform to HSE standards (Regulation 7). The control of exposure can only be treated as adequate if the principles of good practice (discussed under Sections 14.8.2 and 14.9.1) are applied (Schedule 2A).
3. Employers and employees must make proper use of any control measures provided (Regulation 8).
4. Employers must maintain any installed control measures on a regular basis, keep suitable records (Regulation 9) and review systems of work.
5. Monitoring must be undertaken of any employee exposed to items listed in Schedule 5 of the Regulations or in any other case where monitoring is required for the maintenance of adequate control or the protection of employees. Records of this monitoring must be kept for at least 5 years, or 40 years where employees can be identified (Regulation 10).
6. Health surveillance must be provided to any employees who are exposed to any substances listed in Schedule 6. Records of such surveillance must be kept for at least 40 years after the last entry (Regulation 11).
7. Emergency plans and procedures must be prepared to deal with accidents or incidents involving exposure to hazardous substances beyond normal day-to-day risks. This will involve warnings and communication systems to give appropriate response immediately after any incident occurs.
8. Employees who may be exposed to substances hazardous to their health must be given information, instruction and training sufficient for them to know the health risks created by the exposure and the precautions which should be taken (Regulation 12).

Details of a COSHH assessment

Not all hazardous substances are covered by the COSHH Regulations. If there is no warning symbol on the substance container or it is a biological agent which is not directly used in the workplace (such as an influenza virus), then no COSHH assessment is required. The COSHH Regulations do not apply to those hazardous substances which are subject to their own individual Regulations (asbestos, lead or radioactive substances). The COSHH Regulations do apply to the following substances:

➤ substances having occupational exposure limits as listed in the HSE publication EH40 (Occupational Exposure Limits);
➤ substances or combinations of substances listed in the Chemicals (Hazard, Information and Packaging for Supply) Regulations better known as the CHIP Regulations;
➤ biological agents connected with the workplace;
➤ substantial quantities of airborne dust (more than $10\,mg/m^3$ of total inhalable dust or $4\,mg/m^3$ of respirable dust, both 8-hour time-weighted average (TWA), when there is no indication of a lower value);
➤ any substance creating a comparable hazard which for technical reasons may not be covered by CHIP.

14.8.1 Assessment requirements

A COSHH assessment is very similar to a risk assessment but is applied specifically to hazardous substances. The HSE has suggested five steps to COSHH assessments but within these steps there are a number of sub-sections. The steps are as follows.

Step 1. Gather information about the substances, the work and working practices. Assessors should:

➤ identify the hazardous substances present or likely to be present in the workplace, the categories and numbers of persons (e.g. employees and visitors);

➤ gather information about the hazardous substances including the quantity of the substances used;

➤ identify the hazards from these substances by reviewing labels, material safety data sheets, HSE guidance and published literature;

➤ decide who could be affected by the hazardous substances and the possible routes of entry to people exposed (i.e. inhalation, ingestion or absorption). There is a need to look at both the substances and the activities where people could be exposed to hazardous substances.

Step 2. Evaluate the risks to health either individually or collectively. Assessors should:

➤ evaluate the risks to health including the duration and frequency of the exposure of those persons to the substances;

➤ evaluate the level of exposure, for example the concentration and length of exposure to any airborne dusts, gases, fumes or vapours;

➤ consider any WELs (see Section 14.8.2);

➤ decide if existing and potential exposure presents any insignificant risks to health or they pose a significant risk to health.

Step 3. Decide what needs to be done to control the exposure to hazardous substances. Assessors should:

➤ evaluate the existing control measures, including any PPE and RPE, for their effectiveness (using any available records of environmental monitoring) and compliance with relevant legislation;

➤ decide on additional control measures, if any are required (see Section 14.9);

➤ decide what maintenance and supervision of the use of the control measures are needed;

➤ plan what to do in an emergency;

➤ set out how exposure should be monitored;

➤ decide what, if any, health surveillance is necessary;

➤ decide what information, instruction and training is required.

Step 4. Record the assessment. Assessors should:

➤ decide if a record is required (five or more employed, and significant findings);

➤ decide on the format of the record;

➤ decide on storage and how to keep records available to employees, safety representatives, etc.

Step 5. Review the assessment. Assessors should:

➤ decide when a review is necessary (e.g. changes in substances used, processes or people exposed);

➤ decide what needs to be reviewed.

An example of a typical form that can be used for a COSHH assessment is given in Appendix 14.1.

It is important that the assessment is conducted by somebody who is competent to undertake it. Such competence will require some training, the extent of which will depend on the complexities of the workplace. For large organizations with many high-risk operations, a team of competent assessors will be needed. If the assessment is simple and easily repeated, a written record is not necessary. In other cases, a concise and dated record of the assessment together with recommended control measures should be made available to all those likely to be affected by the hazardous substances. The assessment should be reviewed on a regular basis, particularly when there are changes in work process or substances or when adverse ill-health is reported.

14.8.2 Workplace exposure limits

One of the main purposes of a COSHH assessment is to adequately control the exposure of employees and others to hazardous substances. This means that such substances should be reduced to levels which do not pose a health threat to those exposed to them day after day at work. Under the 2006 amendments to the COSHH Regulations 2002, the HSE has assigned WELs to a large number of hazardous substances and publishes any updates in a publication called 'Occupational Exposure limits' EH40. The WEL is related to the concentration of airborne hazardous substances that people breathe over a specified period of time – known as 'time-weighted average'. Before the introduction of WELs, there were two types of exposure limit published – the maximum exposure limit (MEL) and the occupational exposure standard (OES).

The COSHH (Amendment) Regulations 2006 replaced the OES/MEL system with a single WEL. This removed the concern of HSE that the OES was seen as a 'safe' limit rather than a 'likely safe' limit. Hence, the WEL must not be exceeded. Hazardous substances which have been assigned a WEL fall into two groups.

1. Substances which are carcinogenic or mutagenic (having a risk phase R45, R46 or R49) or could cause occupational asthma (having a risk phase R42, or R42/43 or listed in section C of the HSE publication 'Asthmagen? Critical assessment for the agents implicated in occupational asthma' as updated from time to time) or are listed in Schedule 1 of the COSHH Regulations. These are substances which were assigned a MEL before 2005. The level of exposure to these substances should be reduced as far as is reasonably practicable.

2. All other hazardous substances which have been assigned a WEL. Exposure to these substances by inhalation must be controlled adequately to ensure that the WEL is not exceeded. These substances were previously assigned an OES before 2005. For these substances, employers should achieve adequate control of exposure by inhalation by applying the principles of good practice outlined in the Approved Code of Practice and listed in Section 14.9.1. The implication of these principles is discussed later in this chapter.

The WELs are subject to time-weighted averaging. There are two such Time-weighted Averages (TWA): the long-term exposure limit (LTEL) or 8-hour reference period and the short-term exposure limit (STEL) or 15-minute reference period. The 8-hour TWA is the maximum exposure allowed over an 8-hour period so that if the exposure period was less than 8 hours the WEL is increased accordingly with the proviso that exposure above the LTEL value continues for no longer than 1 hour. Table 14.1 shows some typical WELs for various hazardous substances.

For example, if a person was exposed to a hazardous substance with a WEL of 100 mg/m^3 (8-hour TWA) for 4 hours, no action would be required until an exposure level of 200 mg/m^3 was reached (further exposure at levels between 100 and 200 mg/m^3 should be restricted to 1 hour).

If, however, the substance has an STEL of 150 mg/m^3, then action would be required when the exposure level rose above 150 mg/m^3 for more than 15 minutes.

The STEL always takes precedence over the LTEL. When a STEL is not given, it should be assumed that it is three times the LTEL value.

The publication EH40 is a valuable document for the health and safety professional as it contains much additional advice on hazardous substances for use during the assessment of health risks, particularly where new medical information has been made public. The HSE is constantly revising WELs and introducing new ones and it is important to refer to the latest publication of EH40.

It is important to stress that if a WEL from Group 1 is exceeded, the process and use of the substance should cease immediately and employees should be removed from the immediate area until it can be made safe. In the longer term the process and the control and monitoring measures should be reviewed and health surveillance of the affected employees considered.

The over-riding requirement for any hazardous substance which has a WEL from Group 1, is to reduce exposure to as low as is reasonably practicable.

Finally, there are certain limitations on the use of the published WELs:

➤ They are specifically quoted for an 8-hour period (with an additional STEL for many hazardous substances). Adjustments must be made when exposure occurs over a continuous period longer than 8 hours.
➤ They can only be used for exposure in a workplace and not to evaluate or control non-occupational exposure (e.g. to evaluate exposure levels in a neighbourhood close to the workplace, such as a playground).
➤ WELs are only approved where the atmospheric pressure varies from 900 to 1100 millibars. This could exclude their use in mining and tunnelling operations.
➤ They should also not be used when there is a rapid build-up of a hazardous substance due to a serious accident or other emergency. Emergency arrangements should cover these eventualities.

The fact that a substance has not been allocated a WEL does not mean that it is safe. The exposure to these substances should be controlled to a level which nearly all of the working population could experience all the time without any adverse effects to their health. Detailed guidance and references are given for such substances in the HSE ACOP and guidance to the COSHH Regulations (L5).

HSE publishes a revised ACOP and EH40 to include these changes. More information on the latest COSHH Regulations is given in Chapter 17.

Table 14.1 Examples of workplace exposure limits (WELs)		
Group 1 WELs	**LTEL (8 h TWA)**	**STEL (15 min)**
All isocyanates	0.02 mg/m^3	0.07 mg/m^3
Styrene	430 mg/m^3	1080 mg/m^3
Group 2 WELs		
Ammonia	18 mg/m^3	25 mg/m^3
Toluene	191 mg/m^3	574 mg/m^3

14.8.3 Sources of information

There are other important sources of information available for a COSHH assessment in addition to the HSE Guidance Note EH40.

Product labels include details of the hazards associated with the substances contained in the product and any precautions recommended. They may also bear one or more of the CHIP hazard classification symbols.

Material safety data sheets are another very useful source of information for hazard identification and associated advice. Manufacturers of hazardous substances are obliged to supply such sheets to users, giving details of the name, chemical composition and properties of the substance. Information on the nature of the health hazards and any relevant exposure standard (WEL) should also be given together with recommended exposure control measures and personal protective equipment. The sheets contain useful additional information on first aid and fire-fighting measures and handling, storage, transport and disposal information. The data sheets should be stored in a readily accessible and known place for use in the event of an emergency, such as an accidental release.

Other sources of information include trade association publications, industrial codes of practice and specialist reference manuals.

14.8.4 Survey techniques for health risks

An essential part of the COSHH assessment is the measurement of the quantity of the hazardous substance in the atmosphere surrounding the workplace. This is known as air sampling. There are four common types of air-sampling technique used for the measurement of air quality:

1. **Stain tube detectors** use direct reading glass indicator tubes filled with chemical crystals which change colour when a particular hazardous substance passes through them. The method of operation is very similar to the breathalyzer used by the police to check alcohol levels in motorists. The glass tube is opened at each end and fitted into a pumping device (either hand or electrically operated). A specific quantity of contaminated air containing the hazardous substance, is drawn through the tube and the crystals in the tube change colour in the direction of the air flow. The tube is calibrated such that the extent of the colour change along the tube indicates the concentration of the hazardous substance within the air sample (Figure 14.9).

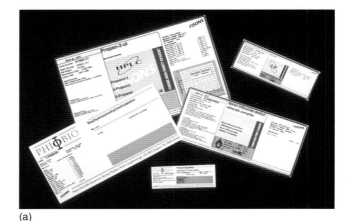

(a)

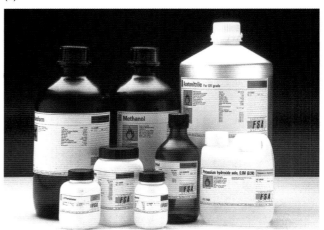

(b)

(c)

Figure 14.8 Typical product labels and material safety data sheets.

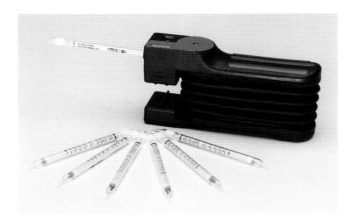

Figure 14.9 Hand pump and stain detectors.

This method can only be effective if there are no leakages within the instrument and the correct volume of sampled air is used. The instrument should be held within 30 cm of the nose of the person whose atmosphere is being tested. A large range of different tubes is available. This technique of sampling is known as **grab** or **spot** sampling since it is taken at one point.

The advantages of the technique are that it is quick, relatively simple to use and inexpensive. There are, however, several disadvantages:

➤ The instrument cannot be used to measure concentrations of dust or fume.
➤ The accuracy of the reading is approximately ±25%; it will yield a false reading if other contaminants present react with the crystals.
➤ The instrument can only give an instantaneous reading, not an average reading over the working period (TWA).
➤ The tubes are very fragile with a limited shelf life.

1. **Passive sampling** is measured over a full working period by the worker wearing a badge containing absorbent material. The material will absorb the contaminant gas and, at the end of the measuring period, the sample is sent to a laboratory for analysis. The advantages of this method over the stain tube are that there is less possibility of instrument errors and it gives a TWA reading.
2. **Sampling pumps and heads** can be used to measure gases and dusts. The worker, whose breathing zone is being monitored, wears a collection head as a badge and a battery-operated pump on his back at waist level. The pump draws air continuously through a filter, fitted in the head, which will either absorb the contaminant gas or trap hazardous dust particles. If this filter is used for dust measurement, it is sometimes called a static dust sampler. Before sampling takes place, the filter is weighed and the pump flow rate preset. After the designated testing period, it is sent to a specialist laboratory for analysis. The quantity of dust present would be determined either by measuring the weight change of the filter or by chemical analysis (e.g. for lead) or by using microscopes to count the number of fibres (e.g. for asbestos). This system is more accurate than stain tubes and gives a TWA result but can be uncomfortable to wear over long periods. Such equipment can only be used by trained personnel.
3. **Direct reading instruments** are available in the form of sophisticated analyzers which can only be used by trained and experienced operatives. Infrared gas analysers are the most common but other types of analysers are also available. They are very accurate and give continuous or TWA readings. They tend to be very expensive and are normally hired or used by specialist consultants.

Other common monitoring instruments include **vane anemometers**, used for measuring air flow speeds and hygrometers, which are used for measuring air humidity.

Qualitative monitoring techniques include smoke tubes and the dust observation lamp. **Smoke tubes** generate a white smoke which may be used to indicate the direction of flow of air – this is particularly useful when the air speed is very low or when testing the effectiveness of ventilation ducting. A **dust observation lamp** enables dust particles which are normally invisible to the human eye to be observed in the light beam. This dust is usually in the respirable range and, although the lamp does not enable any measurements of the dust to be made, it will illustrate the operation of a ventilation system and the presence of such dust.

Regulation 10 requires routine sampling or monitoring of exposure where there could be serious health effects if the controls fail, exposure limits might be exceeded or the control measures may not be working properly. Air monitoring should also be undertaken for any hazardous substances listed in Schedule 5 of the COSHH Regulations. Records of this monitoring should be kept for 5 years unless an employee is identifiable in the records, in which case they should be kept for 40 years.

The control measures required under the COSHH Regulations

14.9.1 The principles of good practice for the control of exposure to substances hazardous to health

The objective of the COSHH Regulations is to prevent ill-health due to the exposure to hazardous substances. Employers are expected to develop suitable and sufficient control measures by:

1. identifying hazards and potentially significant risks;
2. taking action to reduce and control risks;
3. keeping control measures under regular review.

In order to assist employers with these duties, the HSE has produced the following eight principles of good practice:

(a) Design and operate processes and activities to minimize the emission, release and spread of substances hazardous to health.
(b) Take into account all relevant routes of exposure – inhalation, skin absorption and ingestion – when developing control measures.
(c) Control exposure by measures that are proportionate to the health risk.
(d) Choose the most effective and reliable control options which minimize the escape and spread of substances hazardous to health.
(e) When adequate control of exposure cannot be achieved by other means, provide, in combination with other control measures, suitable personal protective equipment.
(f) Check and review regularly all elements of control measures for their continuing effectiveness.
(g) Inform and train all employees on the hazards and risks from the substances with which they work and the use of control measures developed to minimize the risks.
(h) Ensure that the introduction of control measures does not increase the overall risk to health and safety.

All these principles are embodied in the following sections on COSHH control measures.

14.9.2 Hierarchy of control measures

The COSHH Regulations require the prevention or adequate control of exposure by measures other than personal protective equipment, so far as is reasonably practicable, taking into account the degree of exposure and current knowledge of the health risks and associated technical remedies. The hierarchy of control measures is as follows:

➤ elimination;
➤ substitution;
➤ provision of engineering controls;
➤ provision of supervisory (people) controls;
➤ provision of personal protective equipment.

Examples where engineering controls are not reasonably practicable include emergency and maintenance work, short-term and infrequent exposure and where such controls are not technically feasible.

Measures for preventing or controlling exposure to hazardous substances include one or a combination of the following:

➤ elimination of the substance;
➤ substitution of the substance (or the reduction in the quantity used);
➤ total or partial enclosure of the process;
➤ local exhaust ventilation;
➤ dilution or general ventilation;
➤ reduction of the number of employees exposed to a strict minimum;
➤ reduced time exposure by task rotation and the provision of adequate breaks;
➤ good housekeeping;
➤ training and information on the risks involved;
➤ effective supervision to ensure that the control measures are being followed;
➤ personal protective equipment (such as clothing, gloves and masks);
➤ welfare (including first aid);
➤ medical records;
➤ health surveillance.

14.9.3 Preventative control measures

Prevention is the safest and most effective of the control measures and is achieved either by changing the process completely or by substituting for a less hazardous substance (the change from oil-based to water-based paints is an example of this). It may be possible to use a substance in a safer form, such as a brush paint rather than a spray.

The EU has introduced chemical safety regime REACH (Registration, Evaluation and Authorization of Chemicals) Regulations, which restrict the use of high-risk substances or substances of very high concern and require that safer substitutes must be used. Manufacturers and importers of chemicals are responsible for understanding and managing the risks associated with their products.

The Regulations apply to many common items, such as glues, paints, solvents, detergents, plastics, additives, polishes, pens and computers. The three main types of REACH dutyholder are:

➤ **Manufacturers/importers** – businesses that manufacture or import (from outside the EU) 1 tonne or more of any given substance each year are responsible for registering a dossier of information about that substance with the European Chemicals Agency. If substances are not registered, then the data on them will not be available and it will no longer be legal to manufacture or supply them within the EU. Suppliers will be obliged to carry out an inventory and identify where in the supply chain the chemicals come from. Under the REACH system, industry will also have to prepare risk assessments and provide control measures for safe use of the substance by down stream users and get community-wide authorizations for the use of any substances considered to be of high concern.

➤ **Downstream users** – downstream users include any businesses using chemicals, which probably includes most businesses in some way. Companies that use chemicals have a duty to use them in a safe way, and according to the information on risk management measures that should be passed down the supply chain.

➤ **Other users in the supply chain** – However, in order for suppliers to be able to assess these risks they need information from the downstream users about how they are used. REACH provides a framework in which information can be passed both up and down supply chains by using the safety data sheet. This should accompany materials down through the supply chain so that users are provided with the information that they need to ensure chemicals are safely managed. It is envisaged that, in the future, the safety data sheets will include information on safe handling and use.

An important aspect of chemical safety is the need for clear information about any hazardous chemical properties. The classification of different chemicals according to their characteristics (for example, those that are corrosive, or toxic to fish) currently follows an established system, which is reflected in REACH. Over the next few years,

work will be underway to establish in the EU a classification and labelling system based on the United Nations Globally Harmonized System (GHS) [Chapter 17 contains more information on REACH (Section 17.6.7) and GHS (Section 17.6.8)].

14.9.4 Engineering controls

The simplest and most efficient engineering control is the segregation of people from the process; a chemical fume cupboard is an example of this as is the handling of toxic substances in a glove box. Modification of the process is another effective control to reduce human contact with hazardous substances.

More common methods, however, involve the use of forced ventilation – local exhaust ventilation and dilution ventilation.

Local exhaust ventilation

Local exhaust ventilation removes the hazardous gas, vapour or fume at its source before it can contaminate the surrounding atmosphere and harm people working in the vicinity. Such systems are commonly used for the extraction of welding fumes and dust from woodworking machines. All exhaust ventilation systems have the following five basic components (Figure 14.10).

1. A **collection hood and intake** – sometimes this is a nozzle-shaped point which is nearest to the workpiece, while at other times it is simply a hood placed over the workstation. The speed of the air entering the intake nozzle is important; if it is too low then hazardous fume may not be removed (air speeds of up to 1 m/s are normally required).

2. **Ventilation ducting** – this normally acts as a conduit for the contaminated air and transports it to a filter and settling section. It is very important that this section is inspected regularly and any dust deposits removed. It has been known for ventilation ducting attached to a workshop ceiling to collapse under the added weight of metal dust deposits. It has also been known for them to catch fire.

3. **Filter** or **other air cleaning device** – normally located between the hood and the fan, the filter removes the contaminant from the air stream. The filter requires regular attention to remove contaminant and to ensure that it continues to work effectively.

4. **Fan** – this moves the air through the system. It is crucial that the correct type and size of fan is fitted to a given system and it should only be selected by a competent person. It should also be positioned so

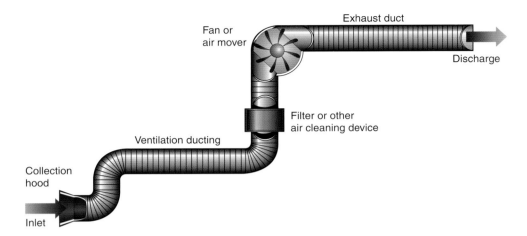

Figure 14.10 Common elements of a simple LEV system.

that it can easily be maintained but does not create a noise hazard to nearby workers.

5. **Exhaust duct** – this exhausts the air to the outside of the building. It should be checked regularly to ensure that the correct volume of air is leaving the system and that there are no leakages. The exhaust duct should also be checked to ensure that there is no corrosion due to adverse weather conditions.

Controlling Airborne Contaminants at Work, HSG258, HSE Books, is a very useful document on ventilation systems.

The COSHH Regulations require that such ventilation systems must be inspected at least every 14 months by a competent person to ensure that they are still operating effectively.

The effectiveness of a ventilation system will be reduced by damaged ducting, blocked or defective filters and poor fan performance. More common problems include the unauthorized extension of the system, poor initial design, poor maintenance, incorrect adjustments and a lack of inspection or testing.

Routine maintenance should include repair of any damaged ducting, checking filters, examination of the fan blades to ensure that there has been no dust accumulation, tightening all drive belts and a general lubrication of moving parts.

The local exhaust ventilation system will have an effect on the outside environment in the form of noise and odour. Both these problems can be reduced by regular routine maintenance of the fan and filter. The waste material from the filter may be hazardous and require the special disposal arrangements described later in this chapter (Section 14.14.4).

Dilution (or general) ventilation

Dilution (or general) ventilation uses either natural ventilation (doors and windows) or fan-assisted forced ventilation system to ventilate the whole working room by inducing a flow of clean air, using extraction fans fitted into the walls and the roof, sometimes assisted by inlet fans. It operates by either removing the contaminant or reducing its concentration to an acceptable level. It is used when airborne contaminants are of low toxicity, low concentration and low vapour density or contamination occurs uniformly across the workroom. Paint-spraying operations often use this form of ventilation as does the glass reinforced plastics (GRP) boat-building industry – these being instances where there are no discrete points of release of the hazardous substances. It is also widely used in kitchens and bathrooms. It is not suitable for dust extraction and where it is reasonably practicable to reduce levels by other means (Figure 14.11).

There are limitations to the use of dilution ventilation. Certain areas of the workroom (e.g. corners and beside cupboards) will not receive the ventilated air and a build-up of hazardous substances occurs. These areas are known as 'dead areas'. The flow patterns are also significantly affected by doors and windows being opened or the rearrangement of furniture or equipment.

14.9.5 Supervisory or people controls

Many of the supervisory controls required for COSHH purposes are part of a good safety culture and were discussed in detail in Chapter 4. These include items such as systems of work, arrangements and procedures, effective

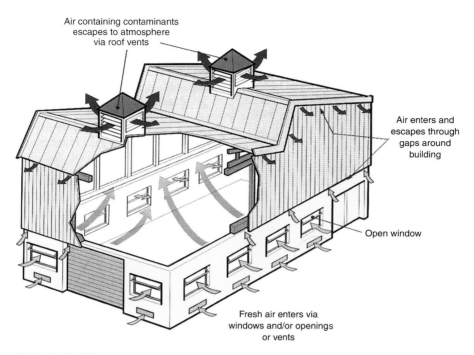

Air containing contaminants
escapes to atmosphere
via roof vents

Air enters and
escapes through
gaps around
building

Open window

Fresh air enters via
windows and/or openings
or vents

Figure 14.11 Natural ventilation in a building.

communications and training. Additional controls when hazardous substances are involved are as follows:

➤ Reduced time exposure – thus ensuring that workers have breaks in their exposure periods. The use of this method of control depends very much on the nature of the hazardous substance and its STEL.

➤ Reduced number of workers exposed – only persons essential to the process should be allowed in the vicinity of the hazardous substance. Walkways and other traffic routes should avoid any area where hazardous substances are in use.

➤ Eating, drinking and smoking must be prohibited in areas where hazardous substances are in use.

➤ Any special rules, such as the use of personal protective equipment, must be strictly enforced.

14.9.6 Personal protective equipment

Personal protective equipment is to be used as a control measure only as a last resort. It does not eliminate the hazard and will present the wearer with the maximum health risk if the equipment fails. Successful use of personal protective equipment relies on good user training, the availability of the correct equipment at all times and good supervision and enforcement (Figure 14.12).

The 'last resort' rule applies in particular to RPE within the context of hazardous substances. There are some working conditions when RPE may be necessary:

➤ during maintenance operations;

➤ as a result of a new assessment, perhaps following the introduction of a new substance;

➤ during emergency situations, such as fire or plant breakdown;

➤ where alternatives are not technically feasible.

All personal protective equipment, except respiratory protective equipment (which is covered by specific Regulations such as COSHH, lead) is controlled by its own set of Regulations – the Personal Protective Equipment at Work Regulations. A detailed summary of these Regulations is given in Chapter 17 (Section 17.25) and needs to be studied in detail with this section. The principal requirements of these Regulations are as follows:

➤ personal protective equipment which is suitable for the wearer and the task;

➤ compatibility and effectiveness of the use of multiple personal protective equipment;

➤ a risk assessment to determine the need and suitability of proposed personal protective equipment;

Figure 14.12 Personal protective equipment at work.

➤ a suitable maintenance programme for the personal protective equipment;

➤ suitable accommodation for the storage of the personal protective equipment when not in use;

➤ information, instruction and training for the user of personal protective equipment;

➤ the supervision of the use of personal protective equipment by employees and a reporting system for defects.

Types of personal protective equipment

There are several types of personal protective equipment such as footwear, hearing protectors and hard hats which are not primarily concerned with protection from hazardous substances; those which are used for such protection include:

➤ respiratory protection PPE;

➤ hand and skin protection PPE;

➤ eye protection PPE;

➤ protective clothing.

For all types of personal protective equipment, there are some basic standards that should be reached. The personal protective equipment should fit well, be comfortable to wear and not interfere with other equipment being worn or present the user with additional hazards (e.g. impaired vision due to scratched eye goggles). Training in the use of particular personal protective equipment is essential, so that it is not only used correctly, but the user knows when to change an air filter or to change a type of glove. Supervision is essential, with disciplinary procedures invoked for non-compliance with personal protective equipment rules.

It is also essential that everyone who enters the proscribed area, particularly senior managers, wear the specified personal protective equipment.

Respiratory protective equipment

Respiratory protective equipment can be subdivided into two categories – respirators (or face masks), which filter and clean the air, and breathing apparatus which supplies breathable air.

Respirators should not be worn in air which is dangerous to health, including oxygen deficient atmospheres. They are available in several different forms but the common ones are:

➤ a **filtering half mask** often called disposable respirator – made of the filtering material. It covers the nose and mouth and removes respirable size dust particles. It is normally replaced after 8–10 hours of use. It offers protection against some vapours and gases;

➤ a **half-mask respirator** – made of rubber or plastic and covering the nose and mouth. Air is drawn through a replaceable filter cartridge. It can be used for vapours, gases or dusts but it is very important that the correct filter is used (a dust filter will not filter vapours);

➤ a **full-face mask respirator** – similar to the half-mask type but covers the eyes with a visor;

➤ a **powered respirator** – a battery-operated fan delivers air through a filter to the face mask, hood, helmet or visor.

Breathing apparatus is used in one of three forms (Figure 14.13):

➤ **self-contained breathing apparatus** – where air is supplied from compressed air in a cylinder and forms a completely sealed system;

➤ **fresh air hose apparatus** – fresh air is delivered through a hose to a sealed face mask from an uncontaminated source. The air may be delivered by the wearer, by natural breathing or mechanically by a fan;

➤ **compressed air line apparatus;** – air is delivered through a hose from a compressed air line. This can be either continuous flow or on demand. The air must be properly filtered to remove oil, excess water and other contaminants and the air pressure must be reduced. Special compressors are normally used.

The selection of appropriate RPE and correct filters for particular hazardous substances is best done by a competent specialist person.

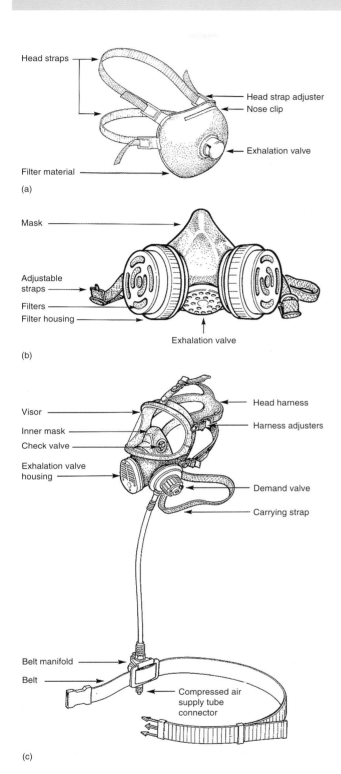

(a)

(b)

(c)

Figure 14.13 Types of respiratory protective equipment. (a) Filtering half mask, (b) half mask re-usable with filters and (c) compressed air line breathing apparatus with full-face mask fitted with demand valve.

There are several important technical standards which must be considered during the selection process. RPE must either be CE marked or HSE approved (HSE approval ceased on 30 June 1995 but such approved equipment may still be used). Other standards include the minimum protection required (MPR) and the assigned protection factor (APF). The CE mark does not indicate that the equipment is suitable for a particular hazard. The following information will be needed before a selection can be made:

➤ details of the hazardous substance, in particular whether it is a gas, vapour or dust or a combination of all three;

➤ presence of a beard or other facial hair which could prevent a good leak-free fit (a simple test to see whether the fit is tight or not is to close off the air supply, breathe in and hold the breath. The respirator should collapse onto the face. It should then be possible to check to see if there is a leak);

➤ the size and shape of the face of the wearer and physical fitness;

➤ compatibility with other personal protective equipment, such as ear defenders;

➤ the nature of the work and agility and mobility required.

Filters and masks should be replaced at the intervals recommended by the supplier or when taste or smell is detected by the wearer.

All RPE should be examined at least once a month except for disposable respirators. A record of the inspection should be kept for at least 5 years. There should be a routine cleaning system in place and proper storage arrangements.

Respiratory Protective Equipment – A Practical Guide for Users, HSG53, HSE Books, contains comprehensive advice and guidance on RPE selection, use, storage, maintenance and training, and should be consulted for more information.

Hand and skin protection

Hand and skin protection is mainly provided by gloves (arm shields are also available). A wide range of safety gloves is available for protection from chemicals, sharp objects, rough working and temperature extremes. Many health and safety catalogues give helpful guidance for the selection of gloves. For protection from chemicals, including paints and solvents, impervious gloves are recommended. These may be made of PVC, nitrile or neoprene. For sharp objects, such as trimming knives, a Kevlar-based glove is the most effective. Gloves should be regularly inspected for tears or holes since this will obviously allow skin contact to take place.

Another effective form of skin protection is the use of barrier creams and these come in two forms – pre-work and after-work. Pre-work creams are designed to provide a barrier between the hazardous substance and the skin. After-work creams are general purpose moisteners which replace the natural skin oils removed either by solvents or by washing.

Eye protection

Eye protection comes in three forms – spectacles (safety glasses), goggles and face visors. Eyes may be damaged by chemical and solvent splashes or vapours, flying particles, molten metals or plastics, non-ionizing radiation (arc welding and lasers) and dust. Spectacles are suitable for low-risk hazards (low-speed particles such as machine swarf). Some protection against scratching of the lenses can be provided but this is the most common reason for replacement. Prescription lenses are also available for people who normally wear spectacles.

Goggles are best to protect the eyes from dust or solvent vapours because they fit tightly around the eyes. Visors offer protection to the face as well as the eyes and do not steam up so readily in hot and humid environments. For protection against very bright lights, special light filtering lenses are used (e.g. in arc welding). Maintenance and regular cleaning are essential for the efficient operation of eye protection (Figure 14.14).

When selecting eye protection, several factors need to be considered. These include the nature of the hazard (the severity of the hazard and its associated risk will determine the quality of protection required), comfort and user acceptability, compatibility with other personal protective equipment, training and maintenance requirements and costs.

Protective clothing

Protective clothing includes aprons, boots and headgear (hard hats and bump caps). Aprons are normally made of PVC and protect against spillages but can become uncomfortable to wear in hot environments. Other lighter fabrics are available for use in these circumstances. Safety footwear protects against falling objects, collision with hard or sharp objects, hot or molten materials, slippery surfaces and chemical spills. It has metal toecaps and comes in the form of shoes, ankle boots or knee-length boots and is made of a variety of materials dependent on the particular hazard (e.g. thermally insulated against cold environments). It must be used with care near live, unprotected-against electricity. Specialist advice is needed for use with flammable liquids.

Appropriate selection of safety footwear involves the matching of the workplace hazards to the performance requirements of the footwear. The key issues are:

➤ the type of hazard (e.g. physical, chemical or thermal)
➤ the type of environment (e.g. indoors or out of doors)
➤ the ergonomics of the job (e.g. standing up, constant movement).

The footwear must have the correct grip for the environment, a hard wearing sole unit and, possibly, a good shock-absorbing capability.

It is important to note that appropriate personal protective equipment should be made available to work-related visitors and other members of the public visiting workplaces where hazardous substances are being used. It is also important to stress that managers and supervisors must lead by example, particularly if there is a legal requirement to wear particular personal protective equipment. Refusals by employees to wear mandatory personal protective equipment must lead to some form of disciplinary action.

In 2002, some extra requirements were added to the Personal Protective Equipment Regulations.

➤ The personal protective equipment must satisfy the basic health and safety requirements which are applicable to that class or type of personal protective equipment.
➤ The appropriate conformity assessment procedures must be carried out.

Figure 14.14 Variety of eye protection goggles.

➤ CE marking must be correctly affixed.

➤ The personal protective equipment must not compromise the safety of individuals, domestic animals or property when properly maintained and used.

14.10 Health surveillance and personal hygiene

Health surveillance enables the identification of those employees most at risk from occupational ill-health. It should not be confused with health monitoring procedures such as pre-employment health checks or drugs and alcohol testing, but it covers a wide range of situations, from a responsible person looking for skin damage on hands to medical surveillance by a medical doctor. Health surveillance detects the start of an ill-health problem and collects data on ill-health occurrences. It also gives an indication of the effectiveness of the control procedures. Health surveillance is needed to protect workers, identify as early as possible any health changes related to exposure and warn of any lapses in control arrangements.

Simple health surveillance is normally sufficient for skin problems and takes the form of skin inspections by a 'responsible person'. A responsible person is a person who has been trained by a competent medical practitioner.

It is required when there appears to be a reasonable chance that ill-health effects are occurring in a particular workplace as a result of reviewing sickness records or when a substance listed in Schedule 6 under Regulation 11 of the COSHH Regulations is being used. Schedule 6 lists the substances and the processes in which they are used. There are a limited number of such substances. The health surveillance includes medical surveillance by an employment medical adviser or appointed doctor at intervals not exceeding 12 months. Records of the health surveillance must contain approved particulars and be kept for 40 years. The need for health surveillance is not common and further advice on the necessary procedures is available from the Employment Medical Advisory Service.

Personal hygiene has already been covered under supervisory controls. It is very important for workers exposed to hazardous substances to wash their hands thoroughly before eating, drinking or smoking. Protection against biological hazards can be increased significantly by vaccination (e.g. tetanus). Finally, contaminated clothing and overalls need to be removed and cleaned on a regular basis.

14.11 Maintenance and emergency controls

Engineering control measures will only remain effective if there is a programme of preventative maintenance available. Indeed the COSHH Regulations require that systems of adequate control are in place and the term 'adequate control' spans normal operations, emergencies and maintenance. Maintenance will involve the cleaning, testing and, possibly, the dismantling of equipment. It could involve the changing of filters in extraction plant or entering confined spaces. It will almost certainly require hazardous substances to be handled and waste material to be safely disposed of. It may also require a permit-to-work procedure to be in place since the control equipment will be inoperative during the maintenance operations. Records of maintenance should be kept for at least 5 years.

Emergencies can range from fairly trivial spillages to major fires involving serious air pollution incidents. The following points should be considered when emergency procedures are being developed:

➤ the possible results of a loss of control (e.g. lack of ventilation);

➤ dealing with spillages and leakages (availability of effective absorbent materials);

➤ raising the alarm for more serious emergencies;

➤ evacuation procedures, including the alerting of neighbours;

➤ fire-fighting procedures and organization;

➤ availability of respiratory protective equipment;

➤ information and training.

The Emergency Services should be informed of the final emergency procedures and, in the case of the Fire Service, consulted for advice during the planning of the procedures. (See Chapter 6 for more details on emergency procedures.)

14.12 The transport of hazardous substances by road

Although this topic is not in the NEBOSH National General Certificate syllabus, a brief mention of the main precautions required to safeguard the health and safety of those directly involved in the transport of hazardous substances and of general members of the public is important.

Data sheets from the manufacturer of the hazardous substance should indicate the safest method of handling it and will give information on emergency procedures (e.g. for spillages and fire). These sheets should be available to all concerned with the transportation of the substance, in particular those responsible for loading/unloading, as well as the driver. The hazardous substance should be loaded correctly on the vehicle in suitable containers and segregated from incompatible materials. There must be adequate emergency information with the substance containers and attached to the vehicle. Drivers of the vehicles must receive special training which covers issues such as emergency procedures and route planning. There should also be emergency provisions for first aid and personal protective equipment on the vehicle. Finally, the Transport of Dangerous Goods (Safety Adviser) Regulations require the appointment of a trained and competent safety adviser to ensure the safe loading, transportation and unloading of the hazardous substance. HSE has produced several guidance publications which offer more detailed advice on this topic.

14.13　An illustrative example using COSHH controls

Organic solvents are widely used throughout industry and commerce in paints, inks, glues and adhesives. The COSHH hierarchy discussed under Section 14.9 should be applied to minimize the health risks from the use of these solvents. The top of this hierarchy is to eliminate or substitute the use of the organic solvent by using a less volatile or water-based alternative. If this is not possible, then some form of engineering control should be applied, such as dilution or local exhaust ventilation. Alternatively, the workplace, where the solvents are being used, could be enclosed or isolated from the main work activities. Other engineering type controls could include the use of properly labelled anti-spill containers, the use of covered disposal units for any used cloths and the transfer of large quantities using pumping/pipe arrangements rather than simply pouring the solvent.

Supervisory controls include the reduction in the length of time any employee is exposed to the solvent and the provision of good housekeeping, such as ensuring that containers are kept closed when not in use and any spills are quickly removed. The provision of barrier creams and personal protective equipment (eye protection, gloves and aprons) and RPE may also be required. Welfare issues will include first-aid provision, washing facilities and the encouragement of high levels of personal hygiene. Smoking and the consumption of food and drink should be prohibited where there might be contamination from organic solvents. All these supervisory items should be reinforced in training sessions and employees given appropriate information on the risks associated with the solvents. Finally, some form of health surveillance will be needed so that employees who show allergies to the solvents can be treated and, possibly, assigned other duties.

14.14　Environmental considerations

Organizations must also be concerned with aspects of the environment. There will be an interaction between the health and safety policy and the environmental policy which many organizations are now developing. Many of these interactions will be concerned with good practice, the reputation of the organization within the wider community and the establishment of a good health and safety culture. The health and safety data sheet, used for a COSHH assessment, also contains information of an environmental nature covering ecological information and disposal considerations.

There are three environmental issues which place statutory duties on employers and are directly related to the health and safety function. These are:

➤ air pollution
➤ water pollution
➤ waste disposal.

The statutory duties are contained in the Environmental Protection Act (EPA) 1990 and several of its subsequent Regulations. The Act is enforced by various state agencies (the Environment Agency, Local Authorities) and these agencies have very similar powers to the HSE (e.g. enforcement, prohibition notices and prosecution). The Act is divided into nine parts but this chapter will only be concerned with Part 1 (the control of pollution into the air, water or land) and Part 2 (waste disposal).

Pollution is a term that covers more than the effect on the environment of atmospheric emissions, effluent discharges and solid waste disposal from industrial processes. It also includes the effect of noise, vibration, heat and light on the environment. This was recognized in the 1996 EU Directive on Integrated Pollution Prevention and Control, and the Pollution Prevention and Control Act 1999 replaced the Integrated Pollution Control Regulations made under Part 1 of the EPA by extending those powers to cover waste minimization, energy efficiency, noise and site restoration.

Thus the Pollution Prevention and Control Act has now replaced Part 1 of the EPA.

The Solvents Emissions Directive (SED) has produced some tougher rules on the use of solvents in industry. Local Authorities and the Environment Agency are beginning to incorporate SED provisions into Integrated Pollution Prevention and Control Permits (IPPCP). Organizations that use solvents on an industrial scale need an IPPCP to operate.

The Control of Pesticides Regulations have been in place since 1986 and the Department of Food and Rural affairs (DEFRA) are proposing to consolidate these Regulations with three Plant Protection Products Regulations to form one set of integrated Regulations (see Section 17.34.10 in Chapter 17 for more information).

14.14.1 Air pollution

The most common airborne pollutants are carbon monoxide, benzene, 1,3-butadiene, sulphur dioxide, nitrogen dioxide and lead (Figure 14.16). Air pollution is monitored by Integrated Pollution Prevention and Control (IPPC). This is a system which extends the Integrated Pollution Control (IPC) system established by Part 1 of the EPA by introducing three tiers of pollution control.

➤ **Regime A1 processes**, which are certain large-scale manufacturing processes with a potential to cause serious environmental damage to the air, water or the land. In England and Wales this is enforced by the Environment Agency. In Scotland there is a parallel system enforced by the Scottish Environmental Protection Agency.

➤ **Regime A2 processes**, which produce emissions to air, water and land with a much smaller potential to pollute than regime A1 processes. The local authority (LA–IPPC) is the regulator for these processes.

➤ **Part B processes**, which may be classified as those from less polluting industries with only emissions released to air being subject to regulatory control. For such processes local authorities are the enforcing body through Environmental Health Officers. The system is known as Local Air Pollution Control (LAPC) in Northern Ireland and Scotland. In England and Wales, it is known as Local Air Pollution Prevention and Control (LAPPC).

This division has led to some anomalies in that some Part A processes create less pollution than some Part B processes. However, the grouping of three pollution destinations under one arrangement tends to a more holistic approach. The aim of IPPC is to control pollution of the

To provide an effective occupational health programme.

Figure 14.15 Health commitment.

Figure 14.16 Example of heavy industrial pollution.

whole environment under a single enforcement system and it offers three principles to prevent and control pollution. These are:

➤ The 'Best Practicable Environmental Option (BPEO)' which considers both the environmental and economic costs and benefits of the possible options available to deal with the pollution problem. BPEO is a legal requirement for Part A processes. It normally requires a technical solution.

➤ The 'Best Available Techniques' are similar to BATNEEC (Best Available Techniques Not Entailing Excessive Cost) introduced by the EPA to minimize the overall environmental impact of a process. Part B processes only need to satisfy the BATNEEC requirement which is not restricted to pollution control technology but can include employee training and competence and building design and maintenance.

➤ 'As low as reasonably practicable' applies the same test to an environmental problem as is applied to a health and safety problem. Any high or unacceptable environmental risk should be reduced to as low as is reasonably practicable.

The EPA proscribes certain listed substances from being released to air, water or land. All proscribed processes must have authorization. An operator of a proscribed process (such as a vehicle spray booth) must apply to the Environment Agency for prior authorization to operate the process. If the application is granted, the operator must monitor emissions and report them to the Environment Agency on a yearly basis. The Agency has the power to revoke the authorization, enforce the terms of the authorization or prohibit the operation of the process. Further information on the authorization process is given in Chapter 17 (Section 17.5).

14.14.2 Water pollution

Pollution of rivers and other water courses can produce very serious effects on the health of plants and animals which rely on that water supply [Figure 14.17(a) and (b)]. The Environment Agency is responsible for coastal waters, inland fresh water and ground waters (known as 'controlled waters'). The EC Groundwater Directive seeks to protect ground water from pollution since this is a source of drinking water. Such sources can become polluted by leakage from industrial soakaways. Discharges to a sewer are controlled by the Water Industry Act that defines trade effluent and those substances which are prohibited from discharge (e.g. petroleum spirit) and the Water Resources Act that covers discharge consent to controlled waters. It is an offence to pollute any controlled waters or sewage system. If hazardous substances are being used by the organization, safety data sheets give advice on the safe disposal of any residues that remain after the particular process has been completed.

The local Water Company has a right to sample discharges into its sewers because it is required to keep a public trade effluent register. There are two lists of proscribed substances which can only be discharged into a public sewer with the permission of the Water Company.

Finally, if oil is stored on the premises, a retaining bund wall should surround the oil store. This will not only ensure that any oil leakage is contained but will also stop the contamination of ground water by fire-fighting foam in the event of a fire. This is a requirement of the Control of Pollution (Oil Storage) Regulations 2001.

(a)

(b)

Figure 14.17 (a) Water pollution from an oil spillage. (b) Water pollution from plastic and other solid waste.

14.14.3 Waste management – environmental permits

The UK produces in excess of 330 million tonnes of waste annually – a quarter of which is from households and business. The remainder derives from construction and demolition, sewage sludge, farm waste and spoils from mines and dredging of rivers. In an attempt to manage this waste, the Environmental Permitting (England and Wales) Regulations have been introduced. These Regulations are one-third of the length of the previous legislation and replace over 40 statutory instruments. They have created one single regulatory system by streamlining and integrating Waste

Management Licensing and Pollution Prevention and Control. Environmental permits will provide industry, regulators and others with a single permitting and compliance system and could include those systems for:

➤ discharge consenting;
➤ ground water authorizations;
➤ water abstraction and impoundment;
➤ radioactive substances Regulation;
➤ licensing of some waste carriers and brokers.

The Environmental Permitting Regime ('the Regime') requires operators to obtain permits for some facilities, the registration of exemptions for other facilities and ongoing supervision by regulators. The regulator may be the Environment Agency or the Local Authority depending on the type of process involved.

The aim of the Regime is to:

➤ protect the environment;
➤ deliver permitting and compliance effectively and efficiently in a way that provides increased clarity and minimizes the administrative burden on both the regulator and the operators of facilities;
➤ encourage regulators to promote best practice in the operation of regulated facilities;
➤ continue to fully implement European legislation.

An Environmental Permit is required for any of the following:

➤ an installation (which carries out the activities listed in Schedule 1 to the Regulations and any activities that are technically linked);
➤ a waste operation; or
➤ a mobile plant (carrying out either one of the Schedule 1 activities or a waste operation).

The collective term used in the EP Regulations for these installations, waste operations and mobile plant is 'regulated facility'. There may be more than one regulated facility on the same site. In such cases there are arrangements in the EP Regulations to allow all such facilities to be regulated by the same regulator and to allow, in many cases, for a single permit. It is an offence under the Regulations to operate a Regulated facility without a permit. More information on environmental permits is available in Chapter 17 (Section 17.5.5).

The single environmental permit will combine and streamline the previous waste management licence (WML) and pollution prevention and control (PPC) systems. All existing WML or PPC permits will automatically become environmental permits.

Figure 14.18 Waste collection.

14.14.4 Waste disposal

The statutory duty of care for the management of waste derives from Part 2 of the EPA (Figure 14.18). The principal requirements are as follows:

➤ to handle waste so as to prevent any unauthorized escape into the environment;
➤ to pass waste only to an authorized person as defined by EPA;
➤ to ensure that a written description accompanies all waste. The Environmental Protection (Duty of Care) Regulations 1991 requires holders or producers of waste to complete a 'Transfer Note' giving full details of the type and quantity of waste for collection and disposal. Copies of the note should be kept for at least 2 years;
➤ to ensure that no person commits an offence under the Act.

The EPA is concerned with controlled waste. Controlled waste comprises household, industrial or commercial waste. It is a criminal offence to deposit controlled waste without a WML and/or in a manner likely to cause environmental pollution or harm to human health.

The EPA also covers 'hazardous wastes' which can only be disposed of using special arrangements. These are sometimes substances which are life threatening (toxic, corrosive or carcinogenic) or highly flammable. Clinical waste falls within this category. A consignment note system accompanies this waste at all the stages to its final destination. Before hazardous waste is removed from the originating premises, a contract should be in place with a licensed carrier. Hazardous waste should be stored securely prior to collection to ensure that the environment is protected.

The Hazardous Waste Regulations, with the exclusion of Scotland, replace the Special Waste Regulations and cover many more substances; for example, computer monitors, fluorescent tubes, end-of-life vehicles and television sets. Hazardous waste is waste which can cause damage to the environment or to human health. The hazardous properties of waste covered by these Regulations are listed in Appendix 14.2 and producers of such waste may need to notify the Environment Agency. The Regulations seek to ensure that hazardous waste is safely managed and its movement is documented. The following points are important for construction sites:

➤ Sites that produce more that 200 kg of hazardous waste each year for removal, treatment or disposal must register with the Environment Agency.

➤ Different types of hazardous waste must not be mixed.

➤ Producers must maintain registers of their hazardous wastes.

Some form of training may be required to ensure that employees segregate hazardous and non-hazardous wastes on site and fully understand the risks and necessary safety precautions which must be taken. Personal protective equipment, including overalls, gloves and eye protection, must be provided and used. The storage site should be protected against trespassers, fire and adverse weather conditions. If flammable or combustible wastes are being stored, adequate fire protection systems must be in place. Finally, in the case of liquid wastes, any drains must be protected and bunds used to restrict spreading of the substance as a result of spills.

A hierarchy for the management of waste streams has been recommended by the Environment Agency.

1. Prevention – by changing the process so that the waste is not produced (e.g. substitution of a particular material).

2. Reduction – by improving the efficiency of the process (e.g. better machine maintenance).

3. Reuse – by recycling the waste back into the process (e.g. using reground waste plastic products as a feed for new products).

4. Recovery – by releasing energy through the combustion, recycling or composting of waste (e.g. the incineration of combustible waste to heat a building).

5. Responsible disposal – by disposal in accordance with regulatory requirements.

In 1998, land disposal accounted for approximately 58% of waste disposal, 26% was recycled and the remainder was incinerated with some of the energy recovered as heat. The Producer Responsibility Obligations (Packaging Waste) Regulations 1997 placed legal obligations on employers to reduce their packaging waste by either recycling or recovery as energy (normally as heat from an incinerator attached to a district heating system). A series of targets have been stipulated which will reduce the amount of waste progressively over the years. These Regulations are enforced by the Environment Agency which has powers of prosecution in the event of non-compliance.

The Landfill (England and Wales) Regulations 2002 are part of the Government's drive to encourage recycling and reduce the amount of rubbish sent to landfill sites, and this means that companies now have to either recycle or treat their waste before it is taken to landfill. The new rules have already come into force in Scotland.

The main changes are that liquid wastes are now banned from landfill and other waste must be treated before it can be passed into landfill. Businesses now need to demonstrate that their waste has been treated in either a physical, thermal, chemical or biological process.

There is more information on the EPA, IPPC and waste disposal in Chapter 17 (Section 17.5). The Environment Agency has similar powers following the introduction of the Waste Electrical and Electronic Equipment (WEEE) Directive, the Restrictions of the use of certain Hazardous Substances in electrical and electronic equipment (RoHS) and the End of Life Vehicle (ELV) Directive. The aim of the WEEE Directive is to minimize the environmental impact of electrical and electronic equipment both during their lifetime and when they are discarded. All electrical and electronic equipment must be returned to the retailer from its end user and reused or reprocessed by the manufacturer. Manufacturers must register with the Environment Agency, who will advise on these obligations. The WEEE Regulations have been amended so that producer-compliance schemes report their activities in a more precise way (by providing evidence of recycling

Figure 14.19 Electronic waste under WEEE.

in kilograms instead of tonnes). It is hoped that this will reduce the delays in recycling and lead to the recycling of smaller amounts at more frequent intervals (Figure 14.19).

14.15 Sources of reference

Step by Step Guide to COSHH Assessment HSG97, HSE Books 2004 ISBN 978 0 7176 2785 1.

Occupational Exposure Limits EH40/2005, HSE Books 2005 ISBN 978 0 7176 2977 0 (As at December 2008 any new editions should be used as they are published.)

Control of Substances Hazardous to Health ACOP, L5 (5th edition), HSE Books 2005 ISBN 978 0 7176 2981 7.

Personal Protective Equipment at Work (Guidance) L25 (2nd edition), HSE Books 2005 ISBN 978 0 7176 6139 8.

An Introduction to Local Exhaust Ventilation HSG37, HSE Books ISBN 0 7176 1001 3 (no longer published).

Respiratory Protective Equipment at Work – A Practical Guide HSG53, HSE Books 2004 ISBN 978 0 7176 2904 6.

Relevant statutory provisions

The Control of Substances Hazardous to Health Regulations 2002 and Amendments.

The Chemicals (Hazard Information and Packaging for Supply) Regulations 2002.

The Personal Protective Equipment at Work Regulations 1992.

The Environmental Protection Act 1990.

The Hazardous Waste (England and Wales) Regulations 2005.

14.16 Practice NEBOSH questions for Chapter 14

1. Health hazards in the workplace may be transported by various chemical and biological agents. **Describe THREE** chemical **AND THREE** biological hazardous agents giving an example in **EACH** case.

2. **(i) Define** the term 'respirable dust'.
    (ii) Outline methods used to measure the levels of airborne dust in the workplace.

3. For **EACH** of the following types of hazardous substance, **give** a typical example **AND state** its primary effect on the body:
    (i) toxic
    (ii) corrosive
    (iii) carcinogenic
    (iv) irritant.

4. **(i) Describe** the differences between acute and chronic health effects.
    (ii) Identify the factors that could affect the level of harm experienced by an employee exposed to a toxic substance.
    (iii) Give TWO acute **AND TWO** chronic health effects on the body from exposure to lead.

5. **(i) Identify** possible routes of entry of biological organisms into the body.
    (ii) Outline control measures that could be used to reduce the risk of infection from biological organisms.

6. **(i) Define** the term 'target organ' within the context of occupational health.
    (ii) Outline the personal hygiene practices that should be followed to reduce the risk of ingestion of a hazardous substance.
    (iii) Describe the respiratory defence mechanisms of the body against atmospheric dust.

7. **(i) Describe** the typical symptoms of occupational dermatitis.
    (ii) Identify the factors that could affect the likelihood of dermatitis occurring in workers handling dermatitic substances.

8. In relation to occupational dermatitis:
    (i) identify TWO common causative agents
    (ii) describe the typical symptoms of the condition

(iii) **state** the sources of information that may help to identify dermatitic substances in the workplace
(iv) **outline** specific measures designed to prevent the occurrence of occupational dermatitis.

9. **(i)** **Explain** the health and safety benefits of restricting smoking in the workplace.
(ii) **Outline** the ways in which an organization could effectively implement a no-smoking policy.

10. For **EACH** of the following agents, **outline** the principal health effects **AND identify** a typical workplace situation in which a person might be exposed:
(i) carbon monoxide
(ii) silica
(iii) legionella bacteria
(iv) hepatitis virus.

11. For **EACH** of the following agents, **outline** the principal health effects **AND identify** a typical workplace situation in which a person might be exposed:
(i) isocyanates
(ii) asbestos
(iii) leptospira bacteria
(iv) lead.

12. **(i)** **Identify TWO** respiratory diseases that may be caused by exposure to asbestos.
(ii) **Identify** the common sources of asbestos in buildings that should be considered when conducting an asbestos survey of work premises.

13. **Outline** the control measures that could be used to minimize the health risks from the use of organic solvents in the workplace.

14. **(i)** **Outline** the main requirements of the Control of Substances Hazardous to Health Regulations.
(ii) Briefly **outline** a typical COSHH assessment of health risks as described by the Regulations.

15. **Outline** the factors to be taken into account when undertaking an assessment of health risks from hazardous substances to be used in the workplace.

16. **(i)** **Explain** the meaning of the term 'workplace exposure limit' (WEL).
(ii) **Outline FOUR** actions management could take when a WEL has been exceeded.

17. An engineering company has noticed a recent increase in work-related ill-health among shop floor workers who use a degreasing solvent for which a workplace exposure limit (WEL) has been assigned.
(i) **Explain** the meaning of the term 'workplace exposure limit' (WEL).
(ii) **Give** reasons for the possible increase in work-related ill-health among the shop floor workers.

18. **Identify** the information that should be included on a manufacturer's safety data sheet supplied with a hazardous substance.

19. The manager of a company is concerned about a substance to be introduced into one of its manufacturing processes. **Outline FOUR** sources of information that might be consulted when assessing the risk from this substance.

20. Hazardous substances in the form of paints, solvents and cleaning chemicals are to be used for the internal re-decoration of a public library.
(i) **Identify FOUR** possible routes of entry of the hazardous substances into the body.
(ii) **Outline** the factors to consider when undertaking an assessment of the risks to health from the hazardous substances as required under the Control of Substances Hazardous to Health Regulations.
(iii) **Outline** the control measures required to prevent employees and members of the public's exposure to the hazardous substances.

21. **(i)** **Describe**, by means of a labelled sketch, a chemical indicator (stain detector) tube suitable for atmospheric monitoring.
(ii) **List** the main limitations of chemical indicator (stain detector) tubes.

22. **(i)** **List FOUR** respiratory diseases that could be caused by exposure to dust at work.
(ii) **Identify** the possible indications of a dust problem in a workplace.
(iii) **Explain** how a 'static' dust sampler is used to assess the level of airborne dust in the workplace.

23. An essential raw material for a process is delivered in powdered form and poured by hand from bags into a mixing vessel. **Outline** the control measures that might be considered in this situation in order to reduce employee exposure to the substance.

24. (i) Identify the possible indications of a dust problem in a workplace.
(ii) Describe how the body may defend itself against the harmful effects of airborne dust.
(iii) Outline, using practical examples where appropriate, the control measures that may be used to reduce levels of dust in a work environment.

25. Outline the factors that need to be considered when managing the risk of exposure to hazardous chemicals at work.

26. An employee is engaged in general cleaning activities in a large veterinary practice.
(i) Identify FOUR specific types of hazard that the cleaner might face when undertaking the cleaning.
(ii) Outline the precautions that could be taken to minimize the risk of harm from these hazards.

27. A furniture factory uses solvent-based adhesives in its manufacturing process.
(i) Identify the possible effects on the health of employees using the adhesives.
(ii) State FOUR control measures to minimize such health effects.

28. An employee is required to install glass-fibre insulation in a loft.
(i) Identify FOUR hazards connected with this activity.
(ii) Outline the precautions that might be taken to minimize harm to the employee carrying out this operation.

29. Outline the precautions to ensure the health and safety of persons engaged in paint spraying in a motor vehicle repair shop.

30. (i) Explain the meaning of the term 'dilution ventilation'.
(ii) Outline the circumstances in which the use of dilution ventilation may be appropriate.

31. A local exhaust ventilation (LEV) system is used to extract welding fume from the working environment in a fabrication workshop.
(i) Outline the factors that might reduce the effectiveness of the LEV system.
(ii) Identify the possible effects that the use of the LEV system may have on the local and wider environment.

32. Local exhaust ventilation (LEV) systems must be thoroughly examined at least every 14 months. **Outline** the routine maintenance that should be carried out between statutory examinations in order to ensure the continuing efficiency of a LEV system.

33. (i) Outline the requirements of the Personal Protective Equipment at Work Regulations.
(ii) Outline the factors which should be considered when selecting personal protective equipment (PPE).

34. An organization uses various types of personal protective equipment (PPE). **Outline** the general issues relating to the selection and use of PPE that will help to ensure its effectiveness in controlling hazards.

35. (i) Explain the difference between breathing apparatus and respirators.
(ii) Outline the main limitations of a half-mask reusable respirator.

36. (i) Identify the types of hazard against which gloves could offer protection.
(ii) Outline the practical limitations of using gloves as a means of protection.

37. (i) Identify THREE types of hazard for which personal eye protection would be required.
(ii) Outline the range of issues that should be addressed when training employees in the use of personal eye protection.

38. (i) Identify ONE advantage **AND ONE** disadvantage of safety goggles compared with safety spectacles.
(ii) Outline SIX factors to be considered in the selection of eye protection for use at work.

39. Identify FOUR different types of hazard that may necessitate the use of special footwear, explaining in **EACH** case how the footwear affords protection.

40. (i) Outline the health and safety risks associated with welding operations.
(ii) Outline the factors to be considered in the selection of respiratory protective equipment for persons carrying out welding activities.

41. Outline the meaning of the term 'health surveillance' as described by the Control of Substances Hazardous to Health Regulations.

42. A large item of process machinery is to be cleaned manually with a flammable solvent before being partially dismantled for repair.

(i) **Identify FOUR** possible health effects from exposure to the solvent.

(ii) **Outline** the safety precautions that should be taken when using such flammable solvents.

(iii) **Outline** further precautions that might be needed in order to ensure the health and safety of those carrying out the maintenance work.

43. **Outline** the main precautions to be taken to ensure the safe transport of hazardous substances by road.

44. In relation to the spillage of a toxic substance from a ruptured drum stored in a warehouse:

(i) **identify THREE** ways in which persons working in close vicinity to the spillage might be harmed

(ii) **outline** a procedure to be adopted in the event of such a spillage.

45. A company produces a range of solid and liquid wastes, both hazardous and non-hazardous. **Outline** the arrangements that should be in place to ensure the safe storage of the wastes prior to their collection and disposal.

46. Absorbent mats and granules have been used to soak up a chemical spillage. **Outline** the issues that will need to be considered in relation to the handling, temporary storage and final disposal of the waste material.

47. **Outline** the issues that should be addressed by an organization when developing a system for the safe collection and disposal of waste.

48. **Outline** the hazards that may be encountered by refuse collectors employed to remove waste from domestic premises and load it into a refuse vehicle.

49. **Identify** the hazards a skip collector could be exposed to when moving a full skip from the ground onto the back of a skip loader vehicle.

50. A full skip is being collected from outside a building. **Outline** the control measures to minimize the risks involved with this activity.

51. **Identify EIGHT** safe practices to be followed when using a skip for the collection and removal of waste from a construction site.

52. Carcinogenic and explosive are two hazardous properties of waste as listed in the Hazardous Waste (England and Wales) Regulations. **Identify EIGHT** other hazardous properties as listed in these Regulations.

Appendix 14.1 A typical set of COSHH assessment forms

COSHH 1 – DETAILS OF SUBSTANCES USED OR STORED

Name of Manager:. . . . .
Name of Department/Area: . . . .

SUBSTANCE DETAILS

1. Information from the label
    Trade name:. . . . .
    Manufacturer's name:. . . . .
    Names of any chemical constituents listed:. . . . .
    . . . .
    Hazard marking – whether corrosive, irritant, harmful, toxic, very toxic: . . . .
    RISKS Phrases noted on label (e.g. harmful in contact with skin): . . . .
    . . . .
    Safety Phrases noted on label (e.g. avoid contact with skin): . . . .
    . . . .
    . . . .

2. Have you got a Health & Safety Data Sheet for this product?   YES/NO

DETAILS OF USE

3. What it is used for? . . . .
    . . . .

4. By whom?. . . . .

5. How often?. . . . .

6. Where? . . . .

7. What CONTROL measures (precautions) are used? (e.g. local ventilation, goggles, respirator, protective gloves). . . . .
    . . . .
    . . . .
    . . . .

8. Is it ABSOLUTELY ESSENTIAL to keep/use this substance?   YES/NO

9. Can it be DISPOSED OF NOW?   YES/NO

COSHH 2 – ASSESSMENT OF A SUBSTANCE

1.   Name of substance:. . . . .

2.   The process or description of job where the substance is used:. . . . .
. . . .

3.   Location of the process where substance is used:. . . . .

4.   Health & safety information on substance:

a)   Hazards to health:. . . . .
. . . .

b)   Precautions required:. . . . .
. . . .
. . . .
. . . .

5.   Number of persons exposed:. . . . .

6.   Frequency and duration of exposure:. . . . .

7.   Control measures that are in use:. . . . .
. . . .
. . . .
. . . .

8.   The assessment, an evaluation of the risks to health:. . . . .
. . . .
. . . .
. . . .

9.   Details of steps to be taken to reduce the exposure:. . . . .
. . . .
. . . .
. . . .

10. Action to be taken by (name):. . . . . (Date): . . . .

11. Date of next assessment/review:. . . . .

12. Name and position of person making this assessment: . . . .

13. Date of assessment: . . . .

Appendix 14.2 Hazardous properties of waste as listed in the Hazardous Waste (England and Wales) Regulations 2005

	Hazard	Description
H1	Explosive	Substances and preparations which may explode under the effect of flame or which are more sensitive to shocks or friction than dinitrobenzene.
H2	Oxidizing	Substances and preparations which exhibit highly exothermic reactions when in contact with other substances, particularly flammable substances.
H3-A	Highly flammable	– liquid substances and preparations having a flash point below 21°C (including extremely flammable liquids), or – substances and preparations which may become hot and finally catch fire in contact with air at ambient temperature without any application of energy, or – solid substances and preparations which may readily catch fire after brief contact with a source of ignition and which continue to burn or to be consumed after removal of the source of ignition, or – gaseous substances and preparations which are flammable in air at normal pressure, or – substances and preparations which, in contact with water or damp air, evolve highly flammable gases in dangerous quantities.
H3-B	Flammable	Liquid substances and preparations having a flash point equal to or greater than 21°C and less than or equal to 55°C.
H4	Irritant	Non-corrosive substances and preparations which, through immediate, prolonged or repeated contact with the skin or mucous membrane, can cause inflammation.
H5	Harmful	Substances and preparations which, if they are inhaled or ingested or if they penetrate the skin, may involve limited health risks.
H6	Toxic	Substances and preparations (including very toxic substances and preparations) which, if they are inhaled or ingested or if they penetrate the skin, may involve serious, acute or chronic health risks and even death.
H7	Carcinogenic	Substances and preparations which, if they are inhaled or ingested or if they penetrate the skin, may induce cancer or increase its incidence.
H8	Corrosive	Substances and preparations which may destroy living tissue on contact.
H9	Infectious	Substances containing viable micro-organisms or their toxins which are known or reliably believed to cause disease in humans or other living organisms.

(Continued)

Appendix 14.2 Continued

	Hazard	Description
H10	Teratogenic	Substances and preparations which, if they are inhaled or ingested or if they penetrate the skin, may induce non-hereditary congenital malformations or increase their incidence.
H11	Mutagenic	Substances and preparations which, if they are inhaled or ingested or if they penetrate the skin, may induce hereditary genetic defects or increase their incidence.
H12		Substances and preparations which release toxic or very toxic gases in contact with water, air or an acid.
H13		Substances and preparations capable by any means, after disposal, of yielding another substance, e.g. a leachate, which possesses any of the characteristics listed above.
H14	Ecotoxic	Substances and preparations which present or may present immediate or delayed risks for one or more sectors of the environment.

Physical and psychological health hazards and control

After reading this chapter you should be able to:

1. identify work processes and practices that may give rise to musculoskeletal health problems (in particular work-related upper limb disorders – WRULD) and suggest practical control measures
2. identify common welfare and work environment requirements in the workplace
3. describe the health effects associated with exposure to noise and suggest appropriate control measures
4. describe the health effects associated with exposure to vibration and suggest appropriate control measures
5. describe the principal health effects associated with ionising and non-ionising radiation and outline basic protection techniques
6. explain the causes and effects of stress at work and suggest appropriate control actions
7. describe the situations that present a risk of violence towards employees and suggest ways of minimizing such risk.

15.1 Introduction

Occupational health is concerned with physical and psychological hazards as well as chemical and biological hazards. The physical occupational hazards have been well known for many years and the recent emphasis has been on the development of lower risk workplace environments. Physical hazards include topics such as electricity and manual handling, which were covered in earlier chapters and noise, display screen equipment (DSE) and radiation, which are discussed in this chapter.

However, it is only really in the last 20 years that psychological hazards have been included among the occupational health hazards faced by many workers. This is now the most rapidly expanding area of occupational health and includes topics such as mental health and workplace stress, violence to staff, passive smoking, drugs and alcohol.

The physical and psychological hazards discussed in this chapter are covered by the following health and safety Regulations:

➤ workplace (Health, Safety and Welfare) Regulations;
➤ health and Safety (Display Screen Equipment) Regulations;
➤ manual Handling Operations Regulations;
➤ noise at Work Regulations;
➤ ionising Radiations Regulations.

15.2 Task and workstation design

15.2.1 The principles and scope of ergonomics

Ergonomics is the study of the interaction between workers and their work in the broadest sense, in that it encompasses the whole system surrounding the work process. It is, therefore, as concerned with the work organization, process and design of the workplace and work methods as it is with work equipment. The common definitions of ergonomics, the 'man–machine interface' or 'fitting the man to the machine rather than vice versa' are far too narrow. It is concerned about the physical and mental capabilities of an individual as well as their understanding of the job under consideration. Ergonomics includes the limitations of the worker in terms of skill level, perception and other personal factors in the overall design of the job and the system supporting and surrounding it. It is the study of the relationship between the worker, the machine and the environment in which it operates and attempts to optimize the whole work system, including the job, to the capabilities of the worker so that maximum output is achieved for minimum effort and discomfort by the worker. Cars, buses and lorries are all ergonomically designed so that all the important controls, such as the steering wheel, brakes, gear stick and instrument panel are easily accessed by most drivers within a wide range of sizes. Ergonomics is sometimes described as human engineering and as working practices become more and more automated, the need for good ergonomic design becomes essential.

The scope of ergonomics and an ergonomic assessment is very wide incorporating the following areas of study:

➤ personal factors of the worker, in particular physical, mental and intellectual abilities, body dimensions and competence in the task required;
➤ the machine and associated equipment under examination;
➤ the interface between the worker and the machine – controls, instrument panel or gauges and any aids including seating arrangements and hand tools;
➤ environmental issues affecting the work process such as lighting, temperature, humidity, noise and atmospheric pollutants;

➤ the interaction between the worker and the task, such as the production rate, posture and system of working;
➤ the task or job itself – the design of a safe system of work, checking that the job is not too strenuous or repetitive and the development of suitable training packages;
➤ the organization of the work, such as shift work, breaks and supervision.

The reduction of the possibility of human error is one of the major aims of ergonomics and an ergonomic assessment. An important part of an ergonomic study is to design the workstation or equipment to fit the worker. For this to be successful, the physical measurement of the human body and an understanding of the variations in these measurements between people are essential. Such a study is known as **anthropometry**, which is defined as the scientific measurement of the human body and its movement. Since there are considerable variations in, for example, the heights of people, it is common for some part of the workstation to be variable (e.g. an adjustable seat) (Figure 15.1).

15.2.2 The ill-health effects of poor ergonomics

Ergonomic hazards are those hazards to health resulting from poor ergonomic design. They generally fall within the physical hazard category and include the manual handling and lifting of loads, pulling and pushing loads, prolonged periods of repetitive activities and work with vibrating tools. The condition of the working environment, such as low lighting levels, can present health hazards to the eyes. It is also possible for psychological conditions, such as occupational stress, to result from ergonomic hazards.

The common ill-health effects of ergonomic hazards are musculoskeletal disorders [back injuries, covered in Chapter 10, and work-related upper limb disorders

Before

After

Figure 15.1 Workstation ergonomic design improvements.

(WRULDs) including repetitive strain injury (RSI) being the main disorders] and deteriorating eyesight.

Work-related upper limb disorders

WRULDs describe a group of conditions which can affect the neck, shoulders, arms, elbows, wrists, hands and fingers. **Tenosynovitis** (affecting the tendons), **carpal tunnel syndrome** (affecting the tendons which pass through the carpal bone in the hand) and **frozen shoulder** are all examples of WRULDs which differ in the manifestation and site of the illness. The term **RSI** is commonly used to describe WRULDs.

WRULDs are caused by repetitive movements of the fingers, hands or arms which involve pulling, pushing, reaching, twisting, lifting, squeezing or hammering. These disorders can occur to workers in offices as well as in factories or on construction sites. Typical occupational groups at risk include painters and decorators, riveters and pneumatic drill operators and desktop computer users.

The main symptoms of WRULDs are aching pain to the back, neck and shoulders, swollen joints and muscle fatigue accompanied by tingling, soft tissue swelling, similar to bruising, and a restriction in joint movement. The sense of touch and movement of fingers may be affected. The condition is normally a chronic one in that it gets worse with time and may lead eventually to permanent damage. The injury occurs to muscle, tendons and/or nerves. If the injury is allowed to heal before being exposed to the repetitive work again, no long-term damage should result. However, if the work is repeated again and again, healing cannot take place and permanent damage can result leading to a restricted blood flow to the arms, hands and fingers.

The risk factors, which can lead to the onset of WRULDs, are repetitive actions of lengthy duration, the application of significant force and unnatural postures, possibly involving twisting and overreaching and the use of vibrating tools. Cold working environments, work organization and worker perception of the work organization have all been shown in studies to be risk factors, as is the involvement of vulnerable workers such as those with pre-existing ill-health conditions and pregnant women.

Ill-health due to vibration

Hand-held vibrating machinery (such as pneumatic drills, sanders and grinders, powered lawn mowers and strimmers and chainsaws) can produce health risks from hand–arm or whole-body vibration (WBV).

Hand–arm vibration syndrome (HAVS)

HAVS describes a group of diseases caused by the exposure of the hand and arm to external vibration. Some of these have been described under WRULDs, such as carpal tunnel syndrome.

However, the best known disease is **vibration white finger (VWF)** in which the circulation of the blood, particularly in the hands, is adversely affected by the vibration. The early symptoms are tingling and numbness felt in the fingers, usually sometime after the end of the working shift. As exposure continues, the tips of the fingers go white and then the whole hand may become affected. This results in

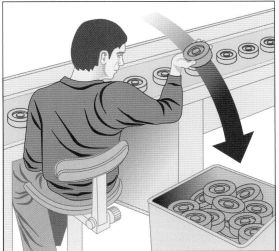 

Figure 15.2 Poor workstation layout may cause WRULDs.

Regular exposure to HAV can cause a range of permanent injuries to hands and arms, collectively known as hand–arm vibration syndrome (HAVS). The injuries can include damage to the:

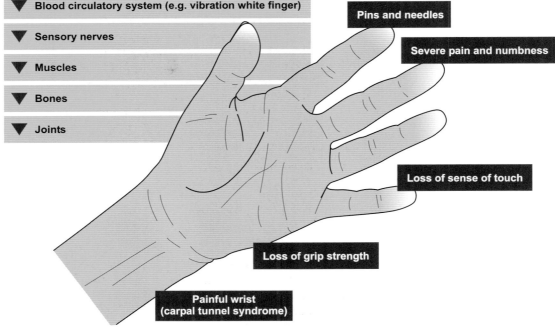

▼ Blood circulatory system (e.g. vibration white finger)

▼ Sensory nerves

▼ Muscles

▼ Bones

▼ Joints

Pins and needles

Severe pain and numbness

Loss of sense of touch

Loss of grip strength

Painful wrist
(carpal tunnel syndrome)

Figure 15.3 Injuries which can be caused by hand–arm vibration.

a loss of grip strength and manual dexterity. Attacks can be triggered by damp and/or cold conditions and, on warming, 'pins and needles' are experienced (Figure 15.3). If the condition is allowed to persist, more serious symptoms become apparent including discoloration and enlargement of the fingers. In very advanced cases, gangrene can develop leading to the amputation of the affected hand or finger. VWF was first detailed as an industrial disease in 1911.

The risk of developing HAVS depends on the frequency of vibration, the length of exposure and the tightness of the grip on the machine or tool. Some typical values of vibration measurements for common items of equipment used in industry are given in Table 15.1.

When assessing the risk of HAVS developing among employees, the source of the vibration, such as reciprocating, rotating and vibrating tools and equipment, needs to be considered first together with the age of the equipment, its maintenance record, its suitability for the job and any information or guidance available from the manufacturer. The number of employees using the tooling or equipment, the duration and frequency of their use and any relevant personal factors, such as a pre-existing circulatory problem, all form part of the assessment. Environmental factors, particularly exposure to cold and/or wet weather and the nature

of the job itself, are also important factors to be considered during such a risk assessment. Finally, an examination of the existing controls and their effectiveness and the frequency, magnitude and direction of the vibration are important elements of the evaluation. Other issues could include the effectiveness of any personal protective equipment provided and any instruction or training given.

Whole-body vibration

WBV is caused by vibration from machinery passing into the body either through the feet of standing workers or the buttocks of sitting workers. The most common ill-health effect is severe back pain which, in severe cases, may result in permanent injury. Other acute affects include reduced visual and manual control, increased heart rate and blood pressure. Chronic or long-term effects include permanent spinal damage, damage to the central nervous system, hearing loss and circulatory and digestive problems.

The most common occupations which generate WBV are driving fork-lift trucks, construction vehicles and agricultural or horticultural machinery and vehicles. There is growing concern throughout the European Union about this problem. Control measures include the proper use of the equipment including correct adjustments of air or hydraulic

Equipment	Condition	Vibration reading (m/s^2)
Road breakers	Typical Modern design and trained operator Worst tools and operating conditions	12 5 20
Demolition hammers	Modern tools Typical Worst tools	8 15 25
Hammer drills	Typical Best tools and operating conditions Worst tools and operating conditions	9 6 25
Large-angle grinders	Modern vibration-reduced designs Other types	4 8
Small-angle grinders	Typical	2–6
Chainsaws	Typical	6
Brush cutter or strimmer	Typical Best	4 2
Sanders (orbital)	Typical	7–10

Table 15.1 Examples of vibration exposure values measured by HSE on work equipment

pressures, seating and, in the case of vehicles, correct suspension, tyre pressures and appropriate speeds to suit the terrain. Other control measures include the selection of suitable equipment with low vibration characteristics, work rotation, good maintenance and fault reporting procedures.

The Health and Safety Executive (HSE) commissioned measurements of WBV on several machines and some of the results are shown in Table 15.2.

Preventative and precautionary measures

The control strategy outlined in Chapter 6 can certainly be applied to ergonomic risks. The common measures used to control ergonomic ill-health effects are to:

➤ implement results of task analysis and identification of repetitive actions;

➤ eliminate vibration-related or hazardous tasks by performing the job in a different way (Figure 15.4);

➤ ensure that the correct equipment (properly adjusted) is always used;

➤ introduce job rotation so that workers have a reduced time exposure to the hazard;

➤ during the design of the job ensure that poor posture is avoided;

➤ undertake a risk assessment;

➤ reports from employees and safety representatives;

➤ ill-health reports and absence records;

➤ introduce a programme of health surveillance;

➤ ensure that employees are given adequate information on the hazards and develop a suitable training programme;

➤ ensure that a programme of preventative maintenance is introduced and include the regular inspection of items such as vibration isolation mountings;

➤ keep up to date with advice from equipment manufacturers, trade associations and health and safety sources (more and more low vibration equipment is becoming available).

Table 15.2 Machines which could produce significant whole-body vibration

Machine	Activity	Vibration reading (m/s^2)
3-tonne articulated site	Removal of spoil	0.78
Dumper truck	Transport of spoil	1.13
25-tonne articulated dumper truck	Transport of spoil	0.91
Bulldozer	Dozing	1.16
4-tonne twin-drum	Finishing tarmac	0.86
80-tonne rigid dumper truck	Transport of spoil	1.03

A very useful and extensive checklist for the identification and reduction of WRULDs is given in Appendix 2 of the HSE Guide to work-related upper limb disorders HSG60. HSE has also produced two very useful guides – Hand–arm vibration, The Control of Vibration at Work Regulations 2005 Guidance on Regulations L140 and Whole-body vibration, The Control of Vibration at Work Regulations 2005 Guidance on Regulations L141. These guidance documents offer detailed advice on the implementation of the Regulations.

15.2.3 The Control of Vibration at Work Regulations 2005

The Control of Vibration at Work Regulations introduce, for both HAV and WBV, a daily exposure limit and action values. These values are as follows:

1. For HAV
   (a) the daily exposure limit value normalized to an 8-hour reference period is 5 m/s^2
   (b) the daily exposure action value normalized to an 8-hour reference period is 2.5 m/s^2.
2. For WBV
   (a) the daily exposure limit value normalized to an 8-hour reference period is 1.15 m/s^2
   (b) the daily exposure action value normalized to an 8-hour reference period is 0.5 m/s^2.

An exposure limit value must not be exceeded. If an exposure action value is exceeded, then action must be taken to reduce the value. The term A(8) is added to the exposure limit or action value to denote that it is an

Figure 15.4 Mounted breaker to reduce vibrations.

average value spread over an 8-hour working day. Thus the daily exposure limit value for HAV is 5 m/s^2 A(8).

Hand–arm vibration
Many machines and the processes used in industry produce HAV. Typical high-risk processes include:

➤ grinding, sanding and polishing wood and stone;
➤ cutting stone, metal and wood;
➤ riveting, caulking and hammering;
➤ compacting sand, concrete and aggregate;
➤ drilling and breaking rock, concrete and road surfaces;
➤ surface preparation, including de-scaling and paint removal.

Table 15.3 The change in exposure times as vibration increase

Value of vibration (m/s^2)	2.5	3.5	5	7	10	14	20
Exposure time to reach action value (hours)	8	4	2	1	30 min	15 min	8 min
Exposure time to reach limit value (hours)	over 24	16	8	4	2	1	30 min

There are several ways to ascertain the size of the vibration generated by equipment and machines. Manufacturers must declare vibration emission values for portable hand-held and hand-guided machines and provide information on risks. Other important sources for vibration information include scientific and technical journals, trade associations and online databases. HSE experience has shown that the vibration level is higher in practice than that quoted by many manufacturers. The reasons for this discrepancy may be that:

➤ the equipment is not well maintained;
➤ the equipment is not suitable for the material being worked;
➤ the tool has not been purchased from a reputable supplier;
➤ the accessories are not appropriate or are badly fitted;
➤ the operative is not using the tool properly.

In view of these problems, it is recommended that the declared value should be doubled when comparisons are made with exposure limits. As the exposure limit or action value is averaged over 8 hours, it is possible to work with higher values for a reduced exposure time. Table 15.3 shows the reduction in exposure time as the size of the vibration increases.

The Guidance (L140) gives very useful advice on the measurement of vibration, undertaking a suitable and sufficient risk assessment, control measures, health surveillance and training of employees. The following points summarize the important measures which should be taken to reduce the risks associated with HAV:

1. avoid, whenever possible, the need for vibration equipment;
2. undertake a risk assessment which includes a soundly based estimate of the employees' exposure to vibration;
3. develop a good maintenance regime for tools and machinery. This may involve ensuring that tools are

regularly sharpened, worn components are replaced or engines are regularly tuned and adjusted;
4. introduce a work pattern that reduces the time exposure to vibration;
5. issue employees with gloves and warm clothing. There is a debate as to whether anti-vibration gloves are really effective but it is agreed that warm clothing helps with blood circulation which reduces the risk of VWF. Care must be taken so that the tool does not cool the hand of the operator;
6. introduce a reporting system for employees to use so that concerns and any symptoms can be recorded and investigated.

It is important that drill bits and tools are kept sharp and used intact – an angle grinder with a chipped cutting disc will lead to a large increase in vibration as well as being dangerous.

Whole-body vibration

WBV in industry arises from driving vehicles, such as tractors or fork-lift trucks, over rough terrain or uneven floors. It is highly unlikely that driving vehicles on smooth roads will produce WBV problems. As explained earlier, the most common health problem associated with WBV is back pain. This pain may well have been caused by other activities but WBV will aggravate it. The reasons for back pain in drivers include:

➤ poor posture while driving;
➤ incorrect adjustment of the driver's seat;
➤ difficulty in reaching all relevant controls due to poor design of the controls layout;
➤ frequent manual handling of loads;
➤ frequent climbing up and down from a high cab.

The Regulations require that where there is a likelihood of WBV, the employer must undertake a risk assessment. The HSE Guidance document, L141, gives detailed advice to help with this risk assessment and on estimating daily exposure levels. WBV risks are low for exposures around the action value and usually only simple control measures are necessary.

The Regulations allow a transitional period for the limit value until July 2010 for vehicles supplied before July 2007.

The measurement of WBV is very difficult and can only be done accurately by a specialist competent person. If the risk assessment has been made and the recommended control actions are in place, there is no need to measure the exposure of employees to vibration. However, HSE has suggested that employers can use the following checklist to estimate whether exposure to WBV is high:

1. there is a warning in the machine manufacturer's handbook that there is a risk of WBV;
2. the task is not suitable for the machine or vehicle being used;
3. operators or drivers are using excessive speeds or operating the machine too aggressively;
4. operators or drivers are working too many hours on machines or vehicles that are prone to WBV;
5. road surfaces are too rough and potholed or floors uneven;
6. drivers are being continual jolted or when going over bumps rising from their seats;
7. vehicles designed to operate on normal roads are used on rough or poorly repaired roads;
8. operators or drivers have reported back problems.

If one or more of the above applies, then exposure to WBV may be high.

The actions for controlling the risks from WBV include the following.

Ensure that:

➤ the driver's seat is correctly adjusted so that all controls can be reached easily and that the driver weight setting on their suspension seat, if available, is correctly adjusted. The seat should have a back rest with lumbar support (Figure 15.5);
➤ anti-fatigue mats are used if the operator has to stand for long periods;
➤ the speed of the vehicle is such that excessive jolting is avoided. Speeding is one of the main causes of excessive WBV;
➤ all vehicle controls and attached equipment are operated smoothly;
➤ only established site roadways are used;
➤ only suitable vehicles and equipment are selected to undertake the work and cope with the ground conditions;

Figure 15.5 Vibrating roller.

➤ the site roadway system is regularly maintained;
➤ all vehicles are regularly maintained with particular attention being paid to tyre condition and pressures, vehicles suspension systems and the driver's seat;
➤ work schedules are regularly reviewed so that long periods of exposure on a given day are avoided and drivers have regular breaks;
➤ prolonged exposure to WBV is avoided for at risk groups (older people, young people, people with a history of back problems and pregnant women);
➤ employees are aware of the health risks from WBV, the results of the risk assessment and the ill-health reporting system. They should also be trained to drive in such a way that excessive vibration is reduced.

A simple health monitoring system that includes a questionnaire checklist should be agreed with employees or their representatives (available on the HSE website) to be completed once a year by employees at risk.

More information on the Control of Vibration at Work Regulations is given in Chapter 17 (Section 17.31).

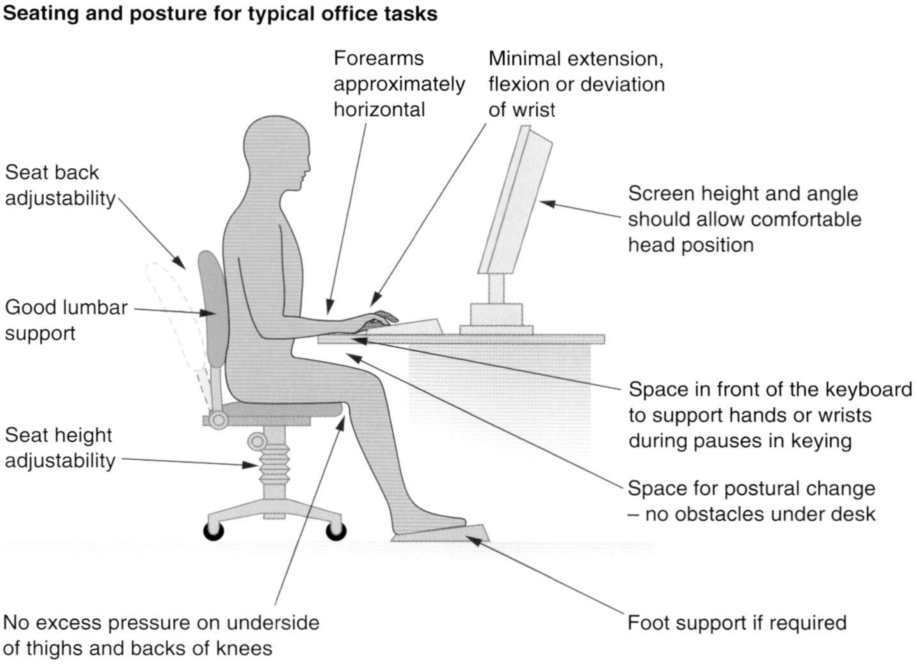

Seating and posture for typical office tasks

Figure 15.6 Workstation design.

15.2.4 Display screen equipment

DSE, which includes visual display units, is a good example of a common work activity which relies on an understanding of ergonomics and the ill-health conditions which can be associated with poor ergonomic design. A recent survey of safety representatives by the TUC found that injuries/illnesses caused by the poor use of DSE and repetitive strain injuries together with stress or overwork were among their major concerns.

Legislation governing DSE is covered by the Health and Safety (Display Screen Equipment) Regulations 1992 and a detailed summary of them is given in Chapter 17. The Regulations apply to a user or operator of DSE and define a user or operator. The definition is not as tight as many employers would like it to be, but usage in excess of approximately 1 hour continuously each day would define a user. The definition is important as users are entitled to free eye tests and, if required, a pair of spectacles specifically for DSE use. The basic requirements of the Regulations are:

➢ a suitable and sufficient risk assessment of the workstation, including the software in use, trip and electrical hazards from trailing cables and the surrounding environment;

➢ workstation compliance with the minimum specifications laid down in the schedules appended to the Regulations;

➢ a plan of the work programme to ensure that there are adequate breaks in the work pattern of workers;

➢ the provision of free eye sight tests and, if required, spectacles to users of DSE;

➢ a suitable programme of training and sufficient information given to all users.

The risk assessment of a DSE work station needs to consider the following factors, many of which are shown in Figure 15.6:

➢ the height and adjustability of the monitor;

➢ the adjustability of the keyboard, the suitability of the mouse and the provision of wrist support;

➢ the stability and adjustability of the DSE user's chair;

➢ the provision of ample foot room and suitable foot support;

➢ the effect of any lighting and window glare at the work station;

➢ the storage of materials around the work station;

➢ the safety of trailing cables, plugs and sockets;

➢ environmental issues – noise, heating, humidity and draughts.

There are three basic ill-health hazards associated with DSE. These are:

➢ musculoskeletal problems;

➢ visual problems;

➢ psychological problems.

A fourth hazard, of radiation, has been shown from several studies to be very small and is now no longer normally considered in the risk assessment.

Similarly, in the past, there have been suggestions that DSE could cause epilepsy and there were concerns about adverse health effects on pregnant women and their unborn children. All these risks have been shown in various studies to be very low.

The provision of DSE training and risk assessments online has become more common. The risk assessment, however, still has to be managed, made appropriate to the particular workplace setting and reviewed from time to time. Various studies have shown that users of any e-learning package lose concentration after 30–40 minutes.

Musculoskeletal problems

Tenosynovitis is the most common and well-known problem which affects the wrist of the user. The symptoms and effects of this condition have already been covered. Suffice it to say that if the condition is ignored, then the tendon and tendon sheath around the wrist will become permanently injured. Tenosynovitis, is caused by the continual use of a keyboard and can be relieved by the use of wrist supports. Other WRULDs are caused by poor posture and can produce pains in the back, shoulders, neck or arms. Less commonly, pain may also be experienced in the thighs, calves and ankles. These problems can be mitigated by the application of ergonomic principles in the selection of working desks, chairs, foot rests and document holders. It is also important to ensure that the desk is at the correct height and the computer screen is tilted at the correct angle to avoid putting too much strain on the neck. (Ideally the user's eyes should be at the same height as the top of the screen.)

The keyboard should be detachable so that it can be positioned anywhere on the desktop and a correct posture adopted while working at the keyboard. The chair should be adjustable in height, stable and have an adjustable backrest. If the knees of the user are lower than the hips when seated, then a footrest should be provided. The surface of the desk should be non-reflecting and uncluttered but ancillary equipment (e.g. telephone and printer) should be easily accessible.

Visual problems

There does not appear to be much medical evidence that DSE causes deterioration in eye sight, but users may suffer from visual fatigue which results in eye strain, sore eyes and headaches. Less common ailments are skin rashes and nausea.

The use of DSE may indicate that reading spectacles are needed and the Regulations make provision for this. It is possible that any prescribed lenses may only be suitable for DSE work as they will be designed to give optimum clarity at the normal distance at which screens are viewed (50–60 cm).

Eye strain is a particular problem for people who spend a large proportion of their working day using DSE. A survey has indicated that up to 90% of DSE users complain of eye fatigue. Eye strain can be reduced by the following steps additional to those already identified in this section:

> Train staff in the correct use of the equipment.
> Ensure that a font size of at least 12 is used on the screen.
> Ensure that users take regular breaks away from the screen (up to 10 minutes every hour).

The screen should be adjustable in tilt angle and screen brightness and contrast. Finally, the lighting around the workstation is important. It should be bright enough to allow documents to be read easily but not too bright such that either headaches are caused or there are reflective glares on the computer screen.

Psychological problems

These are generally stress-related problems. They may have environmental causes, such as noise, heat, humidity or poor lighting, but they are usually due to high-speed working, lack of breaks, poor training and poor workstation design. One of the most common problems is the lack of understanding of all or some of the software packages being used.

There are several other processes and activities where ergonomic considerations are important. These include the assembly of small components (microelectronics assembly lines) and continually moving assembly lines (car assembly plants).

Many of these processes including some of the chemical hazards (such as soldering fumes) are described in the previous chapter.

15.3 Welfare and work environment issues

Welfare and work environment issues are covered by the Workplace (Health, Safety and Welfare) Regulations 1992 together with an Approved Code of Practice and additional guidance.

15.3.1 Welfare

Welfare arrangements include the provision of sanitary conveniences and washing facilities, drinking water, accommodation for clothing, facilities for changing clothing and facilities for rest and eating meals. First-aid provision is also a welfare issue, but is covered in Chapter 6.

Sanitary conveniences and washing facilities must be provided together and in a proportion to the size of the workforce. The Approved Code of Practice provides two tables offering guidance on the requisite number of water closets, wash stations and urinals for varying sizes of workforce (approximately one of each for every 25 employees). Special provision should be made for disabled workers and there should normally be separate facilities for men and women. A single convenience would only be acceptable if it were situated in a separate room whose door could be locked from the inside. There should be adequate protection from the weather and only as a last resort should public conveniences be used. A good supply of warm water, soap and towels must be provided as close to the sanitary facilities as possible. The facilities should be well lit and ventilated and their walls and floors easy to clean. It may be necessary to install a shower for certain types of work. Hand dryers are permitted but there are concerns about their effectiveness in drying hands completely and thus removing all bacteria. In the case of temporary or remote worksites, sufficient chemical closets and sufficient washing water in containers must be provided.

All such facilities should be well ventilated and lit and cleaned regularly.

Drinking water must be readily accessible to all the workforce. The supply of drinking water must be adequate and wholesome. Normally mains water is used and should be marked as 'drinking water' if water not fit for drinking is also available.

Accommodation for clothing and facilities for changing clothing must be provided which is clean, warm, dry, well ventilated and secure. Such accommodation is only necessary when the work activity requires employees to change into specialist clothing. Where workers are required to wear special or protective clothing, arrangements should be such that the workers' own clothing is not contaminated by any hazardous substances.

Facilities for rest and eating meals must be provided so that workers may sit down during break times in areas where they do not need to wear personal protective equipment. From 1 July 2007 smoking was banned inside public premises (see Chapter 17.34.13). Facilities should also be provided for pregnant women and nursing mothers to rest. Arrangements must be in place to ensure that food is not contaminated by hazardous substances.

15.3.2 Workplace environment

The issues governing the workplace environment are ventilation, heating and temperature, lighting, workstations and seating.

Ventilation

Ventilation of the workplace should be effective and sufficient and free of any impurity and air inlets should be sited clear of any potential contaminant (e.g. a chimney flue). Care needs to be taken to ensure that workers are not subject to uncomfortable draughts. The ventilation plant should have an effective visual or audible warning device fitted to indicate any failure of the plant. The plant should be properly maintained and records kept. The supply of fresh air should not normally fall below 5–8 L per second per occupant (see Figure 14.11).

Heating and temperature

During working hours, the temperature in all workplaces inside buildings shall be reasonable (not uncomfortably high or low). 'Reasonable' is defined in the Approved Code of Practice as at least 16°C, unless much of the work involves severe physical effort in which case the temperature should be at least 13°C. These temperatures refer to readings taken close to the workstation at working height and away from windows. The Approved Code of Practice recognizes that these minimum temperatures cannot be maintained where rooms open to the outside or where food or other products have to be kept cold. A heating or cooling method must not be used in the workplace which produces fumes, injurious or offensive to any person. Such equipment needs to be regularly maintained.

A sufficient number of thermometers should be provided and maintained to enable workers to determine the temperature in any workplace inside a building (but need not be provided in every workroom).

Where, despite the provision of local heating or cooling, the temperatures are still unreasonable, suitable protective clothing and rest facilities should be provided.

Lighting

Every workplace shall have suitable and sufficient lighting and this shall be natural lighting so far as is reasonably practicable. Suitable and sufficient emergency lighting must also be provided and maintained in any room where workers are particularly exposed to danger in the event of a failure of artificial lighting (normally due to a power cut and/or a fire). Windows and skylights should be kept clean and free from obstruction so far as is reasonably practicable unless it would prevent the shading of windows or skylights or prevent excessive heat or glare (Figure 15.7).

When deciding on the suitability of a lighting system, the general lighting requirements will be affected by the following factors:

➤ the availability of natural light;
➤ the specific areas and processes, in particular any colour rendition aspects or concerns over stroboscopic effects (associated with fluorescent lights);
➤ the type of equipment to be used and the need for specific local lighting;

Figure 15.7 A well-lit workplace.

➤ the lighting characteristics required (type of lighting, its colour, intensity and local adjustability);
➤ the location of visual display units and any problems of glare;
➤ structural aspects of the workroom, such as the use of screens in open office layout and the reduction of shadows;
➤ the presence of atmospheric dust;
➤ the heating effects of the lighting;
➤ lamp and window cleaning and repair (and disposal issues);
➤ the need and required quantity of emergency lighting.

Light levels are measured in illuminance, having units of lux (lx), using a light meter. A general guide to lighting levels in different workplaces is given in Table 15.4.

Poor lighting levels will increase the risk of accidents such as slips, trips and falls. More information is available on lighting from *Lighting at Work*, HSG38, HSE Books.

Workstations and seating

Workstations should be arranged so that work may be done safely and comfortably. The worker should be at a suitable height relative to the work surface and there should be no need for undue bending and stretching. Workers must not be expected to stand for long periods of time, particularly on solid floors. A suitable seat should be provided when a substantial part of the task can or must be done sitting. The seat should, where possible, provide adequate support for the lower back and a footrest be provided for any worker whose feet cannot be placed flat on the floor. It should be made of materials suitable for the environment, be stable and, possibly, have arm rests.

Table 15.4 Typical workplace lighting levels	
Workplace or type of work	**Illuminance (lx)**
Warehouses and stores	150
General factories or workshops	300
Offices	500
Drawing offices (detailed work)	700
Fine working (ceramics or textiles)	1000
Very fine work (watch repairs or engraving)	1400

It is also worth noting that sitting for prolonged periods can present health risks, such as blood circulation and pressure problems, and vertebral and muscular damage.

Seating at Work, HSG57, HSE Books, provides useful guidance on how to ensure that seating in the workplace is safe and suitable.

Other factors

The Workplace (Health, Safety and Welfare) Regulations cover other important work environment issues, and details of these are given in Chapter 17 (17.32). The condition of floors, stairways and traffic routes should be suitable for the purpose and well maintained and undue space constraints anywhere in the workplace should be avoided. Translucent or transparent doors should be constructed with safety glass and properly marked to warn pedestrians of their presence. Windows and skylights should be designed so that, when they are opened, they do not present an obstruction to passing pedestrians. There must be adequate arrangements in place to ensure the safe cleaning of windows and skylights. Finally, there need to be adequate provisions for the needs of disabled workers.

Noise

There was considerable concern for many years over the increasing cases of occupational deafness and this led to the introduction of the Noise at Work Regulations in 1989 and the revised Noise at Work Regulations in 2005. HSE has estimated that an additional 1.1 million workers will be covered by the revised Regulations. These Regulations, which are summarized in Chapter 17, require the employer to:

➤ assess noise levels and keep records;
➤ reduce the risks from noise exposure by using engineering controls in the first instance and the provision and maintenance of hearing protection as a last resort;
➤ provide employees with information and training;
➤ if a manufacturer or supplier of equipment, provide relevant noise data on that equipment (particularly if any of the three action levels is likely to be reached).

The main purpose of the Noise Regulations is to control noise levels rather than measuring them. This involves the better design of machines, equipment and work processes, and ensuring that personal protective equipment is correctly worn and employees are given adequate training and health surveillance (Figure 15.8).

Sound is transmitted through the air by sound waves which are produced by vibrating objects. The vibrations cause a pressure wave which can be detected by a receiver, such as a microphone or the human ear. The ear may detect vibrations which vary from 20 to 20 000 (typically 50–16 000) cycles each second (or hertz – Hz). Sound travels through air at a finite speed (342 m/s at 20°C and sea level). The existence of this speed is shown by the time lag between lightning and thunder during a thunderstorm. Noise normally describes loud, sudden, harsh or irritating sounds although noise is defined as any audible sound.

Noise may be transmitted directly through the air, by reflection from surrounding walls or buildings or through the structure of a floor or building. In construction work, the noise and vibrations from a pneumatic drill will be transmitted from the drill itself, from the ground being drilled and from the walls of surrounding buildings.

15.4.1 Health effects of noise

The human ear

There are three sections of the ear – the outer (or external) ear, the middle ear and the inner (or internal) ear. The sound pressure wave passes into and through the outer ear and strikes the eardrum causing it to vibrate. The eardrum is situated approximately 25 mm inside the head. The vibration of the eardrum causes the proportional movement of three interconnected small bones in the middle ear, thus passing the sound to the cochlea situated in the inner ear.

Within the cochlea the sound is transmitted to a fluid causing it to vibrate. The motion of the fluid induces a membrane to vibrate which, in turn, causes hair cells attached to the membrane to bend. The movement of the hair cells causes a minute electrical impulse to be transmitted to the brain along the auditory nerve. Those hairs nearest to the middle ear respond to high frequency, while those at the tip of the cochlea respond to lower frequencies (Figure 15.9).

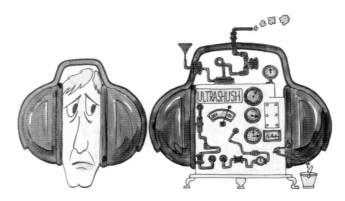

Figure 15.8 Better to control noise at source rather than wear ear protection.

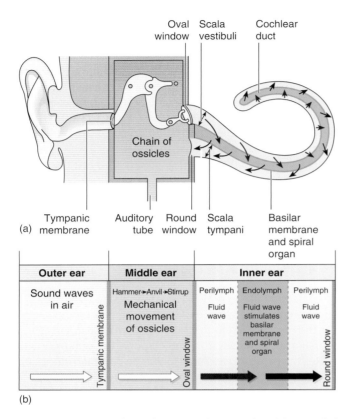

Figure 15.9 Passage of sound waves. (a) The ear with cochlea uncoiled and (b) summary of transmission.

There are about 30 000 hair cells within the ear and noise-induced hearing loss causes irreversible damage to these hair cells.

Ill-health effects of noise

Noise can lead to ear damage on a temporary (acute) or permanent (chronic) basis.

There are three principal **acute** effects:

➤ temporary threshold shift – caused by short excessive noise exposures and affects the cochlea by reducing the flow of nerve impulses to the brain. The result is a slight deafness, which is reversible when the noise is removed;

➤ tinnitus – a ringing in the ears caused by an intense and sustained high noise level. It is caused by the over-stimulation of the hair cells. The ringing sensation continues for up to 24 hours after the noise has ceased;

➤ acute acoustic trauma – caused by a very loud noise such as an explosion. It affects either the eardrum or the bones in the middle ear and is usually reversible. Severe explosive sounds can permanently damage the eardrum.

Occupational noise can also lead to one of the following three **chronic** hearing effects:

➤ noise-induced hearing loss – results from permanent damage to the cochlear hair cells. It affects the ability to hear speech clearly but the ability to hear is not lost completely;

➤ permanent threshold shift – this results from prolonged exposure to loud noise and is irreversible due to the permanent reduction in nerve impulses to the brain. This shift is most marked at the 4000 Hz frequency, which can lead to difficulty in hearing certain consonants and some female voices;

➤ tinnitus – is the same as the acute form but becomes permanent. It is a very unpleasant condition, which can develop without warning.

It is important to note that, if the level of noise exposure remains unchanged, noise-induced hearing loss will lead to a permanent threshold shift affecting an increasing number of frequencies.

Presbycusis is the term used for hearing loss in older people which may have been exacerbated by occupational noise earlier in their lives.

15.4.2 Noise assessments

The Control of Noise at Work Regulations specify action levels at which the hearing of employees must be protected. The conclusion as to whether any of those levels has been breached is reached after an assessment of noise levels has been made. However, before noise assessment can be discussed, noise measurement and the statutory action levels must be described.

Noise measurement

Sound intensity is measured by a unit known as a pascal (Pa – N/m^2), which is a unit of pressure similar to that used when inflating a tyre. If noise was measured in this way, a large scale of numbers would be required ranging from 1 at one end to 1 million at the other. The sound pressure level (SPL) is a more convenient scale because:

➤ it compresses the size of the scale by using a logarithmic scale to the base 10;

➤ it measures the ratio of the measured pressure, p, to a reference standard pressure, p_0, which is the pressure at the threshold of hearing (2×10^{-5} Pa).

The unit is called a decibel (dB) and is defined as:

$$SPL = 20\log_{10}(p/p_0)\,dB$$

Activity or environment	SPL [dB(A)]
Threshold of pain	140
Pneumatic drill	125
Pop group or disco	110
Heavy lorry	93
Street traffic	85
Conversational speech	65
Business office	60
Living room	40
Bedroom	25
Threshold of hearing	0

Table 15.5 Some typical sound pressure levels (SPL) [dB(A) values]

It is important to note that since a logarithmic scale to the base 10 is used, each increase of 3 dB is a doubling in the sound intensity. Thus, if a sound reading changes from 75 dB to 81 dB, the sound intensity or loudness has increased by four times.

Finally, as the human ear tends to distort its sensitivity to the sound it receives by being less sensitive to lower frequencies, the scale used by sound meters is weighted so that readings mimic the ear. This scale is known as the A scale and the readings known as dB(A). There are also three other scales known as B, C and D.

Originally the A scale was used for sound pressure levels (SPLs) up to 55 dB, the B scale for levels between 55 dB and 85 dB and the C scale for values above 85 dB. However today, the A scale is used for nearly all levels except for very high SPLs when the C scale is used. The B scale is rarely used and the D scale is mainly used to monitor jet aircraft engine noise.

Table 15.5 gives some typical decibel readings for common activities.

Noise is measured using a sound level meter which reads SPLs in dB(A) and the **peak sound pressure** in

pascal (Pa), which is the highest noise level reached by the sound. There are two basic types of sound meter – integrated and direct reading meters. Meters which integrate the reading provide an average over a particular time period, which is an essential technique when there are large variations in sound levels. This value is known as the **continuous equivalent noise level (L_{Eq})**, which is normally measured over an 8-hour period.

Direct reading devices, which tend to be much cheaper, can be used successfully when the noise levels are continuous at a near constant value.

Another important noise measurement is the **daily personal exposure level** of the worker, $L_{EP,d}$, which is measured over an 8-hour working day. Hence, if a person was exposed to 87 dB(A) over 4 hours, this would equate to an $L_{EP,d}$ of 84 dB(A) since a reduction of 3 dB(A) represents one half of the noise dose.

The HSE Guidance on the Control of Noise at Work Regulations, L108, offers some very useful advice on the implementation of the Regulations and should be read by anyone who suspects that they have a noise problem at work. The guidance covers 'equipment and procedures for noise surveys' and contains a noise exposure ready reckoner which can be used to evaluate $L_{EP,d}$ when the noise occurs during several short intervals and/or at several different levels during the 8-hour period.

Noise action levels

Regulation 6 of the Control of Noise at Work Regulations places a duty on employers to reduce the risk of damage to the hearing of their employees from exposure to noise to the lowest level reasonably practicable.

The Regulations introduce exposure **action** level values and exposure **limit** values.

An exposure action value is a level of noise at which certain action must be taken.

An exposure limit value is a level of noise at the ear above which an employee must not be exposed. Therefore if the workplace noise levels are above this value, any ear protection provided to the employee must reduce the noise level to the limit value at the ear.

These exposure action and limit values are as follows.

1. The lower exposure action levels are:
   (a) a daily or weekly personal noise exposure of **80 dB(A)**;
   (b) a peak sound pressure of **135 dB(C)**.
2. The upper exposure action levels are:
   (a) a daily or weekly personal noise exposure of **85 dB(A)**;
   (b) a peak sound pressure of **137 dB(C)**.

Observation at the workplace	Likely noise level [dB(A)]	A noise risk assessment must be made if this noise level persists for
The noise is noticeable but does not interfere with normal conversation – equivalent to a domestic vacuum cleaner	80	6 hours
People have to shout to be heard if they are more than 2 m apart	85	2 hours
People have to shout to be heard if they are more than 1 m apart	90	45 minutes

Table 15.6 Simple observations to determine the need for a noise risk assessment

3. The exposure limit values are:
   (a) a daily or weekly personal noise exposure of **87 dB(A)**;
   (b) a peak sound pressure of **140 dB(C)**.

The peak exposure action and limit values are defined because high level peak noise can lead to short-term and long-term hearing loss. Explosives, guns (including nail guns), cartridge tools, hammers and stone chisels can all produce high peak sound pressures.

If the daily noise exposure exceeds the lower exposure action level, then a noise assessment should be carried out and recorded by a competent person. There is a very simple test which can be done in any workplace to determine the need for an assessment. Table 15.6 gives information on the simple test to determine the need for a noise risk assessment.

If there is a marked variation in noise exposure levels during the working week, then the Regulations allow a weekly rather than daily personal exposure level, $L_{EP,w}$, to be used. It is only likely to be significantly different to the daily exposure level if exposure on one or two days in the working week is 5 dB(A) higher than on the other days, or the working week has three or fewer days of exposure. The weekly exposure rate is not a simple arithmetic average of the daily rates. If an organization is considering the use of a weekly exposure level, then the following provisions must be made:

➤ hearing protection must be provided if there are very high noise levels on any one day;
➤ the employees and their representatives must be consulted on whether weekly averaging is appropriate;

➤ an explanation must be given to the employees on the purpose and possible effects of weekly averaging.

Finally, if the working day is 12 hours, then the action levels must be reduced by 3 dB(A) because the action levels assume an 8-hour working day.

The HSE Guidance document L108 gives detailed advice on noise assessments and surveys. The most important points are that the measurements should be taken at the working stations of the employees closest to the source of the noise and over as long a period as possible, particularly if there is a variation in noise levels during the working day. Other points to be included in a noise assessment are:

➤ details of the noise meter used and the date of its last calibration;
➤ the number of employees using the machine, time period of usage and other work activities;
➤ an indication of the condition of the machine and its maintenance schedule;
➤ the work being done on the machine at the time of the assessment;
➤ a schematic plan of the workplace showing the position of the machine being assessed;
➤ other noise sources, such as ventilation systems, that should be considered in the assessment. The control of these sources may help to reduce overall noise levels;
➤ recommendations for future actions, if any.

Other actions which the employer must undertake when the lower exposure action level is exceeded are to:

➤ inform, instruct and train employees on the hearing risks;

Figure 15.10 Typical ear protection zone sign.

➤ supply hearing protection to those employees requesting it;

➤ ensure that any equipment or arrangements provided under the Regulations are correctly used or implemented.

The additional measures which the employer must take if the upper exposure action level is reached are:

➤ reduce and control exposure to noise by means other than hearing protection;

➤ establish hearing protection zones, marked by notices and ensure that anybody entering the zone is wearing hearing protection (Figure 15.10);

➤ supply hearing protection and ensure that it is worn.

The Control of Noise at Work Regulations place a duty on the employer to undertake health surveillance for employees whose exposure regularly exceeds the upper action level irrespective of whether ear protection was worn. The recommended health surveillance is a hearing test at induction, followed by an annual check and review of hearing levels. The checks may be extended to every three years if no adverse effects are found during earlier tests. If exposure continues over a long period, health surveillance of employees is recommended using a more substantial audiometric test. This will indicate whether there has been any deterioration in hearing ability. Where exposure is between the lower and upper exposure action levels or occasionally above the upper action values, health surveillance will only be required if information becomes available, perhaps from medical records, that the employee is particularly sensitive to noise-induced hearing loss.

The Regulations also place the following statutory duties on employees:

➤ for noise levels above the lower exposure action level, they must use any control equipment (other than hearing protection), such as silencers, supplied by the employer and report any defects;

➤ for noise levels above the upper exposure action level, they fulfil the obligations given above and wear the hearing protection provided;

➤ they must take care of any equipment provided under these Regulations and report any defects;

➤ they must see their doctor if they feel that their hearing has become damaged;

➤ present themselves for health surveillance.

The Management of Health and Safety Regulations prohibits the employment of anyone under the age of 18 years where there is a risk to health from noise.

15.4.3 Noise control techniques

In addition to **reduced time exposure** of employees to the noise source, there is a simple hierarchy of control techniques:

➤ reduction of noise at source;

➤ reduction of noise levels received by the employee (known as attenuation);

➤ personal protective equipment, which should only be used when the above two remedies are insufficient.

Reduction of noise at source

There are several means by which noise could be reduced at source (Figure 15.11):

➤ change the process or equipment (e.g. replace solid tyres with rubber tyres or replace diesel engines with electric motors);

➤ change the speed of the machine;

➤ improve the maintenance regime by regular lubrication of bearings, tightening of belt drives.

Attenuation of noise levels

There are many methods of attenuating or reducing noise levels and these are covered in detail in the guide to the Regulations. The more common ones will be summarized here.

Orientation or re-location of the equipment – turn the noisy equipment away from the workforce or locate it away in separate and isolated areas.

Enclosure – surrounding the equipment with a good sound-insulating material can reduce sound levels by up to 30 dB(A). Care will need to be taken to ensure that the machine does not become overheated.

Screens or absorption walls – can be used effectively in areas where the sound is reflected from walls. The

(a)

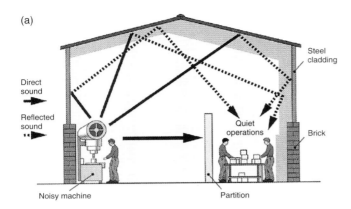

(b)

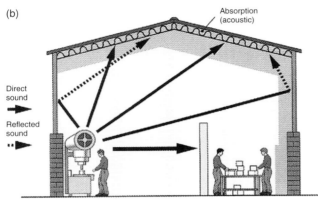

Added absorption to roof 'soaks' up reflected noise

(c)

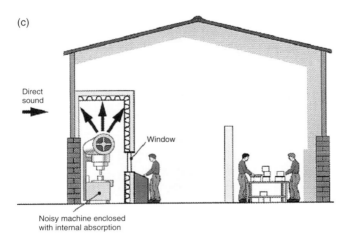

Noisy machine enclosed
with internal absorption

Figure 15.11 (a) Noise paths found in a workplace. The quiet area is subjected to reflected noise from a machine somewhere else in the building. (b) The correct use of absorption in the roof will reduce the reflected noise reaching the quiet area. (c) Segregation of the noisy operation will benefit the whole workplace.

walls of the rooms housing the noisy equipment are lined with sound absorbent material, such as foam or mineral wool, or sound absorbent (acoustic) screens are placed around the equipment.

Damping – the use of insulating floor mountings to remove or reduce the transmission of noise and vibrations through the structure of the building such as girders, wall panels and flooring.

Lagging – the insulation of pipes and other fluid containers to reduce sound transmission (and, incidentally, heat loss).

Silencers – normally fitted to engines which are exhausting gases to atmosphere. Silencers consist of absorbent material or baffles.

Isolation of the workers – the provision of sound-proofed workrooms or enclosures isolated away from noisy equipment (a power station control room is an example of worker isolation).

15.4.4 Personal ear protection

The provision of personal hearing protection should only be considered as a last resort. There is usually resistance from the workforce to use them and they are costly to maintain and replace. They interfere with communications, particularly alarm systems, and they can present hygiene problems.

The following factors should be considered when selecting personal ear protection:

➤ suitability for the range of sound spectrum of frequencies to be encountered;
➤ noise reduction (attenuation) offered by the ear protection;
➤ pattern of the noise exposure;
➤ acceptability and comfort of the wearer, particularly if there are medical problems;
➤ durability;
➤ hygiene considerations;
➤ compatibility with other personal protective equipment;
➤ ease of communication and ability to hear warning alarms;
➤ maintenance and storage arrangements;
➤ cost.

There are two main types of ear protection – earplugs and ear defenders (earmuffs).

Earplugs are made of sound absorbent material and fit into the ear. They can be reusable or disposable, and are able to fit most people and can easily be used with safety glasses and other personal protective equipment. Their effectiveness depends on the quality of the fit in the ear which, in turn, depends on the level of training given to the wearer. Permanent earplugs come in a range of sizes so that a good fit is obtained. The effectiveness of earplugs decreases with age and they should be replaced at the intervals specified by the supplier. A useful simple rule to ensure that the selected ear plug reduces the noise level at the ear to 87 dB(A) is to choose one with a manufacturer's rating of 83 dB(A). This should compensate for any fitting problems. The main disadvantage of ear plugs is that they do not reduce the sound transmitted through the bone structure which surrounds the ear and they often work loose with time.

Ear defenders (earmuffs) offer a far better reduction of all sound frequencies. They are generally more acceptable to workers because they are more comfortable to wear and they are easy to monitor as they are clearly visible. They also reduce the sound intensity transmitted through the bone structure surrounding the ear. A communication system can be built into earmuffs. However, they may be less effective if the user has long hair or is wearing spectacles or large ear rings. They may also be less effective if worn with helmets or face shields and uncomfortable in warm conditions. Maintenance is an important factor with earmuffs and should include checks for wear and tear and general cleanliness.

Selection of suitable ear protection is very important as it should not just reduce sound intensities below the statutory action levels but also reduce those intensities at particular frequencies. Normally advice should be sought from a competent supplier who will be able to advise on ear protection to suit a given spectrum of noise using 'octave band analysis'.

Finally, it is important to stress that the use of ear protection must be well supervised to ensure that not only is it being worn correctly, but that it is, in fact, actually being worn.

15.5 Heat and radiation hazards

15.5.1 Extremes of temperature

The human body is very sensitive to relatively small changes in external temperatures. Food not only provides energy and the build-up of fat reserves, but also generates heat which needs to be dissipated to the surrounding environment. The body also receives heat from its surroundings. The temperature of the body is normally around 37°C and it will attempt to maintain this temperature irrespective of the temperature of the surroundings. Therefore, if the surroundings are hot, sweating will allow heat loss to take place by evaporation caused by air movement over the skin. On the

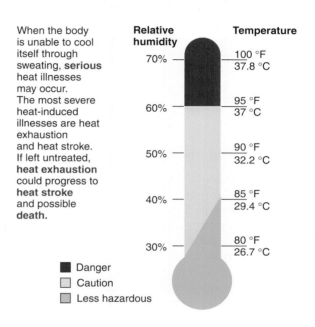

The Heat Equation

HIGH TEMPERATURE + HIGH HUMIDITY + PHYSICAL WORK = HEAT ILLNESS

When the body is unable to cool itself through sweating, **serious** heat illnesses may occur. The most severe heat-induced illnesses are heat exhaustion and heat stroke. If left untreated, **heat exhaustion** could progress to **heat stroke** and possible **death.**

Relative humidity	Temperature
70%	100 °F / 37.8 °C
60%	95 °F / 37 °C
50%	90 °F / 32.2 °C
40%	85 °F / 29.4 °C
30%	80 °F / 26.7 °C

■ Danger
□ Caution
▨ Less hazardous

Figure 15.12 The heat equation.

other hand, if the surroundings are cold, shivering causes internal muscular activity, which generates body heat.

At high temperatures, the body has more and more difficulty in maintaining its natural temperature unless sweating can take place and therefore water must be replaced by drinking. If the surrounding air has high humidity, evaporation of the sweat cannot take place and the body begins to overheat. This leads to heart strain and, in extreme cases, heat stroke. It follows that when working is required at high temperatures, a good supply of drinking water should be available and, further, if the humidity is high, a good supply of ventilation air is also needed. Heat exhaustion is a particular hazard in confined spaces.

At low temperatures, the body will lose heat too rapidly and the extremities of the body will become very cold leading to frostbite and possibly the loss of limbs. Under these conditions, thick, warm (thermal) clothing, the provision of hot drinks and external heating will be required. For those who work in sub-zero temperatures, such as cold store workers, additional precautions will be needed. The store doors must be capable of being unlocked from the inside and an emergency alarm system should be installed. Appropriate equipment selection and a good preventative maintenance system is very important as well as a regular health surveillance programme for the workers, who should be provided with information and training on the hazards associated with working in very low temperatures.

In summary, extremes of temperature require special measures, particularly if accompanied by extremes of humidity. Frequent rest periods will be necessary to allow the body to acclimatize to the conditions. An index called WBGT (Wet Bulb Globe Temperature) is normally used.

15.5.2 Ionising radiation

Ionising radiation is emitted from radioactive materials, either in the form of directly ionising alpha and beta particles or indirectly ionising X-rays and gamma rays or neutrons. It has a high energy potential and an ability to penetrate, ionise and damage body tissue and organs.

All matter consists of atoms within which is a nucleus containing protons and neutrons, and orbiting electrons. The number of electrons within the atom defines the element – hydrogen has 1 electron and lead has 82 electrons. Some atoms are unstable and will change into atoms of another element, emitting ionising radiation in the process. The change is called radioactive decay and the ionising radiations most commonly emitted are alpha and beta particles and gamma rays. X-rays are produced by bombarding a metal target with electrons at very high speeds using very high voltage electrical discharge equipment. Neutrons are released by nuclear fission and are not normally found in manufacturing processes.

Alpha particles consist of two protons and two neutrons and have a positive charge. They have little power to penetrate the skin and can be stopped using very flimsy material, such as paper. Their main route into the body is by ingestion.

Beta particles are high-speed electrons whose power of penetration depends on their speed, but penetration is usually restricted to 2 cm of skin and tissue. They can be stopped using aluminium foil. There are normally two routes of entry into the body – inhalation and ingestion.

Gamma rays, which are similar to **X-rays**, are electromagnetic radiations and have far greater penetrating power than alpha or beta particles. They are produced from nuclear reactions and can pass through the body (Figure 15.13).

There are two principal units of radiation – the becquerel (1 Bq) which measures the amount of radiation

Figure 15.13 Typical ionising warning sign.

Figure 15.14 Warning sign for X-rays.

in a given environment and the millisievert (1 mSv) which measures the ionising radiation dose received by a person.

Ionising radiation occurs naturally from man-made processes and about 87% of all radiation exposure is from natural sources. The Ionising Radiations Regulations specify a range of dose limits, some of which are given in Table 15.7.

Harmful effects of ionising radiation

Ionising radiation attacks the cells of the body by producing chemical changes in the cell DNA by ionising it (thus producing free radicals), which leads to abnormal cell growth. The effects of these ionising attacks depend on the following factors:

➤ the size of the dose – the higher the dose then the more serious will be the effect;
➤ the area or extent of the exposure of the body – the effects may be far less severe if only a part of the body (e.g. an arm) receives the dose;
➤ the duration of the exposure – a long exposure to a low dose is likely to be more harmful than a short exposure to the same quantity of radiation.

Table 15.7 Typical radiation dose limits

	Dose (mSv)	Area of body
Employees aged 18 years+	20	Whole body per year
Trainees 16–18 years	6	Whole body per year
Any other person	1	Whole body per year
Women employees of child-bearing age	13	The abdomen in any consecutive 3-month period
Pregnant employees	1	During the declared term of the pregnancy

Acute exposure can cause, dependent on the size of the dose, blood cell changes, nausea and vomiting, skin burns and blistering, collapse and death. Chronic exposure can lead to anaemia, leukaemia and other forms of cancer. It is also known that ionising radiation can have an adverse effect on the function of human reproductive organs and processes. Increases in the cases of sterility, stillbirths and malformed foetuses have also been observed.

The health effects of ionising radiation may be summarized into two groups – **somatic effects**, which refer to cell damage in the person exposed to the radiation dose and **genetic effects**, which refer to the damage done to the children of the irradiated person.

Sources of ionising radiation

The principal workplaces which could have ionising radiation present are the nuclear industry, medical centres (hospitals and research centres) and educational centres. Radioactive processes are used for the treatment of cancers, and radioactive isotopes are used for many different types of scientific research. X-rays are used extensively in hospitals, but they are also used in industry for non-destructive testing (e.g. crack detection in welds). Smoke detectors, used in most workplaces, also use ionising radiations.

Ionising radiations can also occur naturally – the best example being radon, which is a radioactive gas that occurs mainly at or near granite outcrops where there is a presence of uranium. It is particularly prevalent in Devon and Cornwall. The gas enters buildings normally from the substructure through cracks in flooring or around service inlets. The Ionising Regulations have set two action levels above which remedial action, such as fitting sumps and extraction fans, has to be taken to lower the radon level in the building. The first action level is 400 Bq/m^3 in workplaces and 200 Bq/m^3 in domestic properties. At levels above 1000 Bq/m^3, remedial action should be taken within 1 year.

Personal radiation exposure can be measured using a film badge, which is worn by the employee over a fixed time interval. The badge contains a photographic film which, after the time interval, is developed and an estimate of radiation exposure is made. A similar device, known as a radiation dose meter or detector, can be positioned on a shelf in the workplace for 3 months, so that a mean value of radiation levels may be measured. Instantaneous radiation values can be obtained from portable hand-held instruments, known as geiger counters, which continuously sample the air for radiation levels. Similar devices are available to measure radon levels.

15.5.3 Non-ionising radiation

Non-ionising radiation includes ultraviolet, visible light (this includes lasers which focus or concentrate visible light), infrared and microwave radiations. As the wavelength is relatively long, the energy present is too low to ionise atoms which make up matter. The action of non-ionising radiation is to heat cells rather than change their chemical composition.

Other than the Health and Safety at Work Act there is no specific set of Regulations governing non-ionising radiation. However, the Personal Protective Equipment at Work Regulations are particularly relevant since the greatest hazard is tissue burning of the skin or the eyes.

Ultraviolet radiation occurs with sunlight and with electric arc welding. In both cases, the skin and the eyes are at risk from the effect of burning. The skin will burn (as in sunburn) and repeated exposure can lead to skin cancer. Skin which is exposed to strong sunlight should be protected either by clothing or sun creams. This problem has become more common with the reduction in the ozone layer (which filters out much ultraviolet light). The eyes can be affected by a form of conjunctivitis which feels like grit in the eye, and is called by a variety of names dependent on the activity causing the problem. Arc welders call it 'arc eye' or 'welder's eye' and skiers 'snow blindness'. Cataracts caused by the action of ultraviolet radiation on the eye lens are another possible outcome of exposure.

The most dangerous form of skin cancer, malignant melanoma, has increased by over 40% over the last 10 years making it the fastest growing cancer in the UK. Employers need to be aware of the risks their employees are taking when they work outside without adequate protection from the sun. With growing concern following the rise in skin cancers, the HSE has suggested the following hierarchy of controls for outdoor working:

➤ relocate some jobs inside a building or to a shady location;
➤ undertake some outdoor work earlier or later in the day;
➤ provide personal protection (hats and/or sun cream);
➤ provide suitable education and training for outdoor workers;
➤ provide suitable information and supervision to instigate safe systems of work that protect workers from the sun.

Lasers use visible light and light from the invisible wavelength spectrum (infrared and ultraviolet). As the word laser implies, they produce 'light amplification by stimulated emission of radiation'. This light is highly concentrated and does not diverge or weaken with distance and the output is directly related to the chemical composition of the medium used within the particular laser. The output beam may be pulsating or continuous – the choice being dependent on the task of the laser. Lasers have a large range of applications including bar code reading at a supermarket checkout, the cutting and welding of metals and accurate measurement of distances and elevations required in land and mine surveying. They are also extensively used in surgery for cataract treatment and the sealing of blood vessels.

Lasers are classified into five classes (1, 2, 3a, 3b and 4) in ascending size of power output. Classes 1 and 2 are relatively low hazard and only emit light in the visible band. Classes 3a, 3b and 4 are more hazardous and the appointment of a laser safety officer is recommended. All lasers should carry information stating their class and any precautions required during use.

The main hazards associated with lasers are eye and skin burns, toxic fumes, electricity and fire. The vast majority of accidents with lasers affect the eyes. Retinal damage is the most common and is irreversible. Cataract development and various forms of conjunctivitis can also result from laser accidents. Skin burning and reddening (erythema) are less common and are reversible.

Infrared radiation is generated by fires and hot substances and can cause eye and skin damage similar to that produced by ultraviolet radiation. It is a particular problem to fire fighters and those who work in foundries or near furnaces. Eye and skin protection are essential.

Microwaves are used extensively in cookers and mobile telephones and there are ongoing concerns about associated health hazards (and several inquiries are currently under way). The severity of any hazard is proportional to the power of the microwaves. The principal hazard is the heating of body cells, particularly those with little or no blood supply to dissipate the heat. This means that tissues such as the eye lens are most at risk from injury. However, it must be stressed that any risks are higher for items such as cookers than for low-powered devices such as mobile phones.

The measurement of non-ionising radiation normally involves the determination of the power output being received by the worker. Such surveys are best performed by specialists in the field, as the interpretation of the survey results requires considerable technical knowledge.

15.5.4 Radiation protection strategies

Ionising radiation

Protection is obtained by the application of shielding, time and distance either individually or, more commonly, using a mixture of all three.

Shielding is the best method because it is an 'engineered' solution. It involves the placing of a physical shield, such as a layer of lead, steel and concrete, between the worker and the radioactive source. The thicker the shield the more effective it is.

Time involves the use of the reduced time exposure principle and thus reduces the accumulated dose.

Distance works on the principle that the effect of radiation reduces as the distance between the worker and the source increases.

Other measures include the following:

➤ effective emergency arrangements;
➤ training of employees;
➤ the prohibition of eating, drinking and smoking adjacent to exposed areas;
➤ a high standard of personal cleanliness and first-aid arrangements;
➤ strict adherence to personal protective equipment arrangements, which may include full body protection and respiratory protection equipment;
➤ procedures to deal with spillages and other accidents;
➤ prominent signs and information regarding the radiation hazards;
➤ medical surveillance of employees.

The Ionising Radiations Regulations specify a range of precautions which must be taken, including the appointment of a Radiation Protection Supervisor and a Radiation Protection Adviser.

The **Radiation Protection Supervisor** must be appointed by the employer to advise on the necessary measures for compliance with the Regulations and its Approved Code of Practice. The person appointed, who is normally an employee, must be competent to supervise the arrangements in place and have received relevant training.

The **Radiation Protection Adviser** is appointed by the employer to give advice to the Radiation Protection Supervisor and employer on any aspect of working with ionising radiation including the appointment of the Radiation Protection Supervisor. The Radiation Protection Adviser is often an employee of a national organization with expertise in ionising radiation. A Radiation Protection Adviser is not needed when the only work with ionising radiation is as specified in Schedule 1 of the Regulations.

Non-ionising radiation

For ultraviolet and infrared radiation, eye protection in the form of goggles or a visor is most important, particularly when undertaking arc welding or furnace work. Skin protection is also likely to be necessary for the hands, arms and neck in the form of gloves, sleeves and a collar. For construction and other outdoor workers, protection from sunlight is important, particularly for the head and nose. Sun creams should also be used.

For laser operations, engineering controls such as fixed shielding and the use of non-reflecting surfaces around the workstation are recommended. For laser in the higher class numbers, special eye protection is recommended. A risk assessment should be undertaken before a laser is used.

Engineering controls are primarily used for protection against microwaves. Typical controls include the enclosure of the whole microwave system in a metal surround and the use of an interlocking device that will not allow the system to operate unless the door is closed.

Welding operations

Over 1000 accidents involving welding work are reported to the HSE each year. There are several different types of welding operation. The most common are:

➤ manual metal arc welding;
➤ metal inert-gas welding (MIG);
➤ tungsten inert-gas welding (TIG);
➤ oxy-acetylene welding.

The non-ionising radiation hazards caused by arc welding are not the only hazards associated with welding operations. The hazards from fume inhalation, trailing cables and pipes and the manual handling of cylinders are also present. There have also been serious injuries resulting from explosions and fires during welding processes. Accidents are often caused by lack of training and faulty equipment and they are often made worse by the lack of complete personal protective equipment. Many of these accidents occur on farms where welding equipment is used to make on-the-spot repairs to agricultural machinery.

15.6 The causes and prevention of workplace stress

The HSE has estimated that work-related stress costs society around £3.8bn a year. Approximately 12.8 million working days have been lost due to stress, depression and anxiety in one year alone when 530 000 people reported that they were suffering from work-related stress at a level that was making them ill. This has been accompanied by an increase in civil claims resulting from stress at work. Stress and the related issues of workplace bullying and harassment are issues that simply cannot be ignored.

Stress is not a disease – it is the natural reaction to excessive pressure. It can be defined as the reaction that people have when they are unable to cope with excessive pressures and demands. Stress can lead to an improved performance but is not generally a good thing as it is likely to lead to both physical and mental ill-health, such as high blood pressure, peptic ulcers, skin disorders and depression.

Most people experience stress at some time during their lives, e.g. during an illness or death of a close relative or friend. However, recovery normally occurs after the particular crisis has passed. The position is, however, often different in the workplace because the underlying causes of the stress, known as work-related stressors, are not relieved but continue to build up the stress levels until the employee can no longer cope.

The basic workplace stressors are:

➤ the job itself – boring or repetitive, unrealistic performance targets or insufficient training, job insecurity or fear of redundancy;
➤ individual responsibility – ill-defined roles and too much responsibility with too little power to influence the job outputs;
➤ working conditions – cramped, dirty and untidy workplace; unsafe practices; lack of privacy or security; inadequate welfare facilities; threat of violence; excessive noise, vibration or heat; poor lighting; lack of flexibility in working hours to meet domestic requirements and adverse weather conditions for those working outside;
➤ management attitudes – poor communication, consultation or supervision, negative health and safety culture, lack of support in a crisis;
➤ relationships – unhappy relationship between workers, bullying, sexual and racial harassment.

Possible solutions to all these stressors have been addressed throughout this book and involve the creation of a positive health and safety culture, effective training and consultation procedures and a set of health and safety arrangements which work on a day-to-day basis. The HSE has produced its own generic stress audit survey tool which is available free of charge on its website and advises the following action plan:

➤ identify the problem;
➤ identify the background to the problem and how it was discovered;
➤ identify the remedial action required and give reasons for that action;
➤ identify targets and reasonable target dates;
➤ agree a review date with employees to check that the remedial action is working.

The following additional measures have also been found to be effective by some employers:

➤ take a positive attitude to stress issues by becoming familiar with its causes and controls;
➤ take employees' concerns seriously and develop a counselling system which will allow a frank, honest and confidential discussion of stress-related problems;

➤ develop an effective system of communication and consultation and ensure that periods of uncertainty are kept to a minimum;
➤ set out a simple policy on work-related stress and include stressors in risk assessments;
➤ ensure that employees are given adequate and relevant training and realistic performance targets;
➤ develop an effective employee appraisal system which includes mutually agreed objectives;
➤ discourage employees from working excessive hours and/or missing break periods (this may involve a detailed job evaluation);
➤ introduce job rotation and increase job variety;
➤ develop clear job descriptions and ensure that the individual is matched to them;
➤ encourage employees to improve their lifestyle (e.g. many local health authorities provide smoking cessation advice sessions);
➤ monitor incidents of bullying, sexual and racial harassment and, where necessary, take disciplinary action;
➤ train supervisors to recognize stress symptoms among the workforce;
➤ avoid a blame culture over accidents and incidents of ill-health.

The individual can also take action if he feels that he is becoming over-stressed. Regular exercise, change of job, review of diet and talking to somebody, preferably a trained counsellor, are all possibilities.

Workplace stress has no specific health and safety Regulations but is covered by the duties imposed by the Health and Safety at Work Act and the Management of Health and Safety at Work Regulations 1999 to:

➤ ensure so far as is reasonably practicable that workplaces are safe and without risk to health;
➤ carry out a risk assessment relating to the risks to health;
➤ introduce and maintain appropriate control measures.

The risk assessment will need to relate to the individual, such as absentee record, production performance and which of the stressors are applicable, and to the organization, such as training programmes, communication and appraisal procedures. Many stressful conditions can be reduced by an effective 'whistle-blowing' policy that enables individuals to highlight any concerns to senior managers or directors in a confidential manner.

There have been several successful civil actions for compensatory claims resulting from the effects of workplace stress. However, the Court of Appeal in 2002 redefined the guidelines under which workplace stress compensation claims may be made. Their full guidelines should be consulted and consist of sixteen points. In summary, these guidelines are as follows:

➤ no occupations should be regarded as intrinsically dangerous to mental health;

➤ it is reasonable for the employer to assume that the employee can withstand the normal pressures of the job, unless some particular problem or vulnerability has developed;

➤ the employer is only in breach of duty if they have failed to take the steps that are reasonable in the circumstances;

➤ the size and scope of the employer's operation, its resources and the demands it faces are relevant in deciding what is reasonable; including the interests of other employees and the need to treat them fairly in, for example, the redistribution of duties;

➤ if a confidential counselling advice service is offered to the employee, the employer is unlikely to be found in breach of his duty;

➤ if the only reasonable and effective action is to dismiss or demote the employee, the employer is not in breach of his duty by allowing a willing worker to continue in the same job;

➤ the assessment of damages will take account of any pre-existing disorder or vulnerability and the chance that the claimant would have succumbed to a stress-related disorder in any event.

Stress usually occurs when people feel that they are losing control of a situation. In the workplace, this means that the individual no longer feels that they can cope with the demands made on them. Many such problems can be partly solved by listening to rather than talking at people.

15.7 Causes and prevention of workplace violence

Violence at work, particularly from dissatisfied customers, clients, claimants or patients, causes a lot of stress and in some cases injury. This is not only physical violence as people may face verbal and mental abuse, discrimination, harassment and bullying. Fortunately, physical violence is still rare, but violence of all types has risen significantly in recent years. Violence at work is known to cause pain, suffering, anxiety and stress, leading to financial costs due to absenteeism and higher insurance premiums to cover increased civil claims. It can be very costly to ignore the problem. Violence from members of the public is a higher risk with several occupations – the health and social services, police and fire fighters, various types of enforcement officers, education, benefit services, various service industries and debt collection.

In 1999 the Home Office and the HSE published a comprehensive report entitled *Violence at Work: Findings from the British Crime Survey*. This was updated with a joint report *Violence at Work: New Findings from the 2000 British Crime Survey*, which was published in July 2001. This report shows the extent of violence at work and how it has changed during the period 1991–1999 (Table 15.8).

The report defines violence at work as:

> *All assaults or threats which occurred while the victim was working and were perpetrated by members of the public.*

Physical assaults include the offences of common assault, wounding, robbery and snatch theft. Threats include both verbal threats, made to or against the victim and non-verbal intimidation. These are mainly threats to assault the victim and, in some cases, to damage property.

Excluded from the survey are violent incidents where there was a relationship between the victim and the offender and also where the offender was a work colleague. The latter category was excluded because of the different nature of such incidents.

Figure 15.15 Security-coded access and surveillance CCTV camera.

Table 15.8 Trend in physical assaults and threats at work, 1991–1999 (based on working adults of working age)

Number of incidents (000s)	1991	1993	1995	1997	1999
All violence	947	1275	1507	1226	1288
Assaults	451	652	729	523	634
Threats	495	607	779	703	654
Number of victims (000s)					
All violence	472	530	570	649	604
Assaults	227	287	290	275	304
Threats	264	286	352	395	338

Source: British Crime Survey 1999.

The British Crime Survey shows that the number of incidents and victims rose rapidly between 1991 and 1995 but then declined in 1997. Unfortunately, since 1997 the decline seems to have reversed as the number of incidents has increased by 5%.

It is interesting that almost half of the assaults and a third of the threats happened after 1800 hrs, which suggests that the risks are higher if people work at night or in the late evening. About 16% of the assaults involved offenders under the age of 16 and were mainly against teachers or other education workers.

Violence at work is defined by the HSE as:

> *any incident in which an employee is abused, threatened or assaulted in circumstances relating to their work.*

In recognition of this, the HSE has produced a useful guide to employers which includes a four-stage action plan and some advice on precautionary measures [*Violence at Work: a Guide for Employers*, INDG69 (rev)]. The employer is just as responsible, under health and safety legislation, for protecting employees from violence as they are for any other aspects of their safety.

The Health and Safety at Work Act puts broad general duties on employers and others to protect the health and safety of staff. In particular, section 2 of the HSW Act gives employers a duty to safeguard, so far as is reasonably practicable, the health, safety and welfare at work of their staff.

Employers also have a common law general duty of care towards their staff, which extends to the risk of violence at work. Legal precedents (see *West Bromwich Building Society v Townsend* [1983] IRLR 147 and *Charlton v Forrest Printing Ink Company Limited* [1980] IRLR 331) show that employers have a duty to take reasonable care to see that their staff are not exposed to unnecessary risks at work, including the risk of injury by criminals. In carrying out their duty to provide a safe system of work and a safe working place, employers should, therefore, have regard to, and safeguard their staff against, the risk of injury from violent criminals.

The HSE recommends the following four-point action plan:

1. find out if there is a problem;
2. decide on what action to take;
3. take the appropriate action;
4. check that the action is effective.

15.7.1 Find out if there is a problem

This involves a risk assessment to determine what the real hazards are. It is essential to ask people at the workplace and, in some cases, a short questionnaire may be useful.

Record all incidents to get a picture of what is happening over time, making sure that all relevant detail is recorded. The records should include:

- a description of what happened;
- details of who was attacked, the attacker and any witnesses;
- the outcome, including how people were affected and how much time was lost;
- information on the location of the event.

Owing to the sensitive nature of some aggressive or violent actions, employees may need to be encouraged to report incidents and be protected from future aggression.

All incidents should be classified so that an analysis of the trends can be examined.

Consider the following:

- fatalities;
- major injury;
- less severe injury or shock which requires first-aid treatment, outpatient treatment, time off work or expert counselling;
- threat or feeling of being at risk or in a worried or distressed state.

15.7.2 Decide on what action to take

It is important to evaluate the risks and decide who may be harmed and how this is likely to occur. The threats may be from the public or co-workers at the workplace or may be as a result of visiting the homes of customers. Consultation with employees or other people at risk will improve their commitment to control measures and will make the precautions much more effective. The level of training and information provided, together with the general working environment and the design of the job, all have a significant influence on the level of risk.

Those people at risk could include those working in:

- reception or customer service points;
- enforcement and inspection;
- lone working situations and community-based activities;
- front-line service delivery;
- education and welfare;
- catering and hospitality;
- retail petrol and late-night shopping operations;
- leisure facilities, especially if alcohol is sold;
- healthcare and voluntary roles;
- policing and security;
- mental health units or in contact with disturbed people;
- cash handling or control of high value goods.

Consider the following issues:

- quality of service provided;
- design of the operating environment;
- type of equipment used;
- designing the job.

Some violence may be deterred if measures are taken which suggest that any violence may be recorded. Many public bodies use the following measures:

- informing telephone callers that their calls will be recorded;
- displaying prominent notices that violent behaviour may lead to the withdrawal of services and prosecution;
- using closed circuit television (CCTV) or security personnel.

Quality of service provided

The type and quality of service provision has a significant effect on the likelihood of violence occurring in the workplace. Frustrated people whose expectations have not been met and who are treated in an unprofessional way may believe they have the justification to cause trouble.

Sometimes circumstances are beyond the control of the staff member and potentially violent situations need to be defused. The use of correct skills can turn a dissatisfied customer into a confirmed supporter simply by careful response to their concerns. The perceived lack of or incorrect information can cause significant frustrations.

Design of the operating environment

Personal safety and service delivery are very closely connected and have been widely researched in recent years. This has resulted in many organizations altering their facilities to reduce customer frustration and enhance sales. It is interesting that most service points experience less violence when they remove barriers or screens, but the transition needs to be carefully planned in consultation with staff and other measures adopted to reduce the risks and improve their protection.

The layout, ambience, colours, lighting, type of background music, furnishings including their comfort, information, things to do while waiting and even smell all have a major impact. Queue-jumping causes a lot of anger and frustration and needs effective signs and proper queue management, which can help to reduce the potential for conflict.

Wider desks, raised floors and access for special needs, escape arrangements for staff, carefully arranged furniture and screening for staff areas can all be utilized.

Type of security equipment used

There is a large amount of equipment available and expert advice is necessary to ensure that it is suitable and sufficient for the task. Some measures that could be considered include the following:

- **Access control** to protect people and property. There are many variations from staffed and friendly receptions, barriers with swipe-cards and simple coded security locks. The building layout and design may well partly dictate what is chosen. People inside the premises need access passes so they can be identified easily.
- **Closed circuit television** is one of the most effective security arrangements to deter crime and violence. Because of the high cost of the equipment, it is essential to ensure that proper independent advice is obtained on the type and the extent of the system required.
- **Alarms** – there are three main types:
  - intruder alarms fitted in buildings to protect against unlawful entry, particularly after working hours
  - panic alarms used in areas such as receptions and interview rooms covertly located so that they can be operated by the staff member threatened
  - personal alarms carried by an individual to attract attention and to temporarily distract the attacker.
- **Radios and pagers** can be a great asset to lone workers in particular, but special training is necessary as good radio discipline with a special language and codes are required.
- **Mobile phones** are an effective means of communicating and keeping colleagues informed of people's movements and problems such as travel delays. Key numbers should be inserted for rapid use in an emergency.

Job design

Many things can be done to improve the way in which the job is carried out to improve security and avoid violence. These include:

- using cashless payment methods;
- keeping money on the premises to a minimum;
- careful check of customer or client's credentials;
- careful planning of meetings away from the workplace;
- team work where suspected aggressors may be involved;
- regular contact with workers away from their base. There are special services available to provide contact arrangements;

- avoidance of lone working as far as is reasonably practicable;
- thinking about how staff who have to work shifts or late hours will get home. Safe transport and/or parking areas may be required;
- setting up support services to help victims of violence and, if necessary, other staff who could be affected. They may need debriefing, legal assistance, time off work to recover or counselling by experts.

A busy accident and emergency department of a general hospital, for example, has to balance the protection of staff from violent attack with the need to offer patients a calm and open environment. Protection could be given to staff by the installation of wide counters, coded locks on doors, CCTV systems, panic buttons and alarm systems. The employment of security staff and strict security procedures for the storage and issuing of drugs are two further precautions taken by such departments. Awareness training for staff so that they can recognize early signs of aggressive behaviour and an effective counselling service for those who have suffered from violent behaviour should be provided.

15.7.3 Take the appropriate action

The arrangements for dealing with violence should be included in the safety policy and managed like any other aspect of the health and safety procedures. Action plans should be drawn up and followed through using the consultation arrangements as appropriate. The police should also be consulted to ensure that they are happy with the plan and are prepared to play their part in providing back-up and the like.

15.7.4 Check that the action is effective

Ensure that the records are being maintained and any reported incidents are investigated and suitable action taken. The procedures should be regularly audited and changes made if they are not working properly.

Victims should be provided with help and assistance to overcome their distress, through debriefing, counselling, time off to recover, legal advice and support from colleagues.

15.8 The effects of alcohol and drugs

Alcohol and drug abuse damages health and causes absenteeism and reduced productivity. The HSE is keen to see employers address the problem and offers advice in two separate booklets.

Alcohol abuse is a considerable problem when vehicle driving is part of the job, especially if driving is required on public roads. Misuse of alcohol can reduce productivity, increase absenteeism, increase accidents at work and, in some cases, endanger the public. The HSE has estimated that between 3% and 5% of all absences from work are due to alcohol and result in approximately 14 million working days lost each year. Employers need to adopt an alcohol policy following employee consultation. The following matters need to be considered:

➢ how the organization expects employees to restrict their level of drinking;
➢ how drinking problems can be recognized and help offered; and
➢ at what point and under what circumstances will an employee's drinking be treated as a disciplinary rather than a health problem.

Prevention of the problem is better than remedial action after a problem has occurred. During working hours, there should be no drinking, and drinking during break periods should be discouraged. Induction training should stress this policy and managers should set it. Posters can also help to communicate the message. However, it is important to recognize possible symptoms of an alcohol problem, such as lateness and absenteeism, poor work standards, impaired concentration, memory and judgement, and deteriorating relations with colleagues. It is always better to offer counselling rather than dismissal. The policy should be monitored to check on its effectiveness.

Drug abuse presents similar problems to those found with alcohol abuse – absenteeism, reduced productivity and an increase in the risk of accidents. A recent study undertaken by Cardiff University found that 'although drug use was lower among workers than the unemployed. One in four workers under the age of 30 years reported having used drugs in the previous year.... There are well documented links between drug use and impairments in cognition, perception and motor skills, both at the acute and chronic levels. Associations may therefore exist between drug use and work performance'. The following conclusions were also drawn from the study:

➢ about 13% of working respondents reported drug use in the previous year. The rate varied with age, from 3% of those over 50 years to 29% of those under 30 years;
➢ drug use is strongly linked to smoking and heavy drinking in that order;
➢ there is an association between drug use and minor injuries among those who are also experiencing other minor injury risk factors;

➢ the project has shown that recreational drug use may reduce performance efficiency and safety at work.

A successful drug misuse policy will benefit the organization and employees by reducing absenteeism, poor productivity and the risk of accidents. There is no simple guide to the detection of drug abuse, but the HSE has suggested the following signals:

➢ sudden mood changes;
➢ unusual irritability or aggression;
➢ a tendency to become confused;
➢ abnormal fluctuations in concentration and energy;
➢ impaired job performance;
➢ poor time-keeping;
➢ increased short-term sickness leave;
➢ a deterioration in relationships with colleagues, customers or management;
➢ dishonesty and theft (arising from the need to maintain an expensive habit).

A policy on drug abuse can be established by:

1. **Investigation of the size of the problem**
   Examination of sicknesses, behavioural and productivity changes and accident and disciplinary records is a good starting point.

2. **Planning actions**
   Develop an awareness programme for all staff and a special training programme for managers and supervisors. Employees with a drug problem should be encouraged to seek help in a confidential setting.

3. **Taking action**
   Produce a written policy that includes everyone in the organization and names the person responsible for implementing the policy. It should include details of the safeguards to employees and the confidentiality given to anyone with a drug problem. It should also clearly outline the circumstances in which disciplinary and/or reported action will be taken (the refusal of help, gross misconduct and possession/dealing in drugs).

4. **Monitoring the policy**
   The policy can be monitored by checking for positive changes in the measures made during the initial investigation (improvements in the rates of sickness and accidents). Drug screening and testing is a sensitive issue and should only be considered with the agreement of the workforce (except in the case of pre-employment testing). Screening will only be acceptable if it is seen as part of the health policy of the organization and its purpose is to reduce risks to the misusers and others. Laboratories are accredited

Figure 15.16 Help needed here.

by the United Kingdom Accreditation Service (UKAS) and are able to undertake reliable testing.

It is important to stress that some drugs are prescribed and controlled. The side effects of these can also affect performance and pose risks to colleagues. Employers should encourage employees to inform them of any possible side effects from prescribed medication and be prepared to alter work programmes accordingly.

15.9 Sources of reference

Essentials of Health and Safety at Work, HSE Books 2006 ISBN 978 0 7176 6179 4.

Safe Use of Work Equipment ACOP L22, HSE Books 2008 ISBN 978 0 7176 6295 1.

The Workplace (Health, Safety and Welfare) Regulations 1992 ACOP L24, HSE Books 1992 ISBN 978 0 7176 0413 5.

Personal Protective Equipment at Work (Guidance) L25, HSE Books 2005 ISBN 978 0 7176 6139 8.

Display Screen Equipment Work (Guidance) L26, HSE Books 2003 ISBN 978 0 7176 2582 6.

Controlling Noise at Work L108, HSE Books 2005 ISBN 978 0 7176 6164 0.

Lighting at Work HSG38, HSE Books 1998 ISBN 978 0 7176 1232 1.

Seating at Work HSG57, HSE Books 1997 ISBN 978 0 7176 1231 4.

Work-related Upper Limb Disorders – A Guide HSG60, HSE Books 2002 ISBN 978 0 7176 1978 8.

Hand–arm vibration – Control of Vibration at Work Regulations 2005 L140, HSE Books 1996 ISBN 978 0 7176 0967 3.

Whole-body vibration – Control of Vibration at Work Regulations 2005 L141, HSE Books 1995 ISBN 978 0 7176 1000 6.

The Law on VDUs – An Easy Guide HSG90, HSE Books 2003 ISBN 978 0 7176 2602 1.

Managing the Causes of Work-related Stress HSG218, HSE Books 2007 978 0 7176 6273 9. Plus a CD-ROM.

HSE Stress Management Standards www.hse.gov.uk/stress/standards

Relevant statutory provisions
The Workplace (Health, Safety and Welfare) Regulations 1992.

The Health and Safety (Display Screen Equipment) Regulations 1992.

The Provision and Use of Work Equipment Regulations 1998.

Control of Noise at Work Regulations 2005.

The Personal Protective Equipment at Work Regulations 1992.

The Ionising Radiations Regulations 1999.

The Control of Vibration at Work Regulations 2005.

15.10 Practice NEBOSH questions for Chapter 15

1. **(i)** **Define** the term 'ergonomics'.
   (ii) **List SIX** observations made during an inspection of a machine operation which may suggest that the machine has not been ergonomically designed.

2. In relation to work-related upper limb disorders (WRULDs):
   (i) Identify the typical symptoms that might be experienced by affected individuals.
   (ii) Outline the factors that would increase the risk of developing work-related upper limb disorders (WRULDs).

3. **(i) Identify TWO** work activities that may give rise to work-related upper limb disorders (WRULDs).
   (ii) Outline measures that can be taken to reduce the risk of work-related upper limb disorders (WRULDs).

4. The number of absences due to upper limb disorders in an organization appears to be increasing. **Outline** the possible sources of information that could be consulted when investigating this problem.

5. **Outline** the factors that could contribute towards the development of work-related upper limb disorders (WRULDs) among employees working at a supermarket checkout.

6. An increase in complaints of ill-health effects associated with work-related upper limb disorders (WRULDs) has been reported among bricklayers who have been involved in building work on a long-term construction project.
   (i) Identify the typical symptoms the bricklayers would have shown.

   (ii) Give reasons why the work undertaken by the bricklayers may increase the risk of them developing the condition.
   (iii) Describe the measures that should be taken to minimize the risk to these workers.

7. In relation to the ill-health effects from the use of vibrating hand-held tools:
   (i) identify the typical symptoms that might be shown by affected individuals
   (ii) outline the control measures that may be used to minimize the risk of such effects.

8. A number of employees who are required to work with vibrating hand-held tools for lengthy periods during a work shift have reported symptoms of tingling and numbness in their fingers. Further analysis indicates that the employees concerned could be showing early symptoms of hand–arm vibration syndrome (HAVS).
   (i) Describe further symptoms that might develop should the work continue.
   (ii) Outline factors to consider when assessing the risk of HAVS developing among the employees.
   (iii) Outline precautions that could be taken to minimize the risk of the employees developing the condition.

9. A road worker is operating a hand-operated pneumatic road breaker to split concrete, subjecting the hands and arms to heavy vibration. **Outline** control measures that could be considered to reduce the health effects from vibration.

10. **Outline** the factors to consider when making an assessment of a display screen equipment (DSE) workstation.

11. **(i) Outline** the possible risks to health associated with the use of display screen equipment.
    (ii) Describe the precautionary measures which could be taken to eliminate or reduce the risks outlined in (a) above.
    (iii) Identify the features of a suitable seat for use at a DSE workstation.

12. A computer operator has complained of neck and back pain. **Outline** the features associated with the workstation that might have contributed towards this condition.

13. Employees who work for a computer manufacturer are required to solder components, some of which

are less than 2 mm in size, onto printed circuit boards for up to 8 hours a day.
(i) **Giving** reasons in **EACH** case, **identify** the risks to the health and safety of persons undertaking the work.
(ii) **Outline** the measures to take to minimize the risks identified in (i).

14. **Outline** the requirements of the Workplace (Health, Safety and Welfare) Regulations relating to the provision of welfare facilities.

15. Excluding welfare facilities, **outline** issues associated with the work environment that should be addressed before a new office building is occupied in order to meet the requirements under the Workplace (Health, Safety and Welfare) Regulations.

16. **Outline** the health, safety and welfare issues that a company might need to consider before introducing a night shift to cope with an increased demand for its products.

17. **Outline** the specific factors that should be considered when assessing the risks to employees working on night shifts.

18. An office building is about to be occupied by new owners.
(i) **Identify** the factors that should be considered by the new owners when assessing the suitability of lighting within the building.
(ii) Other than lighting, **outline FOUR** factors associated with the physical working environment that may affect the health and safety of employees.
(iii) **Outline** the requirements with respect to the welfare facilities that should be provided in the building.

19. **Outline** the factors to consider when assessing the adequacy of lighting within an open-plan office.

20. (i) **Outline** the possible health risks associated with working in a seated position for prolonged periods of time.
(ii) **Outline** the features of a suitable seat for office work.

21. (i) **Explain** the following terms in relation to noise exposure at work:
(a) 'noise-induced hearing loss'
(b) 'tinnitus'.

(ii) **Identify FOUR** limitations of personal hearing protection as a means of protecting against the effects of noise.

22. (i) **Outline** measures that can be taken to reduce levels of noise to which workers on a construction site are exposed from the use of cement mixers.
(ii) **Identify FOUR** other noise hazards that may be present on construction sites.

23. A pneumatic drill is to be used during extensive repair work to the floor of a busy warehouse.
(i) **Identify**, by means of a labelled sketch, three possible transmission paths the noise from the drill could take.
(ii) **Outline** appropriate control measures to reduce the noise exposures of the operator and the warehouse staff.

24. Maintenance workers in a factory are required to clean machinery on a regular basis using high pressure compressed air. Noise levels have been measured at 90 dB(A).
(i) **Explain** the meaning of 'dB(A)'.
(ii) **Outline** the options that might be considered in order to reduce the risk of hearing damage **BOTH** to the maintenance staff **AND** to other employees.
(iii) **Outline** the criteria that should be used when selecting suitable hearing protection for the task, **identifying** the limitations of such protective equipment.

25. (i) **Explain** the meaning of the term 'daily personal noise exposure' ($L_{EP,d}$).
(ii) **Outline** the actions required when employees' exposure to noise is found to be in excess of 85 dB(A) $L_{EP,d}$.

26. With respect to the Control of Noise at Work Regulations 2005:
(i) **state**, in dB(A), **BOTH** the lower **AND** the upper exposure action values
(ii) **outline** the measures that the employer is required to take when employees are exposed to noise at or above an upper exposure action level.

27. An industrial washing machine has been installed on the concrete floor of a bakery in order to clean the flour off employees' work clothes after their shift.

When in use, employees are exposed to excessive noise levels that are emitted from the machine.

(i) Identify FOUR possible effects on the health of the bakery employees from long-term exposure to the noise.

(ii) Outline practical measures that can be taken to reduce the levels of noise to which the employees are exposed.

28. **Explain** the meaning of the following terms in relation to noise control:
    (i) silencing
    (ii) absorption
    (iii) damping
    (iv) isolation.

29. **(i) Describe** the **TWO** main types of personal hearing protection.
    (ii) Identify FOUR reasons why personal hearing protection may fail to provide adequate protection against noise.
    (iii) Outline the limitations of **EACH** of the above types of hearing protection.

30. **(i) Identify** the possible effects on health that may be caused by working in a hot environment such as a foundry.
    (ii) Outline measures that may be taken to help prevent the health effects identified in (a).

31. **Outline** the precautions to be taken to minimize the risks to persons working in cold stores operating at sub-zero temperatures.

32. **(i) Identify** the persons that an employer may need to appoint in order to comply with the Ionising Radiations Regulations.
    (ii) Outline the means of controlling exposure to ionising radiation.

33. **(i) Identify TWO** types of ionising radiation.
    (ii) Outline the ways in which exposure to ionising radiation at work may be controlled.

34. **(i) Identify** the possible effects on health arising from exposure to ionising radiation.
    (ii) Outline possible means of ensuring that workers are not exposed to unacceptable levels of ionising radiation.

35. **(i) Identify TWO** types of non-ionising radiation, giving an occupational source of each.
    (i) Outline suitable control measures for **ONE** of the above sources of non-ionising radiation.
    (iii) Outline the health effects associated with exposure to non-ionising radiation.

36. **(i)** For each of the following types of non-ionising radiation, **identify** a source **AND state** the possible ill-health effects on exposed individuals:
    (a) infrared radiation
    (b) ultraviolet radiation.
    (ii) Identify the general methods for protecting people against exposure to non-ionising radiation.

37. **Outline** the factors that may lead to unacceptable levels of occupational stress among employees.

38. Other than those associated with the physical environment:
    (i) outline EIGHT possible causes of increased stress levels among employees
    (ii) outline the options that might be available to an organization to reduce stress levels among its employees.

39. Inadequate lighting in the workplace may affect the level of stress among employees. **Outline EIGHT** other factors associated with the physical environment that may increase stress at work.

40. **Outline** the practical measures that might be taken to reduce the risk of violence to employees who deal with members of the public as part of their work.

41. **(i) Identify FOUR** types of work where employees may be at an increased risk of violence from dealing with members of the public.
    (ii) Outline physical measures that can be taken to protect employees from the risk of violence that may arise out of dealing with members of the public.

42. **Identify** control measures that can be implemented to reduce the risk of violence in a busy accident and emergency department.

43. **Outline FIVE** possible indications that an employee may be suffering from alcohol or drug abuse.

Appendix 15.1 Workstation self-assessment checklist

Name: **Department:**   **Date:**

The completion of this checklist will enable you to carry out a self-assessment of your own workstation. Your views are essential in order to enable us to achieve our objective of ensuring your comfort and safety at work. Please circle the answer that best describes your opinion, for each of the questions listed. The form should be returned to............................... as soon as it has been completed.

Environment

1. *Lighting*			
Describe the lighting at your usual workstation.	About right	Too bright	Too dark
Do you get distracting reflections on your screen?	Never	Sometimes	Constantly
What control do you have over local lighting?	Full control	Some control	No control
2. *Temperature and humidity*			
At your workstation, is it usually:	Comfortable	Too warm	Too cold?
Is the air around your workstation:	Comfortable	Too dry?	
3. *Noise*			
Are you distracted by noise from work equipment?	Never	Occasionally	Constantly
4. *Space*			
Describe the amount of space around your workstation.		Adequate	Inadequate

Furniture

5. *Chair*	
Can you adjust the height of the seat?	Yes/No
Can you adjust the height and angle of the backrest?	Yes/No
Is the chair stable?	Yes/No
Does it allow movement?	Yes/No

Record (continued)

Furniture (continued)

Chair (continued)		
Is the chair in a good state of repair?		Yes/No
If your chair has arms, do they get in the way?		Yes/No

6. Desk		
Is the desk surface large enough to allow you to place all your equipment where you want it?		Yes/No
Is the height of the desk suitable?		Yes Too high Too low
Does the desk have a matt surface (non-reflectant)?		Yes/No

7. Footrest		
If you cannot place your feet flat on the floor whilst keying, has a footrest been supplied?		Yes/No

8. Document holder		
If it would be of benefit to use a document holder, has one been supplied?		Yes/No
If you have a document holder, is it adjustable to suit your needs?		Yes/No

Display Screen Equipment

9. Display screen		
Can you easily adjust the brightness and the contrast between the characters on screen and the background?		Yes/No
Does the screen tilt and swivel freely?		Yes/No
Is the screen image stable and free from flicker?		Yes/No
Is the screen at a height which is comfortable for you?		Yes/No

10. Keyboard		
Is the keyboard separate from the screen?		Yes/No

Record (continued)

Display Screen Equipment (continued)

Keyboard (continued)	
Can you raise and lower the keyboard height?	Yes/No
Can you easily see the symbols on the keys?	Yes/No
Is there enough space to rest your hands in front of the keyboard?	Yes/No
11. Software	
Do you understand how to use the software?	Yes/No
12. Training	
Have you been trained in the use of your workstation?	Yes/No
Have you been trained in the use of software?	Yes/No
If you were to have a problem relating to display screen work, would you know the correct procedures to follow?	Yes/No
Do you understand the arrangements for eye and eyesight tests?	Yes/No

Any other comments?

Appendix 15.2 Example of a noise assessment record form

NOISE ASSESSMENT RECORD

Name of Department:

Date of Survey:

Lower Exposure Action Level: 80 dB(A) daily or weekly		Upper Exposure Action Level: 85 dB(A) daily or weekly			Peak Pressure: 135 dB(C)/137 dB(C)	
Workplace	Number of Persons Exposed	Noise Level (Leq) dB(A)	Daily Exposure Period	LEP'd dB(A)	Peak Pressure dB(C)	Comments

General Comments:

Instrument Used:

Date of Last Calibration:

Signature:

Position:

Date:

Construction activities – hazards and control

16

After reading this chapter you should be able to:

1. identify the main hazards of construction and demolition work and outline the general requirements necessary to control them
2. identify the hazards of work above ground level, outline the general requirements necessary to control them and describe the safe working practices for common forms of access equipment
3. identify the hazards of excavations and outline the general requirements necessary to control them
4. identify the hazards to health commonly encountered in small construction activities and explain how risks might be reduced.

16.1 Introduction

The construction industry covers a wide range of activities from large-scale civil engineering projects to very small house extensions. The construction industry has approximately 200 000 firms, of whom only 12 000 employ more than 7 people – many of these firms are much smaller. The use of sub-contractors is very common at all levels of the industry.

It is most likely that everybody will be aware or involved with some aspect of the construction industry at their place of work – either in terms of the repair and modification of existing buildings or a major new engineering project. It is, therefore, important that the health and safety practitioner has some basic knowledge of the hazards and health and safety legal requirements associated with construction.

Over many years, the construction industry has had a poor health and safety record. In 1966, there were 292 fatalities in the industry and by 1995 this figure had reduced to 62, but by 2000/2001 the figure had increased to 106. These figures include deaths of members of the public, including children playing on construction sites. Most of these fatalities (over 70%) were caused by falls from height.

In 2006/2007, 77 workers died and thousands were injured as a result of construction work. The main causes were:

➤ falling through fragile roofs and roof lights;
➤ falling from ladders, scaffolds and other workplaces;
➤ being struck by excavators, lift trucks or dumpers;
➤ overturning vehicles;
➤ being crushed by collapsing structures.

In 2007/2008, the Health and Safety Executive (HSE) made more than 1000 spot checks on construction refurbishment sites and found that one in three put the lives of workers at risk. They served 395 enforcement notices and stopped work at 30% of the sites. Most of the enforcement action taken was against dangerous work at height.

At a conference organized by the Health and Safety Commission in February 2001 to address the problem, it was noted that at least two construction workers are being killed each week. Targets were set to reduce the number of fatalities and major injuries by 40% over a 4-year period.

Owing to the fragmented nature of the industry and its accident and ill-health record, the recent construction industry legal framework has concentrated on hazards associated with the industry, welfare issues and the need for management and control at all stages of a construction project. In addition to the Health and Safety at Work (HSW) Act 1974 and its associated relevant Regulations, there are three sets of specific construction Regulations which provide this legal framework, as follows:

➤ Construction (Head Protection) Regulations;
➤ Construction (Design and Management) Regulations;
➤ Work at Height Regulations.

A summary of these Regulations is given in Chapter 17.

To protect workers at height from serious injury, the Work at Height Regulations gives the following hierarchy of control:

1. Avoid working at height if possible.
2. Use an existing safe place of work.
3. Provide work equipment to prevent falls.
4. Mitigate distance and consequences of a fall.
5. Provide instruction, training and supervision.

It is important to stress that this control hierarchy is relevant to any work at height and is not limited only to construction work.

16.2 The scope of construction

The scope of the construction industry is very wide. The most common activity is general building work which is domestic, commercial or industrial in nature. This work may be new building work, such as a building extension or, more commonly, the refurbishment, maintenance or repair of existing buildings. Larger civil engineering projects involving road and bridge building, water supply and sewage schemes and river and canal work all come within the scope of construction.

The work could involve hazardous operations, such as demolition or roof work, or contact with hazardous materials, such as asbestos or lead. Construction also includes the use of woodworking workshops together with woodworking machines and their associated hazards, painting and decorating and the use of heavy machinery. It will often require work to take place in confined spaces, such as excavations and underground chambers.

Finally, at any given time, there are many young people receiving training on site in the various construction trades. These trainees need supervision and structured training programmes.

16.3 Construction hazards and controls

The Construction Design and Management (CDM) Regulations 2007 deal with many of the hazards likely to be found on a building site. In addition to these specific hazards, there will be the more general hazards (working at height, manual handling, electricity, noise, etc.) which have been discussed in more detail in earlier chapters. The hazards and controls identified in CDM 2007 and the Working at Height Regulations are as follows.

16.3.1 Safe place of work

Safe access to and egress from the site and the individual places of work on the site are fundamental to a good health and safety environment. This clearly requires that all ladders, scaffolds, gangways, stairways and passenger hoists are safe for use. It further requires that all excavations are fenced, the site is tidy and proper arrangements are in place for the storage of materials and the disposal of waste. The site needs to be adequately lit and secured against intruders, particularly children, when it is unoccupied. Such security will include:

➤ secure and locked gates with appropriate notices posted (Figure 16.1);
➤ a secure and undamaged perimeter fence with appropriate notices posted;
➤ all ladders either stored securely or boarded across their rungs;
➤ all excavations covered;
➤ all mobile plant immobilized and fuel removed, where practicable, and services isolated;
➤ secure storage of all inflammable and hazardous substances;
➤ visits to local schools to explain the dangers present on a construction site. This has been shown to reduce the number of child trespassers;
➤ if unauthorized entry persists, then security patrols and closed circuit television may need to be considered.

16.3.2 Work at height

Work at height accounts for about 50–60 deaths – more than any other workplace activity – and 4000 injuries each year. This is being addressed by the introduction of the Work at Height Regulations, which apply to all operations carried out at height, not just construction work, so

Figure 16.1 Secure site access gate.

they are also relevant to, for example, window cleaning, tree surgery, maintenance work at height and the changing of street lamps. The Work at Height Regulations affect approximately 3 million workers where working at height is essential to their work.

The Work at Height Regulations have no minimum height requirement for work at height. They include all work activities where there is a need to control a risk of falling a distance liable to cause personal injury. This is regardless of the work equipment being used, the duration of the work at height involved or the height at which the work is performed. They include access to and egress from a place of work. They would therefore include:

➤ working on a scaffold or from a mobile elevated work platform (MEWP);
➤ sheeting a lorry or dipping a road tanker;
➤ working on the top of a container in docks or on a ship or storage area;
➤ tree surgery and other forestry work at height;
➤ using cradles or rope for access to a building or other structure like a ship under repair;
➤ climbing permanent structures like a gantry or telephone pole;
➤ working near an excavation area or cellar opening if a person could fall into it and be injured;
➤ painting or pasting and erecting bill posters at height;
➤ work on staging or trestles for example, for filming or events;

➤ using a ladder/stepladder or kick stool for shelf filling, window cleaning and the like;
➤ using manriding harnesses in ship repair, or off-shore or steeple jack work;
➤ working in a mine shaft or chimney;
➤ work carried out at a private house by a person employed for the purpose, for example, painter and decorator (but not if the private individual carries out work on their own home).

However, it would not include:

➤ slips, trips and falls on the same level;
➤ falls on permanent stairs if there is no structural or maintenance work being done;
➤ work in the upper floor of a multi-storey building where there is no risk of falling (except separate activities like using a stepladder).

At the centre of the Regulations (Regulation 6), the employer is expected to apply a three-stage hierarchy to all work which is to be carried out at height. The three steps are the avoidance of work at height, the prevention of workers from falling and the mitigation of the effect on workers of falls should they occur.

The Regulations require that:

➤ work is not carried out at height when it is reasonably practicable to carry the work out safely other than at height (e.g. the assembly of components should be done at ground level);

- when work is carried out at height, the employer shall take suitable and sufficient measures to prevent, so far as is reasonably practicable, any person falling a distance liable to cause injury (e.g. the use of guard rails);
- the employer shall take suitable and sufficient measures to minimize the distance and consequences of a fall (collective measures, for example airbags or safety nets, must take precedence over individual measures, for example safety harnesses).

The risk assessment and action required to control risks from using a kick stool to collect books from a shelf should be simple (not overloading, not overstretching, etc). However, the action required for a complex construction project would involve significantly greater consideration and assessment of risk.

The Work at Height Regulations were amended in 2007 and now apply to those who work at height providing instruction or leadership to one or more people engaged in caving or climbing by way of sport, recreation, team building or similar activities in Great Britain.

A risk assessment for working at height should first consider whether the work could be avoided. If this is not possible, then the risk assessment should consider the following issues:

- the nature and duration of the work;
- the competence level of all those involved with the work and any additional training requirements;
- the required level of supervision;
- use of guard rails, toe boards, working platforms and means of access and egress;
- required personal protective equipment (PPE), such as helmets and harnesses;
- the presence of fall arrest systems, such as netting or soft landing systems;
- the health status of the workers;
- the possible weather conditions;
- compliance with the relevant legislation, in particular the Work at Height Regulations and the Management of Health and Safety at Work Regulations.

A summary of the Regulations is given in Chapter 17 (Section 17.33).

16.3.3 Protection against falls from work at height

The Work at Height Regulations require that guard rails on scaffolds are at a minimum of 950 mm and the maximum unprotected gap between the toe and guard rail of a scaffold is 470 mm. This implies the use of an intermediate

guard rail although other means, such as additional toe boards or screening, may be used. They also specify requirements for personal suspension equipment and means of arresting falls (such as safety nets).

When working at height, a hierarchy of measures should then be followed, to prevent falls from occurring. These measures are:

- avoid working at height, if possible;
- provide of a properly constructed working platform, complete with toe boards and guard rails;
- if this is not practicable or where the work is of short duration, suspension equipment should be used;
- collective fall arrest equipment (airbags or safety nets) may be used;
- where this is not practicable individual fall restrainers (safety harnesses) should be used;
- only when none of the above measures is practicable, should ladders or stepladders be considered.

Much of the work which is done at height could often be done, or partly done, at ground level – thus avoiding the hazards of working at height. The partial erection of scaffolding or edge protection at ground level and the use of cranes to lift it into place at height are examples of this. The manufacture of complete window frames in a workshop and then the final installation of the frame into the building is another. By the use of suitable extension equipment high windows can be cleaned from the ground and high walls can be painted from the ground. However, in most construction work at height, the work cannot be done at ground level and suitable control measures to address the hazards of working at height will be required.

16.3.4 Fragile roofs and surfaces

Work on or near fragile surfaces is also covered by the Work at Height Regulations (see summary in Chapter 17). Roof work, particularly work on pitched roofs, is hazardous and requires a specific risk assessment and method statement (see later under the management of construction activities for a definition) prior to the commencement of work. Particular hazards are fragile roofing materials, including those materials which deteriorate and become more brittle with age and exposure to sunlight, exposed edges, unsafe access equipment and falls from girders, ridges or purlins. There must be suitable means of access such as scaffolding, ladders and crawling boards; suitable barriers, guard rails or covers where people work near to fragile materials and roof lights; and suitable warning signs indicating that a roof is fragile should be on display at ground level (Figure 16.2).

There are other hazards associated with roof work – overhead services and obstructions, the presence of asbestos or other hazardous substances, the use of equipment such as gas cylinders and bitumen boilers and manual handling hazards.

It is essential that only trained and competent persons are allowed to work on roofs and that they wear footwear having a good grip. It is a good practice to ensure that a person does not work alone on a roof.

16.3.5 Protection against falling objects

This is also now covered by the Work at Height Regulations (see summary of Chapter 17). Both construction workers and members of the public need to be protected from the hazards associated with falling objects. Both groups should be protected by the use of covered walkways or suitable netting to catch falling debris. Waste material should be brought to ground level by the use of chutes or hoists. Waste should not be thrown and only minimal quantities of building materials should be stored on working platforms. The Construction (Head Protection) Regulations 1989 virtually mandates employers to supply head protection (hard hats) to employees whenever there is a risk of head injury from falling objects. (Sikhs wearing turbans are exempted from this requirement.) The employer is also responsible for ensuring that hard hats are properly maintained and replaced when they are damaged in any way. Self-employed workers must supply and maintain their own head protection. Visitors to construction sites should always be supplied with head protection and mandatory head protection signs displayed around the site.

16.3.6 Demolition

Demolition is one of the most hazardous construction operations and is responsible for more deaths and major injuries than any other activity. The management of demolition work is controlled by the Construction (Design and Management) Regulations (see the next section of this chapter).

The principal hazards associated with demolition work are:

➤ falls from height or on the same level;
➤ falling debris;
➤ premature collapse of the structure being demolished;
➤ dust and fumes;
➤ the silting up of drainage systems by dust;
➤ the problems arising from spilt fuel oils;
➤ manual handling;
➤ presence of asbestos and other hazardous substances;
➤ noise and vibration from heavy plant and equipment;
➤ electric shock;
➤ fires and explosions from the use of flammable and explosive substances;
➤ smoke from burning waste timber;
➤ pneumatic drills and power tools;
➤ the existence of services, such as electricity, gas and water;
➤ collision with heavy plant;
➤ plant and vehicles overturning.

Figure 16.2 Proper precautions must always be taken when working on or near fragile roofs.

Before any work is started, a full site investigation must be made by a competent person to determine the hazards and associated risks which may affect the demolition workers and members of the public who may pass close to the demolition site. The investigation should cover the following topics:

➤ construction details of the structures or buildings to be demolished and those of neighbouring structures or buildings;

➤ the presence of asbestos, lead or other hazardous substances;

➤ the location of any underground or overhead services (water, electricity, gas, etc.);

➤ the location of any underground cellars, storage tanks or bunkers, particularly if flammable or explosive substances were previously stored;

➤ the location of any public thoroughfares adjacent to the structure or building.

The CDM Co-ordinator, who is responsible for notifying the HSE of the proposed demolition work, must ensure that a written risk assessment is made of the design of the structure to be demolished and the influence of that design on the demolition method proposed. This risk assessment will normally be made by the project designer who will also plan the demolition work. A further risk assessment should then be made by the contractor undertaking the demolition – this risk assessment will be used to draw up a method statement for inclusion in the Health and Safety Plan. A written method statement will be required before demolition takes place. The contents of the method statement will include the following:

➤ details of method of demolition to be used, including the means of preventing premature collapse or the collapse of adjacent buildings and the safe removal of debris from upper levels;

➤ details of equipment, including access equipment, required and any hazardous substances to be used;

➤ arrangements for the protection of the public and the construction workforce, particularly if hazardous substances, such as asbestos or other dust, are likely to be released;

➤ details of the isolation methods of any services which may have been supplied to the site and any temporary services required on the site;

➤ details of PPE which must be worn;

➤ first aid, emergency and accident arrangements;

➤ training and welfare arrangements;

➤ arrangements for waste disposal;

➤ names of site foremen and those with responsibility for health and safety and the monitoring of the work;

➤ control of Substances Hazardous to Health (COSHH) and other risk assessments (PPE, manual handling, etc.) should be appended to the method statement.

There are two forms of demolition:

➤ piecemeal – where the demolition is done using hand and mechanical tools such as pneumatic drills and demolition balls;

➤ deliberate controlled collapse – where explosives are used to demolish the structure. This technique should be used only by trained, specialist competent persons.

A very important element of demolition is the training required by all construction workers involved in the work. Specialist training courses are available for those concerned with the management of the process, from the initial survey to the final demolition. However, induction training, which outlines the hazards and the required control measures, should be given to all workers before the start of the demolition work. The site should be made secure with relevant signs posted to warn members of the public of the dangers.

16.3.7 Excavations

This topic will be covered in more detail later in this chapter (see Section 16.6). Excavations must be constructed so that they are safe environments for construction work to take place. They must also be fenced and suitable notices posted so that neither people nor vehicles fall into them, as required by the Work at Height Regulations (Figure 16.3).

16.3.8 Prevention of drowning

Where construction work takes place over water, steps should be taken to prevent people falling into the water and rescue equipment should be available at all times.

16.3.9 Vehicles and traffic routes

All vehicles used on site should be regularly maintained and records kept. Only trained drivers should be allowed to drive vehicles and the training should be relevant to the particular vehicle (fork-lift truck, dumper truck, etc.). Vehicles should be fitted with reversing warning systems. HSE investigations have shown that in over 30% of dumper truck accidents, the drivers had little experience and no training. Common forms of accident include driving into excavations, overturning while driving up steep inclines and runaway vehicles which have been left unattended with the engine running. Many vehicles such as mobile cranes require regular inspection and test certificates.

Figure 16.3 Barriers around excavation by footpath.

The small dumper truck is widely used on all sizes of construction site. Compact dumper trucks are involved in about 30% of construction transport accidents. The three main causes of such accidents are:

➤ overturning on slopes and at the edges of excavations;
➤ poorly maintained braking systems;
➤ driver error due to lack of training and/or inexperience.

Some of the hazards associated with these vehicles are: collisions with pedestrians, other vehicles or structures such as scaffolding. They can be struck by falling materials and tools or be overloaded. The person driving the truck can be thrown from the vehicle, come into contact with moving parts on the truck, suffer the effects of whole body vibrations due to driving over potholes in the roadway and suffer from the effects of noise and dust. The precautions that can be taken to address these hazards include the use of authorized, trained, competent and supervised drivers only. As with so many other construction operations, risks should be assessed, safe systems of work followed and drivers forbidden from taking shortcuts. The following site controls should also be in place:

➤ designated traffic routes and signs;
➤ speed limits;
➤ stop blocks used when the vehicle is stationary;
➤ proper inspection and maintenance procedures;
➤ procedures for starting, loading and unloading the vehicle;

➤ provision of roll-over protective structures (ROPS) and seat restraints;
➤ provision of falling-object protective structures (FOPS) when there is a risk of being hit by falling materials;
➤ visual and audible warning of approach;
➤ where necessary, hearing protection.

For other forms of mobile construction equipment, such as fork-lift trucks (covered in Chapter 9), the risk to people from the overturning of the equipment must always be safeguarded. This can usually be achieved by the avoidance of working on steep slopes, the provision of stabilizers and ensuring that the load carried does not affect the stability of the equipment/vehicle. Chapter 14 describes the hazards, safeguards and precautions for several pieces of equipment and machinery used in construction work, such as the cement/concrete mixer and the bench-mounted circular saw.

Traffic routes and loading and storage areas need to be well designed with enforced speed limits, good visibility and the separation of vehicles and pedestrians being considered. The use of one-way systems and separate site access gates for vehicles and pedestrians may be required. Finally, the safety of members of the public must be considered, particularly where vehicles cross public footpaths.

16.3.10 Fire and other emergencies

Emergency procedures relevant to the site should be in place to prevent or reduce injury arising from fire, explosions, flooding or structural collapse. These procedures should include the location of fire points and assembly points, extinguisher provision, site evacuation, contact with the emergency services, accident reporting and investigation, and rescue from excavations and confined spaces. There also needs to be training in these procedures at the induction of new workers and ongoing for all workers.

16.3.11 Welfare facilities

The HSE has been concerned for some time at the poor standard of welfare facilities on many construction sites. Sanitary and washing facilities (including showers if necessary) with an adequate supply of drinking water should be provided for everybody working on the site. Accommodation will be required for the changing and storage of clothes and rest facilities for break times. There should be adequate first-aid provision, an accident book and protective clothing against adverse weather conditions.

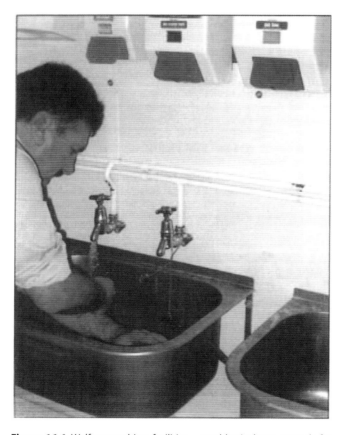

Figure 16.4 Welfare: washing facilities – washbasin large enough for people to wash their forearms.

16.3.12 Electricity

Electrical hazards have been covered in detail earlier in Chapter 12, and all the control measures mentioned apply on a construction site. However, due to the possibility of wet conditions, it is recommended that only 110 V equipment is used on site. Where mains electricity is used (perhaps during the final fitting out of the building), residual current devices should be used with all equipment. Where workers or tall vehicles are working near or under overhead power lines, either the power should be turned off or 'goal posts' or taped markers used to prevent contact with the lines. Similarly, underground supply lines should be located and marked before digging takes place.

16.3.13 Noise

Noisy machinery should be fitted with silencers. When machinery is used in a workshop (such as woodworking machines), a noise survey should be undertaken and, if the noise levels exceed the second action level, the use of ear defenders becomes mandatory.

A cement mixer can be a particularly noisy piece of machinery on a construction site. The levels of noise exposure to workers on the site can be reduced by fixing silencers to diesel-powered mixers, ensuring that the mixer is regularly maintained, minimizing the exposure time of the workers by job rotation and providing ear defenders to those working near to the mixer. A better solution would be to use a mixer with lower noise emissions.

Excessive noise hazards may also be present during demolition work and by the use of pneumatic drills, compressors and vehicles used on site.

16.3.14 Health hazards

Health hazards are present on a construction site. These hazards include vibration, dust (including asbestos), cement, solvents and paints and cleaners. A COSHH assessment is essential before work starts with regular updates as new substances are introduced. Copies of the assessment and the related safety data sheets should be kept in the Site Office for reference after accidents or fires. They will also be required to check that the correct personal protective equipment (PPE) is available. A manual handling assessment should also be made to ensure that the lifting and handling of heavy objects is kept to a minimum. The findings of all assessments must be communicated to the work-force.

Many of the health hazards (both chemical and biological) are covered in Chapter 14 including silica, which is commonly produced during construction activities. Such activities which can expose workers and members of the public to silica dust include:

➤ cutting building blocks and other stone masonry work;
➤ cutting and/or drilling paving slabs and concrete paths;
➤ demolition work;
➤ sand blasting of buildings;
➤ tunnelling.

In general, the use of power tools to cut or dress stone and other silica-containing materials will lead to very high exposure levels while the work is occurring. In most cases, exposure levels are in excess of workplace exposure limits (WELs) by factors greater than 2 and in some cases as high as 12. In response to the growing evidence of ill-health effects of silica inhalation, the HSE has recently revised the WEL down from $0.3\,mg/m^3$ to $0.1\,mg/m^3$.

The inhalation of very fine silica dust can lead to the development of silicosis. HSE has produced a detailed information sheet on silica – CIS No 36 (Rev 1).

In addition to silica, there are three hazardous substances that are particularly relevant to construction activities – cement dust and wet cement, wood dust and the biological hazard tetanus.

Cement dust and wet cement is an important construction and is also a hazardous substance. Contact with wet cement can cause serious burns or ulcers which will take several months to heal and may need a skin graft. Dermatitis, both irritant and allergic, can be caused by skin contact with either wet cement or cement powder. Allergic dermatitis is caused by an allergic reaction to hexavalent chromium (chromate) which is present in cement. Cement powder can also cause inflammation and irritation of the eye, irritation of the nose and throat, and, possibly, chronic lung problems. Research has shown that between 5% and 10% of construction workers are probably allergic to cement. And plasterers, concreters and bricklayers or masons are particularly at risk. A plasterer, who knelt in wet cement for 5 hours while working, required skin grafts to his legs.

Manual handling of wet cement or cement bags can lead to musculoskeletal health problems and cement bags weighing more that 25 kg should not be carried by a single worker. PPE in the form of gloves, overalls with long sleeves and full-length trousers and waterproof boots must be worn on all occasions. If the atmosphere is dusty, goggles and respiratory protection equipment must be worn. An important factor in the possibility of dermatitis occurring is the sensitivity of the worker to the chromate in the cement and the existing condition of the skin including cuts and abrasions. Finally, adequate welfare facilities are essential so that

workers can wash their hands at the end of the job and before eating, drinking or using the toilet. If cement is left on the skin for long periods without being washed off, the risk of an allergic reaction to hexavalent chromium will increase.

A 2005 amendment to the COSHH Regulations prohibits the supply of cement which has a concentration of more than 2 parts per million of chromium VI. This measure is designed to prevent allergic contact dermatitis when wet cement comes into contact with the skin. However, since the strong alkalinity of cement will remain, there is still the potential for skin burns.

The Approved Code of Practice (ACOP) gives useful advice on possible control measures by offering two routes for employers to comply with the amended COSHH Regulations. They can either use the generic advice given in COSHH Essentials HSG193 or design a solution themselves with the help of competent advice. In any event, the controls should be proportionate to the health risk. The Regulations stress the need to provide adequate washing, changing, eating and drinking facilities.

Wood dust can be hazardous, particularly when it is hard wood dust which is known, in rare cases, to lead to nasal cancer. Composite boards, such as medium-density fibreboard (MDF), are hazardous due to the resin bonding material used, which can also be carcinogenic. There are three types of wood-based boards available: laminated board, particle board and fibreboard. The resins used to bond the fibreboard together contain formaldehyde (usually urea formaldehyde). It is generally recognized that formaldehyde is 'probably carcinogenic to humans' and is subject to a WEL. At low exposure levels, it can cause irritation to the eyes, nose and throat and can lead to dermatitis, asthma and rhinitis. The main problems are most likely to occur when the MDF is being machined and dust is produced. A suitable risk assessment should be made and gloves and appropriate masks should always be worn when machining MDF. However, it is important to stress that safer materials are available which do not contain formaldehyde and these should be considered for use in the first instance.

Wood dust is produced whenever wood materials are machined, particularly sawed, sanded, bagged as dust from dust extraction units or during cleaning operations, especially if compressed air is used. The main hazards associated with all wood dusts are skin disorders, nasal problems, such as rhinitis, and asthma. There is also a hazard from fire and explosion. A COSHH assessment is essential to show whether the particular wood dust is hazardous. When the wood dust is created inside a woodworking shop, a well-designed extraction system is essential. PPE in the form of gloves, suitable respiratory protective equipment, overalls and eye protection may also be necessary as a result of the assessment. Finally, good washing and welfare facilities are also essential.

Tetanus is a serious, sometimes fatal, disease caused by a bacterium that lives in the soil. It usually enters the human body through a wound from an infected object, such as a nail, wood splinter or thorn. On entering the wound, it produces a powerful toxin which attacks the nerves that supply muscle tissue. It is commonly known as lockjaw because after an incubation period of approximately a week, stiffness around the jaw area occurs. Later the disease spreads to other muscles including the breathing system and this can be fatal. The disease has been well controlled with anti-tetanus immunization and it is important that all construction workers are so immunized. Booster shots should be obtained every few years. Any flesh wound should be thoroughly cleaned immediately and an anti-septic cream applied.

Other health hazards which could affect construction workers are noise, vibration, asbestos, solvents, fumes, radiation (particularly from sunlight) and biological hazards (such as leptospirosis from sewer work).

Health surveillance may well be required for some of these hazards (see Chapter 14).

16.3.15 Waste disposal

The collection and removal of waste from a construction site is normally accomplished using a skip. The skip should be located on firm, level ground away from the main construction work, particularly excavation work. This will allow clear access to the skip for filling and removal from site. On arrival at site, the integrity of the skip should be checked. It should be filled either by chute or by mechanical means unless items can be placed in by hand. Skips should not be overfilled and be netted or sheeted over when it is full. Any hazardous waste should be segregated as described in Chapter 14, which also mentioned disposal procedures.

Waste skip selection should be made during the site planning process. The selected skip must be suitable for the particular job. The following points should be considered:

➤ sufficient strength to cope with its load;
➤ stability while being filled;
➤ a reasonable uniform load distribution within the skip at all times;
➤ the immediate removal of any damaged skip from service and the skip inspected after repair before it is used again;
➤ sufficient space around the skip to work safely at all times;
➤ the skip should be resting on firm level ground;

➤ the skip should never be overloaded or overfilled;
➤ there must be sufficient headroom for the safe removal of the skip when it is filled.

There are hazards present during the movement of a loaded skip from the ground to the back of a skip loader vehicle. Entanglement with the vehicle lifting mechanisms, such as the hydraulic arms and lifting chains, is a major hazard. Other hazards include contact by the skip with overhead obstructions, movement of the skip contents and skip overload leading to mechanical or structural failure. Slip hazards may be present due to spillages from the skip and the skip contents could be contaminated with biological material, asbestos or syringes. Passing traffic during the loading operation may also present a hazard.

The control measures for these hazards include the use of outriggers to increase the stability of the loader vehicle and the provision of steps for the driver to alight from the cab or the vehicle flatbed. The contents of the skip should be secured using netting or tarpaulin. Adhering to the safe working loads of the skip and lifting equipment and the use of a banksman during the lifting process are additional controls. The area around the vehicle may need to be cordoned off to protect passing pedestrians and road traffic. All workers concerned with the operation should wear suitable PPE, such as high-visibility jackets, gloves and suitable footwear. Finally, all lifting equipment, including chains and shackles, must be subject to a periodic statutory examination.

16.4 The management of construction activities

16.4.1 Introduction

The management of construction work, including the selection and control of contractors, is governed by the CDM Regulations 2007 known as CDM 2007. The CDM Regulations apply to the whole of a construction project, from the initial feasibility study to the handover of the completed structure to the customer.

CDM 2007 came into force in 2007. They replace the CDM Regulations 1994 (CDM 94) and the Construction (Health, Safety and Welfare) Regulations 1996 (CHSW). The new ACOP replaces the ACOP under CDM 94.

The key aim of CDM 2007 is to integrate health and safety into the management of the project and to encourage everyone involved to work as a team.

CDM 2007 is split into several main parts:

➤ Part 2 – covers all construction projects and sets out general management duties;
➤ Part 3 – sets out additional duties where the project is notifiable;
➤ Part 4 – sets out the duties relating to health and safety on construction sites and Schedule 2 covers welfare arrangements (largely in the former CHSW).

More details of CDM 2007 are given in Chapter 17 (Section 17.8).

16.4.2 Key differences between CDM 94 and CDM 2007

The Regulations in CDM 2007 have been re-ordered to group duties together by duty holder and to show whether individual provisions apply to all projects, only notifiable projects or only non-notifiable projects.

The key differences are:

➤ The definition of 'Notifiable' is unchanged, but the application provision relating to fewer than five workers on site has been removed.
➤ The construction health and safety requirements formerly in the Construction (Health, Safety and Welfare) Regulations have now been included within CDM 2007 as Part 4 and Schedule 2.
➤ Work for domestic clients no longer needs to be notified.
➤ A group of clients involved in a project can now elect one or more of its members to be the only client(s).
➤ The 'CDM Co-ordinator' is introduced to support and advise the Client and to co-ordinate design and planning. The role of Planning Supervisor ceases to exist.
➤ The appointment of a CDM Co-ordinator and Principal Contractor, and a written Health and Safety Plan, are only required for notifiable projects (but demolition work requires a written system of work).
➤ Duty holders cannot accept an appointment/engagement unless they are competent to carry it out. Additionally, they cannot arrange for, or instruct anyone, to carry out or manage design or construction work, unless that person is competent (or being supervised by a competent person).
➤ Assessment and demonstration of competence is simplified. The new ACOP (L144 *Managing Health and Safety in Construction*) has specific sections on individual and corporate competence.
➤ Everyone involved in a project has general co-operation and co-ordination duties (relating to others on the same

or adjoining sites). There is a specific requirement to implement any preventive and protective measures, on the basis of the principles of prevention specified in the MHSW Regulations.
➤ Clients now have the duty to take reasonable steps to ensure that managerial arrangements made by duty holders (including time and other resources) enable construction work to be carried out without risk to health or safety.
➤ Clients have the duty to ensure that the arrangements are maintained and reviewed throughout the project.
➤ Clients must tell designers and contractors how much time they have, before work starts on site, for planning and preparing construction work.

16.4.3 Explanation of terms used in the CDM 2007 and ACOP

(a) **The Client** is an organization or individual for whom a construction project is undertaken. Clients only have duties when the project is associated with a business or other undertaking (whether for profit or not). This can include, for example, local authorities, school governors, insurance companies and project originators for private finance initiative (PFI) projects. Domestic clients are a special case and do not have duties under CDM 2007.

(b) **Designers** are those who have a trade or a business which involves them in:

• preparing designs for construction work, including variations. This includes preparing drawings, design details, specifications, bills of quantities and the specification (or prohibition) of articles and substances, as well as all the related analysis, calculations and preparatory work
• arranging for their employees or other people under their control to prepare designs relating to a structure or part of a structure.

It does not matter whether the design is recorded (e.g. on paper or a computer) or not (e.g. it is communicated only orally).

Designers may include:

➤ architects, civil and structural engineers, building surveyors, landscape architects, other consultants, manufacturers and design practices (of whatever discipline) contributing to, or having overall responsibility for, any part of the design, for example drainage engineers;
➤ anyone who specifies or alters a design, or who specifies the use of a particular method of work or material, such as a design manager, quantity surveyor who insists

on specific material, or a client who stipulates a particular layout for a new building;

➤ building service designers, engineering practices or others designing plant which forms part of the permanent structure (including lifts, heating, ventilation and electrical systems), for example a specialist provider of permanent fire extinguishing installations;

➤ those purchasing materials where the choice has been left open, for example those purchasing building blocks and so deciding the weights that a bricklayer must handle;

➤ contractors carrying out design work as part of their contribution to a project, such as an engineering contractor providing design, procurement and construction management services;

➤ temporary works engineers, including those designing auxiliary structures, such as formwork, falsework, facade retention schemes, scaffolding and sheet piling;

➤ interior designers, including shop-fitters who also develop the design;

➤ heritage organizations who specify how work is to be done in detail, for example providing detailed requirements to stabilize existing structures;

➤ those determining how buildings and structures are altered, for example during refurbishment, where this has the potential for partial or complete collapse.

(c) **The CDM Co-ordinator** is the person appointed under Regulation 14(1) for notifiable projects only. The CDM Co-ordinator provides clients with a key project adviser in respect of construction health and safety risk management matters. The main function of the CDM Co-ordinator is to help clients to carry out their duties, to co-ordinate health and safety aspects of the design work and to prepare the Health and Safety File.

(d) **The Principal Contractor** is the contractor appointed by the Client for notifiable projects. The Principal Contractor can be an organization or an individual, and is the main or managing contractor. A principal contractor's key duties are to co-ordinate and manage the construction phase of the project and to ensure the health and safety of everybody involved with the construction work, or who are affected by it.

(e) **The Contractor is** any person (including a client, principal contractor or other person referred to in the Regulations) who, in the course or furtherance of a business, undertakes or manages construction work.

(f) **Pre-construction health and safety information**
The Client must provide designers and contractors and, for notifiable projects, the CDM Co-ordinator with specific health and safety information needed to identify the hazards and risks associated with the design and construction work. The information must be identified, assembled and supplied in good time, so that those who need it during the preparation of a bid for construction work or the planning of such work may estimate the resources required to enable the design, planning and construction work to be properly organized and carried out.

The topics which should be addressed in the pre-construction health and safety information are given in Appendix 16.4.

(g) **Construction Phase Health and Safety Plan**
The Principal Contractor must define the way in which the construction phase will be managed and the key health and safety issues for a particular project must be recorded. The Health and Safety Plan should outline the organization and arrangements required to manage risks and co-ordinate the work on site. It should not be a repository for detailed generic risk assessments, records of how decisions were reached or detailed method statements; but it may, for example, set out when such documents will need to be prepared. It should be well focused, clear and easy for contractors and others to understand – emphasizing key points and avoiding irrelevant material. It is crucial that all relevant parties are involved and co-operate in the development and implementation of the plan as work progresses.

The plan must be tailored to the particular project. Generic plans will not satisfy the requirements of Regulation 23 of CDM 2007 but relevant photographs and sketches will simplify and shorten explanations. The plan should be organized so that relevant sections can easily be made available to designers and contractors. The Construction Phase Health and Safety Plan for the initial phase of the construction work must be prepared before any work begins. Further details may be added as completed designs become available and construction work proceeds.

Appendix 16.5 gives details of what should be included in the Health and Safety Plan.

(h) **Health and Safety File**
This is a record of health and safety information for the Client to retain. It alerts those who are responsible for the structure and equipment within it to the significant health and safety risks that will need to be addressed during subsequent use, construction, maintenance, cleaning, repair, alterations, refurbishment and demolition work. See Section 16.4.6 for more information on the Health and Safety File.

(i) Method statement

This is a formal document that describes the sequences of operations for safe working that will ensure health and safety during the performance of a task. It results from the risk assessment carried out for the task or operation and the identified control measures. If the risk is low, a verbal statement may normally suffice.

(j) Notifiable work

Construction work (except that for a domestic client) is notifiable to the HSE if it lasts longer than 30 working days or will involve more than 500 person days of work (e.g. 50 people working for 10 days). Holidays and weekends are not counted if there is no construction work on those days. See Figure 16.5. The CDM Co-ordinator is responsible for making the notification

as soon as possible after their appointment to the particular project. The notice must be displayed where it can be read by people working on the site. See Chapter 17 (Section 17.8) for details of the information required.

16.4.4 The responsibilities of duty holders for non-notifiable projects

A summary of duties is shown in Table 17.3 in Chapter 17. The flow diagram in Appendix 16.3 gives a summary of the application of CDM 2007 to both notifiable and non-notifiable construction projects. The ACOP for CDM 2007 covers the competence of each duty holder in some depth. This is an essential part of the effective management of construction projects. The ACOP should be consulted for further details on competence requirements.

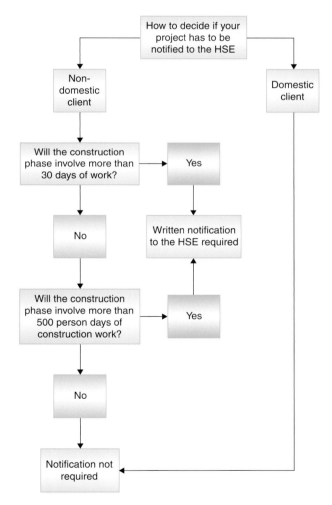

Figure 16.5 HSE notification requirements.

Figure 16.6 (a) Non-notifiable over 30 days but domestic client. (b) Non-notifiable domestic client but only 3 days' work.

Co-ordination and co-operation is a requirement for all members of the project team. However, there is **no** requirement for the appointment of a CDM Co-ordinator, Principal Contractor or a Construction Phase Health and Safety Plan for non-notifiable projects. For low-risk projects a low-key approach will suffice. However, for higher risk projects, for example those involving demolition, a more rigorous approach is required. The architect or lead designer should normally lead the co-ordination of the design work while the main contractor or builder should co-ordinate the construction work.

A brief summary plan may be all that is required in many cases but where the risks are higher, something close to the construction phase plan will be needed. Such higher-risk work includes:

➤ structural alterations;
➤ deep excavations, and those in unstable or contaminated ground;
➤ unusual working methods or safeguards;
➤ ionising radiation or other significant health hazards;
➤ nearby high-voltage power lines;
➤ a risk of falling into water which is, or may become, fast-flowing;
➤ diving;
➤ explosives;
➤ heavy or complex lifting operations.

For all **demolition work**, those in control of the work are required to produce a written plan showing how dangerous hazards will be addressed.

(a) Clients

The role of the Client has been given a higher profile in CDM 2007 to ensure that the construction project has sufficient leadership. Clients are clearly accountable for the effect that their attitudes to health and safety has on those working on, or affected by, the project.

Clients are not expected to manage or plan projects themselves. They are not expected to develop substantial expertise in construction health and safety, unless this is central to their business. However, clients are responsible for ensuring that various actions are taken before and during the construction phase of a project. If advice is needed on client duties, this will be available from the competent person appointed under Regulation 7 of the Management of Health and Safety at Work Regulations. Alternatively clients could seek advice from someone who has acted as a CDM Co-ordinator on other projects. (The appointment of a CDM Co-ordinator is **not** a requirement for non-notifiable projects.)

The Client's duties are to ensure that:

➤ designers, contractors and any others whom they propose to employ are competent or work under the supervision of competent people. They must also be adequately resourced and appointed early enough to fulfil their duties;
➤ they allow sufficient time for each stage of the project;
➤ they co-operate with others who may be involved with any aspect of the project so that they can fulfil their duties;
➤ they co-ordinate their own work with that of others to ensure the safety of those undertaking the construction work;
➤ there are reasonable management arrangements in place throughout the project to ensure the work can be undertaken, so far as is reasonably practicable, safely and without risks to health (this does not mean that they must manage the risk themselves – few clients have the expertise);
➤ contractors have made arrangements for suitable welfare facilities to be provided from the start and throughout the project;
➤ any fixed workplaces which are to be constructed comply with the Workplace (Health Safety and Welfare) Regulations;
➤ relevant information (pre-construction information) likely to be needed by designers, contractors or others to plan and manage the work is passed to them.

The Client will also need to ensure that arrangements are in place to ensure that:

➤ there is clarity as to the roles, functions and responsibilities of members of the project team;
➤ those with duties under CDM 2007 have sufficient time and resource to comply with their duties;
➤ there is good communication, co-ordination and co-operation between members of the project team (e.g. between designers and contractors);
➤ designers are able to confirm that their designs (and any design changes) have taken account of the designers' duties requirements of CDM 2007, and that the different design elements will work together in a way which does not create risks to the health and safety of those constructing, using or maintaining the structure;
➤ that the contractor is provided with the pre-construction information;
➤ contractors are able to confirm that health and safety standards on site will be controlled and monitored, and welfare facilities will be provided by the contractor

from the start of the construction phase through to handover and completion.

Most of these arrangements will be made by others in the project team, such as designers and contractors. Before they start work, a good way of checking that this is the case is to ask the relevant members of the team to explain their arrangements, or to ask for examples of how they will manage these issues during the life of the project. When discussing roles and responsibilities on simple projects, all that may be needed is a simple list of who does what.

Having made these initial checks before work begins, clients should make periodic checks throughout the life of the project to ensure that the arrangements which have been made are maintained. For non-notifiable projects, only simple checks will be needed, for example:

➤ checking that there is adequate protection for the Client's workers and/or members of the public;
➤ checking to make sure that adequate welfare facilities have been provided by the contractor;
➤ checking that there is good co-operation and communication between designers and contractors;
➤ asking for confirmation from the contractor that the arrangements that they agreed to make have been implemented.

Most clients on non-notifiable projects should be able to carry out these checks for themselves.

(b) Designers
CDM 2007 recognizes the key role that designers have for health and safety in construction projects. The aim of the Regulations is to ensure that designers do not produce designs which cannot be constructed, used or maintained safely and with proper consideration for health issues. The amount of effort put in to eliminating hazards and reducing risk should be proportional to the degree of risk.

The duties of designers are to:

➤ make sure that they are competent and adequately resourced to address the health and safety issues likely to be involved in the design of the project;
➤ check that clients are aware of duties;
➤ avoid foreseeable risks, when designing the project, to those involved in the construction and future use of the structure, and in so doing, they should eliminate hazards, so far as is reasonably practicable, and reduce risks associated with the hazards that remain;
➤ provide adequate information about any significant risks associated with the design;

➤ co-ordinate their work with that of others involved with the project in order to improve the way in which risks are managed and controlled.

Designers need to consider risks to those undertaking the construction work, those cleaning and maintaining the structure (particularly windows or translucent panels), those who use the structure for a workplace and those who may be affected by the work (e.g. the public or other customers).

The Designer should not begin work on the project until he/she is sure that the Client is aware of his/her duties. The Designer's duties apply whenever designs are prepared for construction work in Great Britain. This includes concept design, competitions, bids for grants, modifications of existing designs and relevant work carried out as part of feasibility studies. It does not matter whether or not planning permission or funds have been secured, the project is notifiable or high risk, or the Client is a domestic Client.

(c) Contractors
Contractors and their employees are the most at risk of injury and ill health during construction work. They have a key role to play, in co-operation with other duty holders who are planning and managing the work, to ensure that risks are identified and controlled properly. Contractors may also have duties as designers if they are involved in designing elements of their work, such as pre-cast concrete planks or curtain walling.

A contractor defined by the CDM Regulations is anyone who directly employs or engages construction workers, or manages construction work. This includes companies that use their own workforce to carry out construction work on their own premises.

For all projects, contractors must:

➤ check clients are aware of their duties;
➤ satisfy themselves that they and anyone they employ or engage are competent and adequately resourced;
➤ plan, manage and monitor their own work to make sure that workers under their control are safe from the start of their work;
➤ ensure that any contractor whom they appoint or engage to work on the project is informed of the minimum amount of time which will be allowed for them to plan and prepare before starting work;
➤ provide workers under their control (whether employed or self-employed) with any necessary information including relevant aspects of other contractors' work, and site induction which emphasizes the need to work safely, to report problems and to respond appropriately in an emergency;

➤ ensure that any design work they do complies with the requirements on designers in CDM 2007;

➤ comply with any relevant requirement in Part 4 and Schedule 2 of CDM 2007 regarding health and safety on construction sites;

➤ co-operate with others and co-ordinate their own work with others working on the project;

➤ obtain specialist advice where necessary when planning high-risk work (e.g. alterations that could result in a structural collapse or which is on contaminated ground).

Contractors should always plan, manage, supervise and monitor their own work and that of their workers to ensure that it is carried out safely and that health risks are also addressed.

16.4.5 Additional responsibilities of duty holders on notifiable projects

(a) Clients

For notifiable projects, the Client has the following additional duties:

➤ to appoint a CDM Co-ordinator who will advise and assist with the Client's duties and co-ordinate the arrangements for health and safety during the planning phase;

➤ to appoint a Principal Contractor who will plan and manage the construction work – preferably early enough for them to work with the Designer on issues relating to buildability, usability and maintainability;

➤ to ensure that the construction phase does not start until the Principal Contractor has prepared a suitable Health and Safety Plan and made arrangements for suitable welfare facilities to be present from the start of the work;

➤ to ensure that the Health and Safety File is prepared, reviewed or updated and ready for handover at the end of the construction work. The file must be kept available for any future construction work or for transfer to a new owner.

An early appointment of the CDM Co-ordinator is essential as the Client is likely to have to rely on his/her advice on the competence of other appointees and the adequacy of the health and safety arrangements. If a client does not make these appointments, he/she will be liable, under the CDM 2007 Regulations, for the work of the CDM Co-ordinator and the Principal Contractor as well as not making the appointment.

(b) CDM Co-ordinator

A new role of CDM Co-ordinator has been created by CDM 2007. It is a wider role than that of the Planning Supervisor, which it replaces. The CDM Co-ordinator is effectively the Client's health and safety adviser for the project.

The role involves advising and assisting the Client in undertaking the measures needed to comply with CDM 2007, in particular, the Client's duties both at the start of the construction phase and during it.

The CDM Co-ordinator must:

➤ give suitable and sufficient advice and assistance to clients in order to help them to comply with their duties, in particular:

- the duty to appoint competent designers and contractors
- the duty to ensure that adequate arrangements are in place for managing the project;

➤ notify the HSE about the project;

➤ co-ordinate design work, planning and other preparatory work for construction where relevant to health and safety;

➤ identify and collect the pre-construction information and advise the Client if surveys need to be commissioned to fill significant gaps;

➤ promptly provide in a convenient form relevant parts of pre-construction information to those involved with the design of the structure and to every contractor (including the Principal Contractor) who may be or has been appointed by the Client;

➤ manage the flow of health and safety information between clients, designers and contractors;

➤ advise the Client on the suitability of the initial construction phase plan and the arrangements made to ensure that welfare facilities are on site from the start;

➤ produce or update a relevant, user-friendly Health and Safety File suitable for future use at the end of the construction phase.

The CDM Co-ordinator should assist with the development of the health and safety arrangements, and should advise clients on whether or not the arrangements are adequate. They should also advise the Client on the amount of preparatory time that a contractor will need before construction work begins.

(c) Designers

For notifiable projects, designers have the following additional duties:

➤ to ensure that the Client has appointed a CDM Co-ordinator and notified the HSE;

➤ to ensure that design work is not started (other than initial design work) unless a CDM Co-ordinator has been appointed;

➤ to co-operate with the CDM Co-ordinator, Principal Contractor and with any other designers or contractors as necessary so that each of them can comply with their duties – this includes the provision of any information needed for the pre-construction information pack or the Health and Safety File.

If the Client has appointed a CDM Co-ordinator then the Designer can assume that the Client is aware of all other client duties.

(d) The Principal Contractor

The Principal Contractor is the key duty holder, whose main duty is to properly plan, manage and co-ordinate work during the construction phase in order to ensure that hazards are identified and risks are properly controlled. The Principal Contractor has a duty to liaise with all other duty holders and the workforce, in particular to:

➤ consult with the workforce – directly or via their (sub) contractors;

➤ co-operate with the Designer and CDM Co-ordinator, particularly if any changes occur to the design;

➤ ensure that the Client is aware of his/her duties.

Principal Contractors must be competent to carry out their work in a safe manner and ensure that they give proper consideration to the possible effects of their activities on everyone who may be affected by them. Principal Contractors are required to demonstrate to the Client that they have sufficient resources, including properly trained and experienced staff, to carry out the project. It is essential that they are fully aware of the duties of other CDM duty holders.

Principal Contractors must:

➤ satisfy themselves that clients are aware of their duties, that a CDM Co-ordinator has been appointed and HSE notified before work starts;

➤ make sure that they are competent to address the health and safety issues likely to be involved in the management of the construction project phase; attaining a NEBOSH National Certificate in Construction Health and Safety would be a major help;

➤ ensure that the construction phase is properly planned, managed and monitored, with adequately resourced, competent site management appropriate to the risk and activity;

➤ ensure that every contractor who will work on the project is informed of the minimum amount of time

which they will be allowed for planning and preparation before they begin work on site;

➤ ensure that all contractors are provided with the information about the project that they need to enable them to carry out their work safely and without risks to health. Requests from contractors for information should be met promptly;

➤ ensure safe working and co-ordination and co-operation between contractors;

➤ ensure that a suitable Construction Phase Health and Safety Plan is:
  • prepared before construction work begins
  • developed in discussion with, and communicated to, contractors affected by it
  • implemented
  • kept up-to-date as the project progresses;

➤ satisfy themselves that the designers and contractors whom they engage are competent and adequately resourced;

➤ ensure that suitable welfare facilities are provided from the start of the construction phase;

➤ take reasonable steps to prevent unauthorized access to the site;

➤ prepare and enforce any necessary site rules;

➤ provide (copies of or access to) relevant parts of the Health and Safety Plan and other information to contractors including the self-employed in time for them to plan their work;

➤ liaise with the CDM Co-ordinator on design carried out during the construction phase, including design by specialist contractors, and its implications for the Health and Safety Plan;

➤ provide the CDM Co-ordinator promptly with any information relevant to the Health and Safety File;

➤ ensure that all the workers have been provided with suitable health and safety induction, information and training;

➤ ensure that the workforce is consulted about health and safety matters;

➤ display the project notification.

(e) Contractors and the self-employed

In the case of notifiable projects, contractors must also:

➤ check that a CDM Co-ordinator has been appointed and HSE notified before they start work [having a copy of the notification of the project to HSE (form F 10 (rev)) is normally sufficient];

➤ co-operate with the Principal Contractor, CDM Co-ordinator and others working on the project or adjacent sites;

➤ tell the Principal Contractor about risks to others created by their work;

➤ provide details to the Principal Contractor of any contractor whom they engage in connection with carrying out the work;

➤ comply with any reasonable directions from the Principal Contractor, and with any relevant rules in the Construction Phase Health and Safety Plan;

➤ inform the Principal Contractor of any problems with the plan or risks identified during their work that have significant implications for the management of the project;

➤ tell the Principal Contractor about accidents and dangerous occurrences;

➤ provide information for the Health and Safety File.

Contractors must co-operate with the Principal Contractor and assist them in the development of the Construction Phase Health and Safety Plan and its implementation. Where contractors identify shortcomings in the plan, the Contractor should inform the Principal Contractor.

(f) Workers

Workers, along with all others involved in the life of a project, have duties to co-operate and co-ordinate with others. The term 'worker' includes managers and supervisors.

Workers need to be involved as soon as possible and should:

➤ give feedback to their employer via the agreed consultation method;

➤ provide input on risk assessments and developing method statements;

➤ work to the agreed method statement or approach their employer to discuss implementing any change or improvement;

➤ use welfare facilities with respect;

➤ keep tools and PPE in good condition;

➤ be vigilant for hazards and risks and keep management informed of any problem areas;

➤ be aware of the arrangements and actions to take if a dangerous situation arises.

16.4.6 The Health and Safety File

The Health and Safety File is a record of relevant information for the Client or his/her customer and is required for notifiable projects only. It will be developed as the construction progresses with various designers or contractors adding information as it becomes available.

The file should be useful to:

➤ clients who have a duty to provide information about their premises to those who carry out work there;

➤ designers during the development of further designs or alterations;

➤ CDM co-ordinators preparing for construction work;

➤ Principal contractors and contractors preparing to carry out or manage such work.

A number of people have legal duties with respect to the Health and Safety File:

➤ CDM co-ordinators must prepare, review, amend or add to the file as the project progresses and give it to the Client at the end of the project;

➤ clients, designers, principal contractors and other contractors must supply the information necessary for compiling or updating the file;

➤ clients must keep the file to assist with future construction work;

➤ everyone providing information for the file should make sure that it is accurate and provided promptly.

The Client should make sure that the CDM Co-ordinator compiles the file. The contents of the file should be agreed in advance so that designers and contractors can produce the relevant documents as their work progresses.

The file is a key part of the information that a client or their successors will need for future construction projects and so should be kept up-to-date after any relevant surveys or work.

A file must be produced or updated (if one already exists) as part of all notifiable projects. CDM 2007 allows one file to be maintained for a site and then updated after subsequent projects have been completed.

Details of the contents of a Health and Safety File are given in Chapter 17 (Section 17.8.7).

16.4.7 Selection and control of contractors

It is important that health and safety factors are considered as well as technical or professional competence when potential contractors are being shortlisted or employed. The following items will give a guide to health and safety attitudes:

➤ registration with either the HSE or Environmental Health Department of a Local Authority;

➤ a current health and safety policy;

➤ details of any risk assessments made and control measures introduced;

➤ any method statements required to perform the contract;

➤ details of competence certification, particularly when working with gas or electricity may be involved;

➤ details of insurance arrangements in force at the time of the contract;

➤ details of emergency procedures, including fire precautions, for contractor employees;

➤ details of any previous accidents or incidents reported under RIDDOR;

➤ details of accident reporting procedure;

➤ details of previous work undertaken by the contractor;

➤ references from previous employers or main contractors;

➤ details of any health and safety training undertaken by the contractor and his employees.

On being selected, contractors should be expected to:

➤ familiarize themselves with those parts of the Health and Safety Plan which affect them and their employees and/or sub-contractors;

➤ co-operate with the Principal Contractor (or main contact if non-notifiable project) in their health and safety duties to contractors;

➤ comply with their legal health and safety duties.

On arrival at the site, sub-contractors should ensure that:

➤ they report to the Site Office on arrival at the site and report to the Site Manager;

➤ they abide by any site rules, particularly in respect of PPE;

➤ the performance of their work does not place others at risk;

➤ they are familiar with the first aid and accident reporting arrangements of the Principal Contractor;

➤ they are familiar with all emergency procedures on the site;

➤ any materials brought onto the site are safely handled, stored and disposed off in compliance, where appropriate, with the current COSHH Regulations;

➤ they adopt adequate fire precaution and prevention measures when using equipment which could cause fires;

➤ they minimize noise and vibration produced by their equipment and activities;

➤ any ladders, scaffolds and other means of access are erected in conformance with good working practice and the Work at Height Regulations;

➤ any welding or burning equipment brought to the site is in a safe operating condition and used safely with a suitable fire extinguisher to hand;

➤ any lifting equipment brought onto the site complies with the current Lifting Operation and Lifting Equipment Regulations;

➤ all electrical equipment complies with the current Electricity at Work Regulations;

➤ connections to the electricity supply are from a point specified by the Principal Contractor and are by proper cables and connectors. For outside construction work, only 110V equipment should be used;

➤ any restricted access to areas on the site is observed;

➤ welfare facilities provided on site are treated with respect;

➤ any vehicles brought onto the site observe any speed, condition or parking restriction.

The control of sub-contractors can be exercised by monitoring them against the criteria listed above and by regular site inspections. On completion of the contract, the work should be checked to ensure that the agreed standard has been reached and that any waste material has been removed from the site.

16.5 Working above ground level or where there is a risk of falling

16.5.1 Hazards and controls associated with working above ground level

Before any work at height starts, a risk assessment should be undertaken that considers the following simple hierarchy:

➤ work at height should be avoided where possible;

➤ if it is unavoidable, use specific work equipment and control measures to prevent falls;

➤ if it is impossible to eliminate risk of a fall, use work equipment or other control measures to minimize the distance and consequences of a fall if it occurs.

The significance of injuries resulting from falls from height, such as fatalities and other major injuries, has been dealt with earlier in the chapter as has the importance and legal requirements for head protection. Also covered were the many hazards involved in working at height, including fragile roofs and the deterioration of materials, unprotected edges and falling materials. Additional hazards include the weather and unstable or poorly maintained access equipment, such as ladders and various types of scaffold.

The principal means of preventing falls of people or materials includes the use of fencing, guard rails, toe boards, working platforms, access boards, ladder hoops, safety nets and safety harnesses. Safety harnesses arrest the fall by restricting the fall to a given distance due to the fixing of the harness to a point on an adjacent rigid structure. They should only be used when all other possibilities are not practical.

16.5.2 Access equipment

There are many different types of access equipment, but only the following four categories will be considered here:

➤ ladders
➤ fixed scaffold
➤ pre-fabricated mobile scaffold towers
➤ mobile elevated work platforms (MEWPs).

Ladders

The main cause of accidents involving ladders is ladder movement while in use. This occurs when they have not been secured to a fixed point, particularly at the foot. Other causes include over-reaching by the worker, slipping on a rung, ladder defects and, in the case of metal ladders, contact with electricity. The main types of accident are falls from ladders.

There are three common materials used in the construction of ladders – aluminium, timber and glass fibre. Aluminium ladders have the advantage of being light but should not be used in high winds or near live electricity. Timber ladders need regular inspection for damage and should not be painted, as this could hide cracks. Glass fibre ladders can be used near electrical equipment and in food processing areas.

The following factors should be considered when using ladders:

➤ Ensure that the use of a ladder is the safest means of access given the work to be done and the height to be climbed.
➤ The location itself needs to be checked. The supporting wall and supporting ground surface should be dry and slip free. Extra care will be needed if the area is busy with pedestrians or vehicles.
➤ The ladder needs to be stable in use. This means that the inclination should be as near the optimum as possible (1:4 ratio of distance from the wall to distance up the wall).
➤ Wherever possible, a ladder should be tied to prevent it from slipping. This can either be at the top, the bottom or both, making sure both stiles are tied. Never tie a ladder by its rungs.
➤ If the ladder cannot be tied, use an 'effective ladder' or one with an 'effective ladder-stability device'. This means a ladder or ladder-stability device that the

suppliers or manufacturers can confirm will be stable enough to use unsecured in the worst-case scenario.
➤ If the above two precautions are not possible, then the ladder stiles can be wedged against a wall or other similar heavy object; or as a last resort, have a second person 'foot' the ladder.
➤ Weather conditions must be suitable (no high winds or heavy rain).
➤ The proximity of live electricity should also be considered, particularly when ladders are carried near or under power lines (Figure 16.7).
➤ There should be at least 1 m of ladder above the stepping off point.
➤ The work activity must be considered in some detail. Over-reaching must be eliminated and consideration given to the storage of paints or tools which are to be used from the ladder and any loads to be carried up the ladder. The ladder must be matched to work required.
➤ Workers who are to use ladders must be trained in the correct method of use and selection. Such training should include the use of both hands during climbing, clean non-slippery footwear, clean rungs and an undamaged ladder.
➤ Ladders should be inspected (particularly for damaged or missing rungs) and maintained on a regular basis and they should only be repaired by competent persons.
➤ The transportation and storage of ladders is important as much damage can occur at these times. They need to be handled carefully and stored in a dry place.
➤ When a ladder is left secured to a structure during non-working hours, a plank should be tied to the rungs to prevent unauthorized access to the structure.

More information on the safe use of ladders and stepladders is given in INDG402 *Safe Use of Ladders and Stepladders – An Employers' Guide*, HSE Books.

Certain work should not be attempted using ladders. This includes work where:

➤ a secure handhold is not available;
➤ the work is at an excessive height;
➤ the ladder cannot be secured or made stable;
➤ the work is of long duration;
➤ the work area is very large;
➤ the equipment or materials to be used are heavy or bulky;
➤ the weather conditions are adverse;
➤ there is no protection from passing vehicles.

There have been several rumours that the Work at Height Regulations have banned the use of ladders. This is **not** true. Ladders may be used for access and it is legal to work from ladders. Ladders may be used when a risk assessment shows that the risk of injury is low and the task is of short duration or there are unalterable features of the work site and that it is not reasonably practicable to use potentially safer alternative means of access. More information on ladders and their use within the requirements of the Work at Height Regulations is available from the British Ladder Manufacturers' Association (Figure 16.8). Ladders for industrial work in the UK should be marked to:

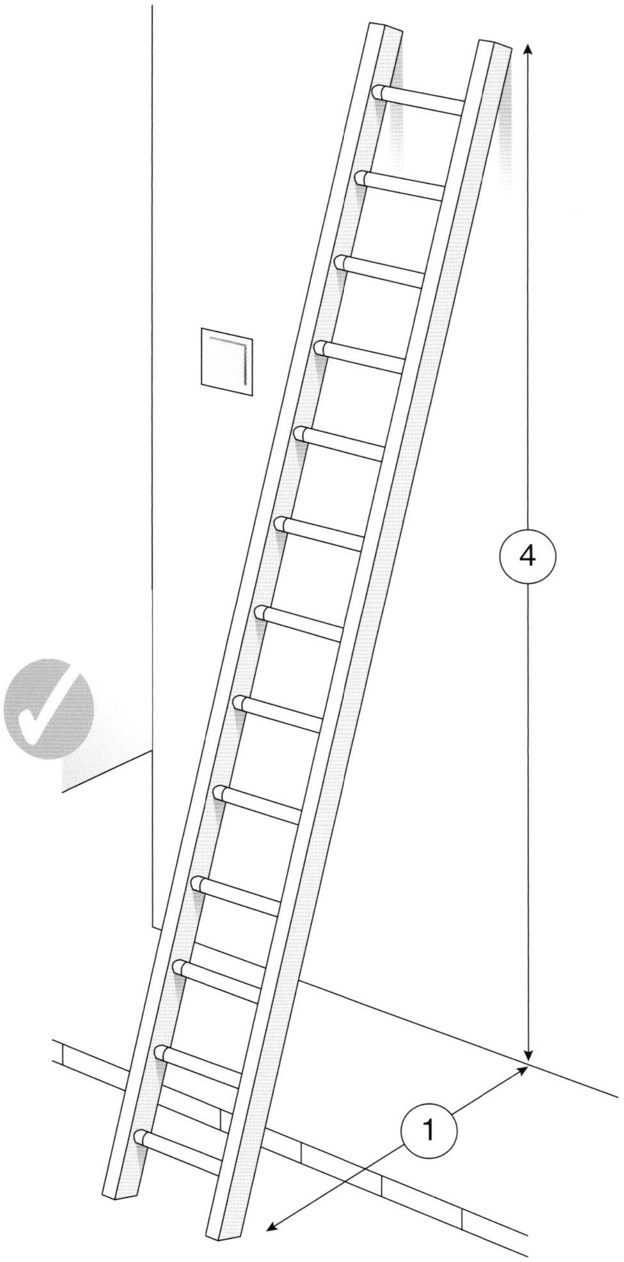

➤ Timber BS1129: Kite marked Class 1 Industrial
➤ Aluminium BS2037: 1994 Kite Marked Class 1 Industrial
➤ Glass fibre BSEN131: 1993 Kite marked Industrial
➤ Step stools BS7377: 1994.

Stepladders, trestles and staging

Many of the points discussed for ladders apply to stepladders and trestles, where stability and over-reaching are the main hazards (Figure 16.9).

All equipment must be checked by the supervisor before use to ensure that there are no defects and must be checked at least weekly whilst in use on site. If a defect is noted, or the equipment is damaged, it must be taken out of use immediately. Any repairs must only be carried out by competent persons.

Supervisors must also check that the equipment is being used correctly and not being used where a safer method should be provided.

Where staging, such as a 'Youngmans' staging platform, is being used in roof areas, supervisors must ensure that only experienced operatives are permitted to carry out this work and that all necessary safety harnesses and anchorage points are provided and used.

The main hazards associated with stepladders, trestles and staging are:

➤ unsuitable base (uneven or loose materials);
➤ unsafe and incorrect use of equipment (e.g. the use of staging for barrow ramps);
➤ overloading;
➤ use of equipment where a safer method should be provided;
➤ overhang of boards or staging at supports ('trap ends');
➤ use of defective equipment.

Stepladders and trestles must be:

➤ manufactured to a recognized industrial specification;
➤ stored and handled with care to prevent damage and deterioration;
➤ subject to a programme of regular inspection (there should be a marking, coding or tagging system to confirm that the inspection has taken place);
➤ checked by the user before use;

Figure 16.7 Ladders showing correct 1 in 4 angle (means of securing omitted for clarity).

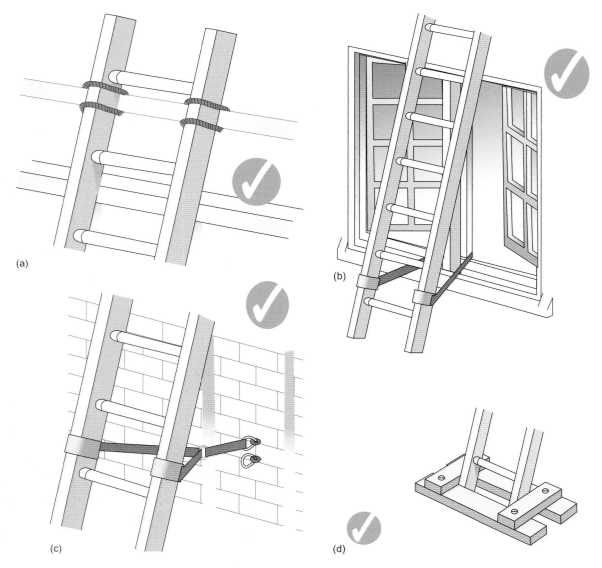

Figure 16.8 (a) Ladder tied at top stiles (correct for working on, not for access). (b) Tying part way down. (c) Tying at base. (d) Securing at the base.

➤ taken out of use if damaged – and destroyed or repaired;

➤ used on a secure surface, and with due regard to ensuring stability at all times;

➤ kept away from overhead cables and similar hazards.

The small platform fitted at the top of many stepladders is designed to support tools, paint pots and other working materials. It should not be used as a working place unless the stepladder has been constructed with a suitable handhold above the platform. Stepladders must not be used if they are too short for the work being undertaken, or if there is not enough space to open them out fully.

Platforms based on trestles should be fully boarded, adequately supported (at least one support for each 1.5 m of board for standard scaffold boards) and provided with edge protection when there is a risk of falling a distance liable to cause injury.

Fixed scaffolds

It is quicker and easier to use a ladder as a means of access, but it is not always the safest. Jobs, such as painting,

✗ Wrong way

Right way ✓

✗ Stepladder too short
✗ Hazard overhead
✗ Over-reaching up and sideways
✗ No grip on ladder
✗ Sideways-on to work
✗ Foot on handrail
✗ Wearing slippers
✗ Loose tools on ladder
✗ Slippery and damaged steps
✗ Uneven soft ground
✗ Damaged stiles
✗ Non-slip rubber foot missing

Steps at right height ✓
No need to over-reach ✓
Good grip on handrail ✓
Working front-on ✓
Wearing good flat shoes ✓
Clean undamaged steps ✓
Firm level base ✓
Undamaged stiles ✓
Rubber non-slip feet all in position ✓
Meets British or European standards ✓

Figure 16.9 Working with stepladders.

gutter repair, demolition work or window replacement, are often easier done using a scaffold. If the work can be completed comfortably using ladders, a scaffold need not be considered. Scaffolds must be capable of supporting building workers, equipment, materials, tools and any accumulated waste. A common cause of scaffold collapse is the 'borrowing' of board s and tubes from the scaffold, thus weakening it. Falls from scaffolds are often caused by badly constructed working platforms, inadequate guard rails or climbing up the outside of a scaffold. Falls also occur during the assembly or dismantling process.

There are two basic types of external scaffold:

➤ **Independent tied** – These are scaffolding structures which are independent of the building but tied to it often using a window or window recess. This is the most common form of scaffolding.
➤ **Putlog** – This form of scaffolding is usually used during the construction of a building. A putlog is a scaffold tube which spans horizontally from the scaffold into the building – the end of the tube is flattened and is usually positioned between two brick courses.

The important components of a scaffold have been defined in a guidance note issued by the HSE as follows (Figure 16.10).

Standard – An upright tube or pole used as a vertical support in a scaffold.

Ledger – A tube spanning horizontally and tying standards longitudinally.

Transom – A tube spanning across ledgers to tie a scaffold transversely. It may also support a working platform.

Bracing – Tubes which span diagonally to strengthen and prevent movement of the scaffold.

Guard rail – A horizontal tube fitted to standards along working platforms to prevent persons from falling.

Toe boards – These are fitted at the base of working platforms to prevent persons, materials or tools falling from the scaffold.

Base plate – A square steel plate fitted to the bottom of a standard at ground level.

Sole board – Normally a timber plank positioned beneath at least two base plates to provide a more uniform distribution of the scaffold load over the ground.

Ties – Used to secure the scaffold by anchoring it to the building. The scaffold in Figure 16.10(a) is tied to the building using a through-tie.

Working platform – An important part of the scaffold, as it is the platform on which the building workers operate and where building materials are stored prior to use.

Other components of a scaffold include access ladders, brick or block guards and chutes to dispose of waste materials.

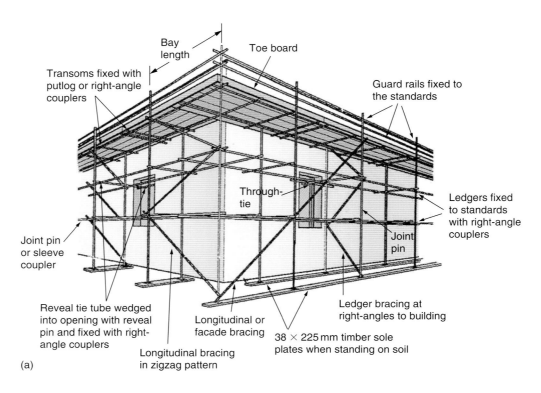

Bay length

Toe board

Transoms fixed with putlog or right-angle couplers

Guard rails fixed to the standards

Through-tie

Ledgers fixed to standards with right-angle couplers

Joint pin

Joint pin or sleeve coupler

Reveal tie tube wedged into opening with reveal pin and fixed with right-angle couplers

Longitudinal or facade bracing

Ledger bracing at right-angles to building

38 × 225 mm timber sole plates when standing on soil

Longitudinal bracing in zigzag pattern

(a)

(b)

Figure 16.10 (a) Typical independent tied scaffold. (b) Fixed scaffold left in place to fit the gutters.

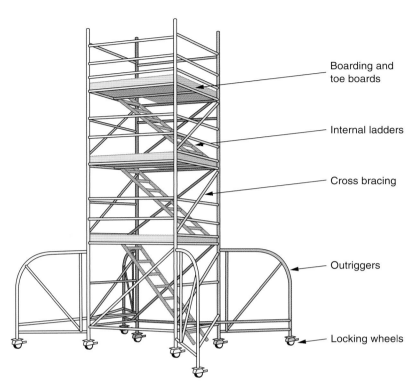

Boarding and
toe boards

Internal ladders

Cross bracing

Outriggers

Locking wheels

Figure 16.11 Typical pre-fabricated tower scaffold.

The following factors must be addressed if a scaffold is being considered for use for construction purposes:

➤ Scaffolding must only be erected by competent people who have attended recognized training courses. Any work carried out on the scaffold must be supervised by a competent person. Any changes to the scaffold must be done by a competent person.

➤ Adequate toe boards, guard rails and intermediate rails must be fitted to prevent people or materials from falling.

➤ The scaffold must rest on a stable surface; uprights should have base plates and timber sole plates if necessary.

➤ The scaffold must have safe access and egress.

➤ Work platforms should be fully boarded with no tipping or tripping hazards.

➤ The scaffold should be sited away from or protected from traffic routes so that it is not damaged by vehicles.

➤ Lower level uprights should be prominently marked in red and white stripes.

➤ The scaffold should be properly braced, secured to the building or structure.

➤ Overloading of the scaffold must be avoided.

➤ The public must be protected at all stages of the work.

➤ Regular inspections of the scaffold must be made and recorded.

Pre-fabricated mobile scaffold towers

Mobile scaffold towers are frequently used throughout industry. It is essential that the workers are trained in their use as recent research has revealed that 75% of lightweight mobile pre-fabricated tower scaffolding is erected, used, moved or dismantled in an unsafe manner (Figure 16.11).

The following points must be considered when mobile scaffold towers are to be used:

➤ The selection, erection and dismantling of mobile scaffold towers must be undertaken by competent and trained persons with maximum height to base ratios not being exceeded.

➤ Diagonal bracing and stabilizers should always be used.

➤ Access ladders must be fitted to the narrowest side of the tower or inside the tower and persons should not climb up the frame of the tower.

- All wheels must be locked while work is in progress and all persons must vacate the tower before it is moved.
- The tower working platform must be boarded, fitted with guard rails and toe boards and not overloaded.
- Towers must be tied to a rigid structure if exposed to windy weather or to be used for work such as jet blasting.
- Persons working from a tower must not over-reach or use ladders from the work platform.
- Safe distances must be maintained between the tower and overhead power lines both during working operations and when the tower is moved.
- The tower should be inspected on a regular basis and a report made.

Mobile elevated work platforms (MEWPs)

MEWPs are very suitable for high-level work such as changing light bulbs in a warehouse (Figure 16.12). The following factors must be considered when using MEWPs:

- The MEWP must only be operated by trained and competent persons.
- It must never be moved in the elevated position.
- It must be operated on level and stable ground with consideration being given for the stability and loading of floors.
- The tyres must be properly inflated and the wheels immobilized.
- Outriggers should be fully extended and locked in position.
- Due care must be exercised with overhead power supplies, obstructions and adverse weather conditions.
- Warning signs should be displayed and barriers erected to avoid collisions.
- It should be maintained regularly and procedures should be in place in the event of machine failure.
- Drivers of MEWPs must be instructed in emergency procedures, particularly to cover instances of power failure.
- All workers on MEWPs should wear safety harnesses.

Workers, using mobile elevating work platforms, have been injured by falling from the platform due to a lack of handrails or inadvertent movement of the equipment because the brakes had not been applied before raising the platform. Injuries have also resulted from the mechanical failure of the lifting mechanism and by workers becoming trapped in the scissor mechanism.

When working on an MEWP, there is a danger that the operator may become trapped against an overhead or adjacent object, preventing him from releasing the controls. There have also been accidents caused when an MEWP is

Figure 16.12 Mobile elevating work platform – scissor lift.

reversed into areas where there is poor pedestrian segregation and the driver has limited visibility. During any manoeuvring operation, a dedicated banksman should be used.

16.5.3 Inspection and maintenance

Inspection

Equipment for work at height needs regular inspection to ensure that it is fit for use. A marking system is probably required to show when the next inspection is due. Formal inspections should not be a substitute for any pre-use checks or routine maintenance. Inspection does not necessarily cover the checks that are made during maintenance although there may be some common features. Inspections need to be recorded but checks do not.

Under the Work at Height Regulations weekly inspections are still required for scaffolding used in construction, as required by the CDM Regulations where a person could fall 2 m or more. The requirements for inspection are set out in the Regulations as follows:

- the name and address of the person for whom the inspection was carried out;

- the location of the work equipment inspected;
- a description of the work equipment inspected;
- the date and time of the inspection;
- details of any matter identified that could give rise to a risk to the health or safety of any person;
- details of any action taken as a result of any matter identified in paragraph 5;
- details of any further action considered necessary;
- the name and position of the person making the report.

Appendix 16.1 shows the inspection format under the CDM Regulations, which still covers the essential items under the Work at Height Regulations and has been slightly adapted to suit.

Maintenance

Inspections and even thorough examinations are not substitutes for properly maintaining equipment. The information gained in the maintenance work, inspections and thorough technical examinations should inform one another. A maintenance log should be kept and be up-to-date. The whole maintenance system will require proper management systems. The frequency will depend on the equipment, the conditions in which it is used and the manufacturers' instructions.

16.6 Excavations

16.6.1 Hazards associated with excavations

There are about seven deaths each year due to work in excavations. Many types of soil, such as clays, are self-supporting but others, such as sands and gravel, are not. Many excavations collapse without any warning, resulting in death or serious injury. Many such accidents occur in shallow workings. It is important to note that, although most of these accidents affect workers, members of the public can also be injured. The specific hazards associated with excavations are as follows (Figure 16.13):

- collapse of the sides;
- materials falling on workers in the excavation;
- falls of people and/or vehicles into the excavation;
- workers being struck by plant;
- specialist equipment such as pneumatic drills;
- hazardous substances, particularly near the site of current or former industrial processes;
- influx of ground or surface water and entrapment in silt or mud;

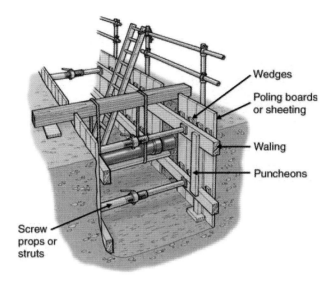

Figure 16.13 Timbered excavation with ladder access and supported services (guard removed on one side for clarity).

- proximity of stored materials, waste materials or plant;
- proximity of adjacent buildings or structures and their stability;
- contact with underground services;
- access and egress to the excavation;
- fumes, lack of oxygen and other health hazards (such as Weil's disease).

Clearly, alongside these specific hazards, more general hazards, such as manual handling, electricity, noise and vibrations, will also be present.

16.6.2 Precautions and controls required for excavations

The following precautions and controls should be adopted:

- At all stages of the excavation, a competent person must supervise the work and the workers must be given clear instructions on working safely in the excavation.
- The sides of the excavation must be prevented from collapsing either by digging them at a safe angle (between 5° and 45° dependent on soil and dryness) or by shoring them up with timber, sheeting or a proprietary support system. Falls of material into the workings can also be prevented by not storing spoil material near the top of the excavation.
- The workers should wear hard hats.

- If the excavation is more than 2 m deep, a substantial barrier consisting of guard rails and toe boards should be provided around the surface of the workings.

- Vehicles should be kept away as far as possible using warning signs and barriers. Where a vehicle is tipping materials into the excavation, stop blocks should be placed behind its wheels.

- It is very important that the excavation site is well lit at night.

- All plant and equipment operators must be competent and non-operators should be kept away from moving plant.

- PPE must be worn by operators of noisy plant.

- Nearby structures and buildings may need to be shored up if the excavation may reduce their stability. Scaffolding could also be de-stabilized by adjacent excavation trenches.

- The influx of water can only be controlled by the use of pumps after the water has been channelled into sumps. The risk of flooding can be reduced by the isolation of the mains water supply.

- The presence of hazardous substances or health hazards should become apparent during the original survey work and, when possible, removed or suitable control measures adopted. Any such hazards found after work has started, must be reported and noted in the inspection report and remedial measures taken. Exhaust fumes can be dangerous and petrol or diesel plant should not be sited near the top of the excavation.

- The presence of buried services is one of the biggest hazards and the position of such services must be ascertained using all available service location drawings before work commences. As these will probably not be accurate, service location equipment should be used by specifically trained people. The area around the excavation should be checked for service boxes. If possible, the supply should be isolated. Only hand tools should be used in the vicinity of underground services. Overhead services may also present risks to cranes and other tall equipment. If the supply cannot be isolated then 'goal posts' beneath the overhead supply together with suitable bunting and signs must be used.

- Safe access by ladders is essential, as are crossing points for pedestrians and vehicles. Whenever possible, the workings should be completely covered outside working hours, particularly if there is a possibility of children entering the site.

- Finally, care is needed during the filling-in process.

Wells and disused mine shafts are found during construction work and must be treated with caution, and in the same way as an excavation. The obvious hazards include falling in and/or drowning and those associated with confined spaces [see Chapter 6 (Section 6.7.4)] – oxygen deficiency, the presence of toxic gases and the possible collapse of the walls. Controls include fencing off the well and covering it until the situation has been reviewed by specialists. Shallow wells would normally be drained and filled with hard core whereas deeper ones would be capped.

16.6.3 Inspection and reporting requirements

The duty to inspect and prepare a report only applies to excavations which need to be supported to prevent accidental fall of material. Only persons with a recognized and relevant competence should carry out the inspection and write the report. Inspections should take place at the following timing and frequency:

- after any event likely to affect the strength or stability of the excavation;
- before work at the start of every shift;
- after an accidental fall of any material.

Although an inspection must be made at the start of every shift, only one report is required of such inspections every 7 days. However, reports must be completed following all other inspections. The report should be completed before the end of the relevant working period and a copy given to the manager responsible for the excavation within 24 hours. The report must be kept on site until the work is completed and then retained for 3 months at an office of the organization which carried out the work.

A suitable form is shown in Appendix 16.1.

16.7 Sources of reference

Essentials of Health and Safety at Work, HSE Books 2006 ISBN 978 0 7176 6179 4.

Health and Safety in Construction, HSG 150 HSE Books 2006 ISBN 978 0 7176 6182 4.

Managing Health and Safety in Construction (ACOP), L144, HSE Books 2007 ISBN 978 0 7176 6223 4.

Safe Use of Work Equipment (ACOP) L22, HSE Books 2008 ISBN 978 0 7176 6295 1.

Work at Height Regulations 2005 As Amended – A Brief Guide INDG 401 HSE Books 2007 ISBN 978 0 7176 6231 9.

Relevant statutory provisions

The Construction (Design and Management) Regulations 2007.

The Construction (Head Protection) Regulations 1989.

The Provision and Use of Work Equipment Regulations 1998.

The Personal Protective Equipment at Work Regulations 1992.

The Work at Height Regulations 2005.

16.8 Practice NEBOSH questions for Chapter 16

1. **Outline** a hierarchy of measures to be considered when a construction worker is likely to fall while working at height.

2. **Outline EIGHT** precautions that should be considered to prevent accidents to children who might be tempted to gain access to a construction site.

3. A contractor has been engaged to undertake building maintenance work in a busy warehouse. **Outline** the issues that should be covered in an induction programme for the contractor's employees.

4. The exterior paintwork of a row of shops in a busy high street is due to be re-painted. **Identify** the hazards associated with the work and **outline** the corresponding precautions to be taken.

5. **Outline** the precautions to be taken when carrying out repairs to the flat roof of a building.

6. **Outline** the precautions to be taken when repair work is to be carried out on the sloping roof of a building.

7. **Outline** the precautions to be taken to ensure the safety of workers required to undertake repair work on a fragile roof.

8. (i) **Explain** the meaning of the term 'hazard'.
   (ii) **Outline** the main hazards that may be present during the demolition of a multi-storey building.

9. (i) **Outline** the possible causes of a dumper truck overturn on a construction site.

(ii) **Identify** the design features of a dumper truck intended to minimize the risk of, or severity of injury from, an overturn.

10. A manufacturing company is to relocate to premises that require refurbishment before equipment and staff can be moved. **Outline** the sanitary and washing facilities that should be considered when planning the refurbishment.

11. **Outline** the health and safety requirements relating to the provision of welfare, fire prevention and first-aid facilities on construction sites.

12. (i) **Outline** the main duties of a CDM Co-ordinator under the Construction (Design and Management) Regulations.
    (ii) **Identify FOUR** items of information in the Health and Safety File that might be needed for an existing building by a contractor carrying out refurbishment work.

13. With reference to the Construction (Design and Management) Regulations:
    (i) **identify** the circumstances under which a construction project must be notified to an enforcing authority
    (ii) **outline** the duties of the Client under the Regulations.

14. **Outline FOUR** duties of each of the following persons under the Construction (Design and Management) Regulations:
    (i) the CDM Co-ordinator
    (ii) the Principal Contractor.

15. (i) **Identify FOUR** hazards associated with work at height above ground level.
    (ii) **Outline** the factors to consider, specifically related to work at height, whilst conducting a risk assessment.
    (iii) **Describe** safe working practices associated with the use of ladders.

16. **Explain** the issues that would need to be addressed if work is to be carried out safely from a ladder.

17. **Outline** the main dangers and the corresponding precautions that should be taken with the use of ladders.

18. **Identify** the ways in which accidents may be prevented when using ladders as a means of access.

19. Outline the precautions that might be taken in order to reduce the risk of injury when using stepladders.

20. Outline THREE causes of a scaffold collapse.

21. List EIGHT components of an independent tied scaffold that has been erected by a competent person.

22. Mobile tower scaffolds should be used on stable, level ground. **List EIGHT** additional points that should be considered to ensure safe use of a mobile tower scaffold.

23. Identify measures that should be adopted in order to protect against the dangers of people and/or materials falling from a mobile tower scaffold.

24. Give reasons that may cause a mobile tower scaffold to become unstable.

25. Outline the precautions to be taken when using a mobile elevating work platform (MEWP) to reach a high point such as a streetlight.

26. Explain how a person may be injured when using a mobile elevated working platform (MEWP) to undertake maintenance work at height.

27. Fluorescent tubes in the roof space of a busy warehouse are to be replaced by maintenance workers using a mobile elevated working platform (MEWP).
(i) Identify the potential hazards associated with the task.
(ii) Outline the measures to be taken to ensure the safety of the maintenance workers and others who may be affected by the work.

28. Identify the main hazards associated with excavation work on construction sites.

29. Outline the main precautions to be taken when carrying out excavation work.

30. The water main supplying a school is to be replaced. The work will be carried out in a 1.5 m deep excavation, which will be supported in order to ensure the safety of the employees working in the excavation.
(i) Identify when the **THREE** statutory inspections of the supported excavation must be carried out by the competent person.
(ii) State the information that should be recorded in the excavation inspection report.
(iii) Other that the provision of supports for the excavation, **outline** additional precautions to be taken during the repair work in order to reduce the risk of injury to the employees and others who may be affected by the work.

31. A well found on a development site needs to be safe before construction work can begin.
(i) Identify the hazards that may be present in these circumstances.
(ii) Outline the control measures that could be used in order to render the well safe.

32. Outline the precautions to protect against electrical contact when:
(i) excavating near underground cables
(ii) working in the vicinity of overhead power lines.

33. Identify EIGHT possible health hazards to which construction workers may be exposed **AND** in **EACH** case **give** an example of a likely source.

34. (i) Identify FOUR possible ill-health effects that can be caused from working with cement.
(ii) Outline ways in which the ill-health effects in (a) can be prevented.

35. (i) Identify the **THREE** types of asbestos commonly found in buildings.
(ii) Explain where asbestos is likely to be encountered in a building during renovation work.

Appendix 16.1 Inspection recording form with timing and frequency chart

Timing and frequency chart (reproduced from HSG150).

Place of work or work equipment	Timing and frequency of checks, inspections and examinations equipment								
	Inspect before work at the start of every shift (see note 1)	Inspect after any event likely to have affected its strength or stability	Inspect after accidental fall of rock or other material	Inspect after installation or assembly in any position (see notes 2 and 3)	Inspect at suitable intervals	Inspect after exceptional circumstances which are liable to jeopardize the safety at work equipment	Inspect at intervals not exceeding 7 days (see note 3)	Check on each occasion before use (REPORT NOT REQUIRED)	LOLER Thorough Examination (if work equipment subject to LOLER) (see note 4)
Excavations which are supported to prevent any person being buried or trapped by an accidental collapse or a fall or dislodgement of material	✓	✓	✓						
Cofferdams and caissons	✓	✓							
The surface and every parapet or permanent rail of every existing place of work at height								✓	
Guard rails, toe boards, barriers and similar collective means of fall protection				✓	✓	✓			
Scaffolds and other working platforms (including tower scaffolds and MEWPs) used for construction work and from which a person could fall more than 2 m				✓		✓	✓		✓
All other working platforms				✓	✓	✓			✓
Collective safeguards for arresting falls (e.g. nets, airbags, soft landing systems)				✓	✓	✓			
Personal fall protection systems (including work positioning, rope access, work restraint and fall arrest systems)				✓	✓	✓			✓
Ladders and stepladders					✓	✓		✓	

Notes:

1. Although an excavation must be inspected at the start of every shift, only one report is needed in any seven-day period. However, if something happens to affect its strength or stability, and/or an additional inspection is carried out, a report must then be completed. A record of this inspection must be made and retained for three months.

2. 'Installation' means putting into position and 'assembly' means putting together. You are not required to inspect and provide a report every time a ladder, tower scaffold or mobile elevated work platform (MEWP) is moved on site or a personal fall protection system is clipped to a new location.

3. An inspection and a report (see image titled Inspection Report) is required for a tower scaffold or MEWP (used for construction work and from which a person could fall 2 metres) after installation or assembly and every seven days thereafter, provided the equipment is being used on the same site. A record of this inspection must be made and retained for three months. If a tower scaffold is reassembled rather than simply moved, then an additional, pre-use inspection and report is required. It is acceptable for this inspection to be carried out by the person responsible for erecting the tower scaffold, provided they are trained and competent. A visible tag system, which supplements inspection records as it is updated following each pre-use inspection, is a way of recording and keeping the results until the next inspection.

4. All work equipment subject to LOLER Regulation 9, thorough examination and inspection requirements, will continue to be subject to LOLER Regulation 9 requirements.

INSPECTION REPORT

1. Name and address of person for whom inspection was carried out.

2. Site address.

3. Date and time of inspection.

4. Location and description of place of work or work equipment inspected.

5. Matters which give rise to any health and safety risks.

6. Can work be carried out safely?

Y/N

7. If not, name of person informed.

8. Details of any other action taken as a result of matters identified in 5 above.

9. Details of any further action considered necessary.

10. Name and position of person making the report.

11. Date and time report handed over.

12. Name and position of person receiving report.

Appendix 16.2 Checklist of typical scaffolding faults

Footings	Standards	Ledgers	Bracing	Putlogs and transoms	Couplings	Bridles	Ties	Boarding	Guard rails and toe boards	Ladders
Soft and uneven	Not plumb	Not level	Some missing	Wrongly spaced	Wrong fitting	Wrong spacing	Some missing	Bad boards	Wrong height	Damaged
No base plates	Jointed at same height	Joints in same bay	Loose	Loose	Loose	Wrong couplings	Loose	Trap boards	Loose	Insufficient length
No sole plates	Wrong spacing	Loose	Wrong fittings	Wrongly supported	Damaged	No check couplers	Not enough	Incomplete	Some missing	Not tied
Undermined	Damaged	Damaged			No check couplers			Insufficient supports		

Appendix 16.3 Summary of application and notification under CDM 2007

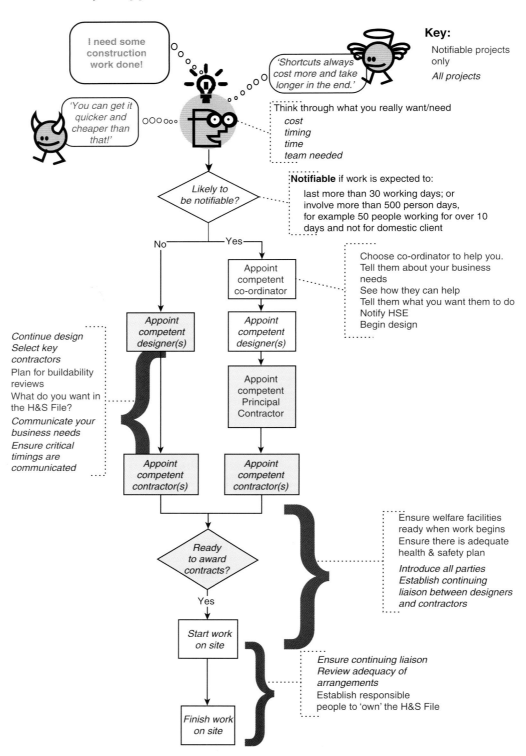

Appendix 16.4 Pre-construction information

This is taken from Appendix 2 of the ACOP *Managing Health and Safety in Construction*, L144.

When drawing up the pre-construction information, each of the following topics should be considered. Information should be included where the topics are relevant to the work proposed. The pre-construction information provides information for those bidding for or planning work, and for the development of the construction phase plan. The level of detail in the information should be proportionate to the risks involved in the project.

Pre-construction information

1. **Description of project**

   (a) Project description and programme details including:
   - key dates (including planned start and finish of the construction phase), and
   - the minimum time to be allowed between appointment of the principal contractor and instruction to commence work on site.

   (b) Details of client, designers, CDM co-ordinator and other consultants

   (c) Whether or not the structure will be used as a workplace – if so, the finished design will need to take account of the relevant requirements of the Workplace (Health, Safety and Welfare) Regulations

   (d) Extent and location of existing records and plans.

2. **Client's considerations and management requirements**

   (a) Arrangements for:
   - planning for and managing the construction work, including any health and safety goals for the project
   - communications and liaison between client and others
   - security of the site
   - welfare provisions.

   (b) Requirements relating to the health and safety of the client's employees or customers or those involved in the project such as:
   - site hoarding requirements
   - site transport arrangements or vehicle movement restrictions
   - client permit-to-work systems
   - fire precautions
   - emergency procedures and means of escape

   - 'no-go' areas or other authorizations requirements for those involved in the project
   - any areas the client has designated as confined spaces
   - smoking and parking restrictions.

3. **Environmental restrictions and existing on-site risks**

   (a) Safety hazards, including:
   - boundaries and access, including temporary access, e.g. narrow streets, lack of parking, turning or storage
   - any restrictions on deliveries or waste collection or storage
   - adjacent land uses, e.g. schools, railway lines or busy roads
   - existing storage of hazardous materials
   - location of existing services particularly those that are concealed – water, electricity, gas etc.
   - ground conditions, underground structures or water courses where these might affect the safe use of plant, e.g. cranes, or the safety of ground work
   - information about existing structures – stability, structural form, fragile or hazardous materials, anchorage points for fall arrest systems (particularly where demolition is involved)
   - previous structural modifications, including weakening or strengthening of the structure (particularly where demolition is involved)
   - fire damage, ground shrinkage, movement or poor maintenance which may have adversely affected the structure
   - any difficulties relating to plant and equipment in the premises, such as overhead gantries whose height restricts access
   - health and safety information contained in earlier design, construction or 'as-built' drawings, such as details of pre-stressed or post-tensioned structures.

   (b) Health hazards, including:
   - asbestos, including results of surveys (particularly where demolition is involved)
   - existing storage of hazardous material
   - contaminated land, including results of surveys
   - existing structures containing hazardous materials
   - health risks arising from client's activities.

4. **Significant design and construction hazards**

   (a) Significant design assumptions and suggested work methods, sequences or other control measures

   (b) Arrangements for co-ordination of ongoing design work and handling design changes

   (c) Information on significant risks identified during design

   (d) Materials requiring particular precautions.

5. **The health and safety file**

   Description of its format and any conditions relating to its content.

Appendix 16.5 Construction phase plan

This is taken from Appendix 3 of the ACOP *Managing Health and Safety in Construction*, L144.

When drawing up the construction phase plan, employers should consider each of the following topics: information should be included in the plan where the topic is relevant to the work proposed; the plan sets out how health and safety is to be managed during the construction phase. The level of detail should be proportionate to the risks involved in the project.

Construction phase plan

1. **Description of project**
   (a) Project description and programme details including key dates
   (b) Details of client, CDM co-ordinator, designers, principal contractor and other consultants
   (c) Extent and location of existing records and plans that are relevant to health and safety on site, including information about existing structures when appropriate.

2. **Management of the work**
   (a) Management structure and responsibilities
   (b) Health and safety goals for the project and arrangements for monitoring and review of health and safety performance
   (c) Arrangements for:
   - regular liaison between parties on site
   - consultation with the workforce
   - exchange of design information between the client, designers, CDM co-ordinator and contractors on site
   - handling design changes during the project
   - the selection and control of contractors
   - the exchange of health and safety information between contractors
   - site security
   - site induction
   - on-site training
   - welfare facilities and first aid
   - the reporting and investigation of accidents and incidents including near misses
   - the production and approval of risk assessments and written systems of work
   - site rules
   - fire and emergency arrangements.

3. **Arrangements for controlling significant site risks**
   (a) Safety risks, including:
   - delivery and removal of materials (including waste) and work equipment taking account of any risks to the public, e.g. during access to and egress from the site
   - dealing with services – water, electricity and gas, including overhead power lines and temporary electrical installations
   - accommodating adjacent land use
   - stability of structures whilst carrying out construction work, including temporary structures and existing unstable structures
   - preventing falls
   - work with or near fragile materials
   - control of lifting operations
   - the maintenance of plant and equipment
   - work on excavations and work where there are poor ground conditions
   - work on well, underground earthworks and tunnels
   - work on or near water where there is a risk of drowning
   - work involving diving
   - work in a caisson or compressed air working
   - work involving explosives
   - traffic routes and segregation of vehicles and pedestrians
   - storage of materials (particularly hazardous materials) and work equipment
   - any other significant safety risks.

(b) health risks, including:

- the removal of asbestos
- dealing with contaminated land
- manual handling
- use of hazardous substances, particularly where there is a need for health monitoring
- reducing noise and vibration
- working with ionizing radiation; any other significant health risks.

4. The health and safety file

(a) layout and format

(b) arrangements for the collation and gathering of information

(c) storage of information.

Summary of the main legal requirements

17

17.1 Introduction

The achievement of an understanding of the basic legal requirements can be a daunting task. However, the health and safety student should not despair because at NEBOSH National Certificate level they do not need to know the full history of UK health and safety legislation, the complete range of regulations made under the HSW Act 1974 or the details of the appropriate European Directives. These summaries, which are correct up to 1 May 2009, cover those acts and regulations which are required for the Certificate student and will provide the essential foundation of knowledge. They will also cover the requirements for similar awards at Certificate level, like the British Safety Council's Certificate in Occupational Safety and Health, and other courses such as IOSH's Managing Safely. Students should check the latest edition of the NEBOSH Certificate guide to ensure that the regulations summarized here are the latest for the course being undertaken.

Many managers will find these summaries a useful quick reference to health and safety legal requirements. However, anyone involved in achieving compliance with legal requirements should ensure they read the regulations themselves, any approved codes of practice and Health and Safety Executive (HSE) guidance.

17.2 The legal framework

17.2.1 General

The HSW Act is the foundation of British health and safety law. It describes the general duties that employers have towards their employees and to members of the public, and also the duties that employees have to themselves and to each other.

The term 'so far as is reasonably practicable' (sfarp) qualifies some of the duties in some sections, notably 2, 3, 4 and 6 of the HSW Act. In other words, the degree of risk in a particular job or workplace needs to be balanced against the time, trouble, cost and physical difficulty of taking measures to avoid or reduce the risk.

The law simply expects employers to behave in a way that demonstrates good management and common sense. They are required to look at what the hazards are and take sensible measures to control/reduce the risks from them.

The Management of Health and Safety at Work Regulations 1999 (MHSWR) (the Management Regulations) give employers specific requirements to manage health and safety under the HSW Act. Like the Act, they apply to every work activity.

The MHSW Regulations requires every employer to carry out *risk assessments*. If there are five or more employees in the workplace, the significant findings of the risk assessments need to be recorded.

In a place like an office, risk assessment should be straightforward; but where there are serious hazards, such as those in a chemical plant, laboratory or an oil rig, it is likely to be more complicated.

The Factories Act 1961 and The Office Shops and Railway Premises Act of 1963, while still partly remaining on the Statute Book for most practical health and safety work, can be ignored. In April 2009, the Repeals and Modification Regulations came into force which removed one of the last requirements of these Acts for premises notification [Forms 9 (factories), OSR1 (offices and Shops) and OSR7 (railway trackside buildings)] and the need to keep a general register in factories. F10 for construction site notification remains.

17.2.2 The relationship between the regulator and industry

In April 2008 the HSC and the HSE merged into a single unified body called the Health and Safety Executive (HSE). The HSE consults widely with those affected by its proposals.

The HSE has many contact points with industry and work through:

➤ the Industry and Subject Advisory Committees, which have members drawn from the areas of work they cover, and focus on health and safety issues in particular industries (such as the textile industry, construction and education or areas such as toxic substances and genetic modification);
➤ intermediaries, such as small firms' organizations providing information and advice to employers and others with responsibilities under the HSW Act;
➤ guidance to enforcers, both the HSE inspectors and those of local authorities (LAs);
➤ the day-to-day contact which inspectors have with people at work.

The HSE consults with small firms through Small Firms Forums. It also seeks views in detail from representatives of small firms about the impact on them of proposed legislation.

17.2.3 Health and safety in Scotland

Although occupational health and safety is a reserved issue [i.e. not devolved to the Scottish Executive (SE)], there are some significant differences with some issues. The following extracts from the concordat between the HSE and the SE outline some of these issues.

Extracts from *The Concordat between the Health and Safety Executive and the Scottish Executive*

Introduction

1. This concordat is an agreement between the Scottish Executive and the Health and Safety Executive. It is intended to provide the framework to guide the future working relationship between the Health and Safety Executive (HSE) and the Scottish Executive (SE). The objective of the concordat is to ensure that the roles and responsibilities of HSE and the SE in the new constitutional structure are effectively translated into practical working arrangements between the two organizations. The aim is to promote the establishment of close and harmonious working relationships and good communications at all levels between HSE and the SE, and in particular to foster constructive co-operation.

Most day to day liaison between the SE and HSE will, in the first instance be handled by HSE's Director for Scotland, based in Edinburgh.

(*Note: where the concordat referred to HSC/E this has been changed to HSE meaning the new combined Authority*)

Nature of concordat

3. This concordat is a voluntary arrangement between HSE and the SE. It is not a binding agreement or contract and so does not create any legally enforceable rights, obligations, or restrictions. It is intended to be binding in honour only. The HSW Act provides for statutory consultation in connection with proposals for regulatory change; this concordat does not create any other right to be consulted or prevent consultation beyond that required by statute.

Any failure to follow the terms of the concordat is not to be taken as invalidating decisions taken by HSE or the Scottish Executive.

Consultation

11. There are a significant number of areas where HSE and Scottish Ministers share or have closely related interests, as set out in the Annex. HSE and the SE recognize that the extent of these areas makes good communications essential in order to assist the process of policy formation and decision-making in each administration and to meet any consultation or other requirements connected with the exercise of a function. They also recognize that there will be mutual benefit from the exchange of information on scientific, technical and, where appropriate, policy matters.

12. Information obtained by use of HSE's statutory powers cannot be made available to SE where the release of that information is not for the purposes of HSE's functions. Subject to that, the HSE and the SE will:

➤ share relevant information, analysis and research;
➤ inform each other of any relevant information which comes to their attention which may require action or have resource consequences for the other party;
➤ seek to involve each other, as and when appropriate, in policy development on all topics where there is a reasonable expectation that a policy initiative might affect the other's responsibilities, or be used or adapted by the other;
➤ inform each other at the earliest practicable stage of any emerging proposal to change primary or secondary legislation which might have an impact upon the other's responsibilities;
➤ inform each other at the earliest practicable stage of substantive new policy announcements which may be relevant to the other's responsibilities; and
➤ co-ordinate activities where appropriate

Local authorities

16. The constitution of Scottish local authorities, which includes their establishment, dissolution, assets, liabilities, funding and receipts, is a devolved matter in which the Scottish Ministers have a policy interest as are most functions conferred upon such authorities. The HSE has an interest in the functions of local authorities under health and safety legislation, and under section 18(4) of HSWA, the Commission may issue guidance to local authorities, as enforcing authorities, which they must

follow. The HSE will consult the SE in advance where proposed changes in health and safety legislation, or directions under section 18(4) might result in new burdens being imposed upon Scottish local authorities.

*ANNEX – areas of common and closely related interest – **extracts only***

This annex lists subjects in which the Scottish Ministers and the HSE share an interest and other areas where there may be a particular need for consultation between HSE and the Scottish Executive.

Building control

This is a devolved matter, but the HSE has an interest in the related matter of construction safety.

Dangerous substances

The HSE have an interest under the HSW Act 1974 and the European Communities Act 1972 in the control of the keeping, notification, supply and use of explosive or highly flammable or otherwise dangerous substances which have the potential to create a major accident and in the prevention of the unlawful acquisition, possession and use of such substances. HSE also has an interest in the carriage by road and rail of dangerous and environmentally hazardous goods. The Scottish Ministers, however, have an interest in such matters for the purposes of protection of the environment and the planning system (by virtue of the Planning (Hazardous Substances) (Scotland) Act 1997). HSE will continue to advise what hazardous substances and in what quantities have significant risk off site. HSE involvement in individual hazardous substances consent/planning applications which are before Scottish Ministers for a decision will be on the basis of established planning and related public inquiry procedures.

Educational facilities; adventure activity centres

The safety of these premises is generally a devolved matter. However, the HSE has interests in the safety of adventure activity centres because of its responsibilities by virtue of the Activity Centres (Young Persons Safety) Act 1995 and subordinate legislation made under that Act. It also has an interest in the health and safety of workers and those affected by the activities of workers in these premises.

Fire safety

General fire safety is a devolved matter and the policy responsibility of the Scottish Ministers. However, HSE has an interest in the reserved matters of process fire precautions, fire precautions in relation to petroleum and petroleum spirit, fire safety on construction sites, ships under construction or repair by persons other than the master or crew, in mines, on offshore installations, and on any other premises which on 1 July 1999 are of a description specified in Part I of Schedule 1 to the Fire Certificates (Special Premises) Regulations 1976.

Food safety

This is generally a devolved matter and the policy responsibility of the Scottish Ministers. However, the HSE has an interest in the safety of workers in the food processing, manufacturing, cooking and food distribution industries and in related risks to the public.

Local authorities

The Scottish Ministers have an interest in the devolved matter of the constitution of Scottish local authorities, which includes their establishment, dissolution, assets, liabilities, funding and receipts. The HSE has an interest in the functions and performance of local authorities in relation to the enforcement of health and safety legislation.

Pesticides (including biocides and plant protection products)

The Scottish Ministers' interest is in the protection of the environment, public health, animal and plant health and food safety. HSE's interest is in the protection of workers and those affected by the activities of workers, protection of the environment, and product approvals.

Places of entertainment (cinemas, theatres, casinos, dance-halls etc), sports facilities, sports events and zoos

The Scottish Ministers' interest is in public safety, fire safety, structural building standards, sanitation, food standards, pest control, security, etc. The HSE has an interest in the safety of workers and the safety of members of the public and others affected by work activity.

Ports, harbours and inland waterways

Scottish Ministers' interest is in ports policy and communications via the various ferry links between mainland Scotland and the islands. Responsibility for safety enforcement for the ships and their crews rests with the Maritime and Coastguard Agency. The HSE has an interest in the safety of workers and the safety of members of the public and others affected by work activity at inland waterways, the safety aspects of new harbour byelaws and the health and safety of shore side workers loading or unloading from berthed ships or the dock side.

Protection of the environment

This is a devolved matter and the policy responsibility of the Scottish Ministers. Responsibility for enforcement of much environmental legislation in Scotland rests with the Scottish Environment Protection Agency. However, the HSE has enforcement responsibilities for certain environmental legislation under agency arrangements under section 13 of HSWA.

Public health, occupational health and health promotion

Public health is generally a devolved matter and the policy responsibility of the Scottish Ministers: Occupational Health is a reserved matter for HSE. General health promotion in Scotland is a devolved matter and the policy responsibility of the Scottish Ministers, but HSE has an interest in 'lifestyle' issues such as misuse of drugs, alcohol and smoking (see also below), when these impinge on the workplace.

Rail safety

This is a reserved matter. However, the Scottish Ministers have an interest because of their executively devolved functions in relation to a number of aspects of the regulation of the rail industry in Scotland. HSE has an interest in any matter which could have an impact on its responsibility for the approval of new and altered works on railways, light rail and tramways.

Smoking

The regulation of smoking and passive smoking in workplaces is a reserved matter. However, the Scottish Ministers have an interest. Other aspects of smoking such as the regulation of tobacco advertising are generally devolved and are the policy responsibility of the Scottish Ministers.

Water and sewerage

This is generally a devolved matter and the policy responsibility of the Scottish Ministers. Water and sewerage services in Scotland are provided by three publicly owned water authorities. The HSE's interest is in matters relating to the safety of workers and those affected by the activities of workers.

17.3 List of Acts, orders and Regulations summarized

The following Acts and Regulations are covered in this summary:

➤ The HSW Act 1974;
➤ The Environmental Protection Act (EPA) 1990.

By far the most important of these is the HSW Act. Most of the relevant regulations covering health and safety at work have been made under this Act since 1974. Those relevant to the Certificate student are listed here and summarized in this chapter. The first list is alphabetical and some titles have been modified to allow an easier search. The second list is chronological with Statutory Instrument (SI) numbers.

The Regulatory Reform Fire Safety Order was made under the Regulatory Reform Act 2001, and the Hazardous Waste Regulations under the EPA.

The regulations/orders covered are as follows.

17.3.1 Alphabetical list of regulations/order summarized

➤ Chemicals (Hazardous Information and Packaging for Supply) Regulations 2002 Plus 2005 amendment. A further amendment is expected in 2009;
➤ Confined Spaces Regulations 1997;
➤ Construction (Design and Management) Regulations 2007;
➤ Construction (Head Protection) Regulations 1989;
➤ Health and Safety (Consultation with Employees) Regulations 1996;
➤ Control of Substances Hazardous to Health Regulations 2002 and 2004 Amendment;
➤ Dangerous Substances and Explosive Atmospheres Regulations 2002;
➤ Health and Safety (Display Screen Equipment) Regulations 1992;
➤ Electricity at Work Regulations 1989;
➤ Employers Liability (Compulsory Insurance) Act 1969 and Regulations 1998;
➤ The Regulatory Reform (Fire Safety) Order 2005;
➤ Fire (Scotland Act) 2005;
➤ Health and Safety (First Aid) Regulations 1981;
➤ Hazardous Waste (England and Wales) Regulations 2005;
➤ Health and Safety (Information for Employees) Regulations 1989;

➤ Ionising Radiations Regulations 1999;
➤ Lifting Operations and Lifting Equipment Regulations 1998;
➤ Management of Health and Safety at Work Regulations 1999;
➤ Manual Handling Operations Regulations 1992;
➤ Noise at Work Regulations 2005;
➤ Personal Protective Equipment at Work Regulations 1992;
➤ Provision and Use of Work Equipment Regulations 1998 (except Part IV – Power Presses);
➤ Reporting of Injuries, Diseases and Dangerous Occurrences Regulations 1995;
➤ Safety Representatives and Safety Committees Regulations 1977;
➤ Health and Safety (Safety Signs and Signals) Regulations 1996;
➤ Supply of Machinery (Safety) Regulations 1992 & New 2008 Regulations;
➤ Vibration at Work Regulations 2005;
➤ Workplace (Health, Safety and Welfare) Regulations 1992;
➤ Work at Height Regulations 2005.

The following acts and regulations have also been included even though they are not in the NEBOSH General Certificate syllabus. Very brief summaries only are given.

Corporate Manslaughter and Corporate Homicide Act 2007;
Asbestos at Work Regulations 2006;
Disability Discrimination Act 1995 and 2005;
Electrical Equipment (Safety) Regulations 1994;
Gas Appliances (Safety) Regulations 1992;
Gas Safety (Installation and Use) Regulations 1998;
Lead at Work Regulations 2002;
Occupiers Liability Acts 1957 and 1984;
Health and Safety (Offences) Act 2008;
Pesticides Regulations 1986;
Personal Protective Equipment Regulations 2002;
Pressure Systems Safety Regulations 2000;
Smoke-free Regulations 2006;
Working Time Regulations 1998, 2001, 2003 and 2007.

17.3.2 Chronological list of regulations/orders summarized

The list below gives the correct titles, year produced and SI number.

Year	SI number	Title
1977	0500	Safety Representatives and Safety Committees Regulations
1981	0917	Health and Safety (First Aid) Regulations
1989	0635	Electricity at Work Regulations
1989	0682	Health and Safety (Information for Employees) Regulations
1989	2209	Construction (Head Protection) Regulations
1992	2792	Health and Safety (Display Screen Equipment) Regulations
1992	2793	Manual Handling Operations Regulations
1992	2966	Personal Protective Equipment at Work Regulations
1992	3004	Workplace (Health, Safety and Welfare) Regulations
1992	3073	Supply of Machinery (Safety) Regulations
1995	3246	Reporting of Injuries, Diseases and Dangerous Occurrences Regulations
1996	0341	Health and Safety (Safety Signs and Signals) Regulations
1996	1513	Health and Safety (Consultation with Employees) Regulations
1997	1713	Confined Spaces Regulations
1998	2306	Provision and Use of Work Equipment Regulations (except Part IV – Power Presses)
1998	2307	Lifting Operations and Lifting Equipment Regulations
1998	2573	Employers Liability (Compulsory Insurance) Regulations
1999	437	Control of Substances Hazardous to Health Regulations
1999	3232	Ionising Radiations Regulations
1999	3242	Management of Health and Safety at Work Regulations
2002	1689	Chemicals (Hazardous Information and Packaging for Supply) Regulations

(Continued)

Year	SI number	Title
2002	2677	Control of Substances Hazardous to Health Regulations
2002	2776	Dangerous Substances and Explosive Atmospheres Regulations
2004	3386	Control of Substances Hazardous to Health (Amendment) Regulations
2005	735	Work at Height Regulations
2005	1541	The Regulatory Reform (Fire Safety) Order
2005	1093	Control of Vibration at Work Regulations
2007	0320	Construction (Design and Management) Regulations
2008	1597	The Supply of Machinery (Safety) Regulations

Very brief summaries only for:

Year	SI number	Title
1957	Ch 31	Occupiers Liability Act
1984	Ch 3	Occupiers Liability Act
1986	1510	Control of Pesticides Regulations
1995	1629	Gas Appliances (Safety) Regulations
1998	2451	Gas Safety (Installation and Use) Regulations
1994	3260	Electrical Equipment (Safety) Regulations
1998	1833	Working Time Regulations
2000	128	Pressure Systems Safety Regulations
2002	1144	Personal Protective Equipment Regulations
2002	2676	The Control of Lead at Work Regulations
2003	1684	Working Time (Amendment) Regulations
2005	Ch 50	Disability Discrimination Act 1995 and 2005

(Continued)

2006	2739	Control of Asbestos Regulations
2006	3368	Smoke-free (Premises and Enforcement) Regulations
2006	090	Scottish SI – The Prohibition of Smoking in Certain Premises (Scotland) Regulations
2007	W.68	The Smoke-free Premises etc. (Wales) Regulations 2007
2007	Ch 19	Corporate Manslaughter and Corporate Homicide Act
2008	Ch 20	Health and Safety (Offences) Act

17.4 HSW Act 1974

The HSW Act was introduced to provide a comprehensive and integrated piece of legislation dealing with the health and safety of people at work and the protection of the public from work activities.

The Act imposes a duty of care on everyone at work related to their roles. This includes employers, employees, owners, occupiers, designers, suppliers and manufacturers of articles and substances for use at work. It also includes self-employed people. The detailed requirements are spelt out in Regulations.

The Act basically consists of four parts.

Part 1 covers:

➤ the health and safety of people at work;
➤ protection of other people affected by work activities;
➤ the control of risks to health and safety from articles and substances used at work.

Part 2 sets up the Employment Medical Advisory Service.
Part 3 makes amendments to the safety aspects of building regulations.
Part 4 consists of general and miscellaneous provisions.

17.4.1 Duties of employers – section 2

The employers' main general duties are to ensure, sfarp, the health, safety and welfare at work of all their employees, in particular:

➤ the provision of safe plant and systems of work;
➤ the safe use, handling, storage and transport of articles and substances;

➤ the provision of any required information, instruction, training and supervision;
➤ a safe place of work including safe access and egress; and
➤ a safe working environment with adequate welfare facilities.

When five or more people [The Employer's Health and Safety Policy Statements (Exception) Regulations 1975 (SI No. 1584) exempted an employer who employs less than 5 people] are employed the employer must:

➤ prepare a written general health and safety policy;
➤ set down the organization and arrangements for putting that policy into effect;
➤ revise and update the policy as necessary;
➤ bring the policy and arrangements to the notice of all employees.

Employers must also:

➤ consult safety representatives appointed by recognized trade unions[1];
➤ consult safety representatives elected by employees[1];
➤ establish a safety committee if requested to do so by recognized safety representatives[1].

17.4.2 Duties of owners/occupiers – sections 3 and 4

Every employer and self-employed person is under a duty to conduct their undertaking in such a way as to ensure,

[1]That these have been enacted by the Safety Representatives and Safety Committees Regulations 1977 and enhanced by the Health and Safety (Consultation with Employees) Regulations 1996 (see later summaries).

sfarp, that persons not in their employment (and themselves for self-employed), who may be affected, are not exposed to risks to their health and safety.

Those in control of non-domestic premises have a duty to ensure, sfarp, that the premises, the means of access and exit, and any plant or substances are safe and without risks to health. The common parts of residential premises are non-domestic.

Note: Section 5 was repealed by the EPA 1990.

17.4.3 Duties of manufacturers/suppliers – section 6

Persons who design, manufacture, import or supply any article or substance for use at work must ensure, sfarp, that:

- it is safe and without risks to health when properly used (i.e. according to manufacturers' instructions);
- they carry out such tests or examinations as are necessary for the performance of their duties;
- they provide adequate information (including revisions) to perform their duties;
- they carry out any necessary research to discover, eliminate or minimize any risks to health or safety;
- the installer or erector has done nothing regarding the way in which the article has been installed or erected to make it unsafe or a risk to health.

17.4.4 Duties of employees – section 7

Two main duties are placed on employees:

- to take reasonable care for the health and safety of themselves and others who may be affected by their acts or omissions at work;
- to co-operate with their employer and others to enable them to fulfil their legal obligations.

17.4.5 Other duties – sections 8 and 9

No person may misuse or interfere with anything provided in the interests of health, safety or welfare in pursuance of any of the relevant statutory provisions.

Employees cannot be charged for anything done, or provided, to comply with the relevant statutory provisions. For example, employees cannot be charged for personal protective equipment (PPE) required by a health and safety regulation.

17.4.6 Powers of inspectors – sections 20–25

Inspectors appointed under this Act have the authorization to enter premises at any reasonable time (or anytime in a dangerous situation), and to:

- take a constable with them if necessary;
- take with them another authorized person and necessary equipment;
- examine and investigate;
- require premises or anything in them to remain undisturbed for purposes of examination or investigation;
- take measurements, photographs and recordings;
- cause an article or substance to be dismantled or subjected to any test;
- take possession of or retain anything for examination or legal proceedings;
- take samples as long as a comparable sample is left behind;
- require any person who can give information to answer questions and sign a statement. Evidence given under this Act cannot be used against that person or their spouse;
- require information, facilities, records or assistance;
- do anything else necessary to enable them to carry out their duties;
- issue an **Improvement Notice(s)**, which is a notice identifying a contravention of the law and specifying a date by which the situation is remedied. There is an appeal procedure which must be triggered within 21 days. The notice is suspended pending the outcome of the appeal;
- issue a **Prohibition Notice(s)**, which is a notice identifying and halting a situation which involves or will involve a risk of serious personal injury to which the relevant statutory provisions apply. A contravention need not have been committed. The notice can have immediate effect or be deferred, for example to allow a process to be shut down safely. Again there is provision to appeal against the notice but the order still stands until altered or rescinded by an employment tribunal;
- initiate prosecutions;
- seize, destroy or render harmless any article or substance which is a source of imminent danger.

17.4.7 Offences – section 33

New legislation, the Health and Safety Offences Act 2008, increases penalties and provides courts with greater sentencing powers for those who flout health and safety legislation.

The Act raises the maximum penalties that can be imposed for breaching health and safety regulations in the lower courts from £5000 to £20 000 and the range of offences for which an individual can be imprisoned has also been broadened.

The Act amends section 33 of the HSW Act 1974, and raises the maximum penalties available to the courts in respect of certain health and safety offences. It received Royal Assent on 16 October 2008 and came into force in January 2009.

Summary of prior and new penalties under the HSW Act

Former maxima (to December 2008):

➤ £5000 (mainly for breaches of Regulations) or £20 000 for summary offence in lower courts, depending on offence; **unlimited fine** for indictable offence in higher courts

➤ Imprisonment not available for most offences (but up to 6 months in Magistrates Court/2 years in Crown Court for few offences, e.g. failing to comply with a prohibition notice or breaching a licencing requirement).

New maxima (from January 2009):

➤ £20 000 fines in lower courts for nearly all summary offences, unlimited fines in higher courts

➤ Imprisonment for nearly all offences – up to 12 months[2] in Magistrates Courts and 2 years in the Crown Court.

The new Act covers Great Britain and Northern Ireland.

There are strict guidelines which are observed by the regulators in their approach to the prosecution of health and safety offences. The HSE Enforcement Policy Statement makes it clear that prosecutions should be in the public interest and where one or more of a list of circumstances apply. These include where:

➤ death was a result of a breach of the legislation;

➤ there has been reckless disregard of health and safety requirements;

[2]*Note:* (1) To be read as a reference to not exceeding 6 months until the coming into force of section 154(1) of the Criminal Justice Act 2003. At present there are no plans to switch on this clause.

➤ there have been repeated breaches which give rise to significant risk, or persistent and significant poor compliance;

➤ false information has been supplied wilfully, or there has been intent to deceive in relation to a matter which gives rise to significant risk.

Prosecutions of individuals by health and safety regulators are not undertaken lightly. Any prosecutions of individuals are subject to the same strict considerations set out above and are only taken if warranted, and not in lieu of a case against their employer (Table 17.1).

If a regulation has been contravened, failure to comply with an Approved Code of Practice is admissible in evidence as failure to comply. Where an offence is committed by a corporate body with the knowledge, connivance or neglect of a responsible person, both that person and the body are guilty of the offence. In proceedings the onus of proving the limits of what is reasonably practicable rests with the accused.

On conviction of directors for indictable offences in connection with the management of a company, the courts may also make a disqualification order.

Lower court maximum	5 years disqualification
Higher court maximum	15 years disqualification

17.4.8 The new merged HSE – section 10

The HSE made the following statement about the new merged HSE. The modernization of health and safety law in Great Britain has its origins in the report 'Safety and Health at Work' (1972). The report was used as the basis for the HSW Act 1974. It proposed the introduction of 'a single authoritative body to facilitate and promote health and safety within the workplace with autonomy, its own budget, executive powers and functions'.

The majority of the proposals set out in Lord Robens' report were adopted in full and formed the basis of the HSW Act. However, contrary to Robens' recommendation, the HSW Act did not provide for a single authority, but two separate Crown Non-Department Public Bodies (NDPBs): the HSC and the HSE.

This recommendation has finally been implemented and the HSC and the HSE merged on 1 April 2008 to form a single national regulatory body responsible for promoting the cause of better health and safety at work. The merged body is called the HSE and will provide greater clarity and transparency whilst maintaining its public accountability. Section 10 of the Act has been revised.

Table 17.1 New Schedule 3A to the Health and Safety at Work etc. Act 1974

SCHEDULE 3A OFFENCES: MODE OF TRIAL AND MAXIMUM PENALTY

Offence	Mode of trial	Penalty on summary conviction	Penalty on conviction on indictment
An offence under section 33(1)(a) consisting of a failure to discharge a duty to which a person is subject by virtue of sections 2–6.	Summarily or on indictment.	Imprisonment for a term not exceeding 12 months, or a fine not exceeding £20 000, or both.	Imprisonment for a term not exceeding 2 years, or a fine, or both.
An offence under section 33(1)(a) consisting of a failure to discharge a duty to which a person is subject by virtue of section 7.	Summarily or on indictment.	Imprisonment for a term not exceeding 12 months, or a fine not exceeding the statutory maximum, or both.	Imprisonment for a term not exceeding 2 years, or a fine, or both.
An offence under section 33(1)(b) consisting of a contravention of section 8.	Summarily or on indictment.	Imprisonment for a term not exceeding 12 months, or a fine not exceeding £20 000, or both.	Imprisonment for a term not exceeding 2 years, or a fine, or both.
An offence under section 33(1)(b) consisting of a contravention of section 9.	Summarily or on indictment.	A fine not exceeding £20 000.	A fine.
An offence under section 33(1)(c).	Summarily or on indictment.	Imprisonment for a term not exceeding 12 months, or a fine not exceeding £20 000, or both.	Imprisonment for a term not exceeding 2 years, or a fine, or both.
An offence under section 33(1)(d).	Summarily only.	A fine not exceeding level 5 on the standard scale.	
An offence under section 33(1)(e), (f) or (g).	Summarily or on indictment.	Imprisonment for a term not exceeding 12 months, or a fine not exceeding £20 000, or both.	Imprisonment for a term not exceeding 2 years, or a fine, or both.
An offence under section 33(1)(h).	Summarily only.	Imprisonment for a term not exceeding 51 weeks (in England and Wales) or 12 months (in Scotland), or a fine not exceeding level 5 on the standard scale, or both.	

(Continued)

Table 17.1 Continued

SCHEDULE 3A OFFENCES: MODE OF TRIAL AND MAXIMUM PENALTY

Offence	Mode of trial	Penalty on summary conviction	Penalty on conviction on indictment
An offence under section 33(1)(i).	Summarily or on indictment.	A fine not exceeding the statutory maximum.	A fine.
An offence under section 33(1)(j).	Summarily or on indictment.	Imprisonment for a term not exceeding 12 months, or a fine not exceeding the statutory maximum, or both.	Imprisonment for a term not exceeding 2 years, or a fine, or both.
An offence under section 33(1)(k), (l) or (m).	Summarily or on indictment.	Imprisonment for a term not exceeding 12 months, or a fine not exceeding £20 000, or both.	Imprisonment for a term not exceeding 2 years, or a fine, or both.
An offence under section 33(1)(n).	Summarily only.	A fine not exceeding level 5 on the standard scale.	
An offence under section 33(1)(o).	Summarily or on indictment.	Imprisonment for a term not exceeding 12 months, or a fine not exceeding £20 000, or both.	Imprisonment for a term not exceeding 2 years, or a fine, or both.
An offence under the existing statutory provisions for which no other penalty is specified.	Summarily or on indictment.	Imprisonment for a term not exceeding 12 months, or a fine not exceeding £20 000, or both.	Imprisonment for a term not exceeding 2 years, or a fine, or both.

(1) This paragraph makes transitional modifications of the table as it applies to England and Wales.

(2) In relation to an offence committed before the commencement of section 154(1) of the Criminal Justice Act 2003 (general limit on magistrates' courts' powers to imprison), a reference to imprisonment for a term not exceeding 12 months is to be read as a reference to imprisonment for a term not exceeding 6 months.

(3) In relation to an offence committed before the commencement of section 281(5) of that Act (alteration of penalties for summary offences), a reference to imprisonment for a term not exceeding 51 weeks is to be read as a reference to imprisonment for a term not exceeding 6 months.

The decision to merge the HSC and the HSE was reached after extensive consultation with stakeholders and through the process determined by the Legislative and Regulatory Reform Act 2006.

The merger does not fundamentally change the day-to-day operations but will set the tone for closer working throughout the organization.

HSE will build on good relationships with stakeholders and in particular relationships with LAs to develop a revised strategy for health and safety in Great Britain.

HSE retains its independence, reflecting the interests of employers, employees and LAs and is committed to maintaining its service delivery. The Board of the new Executive assumed responsibility for running all aspects of the organization, including setting the overall strategic direction, financial and performance management and prioritization of resources.

The merger means:

➤ there is a single national regulatory body responsible for promoting the cause of better health and safety at work;

➤ the existing Chair of the HSC becomes Chair of the Board of the new Executive;

➤ existing Commissioners are appointed as non-executive directors of the new Executive for the remainder of their term of office with the relevant responsibilities of the new roles;

➤ the potential size of the Board of the new Executive will be no more than 11 members plus the Chair, and members will continue to be appointed by the Secretary of State;

➤ all the fundamental contents of the HSW Act remain;

➤ none of the statutory functions of the previous Commission and Executive has been removed;

➤ there is no change in health and safety requirements, how they are enforced or how stakeholders relate to the health and safety regulator – no health and safety protections have been removed.

17.4.9 Sources of References

An Introduction to Health and Safety: Guidance on Health and Safety for Small Firms INDG259 (rev 1), reprinted 2006, HSE Books ISBN 978 0 7176 2685 4.

Health and Safety at Work etc. Act 1974, chapter 37, London, The Stationery Office www.opsi.gov.uk/stst.htm

The Health and Safety System in Great Britain HSC, 2002, HSE Books, ISBN 0 7176 2243 6 (no longer published).

17.5 Environmental Protection Act 1990

17.5.1 Introduction

The EPA 1990 is still the centrepiece of current UK legislation on environmental protection. It is divided into nine parts, corresponding to the wide range of subjects dealt with by the Act.

Integrated Pollution Control (IPC) was a system established by Part 1 of the Act. Part 1 introduced *Part A Processes*, which are the most potentially polluting or technologically complex processes. In England and Wales these are enforced by the Environment Agency. In Scotland there is a parallel system enforced by the Scottish Environment Protection Agency.

Less polluting industries were classified as Part B, with only emissions released to air being subject to regulatory control. For such processes LAs are the enforcing body and the system is known as Local Air Pollution Control (LAPC).

Both IPC and LAPC have now been replaced by an Integrated Pollution Prevention and Control (IPPC) regime that implements the requirements of the European Community (EC) Directive 96/61 on IPPC. This was introduced under the Pollution Prevention and Control Act of 1999 (1999, Chapter 24), which repealed Part 1 of the EPA.

The change is outlined in Box 17.1.

17.5.2 Integrated pollution prevention and control

The system of IPPC applies an integrated environmental approach to the regulation of certain industrial activities. This means that emissions to air, water (including discharges to sewer) and land, plus a range of other environmental effects, must be considered together. It also means that regulators must set permit conditions so as to achieve a high level of protection for the environment as a whole. These conditions are based on the use of the 'Best Available Techniques' (BAT), which balances the costs to the operator against the benefits to the environment (see Box 17.2). IPPC aims to prevent emissions and waste production and where that is not practicable, reduce them to acceptable levels. IPPC also takes the integrated approach beyond the initial task of permitting, through to the restoration of sites when industrial activities cease.

17.5.3 Setting the legal framework

The PPC Regulations implement the EC Directive 96/61/EC on IPPC ('the IPPC Directive'), insofar as it relates to installations in England and Wales. Separate regulations apply the IPPC Directive in Scotland and Northern Ireland and to the offshore oil and gas industries.

➤ Prior to the PPC Regulations coming into force, many industrial sectors covered by the IPPC Directive were regulated under Part I of the EPA 1990. This introduced the systems of IPC, which controlled releases to all environmental media, and LAPC, which controlled releases to air only. Other industrial sectors new to integrated permitting, such as the landfill, intensive farming and food and drink sectors, were regulated, where appropriate, by separate waste management licences issued under Part II of the EPA and/or water discharge consents under the Water Resources Act 1991 or Water Industry Act 1991.

Box 17.1   Pollution prevention and control regimes

PPC REGIME
Pollution prevention and control

IPC (regime A)	IPPC (regime A1 and A2)
Integrated pollution control	Integrated pollution prevention and control
LAPC (regime B)	**LAPPC (regime B)**
Local air pollution control	Local air pollution prevention and control

Under EPA 1990 Part 1

Regime A – This is an integrated permitting regime. Emissions to the air, land and water of the potentially more polluting processes are regulated. The Environment Agency is the regulator.

Regime B – This regime permits processes with a lesser potential for polluting emissions. Only emissions to the air are regulated. The Local Authority is the regulator.

Under Pollution Prevention and Control Act 1999

Regime A1 – This is an integrated permitting regime. Emissions to the air, land and water of potentially more polluting processes are regulated. The Environment Agency is the regulator.

Regime A2 – This is an integrated permitting regime. Emissions to the air, land and water of processes with a lesser potential to pollute are regulated. The Local Authority Agency is the regulator.

Regime B – This is the permitting of processes with a lesser potential to pollute. Only emissions to the air are regulated. The Local Authority is the regulator.

Box 17.2   Best available techniques (BAT)

This term is defined as:

The most effective and advanced stage in the development of activities and their methods of operation which indicates the practicable suitability of particular techniques for providing the basis for emission limit values designed to prevent, and where that is not practicable, generally to reduce the emissions and the impact on the environment as a whole.

This definition implies that BAT not only covers the technology used but also the way in which the installation is operated, to ensure a high level of environmental protection as a whole. BAT takes into account the balance between the costs and environmental benefits (i.e. the greater the environmental damage that can be prevented, the greater the cost for the techniques).

➤ The PPC Regulations create a coherent new framework to prevent and control pollution, with two parallel systems similar to the old regimes of IPC and LAPC. The first of these – the 'Part A' regime of IPPC – applies a similar integrated approach to IPC while delivering the additional requirements of the IPPC Directive. 'Part A' extends the issues that regulators must consider alongside emissions into areas such as energy use and site restoration. The main provisions of IPPC apply equally to the ex-IPC processes and the other sectors new to integrated permitting. There are also some further requirements that apply solely to waste management activities falling under IPPC.

➤ The IPPC Directive applies to those landfills receiving more than 10 tonnes per day or with a total capacity exceeding 25000 tonnes (but excluding landfills taking only inert waste), the landfill Directive applies to all landfills. The PPC Regulations have been amended to include all landfills. For landfills the technical requirements are met through the Landfill Regulations. Department of Environment, Food and Rural Affairs (Defra) issued separate guidance on the Landfill Regulations in 2004.

➤ The Environment Agency regulates Part A(1) installations. Part A(2) installations are regulated by the relevant local authority – usually the district, London or metropolitan borough council in England and the county or borough council in Wales. However, the local authority will always be a statutory consultee where the Environment Agency is the regulator, and vice versa. Moreover, the local authority and the Environment Agency will work together in the permitting process. LAs have expertise in setting standards for noise control, while the Environment Agency will ensure that permit conditions protect water adequately. Annex I describes how IPPC installations are classified into either Part A(1) or Part A(2) installations depending on what activities take place within them.

➤ The second new regime – the 'Part B' regime of Local Air Pollution Prevention and Control (LAPPC) – represents a continuation of the old LAPC regime. LAPPC is similar to IPPC from a procedural perspective, but it still focuses on controlling emissions to air only. Defra provides separate guidance on local authority air pollution control.

17.5.4 Overview of the regulatory process

The basic purpose of the IPPC regime is to introduce a more integrated approach to controlling pollution from industrial sources. It aims to achieve 'a high level of protection of the environment taken as a whole by, in particular, preventing or, where that is not practicable, reducing emissions into the air, water and land'. The main way of doing that is by determining and enforcing permit conditions based on BAT.

➤ The entire regulatory process for IPPC consists of a number of elements. These are outlined below. IPPC applies to specified 'installations' – both 'existing' and 'new' – requiring each 'operator' to obtain a permit (from April 2008 an environmental permit) from the regulator – either the Environment Agency or the Local Authority.

17.5.5 Environmental permitting

(a) Introduction
The Environment Agency provides guidance on environmental permitting which aims to provide comprehensive help for those operating, regulating or interested in facilities that are covered by the Environmental Permitting (England and Wales) Regulations 2007 SI 2007 No. 3538 (EP Regulations). Environment Agency guidance on the Regime can be found on its website at www.environment-agency.gov.uk/epr.

These Regulations replace the system of waste management licencing in Part II of the EPA 1990 (c. 43) and the Waste Management Licensing Regulations 1994 (SI 1994/1056, as amended), and the system of permitting in the Pollution Prevention and Control (England and Wales) Regulations

2000 (SI 2000/1973, as amended), with a new system of environmental permitting in England and Wales.

For local authority-regulated facilities, guidance can be found on the Defra website: www.defra.gov.uk/environment/ppc/localauth/pubs/guidance/manuals.htm.

(b) What is environmental permitting?
Some activities could harm the environment or human health unless they are controlled. The Environmental Permitting Regime ('the Regime') requires operators to obtain permits for some facilities, the registration of exemptions for other facilities and ongoing supervision by regulators. The Regime is operated under the EP Regulations which came into force on 1 April 2008.

The aim of the Regime is to:

➤ protect the environment;
➤ deliver permitting and compliance effectively and efficiently in a way that provides increased clarity and minimizes the administrative burden on both the regulator and the operators of facilities;
➤ encourage regulators to promote best practice in the operation of regulated facilities;
➤ continue to fully implement European legislation.

The Regime covers facilities previously regulated under the PPC Regulations and Waste Management Licensing and exemptions. The Regime extends to England and Wales. It also covers the adjacent sea as far as the seaward boundary of the territorial sea.

(c) The legal framework
The Regulations set out the following:

➤ the facilities that need environmental permits or need to be registered as exempt;
➤ the process for registering exemptions;
➤ how to apply for and determine permit applications;
➤ requirements that environmental permits contain conditions to protect the environment as required by Directives and national policy;
➤ how environmental permits can be changed and ultimately be surrendered;
➤ a simplified permitting system called standard rules;
➤ compliance obligations backed up by enforcement powers and offences;
➤ provisions for public participation in the permitting process;
➤ the powers and functions of regulators, the Secretary of State and the Welsh Assembly Government (WAG);
➤ a simple transition to the new regime;
➤ provisions for appeals against permitting decisions.

17.5.6 What facilities require an environmental permit?

The EP Regulations specify which activities and waste operations require an Environmental Permit and provide that some waste operations can be exempt from those requirements. Certain waste operations covered by other legislation are excluded from permitting.

(a) Regulated facilities

An Environmental Permit is required for any of the following:

➤ an installation (which carries out the activities listed in Schedule 1 to the Regulations and any activities that are technically linked);
➤ a waste operation; or
➤ a mobile plant (carrying out either one of the Schedule 1 activities or a waste operation).

The collective term used in the EP Regulations for these installations, waste operations and mobile plant is 'regulated facility'.

There may be more than one regulated facility on the same site. In such cases there are arrangements in the EP Regulations to allow all such facilities to be regulated by the same regulator and to allow, in many cases, for a single permit.

It is an offence under Regulation 12 to operate a regulated facility without a permit.

(b) A single permit

A regulated facility will be either an installation, a waste operation (other than as part of an installation) or a mobile plant. An Environmental Permit can, however, cover more than one regulated facility (Regulation 17).

(c) When can a single permit be granted?

A single Environmental Permit can only be granted for more than one regulated facility where:

➤ the regulator is the same for each facility;
➤ the operator is the same for each facility;
➤ all the facilities are on the same site (the exceptions to this are set out as follows).

Where the regulator and operator are the same, a single Environmental Permit can be granted to an operator for more than one mobile plant. Mobile plants do not have to be operating on the same site in order to be included in a single permit.

Where the regulator and operator are the same, a single Environmental Permit can be granted to an operator for more than one regulated facility to which standard rules apply. Standard facilities do not have to be on the same site in order to be included in a single permit. However, standard facilities on different sites cannot be combined in a single permit where the IPPC Directive applies to any of the facilities.

Regulated facilities have to be on the same site in order to be covered by the same permit (with the exceptions of mobile plant and standard facilities set out previously). The regulator should consider the following factors in determining whether the facilities are on the same site.

➤ Proximity – there should, however, be no simple 'cut off' distance since some industrial complexes cover very large areas but still can be regarded as one site for permitting purposes.
➤ Coherence of a site – some regulated facilities will be within a single fenced area or may share security or emergency systems.
➤ Management systems – the extent to which the regulated facilities share a common management system is a relevant consideration.

It is expected that a regulator will adopt a common-sense approach to determining when facilities should be regulated under one permit. This consideration should be based on achieving protection of the environment in the most efficient regulatory manner.

17.5.7 Exempt and excluded waste operations

(a) Exempt waste operations

A waste operation (including those listed in Chapter 5 of Part 2 of Schedule 1 of the EP Regulations) which is exempt does not require an Environmental Permit. Exempt waste operations are not included in the definition of a regulated facility).

An exempt waste operation must (Regulation 5):

➤ meet the criteria of one of the paragraphs in Part 1 of Schedule 3;
➤ be consistent with the need to attain the objectives in Article 4 of the Waste Framework Directive 14;
➤ be registered;
➤ not involve hazardous waste or the storage or treatment of WEEE (Waste Electrical and Electronic Equipment), unless allowed by the relevant paragraph of Schedule 3.

The permitting requirements set out here are not relevant to exempt waste operations.

(b) Excluded activities

Some waste operations are excluded from the regime because there is already appropriate environmental regulation under other regimes.

These excluded operations are those that are carried out under a FEPA permit (Food and Environmental Protection Act 1985); under a Water Resources Act (chapter II of Part II of the Water Resources Act 1991) consent for liquid waste discharge; under a Groundwater Regulations authorization (Regulation 18 of the Groundwater Regulations 1998); for agricultural waste disposal (Regulation 4) or the disposal or recovery of sludges which are not to be treated as industrial or commercial waste under the Controlled Waste Regulations [Regulation 17(1) of the Controlled Waste Regulations 1992].

17.5.8 The regulator

The regulator for each category of regulated facility is identified in Regulation 32.

The Environment Agency regulates:

➤ Part A(1) installations;
➤ Part A(1) mobile plant;
➤ waste operations (though see below).

The relevant local authority regulates:

➤ Part A(2) installations;
➤ the 'Part B' regime of LAPPC;
➤ Part A(2) and Part B mobile plant;
➤ waste operations which are carried out as part of the Part A(2) or Part B installations or Part A(2) and Part B mobile plant.

LAPPC focuses on controlling emissions to air only for Part B installations. Defra and WAG jointly provide guidance on local authority air pollution and control. Available at: http://www.defra.gov.uk/environment/ppc/localauth/pubs/guidance/manuals.htm.

Rules for determining whether an installation is Part A(1), Part A(2) or Part B can be found in Part 1 of Schedule 1 to the Regulations. Guidance on Part A(1) and Part A(2) installations can be found in the Guidance on Part A installations. Available at http://www.defra.gov.uk/environment/epp/guidance.htm.

17.5.9 Environmental permit applications

(a) The operator

Only a person who is in control of the facility may obtain or hold an Environmental Permit. This person is the 'operator' (Regulation 7).

Box 17.3 Definition of operator

'Operator' is defined in Regulation 7 as the person who has control over the operation of a regulated facility.

Regulation 7(b) provides that a regulated facility need not be in operation for there to be an operator. Legal obligations may also be imposed on an operator during the pre- and post-operational phases.

The operator must demonstrably have the authority and ability to ensure the environmental permit is complied with.

The meaning of 'operator' is set out in Box 17.3. In most cases a single operator will have to obtain a single Environmental Permit for each regulated facility.

However, in some circumstances different operators run different parts of a regulated facility. This does not affect the regulator's determination of what actually constitutes the regulated facility.

Where two or more operators run different parts of a single regulated facility, they will each need a separate Environmental Permit and be responsible for complying with their permit conditions. In such cases, there should be no ambiguity over which operator has responsibility for which part of the regulated facility.

The requirement that the permit holder must be the operator of the facility is different from the previous system under waste management licencing. The transitional provisions therefore allow the holder of a waste management licence to be treated as the operator for the purposes of the Regulations [see Regulation 69(2)]. However, when permits are transferred over time they must be transferred to the person who will have control of the regulated facility.

(b) Pre-application discussions

Pre-application discussions between operators and regulators can help in improving the quality of the formal application and are therefore encouraged. In order for such discussions to make the best use of time, the operator is expected to have read the relevant published guidance. The regulator should not be expected to provide advice that might prejudice its determination of an application.

Operators and regulators may use the discussions to clarify whether a permit is likely to be needed. The regulator may also give operators general advice on how to prepare their applications, focus on the key issues and tell them what additional guidance is available. Other parties may be invited to join these discussions – for example, a public consultor.

Operators should bear in mind that, especially for controversial cases, good engagement with local or national interested parties at the pre-application stage can be beneficial to all sides.

(c) Using existing data

Operators may, where relevant, draw upon or attach other sources of information in their applications. These might include:

➤ environmental impact assessments;
➤ documents relating to an installation's regulation under the Control of Major Accident Hazards (COMAH) Regulations;
➤ prior investigations for compliance with the Groundwater Regulations;
➤ externally certified environmental management systems (EMSs);
➤ site reports prepared for planning purposes

They should make clear which parts of any attachments are relevant to their environmental permit applications and should demonstrate how they relate to the relevant Directive requirements.

(d) Timing of applications

Operators should normally make an application when they have drawn up full designs but before construction work commences (whether on a new regulated facility or when making changes to an existing one). Where facilities are not particularly complex or novel, the operator should usually be able to submit an application at the design stage containing all information the regulator needs. If, in the course of construction or commissioning, the operator wants to make any changes which mean that the permit conditions have to be varied, the operator may apply for this in the normal way.

There is nothing in the regulations to stop an operator from beginning construction before a permit has been issued. However, regulators may not agree with the design and infrastructure put in place. Therefore, to avoid any expensive delays and re-work, it is in the operator's interest to submit applications at the design stages. Any investment or construction work that an operator carries out before they have got an Environmental Permit will be at their own risk and will in no way affect the regulator's decision.

(e) Planning and environmental permit applications

If a regulated facility also needs planning permission, it is recommended that the operator should make both applications in parallel whenever possible. This will allow the pollution control regulator to start its formal consideration early on, thus allowing it to have a more informed input to the planning process. For certain waste operations, where planning permission is required, this must be in force before an Environmental Permit can be granted (paragraph 2 Schedule 9).

17.5.10 Transitional arrangements

(a) Existing permits and licences

The EP Regulations provide transitional arrangements for existing PPC permits and waste management licences so new applications for environmental permits are not required.

The regulator for these permits remains the same and will not change unless there is a subsequent direction from the Secretary of State or the WAG.

Except where there is an outstanding application (see below) these existing permits and licences automatically become environmental permits from the date the Regulations come into force [Regulation 69(1)].

(b) Existing permits and licences with an outstanding application to transfer, surrender, vary or modify

PPC permits and waste management licences do not become environmental permits on the date the EP Regulations come into force where there is an outstanding, duly-made, application in relation to the existing permit or licence. The relevant applications are those to transfer, surrender, vary or modify the existing permit or licence.

The existing permit or licence will not become an Environmental Permit until the outstanding application has been determined and the appeal period has elapsed. The existing permit or licence will become an Environmental Permit the day after the last day that an appeal could be made against the regulator's decision [Regulation 70(2)(c)].

If an appeal is made against the regulator's decision regarding the outstanding application, the existing permit or licence will become an Environmental Permit on the date the appeal is determined or withdrawn.

The provisions in Regulation 72 ensure that enforcement of the waste management licence or PPC permit can continue under the original legislation.

17.5.11 Application procedures

(a) Applications

The requirements for applications are set out in Schedule 5 to the Regulations. The application must:

➤ be made by the operator (though it may be made by an agent acting on behalf of the operator);
➤ in the case of a transfer application, be made by the current operator and the future operator;
➤ be made on the form provided by the regulator;
➤ include the information required by the application form;
➤ include the relevant fee.

An applicant can withdraw an application at any time before it is determined but the regulator is not obliged to return any of the application fee.

(b) Specific procedures for different types of applications

Variation applications

Once an operator has an Environmental Permit, changes in the operation of the facility may require the operator to apply to vary the permit.

The operator must apply to the regulator to vary the permit conditions when proposing a change that would mean that a permit condition could no longer be complied with. Other aspects of the Environmental Permit may also require a variation application – for example, to change the name of the operator on the permit (though not the operator's identity, which would require a transfer application).

A variation application may include an increase to the area over which the regulated facility operates. Where this occurs, issues such as the protection of the land and, where relevant, land use planning must be addressed.

(c) Surrender applications and notifications

There are two separate methods for surrender. Some regulated facilities may simply notify the regulator but others must make an application to the regulator (Regulations 24 and 25).

Surrender of the Environmental Permit by notifying the regulator is restricted to environmental permits for Part B installations and for mobile plant.

The pollution control measures for Part B installations relate solely to emissions to air. There is therefore no requirement to consider the condition of the land prior to surrendering the permit.

For mobile plant the position is similar in that there is no geographical site associated with the environmental permit. There cannot therefore be a consideration of the condition of the land before a surrender takes place. It should be noted that, where relevant, the permit conditions for mobile plant should be in place to ensure the protection of the land on which they operate.

It is possible to surrender part of the Environmental Permit for a regulated facility. This is the only method of reducing the extent of the site of the regulated facility covered by the permit. Where there is a partial surrender, the regulator may need to vary the permit conditions to reflect this.

17.5.12 Permit conditions

If the regulator grants a permit it can include any conditions it sees fit [paragraph 12(2) Part 1 Schedule 5]. It has a duty to impose conditions in order to secure the objectives that apply to the category of facility.

Where the regulator grants an application for the variation, transfer or partial surrender of an Environmental Permit and there are additional variations needed as a consequence of the application, the regulator should make those necessary variations to the Environmental Permit [paragraph 12(3)(a) Part 1 Schedule 5].

All permit conditions should be both necessary and enforceable. 'Necessary' means that the regulator should be able to justify – at appeal if necessary – the permit conditions it attaches. To be enforceable, conditions should clearly state the objective, standard or desired outcome of the condition so that the operator can understand what is required.

Permit conditions may comprise some or all of the following:

➤ conditions stipulating objectives or outcomes;
➤ standards to mitigate a particular hazard/risk;
➤ conditions addressing particular legislative requirements.

The regulator can include conditions in the permit setting out steps to be taken during, prior to and after the operation of the regulated facility.

17.5.13 Standard rules

(a) Standard rules

The Secretary of State, WAG and the Environment Agency can make standard rules (Regulation 26).

These rules consist of requirements common to the class of facilities subject to them (standard facilities – Regulation 2) and can be used instead of site-specific permit conditions. Standard rules would be suitable for industry sectors where a number of regulated facilities share similar characteristics in relation to environmental hazards.

The standard rules must achieve the same high level of environmental protection as site-specific conditions.

The rules are the conditions of the standard permit for all purposes other than for appeals. The standard rules cannot be appealed [Regulation 27(3)] as applying for a permit subject to the rules is voluntary.

(b) Developing standard rules

In preparing standard rules, it is necessary to consult widely with those who may be affected by or have an interest in the rules [Regulation 26(2)], including relevant statutory bodies. The standard nature of the facilities for which standard rules will be produced allows a general consideration of the requirements and standards for all such facilities.

It is expected that standard rules will be developed in consultation with the relevant industry.

Assessments of risk can be carried out nationally for common generic activities. This understanding of the hazards and risks posed by these activities would form the basis for the development of standard rules for standard facilities.

17.5.14 Standard permits

It is the operator's decision as to whether they wish to operate under standard rules. Where standard rules have been made, operators of standard facilities can, if they so wish, request that their facility be subject to the relevant rules. This request may be made in an application for a new permit or an application to vary an Environmental Permit.

From April 2008, operators may apply to operate under a set of standard rules.

The generic assessments of risk for standard facilities should be made available to applicants to assist them in determining whether their activity is within the scope of the standard rules and, if they apply for a standard permit, in the adoption of suitable control measures to meet those rules.

One important difference from other regulated facilities is that any additional site-specific assessment of risk is not necessary for a standard facility. Regulated facilities that require a location-specific assessment of impact and risk are not suitable for standard rules.

Public consultation on applications for individual standard facilities is not required (other than for Part A installations).

17.5.15 Management systems

In order to ensure a high level of environmental protection, operators should have effective management systems in place. The nature of the required management system depends upon the complexity of the regulated facility.

Complex regulated facilities are encouraged to put in place a formal EMS externally certified to the international standard ISO 14001 by the UKAS accredited certification body or other European equivalent and to register for the EU's Eco Management and Audit Scheme (EMAS). These standards require that the management system include safeguards for legal compliance and a commitment to continuous improvement of environmental performance. Additionally EMAS requires organizations to produce an independently validated public report about their environmental performance and progress against targets and objectives. Where relevant the performance should be benchmarked against European legislation, e.g. BAT under IPPC. EMAS and ISO 14001 are also recognized by the Environment Agency's risk rating scheme OPRA (Operator and Pollution Risk Appraisal scheme). OPRA scores are linked to fees and charges. Organizations which have implemented an EMS may achieve a better OPRA score and can pay lower fees and charges.

For simpler regulated facilities, externally certified schemes or a full EMS may be less appropriate but should still be carefully considered by operators and, where appropriate, encouraged by regulators. The stepwise approach provided by BS8555 is particularly appropriate for smaller facilities and can make EMS implementation much simpler. Organizations can achieve UKAS accredited certification to one or more stages of BS8555 under the IEMA Acorn or BSI Stems schemes. The European Commission has also developed a simplified implementation guide, EMAS 'easy', which aims to help small and medium enterprises (SMEs) achieve registration for EMAS. There is also specific guidance on management systems for some industry sectors on the website of the Institute of Environmental Management and Assessment.

EMSs have relevance to other aspects of regulation, such as determining risk-based inspection frequencies. Recognized quality assurance schemes may also be relevant, and regulators may also take account of non-certified systems where these can be demonstrated to provide an equivalent role in safeguarding compliance and continual improvement of environmental performance.

17.5.16 Enforcement

(a) Enforcement notices

Regulation 36 of the EP Regulations allows the regulator to serve an 'enforcement notice' if it believes an operator has contravened, is contravening or is likely to contravene any permit conditions.

Enforcement notices will specify the steps required to remedy the problem and the timescale in which they must be taken. Enforcement notices may include steps to remedy the effects of any harm and to bring a regulated facility back into compliance.

(b) Suspension notices

If the operation of a regulated facility involves a risk of serious pollution, the regulator may serve a 'suspension notice' under regulation 37 of the Regulations. This applies whether or not the operator has breached a permit condition.

The suspension notice must describe the nature of the risk of pollution and the actions necessary to remove that risk. The notice must specify the deadline for taking actions.

When the regulator serves a suspension notice, the permit ceases to authorize the operation of the entire facility or specified activities depending upon what is specified in the notice.

A suspension notice should allow activities to continue unless their cessation is necessary to address the risk of pollution. While the suspension notice is in force, additional restrictions may be necessary for any activities that are allowed to continue. Where this is the case the suspension notice must set out these additional steps.

When the operator has taken the remedial steps required by the notice, the regulator must withdraw the notice.

(c) Prosecutions

If an operator has committed a criminal offence under the Regulations, regulators should consider a prosecution. Conviction in a magistrates' court carries a fine of up to £50 000 and up to 12 months' imprisonment for the most serious offences under the Regulations. Conviction in the Crown Court for those offences may lead to an unlimited fine and imprisonment for up to 5 years.

17.5.17 Wider scope of IPPC

IPPC takes a wider range of environmental impacts into account than IPC. The current system of IPC regulates emissions to land, water and air. The IPPC regime will additionally take into account: waste avoidance or minimization, energy efficiency, accident avoidance, and minimization of noise, heat and vibrations. These aims will achieve a higher level of protection as a whole.

IPPC also applies to a wider range of industries than IPC. These industries include all installations that are currently regulated under IPC, some installations currently under LAPC and some installations that are not currently under either regime, such as landfill sites, intensive agriculture, large pig and poultry units, and food and drink manufacturers.

Under IPPC, regulated industries are referred to as 'installations' as opposed to 'processes', which is the term used for IPC.

This change in terminology enables a more integrated approach to regulation; a whole installation must be permitted rather than just individual processes within the installation.

Guidelines to establish which techniques are BAT are published by the European Commission's IPPC Bureau. These reference notes are known as BREF notes and provide the basis for national sectoral guidance. The Environment Agency has supplementary guidance to cover many issues; some, for example, energy efficiency, site remediation and noise, are new issues under IPPC. Industry sectors not previously regulated under the EPA 1990 such as intensive farming and food and drink installations are also covered by guidance.

17.5.18 Duty of care

Waste and the duty of care

The duty of care is covered in Part II of the EPA 1990. The duty of care applies to anyone who produces or imports, keeps or stores, transports, treats or disposes of waste. It also applies if they act as a broker and arrange these things.

The duty holder is required to take all reasonable steps to keep waste safe. If they give waste to someone else, the duty holder must be sure they are authorized to take it and can transport, recycle or dispose of it safely.

The penalty for breach of this law is an unlimited fine.

Waste can be anything owned, or a business produces, which a duty holder wants to get rid of. Controlled waste is defined in Box 17.4.

If the waste comes from a person's own home, the duty of care **does not** apply to them. But if the waste is not from the house they live in – for example, if it is waste from their workplace or waste from someone else's house – the duty of care **does** apply.

Animal waste collected and transported under the Animal By-Products Order 1992 is not subject to the duty of care.

Duty holders must take all reasonable steps to fulfil the duty and complete some paperwork. What is reasonable depends on what is done with the waste.

Box 17.4 Definition of controlled waste

Controlled waste means household, commercial or industrial waste. It includes any waste from a house, school, university, hospital, residential or nursing home, shop, office, factory or any other trade or business premises. It is controlled waste whether it is solid or liquid and even if it is not hazardous or toxic.

Box 17.5 Who has authority to take waste?

Council waste collectors

The duty holder does not have to do any checking, but if they are not a householder, there is some paperwork to complete. This is explained in Box 17.6.

Registered waste carriers

Most carriers of waste have to be registered with the Environment Agency or the Scottish Environment Protection Agency. Look at the carrier's certificate of registration or check with the Agencies.

Exempt waste carriers

The main people who are exempt are charities and voluntary organizations. Most exempt carriers need to register their exemption with the Environment Agency or the Scottish Environment Protection Agency. If someone says they are exempt, ask them why. Check with the Agencies that their exemption is registered.

Holders of environmental permits (from April 2008 – formerly Licences)

Some environmental permits are valid only for certain kinds of waste or certain activities. Ask to see the permit. Check that it covers the kind of waste being consigned.

Businesses exempt from environmental permits

There are exemptions from permitting for certain activities and kinds of waste. Most exempt businesses need to register their exemption with the Environment Agency

or the Scottish Environment Protection Agency. Check with the Agencies that their exemption is registered.

Authorized transport purposes

Waste can also be transferred to someone for 'Authorized transport purposes'. This means:

the transfer of controlled waste between different places within the same premises

the transport of controlled waste into Great Britain from outside Great Britain

the transport by air or sea of controlled waste from a place in Great Britain to a place outside Great Britain.

Registered waste brokers

Anyone who arranges the recycling or disposal of waste, on behalf of someone else, must be registered as a waste broker. Check with the Environment Agency or the Scottish Environment Protection Agency that the broker is registered.

Exempt waste brokers

Most exempt waste brokers need to register with the Environment Agency or the Scottish Environment Protection Agency. Those who are exempt are mainly charities and voluntary organizations. If someone tells you they are exempt, ask them why. You can check with the Environment Agencies that their exemption is registered.

Steps to take if the duty of care applies

When a duty holder has waste, they must:

➤ stop it escaping from their control and store it safely and securely. They must prevent it causing pollution or harming anyone;

➤ keep it in a suitable container. Loose waste in a skip or on a lorry must be covered;

➤ if the duty holder gives waste to someone else, check that they have authority to take it. The law says the person to whom they give the waste must be authorized to take it. Box 17.5 shows who is allowed to take waste and how the duty holder can check;

➤ describe the waste in writing. The duty holder must fill in and sign a transfer note for it and keep a copy. To save on paperwork, the description of the waste can be written on the transfer note (see Box 17.6).

When a person takes waste from someone else they must:

➤ be sure the law allows them to take it. Box 17.5 shows who is allowed to take waste;

➤ make sure the person giving them the waste describes it in writing. The waste receiver must fill in and sign a transfer note and keep a copy (see Box 17.6).

17.5.19 Hazardous waste

On 16 July 2005, the Hazardous Waste (England and Wales) Regulations and the List of Wastes (England) Regulations, replacing the Special Waste Regulations, came into force (Scotland has retained The Special Waste Regulations). The Regulations can be found on the Office of Public

Box 17.6 Filling in paperwork

When waste is passed from one person to another the person taking the waste must have a written description of it. A transfer note must also be filled in and signed by both persons involved in the transfer.

The duty holder can write the description of the waste on the transfer note. Who provides the transfer note is not important as long as it contains the right information. The Government has published a model transfer note with the Code of Practice which can be used if desired.

Repeated transfers of the same kind of waste between the same parties can be covered by one transfer note for up to a year, for example weekly collections from shops.

The transfer note to be completed and signed by both persons involved in the transfer must include:

➤ what the waste is and how much there is
➤ what sort of containers it is in
➤ the time and date the waste was transferred
➤ where the transfer took place
➤ the names and addresses of both persons involved in the transfer
➤ whether the person transferring the waste is the importer or the producer of the waste
➤ the details of which category of authorized person each one is. If the waste is passed to someone for

authorized transport purposes, you must say which of those purposes applies
➤ if either or both persons is a registered waste carrier, the certificate number and the name of the Environment Agency which issued it
➤ if either or both persons have an environmental permit (or an old waste management licence), the permit number (or old licence number) and the name of the Environment Agency which issued it
➤ the reasons for any exemption from the requirement to register or have a an environmental permit
➤ where appropriate, the name and address of any broker involved in the transfer of waste.

The written description – The written description must provide as much information as someone else might need to handle the waste safely.

Keeping the papers – Both persons involved in the transfer must keep copies of the transfer note and the description of the waste for 2 years. They may have to prove in court where the waste came from and what they did with it. A copy of the transfer note must also be made available to the Environment Agency or the Scottish Environment Protection Agency if they ask to see it.

Sector Information (OPSI) website at: http://www.opsi.gov.uk/legislation.

The regime includes a requirement for most producers of hazardous waste to notify their premises to the Environment Agency. The facility to notify premises has been available since April 2005. Preliminary guidance on notification was published by Defra in January 2005. Guidance on notification, including the online notification facility, and more general guidance on the new regime, can be found on the Environment Agency's *hazardous waste pages*. See a brief summary of these Regulations in Section 17.19.

17.5.20 Applying for a waste management licence

From April 2008 this has now been replaced by the Environmental Permitting regime. See Section 17.5.5.

17.5.21 Sources of references

The law
The Environmental Protection Act 1990 ISBN 0 10 544390 5.

The Pollution Prevention and Control Act 1999, Stationery Office 1999 ISBN 0 10 542499.

Hazardous Waste (England and Wales) Regulations 2005, The Stationery Office SI 2005 No. 894.

Government guidance
Environmental Permitting Core Guidance for the Environmental Permitting (England and Wales) Regulations 2007, Defra, 2008, http://www.defra.gov.uk.

Environment Agency HWR01. A guide to the Hazardous Waste Management Regulations and the List of Waste Regulations in England and Wales.

Integrated Pollution Prevention and Control Practical Guidance, 4th edition, Defra, 2005.

17.6 Chemicals (Hazard Information and Packaging for Supply) Regulations 2002 and Amendment Regulations

17.6.1 Introduction

CHIP is the Chemicals (Hazard Information and Packaging for Supply) Regulations 2002, sometimes known as CHIP 3. The aim of CHIP 3 is to ensure that people who are supplied with chemicals receive the information they need to protect themselves, others and the environment.

To achieve this, CHIP 3 obliges suppliers of chemicals to identify their hazards (flammability, toxicity, etc.) and to pass on this information together with advice on safe use to the people they supply the chemicals to. This is usually done by means of package labels and safety data sheets (SDSs).

CHIP applies to most chemicals but not all. The exceptions (which generally have Regulations of their own) are set out in Regulation 3 of CHIP and include cosmetic products, medicinal products, foods, etc. CHIP 3, a new set of Regulations which consolidates and extends the previous ones, came into force on 24 July 2003.

CHIP 3 is intended to protect people and the environment from the harmful effects of dangerous chemicals by making sure users are supplied with information about the dangers. Many chemicals such as cosmetics and medicines are outside the scope of CHIP and have their own specific regulatory regimes. However, biocides and plant protection products, which have their own specific laws, have had to be classified and labelled according to CHIP as of July 2004.

CHIP 3 requires the supplier of a dangerous chemical to:

➤ **identify the hazards (dangers)** of the chemical (this is known as 'classification');
➤ **package the chemical safely**;
➤ **give information about the hazards** to their customers (usually by means of information on the package (e.g. a label) and, if supplied for use at work, an SDS).

These are known as supply requirements. 'Supply' is defined as making a chemical available to another person. Manufacturers, importers, distributors, wholesalers and retailers are examples of suppliers.

The CHIP (Amendment) Regulations 2005 entered into force on 31 October 2005. The regulations are known as CHIP 3.1. The regulations bring into legal effect all the new entries, revisions, deletions and amendments to the classification and labelling requirements of hazardous substances set out in the 29th Adaptation to Technical Progress (29th ATP) to the Dangerous Substances Directive (European Commission Directive 2004/73/2004). (A further amendment CHIP3.2 is expected in 2008/2009).

Table 17.2 outlines a description of commonly used terms, and Figure 17.1 shows a summary of what needs to be done for compliance.

17.6.2 Classification – Regulation 4

The basic requirement for CHIP 3 is for the supplier to decide whether the chemical is hazardous. CHIP 3 with its Approved Classification and Labelling Guide (ACLG) sets out the rules for this. It tells the supplier how to:

➤ decide what kind of hazard the chemical has;
➤ explain the hazard by assigning a simple sentence that describes it (known as a 'risk phrase' or R-Phrase). This is known as classification. Many commonly used substances have already been classified. They are contained in the CHIP Approved Supply List (ASL) which must be used.

17.6.3 Information and labelling – Regulations 5–10

Information has to be supplied for the customers via a data sheet and a label on the package (unless the substance is provided in bulk such as a tanker or by pipeline). If the chemical is supplied for use at work an SDS must be provided. CHIP gives 16 headings for the SDS to set a standard for their quality.

CHIP 3 specifies what has to go on to the label including where it must be displayed, the size of the label, name and address of supplier, name of the substance, risk and safety phrases and indications of danger with symbols.

17.6.4 Packaging of Dangerous Substance – Regulation 7

Packaging used for a chemical must be suitable. That means:

➤ the receptacle containing the dangerous substance or dangerous preparation is designed and constructed so that its contents cannot escape;
➤ the materials constituting the packaging and fastenings are not susceptible to adverse attack by the contents or liable to form dangerous compounds with the contents;
➤ the packaging and fastenings are strong and solid throughout to ensure that they will not loosen and will meet the normal stresses and strains of handling;

Table 17.2 Commonly used terms

Category of danger	A description of hazard type
Classification	Precise identification of the hazard of a chemical by assigning a category of danger and a risk phrase using set criteria
Risk phrase (R-phrase)	A standard phrase which gives simple information about the hazards of a chemical in normal use
Safety phrase (S-phrase)	A standard phrase which gives advice on safety precautions which may be appropriate when using a chemical
Substance	A chemical element or one of its compounds, including any impurities
Preparation	A mixture of substances
Chemical	A generic term for substances and preparations
Tactile warning devices (TWDs)	A small raised triangle applied to a package intended to alert the blind and visually impaired to the fact that they are handling a container of a dangerous chemical
Child-resistant fastenings (CRFs)	A closure which meets certain standards intended to protect young children from accessing the hazardous contents of a package
Chain of supply	The successive ownership of a chemical as it passes from manufacturer to its ultimate user

➤ any replaceable fastening fitted to the receptacle containing the dangerous substance or dangerous preparation is designed so that the receptacle can be repeatedly refastened without the contents of the receptacle escaping.

17.6.5 Child-resistant fastenings, tactile warnings and other consumer protection measures – Regulation 11

CHIP 3 sets out special requirements for chemicals that are sold to the public.

Some containers have to be fitted with a child-resistant closure to a laid-down standard to prevent young children opening them and swallowing the contents.

Some must have tactile danger warning to alert the blind and partially sighted to the danger. This is often a raised triangle.

17.6.6 Retention of Data – Regulation 12

Data used for classification, labelling, child-resistant fasteners and for preparing the SDS must be kept for at least 3 years after the dangerous chemical is supplied for the last time.

17.6.7 Chemical manufacture and storage – the European Commission's new chemical strategy (REACH)

(a) What is REACH?
REACH is a new EC regulation concerning the Registration, Evaluation, Authorization and restriction of CHemicals. It came into force on 1 June 2007 and replaces a patchwork of European Directives and Regulations with a single system.

Defra is responsible for implementing this European Regulation in the UK.

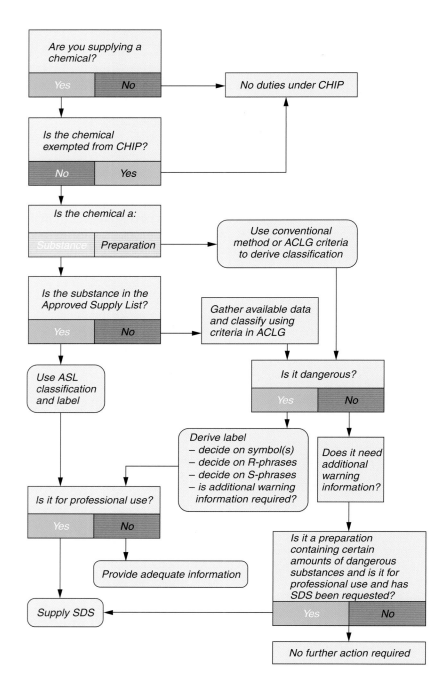

Figure 17.1 Summary of what you need to do for compliance.

(b) Competent authority

Many European regulatory systems are operated at the national ('member state') level by a Competent Authority in each member state. In October 2006, Defra nominated HSE to be the UK Competent Authority for REACH, working closely with the Environment Agency and other partners to manage certain key aspects of the REACH system in the UK.

The Competent Authority's responsibilities under REACH will be to:

➤ provide advice to manufacturers, importers, downstream users and other interested parties on their respective responsibilities and obligations under REACH (Competent Authorities' helpdesks);

➤ conduct substance evaluation of prioritized substances and prepare draft decisions;

➤ propose harmonized Classification and Labelling for carcinogens, mutagens or reproductive toxicants (CMRs) and respiratory sensitizers;

➤ identify substances of very high concern for authorization;

➤ propose restrictions;

➤ nominate candidates to membership of ECA committees on Risk Assessment and Socio-economic Analysis;

➤ appoint members for the Member State Committee to resolve differences of opinion on evaluation decisions;

➤ appoint a member to the Forum for Information Exchange and meet to discuss enforcement matters;

➤ provide adequate scientific and technical resources to the members of the Committees that they have nominated;

➤ work closely with the European Chemical Agency in Helsinki.

(c) Aims

REACH has several aims:

➤ to provide a high level of protection of human health and the environment from the use of chemicals;

➤ to allow the free movement of substances on the EU market;

➤ to make the people who place chemicals on the market (manufacturers and importers) responsible for understanding and managing the risks associated with their use;

➤ to promote the use of alternative methods for the assessment of the hazardous properties of substances;

➤ to enhance innovation in and the competitiveness of the EU chemicals industry.

(d) No data, no market

A major part of REACH is the requirement for manufacturers or importers of substances to register them with a central European Chemicals Agency (ECHA). A registration package will be supported by a standard set of data on that substance. The amount of data required is proportionate to the amount of substance manufactured or supplied.

If you do not register your substances, then the data on them will not be available and as a result, you will no longer be able to manufacture or supply them legally, that is no data, no market.

(e) Scope and exemptions

REACH applies to substances manufactured or imported into the EU in quantities of 1 tonne per year or more. Generally, it applies to all individual chemical substances on their own, in preparations or in articles (if the substance is intended to be released during normal and reasonably foreseeable conditions of use from an article).

Some substances are specifically excluded.

➤ radioactive substances;

➤ substances under customs supervision;

➤ the transport of substances;

➤ non-isolated intermediates;

➤ waste;

➤ some naturally occurring low-hazard substances;

Some substances, covered by more specific legislation, have tailored provisions, including:

➤ human and veterinary medicines;

➤ food and foodstuff additives;

➤ plant protection products and biocides;

➤ isolated intermediates;

➤ substances used for research and development;

(f) Pre-registration

It is estimated that there are around 30 000 substances on the European Market in quantities of 1 tonne or more per year. Registering all of these at once would be a huge task for both industry and regulators. To overcome this, the registration of those substances already being manufactured or supplied is to take place in three phases. These phases are spread over 11 years. To benefit from these provisions, manufacturers or suppliers had to pre-register their substances between 1 June and 30 November 2008.

Once pre-registered, the ECHA will identify who is intending to register the same substance and put them in contact with each other. The potential registrants can then come together and form a 'Substance Information Exchange Forum' (SIEF) where they can negotiate sharing their available data and the costs of generating any new data.

(g) One substance, one registration

This is the principle that, for any one substance, a single set of data is produced that is shared by all those companies that manufacture or supply that substance. The details of how this sharing is arranged (costs, etc.); business specific (e.g. company name) and business sensitive (e.g. how it is used) information; is submitted separately by each company.

(h) Registration

Registration is a requirement on industry (manufacturers/suppliers/importers) to collect and collate specified sets of information on the properties of those substances they manufacture or supply. This information is used to perform an assessment of the hazards and risks that substance may pose and how those risks can be controlled. This information and its assessment is submitted to the ECHA in Helsinki.

Chemicals already existing (those on EINECS (European Inventory of Existing Commercial Chemical Substances) or manufactured in the EU prior to entry into force of REACH) are known as 'phase-in' substances under REACH. These will be registered in three phases according to their tonnage and/or hazardous properties.

➤ Phase 1 – substances supplied at ≥1000 tonnes per year; substances classified under CHIP as very toxic to aquatic organisms supplied at ≥100 tonnes per year; substance classified under CHIP as Category 1 or 2 CMRs supplied at ≥1 tonne per year; substances classified as very toxic to aquatic organisms must be registered in the first 3 years (by 1 December 2010).

➤ Phase 2 – substances supplied at ≥100 tonnes per year must be registered in the first 6 years (by 1 June 2013).

➤ Phase 3 – substances supplied at ≥1 tonne per year must be registered in the first 11 years (by 1 June 2018).

A substance can be registered at any time prior to these deadlines.

Non-phase-in substances (i.e. those not on EINECS, or those which have never been manufactured previously or not pre-registered) will be subject to registration 1 June 2008. Until then, the current Notification of New Substances Regulations (NONS) continue to apply. For further details on registration, please see the specific ECHA webpage or detailed guidance document.

(i) Evaluation

Registration packages (dossiers) submitted under REACH can be evaluated for:

➤ **Compliance check:** A quality/accuracy check of the information submitted by industry.

➤ **Dossier evaluation:** A check that an appropriate testing plan has been proposed for substances registered at the higher tonnage levels (≥100 tonnes/annum).

➤ **Substance evaluation:** An evaluation of all the available data on a substance, from all registration dossiers. This is done by national Competent Authorities on substances that have been prioritized for potential regulatory action because of concerns about their properties or uses.

(j) Substances of very high concern

Some substances have hazards that have serious consequences; for example they cause cancer, or they have other harmful properties and remain in the environment for a long time and gradually build up in animals. These are 'substances of high concern'. One of the aims of REACH is to control the use of such substances.

(k) Authorization

Authorization is a feature of REACH that is new to the area of general chemicals management. As REACH progresses, a list of 'substances of very high concern' will be created. Substances on this list cannot be supplied or used unless an authorization has been granted. A company wishing to market or use such a substance must apply to the ECHA in Helsinki for an authorization, which may be granted or refused.

(l) Restrictions

Any substance that poses a particular threat can be restricted. Restrictions take many forms, for example, from a total ban to not being allowed to supply it to the general public. Restrictions can be applied to any substance, including those that do not require registration. This part of REACH takes over the provisions of the Marketing & Use Directive.

(m) Classification and labelling

An important part of chemical safety is clear information about any hazardous properties a chemical has. The classification of different chemicals according to their characteristics (e.g. those that are corrosive, or toxic to fish) currently follows an established system, which is reflected in REACH. Over the next few years, work will be underway to establish in the EU a classification and labelling system based on the United Nations (UN) Globally Harmonized System, or GHS. REACH has been written with GHS in mind (See Section 17.6.8 for more information on GHS).

(n) Information in the supply chain

The passage of information up and down the supply chain is a key feature of REACH. Users should be able to understand what manufacturers and importers know about the dangers involved in using chemicals and how to control risks. However, in order for suppliers to be able to assess these risks they need information from the downstream users about how the chemicals are used. REACH provides a framework in which information can be passed both up and down supply chains.

REACH adopts and builds on the previous system for passing information – the SDS. This should accompany materials down through the supply chain, providing the information users need to ensure chemicals are safely managed. In time these SDSs will include information on safe handling and use.

More detailed information about REACH can be found at the Defra REACH website, and HSE website, including impact assessment work at the national and community levels, and consultations. A wide range of industry and commercial sources also offer commentary and advice regarding REACH.

(o) Enforcement

Enforcing this very wide ranging new system presents new challenges to regulators across Europe. REACH places new duties on a range of different businesses. Mostly, the new duties will be on manufacturers and importers of chemicals, but there are also requirements for downstream users of chemicals to share information with their suppliers. Although HSE will play a key role in enforcing REACH, both as the UK Competent Authority and more generally as the UK occupational health and safety regulator, enforcing REACH will fall to a number of regulatory bodies. HSE is working closely with other regulators to support Defra in setting up the framework for enforcing REACH in the UK.

17.6.8 Globally Harmonized System of Classification and labelling of Chemicals

Worldwide, there are many different laws on how to identify hazardous chemicals (classification) and how this information should be communicated to users (via labels and SDSs).

This is often confusing as the same chemical can have different hazard descriptions in different countries. The UN brought together experts from different countries to create the GHS.

The aim of GHS is to have, throughout the world, the same:

➤ criteria for classifying chemicals according to their health, environmental and physical hazards;
➤ hazard communication requirements for labelling and SDSs.

The UN GHS is not a formal treaty, but instead is a non-legally binding international agreement. This means that countries (or trading blocks like the EU) must create local or national legislation to implement the GHS.

The EU has proposed new Regulation which will apply directly in all member states. This means that member states do not need local legislation to implement the new Regulations and existing legislation that implements classification and labelling will need to be repealed. In

Figure 17.2 New harmonized hazard warning symbols for labels.

Great Britain this is the CHIP Regulations which will be repealed in full when the new European Regulation is fully in force (anticipated in June 2015). Amendments to CHIP are likely as the transitional period progresses. The ASL (an integral part of CHIP) will also be repealed at the end of the transitional period.

Under the proposed new Regulation there will be:

➤ new scientific criteria to assess hazardous properties of chemicals;
➤ two new harmonized hazard warning symbols for labels (known as 'pictograms'; Figure 17.2);
➤ a new design for existing symbols (Figure 17.3);
➤ new harmonized warning and precautionary statements for labels, which will replace the existing risk and safety phrases.

17.6.9 Sources of references

Approved Classification and Labeling Guide (fifth edition) L131, HSE Books, 2002, ISBN 9780 7176 2369 3.

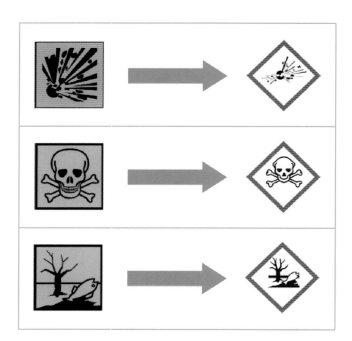

Figure 17.3 New design for existing symbols.

Approved Supply List. Information Approved for the Classification and Labeling of Dangerous Substances and Preparations for Supply (eighth edition) L142, HSE Books, 2005, ISBN 9780 7176 6138 1.

Chemical (Hazard Information and Packaging for Supply) Regulations 2002, SI No. 2002, 1689, The Stationery Office.

Chip for Everyone HSG 228, HSE Books, 2002, ISBN 9780 7176 2370 9.

The Compilation of Safety Data Sheets. Approved Code of Practice (third edition) L130, HSE Books, 2002, ISBN 9780 7176 2371 6.

The Idiot's Guide to CHIP INDG 350, HSE Books, 2002, ISBN 9780 7176 2333 4.

Why Do I Need a Safety Data Sheet? INDG 353, HSE Books, 2002, ISBN 9780 7176 2367 9.

HSE's CHIP website: www.hse.gov.uk/chip

HSE's REACH website: www.hse.gov.uk/reach

HSE's GHS website: www.hse.gov.uk/ghs/index.htm

17.7 Confined Spaces Regulations 1997

17.7.1 Introduction

These Regulations concern any work that is carried on in a place which is substantially (but not always entirely) enclosed, where there is a reasonably foreseeable risk of serious injury from conditions and/or hazardous substances in the space or nearby. Every year about 15 people are killed and a number seriously injured across a wide range of industries (e.g. from simple open top pits to complex chemical plants). Rescuers without proper training and equipment, often become the victims.

17.7.2 Definitions

Confined space – means any place, including any chamber, tank, vat, silo, pit, trench, pipe, sewer, flue, well or similar space in which, by virtue of its enclosed nature, there arises a reasonably foreseeable specified risk.

Specified risk – means a risk to any person at work of:

➤ serious injury arising from a fire or explosion;
➤ loss of consciousness arising from an increase in body temperature;
➤ loss of consciousness or asphyxiation arising from gas, fume, vapour or the lack of oxygen;
➤ drowning arising from the an increase in the level of liquid;
➤ asphyxiation arising from a free-flowing solid or because of entrapment by it.

17.7.3 Employers' duties – Regulation 3

Duties are placed on employers to:

➤ comply regarding any work carried out by employees;
➤ ensure, sfarp, that other persons (e.g. use competent contractors) comply regarding work in the employer's control.

The self-employed also have duties to comply.

17.7.4 Work in confined space – Regulation 4

1. No person at work shall enter a confined space for any purpose unless it is not reasonably practicable to achieve that purpose without such entry.
2. Other than in an emergency, no person shall enter, carry out work or leave a confined space otherwise than in accordance with a safe system of work, relevant to the specified risks.

17.7.5 Risk assessment

Risk assessment is an essential part of complying with these Regulations and must be done (under the MHSWR 1999) to determine a safe system of work. The risk assessment needs to follow a hierarchy of controls for compliance. This should start with the measures – both in design and procedures – that can be adopted to enable any work to be carried out outside the confined space.

The assessment must be done by a competent person and will form the basis of a safe system of work. This will normally be formalized into a specific permit to work, applicable to a particular task. The assessment will involve the following.

(a) The general conditions to assess what may or may not be present. Consider:
  - what have been the previous contents of the space
  - residues that have been left in the space, for example, sludge, rust, scale and what may be given off if they are disturbed
  - contamination which may arise from adjacent plant, processes, services, pipes or surrounding land, soil or strata
  - oxygen deficiency and enrichment. There are very high risks if the oxygen content differs significantly from the normal level of 20.8%; if it is above this level increased flammability exists; if it is below then impaired mental ability occurs, with loss of consciousness under 16%
  - that physical dimensions and layout of the space can affect air quality.

(b) Hazards arising directly from the work being undertaken. Consider:
  - the use of cleaning chemicals and their direct effect or interaction with other substances
  - sources of ignition for flammable dusts, gases, vapours, plastics and the like.

(c) The need to isolate the confined space from outside services or from substances held inside, such as liquids, gases, steam, water, fire extinguishing media, exhaust gases, raw materials and energy sources.

(d) The requirement for emergency rescue arrangements including trained people and equipment.

17.7.6 Safe system of work

The detailed precautions required will depend on the nature of the confined space and the actual work being carried out. The main elements of a safe system of work which may form the basis of a 'permit-to-work' are:

➤ the type and extent of supervision;
➤ competence and training of people doing the work;
➤ communications between people inside, and from inside to outside and to summon help;
➤ testing and monitoring the atmosphere for hazardous gas, fume, vapour, dust, etc. and for concentration of oxygen;

➤ gas purging of toxic or flammable substances with air or inert gas such as nitrogen;
➤ good ventilation, sometimes by mechanical means;
➤ careful removal of residues using equipment which does not cause additional hazards;
➤ effective isolation from gases, liquids and other flowing materials by removal of pieces of pipe, blanked off pipes, locked off valves;
➤ effective isolation from electrical and mechanical equipment to ensure complete isolation with lock off and a tag system with key security. Need to secure against stored energy or gravity fall of heavy presses etc.;
➤ if it is not possible to make the confined space safe, the provision of personal and respiratory protective equipment;
➤ supply of gas via pipes and hoses carefully controlled;
➤ access and egress to give quick unobstructed and ready access and escape;
➤ fire prevention;
➤ lighting, including emergency lighting;
➤ prohibition of smoking;
➤ emergencies and rescue;
➤ limiting of working periods and the suitability of individuals.

17.7.7 Emergency arrangements – Regulation 5

Before people enter a confined space, suitable and sufficient rescue arrangements must be set up. These must:

➤ reduce the risks to rescuers sfarp;
➤ include the provision and maintenance of suitable resuscitation equipment designed to meet the specified risks.

To be suitable and sufficient arrangements will need to cover:

➤ rescue and resuscitation equipment;
➤ raising the alarm, alerting rescue and watch keeping;
➤ safeguarding the rescuers;
➤ fire safety precautions and procedures;
➤ control of adjacent plant;
➤ first-aid arrangements;
➤ notification and consultation with emergency services;
➤ training of rescuers and simulations of emergencies;
➤ size of access openings to permit rescue with full breathing apparatus, harnesses, fall arrest gear and lifelines, which is the normal suitable respiratory protection and rescue equipment for confined spaces.

17.7.8 Training

Specific, detailed and frequent training is necessary for all people concerned with confined spaces, whether they are acting as rescuers, watchers or those carrying out the actual work inside the confined space. The training will need to cover all procedures and the use of equipment under realistic simulated conditions.

17.7.9 References

Guidance on Permit-to-Work Systems: A Guide for the Petroleum, Chemical and Allied Industries. HSG250, HSE Books, 2005, ISBN 978 0 7176 2943 5.

Respiratory Protective Equipment at Work: A Practical Guide HSG53, 2004, HSE Books, ISBN 978 0 7176 2904 6.

Safe Work in Confined Spaces. Confined Spaces Regulations 1997. Approved Code of Practice, Regulations and Guidance. L101, 1997, HSE Books, ISBN 978 0 7176 1405 9.

17.8 Construction (Design and Management) (CDM) Regulations 2007

17.8.1 Background and introduction

The following is extracted from the new Approved Code of Practice (ACOP) L144.

The revised Construction (Design and Management) Regulations 2007 (CDM 2007) came into force on 6 April 2007. They replace the Construction (Design and Management) Regulations 1994 (CDM94) and the Construction (Health, Safety and Welfare) Regulations 1996 (CHSW). The new Approved Code of Practice replaces the ACOP under CDM94.

The key aim of the CDM 2007 is to integrate health and safety into the management of the project and to encourage everyone involved to work together to:

➤ improve the planning and management of projects from the very start;
➤ identify risks early on so they can be eliminated or reduced at the design or planning stage and the remaining risks can be properly managed;
➤ target effort where it can do the most good in terms of health and safety;
➤ discourage unnecessary bureaucracy.

The regulations are intended to focus attention on planning and management throughout construction projects, from design concepts onwards. The aim is for health and safety considerations to be treated as an essential, but normal part of a projects development – not an afterthought or bolt-on extra.

The effort devoted to planning and managing health and safety should be in proportion to the risks and complexity associated with the project. The focus should always be on action necessary to reduce and manage risks. Any paperwork produced should help with communication and risk management. Paperwork which adds little to the management of risk is a waste of effort, and can be a dangerous distraction from the real business of risk reduction and management.

Time and thought invested at the start of the project will pay dividends not only in improved health and safety, but also in:

➤ reductions in the overall cost of ownership, because the structure is designed for safe and easy maintenance and cleaning work, and because key information is available in the health and safety file;
➤ reduced delays;
➤ more reliable costings and completion dates;
➤ improved communication and co-operation between key parties;
➤ improved quality of the finished product.

Note:
Typical operating and owning cost of a building are in the ratio:
➤ 1 for construction costs
➤ 5 for maintenance and building operating costs
➤ 200 for business operating costs

Source: Royal Academy of Engineering report on 'Long Term costs of Owning and Using Buildings'.

17.8.2 Outline of regulations, application and notification

The Regulations are divided into five parts. Part 1 of the Regulations deals with matters of interpretation and application. The Regulations apply to all construction work in Great Britain and its territorial waters, and apply to both employers and self-employed without distinction.

Part 2 covers general management duties which apply to all construction projects, including those which are non-notifiable.

Part 3 sets out additional management duties which apply to projects above the notification threshold (projects lasting more than 3 days, or involving more than 500 person days of construction work). These additional duties require particular appointments or particular documents which will assist with the management of health and safety from concept to completion.

Part 4 of the Regulations apply to all construction work carried out on construction sites, and covers physical safeguards which need to be provided to prevent danger. Duties to achieve these standards are held by contractors who actually carry out the work, irrespective of whether they are employers or are self-employed. Duties are also held by those who do not do construction work themselves, but control the way in which the work is done. In each case, the extent of the duty is in proportion to the degree of control which the individual or organization has over the work in question.

This does not mean that everyone involved with design, planning or management of the project legally must ensure that all the specific requirements in Part 4 are complied with. They only have duties if, in practice, they exercise sufficient control over the actual working methods, safeguards and site conditions. If, for example, a client specifies that a particular job is done in a particular way then the client will have a duty to make sure that their instructions comply with the requirements.

Contractors must not allow work to start or continue unless the necessary safeguards are in place, whether they are providing the safeguards, or using something, for example a scaffold supplied by someone else.

Part 5 of the Regulations covers issues of civil liability; transitional provisions which will apply during the period when the regulations come into force; and amendments and revocations of other legislation.

The distinction in the CDM94 Regulations between their application and notification of projects was confusing. For the purposes of CDM 2007 Regulations, there should only be two types of construction project: notifiable and non-notifiable. All of the requirements apply to notifiable projects, but the requirements relating to appointments, plans and other paperwork would not apply to non-notifiable projects.

Except where the project is for domestic clients, the HSE must be notified of projects if the construction phase is likely to involve more than (a) 30 days, or (b) 500 person days, of construction work. An F10 (Rev) may be used and is available from local HSE offices or can be completed online. This form does not have to be used as long as all the particulars shown in Schedule 1 (see Figure 17.4) are provided.

SCHEDULE 1
PARTICULARS TO BE NOTIFIED TO THE EXECUTIVE

1. Date of forwarding.
2. Exact address of the construction site.
3. The name of the local authority where the site is located.
4. A brief description of the project and the construction work which it includes.
5. Contact details of the client (name, address, telephone number and any e-mail address).
6. Contact details of the CDM co-ordinator (name, address, telephone number and any e-mail address).
7. Contact details of the principal contractor (name, address, telephone number and any e-mail address).
8. Date planned for the start of the construction phase.
9. The time allowed by the client to the principal contractor referred to in Regulation 15(b) for planning and preparation for construction work.
10. Planned duration of the construction phase.
11. Estimated maximum number of people at work on the construction site.
12. Planned number of contractors on the construction site.
13. Name and address of any contractor already appointed.
14. Name and address of any designer already engaged.
15. A declaration signed by or on behalf of the client that he is aware of his duties under these Regulations.

Figure 17.4 Schedule 1 of CDM 2007 – Particulars to be notified to the HSE.

17.8.3 Definition of Construction – Regulation 2

Regulation 2 defines construction work as the following; 'construction work' means the carrying out of any building, civil engineering or engineering construction work and includes:

(a) the construction, alteration, conversion, fitting-out, commissioning, renovation, repair, upkeep, redecoration or other maintenance (including cleaning which involves the use of water or an abrasive at high pressure or the use of corrosive or toxic substances), decommissioning, demolition or dismantling of a structure;

(b) the preparation for an intended structure, including site clearance, exploration, investigation (but not site survey) and excavation, and the clearance or preparation of the site or structure for use or occupation at its conclusion;

(c) the assembly on site of prefabricated elements to form a structure or the disassembly on site of prefabricated elements which, immediately before such disassembly, formed a structure;

(d) the removal of a structure or of any product or waste resulting from demolition or dismantling of a structure or from disassembly of prefabricated elements which immediately before such disassembly formed such a structure;

(e) the installation, commissioning, maintenance, repair or removal of mechanical, electrical, gas, compressed air, hydraulic, telecommunications, computer or similar services which are normally fixed within or to a structure.

This means that there are some things which may take place on a construction site, like tree planting and horticulture work, putting up marquees, positioning lightweight panels in an open plan office and surveying, which are not construction work.

Domestic clients are people who have work done on their own home or the home of a family member, that does not relate to a trade or business whether for profit or not. Where an insurance company arranges construction work to be done on a domestic dwelling, the insurance company becomes the client for the purposes of CDM. If, however, the insured arranges the work and is reimbursed by the insurance company then the insured is the client.

As with all projects, designers and contractors working for domestic clients will have to be competent and take reasonable steps to ensure that anyone they arrange for, or instruct, to manage, design or construct work is also competent. They will also have to co-operate with others involved in the project, to safeguard the health and safety of everyone involved. When preparing or modifying a design, designers will have to avoid risks to the health and safety of anyone constructing, maintaining, using or demolishing the structures concerned, by removing the hazards (and reducing the risks arising from any that remain). Contractors will have to plan, manage and monitor their own work and ensure that there are suitable welfare facilities.

17.8.4 General duties of clients, CDM co-ordinators, principal contractors and contractors – Part 2

The Regulations require that appointees are competent and state that:

No person on whom these Regulations place a duty shall:

(a) *appoint or engage a CDM co-ordinator, designer, principal contractor or contractor unless he has taken reasonable steps to ensure that the person to be appointed or engaged is competent*

(b) *accept such an appointment or engagement unless he is competent*

(c) *arrange for or instruct a worker to carry out or manage design or construction work unless the worker is:*
  (i) *competent*
  (ii) *under the supervision of a competent person.*

The ACOP goes into considerable detail about the assessment of organizational and individual competence. In both cases it lays down a two-stage approach.

➤ Stage 1 – assessing a company's organization and arrangements for health and safety and an individual's task knowledge to assess whether they appreciate the risks.

➤ Stage 2 – assessing the company's and individual's experience and track record to see if they are capable of doing the work involved.

To be competent, an organization or individual must have:

➤ sufficient knowledge of the specific tasks to be undertaken and the risks which the work will entail;

➤ sufficient experience and ability to carry out their duties in relation to the project; to recognize their limitations; and take appropriate action in order to prevent harm to those carrying out construction work, or those affected by the work.

The Regulations require all duty holders to co-operate and co-ordinate their work with one another to enable people to carry out their duties effectively.

Duties are placed on a number of people, which are summarized in Table 17.3.

All projects require:

➤ non-domestic clients to check the competence of all their appointees; ensure there are suitable management arrangements for the project; allow sufficient time and resources for all stages; provide pre-construction information to designers and contractors;

➤ designers to eliminate hazards and reduce risks during design; and provide information about remaining risks;

➤ contractors to plan, manage and monitor their own work and that of workers; check the competence of all their appointees and workers; train their own employees; provide information to their workers; comply with the requirements for health and safety on site detailed in Part 4 of the Regulations and other regulations such as the Work at Height Regulations; and ensure there are adequate welfare facilities for their workers;

➤ everyone to assure their own competence; co-operate with others and co-ordinate work so as to ensure the health and safety of construction workers and others who may be affected by the work; report obvious risks; take account of the general principles of prevention in planning or carrying out construction work; and comply with the requirements in Schedule 3, Part 4 of CDM 2007 and other regulations for any work under their control.

Table 17.3 Summary of duties under CDM 2007

	All construction projects (Part 2 of the Regulations)	Additional duties for notifiable projects (Part 3 of Regulations)

Clients (excluding domestic clients)	• Check competence and resources of all appointees
	• Ensure there are suitable management arrangements for the project including welfare facilities
	• Allow sufficient time and resources for all stages
	• Provide pre-construction information to designers and contractors
		• Appoint CDM co-ordinator[*]
		• Appoint principal contractor[*]
		• Provide pre-construction information to CDM co-ordinator
		• Make sure that the construction phase does not start unless there are suitable:
		  – welfare facilities; and
		  – construction phase plan in place
		• Provide information relating to the health and safety file to the CDM co-ordinator
		• Retain and provide access to the health and safety file

CDM co-ordinators		• Advise and assist the client with his/her duties • Notify HSE • Co-ordinate health and safety aspects of design work and co-operate with others involved in the project • Facilitate good communication between client, designers and contractors • Liaise with principal contractor regarding ongoing design • Identify, collect and pass on pre-construction information • Prepare/update health and safety file
Designers	• Eliminate hazards and reduce risks due to design • Provide information about remaining risks	• Check client is aware of duties and CDM co-ordinator has been appointed • Check HSE has been notified • Provide any information needed for the health and safety file
Principal contractors		• Plan, manage and monitor construction phase in liaison with contractors • Prepare, develop and implement a written plan and site rules (initial plan completed before the construction phase begins) • Give contractors relevant parts of the plan • Make sure suitable welfare facilities are provided from the start and maintained throughout the construction phase

(Continued)

Table 17.3 Continued

	All construction projects (Part 2 of the Regulations)	Additional duties for notifiable projects (Part 3 of Regulations)
		• Check competence of all their appointees • Ensure all workers have site inductions and any further information and training needed for the work • Consult with the workers • Liaise with CDM co-ordinator regarding ongoing design • Secure the site
Contractors	• Plan, manage and monitor own work and that of workers • Check competence of all their appointees and workers • Train own employees • Provide information to their workers • Comply with requirements in Part 4 of the Regulations • Ensure there are adequate welfare facilities for their workers	• Check client is aware of duties and a CDM co-ordinator has been appointed and HSE notified before starting work • Co-operate with principal contractor in planning and managing work, including reasonable directions and site rules • Provide details to the principal contractor of any contractor whom he engages in connection with carrying out the work • Provide any information needed for the health and safety file • Inform principal contractor of problems with the plan • Inform principal contractor of reportable accidents, diseases and dangerous occurrences
Everyone	• Check own competence • Co-operate with others and co-ordinate work so as to ensure the health and safety of construction workers and others who may be affected by the work • Report obvious risks • Comply with requirements in Schedule 2 (ACOP says 3 but believed a mistake) and Part 4 of the Regulations for any work under their control.	

Source: From L144 *Managing H&S in Construction – Construction (Design and Management) Regulations 2007 (CDM). Approved Code of Practice* (HSE Books, 2007) ISBN 9780717662234. © Crown copyright material is reproduced with the permission of the Controller of HMSO and Queen's Printer. *There must be a CDM co-ordinator and principal contractor until the end of the construction phase.

To ensure the revised Regulations are proportionate to risk and the needs of small businesses, and to minimize bureaucracy, the CDM94 requirement for appointment of a planning supervisor (PS) and principal contractor (PC) and written plans for projects involving five or more workers has been withdrawn. This does not mean any lessening in the health and safety standards required by the CDM 2007 Regulations, as they have strengthened or introduced other requirements. These place the emphasis on risk management, while avoiding disproportionate bureaucracy for smaller projects.

17.8.5 Additional duties where the project is notifiable

As well as the above requirements under Section 17.8.3, a notifiable project requires:

➤ non-domestic clients to appoint a CDM co-ordinator; appoint a PC; provide pre-construction information to CDM co-ordinator;
➤ check (before construction work starts) that there is a construction phase plan and suitable welfare facilities; and retain and provide access to the health and safety file;
➤ CDM co-ordinators to advise and assist clients with their duties; notify HSE; co-ordinate health and safety aspects of design work; provide pre-construction information to designers and contractors; facilitate good communications between client, designers and contractors; liaise with the PC on ongoing design issues; prepare and update the health and safety file;
➤ designers to check, before they start work, that clients are aware of their duties and a CDM co-ordinator has been appointed; check HSE has been notified; and provide any information needed for the health and safety file;
➤ PCs to plan, manage and monitor the construction phase in liaison with contractors; prepare, develop and implement a written plan (the initial plan to be completed before the construction phase begins); give contractors relevant parts of the plan; make sure suitable welfare facilities are provided from the start and maintained throughout the construction phase; check the competence of all their appointees; provide site inductions; consult with the workers; liaise with the CDM co-ordinator on ongoing design issues; and secure the site;
➤ contractors to confirm clients are aware of their duties and a CDM co-ordinator has been appointed; co-operate with the PC in planning and managing

work; check HSE has been notified, and provide any information needed for the health and safety file; inform PC of problems with plan; inform PC of any reportable accidents, diseases and dangerous occurrences.

17.8.6 Roles of duty holders under CDM 2007

(a) Clients

The role of clients is one of the most difficult areas to cover in law, because of the vast range of interest and expertise in construction, and clients have real questions as to why they should be involved. The HSC is very conscious of the substantial influence and control that clients exert over construction projects in practice. For example, they:

➤ set the tone for projects;
➤ control contractual arrangements;
➤ make crucial decisions (e.g. budget, time, suitability of designs); select procurement method and construction team/supply chain;
➤ may have essential information about site/building.

There is a new duty on clients to ensure that suitable project management arrangements for health and safety are in place.

Clients are not expected to develop these arrangements themselves and few have the expertise to do so. They should be able to rely on the advice and support of their construction team and, in particular, the CDM co-ordinator. What the HSC expects is for them to exercise their influence and control responsibly and with due regard for those who will construct, maintain and demolish the structure.

(b) CDM co-ordinator

There is widespread agreement that the role of PS, as developed under CDM94, has not proved as effective as intended. However, views on this tend to be highly polarized. The main problems are that PSs:

➤ are not seen as a natural part of the construction team. To be effective they need to be better integrated with the rest of the design and construction team;
➤ are often appointed too late in the project so that they cannot do their job;
➤ frequently have to operate at a disadvantage, due to insufficient allocation of resources by the client, in terms of both money and time;
➤ have no authority to carry out their duties unless the client effectively empowers them and others to co-operate;

➤ have, fairly or not, become the scapegoat for the bureaucracy linked to CDM.

To address these points, CDM 2007:

➤ creates a new function – the CDM co-ordinator – to advise and assist the client;

➤ places responsibility on clients to ensure that the co-ordinator's duties are carried out – only they have the information and authority to empower the CDM co-ordinator;

➤ explicitly requires the CDM co-ordinator to be appointed before design work starts, with corresponding duties on designers and contractors not to begin work unless a CDM co-ordinator has been appointed;

➤ requires the client to ensure that the arrangements for managing projects include the allocation of adequate resources (including time).

(c) Designers

The HSE fully recognizes that, as well as health and safety considerations, designers need to take into account issues such as aesthetics, buildability and cost. The challenge is to ensure that health and safety considerations are not outweighed by aesthetic and commercial priorities and, conversely, that health and safety does not inhibit aesthetics. However, it is a truth, almost universally acknowledged, that designers have considerable potential to eliminate hazards and reduce risks associated with construction work, as well as those associated with building use, maintenance, cleaning and eventual demolition.

As part of balancing their design priorities, designers must take positive steps to use that potential and pay sufficient regard to health and safety in their designs to ensure that in the construction, use, maintenance and demolition of the resulting structures, hazards are removed **where possible** and any remaining risks reduced. Although this is already stated in the draft guidance, the HSE has explicitly acknowledged the need for such balanced decisions in CDM 2007.

The HSE also wants designers to focus on how their decisions are likely to affect those constructing, maintaining, using or demolishing the structures that they have designed and what they can do, **in the design**, to remove the hazards, for example by not specifying hazardous materials and avoiding the need for processes that create hazardous fumes, vapours, dust, noise or vibration, and reduce the resulting risks where the hazard cannot be removed. The ACOP under CDM94 and guidance already sets out most of this and there is no plan to change these standards.

CDM 2007 is intended to require designers to eliminate hazards where they can, and then reduce those risks which remain. It does not ask designers to minimize all risks, as the HSC does not expect structures to be restricted to a height of 1 m. Also, there are often too many variables and no obvious *safest* design.

The duty regarding maintenance is limited in CDM94 to structural matters, but it is important that designers also consider safety during routine maintenance that is affected by their designs – for example, how are high-level lights and ventilation systems to be maintained?

Designers had no duty under CDM94 to ensure that their designs are safe to use. However, occupiers of workplaces have to ensure that the finished structure complies with other health and safety law, particularly the Workplace Regulations. To ensure that these issues are addressed at the design stage, the CDM 2007 extends designers' duties for fixed workplaces (e.g. offices, shops, schools, hospitals and factories) to cover safe use. Competent designers should be doing this already – so this is likely to require minimal additional work in practice.

(d) Principal contractor

The role of PC, introduced when CDM94 came into force, was built on the longstanding role of main or managing contractor and did not, therefore, require any substantial changes in industry practice. Because of this, as a role it has worked fairly well since CDM94 came into force, and has not changed significantly in CDM 2007.

The only substantial change is to make explicit, in the Regulations, the PC's key role in managing the construction phase, to ensure that it is carried out, sfarp, safely and without risk to health. This does not mean that the PC has to manage the work of contractors in detail – that is the contractors' own responsibility. They do have to make sure that they themselves are competent to address the health and safety issues likely to be involved in the management of the construction phase; satisfy themselves that the designers and contractors that they engage are competent and adequately resourced; and ensure that the construction phase is properly planned, managed and monitored, with adequately resourced, competent site management appropriate to the risk and activity.

The HSE did not feel it was necessary, legally or otherwise, to specify in CDM 2007 that the PC must be a contractor. In over 90% of projects, contractors discharge the role of PC and those with contractor's experience and expertise are most likely to have the competence and resources to manage the work. The HSE believed that few clients have the competence or resources to manage significant construction work and does not want to encourage them to do so, although there is nothing to prevent this if they are competent – which is most likely in simple, low-risk projects.

(e) Contractors

The only substantive change regarding contractors is to make explicit their own duty to plan, manage and monitor their own work. The intention is that the management duties on PCs and contractors should complement one another, with the contractor's duty focusing on their own work and the PC's on the co-ordination of the work of the various contractors.

17.8.7 Health and safety file

Under the CDM94, a separate health and safety file is required for each project. The HSE believes that it would be more useful to have one file for each site, structure or, occasionally, group of structures – for example, bridges along a road. The file can then be developed over time as information is added from different projects. The CDM co-ordinator therefore has the responsibility to prepare a suitable health and safety file or update it if one already exists.

There are also opportunities to link the health and safety file with other documents such as the Buildings Regulations Log Book. The potential practical value of the information contained in the file is also likely to increase as more clients make use of the Internet to share this information with designers and contractors (for example, maintenance contractors could check what access equipment and parts they are likely to need to repair a fault before leaving for the site, saving them and their clients inconvenience, time and money).

The contents of the Health and Safety File are given in the ACOP as follows:

➤ a brief description of the work carried out;
➤ any residual hazards which remain and how they have been dealt with (e.g. surveys or other information concerning asbestos, contaminated land, water-bearing strata and buried services);
➤ key structural principles (e.g. bracing, sources of substantial stored energy – including pre- or post-tensioned members) and safe working loads (SWL) for floors and roofs, particularly where these may preclude placing scaffolding or heavy machinery there;
➤ hazardous material used (e.g. lead paint, pesticides and special coatings which should not be burnt off);
➤ information regarding the removal or dismantling of installed plant and equipment (e.g. any special

arrangements for lifting, order of or other special instructions for dismantling);
➤ health and safety information about equipment provided for cleaning or maintaining the structure;
➤ the nature, location and markings of significant services, including underground cables, gas supply equipment and firefighting services;
➤ information and as-built drawings of the structure, its plant and equipment (e.g. the means of safe access to and from service voids, fire doors and compartmentalization).

17.8.8 Part 4 – health and safety on construction sites

The CHSW Regulations 1996 have been revoked by CDM 2007. Without the work at height provisions (Regulations 6, 7 and 8), their requirements form the basis of Part 4 and Schedule 2 (Welfare facilities) of the CDM 2007 Regulations. The revision is mainly intended to simplify and clarify the wording of the Regulations, without making substantive changes to what is expected in practice. One substantive change, however, has been to broaden the duty regarding explosives to cover the important issues of security and safety of storage and transport, as well as safety in use.

However, some issues already covered by the Workplace Regulations are particularly important in construction. These include traffic management and lighting. Because these are so important, CDM 2007 duplicates aspects of the Workplace Regulations requirements, although that was not legally necessary.

The issues covered by Part 4 apply to all construction projects and are as follows.

(a) General – Regulation 25

Duties to comply with this part are placed on contractor or other person who controls the way in which construction work is carried on.

Duty placed on workers to report any defect they are aware of which may endanger the health and safety of themselves or another person.

(b) Safe place of work – Regulation 26

This regulation requires sfarp:

➤ safe access and egress to places of work;
➤ maintenance of access and egress;
➤ that people are prevented from gaining access to unsafe access or workplaces;
➤ provision of safe places of work with adequate space and suitable for workers.

(c) Good order and site security – Regulation 27

➤ Requires that all parts of a construction be kept, sfarp, in good order and in a reasonable state of cleanliness.

➤ Where necessary in the interests of health and safety the site's perimeter should be identified, signed and fenced off.

➤ No timber or other materials with projecting nails or similar sharp objects shall be used or allowed to remain if they could be a source of danger.

(d) Stability of structures – Regulation 28

All practical steps must be taken to ensure that:

➤ any new or existing structure which may be, or become, weak does not collapse accidentally;

➤ temporary supporting structures are designed, installed and maintained so as to withstand all foreseeable loads;

➤ a structure is not overloaded so as to be a source of danger.

(e) Demolition or dismantling – Regulation 29

➤ Demolition or dismantling must be planned and carried out so as to prevent danger or reduce it to as low a level as is reasonably possible.

➤ Arrangements must be recorded before the work takes place.

(f) Explosives – Regulation 30

➤ Explosives, sfarp, shall be stored, transported and used safely and securely.

➤ Explosives may only be fired if suitable and sufficient steps have been taken to ensure no one is exposed to risk of injury.

(g) Excavations – Regulation 31

All practicable steps shall be taken to prevent danger to people including, where necessary, supports or battering to ensure that:

➤ excavations do not collapse accidentally;

➤ no material from side, roof or adjacent is dislodged or falls;

➤ a person is prevented from being buried or trapped by a fall of material;

➤ persons, work equipment or any accumulation of material are prevented from falling into the excavation;

➤ any part of an excavation or ground adjacent is not overloaded by work equipment or material.

No construction work may be carried out in an excavation where any supports or battering has been provided unless:

➤ it has been inspected by a competent person:
(i) at the start of the shift (only one report required every 7 days)
(ii) after any event likely to affect the strength or stability of the excavation
(iii) after any material accidentally falls or is dislodged
(iv) the person inspecting is satisfied that it is safe.

Where the person who carries out the inspection has informed the person on whose behalf the inspection was carried out, work must cease until the relevant matters have been corrected.

(h) Cofferdams and caissons – Regulation 32

Every cofferdam or caisson must be:

➤ suitably designed and constructed;

➤ appropriately equipped so that people can gain shelter or escape if water or material enters;

➤ properly maintained.

A cofferdam or caisson may be used to carry out construction only if:

➤ it and any work equipment and materials which affect its safety have been inspected by a competent person:
• at the start of the shift (only one report required every 7 days)
• after any event likely to affect its strength or stability

➤ the person inspecting is satisfied that it is safe.

Where the person who carries out the inspection has informed the person on whose behalf the inspection was carried out, work must cease until the relevant matters have been corrected.

(i) Reports of inspection – Regulation 33

Before the end of the shift in which the report was completed, the person who carries out the inspection under Regulations 31 or 32 must:

➤ inform the person for whom they carried out the inspection if they are not satisfied that construction work can be carried out safely;

➤ prepare a report with the particulars set out in Schedule 3 to the Regulations as shown in Figure 17.5.

A copy of the report must be provided within 24 hours of the relevant inspection. If the inspector is an employee or works under someone else's control, that person must ensure that these duties are performed.

SCHEDULE 3 Regulation 33(1)(b)

PARTICULARS TO BE INCLUDED IN A REPORT OF INSPECTION

1. Name and address of the person on whose behalf the inspection was carried out.
2. Location of the place of work inspected.
3. Description of the place of work or part of that place inspected (including any work equipment and materials).
4. Date and time of the inspection.
5. Details of any matter identified that could give rise to a risk to the health or safety of any person.
6. Details of any action taken as a result of any matter identified in paragraph 5 above.
7. Details of any further action considered necessary.
8. Name and position of the person making the report.

Figure 17.5 Schedule 3 Regulation 33(1)(b).

Records must be kept at the site where the inspection was carried out until the work is complete and then for 3 months. Extracts must be provided for an inspector as and when they require.

(j) Energy distribution installations – Regulation 34

Energy distribution installations must be suitably located, checked and clearly indicated.

Where there is a risk from electric power cables:

➤ they should be routed away from the risk areas, or
➤ the power should be cut off, or
➤ if the above safety measures are not reasonably practicable suitable warning notices, with barriers to exclude work equipment and suspended protections where vehicles have to pass underneath, should be provided, or
➤ something equally as safe must be done.

No construction work which is liable to create a risk to health or safety from an underground service, or from damage to or disturbance of it, shall be carried out unless suitable and sufficient steps have been taken to prevent such risk, sfarp.

(k) Prevention of drowning – Regulation 14

Where persons could fall into water or other liquid, steps must be taken to:

➤ prevent, sfarp, people from falling;
➤ minimize the risk of drowning if people fall;
➤ provide, use and maintain suitable rescue equipment to ensure prompt rescue.

Transport to work by water must be suitable and sufficient and any vessel must not be overcrowded or overloaded.

(l) Traffic routes – Regulation 15

Every construction site shall be organized in such a way that, sfarp, pedestrians and vehicles can move safely.

Traffic routes shall be suitable for the persons or vehicles using them, sufficient in number, in suitable positions and of sufficient size.

Steps must be taken to ensure that:

➤ pedestrians or vehicles may use them without causing danger to the health or safety of persons near them;
➤ any door or gate for pedestrians which leads onto a traffic route is sufficiently separated from it to enable them from a place of safety to see any approaching vehicle;
➤ there is sufficient separation between vehicles and pedestrians to ensure safety or, where this is not reasonably practicable:
  (i) there are provided other means for the protection of pedestrians
  (ii) there are effective arrangements for warning any person liable to be crushed or trapped by any vehicle of its approach;
➤ any loading bay has at least one exit point for the exclusive use of pedestrians;
➤ where it is unsafe for pedestrians to use a gate intended primarily for vehicles, one or more doors for pedestrians provided in the immediate vicinity of the gate, is clearly marked and is kept free from obstruction.

Every traffic route shall be:

➤ indicated by suitable signs;
➤ regularly checked;
➤ properly maintained.

No vehicle shall be driven on a traffic route unless, sfarp, that traffic route is free from obstruction and permits sufficient clearance.

(m) Vehicles – Regulation 37

The unintended movement of any vehicle must be prevented or controlled.

Suitable and sufficient steps shall be taken to ensure that, where a person may be endangered by the movement of any vehicle, the person having effective control of the vehicle shall give suitable and sufficient warning.

Any vehicle being used for the purposes of construction work, shall, when being driven, operated or towed:

➤ be driven, operated or towed in a safe manner;
➤ be loaded so that it can be driven, operated or towed safely.

No person shall ride on any vehicle being used for the purposes of construction work except in a safe place provided.

No person shall remain on any vehicle during the loading or unloading of any loose material unless a safe place of work is provided and maintained.

Suitable measures must be taken to prevent any vehicle from falling into any excavation or pit, or into water, or overrunning the edge of any embankment or earthwork.

(n) Prevention of risk from fire – Regulation 38

Steps must be taken to prevent injury arising from:

➤ fires or explosions;
➤ flooding;
➤ substances liable to cause asphyxiation.

(o) Emergency procedures – Regulation 39

Where necessary, emergency procedures must be prepared and, where necessary, implemented to deal with any foreseeable emergency. They must include procedures for any necessary evacuation of the site or any part thereof.

The arrangements must take account of:

(a) the type of work for which the construction site is being used;
(b) the characteristics and size of the construction site and the number and location of places of work on that site;
(c) the work equipment being used;
(d) the number of persons likely to be present on the site at any one time;
(e) the physical and chemical properties of any substances or materials on or likely to be on the site.

Steps must be taken to make people on site familiar with the procedures and test the procedures at suitable intervals.

(p) Emergency routes and exits – Regulation 40

A sufficient number of suitable emergency routes and exits shall be provided to enable any person to reach a place of safety quickly in the event of danger.

An emergency route or exit (or, as appropriate, traffic route) provided must:

➤ lead as directly as possible to an identified safe area;
➤ be kept clear and free from obstruction and, where necessary, provided with emergency lighting so that such emergency route or exit may be used at any time;
➤ be indicated by suitable signs.

(q) Fire detection and firefighting – Regulation 41

Where necessary, duty holders must provide suitable and sufficient:

➤ firefighting equipment, and
➤ fire detection and alarm systems

which shall be suitably located.

The equipment must be examined and tested at suitable intervals, properly maintained and suitably indicated by signs. If non-automatic, it must be easily accessible.

Every person at work on a construction site shall, sfarp, be instructed in the correct use of any firefighting equipment which it may be necessary for them to use. Where a work activity may give rise to a particular risk of fire, a person shall not carry out such work unless suitably instructed.

(r) Fresh air – Regulation 42

Steps must be taken to ensure, sfarp, that every place of work or approach has sufficient fresh or purified air to ensure that the place or approach is safe and without risks to health. Any plant used for the provision of fresh air must include, where necessary, an effective device to give visible or audible warning of any failure of the plant.

(s) Temperature and weather protection – Regulation 43

Steps must be taken to ensure, that sfarp, during working hours the temperature at any place of work indoors is reasonable having regard to its purpose.

Every place of work outdoors must, where necessary, sfarp, and having regard to the purpose for which that place is used and any protective clothing or work equipment provided, be so arranged that it provides protection from adverse weather.

(t) Lighting – Regulation 44

Every place of work and approach thereto and every traffic route shall be provided with suitable and sufficient lighting, which shall be, sfarp, by natural light.

The colour of any artificial lighting provided shall not adversely affect or change the perception of any sign or signal provided for the purposes of health and safety.

Suitable and sufficient secondary lighting shall be provided in any place where there would be a risk to the health or safety of any person in the event of failure of primary artificial lighting.

17.8.9 Welfare facilities – Schedule 2

(a) Sanitary conveniences

1. Suitable and sufficient sanitary conveniences shall be provided or made available at readily accessible places. So far as is reasonably practicable, rooms containing sanitary conveniences shall be adequately ventilated and lit.

2. So far as is reasonably practicable, sanitary conveniences and the rooms containing them shall be kept in a clean and orderly condition.

3. Separate rooms containing sanitary conveniences shall be provided for men and women, except where and so far as each convenience is in a separate room the door of which is capable of being secured from the inside.

(b) Washing facilities

4. Suitable and sufficient washing facilities, including showers if required by the nature of the work or for health reasons, shall sfarp be provided or made available at readily accessible places.

5. Washing facilities shall be provided:
(a) in the immediate vicinity of every sanitary convenience, whether or not provided elsewhere;
(b) in the vicinity of any changing rooms required by paragraph 15 whether or not provided elsewhere.

6. Washing facilities shall include:
(a) a supply of clean hot and cold, or warm, water (which shall be running water sfarp);
(b) soap or other suitable means of cleaning;
(c) towels or other suitable means of drying.

7. Rooms containing washing facilities shall be sufficiently ventilated and lit.

8. Washing facilities and the rooms containing them shall be kept in a clean and orderly condition.

9. Subject to paragraph 10 below, separate washing facilities shall be provided for men and women, except where and so far as they are provided in a room the door of which is capable of being secured from inside and the facilities in each such room are intended to be used by only one person at a time.

10. Paragraph 9 above shall not apply to facilities which are provided for washing hands, forearms and face only.

(c) Drinking water

11. An adequate supply of wholesome drinking water shall be provided or made available at readily accessible and suitable places.

12. Every supply of drinking water shall be conspicuously marked by an appropriate sign where necessary for reasons of health and safety.

13. Where a supply of drinking water is provided, there shall also be provided a sufficient number of suitable cups or other drinking vessels unless the supply of drinking water is in a jet from which persons can drink easily.

(d) Changing rooms and lockers

14. (1) Suitable and sufficient changing rooms shall be provided or made available at readily accessible places if:
(a) a worker has to wear special clothing for the purposes of their work;
(b) they cannot, for reasons of health or propriety, be expected to change elsewhere being separate rooms for, or separate use of rooms by, men and women where necessary for reasons of propriety.

14. (2) Changing rooms shall:
(a) be provided with seating;
(b) include, where necessary, facilities to enable a person to dry any such special clothing and their own clothing and personal effects.

14. (3) Suitable and sufficient facilities shall, where necessary, be provided or made available at readily accessible places to enable persons to lock away:
(a) any such special clothing which is not taken home;
(b) their own clothing which is not worn during working hours;
(c) their personal effects.

(e) Facilities for rest

15. (1) Suitable and sufficient rest rooms or rest areas shall be provided or made available at readily accessible places.

15. (2) Rest rooms and rest areas shall:
(a) include suitable arrangements to protect non-smokers from discomfort caused by tobacco smoke;
(b) be equipped with an adequate number of tables and adequate seating with backs for the number of persons at work likely to use them at any one time;
(c) where necessary, include suitable facilities for any person at work who is a pregnant woman or nursing mother to rest lying down;
(d) include suitable arrangements to ensure that meals can be prepared and eaten;
(e) include the means for boiling water;
(f) be maintained at an appropriate temperature.

17.8.10 Civil liability – Regulation 45

The Management Regulations (MHSWR 1999) were amended in 2003 to provide employees with a right of action in civil proceedings, in relation to breach of duties by their employer. To maintain consistency, CDM 2007 carries forward with three small exceptions, the rights of civil action in CDM and CHSW and also allows employees to take action in the civil courts for injuries resulting from a failure to comply with duties under these Regulations.

17.8.11 Enforcement

General

The enforcement demarcation between HSE and LAs in respect of construction work is currently set out in the Health and Safety (Enforcing Authority) Regulations 1998, and in Regulations 3(4) and 22 of CDM94. Interpretation of these requirements is not straightforward, but the practical effect is that LAs were prevented from enforcing CDM.

The HSE wished to simplify this by omitting Regulations 3(4) and 22 from CDM 2007.

The effect of this that under CDM 2007 HSE would then be the enforcing authority for:

➤ all notifiable construction work, except that undertaken by people in LA-enforced premises who normally work on the premises;
➤ work done to the exterior of the premises;
➤ work done in segregated areas.

Enforcement in respect of fire – Regulation 46

This regulation of CDM 2007 is given in full.

(1) Subject to paragraphs (2) and (3):
(a) in England and Wales the enforcing authority within the meaning of article 25 of the Regulatory Reform (Fire Safety) Order 2005.
(b) in Scotland the enforcing authority within the meaning of section 61 of the Fire (Scotland) Act 2005, shall be the enforcing authority in respect of a construction site which is contained within, or forms part of, premises which are occupied by persons other than those carrying out the construction work or any activity arising from such work as regards Regulations 39 and 40, in so far as those regulations relate to fire, and Regulation 41.
(2) In England and Wales, paragraph (1) only applies in respect of premises to which the Regulatory Reform (Fire Safety) Order 2005 applies.
(3) In Scotland, paragraph (1) only applies in respect of premises to which Part 3 of the Fire (Scotland) Act 2005 applies.

17.8.12 Sources of references

Backs for the Future HSG 149, HSE Books, ISBN 978 0 7176 1122 5.

Construction (Design and Management) Regulations 2007, Approved Code of Practice L144, 2007, HSE Books, ISBN 978 0 7176 6223 4.
Electrical Safety on Construction Sites HSG141, 1995, HSE Books, ISBN 978 0 7176 1000 6.
Fire Safety in Construction Work HSG 168, 1997, HSE Books, ISBN 0 7176 1332 1 (being revised).
Health and Safety in Construction HSG150 (rev 3), 2006, HSE Books, ISBN 9780 7176 6182 4 (due for revision).
Health and Safety in Excavations HSG 185, 1999, HSE Books, ISBN 9780 0 7176 1563 6.
Health and Safety in Roof Work, revised HSG33, 1999, HSE Books, ISBN 0 7176 1425 5 (being revised).
Safe Use of Vehicles on Construction Sites HSG 144, 1998, HSE Books, ISBN 0 7176 1610 X (being revised).

17.9 Construction (Head Protection) Regulations 1989

17.9.1 Application

These Regulations apply to 'building operations' and 'works of engineering construction' as defined originally by the Factories Act 1961.

17.9.2 Provision and maintenance

Every employer must provide suitable head protection for each employee and shall maintain it and replace it whenever necessary. A similar duty is placed on self-employed people. The employer must make an assessment and review it as necessary, to determine whether the head protection is suitable. Accommodation must be provided for head protection when it is not in use.

17.9.3 Ensuring that head protection is worn – Regulation 4

Every employer shall ensure, sfarp, that each employee (and any other person over whom they have control) at work wears suitable head protection, unless there is no foreseeable risk of injury to their head other than through falling.

17.9.4 Rules and directions – Regulation 5

The person in control of a site **may** make rules regulating the wearing of suitable head protection. These must be in writing. An employer may give directions requiring their employees to wear head protection.

17.9.5 Wearing of suitable head protection, reporting loss – Regulations 6, 7 and 9

Every employee who has been provided with suitable head protection shall wear it when required to do so by rules made or directions given under Regulation 5 of these Regulations. They must make full and proper use of it and return it to the accommodation.

Employees must take reasonable care of head protection, and report any loss or obvious defect.

Exemption certificates, issued by the HSE, are allowed.

17.9.6 Sources of references

Construction (Head Protection) Regulations 1989 Guidance on Regulations L 102 (rev), 1998, HSE Books, ISBN 9780 0 7176 1478 3.
Head Protection for Sikhs Wearing Turbans INDG262.

17.10 Health and Safety (Consultation with Employees) Regulations 1996

17.10.1 Application

These Regulations apply to all employers and employees in Great Britain except:

➤ where employees are covered by safety representatives appointed by recognized trade unions under the Safety Representatives and Safety Committees Regulations 1977 [note in workplaces where there is trade union recognition but either the trade union has not appointed a safety representative, or the union safety representative does not cover the whole workforce, the Health and Safety (Consultation with Employees) Regulations will apply];
➤ domestic staff employed in private households;
➤ crew of a ship under the direction of the master.

17.10.2 Employers' duty to consult – Regulation 3

The employer must consult relevant employees in good time with regard to:

➤ the introduction of any measure which may substantially affect their health and safety;

➤ the employer's arrangements for appointing or nominating competent persons under the Management of Health and Safety at Work Regulations 1999;
➤ any information required to be provided by legislation;
➤ the planning and organization of any health and safety training required by legislation;
➤ the health and safety consequences to employees of the introduction of new technologies into the workplace.

The guidance emphasizes the difference between informing and consulting. Consultation involves listening to employees' views and taking account of what they say before any decision is taken.

17.10.3 Persons to be consulted – Regulation 4

Employers must consult with either:

➤ the employees directly; or
➤ one or more persons from a group of employees, who were elected by employees in that group to represent them under these regulations. They are known as 'Representative(s) of Employee Safety' (RES).

Where RES are consulted, all employees represented must be informed of:

➤ the names of RES;
➤ the group of employees represented.

Employers shall not consult an RES if:

➤ the RES does not wish to represent the group;
➤ the RES has ceased to be employed in that group;
➤ the election period has expired;
➤ the RES has become incapacitated from carrying out the necessary functions.

If an employer decides to consult directly with employees, they must inform them and RES of that fact. Where no RES is elected, employers will have to consult directly.

17.10.4 Duty to provide information – Regulation 5

Employers must provide enough information to enable RES to participate fully and carry out their functions. This will include:

➤ what the likely risks and hazards arising from their work may be;

➤ reported accidents and diseases etc. under the Reporting of Injuries, Diseases and Dangerous Occurrences Regulations (RIDDOR) 95;

➤ the measures in place, or which will be introduced, to eliminate or reduce the risks;

➤ what employees ought to do when encountering risks and hazards.

An employer need not disclose information which:

➤ could endanger national security;

➤ violates a legal prohibition;

➤ relates specifically to an individual without their consent;

➤ could substantially hurt the employer's undertaking or infringe commercial security;

➤ was obtained in connection with legal proceedings.

17.10.5 Functions of RES – Regulation 6

ROES have the following functions (but no legal duties):

➤ to make representations to the employer on potential hazards and dangerous occurrences related to the group of employees represented;

➤ to make representations to the employer on matters affecting the general health and safety of relevant employees;

➤ to represent the group of employees for which they are the RES in consultation with inspectors.

17.10.6 Training, time off and facilities – Regulation 7

Where an employer consults RES, they shall:

➤ ensure that each RES receives reasonable training at the employer's expense;

➤ allow time off with pay during RES working hours to perform the duties of a RES and while the RES is a candidate for election;

➤ provide such other facilities and assistance that RES may reasonably require.

17.10.7 Civil liability and complaints – Regulation 9 and Schedule 2

Breach of these Regulations d oes not confer any right of action in civil proceedings subject to Regulation 7(3) and Schedule 2 relating to complaints to industrial tribunals.

A RES can complain to an industrial tribunal that:

➤ their employer has failed to permit time off for training or to be a candidate for election;

➤ their employer has failed to pay them as set out in Schedule 1 to the Regulations.

17.10.8 Elections

The guidance lays down some ideas for the elections, although there are no strict rules.

RES do not need to be confined to consultation related to these Regulations. Some employers have RES sitting on safety committees and taking part in accident investigation similarly to union Safety Representatives.

17.10.9 Sources of references

A Guide to the Health and Safety (Consultation with Employees) Regulations 1996. L95, HSE Books, ISBN 978 0 7176 1234 5.

Consulting Workers on Health and Safety. Safety Representatives and Safety Committees Regulations 1977 (as amended) and Health and Safety (Consultation with Employees) Regulations 1996 (As amended). Approved Code of Practice and Guidance L146, HSE Books, 2008, ISBN 978 0 7176 6311 8.

Consulting Employees on Health and Safety. A Brief Guide to the Law INDG 232 (rev 1), HSE Books, ISBN 978 0 7176 6312 5.

Involving Your Workforce in Health and Safety: Good Practice for All Workplaces HSG263, HSE Books, 2008, ISBN 978 0 7176 6227 2.

Involving Your Workers in Health and Safety: A Guide for Small Businesses. Available on HSE's website www.hse.gov.uk/pubns/web35.pdf

17.11 Control of Substances Hazardous to Health Regulations (COSHH) 2002 and 2005 Amendment

17.11.1 Introduction

The 2002 COSHH Regulations have updated the 1999 Regulations with a few changes which include: additional definitions like 'inhalable dust' and 'health surveillance'; clarify and extend the steps required under risk assessment; introduce a duty to deal with accidents and emergencies.

The 2005 Amendment Regulations replace Regulation 7(7) and (8) by substituting new requirements to observe principles of good practice for the control of exposure to substances hazardous to health introduced by Schedule 2A, to ensure that workplace exposure limits are not exceeded, and to ensure in respect of carcinogens and asthmagens that exposure is reduced to as low a level as is reasonably practicable. They also introduce a single new Workplace Exposure Limit for substances hazardous to health which replaces occupational exposure standards and maximum exposure limits. The amendment regulations introduce a duty to review control measures other than the provision of plant and equipment, including systems of work and supervision, at suitable intervals.

COSHH covers most substances hazardous to health found in workplaces of all types. The substances covered by COSHH include:

➤ substances used directly in work activities (e.g. solvents, paints, adhesives and cleaners);
➤ substances generated during processes or work activities (e.g. dust from sanding and fumes from welding);
➤ naturally occurring substances (e.g. grain dust).

But COSHH does **not** include:

➤ asbestos and lead, which have specific regulations;
➤ substances which are hazardous only because they:

➤ are radioactive;
➤ are simple asphyxiants;
➤ are at high pressure;
➤ are at extreme temperatures;
➤ have explosive or flammable properties (separate Dangerous Substances and Explosive Atmospheres Regulations, DSEAR, cover these);
➤ are biological agents if they are not directly connected with work and are not in the employer's control, such as catching flu from a workmate.

17.11.2 Definition of substance hazardous to health – Regulation 2

The range of substances regarded as hazardous under COSHH are:

➤ substances or mixtures of substances classified as dangerous to health under the Chemicals (Hazard, Information and Packaging for Supply) Regulations 2002 (CHIP). These have COSHH warning labels and manufacturers must supply data sheets. They cover substances that are: very toxic, toxic, harmful, corrosive or irritant under CHIP;
➤ substances with a workplace exposure limit as listed in EH40 published by the HSE;
➤ biological agents (bacteria and other microorganisms) if they are directly connected with the work;

➤ any kind of dust in a concentration specified in the regulations, that is
  (i) $10\,mg/m^3$, as a time-weighted average over an 8-hour period, of total inhalable dust
  (ii) $4\,mg/m^3$, as a time-weighted average over an 8-hour period, of respirable dust;
➤ any other substance which has comparable hazards to people's health but which, for technical reasons, is not covered by CHIP.

Workplace exposure limit for a substance hazardous to health means the exposure limit approved by the HSC for that substance in relation to the specified reference period when calculated by a method approved by the HSC, as contained in HSE publication 'EH/40 Workplace Exposure Limits 2005' as updated from time to time.

17.11.3 Duties under COSHH – Regulation 3

The duties placed on employers under these Regulations are extended to (except for health surveillance, monitoring and information and training) persons who may be on the premises but not employed whether they are at work or not, for example visitors and contractors.

17.11.4 General requirements

There are eight basic steps to comply with COSHH. They are:

1. Assess the risks to health;
2. Decide what precautions are needed;
3. Prevent or adequately control exposure;
4. Ensure that control measures are used and maintained;
5. Monitor the exposure of employees to hazardous substances;
6. Carry out appropriate health surveillance where necessary;
7. Prepare plans and procedures to deal with accidents, incidents and emergencies;
8. Ensure employees are properly informed, trained and supervised.

17.11.5 Steps 1 and 2: assessment of health risk – Regulation 6

No work may be carried where employees are liable to be exposed to substances hazardous to health unless:

➤ a suitable and sufficient risk assessment, including the steps need to meet COSHH, has been made.

The assessment must be reviewed and changes made regularly and immediately if:

➤ it is suspected that it is no longer valid;
➤ there has been a significant change in the work to which it relates;
➤ the results of any monitoring carried out in accordance with Regulation 10 shows it to be necessary.

The assessment involves identifying the substance present in the workplace; assessing the risk it pre-sents in the way it is used; and deciding what precautions (and health surveillance) are needed. Where more than four employees are employed, the significant findings must be recorded and the steps taken to comply with Regulation 7.

17.11.6 Step 3: prevention or control of exposure – Regulation 7

Every employer must ensure that exposure to substances hazardous to health is either:

➤ prevented or, where this is not reasonably practicable
➤ adequately controlled.

Preference must be given to substituting with a safer substance. Where it is not reasonably practicable to prevent exposure, protection measures must be adopted in the following order of priority.

➤ the design and use of appropriate work processes, systems and engineering controls and the provision and use of suitable work equipment;

➤ the control of exposure at source, including adequate ventilation systems and appropriate organizational measures;
➤ where adequate control of exposure cannot be achieved by other means, the additional provision of suitable PPE.

The measures must include:

➤ safe handling, storage and transport (plus waste);
➤ reducing to a minimum required for the work the number of employees exposed, the level and duration of exposure, and the quantity of substance present at the workplace;
➤ the control of the working environment including general ventilation;
➤ appropriate hygiene measures.

Control of exposure shall only be treated as adequate if:

➤ the principles of good practice as set out in Schedule 2A are applied;
➤ any workplace exposure limit approved for that substance is not exceeded;
➤ for a carcinogen (in Schedule 1 or with risk phrase R45, R46 or R49) or sensitizer (Risk phrases R42 or R42/43) or asthmagen (section C of HSE publication on Asthmagens ISBN 0 7176 1465 4) exposure is reduced to as low a level as is reasonably practicable.

PPE must conform to the Personal Protective Equipment (EC Directive) Regulations 2002.

Additionally, respiratory protective equipment must:

➤ be suitable for the purpose;
➤ comply with the EC Directive Regulations; or

Principles of good practice for the control of exposure to substances hazardous to health (Schedule 2A)

➤ Design and operate processes and activities to minimize emission, release and spread of substances hazardous to health.
➤ Take into account all relevant routes of exposure – inhalation, skin absorption and ingestion – when developing control measures.
➤ Control exposure by measures that are proportionate to the health risk.
➤ Choose the most effective and reliable control options which minimize the escape and spread of substances hazardous to health.

➤ Where adequate control of exposure cannot be achieved by other means, provide, in combination with other control measures, suitable PPE.
➤ Check and review regularly all elements of control measures for their continuing effectiveness.
➤ Inform and train all employees on the hazards and risks from the substances with which they work and the use of control measures developed to minimize the risks.
➤ Ensure that the introduction of control measures does not increase the overall risk to health and safety.

➤ if there is no requirement imposed by these regulations, of a type that shall conform to a standard, approved by the HSE.

Where exposure to a carcinogen or biological agent is involved, additional precautions are laid down in Regulation 7.

17.11.7 Step 4 – use, maintenance, examination and test of control measures – Regulations 8 and 9

➤ Every employer must take all reasonable steps to ensure that control measures, PPE or anything else provided under COSHH, are properly used or applied.

➤ Employees must make full and proper use of control measures, PPE or anything else provided. Employees must, as far as is reasonable, return items to accommodation and report defects immediately.

Employers shall also:

➤ properly maintain and keep clean plant and equipment, other engineering controls and PPE;

➤ in the case of the provision of systems of work and supervision or similar, review at suitable intervals and revise as necessary;

➤ carry out thorough examination and tests on engineering controls:

  • in the case of local exhaust ventilation (LEV) at least once every 14 months (except those in Schedule 4 which covers blasting of castings – monthly; dry grinding, polishing or abrading of metal for more than 12 hours per week – 6 months; and jute manufacture – 6 months)

  • in any other case, at suitable intervals;

➤ carry out thorough examination and tests where appropriate (except on disposable items) on respiratory protective equipment at suitable intervals;

➤ keep a record of examination and tests for at least 5 years;

➤ check, properly store, repair/replace and safely dispose of contaminated PPE as necessary.

17.11.8 Step 5 – monitoring exposure – Regulation 10

Employers must monitor exposure:

➤ where this is necessary to ensure maintenance of control measures or the protection of health;

➤ specifically for vinyl chloride monomer or chromium plating as required by Schedule 5; and

➤ keep a record of identifiable personal exposures for 40 years and any other exposures for 5 years.

17.11.9 Step 6: health surveillance – Regulation 11

Employers must ensure employees, where appropriate, who are exposed or liable to be exposed, are under health surveillance. It is considered appropriate when:

➤ an employee is exposed (if significant) to substances or processes in Schedule 6;

➤ an identifiable disease or adverse health effect may be related to the exposure, and there is a reasonable likelihood that disease may occur and there are valid disease indication or effect detection methods.

Records of health surveillance containing approved particulars must be kept for 40 years or offered to HSE if trading ceases. If a person is exposed to a substance and/or process in Schedule 6, the health surveillance shall include medical surveillance by an employment medical adviser or appointed doctor at intervals not exceeding 12 months.

If a medical adviser certifies that a person should not be engaged in particular work they must not be permitted to carry out that work except under specified conditions. Health records must be available for the individual employee to see after reasonable notice.

Employees must present themselves, during working hours at the employer's expense, for appropriate health surveillance.

An employment medical adviser or appointed doctor has the power to inspect the workplace or look at records for the purpose of carrying out functions under COSHH.

17.11.10 Step 7: accidents and emergencies – Regulation 13

Prepare plans and procedures to deal with accidents, incidents and emergencies involving hazardous substances beyond the risks of day-to-day use. This involves:

➤ first-aid provision and safety drills;

➤ information on emergency arrangements and specific hazards;

➤ suitable warning and other communications to enable appropriate response;

➤ ensuring suitable PPE and other equipment is available;

➤ making sure that the procedure is put into practice in an incident.

17.11.11   Step 8: information, instruction and training – Regulation 12

Where employees are likely to be exposed to substances hazardous to health, employers must provide:

➤ information, instruction and training on the risks to health and the precautions which should be taken (this duty is extended to anyone who may be affected);

➤ information on any monitoring of exposure (particularly if there is a workplace exposure limit where the employee or their representative must be informed immediately);

➤ information on collective results of health surveillance (designed so that individuals cannot be identified).

17.11.12   Defence – Regulation 16

It is a defence under these regulations for a person to show that they have taken all reasonable precautions and exercised all due diligence to avoid the commission of an offence.

17.11.13   Sources of references

A Step by Step Guide to COSHH Assessment HSG97, HSE Books, 2002, ISBN 978 0 7176 2785 1.

The Compilation of Safety Data Sheets L130, HSE Books, 2002, ISBN 978 0 7176 2371 6.

Control of Substances Hazardous to Health Approved Code of Practice and Guidance L5 (fifth edition), 2005, HSE Books, ISBN 9780 0 7176 2981 7.

The Control of Substances Hazardous to Health Regulations 2002, SI 2002, 2677, 2002 Stationery Office/http://www.opsi.gov.uk/si/si2002/20022677.htm

COSHH: A Brief Guide to the Regulations INDG136, 2005, HSE Web only.

COSHH Essentials: Easy Steps to Control Chemicals HSG193, 2007, HSE's Web only.

COSHH HSE Micro website: http://www.hse.gov.uk/coshh/index.htm

EH40/2005 Workplace Exposure Limits, HSE Books, 2005, (revised frequently) ISBN 978 0 7176 2977 0.

Respiratory Protective Equipment at Work HSG53, HSE Books, 2004, ISBN 978 0 7176 2904 6.

Why Do I Need a Safety Data Sheet? INDG353, 2002, HSE Books, ISBN 978 0 7176 2367 9.

17.12   Dangerous Substances and Explosive Atmospheres Regulations (DSEAR) 2002

17.12.1   Introduction

These Regulations are designed to implement the safety requirements of the EU Chemical Agents and Explosive Atmospheres Directives. DSEAR deals with the prevention of fires, explosions and similar energy-releasing events arising from dangerous substances. Following the introduction of DSEAR the opportunity will also be taken to modernize all the existing laws on the use and storage of petrol.

In summary the DSEAR Regulations require employers and the self-employed to:

➤ carry out an assessment of the fire and explosion risks of any work activities involving dangerous substances;

➤ provide measures to eliminate, or reduce sfarp, the identified fire and explosion risks;

➤ apply measures, sfarp, to control risks and to mitigate the detrimental effects of a fire or explosion;

➤ provide equipment and procedures to deal with accidents and emergencies;

➤ provide employees with information and precautionary training.

Where explosive atmospheres may occur:

➤ the workplaces should be classified into hazardous and non-hazardous places; and any hazardous places classified into zones on the basis of the frequency and duration of an explosive atmosphere, and where necessary marked with a sign;

➤ the equipment in classified zones should be safe so as to satisfy the requirements of the Equipment and Protective Systems Intended for Use in Potentially Explosive Atmospheres Regulations 1996;

➤ the workplaces should be verified by a competent person, as meeting the requirements of DSEAR.

17.12.2   Scope of regulations

Dangerous substance

The Regulations give a detailed interpretation of *dangerous substance*, which should be consulted. In summary it means:

➤ a substance or preparation that because of its chemical and sometimes physical properties and the way it is present and/or used at work, creates a fire or explosion risk to people; for example, substances like petrol, LPG, paints, cleaners, solvents and flammable gases;

➤ any dusts which could form an explosive mixture in air (not included in a substance or preparation); for example, many dusts from grinding, milling or sanding.

Explosive atmosphere

The Regulations give the definition as:

> *A mixture under atmospheric conditions, of air and one or more dangerous substances in the form of gases, vapours, mists or dusts, in which, after ignition has occurred, combustion spreads to the entire unburned mixture.*

17.12.3 Application

The Regulations do not in general apply to:

➤ ships under the control of a master;
➤ areas used for medical treatment of patients;
➤ many gas appliances used for cooking, heating, hot water, refrigeration, etc. (except an appliance specifically designed for an industrial process), gas fittings;
➤ manufacture, use, transport of explosives or chemically unstable substances;
➤ mine, quarry or borehole activities;
➤ activity at an offshore installation;
➤ the use of means of transport (but the Regulations do cover means of transport intended for use in a potentially explosive atmosphere).

17.12.4 Risk assessment – Regulation 5

The risk assessment required by Regulation 5 must include:

➤ the hazardous properties of the dangerous substance;
➤ supplier information and SDS;
➤ the circumstances of the work including:
  • work processes and substances used and their possible interactions

• the amount of substance involved
• risks of substances in combination
• arrangements for safe handling, storage and transport and any waste which might contain dangerous substances;

➤ high risk maintenance activities;
➤ likelihood of an explosive atmosphere;
➤ likelihood of ignition sources, including electrostatic discharges being present;
➤ scale of any possible fire or explosion;
➤ any places connected by openings to areas where there could be an explosive atmosphere;
➤ any additional information which may be needed.

17.12.5 Elimination or reduction of risks – Regulation 6

Regulation 6 concerns the reduction of risks and tracks the normal hierarchy as follows:

➤ Substitute a dangerous substance by a substance or process which eliminates or reduces the risk, for example the use of water-based paints, or using a totally enclosed continuous process.
➤ Reduce the quantity of dangerous substance to a minimum; for example, only a half day supply in the workroom.
➤ Avoid releasing of a dangerous substance or minimize releases, for example keeping them in special closed containers.
➤ Control releases at source.
➤ Prevent the formation of explosive atmospheres, including the provision of sufficient ventilation.
➤ Ensure that any releases are suitably collected, contained and removed; suitable LEV in a paint spray booth is an example.
➤ Avoid ignition sources and adverse conditions, for example keep electrical equipment outside the area.
➤ Segregate incompatible dangerous substances, for example oxidizing substances and other flammable substances.

Steps must also be taken to mitigate the detrimental effects of a fire or explosion by:

➤ keeping the number of people exposed to a minimum;
➤ avoidance of fire and explosion propagation;
➤ provision of explosion relief systems;
➤ provision of explosion suppression equipment;
➤ provision of very strong plant which can withstand an explosion;
➤ the provision of suitable PPE.

17.12.6 Classification of workplaces – Schedule 2

Where an explosive atmosphere may occur, workplaces must be classified into hazardous and non-hazardous places. Table 17.4 shows the zones specified by Schedule 2 of the Regulations.

17.12.7 Accidents, incidents and emergencies – Regulation 8

In addition to any normal fire prevention requirements employers must under these Regulations:

➤ ensure that procedures and first aid are in place with tested relevant safety drills;
➤ provide information on emergency arrangements including work hazards and those that are likely to arise at the time of an accident;
➤ provide suitable warning and other communications systems to enable an appropriate response, remedial actions and rescue operations to be made;
➤ where necessary, before any explosion condition is reached, provide visual or audible warnings and withdraw employees;
➤ provide escape facilities where the risk assessment indicates it is necessary.

In the event of an accident, immediate steps must be taken to:

➤ mitigate the effects of the event;
➤ restore the situation to normal;
➤ inform employees who may be affected.

Only essential persons may be permitted in the affected area. They must be provided with PPE, protective clothing and any necessary specialized safety equipment and plant.

17.12.8 Information instruction and training – Regulation 9

Under Regulation 9 where a dangerous substance is present, an employer must provide:

➤ suitable and sufficient information, instruction and training on the appropriate precautions and actions;
➤ details of the substances, any relevant data sheets and legal provisions;
➤ the significant findings of the risk assessment.

17.12.9 Contents of containers and pipes – Regulation 10

Regulation 10 requires that for containers and pipes (except where they are marked under legislation contained in

Table 17.4 Classification zones	
Zone 0	A place in which an explosive atmosphere consisting of a mixture with air of dangerous substances in the form of gas, vapour or mist is present continuously or for long periods.
Zone 1	A place in which an explosive atmosphere consisting of a mixture with air of dangerous substances in the form of gas, vapour or mist is likely to occur in normal operations occasionally.
Zone 2	A place in which an explosive atmosphere consisting of a mixture with air of dangerous substances in the form of gas, vapour or mist is not likely to occur in normal operations but, if it does, will persist for a short period only.
Zone 20	A place in which an explosive atmosphere in the form of a cloud of combustible dust in air is present continuously, or for long periods.
Zone 21	A place in which an explosive atmosphere in the form of a cloud of combustible dust in air is likely to occur in normal operations occasionally.
Zone 22	A place in which an explosive atmosphere in the form of a cloud of combustible dust in air is not likely to occur in normal operations but, if it does occur, will persist for a short period only.

Schedule 5 to the Regulations) the content and the nature of those contents and any associated hazards be clearly identified.

17.12.10 Sources of references

Control and Mitigation Measures, L136. Dangerous Substances and Explosive Atmospheres Regulations 2002. Approved Code of Practice and Guidance, 2003, HSE Books, ISBN 978 0 7176 2201 6.

The Dangerous Substances and Explosive Atmosphere Regulations 2002, SI 2002, No. 2776, ISBN 9780 11 042957 5.

Dangerous Substances and Explosive Atmospheres, L138. Dangerous Substances and Explosive Atmospheres Regulations 2002. Approved Code of Practice, 2003, HSE Books, ISBN 978 0 7176 2203 2.

Fire and Explosions – How Safe is Your Workshop. A Short Guide to DSEAR, HSE INDG370, 2002, HSE Books, ISBN 978 0 7176 2589 5.

Safe Handling of Combustible Dusts; Precautions Against Explosion HSG 103, HSE Books, 2003, ISBN 978 0 7176 2726 4.

Safe Maintenance, Repair and Cleaning Procedures, L137. Dangerous Substances and Explosive Atmospheres Regulations 2002. Approved Code of Practice and Guidance, 2003, HSE Books, ISBN 978 0 7176 2202 3.

Safe Use and Handling of Flammable Liquids HSG 140, HSE Books, ISBN 978 0 7176 0967 3.

Safe Working with Flammable Substances INDG227 HSE Books, 1996, ISBN 978 0 7176 1154 6.

Storage of Dangerous Substances, L135 Dangerous Substances and Explosive Atmospheres Regulations 2002. Approved Code of Practice and Guidance, 2003, HSE Books, ISBN 978 0 7176 2200 9.

17.13 Health and Safety (Display Screen Equipment) Regulations 1992

17.13.1 General

These Regulations cover the minimum health and safety requirements for the use of display screen equipment (DSE) and are accompanied by a guidance note. They typically apply to computer equipment with either a cathode ray tube or liquid crystal monitors. But any type of display is covered with some exceptions, for example, on board a means of transport, or where the main purpose is for screening a film or for a television. Multimedia equipment would generally be covered.

Equipment not specifically covered by these Regulations or where it is not being used by a defined 'user' is, nevertheless, covered by other requirements under the MHSWR and the Provision and Use of Work Equipment Regulations (PUWER).

17.13.2 Definitions – Regulation 1

(a) *Display Screen Equipment* (DSE) refers to any alphanumeric or graphic display screen, regardless of the display process involved.

(b) A *user* is an employee and an *operator* is a self-employed person, both of whom habitually use DSE as a significant part of their normal work. Both would be people to whom most or all of the following apply. A person:

➤ who depends on the DSE to do their job;
➤ who has no discretion as to use or non-use;
➤ who needs particular training and/or skills in the use of DSE to do their job;
➤ who uses DSE for continuous spell of an hour or more at a time;
➤ who does so on a more or less daily basis;
➤ for whom fast transfer of data is important for the job;
➤ of whom a high level of attention and concentration is required, in particular to prevent critical errors.

(c) A *workstation* is an assembly comprising:

➤ DSE with or without a keyboard, software or input device;
➤ optional accessories;
➤ disk drive, telephone, modem, printer, document holder, chair, desk, work surface, etc.;
➤ the immediate working environment.

17.13.3 Exemptions – Regulation 1(4)

Exemptions include DSE used in connection with:

➤ drivers' cabs or control cabs for vehicles or machinery;
➤ on board a means of transport;
➤ mainly intended for public operation;
➤ portable systems not in prolonged use;
➤ calculators, cash registers and small displays related to the direct use of this type of equipment;
➤ window typewriters.

17.13.4 Assessment of risk – Regulation 2

Possible hazards associated with DSE are physical (musculoskeletal) problems, visual fatigue and mental stress. They are not unique to display screen work, nor an inevitable consequence of it, and indeed research shows that the risk to the individual from typical display screen work is low. However, as in other types of work, ill health can result from poor work organization, working environment, job design and posture, and from inappropriate working methods.

Employers must carry out a suitable and sufficient analysis of users' (regardless of who has provided them) and operators' (provided by the employer) workstations in order to assess the risks to health and safety. The guidance gives detailed information on workstation minimum standards and possible effects on health.

The assessment should be reviewed when major changes are made to software, hardware, furniture, environment or work requirements.

17.13.5 Workstations – Regulation 3

All workstations must meet the requirements laid down in the schedule to the Regulations. This schedule lays down minimum requirements for display screen workstations, covering the equipment, the working environment and the interface between computer and user/operator.

17.13.6 Daily work routine of users – Regulation 4

The activities of users should be organized so that their daily work on DSE is periodically interrupted by breaks or changes of activity that reduce their workload at the equipment.

In most tasks, natural breaks or pauses occur from time to time during the day. If such breaks do not occur, deliberate breaks or pauses must be introduced. The guidance requires that breaks should be taken before the onset of fatigue and must be included in the working time. Short breaks are better than occasional long ones, for example, a 5- to 10-minute break after 50 to 60 minutes' continuous screen and/or keyboard work is likely to be better than a 15-minute break every 2 hours. If possible, breaks should be taken **away** from the screen. Informal breaks, with time on other tasks, appear to be more effective in relieving visual fatigue than formal rest breaks.

17.13.7 Eyes and eyesight – Regulation 5

Initially on request, employees have the right to a free eye and eyesight test conducted by a competent person where they are:

➤ already users (as soon as practicable after the request);
➤ to become a user (before they become a user).

The employer must provide a further eye and eyesight test at regular intervals thereafter or when a user is experiencing visual difficulties which could be caused by working with DSE.

There is no reliable evidence that work with DSE causes any permanent damage to eyes or eyesight, but it may make users with pre-existing vision defects more aware of them.

An *eye and eyesight test* means a *sight test* as defined in the Opticians Act 1989. This should be carried out by a registered ophthalmic optician or medical practitioner (normally only those with an ophthalmic qualification do so).

Employers shall provide special corrective appliances to users where:

➤ normal corrective appliances cannot be used;
➤ the result of the eye and eyesight test shows that such provision is necessary.

The guidance indicates that the liability of the employer extends only to the provision of corrective appliances which are of a style and quality adequate for their function. If an employee chooses a more expensive design or multi-function correction appliances, the employer need only pay a proportion of the cost.

Employers are free to specify that users' eye and eyesight tests and correction appliances are provided by a nominated company or optician.

The confidential clinical information from the tests can only be supplied to the employer with the employee's consent.

Vision screen tests can be used to identify people with defects but they are not a substitute for the full eyesight test, and employees have the right to opt for the full test from the outset.

17.13.8 Training – Regulation 6

Employers shall ensure that adequate health and safety training is provided to users and potential users in the use of any workstation and refresh the training following any re-organization. The guidance suggests a range of topics to be covered in the training. In summary these involve:

➤ the recognition of hazards and risks, including the absence of desirable features and the presence of undesirable ones;
➤ causes of risk and how harm may occur;
➤ what the user can do to correct them;
➤ how problems can be communicated to management;
➤ information on the regulations;
➤ the user's contribution to assessments.

17.13.9 Information – Regulation 7

Operators and users shall be provided with adequate information on all aspects of health and safety relating to their workstation and what steps the employer has taken to comply with the Regulations (insofar as the action taken relates to that operator or user and their work). Under Regulation 7, specific information should be provided as outlined in Table 17.5.

17.13.10 Sources of references

The Law on VDUs: An Easy Guide HSG90, 2003, HSE Books, ISBN 978 0 7176 2602 1.

VDU Workstation Checklist, 2003, HSE Books, ISBN 978 0 7176 2617 5.

Work with Display Screen Equipment, Health and Safety (Display Screen Equipment) Regulations 1992 as Amended by the Health and Safety (Miscellaneous Amendments) Regulations 2002 – Guidance on Regulations L26 (second edition), 2003, HSE Books, ISBN 978 0 7176 2582 6.

Working with VDUs INDG36 (rev 3), 2006, HSE Books, ISBN 978 0 7176 6222 7.

17.14 Electricity at Work Regulations 1989

The purpose of these Regulations is to require precautions to be taken against the risk of death or personal injury from electricity in work activities. The Regulations impose duties on persons ('duty holders') in respect of systems, electrical equipment and conductors and in respect of work activities on or near electrical equipment. They apply to almost all places of work and electrical systems at all voltages.

Table 17.5 Provision of information under Regulation 7

		Information on					
		Risks from DSE workstation	**Risk assessment and measures to reduce the risks Regs 2 and 3**	**Breaks and activity changes Reg. 4**	**Eye and eyesight texts Reg. 5**	**Initial training Reg. 6(1)**	**Training when workstation modified Reg. 6(2)**
Does employer have to provide information to display screen workers who are:	Users employed by the undertaking	Yes	Yes	Yes	Yes	Yes	Yes
	Users employed by other employer	Yes	Yes	Yes	No	No	Yes
	Operators in the undertaking	Yes	Yes	No	No	No	No

Source: HSE.

Guidance on the Regulations is contained in the *Memorandum of Guidance on the Electricity at Work Regulations 1989*, published by the HSE.

17.14.1 Definitions

(a) **Electrical equipment** – includes anything used to generate, provide, transmit, rectify, convert, conduct, distribute, control, store, measure or use electrical energy.

(b) **Conductor** – means a conductor of electrical energy. It means any material (solid, liquid or gas), capable of conducting electricity.

(c) **System** – means an electrical system in which all the electrical equipment is, or may be, electrically connected to a common source of electrical energy, and includes the source and equipment. It includes portable generators and systems on vehicles.

(d) **Circuit conductor** – term used in Regulations 8 and 9 only, means a conductor in a system which is intended to carry electric current in normal conditions. It would include a combined neutral and earth conductor, but does not include a conductor provided solely to perform a protective connection to earth or other reference point and energized only during abnormal conditions.

(e) **Danger** – in the context of these Regulations means a risk of injury from any electrical hazard.

Every year about 30 people die from electric shock or electric burns at work. Each year several hundred serious burns are caused by arcing where the heat generated can be very intense. In addition, intense ultraviolet radiation from an electric arc can cause damage to the eyes – known as arc-eye. These hazards are all included in the definition of **Danger**.

17.14.2 Duties – Regulation 3

Duties are imposed on employers, self-employed and employees. The particular duties on employees are intended to emphasize the level of responsibility which many employees in the electrical trades and professions are expected to take on as part of their job. Employees are:

➤ to co-operate with their employer so far as is necessary to enable any duty placed on the employer to be complied with (this reiterates section 7(b) of the HSW Act);

➤ to comply with the provisions of these Regulations insofar as they relate to matters which are within

their control. (This is equivalent to duties placed on employers and self-employed where these matters are within their control.)

17.14.3 Systems, work activities and protective equipment – Regulation 4

Systems must, at all times, be of such construction as to prevent danger. Construction covers the physical condition, arrangement of components and design of the system and equipment.

All systems must be maintained so as to prevent danger.

Every work activity, including operation, use and maintenance or work near a system, shall be carried out in a way, which prevents danger.

Protective equipment shall be suitable, suitably maintained and used properly.

17.14.4 Strength and capability of equipment – Regulation 5

No electrical equipment may be put into use where its strength and capability may be exceeded in such a way as may give rise to danger, in normal transient or fault conditions.

17.14.5 Adverse or hazardous environments – Regulation 6

Electrical equipment which may be exposed to:

➤ mechanical damage;
➤ the effects of weather, natural hazards, temperature or pressure;
➤ the effects of wet, dirty, dusty or corrosive conditions;
➤ any flammable or explosive substances including dusts, vapours or gases

shall be so constructed and protected that it prevents danger.

17.14.6 Insulation, protection and placing of conductors – Regulation 7

All conductors in a system which may give rise to danger shall either be suitably covered with insulating material and further protected as necessary, for example, against mechanical damage, using trunking or sheathing; or have precautions taken that will prevent danger, for example, being suitably placed like overhead electric power cables, or by having strictly controlled working practices.

17.14.7 Earthing, integrity and other suitable precautions – Regulations 8 and 9

Precautions shall be taken, either by earthing or by other suitable means, for example, double insulation, use of safe voltages and earth-free non-conducting environments, where a conductor, other than a circuit conductor, could become charged as a result of either the use of, or a fault in, a system.

An earth conductor shall be of sufficient strength and capability to discharge electrical energy to earth. The conductive part of equipment, which is not normally live but energized in a fault condition, could be a conductor.

If a circuit conductor is connected to earth or to any other reference point nothing which could break electrical continuity or introduce high impedance, for example, fuse, thyristor or transistor, is allowed in the conductor unless suitable precautions are taken. Permitted devices would include a joint or bolted link, but not a removable link or manually operated knife switch without bonding of all exposed metal work and multiple earthing.

17.14.8 Connections – Regulation 10

Every joint and connection in a system shall be mechanically and electrically suitable for its use. This includes terminals, plugs and sockets.

17.14.9 Excess current protection – Regulation 11

Every part of a system shall be protected from excess current, for example, short circuit or overload, by a suitably located efficient means such as a fuse or circuit breaker.

17.14.10 Cutting off supply and isolation – Regulation 12

There should be suitably located and identified means of cutting off (switch) the supply of electricity to any electrical equipment and also isolating any electrical equipment. Although these are separate requirements, they could be affected by a single means. The isolator should be capable of being locked off to allow maintenance to be done safely.

Sources of electrical energy (accumulators, capacitors and generators) are exempt from this requirement, but precautions must be taken to prevent danger.

17.14.11 Work on equipment made dead – Regulation 13

Adequate precautions shall be taken to prevent electrical equipment that has been made dead from becoming live while work is carried out on or near the equipment. This will include means of locking off isolators, tagging equipment, permits to work and removing fuses.

17.14.12 Work on or near live conductors – Regulation 14

No persons shall work near a live conductor, except if it is insulated, unless:

➤ it is unreasonable in all the circumstances for it to be dead;
➤ it is reasonable in **all** the circumstances for them to be at work on or near it while it is live;
➤ suitable precautions (including where necessary the provision of suitable protective equipment) are taken to prevent injury.

17.14.13 Working space access and lighting – Regulation 15

Adequate working space means of access and lighting shall be provided for all electrical equipment at which or near which work is being done in circumstances which may give rise to danger. This covers work of any kind. However, when the work is on live conductors, the access space must be sufficient for a person to fall back out of danger and if needed for persons to pass one another with ease and without hazard.

17.14.14 Competence – Regulation 16

Where technical knowledge or experience is necessary to prevent danger, all persons must possess such knowledge or experience or be under appropriate supervision.

17.14.15 Sources of references

Electrical Safety and You INDG231, 1996, HSE Books, ISBN 978 0 7176 1207 9.

Electrical Switch Gear and Safety. A Concise Guide for Users INDG372, HSE Books, 2003, ISBN 978 0 7176 2187 3.

Electricity at Work: Safe Working Practices HSG85, 2003, HSE Books, ISBN 978 0 7176 2164 4.

Guidance on Safe Isolation Procedures, SELECT, http://www.select.org.uk ISBN 978 0 7176 1272 4.

Maintaining Portable and Transportable Electrical Equipment HSG107 (second edition), 2004, HSE Books, ISBN 978 0 7176 2805 6.

Maintaining Portable Electrical Equipment in Hotels and Tourist Accommodation INDG237, 1997, HSE Books, ISBN 978 0 7176 1273 4.

Maintaining Portable Electrical Equipment in Offices and Other Low-risk Environments INDG236, 1996, HSE Books, ISBN 978 0 7176 1272 7.

Memorandum of Guidance on the Electricity at Work Regulations, Guidance on Regulations HSR25, 2007, HSE Books, ISBN 978 0 7176 6228 9.

Safety in Electrical Testing at Work, General Guidance INDG354, 2002, HSE Books, ISBN 978 0 7176 2296 2.

17.15 Employers' Liability (Compulsory Insurance) Act 1969 and Regulations 1998 amended in 2002

17.15.1 Introduction

Employers are responsible for the health and safety of employees while they are at work. Employees may be injured at work, or they or former employees may become ill as a result of their work while employed. They may try to claim compensation from the employer if they believe them to be responsible. The Employers' Liability Compulsory Insurance Act 1969 ensures that an employer has at least a minimum level of insurance cover against any such claims.

Employers' liability insurance will enable employers to meet the cost of compensation for employees' injuries or illnesses whether they are caused on or off site. However, any injuries or illnesses relating to motor accidents that occur while employees are working for them may be covered separately by motor insurance.

Public liability insurance is different. It covers for claims made against a person/company by members of the public or other businesses, but not for claims made by employees. While public liability insurance is generally voluntary, employers' liability insurance is compulsory. Employers can be fined if they do not hold a current employers' liability insurance policy which complies with the law.

17.15.2 Application

An employer needs employers' liability insurance unless they are exempt from the Employers' Liability Compulsory Insurance Act. The following employers are exempt:

➤ most public organizations including government departments and agencies, LAs, police authorities and nationalized industries;

➤ health service bodies, including National Health Service trusts, health authorities, Family Health Services Authorities and Scottish Health Boards and State Hospital Management Committees;

➤ some other organizations which are financed through public funds, such as passenger transport executives and magistrates' courts committees;

➤ family businesses, i.e. if employees are closely related to the employer (as husband, wife, father, mother, grandfather, grandmother, stepfather, stepmother, son, daughter, grandson, granddaughter, stepson, stepdaughter, brother, sister, half-brother or half-sister). However, this exemption does not apply to family businesses that are incorporated as limited companies except any employer which is a company that has only one employee who owns 50% or more of the share capital.

A full list of employers who are exempt from the need to have employers' liability insurance is shown at Schedule 2 of the Employers' Liability (Compulsory Insurance) Regulations 1998.

17.15.3 Coverage

Employers are only required by law to have employers' liability insurance for people whom they employ. However, people who are normally thought of as self-employed may be considered to be employees for the purposes of employers' liability insurance.

Whether or not an employer needs employers' liability insurance for someone who works for them depends on the terms of the contract. This contract can be spoken, written or implied. It does not matter whether someone is usually called an employee or self-employed or what their tax status is. Whether the contract is called a contract of employment or a contract for services is largely irrelevant. What matters is the real nature of the employer/employee relationship and the degree of control the employer has over the work employees do.

There are no hard and fast rules about who counts as employee for the purposes of Employers' liability insurance. The following paragraphs may help to give some indication.

In general, employers' liability insurance may be needed for a worker if:

➤ national insurance and income tax is deducted from the money paid to them;
➤ the employer has the right to control where and when they work and how they do it;
➤ most materials and equipment are supplied by the employer;
➤ the employer has a right to any profit workers make even though the employer may choose to share this with them through commission, performance pay or shares in the company; similarly, the employer will be responsible for any losses;
➤ that person is required to deliver the service personally and they cannot employ a substitute if they are unable to do the work;
➤ they are treated in the same way as other employees, for example if they do the same work under the same conditions as some other employee.

In most cases, employers' liability insurance is needed for volunteers. However, in general, the law may not require an employer to have insurance for:

➤ students who work unpaid;
➤ people who are not employed but are taking part in youth or adult training programmes;
➤ school children on work experience programmes.

However, in certain cases, these groups might be classed as employees. In practice, many insurance companies will provide cover for people in these situations.

One difficult area is domestic help. In general, an employer will probably not need employers' liability insurance for people such as cleaners or gardeners if they work for more than one person. However, if they only work for one employer, that employer may be required to take out insurance to protect them.

17.15.4 Display of certificate

Under the Regulations, employers must display, in a suitable convenient location, a current copy of the certificate of insurance at each place of business where they employ relevant people.

17.15.5 Retention of certificates

An employer must retain for at least 40 years copies of certificates of insurance which have expired. This is because claims for diseases can be made many years after the disease is caused. Copies can be kept electronically if this is more convenient than paper. An employer must make these available to health and safety inspectors on request.

These requirements do not apply to policies which expired before 1 January 1999. However, it is still very important to keep full records of previous insurance policies for the employer's own protection.

17.15.6 Penalties

The HSE enforces the law on employers' liability insurance and HSE inspectors will check that employers have employers' liability insurance with an approved insurer for at least £5 m. They will ask to see the certificate of insurance and other insurance details.

Employers can be fined up to £2500 for any day they are without suitable insurance. If they do not display the certificate of insurance or refuse to make it available to HSE inspectors when they ask, employers can be fined up to £1000.

17.15.7 Sources of references

Employers' Liability (Compulsory Insurance) Act 1969: A Guide for Employers HSE40 (rev 2), 2008, HSE website.

17.16 **Regulatory Reform (Fire Safety) Order 2005**

17.16.1 Introduction

This Order, made under the Regulatory Reform Act 2001, reforms the law relating to fire safety in non-domestic premises. It replaces fire certification under the Fire Precautions Act 1971 (which it repeals) with a general duty to ensure, sfarp, the safety of employees, a general duty, in relation to non-employees to take such fire precautions as may reasonably be required in the circumstances to ensure that premises are safe and a duty to carry out a risk assessment.

The Fire Certificate (Special Premises) Regulations 1976, The Fire Precautions (Workplace) Regulations 1997 and Amendment Regulations 1999 are all revoked.

The Fire Safety Order is the responsibility of the Department for Communities and Local Government and is enforced by the local fire and rescue authorities, with some exceptions which include:

➤ the HSE for:
   (i) nuclear installations;
   (ii) ships in construction or repair;
   (iii) construction sites other than a construction site which is contained within, or forms part of, premises which are occupied by persons other than those carrying out the construction work or any activity arising from such work.
➤ In Crown-occupied and Crown-owned buildings, enforcement is carried out by the Fire Services Inspectorates appointed under the Fire Services Act 1947.
➤ LAs for Certificated Sports Grounds.

The order came into force on 1 October 2006 with a number of guidance documents.

In Scotland the fire safety legislation is enacted through the Fire (Scotland) Act 2005 which, in brief, covers the following.

The Fire (Scotland) Act 2005 received Royal Assent on 1 April 2005. Parts 1, 2, 4 and 5 of the Act commenced in August 2005. Part 3 introduced a new fire safety regime for non-domestic premises and came into force on 1 October 2006 and replaced the Fire Precautions Act 1971 and the Fire Precautions (Workplace) Regulations 1997, as amended. Fire certificates will no longer be required after 1 October 2006 and the new fire safety regime will be based on the principle of risk assessment.

Ten guidance documents to complement the new legislation and help those with fire safety responsibilities to understand their duties are available.

PART 1 General

17.16.2 Interpretation – Articles 1–7

Here are a few of the more important definitions from the articles. For a full list consult the Order directly.

(a) 'Premises' includes any place and, in particular, includes:
➤ any workplace;
➤ any vehicle, vessel, aircraft or hovercraft;
➤ any installation on land (including the foreshore and other land intermittently covered by water), and any other installation (whether floating, or resting on the seabed or the subsoil thereof, or resting on other land covered with water or the subsoil thereof);
➤ any tent or movable structure.

(b) 'Risk' means the risks to the safety of persons from fire. 'Safety' means the safety of persons in the event of fire. 'Workplace' means any premises or parts of premises, not being domestic premises, used for the purposes of an employer's undertaking and which are made available to an employee of the employer as a place of work and includes:
➤ any place within the premises to which such employee has access while at work;
➤ any room, lobby, corridor, staircase, road or other place:
   ● used as a means of access to or egress from that place of work; or
   ● where facilities are provided for use in connection with that place of work, other than a public road.
(c) 'Responsible person' means:
➤ in relation to a workplace, the employer, if the workplace is to any extent under their control
➤ in relation to any premises not falling within paragraph (a):
   ● the person who has control of the premises (as occupier or otherwise) in connection with the carrying on by them of a trade, business or other undertaking (for profit or not); or
   ● the owner, where the person in control of the premises does not have control in connection with the carrying on by that person of a trade, business or other undertaking.
(d) 'General fire precautions' in relation to premises means:
➤ measures to reduce the risk of fire and the risk of the spread of fire;
➤ the means of escape from the premises;
➤ measures for securing that, at all material times, the means of escape can be safely and effectively used;
➤ measures in relation to fighting fires;
➤ the means for detecting fires and giving warning in case of fire;
➤ action to be taken in the event of fire, including
   ● instruction and training of employees
   ● measures to mitigate the effects of the fire.
These issues do not cover process-related fire risks including:

➤ the use of plant or machinery; or
➤ the use or storage of any dangerous substance.

Duties are placed on responsible persons in a workplace and in premises which are not workplaces to the extent that they have control over the premises.

The order does not apply (Article 8) to:

➤ domestic premises;

➤ an offshore installation;

➤ a ship, in respect of the normal ship-board activities of a ship's crew which are carried out solely by the crew under the direction of the master;

➤ fields, woods or other land forming part of an agricultural or forestry undertaking but which is not inside a building and is situated away from the undertaking's main buildings;

➤ an aircraft, locomotive or rolling stock, trailer or semi-trailer used as a means of transport or a vehicle;

➤ a mine other than any building on the surface at a mine;

➤ a borehole site.

In addition certain provisions of the Order do not apply to groups of workers such as:

➤ occasional work which is not harmful to young people in a family undertaking;

➤ armed forces;

➤ members of police forces;

➤ emergency services.

PART 2 Fire Safety Duties

17.16.3 Duty to take general fire precautions – Article 8

The responsible person must:

(a) take such general fire precautions as will ensure, sfarp, the safety of any of his employees; and

(b) in relation to relevant persons who are not their employees, take such general fire precautions as may reasonably be required in the circumstances of the case to ensure that the premises are safe.

17.16.4 Risk assessment and fire safety arrangements – Articles 9 and 11

The responsible person must make a suitable and sufficient assessment of the risks to identify the general fire precautions needs to take.

Where a dangerous substance is or is liable to be present in or on the premises, the risk assessment must include consideration of the matters set out in Part 1 of Schedule 1, reproduced in Figure 17.6.

Risk assessments must be reviewed by the responsible person regularly and if it is no longer valid or there has been significant changes. The responsible person must not employ a young person unless risks to young persons

---

MATTERS TO BE CONSIDERED IN RISK ASSESSMENT IN RESPECT OF DANGEROUS SUBSTANCES

The matters are:

(a) the hazardous properties of the substance
(b) information on safety provided by the supplier, including information contained in any relevant safety data sheet
(c) the circumstances of the work including:
(i) the special, technical and organizational measures and the substances used and their possible interactions
(ii) the amount of the substance involved
(iii) where the work will involve more than one dangerous substance, the risk presented by such substances in combination
(iv) the arrangements for the safe handling, storage and transport of dangerous substances and of waste containing dangerous substances.
(d) activities, such as maintenance, where there is the potential for a high level of risk
(e) the effect of measures which have been or will be taken pursuant to this Order
(f) the likelihood that an explosive atmosphere will occur and its persistence
(g) the likelihood that ignition sources, including electrostatic discharges, will be present and become active and effective
(h) the scale of the anticipated effects
(i) any places which are, or can be connected via openings to, places in which explosive atmospheres may occur
(j) such additional safety information as the responsible person may need in order to complete the assessment.

Figure 17.6 Matters to be considered in risk assessment in respect of dangerous substances.

have been considered in an assessment covering the following, which is Part 2 of Schedule 1.

Matters to be taken into particular account in risk assessment in respect of young persons
The matters are:

(a) the inexperience, lack of awareness of risks and immaturity of young persons;

(b) the fitting-out and layout of the premises;

(c) the nature, degree and duration of exposure to physical and chemical agents;

(d) the form, range and use of work equipment and the way in which it is handled;

(e) the organization of processes and activities;

(f) the extent of the safety training provided or to be provided to young persons;

(g) risks from agents, processes and work listed in the Annex to Council Directive 94/33/EC(a) on the protection of young people at work.

As soon as practicable after the assessment is made or reviewed, the responsible person must record the information prescribed where:

(a) the employer employs five or more employees;
(b) a licence under an enactment is in force in relation to the premises; or
(c) an **Alterations Notice** requiring this is in force in relation to the premises.

The prescribed information is:

(a) the significant findings of the assessment, including the measures which have been or will be taken by the responsible person;
(b) any group of persons identified by the assessment as being especially at risk.

The responsible person must make and record (as per risk assessments) arrangements as are appropriate, for the effective planning, organization, control, monitoring and review of the preventive and protective measures.

17.16.5 Principles of prevention to be applied and fire safety arrangements – Articles 10 and 11

Preventive and protective measures must be implemented on the basis of the principles specified in Part 3 of Schedule 1 as follows, which are broadly the same as those in the Management Regulations.

Principles of prevention
The principles are:

(a) avoiding risks;
(b) evaluating the risks which cannot be avoided;
(c) combating the risks at source;
(d) adapting to technical progress;
(e) replacing the dangerous substances by the non-dangerous or the less-dangerous substances;
(f) developing a coherent overall prevention policy which covers technology, organization of work and the influence of factors relating to the working environment;
(g) giving collective protective measures priority over individual protective measures;
(h) giving appropriate instructions to employees.

Fire safety arrangements must be made and put into effect by the responsible person where five or more people are employed; there is a licence for the premises; or an Alterations Notice requires them to be recorded. The arrangements must cover effective planning, organization,

control, monitoring and review of the preventative and protective measures.

17.16.6 Elimination or reduction of risks from dangerous substances – Article 12

Where a dangerous substance is present in or on the premises, the responsible person must ensure that risk of the substance is either eliminated or reduced, sfarp, by first replacing the dangerous substance with a safer alternative.

Where it is not reasonably practicable to eliminate risk the responsible person must, sfarp, apply measures including the measures specified in Part 4 of Schedule 1 to the Order (Figure 17.7).

The responsible person must:

(a) arrange for the safe handling, storage and transport of dangerous substances and waste containing dangerous substances;
(b) ensure that any conditions necessary for ensuring the elimination or reduction of risk are maintained.

17.16.7 Firefighting and fire detection – Article 13

The premises must be equipped with appropriate firefighting equipment and with fire detectors and alarms; any non-automatic firefighting equipment provided must be easily accessible, simple to use and indicated by signs.

Appropriate measures must be taken for firefighting in the premises: to nominate and train competent persons to implement the measures; and arrange any necessary contacts with external emergency services.

A person is to be regarded as competent where they have sufficient training and experience or knowledge and other qualities to enable them to properly implement the measures referred to.

17.16.8 Emergency routes and exits – Article 14

Where necessary, routes to emergency exits from premises and the exits themselves must be kept clear at all times.

The following requirements must be complied with:

(a) emergency routes and exits must lead as directly as possible to a place of safety;
(b) in the event of danger, it must be possible for persons to evacuate the premises as quickly and as safely as possible;
(c) the number, distribution and dimensions of emergency routes and exits must be adequate having regard to the use, equipment and dimensions of the

<table>
<tr><td>

Measures to be taken in respect of dangerous substances

1. In applying measures to control risks the responsible person must, in order of priority:
   (a) reduce the quantity of dangerous substances to a minimum
   (b) avoid or minimize the release of a dangerous substance
   (c) control the release of a dangerous substance at source
   (d) prevent the formation of an explosive atmosphere, including the application of appropriate ventilation
   (e) ensure that any release of a dangerous substance which may give rise to risk is suitably collected, safely contained, removed to a safe place, or otherwise rendered safe, as appropriate
   (f) avoid:
      (i) ignition sources including electrostatic discharges
      (ii) such other adverse conditions as could result in harmful physical effects from a dangerous substance
   (g) segregate incompatible dangerous substances.
2. The responsible person must ensure that mitigation measures applied in accordance with Article 12(3)(b) include:
   (a) reducing to a minimum the number of persons exposed
   (b) measures to avoid the propagation of fires or explosions
   (c) providing explosion pressure relief arrangements
   (d) providing explosion suppression equipment
   (e) providing plant which is constructed so as to withstand the pressure likely to be produced by an explosion
   (f) providing suitable personal protective equipment.
3. The responsible person must:
   (a) ensure that the premises are designed, constructed and maintained so as to reduce risk
   (b) ensure that suitable special, technical and organizational measures are designed, constructed, assembled, installed, provided and used so as to reduce risk
   (c) ensure that special, technical and organisational measures are maintained in an efficient state, in efficient working order and in good repair
   (d) ensure that equipment and protective systems meet the following requirements:
      (i) where power failure can give rise to the spread of additional risk, equipment and protective systems must be able to be maintained in a safe state of operation independently of the rest of the plant in the event of power failure
      (ii) means for manual override must be possible, operated by employees competent to do so, for shutting down equipment and protective systems incorporated within automatic processes which deviate from the intended operating conditions, provided that the provision or use of such means does not compromise safety
      (iii) on operation of emergency shutdown, accumulated energy must be dissipated as quickly and as safely as possible or isolated so that it no longer constitutes a hazard
      (iv) necessary measures must be taken to prevent confusion between connecting devices.
   (e) where the work is carried out in hazardous places or involves hazardous activities, ensure that appropriate systems of work are applied including:
      (i) the issuing of written instructions for the carrying out of work
      (ii) system of permits to work, with such permits being issued by a person with responsibility for this function prior to the commencement of the work concerned.
</td></tr>
</table>

Figure 17.7 Measures to be taken in respect of dangerous substances.

premises and the maximum number of persons who may be present there at any one time;

(d) emergency doors must open in the direction of escape;

(e) sliding or revolving doors must not be used for exits specifically intended as emergency exits;

(f) emergency doors must not be so locked or fastened that they cannot be easily and immediately opened by any person who may require to use them in an emergency;

(g) emergency routes and exits must be indicated by signs;

(h) emergency routes and exits requiring illumination must be provided with emergency lighting of adequate intensity in the case of failure of their normal lighting.

17.16.9 Procedures for serious and imminent danger and for danger areas – Article 15

The responsible person must:

(a) establish appropriate procedures, including safety drills;

(b) nominate a sufficient number of competent persons to implement evacuation procedures;

(c) provide adequate safety instruction for restricted areas.

Persons who are exposed to serious and imminent danger must be informed of the nature of the hazard and of the steps taken or to be taken to protect them from it. They must be able to stop work and immediately proceed to a place of safety in the event of their being exposed to serious, imminent and unavoidable danger; and procedures must require the persons concerned to be prevented from resuming work in any situation where there is still a serious and imminent danger.

17.16.10 Additional emergency measures in respect of dangerous substances – Article 16

In order to safeguard persons from an accident, incident or emergency related to the presence of a dangerous substance, the responsible person must (unless the risk assessment shows it is unnecessary) ensure that:

(a) information on emergency arrangements is available, including:
   (i) details of relevant work hazards and hazard identification arrangements;

(ii) specific hazards likely to arise at the time of an accident, incident or emergency.

(b) suitable warning and other communication systems are established to enable an appropriate response, including remedial actions and rescue operations, to be made immediately when such an event occurs;

(c) where necessary, before any explosion conditions are reached, visual or audible warnings are given and relevant persons withdrawn;

(d) where the risk assessment indicates it is necessary, escape facilities are provided and maintained to ensure that, in the event of danger, persons can leave endangered places promptly and safely.

The information required must be:

(a) made available to accident and emergency services;

(b) displayed at the premises, unless the results of the risk assessment make this unnecessary.

In the event of a fire arising from an accident, incident or emergency related to the presence of a dangerous substance in or on the premises, the responsible person must ensure that:

(a) immediate steps are taken to:
    (i) mitigate the effects of the fire;
    (ii) restore the situation to normal;
    (iii) inform those persons who may be affected.

(b) only those persons who are essential for the carrying out of repairs and other necessary work are permitted in the affected area and they are provided with:
    (i) appropriate PPE and protective clothing; and
    (ii) any necessary specialized safety equipment and plant

which must be used until the situation is restored to normal.

17.16.11 Maintenance – Article 17

Any facilities, equipment and devices provided must be subject to a suitable system of maintenance and maintained in an efficient state, in efficient working order and in good repair.

17.16.12 Safety assistance – Article 18

The responsible person must (except a competent self-employed person) appoint one or more competent persons to assist them in undertaking the preventive and protective measures. If more than one person is appointed, they must make arrangements for ensuring adequate co-operation between them.

The number of persons appointed, the time available for them to fulfil their functions and the means at their disposal must be adequate having regard to the size of the premises, the risks to which relevant persons are exposed and the distribution of those risks throughout the premises.

17.16.13 Provision of information to employees and others – Articles 19 and 20

The responsible person must provide their employees with comprehensible and relevant information on:

➤ the risks to them identified by the risk assessment;
➤ the preventive and protective measures;
➤ the procedures for fire drills;
➤ the identities of persons nominated for fire fighting or fire drills;
➤ the risks notified to them regarding shared premises.

Before employing a child, a parent (or guardian) of the child must be provided with comprehensible and relevant information on the risks to that child identified by the risk assessment; the preventive and protective measures; and the risks notified to them regarding shared premises.

Where a dangerous substance is present in or on the premises, additional information must be provided for employees as follows:

(a) the details of any such substance including:
    (i) the name of the substance and the risk which it presents;
    (ii) access to any relevant SDS;
    (iii) legislative provisions which apply to the substance.
(b) the significant findings of the risk assessment.

The responsible person must ensure that the employer of any employees from an outside undertaking who are working in or on the premises is provided with comprehensible and relevant information on:

(a) the risks to those employees;
(b) the preventive and protective measures taken by the responsible person.

The responsible person must ensure that any person working in their undertaking who is not their employee is provided with appropriate instructions and comprehensible and relevant information regarding any risks to that person.

17.16.14 Capabilities and training – Article 21

The responsible person must ensure that all employees are provided with adequate safety training:

(a) at the time when they are first employed;

(b) on their being exposed to new or increased risks because of:

(i) their being transferred or given a change of responsibilities;

(ii) the introduction of new work equipment or a change respecting work equipment already in use;

(iii) the introduction of new technology; or

(iv) the introduction of a new system of work, or a change respecting a system of work already in use.

The training must:

(a) include suitable and sufficient instruction and training on the appropriate precautions and actions to be taken by the employee;

(b) be repeated periodically where appropriate;

(c) be adapted to take account of any new or changed risks to the safety of the employees concerned;

(d) be provided in a manner appropriate to the risk identified by the risk assessment;

(e) take place during working hours.

17.16.15 Co-operation and co-ordination – Article 22

➤ Where two or more responsible persons share, or have duties in respect of, premises (whether on a temporary or a permanent basis) each such person must co-operate with, co-ordinate safety measures and inform the other responsible person concerned so far as is necessary to enable them to comply with the requirements and prohibitions imposed on them.

➤ Where two or more responsible persons share premises (whether on a temporary or a permanent basis) where an explosive atmosphere may occur, the responsible person who has overall responsibility for the premises must co-ordinate the implementation of all the measures required.

17.16.16 General duties of employees at work – Article 23

Every employee must, while at work:

(a) take reasonable care for the safety of himself or herself and of other relevant persons who may be affected by acts or omissions at work;

(b) as regards any duty or requirement imposed on an employer by or under any provision of this Order, the employee must co-operate so far as is necessary to enable that duty or requirement to be performed or complied with;

(c) inform their employer or any other employee with specific responsibility for the safety of fellow employees:

(i) of any work situation which a person with the first-mentioned employee's training and instruction would reasonably consider represented a serious and immediate danger to safety;

(ii) of any matter which a person with the first-mentioned employee's training and instruction would reasonably consider represented a shortcoming in the employer's protection arrangements for safety.

PART 3

17.16.17 Enforcement – Articles 25–31

Part 3 of the Order is about enforcement and penalties. An enforcing authority may be the fire and rescue authority of the area (most cases), the HSE (nuclear sites, ships and construction sites without other operations) and fire Inspectors maintained by the Secretary of State (armed forces, UK Atomic Energy Authority and Crown premises).

Fire Inspectors and an officer of the fire brigade maintained by the fire and rescue authority, have similar powers (Articles 27 and 28) to inspectors under the HSW Act.

There are differences between the HSW Act and the Fire order in the notices which can be issued. Appeals can be made against these notices within 21 days and the notices are suspended until the appeal is heard. However, a Prohibition notice stands until confirmed or altered by the court. Fire Inspectors can issue the following notices.

Alterations Notices – Article 29

These may be issued where the premises concerned constitute a serious risk to relevant people or may constitute a serious risk if any change is made to the premises or its use.

Where an Alterations Notice has been served the responsible person must notify the enforcing authority before making any specified changes, which are as follows:

➤ a change to the premises;

➤ a change to the services, fittings or equipment in or on the premises;

➤ an increase in the quantities of dangerous substances which are present in or on the premises;

➤ a change in the use of the premises.

In addition, the Alterations Notice may also include the requirement for the responsible person to:

➤ record the significant findings of the risk assessment as per Article 9(7) and 9(6);

➤ record the fire safety arrangements as per Article 11(1) and 11(2);

➤ send a copy of the risk assessment to the enforcing authority, before making the above changes, and a summary of the proposed changes to the general fire precautions.

Enforcement Notices – Article 30

Where the enforcing authority is of the opinion that the responsible person has failed to comply with the requirements of this Order or any regulations made under it, they can issue an Enforcement Notice which must:

➤ state that the enforcing authority is of this opinion;

➤ specify the provisions which have not been complied with; and require that person to take steps to remedy the failure within a period (not less than 28 days) from the date of service of the notice.

An Enforcement Notice may include directions on the measures needed to remedy the failures. Choices of remedial action must be left open.

Before issuing an Enforcement Notice the enforcing authority must consult the relevant enforcing authorities including those under HSW Act and the Building Regulations.

Prohibition Notices – Article 31

If the enforcing authority is of the opinion that the risks, relating to escape from the premises, are so serious that the use of the premises ought to be prohibited or restricted they may issue a Prohibition Notice. The notice must:

➤ state that the enforcing authority is of this opinion;

➤ specify the provisions which give or may give rise to that risk;

➤ direct that the use to which the notice relates is prohibited or restricted as may be specified until the specified matters have been remedied.

A Prohibition Notice may include directions on the measures needed to remedy the failures. Choices of remedial action must be left open.

A Prohibition Notice takes effect immediately it is served.

PART 4

17.16.18 Offences and Appeals – Articles 32–36

Cases can be tried in a magistrates' court or on indictment in the Crown Court.

The responsible person can be liable on conviction on indictment to a fine (not limited), or to imprisonment for a term not exceeding 2 years or both.

Any person can be liable to:

➤ on conviction on indictment to a fine (not limited) for an offence where that failure places one or more relevant people at risk of death or serious injury in the case of fire;

➤ on summary conviction to a fine at standard levels 3 or 5 depending on the particular offence.

In general, where an offence committed by a body corporate is proved to have been committed with the consent or connivance of any director, manager, secretary or other similar officer of the body corporate they as well as the body corporate is guilty of that offence – Article 32(9).

17.16.19 Sources of references

Regulatory Reform, England and Wales. The Regulatory Reform (Fire Safety) Order 2005, SI 2005 No. 1541.

Regulatory Reform (Fire Safety) Order 2005 – A short guide to making your premises safe from fire. Code 05 FRSD 03546. Free download http://www.communities.gov.uk/publications/fire/regulatoryreformfire

There are a number of guides produced by the Department of Communities and Local Government including the following subjects:

Fire Risk Assessment – Offices and Shops ISBN 978 1 85112 815 0.

Fire Risk Assessment – Sleeping Accommodation ISBN 978 1 85112 817 4.

Fire Risk Assessment – Residential Care Premises ISBN 978 1 85112 818 1.

Fire Risk Assessment – Small and Medium Places of Assembly ISBN 978 1 85112 820 4.

Fire Risk Assessment – Large Places of Assembly ISBN 978 1 85112 821 1.

Fire Risk Assessment – Factories and Warehouses ISBN 978 1 85112 816 7.

Fire Risk Assessment – Theatres, Cinemas and Similar Premises ISBN 978 1 85112 822 8.

Fire Risk Assessment – Educational Premises ISBN 978 1 85112 819 8.

Fire Risk Assessment – Healthcare Premises ISBN 978 1 85112 824 2.

Fire Risk Assessment – Transport Premises and Facilities ISBN 978 1 85112 825 9.

Fire Risk Assessment – Open Air Events and Venues ISBN 978 1 85112 823 5.

Fire Risk Assessment – Animal Premises and Stables ISBN 978 1 85112 884 6.

See website: http://www.communities.gov.uk/fire/publications/manuals-and-booklets

17.17 Health and Safety (First aid) Regulations 1981 as amended in 2002

These Regulations set out employers' duties to provide adequate first-aid facilities. They define first aid as:

➤ treatment for the purposes of preserving life and minimizing the consequences of injury and illness until medical help is obtained;

➤ treatment of minor injuries which would otherwise receive no treatment or which do not need treatment by a medical practitioner or nurse.

17.17.1 Duty of the employer – Regulation 3

An employer shall provide, or ensure that there are provided:

➤ adequate and appropriate facilities and equipment;

➤ qualified first aiders to render first aid;

➤ an appointed person, being someone to take charge of situations as well as first-aid equipment and facilities, where medical aid needs to be summoned. An appointed person will suffice where:

• the nature of the work is such that there are no specific serious hazards (offices, libraries, etc.); the workforce is small; the location makes further provision unnecessary;

• there is temporary (not planned holidays) or exceptional absence of the first aider.

There must always be at least an appointed person in every workplace during working hours.

Employers must make an assessment of the first-aid requirements that are appropriate for each workplace.

Any first-aid room provided under this Regulation must be easily accessible to stretchers and to any other equipment needed to convey patients to and from the room. It must be sign-posted according to the safety signs and signals regulations.

17.17.2 Employees information – Regulation 4

Employees must be informed of the arrangements for first aid, including the location of facilities, equipment and people.

17.17.3 Self-employed – Regulation 5

The self-employed shall provide such first-aid equipment as is appropriate to render first aid to themselves.

17.17.4 Sources of references

Basic Advice on First Aid at Work INDG347 (revised), 2006, ISBN 9780 7176 6193 0.

First Aid at Work. Health and Safety (First Aid) Regulations 1981. Approved Code of Practice and Guidance L74, HSE Books, 1997, ISBN 978 0 7176 1050 1 (revision due in October 2009).

First Aid at Work INDG214, HSE Books, 1997, ISBN 978 0 7176 1074 7.

HSE First Aid Micro Site for General Information, http://www.hse.gov.uk/firstaid/index.htm

17.18 Health and Safety (Information for Employees) Regulations 1989

17.18.1 General requirements

These Regulations require that the Approved Poster entitled, *Health and safety – what you should know*, is displayed or the Approved Leaflet is distributed.

This information tells employees in general terms about the requirements of health and safety law.

A new modern and easy to read version of the poster and pocket cards were introduced on 6th April 2009. Employers may continue to use the existing posters up to 5th April 2014. However only the new posters and leaflets will be available to purchase from April 2009.

The new poster has three sections for completion by the employer. These are 1. Employee representative; 2. Management representative (e.g. appointed competent person to assist managers health and safety appointee); 3. details of how the Enforcing authority and the Employment Medical Advisory Service (EMAS) can be contacted.

17.18.2 Sources of references

Health and Safety (Information for Employees) Regulations 1989 (SI No. 682).
Health and Safety Information for Employees (Amendment) Regulations 2009, SI 2009/606.

17.19 Hazardous Waste (England and Wales) Regulations 2005

17.19.1 Introduction

The Hazardous Waste (England and Wales) Regulations 2005 were implemented on 16 July 2005. These are made under the EPA 1990. There were many changes to the previous Special Waste Regulations, but the two key ones are that hazardous waste producers are now required to pre-register before any hazardous waste can be collected from their premises and the Regulations apply the European Waste Catalogue codes of hazardous wastes that will affect a much wider range of producers.

17.19.2 Summary

The list below is not exhaustive, but summarizes the main requirements. There is also a range of web links that will help to obtain more detail.

➤ From 16 July 2005, it is an offence for hazardous waste to be collected from a site that has not been registered or is exempt.
➤ All non-exempt sites that produce hazardous waste must be registered even if they are unlikely to have that waste collected for some time. Recent EA (Environment Agency) Guidance has clarified that it is an offence to produce hazardous waste on site and not be registered.
➤ The Regulations implement, through the List of Wastes (England) Regulations 2005, the European Waste Catalogue list of Hazardous wastes for the purposes of collection. This can be found at www.hmso.gov.uk

and will mean that things like PC monitors, PC base units, fridges, TVs, oily rags and separately collected fluorescent tubes require collection under the new hazardous waste notification and documentation procedures.

➤ The EA will accept postal registrations and registrations can be made online through the following website: www.environment-agency.gov.uk
➤ Each site producing hazardous waste has to have a separate registration although multiple sites can be registered on the same notification. Therefore, a head office could register all its sites centrally, but each site would have a separate unique registration number and require a separate fee.
➤ Some sites are exempt if they expect to produce less than 200 kg of hazardous waste a year – although they would then have to register if, part way through the year, they went over the threshold. These include agricultural premises, office premises, shops, premises where WEEE is collected, dental, veterinary and medical practices and ships. The EA indicates that 200 kg is approximately 10 small TVs, 500 fluorescent tubes or 5 small domestic fridges.
➤ Domestic waste is excluded from the Hazardous Waste Regulations on collection from the domestic property, but is then subject to the Regulations if it is separately collected or if it consists of asbestos. This includes Prescription Only Medicines (other than cytotoxic and cytostatic medicines) which will be hazardous waste.
➤ The Regulations require a new consignment note to be used from 16 July 2005 in place of the former section 62. Each consignment will require a fee to be paid to the EA by the consignee with their quarterly returns to the Agency. Clearly, this will be charged back to the collector, but a consignment might well attract more than one consignment fee if, for instance, it goes through a transfer station and the collector would have to ensure that this was considered in the price.
➤ Collection rounds will be possible but again, each site where the waste is collected would have to be left a copy of the consignment note, so the process of tracking and the paper trail will get quite complex especially as each collection will count as a consignment from the fee point of view.
➤ The Regulations ban the mixing of hazardous waste and state that it must be stored separately on site. However, clarification on the interpretation of this is still awaited from the EA as it would, for instance, preclude the collection of computer systems that included a base unit and screen.

➤ Registration as a Hazardous Waste producer places a statutory duty on the EA to inspect the site where the hazardous waste arises.

17.19.3 Sources of references

A guide to the Hazardous Waste Regulations and the List of Waste Regulations in England and Wales, Environment Agency, HWR01. See www.environment-agency.gov.uk.

17.20 Ionising Radiation Regulations 1999

17.20.1 Introduction

The Ionising Radiations Regulations 1999 (IRR99) implement the majority of the Basic Safety Standards Directive 96129/Euratom (BSS Directive). From 1 January 2000, they replaced the Ionising Radiations Regulations 1985 (IRR85) [except for Regulation 26 (special hazard assessments)].

The main aim of the Regulations and the supporting ACOP is to establish a framework for ensuring that exposure to ionizing radiation arising from work activities, whether from man-made or natural radiation and from external radiation (e.g. X-ray set) or internal radiation (e.g. inhalation of a radioactive substance), is kept as low as reasonably practicable and does not exceed dose limits specified for individuals. IRR99 also:

(a) replaces the Ionising Radiations (Outside Workers) Regulations 1993 (OWR93), which were made to implement the Outside Workers Directive 90/641/Euratom;

(b) implements a part of the Medical Exposures Directive 97143/Euratom in relation to equipment used in connection with medical exposures.

The guidance which accompanies the Regulations and ACOP gives detailed advice about the scope and duties of the requirements imposed by IRR99. It is aimed at employers with duties under the Regulations but should also be useful to others such as radiation protection advisers, health and safety officers, radiation protection supervisors and safety representatives.

17.20.2 Working with ionising radiation

Essentially, work with ionising radiation means:

(a) a practice, which involves the production, processing, handling, use, holding, storage, transport or disposal of artificial radioactive substances and some naturally occurring sources, or the use of electrical equipment emitting ionizing radiation at more than 5 kV [see definition of practice in Regulation 2(1)];

(b) work in places where the radon gas concentration exceeds the values in Regulation 3(1)(b); or

(c) work with radioactive substances containing naturally occurring radionuclides not covered by the definition of a practice.

17.20.3 Radiation employers

Radiation employers are essentially those employers who work with ionising radiation, that is they carry out:

(a) a practice [see definition in Regulation 2(1)]; or

(b) work in places where the radon gas concentration exceeds the values in Regulation 3(1)(b); or

(c) work with radioactive substances containing naturally occurring radionuclides not covered by the definition of a practice.

17.20.4 Duties of self-employed people

A self-employed person who works with ionizing radiation will simultaneously have certain duties under these Regulations, both as an employer and as an employee.

For example, self-employed persons may need to take such steps as:

➤ carrying out assessments under Regulation 7;

➤ providing control measures under Regulation 8 to restrict exposure;

➤ designating themselves as classified persons under Regulation 20;

➤ making suitable arrangements under Regulation 21 with one or more approved dosimetry services (ADS) for assessment and recording of doses they receive;

➤ obtaining a radiation passbook and keeping it up to date in accordance with Regulation 2;

➤ if they carry out services as an outside worker, making arrangements for their own training as required by Regulation 14;

➤ ensuring they use properly any dose meters provided by an ADS as required by regulation.

17.20.5 General requirements

Some of the major considerations are:

➤ notification to HSE of specific work unless specified in Schedule 1 to the Regulations (Regulation 6);

➤ carrying out prior risk assessment by a radiation employer before commencing a new activity (Regulation 7);

➤ use of PPE (Regulation 9);

➤ maintenance and examination of engineering controls and PPE (Regulation 10);

➤ dose limitations (Regulation 11);

➤ contingency plans for emergencies (Regulation 12);

➤ radiation protection adviser appointment (Regulation 13);

➤ information instruction and training (Regulation 14).

17.20.6 Prior risk assessment

Where a radiation employer is required to undertake a prior risk assessment, the following matters need to be considered, where they are relevant:

➤ the nature of the sources of ionising radiation to be used, or likely to be present, including accumulation of radon in the working environment;

➤ estimated radiation dose rates to which anyone can be exposed;

➤ the likelihood of contamination arising and being spread;

➤ the results of any previous personal dosimetry or area monitoring relevant to the proposed work;

➤ advice from the manufacturer or supplier of equipment about its safe use and maintenance;

➤ engineering control measures and design features already in place or planned;

➤ any planned systems of work; estimated levels of airborne and surface contamination likely to be encountered;

➤ the effectiveness and the suitability of PPE to be provided;

➤ the extent of unrestricted access to working areas where dose rates or contamination levels are likely to be significant; possible accident situations, their likelihood and potential severity;

➤ the consequences of possible failures of control measures – such as electrical interlocks, ventilation systems and warning devices – or systems of work; steps to prevent identified accident situations, or limit their consequences.

This prior risk assessment should enable the employer to determine:

➤ what action is needed to ensure that the radiation exposure of all persons is kept as low as reasonably practicable [Regulation 8 (1)];

➤ what steps are necessary to achieve this control of exposure by the use of engineering controls, design features, safety devices and warning devices [Regulation 8(2)(a)] and, in addition, by the development of systems of work [Regulation 8 (2) (b)];

➤ whether it is appropriate to provide PPE and if so what type would be adequate and suitable [Regulation 8(2)(c)];

➤ whether it is appropriate to establish any dose constraints for planning or design purposes, and if so what values should be used [Regulation 8(3)];

➤ the need to alter the working conditions of any female employee who declares she is pregnant or is breastfeeding [Regulation 8(5)];

➤ an appropriate investigation level to check that exposures are being restricted as far as are reasonably practicable [Regulation 8(7)];

➤ what maintenance and testing schedules are required for the control measures selected (Regulation 10);

➤ what contingency plans are necessary to address reasonably foreseeable accidents (Regulation 12);

➤ the training needs of classified and non-classified employees (Regulation 14);

➤ the need to designate specific areas as controlled or supervised areas and to specify local rules (Regulations 16 and 17);

➤ the actions needed to ensure restriction of access and other specific measures in controlled or supervised areas (Regulation 18);

➤ the need to designate certain employees as classified persons (Regulation 20);

➤ the content of a suitable programme of dose assessment for employees designated as classified persons and for others who enter controlled areas (Regulations 18 and 21);

➤ the responsibilities of managers for ensuring compliance with these Regulations;

➤ an appropriate programme of monitoring or auditing of arrangements to check that the requirements of these Regulations are being met.

17.20.7 Sources of references

Control of Radioactive Substances IRIS8, HSE Books, 2001, HSE Books website: www.hsebooks.co.uk

HSC, L121, 2000, HSE Books, ISBN 978 0 7176 1746 3. This is a large document and need only be studied by those with a specific need to control IR. The HSE have also produced a number of information sheets, which are available free on the Internet at their website.

Industrial Radiography Managing Radiation Risk IRIS1, 2000, HSE Books website: www.hsebooks.co.uk

Protection of Outsider Workers Against Radiation IRIS4, 2000, HSE Books website: www.hsebooks.co.uk

HSE Radiation Micro site: http://www.hse.gov.uk/radiation/index.htm

17.21 Lifting Operations and Lifting Equipment Regulations (LOLER) 1998 as amended in 2002

17.21.1 Introduction

This summary gives information about the LOLER 1998 which came into force on 5 December 1998.

In the main, LOLER replaced existing legal requirements relating to the use of lifting equipment, for example the Construction (Lifting Operations) Regulations 1961, the Docks Regulations 1988 and the Lifting Plant and Equipment (Records of Test and Examination, etc.) Regulations 1992.

The Regulations aim to reduce risks to people's health and safety from lifting equipment provided for use at work. In addition to the requirements of LOLER, lifting equipment is also subject to the requirements of the PUWER 1998.

Generally, the Regulations require that lifting equipment provided for use at work is:

➤ strong and stable enough for the particular use and marked to indicate SWL;
➤ positioned and installed to minimize any risks;
➤ used safely, i.e. the work is planned, organized and performed by competent people;
➤ subject to ongoing thorough examination and, where appropriate, inspection by competent people.

17.21.2 Definition

Lifting equipment includes **any equipment used at work for lifting or lowering loads**, including attachments used for anchoring, fixing or supporting it. The Regulations cover a wide range of equipment including cranes, fork-lift trucks, lifts, hoists, mobile elevating work platforms and vehicle inspection platform hoists. The definition also includes lifting accessories such as chains, slings, eyebolts, etc. LOLER **does not** apply to escalators; these are covered

by more specific legislation, i.e. the Workplace (Health, Safety and Welfare) Regulations 1992.

If employees are allowed to provide their own lifting equipment, then this too is covered by the Regulations.

17.21.3 Application

The Regulations apply to an employer or self-employed person providing lifting equipment for use at work, or who has control of the use of lifting equipment. They do not apply to equipment to be used primarily by members of the public, for example, lifts in a shopping centre. However, such circumstances are covered by the HSW Act 1974.

LOLER applies to the way lifting equipment is used in industry and commerce. LOLER applies only to work activities, for example:

➤ a crane on hire to a construction site;
➤ a contract lift;
➤ a passenger lift provided for use of workers in an office block;
➤ refuse collecting vehicles lifting on a public road;
➤ patient hoist;
➤ fork-lift truck.

These Regulations add to the requirements of PUWER 98 and should be interpreted with them. For example, when selecting lifting equipment, PUWER Regulation 4, regarding suitability, should be considered in connection with:

➤ ergonomics;
➤ the conditions in which the equipment is to be used;
➤ safe access and egress;
➤ preventing slips, trips and falls;
➤ protecting the operator.

While employees do not have duties under LOLER, they do have general duties under the HSW Act and the MHSWR 1992, for example to take reasonable care of themselves and others who may be affected by their actions and to co-operate with others.

The Regulations cover places where the HSW Act applies – these include factories, offshore installations, agricultural premises, offices, shops, hospitals, hotels, places of entertainment, etc.

17.21.4 Strength and stability – Regulation 4

Lifting equipment shall be of adequate strength and stability for each load, having regard in particular to the stress induced at its mounting or fixing point.

Every part of a load and anything attached to it and used in lifting it shall be of adequate strength.

Account must be taken of the combination of forces to which the lifting equipment will be subjected, as well as the weight of any lifting accessories. The equipment should include an appropriate factor of safety against failure.

Stability needs to take into account the nature, load-bearing strength, stability, adjacent excavations and slope of the surface. For mobile equipment, keeping rails free of obstruction and tyres correctly inflated must be considered.

17.21.5 Lifting equipment for lifting persons – Regulation 5

To ensure safety of people being lifted, there are additional requirements for such equipment. The use of equipment not specifically designed for raising and lowering people should only be used in exceptional circumstances.

The Regulation applies to all lifting equipment used for raising and lowering people and requires that lifting equipment for lifting persons shall:

➤ prevent a person using it from being crushed, trapped or struck, or falling from the carrier;
➤ prevent, sfarp, persons using it while carrying out work from the carrier being crushed, trapped or struck or falling from the carrier;
➤ have suitable devices to prevent the risk of the carrier falling. If a device cannot be fitted, the carrier must have:
  • an enhanced safety coefficient suspension rope or chain;
  • the rope or chain inspected every working day by a competent person.
➤ be such that a person trapped in any carrier is not thereby exposed to any danger and can be freed.

17.21.6 Positioning and installation – Regulation 6

Lifting equipment must be positioned and installed so as to reduce the risks, sfarp, from:

➤ equipment or a load striking another person;
➤ a load drifting, falling freely or being released unintentionally;

and it is otherwise safe.

Lifting equipment should be positioned and installed to minimize the need to lift loads over people and to prevent crushing in extreme positions. It should be designed to stop safely in the event of a power failure and not release its load. Lifting equipment, which follows a fixed path, should be enclosed with suitable and substantial interlocked gates and any necessary protection in the event of power failure.

17.21.7 Marking of lifting equipment – Regulation 7

Machinery and accessories for lifting loads shall be clearly marked to indicate their SWL, and:

➤ where the SWL depends on the configuration of the lifting equipment:
  • the machinery should be clearly marked to indicate its SWL for each configuration;
  • information which clearly indicates its SWL for each configuration should be kept with the machinery.
➤ accessories for lifting (e.g. hooks, slings) are also marked in such a way that it is possible to identify the characteristics necessary for their safe use (e.g. if they are part of an assembly);
➤ lifting equipment which is designed for lifting people is appropriately and clearly marked;
➤ lifting equipment not designed for lifting people, but which might be used in error, should be clearly marked to show it is not for lifting people.

17.21.8 Organization of lifting operations – Regulation 8

Every lifting operation, that is lifting or lowering of a load, shall be:

➤ properly planned by a competent person;
➤ appropriately supervised;
➤ carried out in a safe manner.

The person planning the operation should have adequate practical and theoretical knowledge and experience of planning lifting operations. The plan will need to address the risks identified by the risk assessment and identify the resources, the procedures and the responsibilities required so that any lifting operation is carried out safely. For routine simple lifts a plan will normally be left to the people using the lifting equipment. For complex lifting operations, for example where two cranes are used to lift one load, a written plan may need to be produced each time.

The planning should take account of avoiding suspending loads over occupied areas, visibility, attaching/detaching and securing loads, the environment, location, overturning, proximity to other objects, lifting of people and pre-use checks of the equipment.

17.21.9 Thorough examination and inspection – Regulations 9 and 11

Before using lifting equipment for the first time by an employer, it must be thoroughly examined for any defect unless:

➤ the lifting equipment has not been used before;

➤ an EC declaration of conformity (where one should have been drawn up) has been received or made not more than 12 months before the lifting equipment is put into service;

➤ if it is obtained from another undertaking, it is accompanied by physical evidence of an examination.

A copy of this thorough examination report shall be kept for as long as the lifting equipment is used (or, for a lifting accessory, 2 years after the report is made) (Regulation 11).

Where safety depends on the installation conditions, the equipment shall be thoroughly examined:

➤ after installation and before being put into service,

➤ after assembly and before being put into service at a new site or in new location

to ensure that it has been installed correctly and is safe to operate.

A copy of the thorough examination report shall be kept for as long as the lifting equipment is used at the place it was installed or assembled (Regulation 11).

Lifting equipment which is exposed to conditions causing deterioration that may result in dangerous situations, shall be:

➤ thoroughly examined at least every 6 months (for lifting equipment for lifting persons, or a lifting accessory); at least every 12 months (for other lifting equipment); or in accordance with an examination scheme; and each time that exceptional circumstances, liable to jeopardize the safety of the lifting equipment, have occurred and a copy of the report kept until the next report is made, or for 2 years (whichever is longer);

➤ inspected, if appropriate, by a competent person at suitable intervals between 'thorough examinations' (and a copy of the record kept until the next record is made).

All lifting equipment shall be accompanied by physical evidence that the last 'thorough examination' has been carried out before it leaves an employer's undertaking (or before it is used after leaving another undertaking).

The user, owner, manufacturer or some other independent party may draw up examination schemes, provided they have the necessary competence. Schemes should specify the intervals at which lifting equipment should be thoroughly examined and, where appropriate, those parts that need to be tested. The scheme should take account, for example, of its condition, the environment in which it is used, the number of lifting operations and the loads lifted.

The 'competent person' carrying out a thorough examination should have appropriate practical and theoretical knowledge and experience of the lifting equipment to be examined to enable them to detect defects or weaknesses and to assess their importance in relation to the safety and continued use of the lifting equipment. They should also determine whether a test is necessary and the most appropriate method for carrying it out.

17.21.10 Reports and defects – Regulation 10

The person making a 'thorough examination' of lifting equipment shall:

➤ notify the employer forthwith of any defect which, in their opinion, is or could become, dangerous;

➤ as soon as is practicable (within 28 days) write an authenticated report to:
  • the employer;
  • any person who hired or leased the lifting equipment, containing the information specified in Schedule 1.

➤ send a copy (as soon as is practicable) to the relevant enforcing authority where there is, in their opinion, a defect with an existing or imminent risk of serious personal injury (this will always be HSE if the lifting equipment has been hired or leased).

Every employer notified of a defect following a 'thorough examination' of lifting equipment should ensure that it is not used:

➤ before the defect is rectified;

➤ after a time specified in the schedule accompanying the report.

The person making an 'inspection' shall also notify the employer when, in his or her opinion, a defect is, or could become, dangerous and, as soon as is practicable, make a record of the inspection in writing.

17.21.11 Reports – Schedule 1

Schedule 1 lists the information to be contained in a report of a thorough examination. For example, name and address; identity of equipment; date of last thorough examination; SWL; appropriate interval; any dangerous or potentially dangerous defects; repairs required; date of next examination and test; details of the competent person.

17.21.12 Sources of references

Lifting Equipment and Lifting Operations Regulations 1998, SI No., OPSI website: http://www.opsi.gov.uk/si/si1998/19982307.htm

LOLER 1998 Lifting Operations and Lifting Equipment Regulations 1998, Open Learning Guidance. 1999, HSE Books, ISBN 978 0 7176 2464 5.

LOLER: How the Regulations apply to agriculture AIS28, HSE Books, 1998.

LOLER: How the Regulations apply to forestry AIS29, HSE Books, 1998.

LOLER: How the Regulations apply to Arboriculture AIS30, HSE Books, 1998.

Safe Use of Lifting Equipment, Lifting Operations and Lifting Equipment Regulations, 1998, Approved Code of Practice and Guidance, L113, 1998, HSE Books, ISBN 978 0 7176 1628 2.

Simple Guide to the LOLER, 1999 HSE, INDG290, HSE Web only.

Thorough Examination and Testing of Lifts. Simple Guidance for Lift Owners INDG339 (rev), HSE Books, 2007, ISBN 978 0 7176 6255 5.

17.22 Management of Health and Safety at Work Regulations 1999 as amended in 2003 and 2006

17.22.1 General

These Regulations give effect to the European Framework Directive on health and safety. They supplement the requirements of the HSW Act 1974 and specify a range of management issues, most of which must be carried out in all workplaces. The aim is to map out the organization of precautionary measures in a systematic way, and to make sure that all staff are familiar with the measures and their own responsibilities. They were amended slightly in 2003, particularly in relation to civil liability (Regulation 22).

17.22.2 Risk assessment – Regulation 3

Every employer is required to make a 'suitable and sufficient' assessment of risks to employees, and risks to other people who might be affected by the organization, such as visiting contractors and members of the public.

A systematic investigation of risks involved in all areas and operations is required, together with identification of the persons affected, a description of the controls in place and any further action required to reduce risks.

The risk assessments must take into account risks to new and expectant mothers and young people.

Significant findings from the assessments must be written down (or recorded by other means, such as on a computer) when there are five or more employees. The assessments need to be reviewed regularly and if necessary, when there have been significant changes, they should be modified.

17.22.3 Principles of prevention – Regulation 4

The following principles must be adopted when implementing any preventative and protective measures:

➤ avoiding risks;
➤ evaluating the risks which cannot be avoided;
➤ combating the risks at source;
➤ adapting the work to the individual, especially as regards the design of workplaces, the choice of work equipment and the choice of working and production methods, with a view, in particular, to alleviating monotonous work and work at a predetermined work-rate and to reducing their effect on health;
➤ adapting to technical progress;
➤ replacing the dangerous substances by non-dangerous or less-dangerous substances;
➤ developing a coherent overall prevention policy which covers technology, organization of work, working conditions, social relationships and the influence of factors relating to the working environment;
➤ giving collective protective measures priority over individual protective measures;
➤ giving appropriate instruction to employees.

17.22.4 Effective arrangements for health and safety – Regulation 5

Formal arrangements must be devised (and recorded) for effective planning, organization, control, monitoring and review of safety measures. This will involve an effective health and safety management system to implement the policy. Where there are five or more employees the arrangements should be recorded.

Planning involves a systematic approach to risk assessment, the selection of appropriate risk controls and establishing priorities with performance standards.

Organization involves consultation and communication with employees; employee involvement in risk assessment; the provision of information; and securing competence with suitable instruction and training. Control involves clarifying responsibilities and making sure people adequately fulfil their responsibilities. It involves adequate and appropriate supervision.

Monitoring should include the measurement of how well the policy is being implemented and whether hazards are being controlled properly. It covers inspections of the workplace and management systems; and the investigation of incidents and accidents to ascertain the underlying causes and effect a remedy.

Review is essential to look at the whole of the health and safety management system to ensure that it is effective and achieving the correct standard of risk control.

17.22.5 Health surveillance – Regulation 6

In appropriate circumstances health surveillance of staff may be required – the ACOP describes more fully when this duty will arise. Health surveillance is considered relevant when: there is an identifiable disease or poor health condition; there are techniques to detect the disease; there is a reasonable likelihood that the disease will occur; and surveillance is likely to enhance the protection of the workers concerned. A competent person, who will range from a manager, in some cases, to a fully qualified occupational medical practitioner in others, should assess the extent of the surveillance.

17.22.6 Competent assistance – Regulation 7

Every employer is obliged to appoint one or more 'competent person(s)' to advise and assist in undertaking the necessary measures to comply with the relevant statutory requirements. They may be employees or outside consultants. The purpose is to make sure that all employers have access to health and safety expertise. Preference should be given to an in-house appointee who may be backed up by external expertise.

The competence of the person(s) appointed is to be judged in terms of their training, knowledge and experience of the work involved; it is not necessarily dependent upon particular qualifications. In simple situations, it may involve knowledge of relevant best practice, knowing one's limitations and taking external advice when necessary. In more complex situations or risks, fully qualified and appropriately experienced practitioners will be required.

Appointed competent persons must be provided with adequate information, time and resources to do their job.

17.22.7 Procedures for serious and imminent danger and contact with external services – Regulations 8 and 9

Procedures must be established for dealing with serious and imminent dangers, including fire evacuation plans and arrangements for other emergencies. A sufficient number of competent persons must be appointed to evacuate the premises in the event of an emergency. The procedures should allow for persons at risk to be informed of the hazards and how and when to evacuate to avoid danger. In shared workplaces employers must co-operate. Access to dangerous areas should be restricted to authorized and properly trained staff. Any necessary contact arrangements with external services for first aid, emergency medical care and rescue work must be set up.

17.22.8 Information for employees – Regulation 10

Information must be provided to staff on the risk assessment, risk controls, emergency procedures, the identity of the people appointed to assist on health and safety matters and risks notified by others.

The information provided must take into account the level of training, knowledge and experience of the employees. It must take account of language difficulties and be provided in a form that can be understood by everyone. The use of translations, symbols and diagrams should be considered. Where children under school-leaving age are at work, information on the risk assessments and control measures must be provided to the child's parent or guardians of children at work before the child starts work. It can be provided verbally or directly to the parent, guardians or school.

17.22.9 Co-operation and co-ordination – Regulations 11, 12 and 15

Where two or more employers share a workplace, each must:

➤ co-operate with other employers in health and safety matters;

➤ take reasonable steps to co-ordinate their safety precautions;

➤ inform the other employers of the risks to their employees, i.e. risks to neighbours' employees.

Where people from outside organizations are present to do work, they – and their employers – have to be provided with appropriate information on risks, health and the necessary precautions to be taken.

Temporary staff and staff with fixed-term contracts as well as permanent employees must be supplied with health and safety information before starting work (Regulations 12 and 15).

Regulation 11 does not apply to multi-occupied premises or sites where each unit, under the control of an individual tenant employer or self-employed person, is regarded as a separate workplace. In other cases, common areas may be shared workplaces, such as a reception area or canteen or they may be under the control of a person to whom section 4 of HSW Act applies. Suitable arrangements may need to be put in place for these areas.

17.22.10 Capabilities and training – Regulation 13

When giving tasks to employees, their capabilities with regard to health and safety must be taken into account.

Employees must be provided with adequate health and safety training:

➤ on recruitment;
➤ on being exposed to new or increased risks;
➤ on the introduction of new procedures, systems or technology. Training must be repeated periodically and take place in working hours (or while being paid).

17.22.11 Duties on employees – Regulation 14

Equipment and materials must be used properly in accordance with instructions and training. Obligations on employees are extended to include certain requirements to report serious and immediate dangers and any shortcomings in the employer's protection arrangements.

17.22.12 New or expectant mothers – Regulations 16–18

Where work is of a kind that could present a risk to new or expectant mothers working there or their babies, the risk assessments must include an assessment of such risks. When the risks cannot be avoided the employer must alter a women's working conditions or hours to avoid the risks; offer suitable alternative work; or suspend from work on full pay. The woman must notify the employer in writing of her pregnancy, that she has given birth within the last 6 months or she is breastfeeding.

17.22.13 Young persons – Regulation 19

Employers must protect young persons at work from risks to their health and safety which are the result of lack of experience, or absence of awareness of existing or potential risks or which arise because they have not yet fully matured. Young persons may not be employed in a variety of situations which pose a significant risk to their health and safety that are a consequence of the following factors:

➤ physical or psychological capacity;
➤ pace of work;
➤ temperature extremes, noise or vibration;
➤ radiation;
➤ compressed air and diving;
➤ hazardous substances;
➤ lack of training and experience.

The exception to this is young persons over school-leaving age:

➤ where the work is necessary for their training;
➤ where they will be supervised by a competent person;
➤ where the risk will be reduced to the lowest level that is reasonably practicable.

The risk assessment must take the following specific factors into account:

➤ the fitting-out and layout of the workplace and the particular site where they will work;
➤ the nature of any physical, biological and chemical agents they will be exposed to, for how long and to what extent;
➤ what types of work equipment will be used and how this will be handled;
➤ how the work and processes involved are organized;
➤ the need to assess and provide health and safety training;
➤ risks from the particular agents, processes and work.

17.22.14 Provisions as to liability – Regulation 21

This provision is to prevent a defence for an employer by reason of any act or default by an employee or a competent person appointed under Regulation 7. Employees' duties to take reasonable care of their own health and safety and that of others affected by their work activities are unaffected by this provision.

17.22.15 Restriction of Civil Liability for Breach of Statutory Duty – Regulation 22

The MHSWR 1999 were amended in 2003, by the 2003 Amendment Regulations (SI 2003 No. 2457), to enable employees to claim damages from their employer in a civil action where they suffered injury or illness as a result of the employer being in breach of those Regulations. They were also intended to enable civil claims to be brought against employees for a breach of their duties under those Regulations that resulted in injury or illness. Employees have duties under those Regulations to use any equipment, dangerous substance, etc. in accordance with any training and instruction provided by the employer. Employees are also required to alert their employer of serious and imminent danger in the workplace or any shortcomings in the health and safety arrangements.

In April 2006 the Regulations were further amended by the Management of Health and Safety at Work (Amendment) Regulations 2006.

The amendment changes the civil liability provisions in the Regulations so as to exclude the right of third parties to take legal action against employees for contraventions of their duties under these Regulations. This extends to employees the same protection against third party action as that provided for employers.

The amendment neither creates any new duties nor removes any. The practical effect will be to reduce the likelihood of claims against employees by third parties. Therefore, it is expected that there will be no additional burdens on businesses.

The wording of the 2003 amendment produced the unintended consequence of allowing claims to be brought against employees by third parties who were affected by their work activity, for example members of the public. This had not been the intention. One concerned group raised this unintended consequence of the 2003 amendment. HSE sought independent advice, which also concluded that there was potential for third parties to make claims against employees.

17.22.16 Sources of references

An Introduction to Health and Safety: Health and Safety in Small Businesses INDG259 (rev 1), 2003, HSE Books, ISBN 978 0 7176 2685 4.
Five Steps to Risk Assessment HSE INDG163 (rev 2), 2006, HSE Books ISBN 978 0 7176 6189 3.

Leading Health and Safety at Work. Leadership Action for Directors and Board Members INDG417, HSE Books, 2007, ISBN 978 0 7176 6267 8.
Management of Health and Safety at Work: Approved Code of Practice and Guidance HSC L21 (second edition), 2000, HSE Books, ISBN 978 0 7176 2488 1.
The Management of Health and Safety at Work Regulations 1999, ISBN 9780 11 025051 6, OPSI website: http://www.opsi.gov.uk/si/si1999/uksi_19993242_en.pdf

17.23 Manual Handling Operations Regulations (MHO) 1992 as amended in 2002

17.23.1 General

The Regulations apply to the manual handling (any transporting or supporting) of loads, that is by human effort, as opposed to mechanical handling by fork-lift truck, crane, etc. Manual handling includes lifting, putting down, pushing, pulling, carrying or moving. The human effort may be applied directly to the load, or indirectly by pulling on a rope, chain or lever. Introducing mechanical assistance, like a hoist or sack truck, may reduce but not eliminate manual handling, as human effort is still required to move, steady or position the load.

The application of human effort for purposes other than transporting or supporting a load, for example pulling on a rope to lash down a load or moving a machine control, is not a manual handling operation. A load is a discrete movable object, but it does not include an implement, tool or machine while being used.

Injury in the context of these Regulations means to any part of the body. It should take account of the physical features of the load which might affect grip or cause direct injury, for example, slipperiness, sharp edges and extremes of temperature. It does not include injury caused by any toxic or corrosive substance which has leaked from a load, is on its surface or is part of the load.

See Figure 17.8, which shows a flow chart for the Regulations.

17.23.2 Duties of employers: avoidance of manual handling – Regulation 4(1)(a)

Employers should take steps to avoid the need for employees to carry out MHO which involves a risk of their being injured.

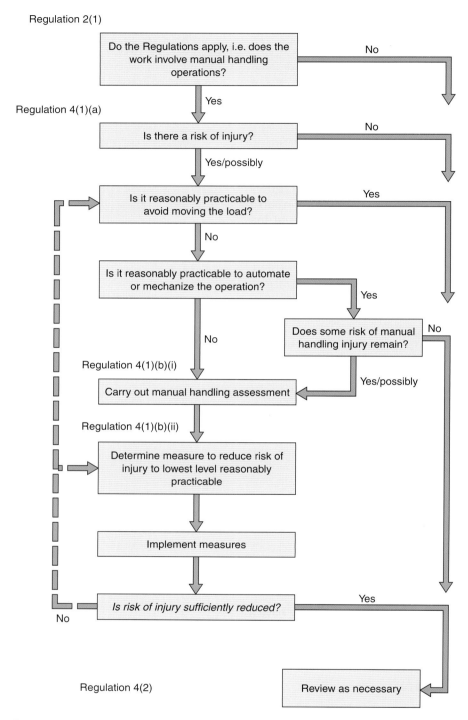

Regulation 2(1)

Regulation 4(1)(a)

Regulation 4(1)(b)(i)

Regulation 4(1)(b)(ii)

Regulation 4(2)

Figure 17.8 Manual Handling Operations Regulations – flow chart.

The guidance suggests that a preliminary assessment should be carried out when making a general risk assessment under the MHSWR 1999. Employers should consider whether the operation can be eliminated, automated or mechanized.

17.23.3 Duties of employers assessment of risk – Regulation 4(1)(b)(i)

Where it is not reasonably practicable to avoid MHO, employers must make a suitable and sufficient risk assessment of all such MHO in accordance with the requirements of Schedule 1 to the Regulations (shown in Table 17.6). This duty to assess the risk takes into account the **task**, the **load**, the **working environment** and **individual capability**.

17.23.4 Duties of employers: reducing the risk of injury – Regulation 4(1)(b)(ii)

Where it is not reasonably practicable to avoid the MHO with which there is a risk of injury, employers must take steps to reduce the risk of injury to the lowest level reasonably practicable.

The structured approach (considering the task, the load, the working environment and the individual capability) is recommended in the guidance. The steps taken will involve ergonomics, changing the load, mechanical assistance, task layout, work routines, PPE, team working and training.

17.23.5 Duties of employers: additional information on the load – Regulation 4(1)(b)(iii)

Employers must take appropriate steps where manual handling cannot be avoided to provide general indications and, where practicable, precise information on:

➤ the weight of each load;
➤ the heaviest side of any load which does not have a central centre of gravity.

The information is probably best marked on the load. Sections 3 and 6 of the HSW Act may place duties on originators of loads, like manufacturers or packers.

17.23.6 Duties of employers – reviewing assessment – Regulation 4(2)

The assessment should be reviewed if there is reason to suspect that it is no longer valid or there have been significant changes in the particular MHO.

Table 17.6 Schedule 1 to the Manual Handling Operations Regulations	
Factors to which the employer must have regard and questions they must consider when making an assessment of manual handling operations	
Factors	**Questions**

1. The tasks	Do they involve:
	    Holding or manipulating loads at distance from trunk?
	Unsatisfactory bodily movement or posture, especially:
	    Twisting the trunk?
	    Stooping?
	    Reaching upwards?
	Excessive movement of loads especially:
	    Excessive lifting or lowering distances?
	    Excessive carrying distances?
	    Excessive pulling or pushing of loads?
	    Risk of sudden movement of loads?
	    Frequent or prolonged physical effort?
	    Insufficient rest or recovery periods?
	    A rate of work imposed by a process?

2. The loads	Are they:
	    Heavy?
	    Bulky or unwieldy?
	    Difficult to grasp?
	    Unstable, or with contents likely to shift?
	    Sharp, hot or otherwise potentially damaging?

3. The working environment	Are there:
	    Space constraints preventing good posture?
	    Uneven, slippery or unstable floors?

	Variations in level of floors or work surfaces? Extremes of temperature or humidity? Conditions causing ventilation problems or gusts of wind? Poor lighting conditions?

4. Individual capability	Does the job:
	    Require unusual strength, height, etc.?
	    Create a hazard to those who might reasonably be considered to be pregnant or have a health problem?
	    Require special information or training for its safe performance?

5. Other factors	Is movement or posture hindered by personal protective equipment or by clothing?

17.23.7 Individual Capability Regulation 4(3)(a)–(f)

A new requirement was added in 2002 by Amendment Regulations concerning the individual capabilities of people undertaking manual handling. Regard must be had in particular to:

(a) the physical suitability of the employee to carry out the operation;

(b) the clothing, footwear or other personal effects they are wearing;

(c) their knowledge and training;

(d) results of any relevant risk assessments carried out under Regulation 3 of the Management Regulations 1999;

(e) whether the employee is within a group of employees identified by that assessment as being especially at risk;

(f) the results of any health surveillance provided pursuant to Regulation 6 of the Management Regulations 1999.

17.23.8 Duty of employees – Regulation 5

Each employee, while at work, has to make proper use of any system of work provided for their use. This is in addition to other responsibilities under the HSW Act and the MHSWR.

The provisions do not include well-intentioned improvisation in an emergency, for example rescuing a casualty or fighting a fire.

17.23.9 Sources of references

A Pain in Your Workplace: Ergonomic Problems and Solutions HSG121, HSE Books, 1994, ISBN 0 7176 06668 6 (no longer published).

Are You Making the Best Use of Lifting and Handling Aids? INDG398, HSE Books, 2004, ISBN 978 0 7176 2900 8.

Backs for the Future: Safe Manual Handling in Construction HSG149, HSE Books, 2000, ISBN 978 0 7176 1122 5.

Manual Handling Assessment Chart INDG383, HSE Books, 2003, ISBN 978 0 7176 2741 7.

Manual Handling, Manual Handling Operations Regulations 1992 (as amended), Guidance on Regulations Revised, L23, HSE Books, 2004, ISBN 978 0 7176 2823 0.

Manual Handling, Solutions You Can Handle HSG 115, 1994, HSE Books, ISBN 978 0 7176 0693 1.

17.24 Control of Noise at Work Regulations 2005

17.24.1 Introduction

The Control of Noise at Work Regulations 2005 require employers to prevent or reduce risks to health and safety from exposure to noise at work. Employees also have duties under the Regulations.

The Regulations require employers to:

- assess the risks to their employees from noise at work;
- take action to reduce the noise exposure that produces those risks;
- provide their employees with hearing protection if they cannot reduce the noise exposure enough by using other methods;
- make sure the legal limits on noise exposure are not exceeded;
- provide their employees with information, instruction and training;
- carry out health surveillance where there is a risk to health.

The Regulations do not apply to:

- members of the public exposed to noise from their non-work activities, or making an informed choice to go to noisy places;
- low-level noise which is a nuisance but causes no risk of hearing damage.

The Noise Regulations came into force for the music and entertainment sectors on 6th April 2008 bringing them in line with all other sectors where the regulations have been in force since April 2006. The noise regulations will now apply to pubs and clubs, amplified live music events, orchestras and other premises where live music or recorded music is played.

17.24.2 Exposure limit values and action levels – Regulation 4

The Noise at Work Regulations require employers to take specific action at certain action values. These relate to:

- the levels of exposure to noise of their employees averaged over a working day or week;
- the maximum noise (peak sound pressure) to which employees are exposed in a working day.

The values are:

➤ lower exposure action values:
- daily or weekly exposure of 80 dB (A-weighted)
- peak sound pressure of 135 dB (C-weighted).

➤ upper exposure action values:
- daily or weekly exposure of 85 dB (A-weighted)
- peak sound pressure of 137 dB (C-weighted).

Figure 17.9 helps decide what needs to be done.

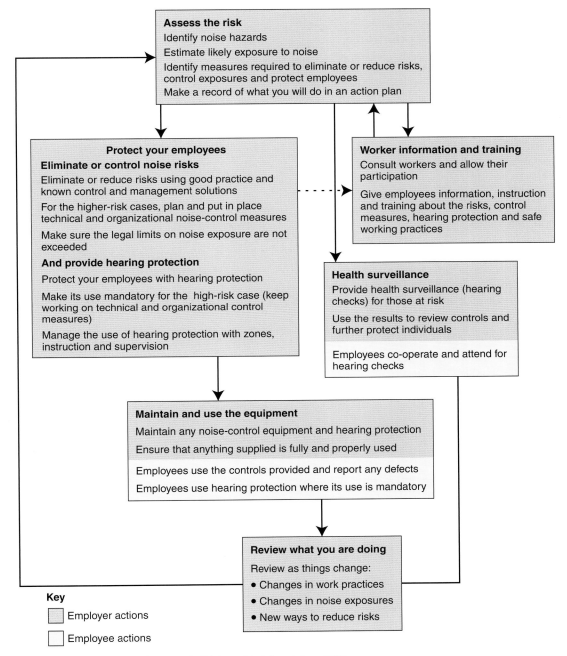

Figure 17.9 What needs to be done under the Control of Noise at Work Regulations 2005.

There are also levels of noise exposure which must not be exceeded:

➤ exposure limit values:
  • daily or weekly exposure of 87 dB (A-weighted)
  • peak sound pressure of 140 dB (C-weighted).

These exposure limit values take account of any reduction in exposure provided by hearing protection.

17.24.3 Risk assessment – Regulation 5

If employees are likely to be exposed to noise at or above the lower exposure value, a suitable and sufficient assessment of the risks must be made. Employers need to decide whether any further action is needed, and plan how to do it.

The risk assessment should:

➤ identify where there may be a risk from noise and who is likely to be affected;
➤ contain a reliable estimate of employees' exposures, and compare the exposure with the exposure action values and limit values;
➤ identify what needs to be done to comply with the law, e.g. whether noise-control measures or hearing protection are needed, and, if so, where and what type;
➤ identify any employees who need to be provided with health surveillance and whether any are at particular risk.

The risk assessment should include consideration of:

➤ the level, type and duration of exposure, including any exposure to peak sound pressure;
➤ the effects of exposure to noise on employees or groups of employees whose health is at particular risk;
➤ sfarp any effects on the health and safety of employees resulting from interaction, e.g. between noise and vibration;
➤ any indirect effects from the interaction between noise and audible warnings;
➤ manufacturers' information;
➤ availability of alternative equipment designed to reduce noise emissions;
➤ any extension to noise exposure due to extended hours or in supervised rest facilities;
➤ information following health surveillance;
➤ availability of personal hearing protectors with adequate attenuation characteristics.

It is essential that employers can show that their estimate of employees' exposure is representative of the work that they do. It needs to take account of:

➤ the work they do or are likely to do;
➤ the ways in which they do the work;
➤ how it might vary from one day to the next.

The estimate must be based on reliable information, for example measurements in their own workplace, information from other similar workplaces or data from suppliers of machinery.

Employers must record the significant findings of their risk assessment. They need to record in an action plan anything identified as being necessary to comply with the law, setting out what they have done and what they are going to do, with a timetable and saying who will be responsible for the work.

The risk assessment should be reviewed if circumstances in the workplace change and affect noise exposures. Also it should be reviewed regularly to make sure that the employer continues to do all that is reasonably practicable to control the noise risks. Even if it appears that nothing has changed, employers should not leave it for more than about 2 years without checking whether a review is needed.

17.24.4 Elimination or control of exposure – Regulation 6

The purpose of the Noise at Work Regulations 2005 is to make sure that people do not suffer damage to their hearing – so efforts to controlling noise risks and noise exposure should be concentrated.

Wherever there is noise at work employers should be looking for alternative processes, equipment and/or working methods which would make the work quieter or mean people are exposed for shorter times. They should also be keeping up with what is good practice or the standard for noise control within their industry.

Where there are things that can be done to reduce risks from noise that are reasonably practicable, they should be done. However, where noise exposures are below the lower exposure action values, risks are low and so employers would only be expected to take actions which are relatively inexpensive and simple to carry out.

Where the assessment shows that employees are likely to be exposed at or above the upper exposure action values, employers must put in place a planned programme of noise control.

The risk assessment will have produced information on the risks and an action plan for controlling noise. Employers should use this information to:

➤ tackle the immediate risk, e.g. by providing hearing protection;

- identify what is possible to control noise, how much reduction could be achieved and what is reasonably practicable;
- establish priorities for action and a timetable (e.g. consider where there could be immediate benefits, what changes may need to be phased in over a longer period of time and the number of people exposed to the noise in each case);
- assign responsibilities to people to deliver the various parts of the plan;
- ensure the work on noise control is carried out;
- check that what has been done has worked.

Actions taken should be based on the general principles set out in the Management Regulations and should include consideration of:

- other working methods;
- choice of appropriate work equipment emitting the least possible noise;
- the design and layout of workplaces, work stations and rest facilities;
- suitable and sufficient information and training;
- reduction of noise by technical means;
- appropriate maintenance programmes;
- limitation of the duration and intensity of exposure;
- appropriate work schedules with adequate rest periods.

17.24.5 Hearing protection – Regulation 7

Hearing protection should be issued to employees:

- where extra protection is needed above what can been achieved using noise control;
- as a short-term measure while other methods of controlling noise are being developed.

Hearing protection should not be used as an alternative to controlling noise by technical and organizational means. Employers should consult with their employees or their representatives on the type of hearing protection to be used.

Employers are required to:

- provide employees with hearing protectors if they ask for it and their noise exposure is between the lower and upper exposure action values;
- provide employees with hearing protectors and make sure they use them properly when their noise exposure exceeds the upper exposure action values;
- identify hearing protection zones, i.e. areas where the use of hearing protection is compulsory, and mark them with signs if possible. Restrict access to hearing protection zones where this is practicable and the noise exposure justifies it;
- provide employees with training and information on how to use and care for the hearing protectors;
- ensure that the hearing protectors are properly used and maintained.

17.24.6 Maintenance and use of equipment – Regulation 8

Employers need to make sure that hearing protection works effectively and to:

- check that all equipment provided in compliance with the regulations remains in good, clean condition and that there are no unofficial modifications;
- check that hearing protection is fully and properly used (except where it is provided for employees who are exposed at or above the lower exposure but below the upper exposure level). This is likely to mean that an employer needs to:
  - put someone in authority in overall charge of issuing it and making sure replacements are readily available;
  - carry out spot checks to see that the rules are being followed and that hearing protection is being used properly;
  - ensure all managers and supervisors set a good example and wear hearing protection at all times when in hearing protection zones;
  - ensure only people who need to be there enter hearing protection zones and do not stay longer than they need to.

17.24.7 Health surveillance – Regulation 9

Employers must provide health surveillance (hearing checks) for all their employees who are likely to be regularly exposed above the upper exposure action values, or are at risk for any reason – for example they already suffer from hearing loss or are particularly sensitive to damage.

The purpose of health surveillance is to:

- warn when employees might be suffering from early signs of hearing damage;
- give employers an opportunity to do something to prevent the damage getting worse;
- check that control measures are working.

The trade union safety representative, or employee representative and the employees concerned, should be consulted before introducing health surveillance. It is important that employees understand that the aim of health surveillance is to protect their hearing. Employers will need their understanding and co-operation if health surveillance is to be effective.

Health surveillance for hearing damage usually means:

➤ regular hearing checks in controlled conditions;
➤ telling employees about the results of their hearing checks (required after reasonable notice by employee);
➤ keeping health records;
➤ ensuring employees are examined by a doctor where hearing damage is identified.

If the doctor considers the hearing damage is likely to be the result of exposure, the employer must:

➤ ensure that the employee is informed by a suitably qualified person;
➤ review the risk assessment and control measures;
➤ consider assigning the employee to alternative work;
➤ ensure continued surveillance.

17.24.8 Information, instruction and training – Regulation 10

Where employees are exposed to noise which is likely to be at or above the lower exposure level, suitable and sufficient information, instruction and training must be provided and kept up to date. This includes the nature of the risks, compliance action taken, significant findings of the risk assessment, hearing protection, how to detect and report signs of hearing damage, entitlement to health surveillance, safe working practices and collective results of health surveillance.

17.24.9 Sources of references

Control of Noise at Work Regulations 2005 (SI No. 1643) http://www.opsi.gov.uk/si/si2005/uksi_20051643_en.pdf
Controlling Noise at Work. The Control of Noise At Work Regulations 2005, L108 (second edition), 2005, HSE Books, ISBN 978 0 7176 6164 0.
Noise at Work. Guidance for Employers on the Control of Noise at Work Regulations 2005 INDG362 (rev 1), 2005, HSE Books, ISBN 978 0 7176 6165 7.
Protect Your Hearing or Lose It! INDG363 (rev 1) 2005, HSE Books, ISBN 978 0 7176 6166 4.

17.25 Personal Protective Equipment at Work Regulations 1992 as amended in 2002

17.25.1 Introduction

The effect of the PPE at Work regulations is to ensure that certain basic duties governing the provision and use of PPE apply to all situations where PPE is required. The Regulations follow sound principles for the effective and economical use of PPE, which all employers should follow.

PPE, as defined, includes all equipment (including clothing affording protection against the weather) which is intended to be worn or held by a person at work and which protects them against one or more risks to their health and safety. Waterproof, weatherproof or insulated clothing is covered only if its use is necessary to protect against adverse climatic conditions.

Ordinary working clothes and uniforms, which do not specifically protect against risks to health and safety, and protective equipment worn in sports' competitions, are not covered.

Where there is overlap in the duties in these Regulations and those covering lead, ionizing radiations, asbestos, hazardous substances (COSHH), noise and construction head protection then the specific legislative requirements should prevail.

17.25.2 Provision of PPE – Regulation 4

Every employer shall ensure that suitable PPE is provided to their employees who may be exposed to risks to their health and safety except where it has been adequately or more effectively controlled by other means. (Management Regulations require PPE to be the last choice in the principles of protection.)

PPE shall not be suitable unless:

➤ it is appropriate for the risks and the conditions of use including the period for which it is worn;
➤ it takes account of ergonomic requirements and the state of health of the wearer and the characteristics of the workstation of each person;
➤ it is capable of fitting the wearer correctly, by adjustments if necessary;
➤ it is, so far as is practicable, able to combat the risks without increasing overall risks;
➤ it complies with UK legislation on design or manufacture, i.e. it has a CE marking.

Where it is necessary that PPE is hygienic and otherwise free of risk to health, PPE must be provided to a person solely for their individual use.

17.25.3 Compatibility – Regulation 5

Where more than one health and safety risk necessitates the wearing of multiple PPE simultaneously then they shall be compatible and remain effective.

17.25.4 Assessment – Regulation 6

Before choosing any PPE, employers must ensure that an assessment is made to determine whether the PPE is suitable.
The assessment shall include:

➤ assessing risks which have not been avoided by other means;
➤ a definition of the characteristics that PPE must have to be effective, taking into account any risks created by the PPE itself;
➤ a comparison of available PPE with the required characteristics;
➤ information on whether the PPE is compatible with any other PPE which is in use, and which an employee would be required to wear simultaneously.

The assessment should be reviewed if it is no longer valid or there have been significant changes. In simple cases it will not be necessary to record the assessment but, in more complex cases, written records should be made and kept readily available for future reference.

17.25.5 Maintenance – Regulation 7

Every employer (and self-employed person) shall ensure that any PPE provided is maintained, including being replaced and cleaned, in an efficient state, in efficient working order and in good repair.
The guide emphasizes the need to set up an effective system of maintenance for PPE. This should be proportionate to the risks and appropriate to the particular PPE. It could include, where appropriate, cleaning, disinfection, examination, replacement, repair and testing. For example, mechanical fall arrestor equipment or sub-aqua breathing apparatus will require planned preventative maintenance with examination, testing and overhaul. Records should be kept of the maintenance work. Gloves may only require periodic inspection by the user as necessary, depending on their use.

Spare parts must be compatible and be the proper part, suitably CE-marked where applicable. Manufacturers' maintenance schedules and instructions should be followed unless alternative schemes are agreed with the manufacturer or agent.

In some cases these requirements can be fulfilled by using disposable PPE which can be discarded after use or when their life has expired. Users should know when to discard and replace disposable PPE.

17.25.6 Accommodation – Regulation 8

When an employer or self-employed person has to provide PPE they must ensure that appropriate accommodation is provided to store it when not in use.

The type of accommodation will vary and may just be suitable hooks for special clothing and small portable cases for goggles. It should be separate from normal outer clothing storage arrangements and protect the PPE from contamination or deterioration.

17.25.7 Information, instruction and training – Regulation 9

Employers shall provide employees with adequate and appropriate information, instruction and training on:

➤ the risks which the PPE will avoid or limit;
➤ the purpose for which and the manner in which PPE should be used;
➤ any action required of the employee to maintain the PPE.

Employers must ensure that the information is kept available. The employer shall where appropriate and at suitable intervals, organize demonstrations of the wearing of PPE.
The guidance suggests the training should include:

➤ an explanation of the risks and why PPE is needed;
➤ the operation, performance and limitations of the equipment;
➤ instructions on the selection, use and storage of PPE;
➤ problems that can affect PPE relating to other equipment, working conditions, defective equipment, hygiene factors and poor fit;
➤ the recognition of defects and how to report problems with PPE;
➤ practice in putting on, wearing and removing PPE;
➤ practice in user cleaning and maintenance;
➤ how to store safely.

17.25.8 Use and reporting of defects – Regulations 10 and 11

Every employer shall take all reasonable steps to ensure that PPE is properly used.

Every employee shall:

➤ use PPE provided in accordance with training and instructions;
➤ return it to the accommodation provided after use;
➤ report any loss or obvious defect.

17.25.9 Sources of references

A Short Guide to the Personal Protective Equipment at Work Regulations 1992, INDG174, HSE Books, 2005, ISBN 978 0 7176 6141 1.

Personal Protective Equipment (EC Directive) Regulations 1992. SI 1992 No. 3139, http://www.opsi.gov.uk/si/ si1992/ Uksi_19922966_en_6.htm

Personal Protective Equipment at Work. Personal Protective Equipment at Work Regulations 1992. Guidance on Regulations. L25 (second edition), HSE Books, 2005, ISBN 978 0 7176 6139 8.

Selecting Protective Gloves for Work with Chemicals. Guidance for Employers and Health and Safety Specialists INDG330, HSE Books, 2000. Web only: http://www.hse.gov.uk/pubns/ppeindex.htm

Sun Protection Advice for Employers of Outdoor Workers INDG337, HSE Books, 2001, ISBN 978 0 7176 1982 5.

17.26 Provision and Use of Work Equipment Regulations 1998 (except Part IV) as amended in 2002

17.26.1 Introduction

The PUWER Regulations 1998 are made under the HSW Act and their primary aim is to ensure that work equipment is used without risks to health and safety, regardless of its age, condition or origin. The requirements of PUWER that are relevant to woodworking machinery are set out in the *Safe use of woodworking machinery Approved Code of Practice*. PUWER has specific requirements for risk assessment which are covered under the Health and Safety Management Regulations 1999.

Part IV of PUWER is concerned with power presses and is not part of the Certificate syllabus, and is therefore not covered in this summary.

17.26.2 Definitions

Work equipment means any machinery, appliance, apparatus, tool or installation for use at work.

Use in relation to work equipment means any activity involving work equipment and includes starting, stopping, programming, setting, transporting, repairing, modifying, maintaining, servicing and cleaning.

17.26.3 Duty holders – Regulation 3

Under PUWER the following groups of people have duties placed on them:

➤ employers;
➤ the self-employed;
➤ people who have control of work equipment, for example plant hire companies.

In addition to all places of work, the Regulations apply to common parts of shared buildings, industrial estates and business parks; to temporary works sites including construction; to home working (but not to domestic work in a private household); to hotels, hostels and sheltered accommodation.

17.26.4 Suitability of work equipment – Regulation 4

Work equipment:

➤ has to be constructed or adapted so that it is suitable for its purpose;
➤ has to be selected with the conditions of use and the users' health and safety in mind;
➤ may only be used for operations for which, and under conditions for which, it is suitable.

This covers all types of use and conditions and must be considered for each particular use or condition. For example, scissors may be safer than knives with unprotected blades and should therefore be used for cutting operations where practicable; risks imposed by wet, hot or cold conditions must be considered.

17.26.5 Maintenance – Regulation 5

The Regulation sets out the general requirement to keep work equipment maintained in:

➤ an efficient state;
➤ efficient working order;
➤ good repair.

Compliance involves all three criteria. In addition, where there are maintenance logs for machinery, they must be kept up to date.

In many cases this will require routine and planned preventive maintenance of work equipment. When checks are made priority must be given to:

➤ safety;
➤ operating efficiency and performance;
➤ the equipment's general condition.

17.26.6 Inspection – Regulation 6

Where the safety of work equipment depends on the installation conditions, it must be inspected:

➤ after installation and before being put into service for the first time;
➤ after assembly at a new site or in a new location;
➤ to ensure that it has been installed correctly and is safe to operate.

Where work equipment is exposed to conditions causing deterioration which is liable to result in dangerous situations, it must be inspected:

➤ at suitable intervals;
➤ when exceptional circumstances occur.

Inspections must be determined and carried out by competent persons. An inspection will vary from a simple visual external inspection to a detailed comprehensive inspection which may include some dismantling and/or testing. However, the level of inspection would normally be less than that required for a thorough examination under, for example, LOLER for certain items of lifting equipment.

Records of inspections must be kept with sufficient information to properly identify the equipment, its normal location, dates, faults found, action taken, to whom faults were reported, who carried out the inspection, when repairs were made, date of the next inspection.

When equipment leaves an employer's undertaking it must be accompanied by physical evidence that the last inspection has been carried out.

17.26.7 Specific risks – Regulation 7

Where the use of work equipment involves specific hazards, its use must be restricted to those persons given the specific task of using it, and repairs, etc. must be restricted to designated persons.

Designated persons must be properly trained to fulfil their designated task.

The ACOP requires that hazards must be controlled using a hierarchy of control measures, starting with elimination where this is possible, then considering hardware

measures such as physical barriers and, lastly, software measures such as a safe system of work.

17.26.8 Information, instruction and training – Regulations 8 and 9

Persons who use work equipment must have adequate:

➤ health and safety information;
➤ where appropriate, written instructions about the use of the equipment;
➤ training for health and safety in methods which should be adopted when using the equipment, and for any hazards and precaution which should be taken to reduce risks.

Any persons who supervise the use of work equipment should also receive information, instruction and training. The training of young persons is especially important with the need for special risk assessments under the Management of Health and Safety at Work Regulations.

Health and safety training should take place within working hours.

17.26.9 Conformity with community requirements – Regulation 10

The intention of this Regulation is to require that employers ensure that equipment, provided for use after 31 December 1992, conforms at all times with the relevant essential requirements in various European Directives made under Article 100A (now 95) of the Treaty of Rome. The requirements of PUWER 98 Regulations 11–19 and 22–29 only apply if the essential requirements do not apply to a particular piece of equipment.

However, PUWER Regulations 11–19 and 22–29 **will apply** if:

➤ they include requirements which were not included in the relevant product legislation;
➤ the relevant product legislation has not been complied with (e.g. the guards fitted on a machine when supplied were not adequate).

Employers using work equipment need to check that any new equipment has been made to the requirements of the relevant Directive, and has a CE marking, suitable instructions and a Certificate of Conformity.

The Machinery Directive was brought into UK law by the Supply of Machinery (Safety) Regulations 1992 as amended, which duplicate PUWER Regulations 11–19 and 22–29. On 29th December 2009 these regulations are replaced by the 2008 Regulations.

The employer still retains the duty to ensure that the equipment is safe to use.

17.26.10 Dangerous parts of machinery – Regulation 11

Measures have to be taken which:

➤ prevent access to any dangerous part of machinery or to any rotating stock-bar;
➤ stop the movement of any dangerous part of machinery or rotating stock-bar before any person enters a danger zone.

The measures required follow the normal hierarchy and should be considered to the extent that they are practicable in each case. They consist of:

➤ the provision of fixed guards enclosing every dangerous part of machinery then;
➤ the provision of other guards or protection devices then;
➤ the provision of jigs, holders, push-sticks or similar protection appliances used in conjunction with the machinery then;
➤ the provision of information, instruction, training and supervision as is necessary.

All guards and protection devices shall:

➤ be suitable for their purpose;
➤ be of good construction, sound material and adequate strength;
➤ be maintained in an efficient state, in efficient working order and in good repair;
➤ not give rise to increased risks to health and safety;
➤ not be easily bypassed or disabled;
➤ be situated at sufficient distance from the danger zone;
➤ not unduly restrict the view of the operating cycle of the machine where this is relevant;
➤ be so constructed or adapted that they allow operations necessary to fit or replace parts and for maintenance work, if possible without having to dismantle the guard or protection device.

17.26.11 Protection against specified hazards – Regulation 12

Exposure to health and safety risks from the following hazards must be prevented or adequately controlled:

➤ any article falling or being ejected from work equipment;
➤ rupture or disintegration of work equipment;
➤ work equipment catching fire or overheating;
➤ the unintended or premature discharge of any article, or of any gas, dust, liquid, vapour or other substance which is produced, used or stored in the work equipment;
➤ the unintended or premature explosion of the work equipment or any article or substance produced, used or stored in it.

17.26.12 High or very low temperature – Regulation 13

Work equipment and any article or substance produced, used or stored in work equipment which is at a high or very low temperature must have protection to prevent injury by burn, scald or sear.

This does not cover risks such as from radiant heat or glare.

Engineering methods of control such as insulation, doors, temperature control guards and should be used where practicable but there are some cases, like cooker hot plates, where this is not possible.

17.26.13 Controls – Regulations 14–18

Where work equipment is provided with (Regulation 14):

➤ starting controls (including restarting after a stoppage), and
➤ controls which change speed, pressure or other operating condition which would affect health and safety, it should not be possible to perform any operation except by a deliberate action on the control. This does not apply to the normal operating cycle of an automatic device.

Where appropriate, one or more readily accessible **Stop controls** shall be provided to bring the work equipment to a safe condition in a safe manner (Regulation 15). They must:

➤ bring the work equipment to a complete stop where necessary;
➤ if necessary, switch off all sources of energy after stopping the functioning of the equipment;
➤ operate in priority to starting or operating controls.

Where appropriate, one or more readily accessible **emergency stop controls** (Regulation 16) must be provided unless it is not necessary:

➤ by the nature of the hazard;
➤ by the time taken for the stop controls to bring the equipment to a complete stop.

Emergency stop controls must have priority over stop controls. They should be provided where other safeguards are not adequate to prevent risk when something irregular happens. They should not be used as a substitute for safeguards or the normal method of stopping the equipment.

All **controls** for work equipment shall (Regulation 17):

➤ be clearly visible and identifiable including appropriate marking where necessary;

➤ not expose any person to a risk to their health and safety except where necessary.

Where appropriate, employers shall ensure that:

➤ controls are located in a safe place;

➤ systems of work are effective in preventing any person being in a danger zone when equipment is started;

➤ an audible, visible or other suitable warning is given whenever work equipment is about to start.

Persons in a danger zone as a result of starting or stopping equipment must have sufficient time and suitable means to avoid any risks.

Control systems (Regulation 18) must be safe and chosen so as to allow for failures, faults and constraints. They must:

➤ not create any increased risk to health and safety;

➤ not result in additional or increased risks when failure occurs;

➤ not impede the operation of any stop or emergency stop controls.

17.26.14 Isolation from sources of energy – Regulation 19

Work equipment must be provided with readily accessible and clearly identified means to isolate it from all sources of energy.

Re-connection must not expose any person using the equipment to any risks.

The main purpose is to allow equipment to be made safe under particular circumstances, such as maintenance, when unsafe conditions occur, or when adverse conditions such as electrical equipment in a flammable atmosphere or wet conditions occur.

If isolation may cause a risk in itself, special precautions must be taken – for example a support for a hydraulic press tool which could fall under gravity if the system is isolated.

17.26.15 Stability – Regulation 20

Work equipment must be stabilized by clamping or otherwise as necessary to ensure health and safety.

Most machines used in a fixed position should be bolted or fastened so that they do not move or rock in use.

17.26.16 Lighting – Regulation 21

Suitable and sufficient lighting, taking account of the operations being carried out, must be provided where people use work equipment.

This will involve general lighting and in many cases local lighting, such as on a sewing machine. If access for maintenance is required regularly, permanent lighting should be considered.

17.26.17 Maintenance operations – Regulation 22

So far as is reasonably practicable, work equipment should be constructed or adapted to allow maintenance operations to be:

➤ conducted while it is shut down;
➤ undertaken without exposing people to risk;
➤ carried out after appropriate protection measures have been taken.

17.26.18 Markings and warnings – Regulations 23 and 24

Work equipment should have all appropriate **markings** for reasons of health and safety made in a clearly visible manner – for example, the maximum safe working load, stop and start controls, or the maximum rotation speed of an abrasive wheel.

Work equipment must incorporate **warnings** or **warning devices** as appropriate, which are unambiguous, easily perceived and easily understood.

They may be incorporated in systems of work, a notice, a flashing light or an audible warning. They are an active instruction or warning to take specific precautions or actions when a hazard exists.

17.26.19 Part III: mobile work equipment – Regulations 25–30

The main purpose of this section is to require additional precautions relating to work equipment while it is travelling from one location to another or where it does work while moving. If the equipment is designed primarily for travel on public roads, the Road Vehicles (Construction and Use) Regulations 1986 will normally be sufficient to comply with PUWER 98.

Mobile equipment would normally move on wheels, tracks, rollers, skids, etc. Mobile equipment may be self-propelled, towed or remote controlled and may incorporate attachments. Pedestrian-controlled work equipment such as lawn mowers is not covered by Part III.

Employees carried on mobile work equipment – Regulation 25

No employee may be carried on mobile work equipment unless:

➤ it is suitable for carrying persons;
➤ it incorporates features to reduce risks as low as is reasonably practicable, including risks from wheels and tracks.

Rolling over of mobile work equipment – Regulation 26

Where there is a risk of overturning it must be minimized by:

➤ stabilizing the equipment;
➤ fitting a structure so that it only falls on its side (tip over protection structure such as the vertical mast of a fork-lift truck);
➤ fitting a structure which gives sufficient clearance for anyone being carried if it turns over further (ROPS rollover protection structures);
➤ a device giving comparable protection;
➤ fitting a suitable restraining system for people if there is a risk of being crushed by rolling over.

This Regulation does not apply:

➤ to a fork-lift truck fitted with ROPS or vertical mast which prevents it turning on its side;
➤ where it would increase the overall risks;
➤ where it would not be reasonably practicable to operate equipment;
➤ to any equipment provided for use before 5 December 1998.

Overturning of fork-lift trucks – Regulation 27

Fork-lift trucks which carry an employee must be adapted or equipped to reduce the risk to safety from overturning to as low as is reasonably practicable.

Self-propelled work equipment – Regulation 28

Where self-propelled work equipment may involve risks while in motion, it shall have:

➤ facilities to prevent unauthorized starting;
➤ (with multiple rail-mounted equipment) facilities to minimize the consequences of collision;
➤ a device for braking and stopping;
➤ (where safety constraints so require) emergency facilities for braking and stopping, in the event of failure of the main facility, which have readily accessible or automatic controls;
➤ (where the driver's vision is inadequate) devices fitted to improve vision;
➤ (if used at night or in dark places) appropriate lighting fitted or otherwise it shall be made sufficiently safe for its use;
➤ if there is anything carried or towed that constitutes a fire hazard liable to endanger employees (particularly if escape is difficult such as from a tower crane), appropriate firefighting equipment carried, unless it is sufficiently close by.

Remote-controlled self-propelled work equipment – Regulation 29

Where remote-controlled self-propelled work equipment involves a risk while in motion it shall:

➤ stop automatically once it leaves its control range;
➤ have features or devices to guard against the risk of crushing or impact.

Drive shafts – Regulation 30

Where seizure of the drive shaft between mobile work equipment and its accessories or anything towed is likely to involve a risk to safety:

➤ the equipment must have means to prevent a seizure;
➤ where it cannot be avoided, every possible measures should be taken to avoid risks;
➤ the shaft should be safeguarded from contacting the ground and becoming soiled or damaged.

17.26.20 Part IV – power presses

Regulations 31–35 relate to power presses and are not included here as they are excluded from the NEBOSH Certificate syllabus. Details can be found in the Power Press ACOP.

17.26.21 Sources of references

Hiring and Leasing Out of Plant: Application of PUWER 98, Regulations 26 and 27, HSE MISC156, 1998, HSE Books.

Provision and Use of Work Equipment Regulations 1992, SI 1992 No. 2932. http://www.opsi.gov.uk/si/si1992/Uksi_19922932_en_5.htm

PUWER 1998, Provision and Use of Work Equipment Regulations 1998: Open Learning Guidance, HSE Books, 2008, ISBN 978 0 7176 6285 2.

Retrofitting of Roll-over Protective Structures, Restraining Systems and Their Attachment Points to Mobile Work Equipment MISC175, HSE Books, 1999.

Safe Use of Work Equipment, Provision and Use of Work Equipment Regulations 1998, Approved Code of Practice and Guidance, HSC L22, HSE Books, 2008, ISBN 978 0 7176 6295 1.

Simple Guide to the Provision and Use of Work Equipment Regulations 1998, INDG291, HSE Books, 1999, ISBN 978 0 7176 2429 4.

The selection and management of mobile elevating work platforms, HSE information sheet, 2008, Construction Information Sheet No. 58.

Using Work Equipment Safely INDG229, 2002, HSE Books, ISBN 978 0 7176 2389 1.

17.27 The Reporting of Injuries, Diseases and Dangerous Occurrences Regulations 1995

17.27.1 Introduction

These Regulations require the reporting of specified accidents, ill health and dangerous occurrences to the enforcing authorities. The events all arise out of or in connection with work activities covered by the HSW Act. They include death, major injury and more than three-day lost-time accidents. Schedules to the Regulations specify the details of cases of ill health and dangerous occurrences.

For most businesses reportable events will be quite rare and so there is little for them to do under these Regulations apart from keeping the guidance and forms available and being aware of the general requirements.

17.27.2 Definitions – Regulation 2

Accident includes an act of non-consensual physical violence done to a person at work.

This means that injuries, through physical violence to people not at work, are not reportable. Neither is any injury that occurs between workers over a personal matter or is carried out by a visiting relative of a person at work to that person. However, if a member of the public causes injury to a person at work through physical violence, that is reportable.

Incidents involving acts of violence may well be reportable to the police, which is outside the requirements of these Regulations.

17.27.3 Notification and reporting of major injuries and dangerous occurrences – Regulations 3(1) and 4 and Schedules 1 and 2

The HSE or local authority shall be notified immediately by the quickest practicable means and sent a report form F2508 (or other approved means) within 10 days following:

(a) **death** of a person as a result of an accident arising out of or in connection with work. Also if death occurs within 1 year of an accident, the authorities must be informed;

(b) **major injury** to a person as a result of an accident arising out of or in connection with work;

(c) a person not at work suffering an injury as a result of or in connection with work, and that person being taken from the site to hospital for treatment;

(d) a person not at work suffering a major injury as a result of an accident arising out of or in connection with work at a hospital

where there is a dangerous occurrence.

17.27.4 Reporting of three-day plus accidents – Regulation 3(2)

Where a person at work is incapacitated for work of a kind which they might reasonably be expected to do, either under their own contract of employment or in the normal course of employment, for more than three consecutive days (excluding the day of the accident, but including any days which would not have been working days) because of an injury resulting from an accident at work, the responsible person shall within 10 days send a report on form F2508 or other approved form, unless it has been reported under Regulation 3(1) as a major injury, etc.

17.27.5 Reporting of cases of disease – Regulation 5

Where a medical practitioner notifies the employer's responsible person that an employee suffers from a reportable work-related disease, a completed disease report form (F2508A) should be sent to the enforcing authority. The full list is contained in Schedule 3 to the Regulations, which is summarized in this guide.

17.27.6 Which enforcing authority?

LAs are responsible for retailing, some warehouses, most offices, hotels and catering, sports, leisure, consumer services and places of worship.

The HSE is responsible for all other places of work. Reports to the Incident Control Centre are automatically transferred to the relevant local authority.

17.27.7 Records

A record of each incident reported must be kept at the place of business for at least 3 years.

Major injuries Schedule 1

1. any fracture, other than to the fingers, thumbs or toes;
2. any amputation;
3. dislocation of the shoulder, hip, knee or spine;
4. loss of sight (whether temporary or permanent);
5. a chemical or hot metal burn to the eye or any penetrating injury to the eye;
6. any injury resulting from an electric shock or electrical burn (including any electrical burn caused by arcing or arcing products) leading to unconsciousness or requiring resuscitation or admittance to hospital for more than 24 hours;
7. any other injury
    (a) leading to hypothermia, heat-induced illness or to unconsciousness;
    (b) requiring resuscitation;
    (c) requiring admittance to hospital for more than 24 hours.
8. loss of consciousness caused by asphyxia or by exposure to a harmful substance or biological agent;
9. either of the following conditions which result from the absorption of any substance by inhalation, ingestion or through the skin
    (a) acute illness requiring medical treatment;
    (b) loss of consciousness.
10. acute illness which requires medical treatment where there is reason to believe that this resulted from exposure to a biological agent or its toxins or infected material.

17.27.8 Dangerous occurrences – Schedule 2 summary

PART I General

1. **Lifting machinery, etc.**
   The collapse, overturning, or the failure of any load-bearing part of lifts and lifting equipment.
2. **Pressure systems**
   The failure of any closed vessel or of any associated pipeline work, in which the internal pressure was above or below atmospheric pressure.
3. **Freight containers**
   The failure of any freight container in any of its load-bearing parts.
4. **Overhead electric lines**
   Any unintentional incident in which plant or equipment comes into contact with overhead power lines.
5. **Electrical short circuit**
   Electrical short circuit or overload attended by fire or explosion resulting in stoppage of the plant for over 24 hrs or has the potential to cause death.
6. **Explosives**
   The unintentional explosion or ignition of explosives, misfire, the failure of the shots in any demolition operation to cause the intended extent of collapse, the projection of material beyond the boundary of the site, any injury to a person resulting from the explosion or discharge of any explosives or detonator.
7. **Biological agents**
   Any accident or incident which resulted, or could have resulted, in the release or escape of a biological agent likely to cause severe human infection or illness.
8. **Malfunction of radiation generators, etc.**
   Any incident in which the malfunction of a radiation generator or its ancillary equipment used in fixed or mobile industrial radiography, the irradiation of food or the processing of products by irradiation, causes it to fail to de-energize at the end of the intended exposure period; or to fail to return to its safe position at the end of the intended exposure period.
9. **Breathing apparatus**
   Any incident in which breathing apparatus malfunctions while in use, or during testing immediately prior to use.
10. **Diving operations**
    In relation to a diving project, the failure or the endangering of diving equipment, the trapping of a diver, any explosion in the vicinity of a diver, any uncontrolled ascent or any omitted decompression.

11. Collapse of scaffolding

The complete or partial collapse of any scaffold which is more than 5 m in height, erected over or adjacent to water, in circumstances such that there would be a risk of drowning to a person falling from the scaffold into the water; or the suspension arrangements of any slung or suspended scaffold which causes a working platform or cradle to fall.

12. Train collisions

Any unintended collision of a train with any other train or vehicle.

13. Wells

Dangerous occurrence at a well other than a well sunk for the purpose of the abstraction of water.

14. Pipelines or pipeline works

A dangerous occurrence in respect of a pipeline or pipeline works in specific circumstances.

15. Fairground equipment

The failure of any load-bearing part or the failure of any part designed to support or restrain passengers or the derailment or the unintended collision of cars or trains.

16. Carriage of dangerous substances by road

Any incident involving a road tanker or tank container used for the carriage of dangerous goods in which the tanker overturns, is seriously damaged, there is an uncontrolled release or there is a fire.

Dangerous occurrences which are reportable except in relation to offshore workplaces.

17. Collapse of building or structure

Any unintended collapse or partial collapse of any building or structure under construction, reconstruction, alteration or demolition which involves a fall of more than 5 tonnes of material; any floor or wall of any building; or any false work.

18. Explosion or fire

An explosion or fire occurring in any plant or premises which results in the suspension of normal work for more than 24 hours where the explosion or fire is due to the ignition of any material.

19. Escape of flammable substances

(a) The sudden, uncontrolled release inside a building:
    (i) of 100 kg or more of a flammable liquid
    (ii) of 10 kg or more of a flammable liquid above its boiling point
    (iii) of 10 kg or more of a flammable gas
(b) 500 kg or more of any of the substances, in (a) above if released in the open air.

20. Escape of substances

The accidental release or escape of any substance in a quantity sufficient to cause the death, major injury or any other damage to the health of any person.

17.27.9 Reportable diseases – Schedule 3, brief summary

These include when certain specified activities are carried out:

➤ certain poisonings;
➤ some skin diseases such as occupational dermatitis, skin cancer, chrome ulcer, oil folliculitis/acne;
➤ lung diseases, such as occupational asthma, farmer's lung, pneumoconiosis, asbestosis, mesothelioma;
➤ infections such as leptospirosis, hepatitis, tuberculosis, anthrax, legionellosis and tetanus;
➤ other conditions, such as occupational cancer, certain musculoskeletal disorders, decompression illness and hand–arm vibration syndrome.

The details can be checked by consulting the guide, looking at the pad of report forms, checking the HSE's website, or ringing the HSE's Infoline or the Incident Control Centre (ICC).

17.27.10 Sources of references

A Guide to the Reporting of Injuries, Diseases and Dangerous Occurrences Regulations (third edition), 2008, L73, HSE Books, ISBN 9780 7176 6290 6.
RIDDOR Information for Doctors HSE 32, 1996, HSE Books.
Incident at Work? MISC769, HSE, 2007. See also: www.hse.gov.uk.riddor.

17.28   Safety Representatives and Safety Committees Regulations 1977

These Regulations, made under the HSW ACT section 2(4), prescribe the cases in which recognized trade unions may appoint safety representatives, specify the functions of such representatives and set out the obligations of employers towards them.

Employers' obligations to consult non-union employees are contained in the Health and Safety (Consultation with Employees) Regulations 1996.

17.28.1 Appointment – Regulation 3

A recognized trade union may appoint safety representatives from among employees in all cases where one or more employees are employed. When the employer is notified in writing the safety representatives have the functions set out in Regulation 4.

A person ceases to be a safety representative when:

➤ the appointment is terminated by the trade union;
➤ they resign;
➤ employment ceases.

A safety representative should have been with the employer for 2 years or have worked in similar employment for at least 2 years.

17.28.2 Functions – Regulation 4

These are functions and not duties. They include:

➤ representing employees in consultation with the employer;
➤ investigating potential hazards and dangerous occurrences;
➤ investigating the causes of accidents (Note re accident book BI510 and Data Protection – if the injured person has ticked the box and signed the form the safety representative may see all details. If not the employer should conceal the injured person's identity and details when giving access to the safety representatives);
➤ investigating employee complaints relating to health, safety and welfare;
➤ making representations to the employer on health, safety and welfare matters;
➤ carrying out inspections;
➤ representing employees at the workplace in consultation with enforcing inspectors;
➤ receiving information;
➤ attending safety committee meetings.

Safety representatives must be afforded time off with pay to fulfil these functions and to undergo training.

17.28.3 Employers' duties – Regulation 4a

Every employer shall consult safety representatives in good time regarding:

➤ the introduction of any measure which may affect health and safety;
➤ the arrangements for appointing or nominating competent person(s) under the Management Regulations;

➤ any health and safety information required for employees;
➤ the planning and organizing of any health and safety training for employees;
➤ the health and safety consequences of introducing new technology.

Employers must provide such facilities and assistance as safety representatives may reasonably require to carry out their functions.

17.28.4 Inspections – Regulations 5 and 6

Following reasonable notice in writing, safety representatives may inspect the workplace every quarter (or more frequently by agreement with the employer). They may inspect the workplace at any time, after consultation, when there have been substantial changes in the conditions of work or there is new information on workplace hazards published by the HSE.

Following an injury, disease or dangerous occurrence subject to Reporting of Injuries Diseases and Dangerous Occurrences Regulations, and after notifying the employer, where it is reasonably practicable to do so, safety representatives may inspect the workplace if it is safe.

Employers must provide reasonable assistance and facilities, including provision for independent investigation and private discussion with employees. The employer may be present in the workplace during inspections.

17.28.5 Information – Regulation 7

Having given reasonable notice to the employer, safety representatives are entitled to inspect and take copies of any relevant statutory documents (except any health record of an identified person).

An exempt document is one:

➤ which could endanger national security;
➤ which could cause substantial commercial injury on the employer;
➤ that contravenes a prohibition;
➤ that relates to an individual without their consent;
➤ which has been obtained specifically for legal proceedings.

17.28.6 Safety committees – Regulation 9

When at least two safety representatives have requested in writing that a safety committee is set up, the employer has 3 months to comply. The employer must consult with the safety representatives and post a notice stating its composition and the workplaces covered by it, in a place where it can be easily read by employees. The guidance gives details on the composition and running of safety committees.

17.28.7 Complaints – Regulation 11

If the employer fails to permit safety representatives time off or fails to pay for time off, complaints can be made to an industrial tribunal within 3 months of the incident.

17.28.8 Sources of references

HSC, Safety Representatives and Safety Committees (The Brown Book) Approved Code of Practice and Guidance on the Regulations (third edition), L87, 1996, HSE Books, ISBN 978 0 7176 1220 8.

17.29 Health and Safety (Safety Signs and Signals) Regulations 1996

17.29.1 Introduction

These Regulations came into force in April 1996. The results of the risk assessment made under the Management of Health and Safety at Work Regulations will have identified situations where there may be a residual risk when warnings or further information are necessary. If there is no significant risk there is no need to provide a sign.

17.29.2 Definitions – Regulation 2

➤ 'Safety sign' means a sign referring to a specific object, activity or situation and providing information or instruction about health and safety at work by means of a signboard, a safety colour, an illuminated sign, an acoustic sign, a verbal communication or a hand signal.
➤ 'Signboard' means a sign which provides information or instructions by a combination of geometric shape, colour and a symbol or pictogram and which is rendered visible by lighting of sufficient intensity.
➤ 'Hand signal' means a movement or position of the arms or hands or a combination, in coded form, for guiding persons who are carrying out manoeuvres

which create a risk to the health and safety of people at work.
➤ 'Acoustic signal' means a coded sound signal which is released and transmitted by a device designed for that purpose, without the use of a human or artificial voice.
➤ 'Verbal communication' means a predetermined spoken message communicated by a human or artificial voice.

17.29.3 Provision and maintenance of safety signs – Regulation 4

The Regulations require employers to use and maintain a safety sign where there is a significant risk to health and safety that has not been avoided or controlled by other means, like engineering controls or safe systems of work, and where the use of a sign can help reduce the risk.

They apply to all workplaces and to all activities where people are employed, but exclude signs used in connection with transport or the supply and marketing of dangerous substances, products and equipment.

The Regulations require, where necessary, the use of road traffic signs in workplaces to regulate road traffic.

17.29.4 Information, instruction and training – Regulation 5

Every employer shall ensure that:

➤ comprehensible and relevant information on the measures to be taken in connection with safety signs is provided to each employee;
➤ each employee receives suitable and sufficient instruction and training in the meaning of safety signs.

17.29.5 Functions of colours, shapes and symbols in safety signs

Safety colours
Red

Red is a safety colour and must be used for any:

➤ prohibition sign concerning dangerous behaviour (e.g. the safety colour on a 'No-Smoking' sign). Prohibition signs must be round, with a black pictogram on a white background with red edging and a red diagonal line (top left, bottom right). The red part must take up at least 35% of the area of the sign;
➤ danger alarm concerning stop, shutdown, emergency cut-out devices, evacuate (e.g. the safety colour of an emergency stop button on equipment);

➤ firefighting equipment. Rectangular or square shape with white pictogram on a red background (red part must be at least 50% of the area of the sign). See Chapter 6, Figure 6.9.

No fork-lift trucks   No smoking

Red and white alternating stripes may be used for marking surface areas to show obstacles or dangerous locations.

Yellow

Yellow (or amber) is a safety colour and must be used for any warning sign concerning the need to be careful, take precautions, examine or the like (e.g. the safety colour on hazard signs, such as for flammable material, electrical danger). Warning signs must be triangular, with a black pictogram on a yellow (or amber) background with black edging. The yellow (or amber) part must take up at least 50% of the area of the sign. Stripes should be of equal size and at 45 degrees.

General danger   Explosive

Yellow and black alternating stripes may be used for marking surface areas to show obstacles or dangerous locations.

Yellow may be used in continuous lines showing traffic routes.

Blue

Blue is a safety colour and must be used for any mandatory sign requiring specific behaviour or action (e.g. the safety colour on a 'Safety Helmet Must Be Worn' sign or a 'Pedestrians Must Use This Route' sign). Mandatory signs

must be round, with a white pictogram on a blue background. The blue part must take up at least 50% of the area of the sign.

Ear protection must be worn   Eye protection must be worn

Green

Green is a safety colour and must be used for emergency escape signs (e.g. showing emergency doors, exits and routes) and first-aid signs (e.g. showing location of first-aid equipment and facilities). Escape and first-aid signs must be rectangular or square, with a white pictogram on a green background. The green part must take up at least 50% of the area of the sign. So long as the green takes up at least 50% of the area, it is sometimes permitted to use a green pictogram on a white background, for example where there is a green wall and the reversal provides a more effective sign than one with a green background and white border – no danger (e.g. for 'return to normal').

Means of Escape   First Aid

Other colours
White

White is *not* a safety colour but is used for: pictograms or other symbols on blue and green signs; in alternating red and white stripes to show obstacles or dangerous locations; in continuous lines showing traffic routes.

Black

Black is *not* a safety colour but is used: for pictograms or other symbols on yellow (or amber) signs and, except for fire signs, red signs; in alternating yellow and black stripes to show obstacles or dangerous locations.

Shapes

Round signs must be used for: any prohibition (red) sign; mandatory (blue) sign.

Triangular signs must be used for any warning (yellow or amber) sign.

Square or rectangular signs must be used for any emergency escape sign and any first-aid sign.

Pictograms and other symbols

The meaning of a sign (other than verbal communication) must not rely on words. However, a sign may be supplemented with words to reinforce the message provided the words do not in fact distract from the message or create a danger.

A sign (other than verbal communication, acoustic signals or hand signals) should use a simple pictogram and/or other symbol (such as directional arrows, exclamation mark) to effectively communicate its message and so overcome language barriers.

Pictograms and symbols are included in the Regulations. Employee training is needed to understand the meaning of these as many are not inherently clear, some are meaningless to anyone who has not had their meaning explained and some can even be interpreted with their opposite meaning.

Pictograms and symbols included in the Regulations do not cover all situations for which graphic representation of a hazard or other detail may be needed. Any sign used for a situation not covered in the Regulations should include:

➤ the international symbol for general danger (exclamation mark!) if the sign is a warning sign and tests show that the sign is effective;

➤ in any other case a pictogram or symbol which has been tested and shown to be effective.

The text of any words used to supplement a sign must convey the same meaning. For example, a round blue sign with a pictogram showing the white outline of a face with a solid white helmet on the head means 'Safety Helmet Must Be Worn' and so any text used must maintain the obligatory nature of the message.

17.29.6 References

The Health and Safety (Safety Signs and Signals) Regulations 1996, SI 1996/341. Stationary Office www.opsi.gov.uk/si

Safety Signs and Signals Guidance on the Regulations L64, HSE Books, ISBN 978 0 7176 0870 6.

Signpost to the Health and Safety (Safety Signs and Signals) Regulations 1996, INDG 184L, HSE Books (see HSE website).

17.30 Supply of Machinery (Safety) Regulations 1992 and amendments

17.30.1 Introduction

The Supply of Machinery (Safety) Regulations 1992 entered into force on 1 January.

The Supply of Machinery (Safety) (Amendment) Regulations 1994 made a number of changes to the 1992 Regulations, in particular, to widen the scope to include machinery for lifting persons and safety components for machinery. The main provisions of the amending Regulations entered into force on 1 January 1995.

Therefore from **1 January 1995**:

➤ most machinery supplied in the UK, including imports, must:
  - satisfy wide-ranging health and safety requirements, for example, on construction, moving parts and stability
  - in some cases, have been subjected to type-examination by an approved body
  - carry CE marking and other information;

➤ the manufacturer or the importer will generally have to be able to assemble a file containing technical information relating to the machine (as from 29th December 2009 the Supply of Machinery (Safety) Regulations 2008 will come into force and the 1992 Regulations and amendments will be revoked.

17.30.2 Other relevant legislation to the supply of machinery

The two sets of Regulations that will often apply are the:

➤ Electrical Equipment (Safety) Regulations 1994, which apply to most electrically powered machinery used in workplaces;

➤ Electromagnetic Compatibility Regulations 2005, which cover equipment likely to cause electromagnetic disturbance, or whose performance is likely to be affected by electromagnetic disturbance.

In some cases, other laws may apply such as the Simple Pressure Vessels (Safety) Regulations 1991 or the Gas Appliances (Safety) Regulations 1995. All these Regulations implement European Directives and contain various requirements. The existence of CE marking on machinery should indicate that the manufacturer has met **all** of the requirements that are relevant.

Special transitional arrangements remain for products covered by existing Directives on rollover and falling-object protective structures and industrial trucks and for safety components and machinery for lifting persons.

Failure to comply with these requirements:

➤ will mean that the machinery cannot legally be supplied in the UK;

➤ could result in prosecution and penalties, on conviction, of a fine up to £5000 or, in some cases, imprisonment for up to 3 months, or both.

The same rules apply everywhere in the European Economic Area (EEA), so machinery complying with the community regime may be supplied in any EEA State.

17.30.3 Coverage

Machinery, described as:

➤ an assembly of linked parts or components, at least one of which moves, with the appropriate actuators, control and power circuits, joined together for a specific application, in particular for the processing, treatment, moving or packaging of a material;

➤ an assembly of machines which, in order to achieve the same end, are arranged and controlled so that they function as an integral whole;

➤ interchangeable equipment modifying the function of a machine which is supplied for the purpose of being assembled with a machine (or a series of different machines or with a tractor) by the operator himself in so far as this equipment is not a spare part or a tool.

Safety components for machinery, described as:

➤ components which are supplied separately to fulfil a safety function when in use and the failure or malfunctioning of which endangers the safety or health of exposed persons.

17.30.4 Exceptions

The Regulations do not apply to machinery or safety components:

➤ listed in Schedule 5;

➤ previously used in the EC or, since 1 January 1994 the EEA (e.g. secondhand);

➤ for use outside the EEA which do not carry CE marking;

➤ where the hazards are mainly of electrical origin (such machinery is covered by the Electrical Equipment (Safety) Regulations 1994);

➤ to the extent that the hazards are wholly or partly covered by other Directives, from the date those other Directives are implemented as UK law;

➤ machinery first supplied in the EC before 1 January 1993.

The Regulations do not apply to safety components and machinery for lifting persons first supplied in the EEA before 1 January 1995. Such products first supplied on or after 1 January 1995 must comply **either** with the Supply of Machinery (Safety) Regulations or the UK health and safety legislation in force relating to these items on 14 June 1993. All such products first supplied after 1 January 1997 must comply with the Supply of Machinery (Safety) Regulations.

17.30.5 Machinery excluded – Schedule 5 to the Regulations

➤ machinery whose only power source is directly applied manual effort, unless it is a machine used for lifting or lowering loads;

➤ machinery for medical use used in direct contact with patients;

➤ special equipment for use in fairgrounds and/or amusement parks;

➤ steam boilers, tanks and pressure vessels;

➤ machinery specially designed or put into service for nuclear purposes, which, in the event of failure, may result in an emission of radioactivity;

➤ radioactive sources forming part of a machine;

➤ firearms.

➤ storage tanks and pipelines for petrol, diesel fuel, highly flammable liquids and dangerous substances;

➤ means of transport, i.e. vehicles and their trailers intended solely for transporting passengers by air or on road, rail or water networks. Also, transport, which is designed for transporting goods by air, on public road or rail networks or on water. Vehicles used in the mineral extraction industry are not excluded;

➤ seagoing vessels and mobile offshore units together with equipment on board, such as vessels or units;

➤ cableways, including funicular railways, for the public or private transportation of people;

- agriculture and forestry tractors, as defined by certain European Directives;
- machines specially designed and constructed for military or police purposes;
- certain goods and passenger lifts;
- means of transport of people using rack and pinion rail-mounted vehicles;
- mine winding gear;
- theatre elevators;
- Construction site hoists intended for lifting individuals or people and goods.

17.30.6 General requirements

Subject to the exceptions and transitional arrangements described above, the Regulations make it an offence for a 'responsible person' to supply machinery or a safety component unless:

- it satisfies the essential health and safety requirements;
- the appropriate conformity assessment procedure has been carried out;
- an EC declaration of conformity or declaration of incorporation has been issued;
- CE marking has been properly affixed (unless a declaration of incorporation has been issued);
- it is, in fact, safe.

17.30.7 Essential health and safety requirements

To comply with the Regulations, machinery and safety components must satisfy the essential health and safety requirements (set out in Schedule 3 to the Regulations) which apply to them. The requirements are wide ranging, and take into account potential dangers to operators and other exposed persons within a 'danger zone'. Aspects covered in Section 1 include: the materials used in the construction of the machinery; lighting; controls; stability; fire; noise; vibration; radiation; emission of dust, gases, etc.; and maintenance. Section 2 has additional requirements for agri-foodstuffs machinery, portable hand-held machinery, and machinery for working wood and analogous materials. Section 3 deals with particular hazards associated with mobility; Section 4 with those associated with lifting; Section 5 with those relating to underground working and Section 6 with those associated with the lifting or moving of persons. The requirements also comment on instructions (including translation requirements) and marking.

When applying the essential health and safety requirements, technical and economic limitations at the time of construction may be taken into account.

17.30.8 Standards

Machinery and safety components manufactured in conformity with specified, published European standards which have also been published as identically worded national standards ('transposed harmonized standards'), will be presumed to comply with the essential health and safety requirements covered by those standards.

The European Committee for Standardization (CEN) is working to produce a complex of European standards at three levels in support of the Machinery Directive. The first (A) level comprises general principles for the design of machinery. The second (B) level covers specific safety devices and ergonomic aspects. The third (C) level deals with specific classes of machinery by calling up the appropriate standards from the first two levels and addressing requirements specific to the class.

17.30.9 Step 1 – conformity assessment ('attestation')

The responsibility for demonstrating that the machinery or safety component satisfies the essential health and safety requirements rests on the 'responsible person'.

For most machinery or safety components (other than those listed in Schedule 4): the 'responsible person' must be able to assemble the technical file described in Regulation 14 and Schedule 4.

For machinery or safety components listed in Schedule 4: the 'responsible person' must follow the special procedures described below.

17.30.10 Step 2 – Declaration procedure

Declaration of conformity: the 'responsible person' must then draw up an EC declaration of conformity, described in Regulation 22 and Schedule 2, for each machine or safety component supplied. This declaration is intended to be issued with the machine or safety component and declares that it complies with the relevant essential health and safety requirements or with the example that underwent type-examination.

Declaration of incorporation: alternatively, where the machinery is intended for incorporation into other machinery or assembly with other machinery to constitute machinery covered by the Regulations, the 'responsible person' may draw up a declaration of incorporation, described in Regulation 23.3, for each machine.

17.30.11 Step 3 – marking

Once a declaration of conformity has been issued, the 'responsible person' must affix the CE marking to the machinery.

CE markings must be affixed in a distinct, visible, legible and indelible manner.

The CE marking should not be affixed to safety components or for machinery for which a declaration of incorporation has been issued.

The Regulations make it an offence to affix a mark to machinery which may be confused with CE marking.

17.30.12 Enforcement

In Great Britain, the HSE is responsible for enforcing the Regulations in relation to machinery and safety components for use at work. Local Authority Trading Standards Officers are responsible for enforcement in relation to machinery and safety components for private use.

In Northern Ireland, the Department of Economic Development and the Department of Agriculture are responsible for enforcing the Regulations in relation to machinery and safety components for use at work. District councils are responsible for enforcement in relation to machinery and safety components for private use.

The enforcement authorities have available to them various powers under the HSW Act 1974, the Health and Safety at Work (Northern Ireland) Order 1978 and the Consumer Protection Act 1987, for example relating to suspension, prohibition and prosecution.

Where machinery bearing the CE marking is safe, but there are breaches of other obligations, the 'responsible person' will be given the opportunity to correct the breach before further enforcement action is taken.

The Machinery Directive, as amended, requires member states to inform the European Commission of any specific enforcement action taken. The Commission will consider whether the action is justified and advise the parties concerned accordingly.

17.30.13 Penalties

The maximum penalty for contravening the prohibition on supply of non-compliant machinery and safety components is imprisonment for up to 3 months or a fine of up to £5000 or both. The penalty for other contraventions

of the Regulations is a fine up to the same amounts. It is for the courts to decide the penalty in any given case, taking into account the severity of the offence.

The Regulations provide a defence of due diligence. They also provide for proceedings to be taken against a person other than the principal offender, if it is the other person's fault, and against officers of a company or other body corporate.

17.30.14 The Supply of Machinery (Safety) Regulations 2008

(a) Introduction

The Supply of Machinery (Safety) Regulations 2008 comes into force on 29 December 2009. They make a number of changes to existing law intended to streamline the process of CE marking.

The new Machinery Regulations implement Directive 2006/42/EC, which also amends the EU Lifts Directive (95/16/EC).

Directive 2006/42/EC, the new Machinery Directive, aims to rationalize and modernize the existing Machinery Directive (98/37/EC). It aims to bring greater clarity to the legislative framework applying to the production and use of machinery. This is in terms of, for example re-drafting the scope and applications of machinery covered by the Directive, and in terms of making a clearer distinction between the Machinery Directive and other relevant legislation, such as the Lifts Directive and the Low Voltage Directive (LVD).

The Machinery Directive is based on the principles of the EU's 'New Approach'.

The 'New Approach' aims to maximize the free movement of goods within the internal market of the EU by removing barriers to trade resulting from the adoption of diverging national technical standards and regulations.

The 'New Approach' introduces 'harmonized legislation' limited to the essential requirements that products must meet to be placed on the EU market, and to benefit from free movement within the community. 'New Approach' Directives are based on Article 95 of the EC Treaty.

The first Directive in relation to machinery entered into force on 1 January 1993. This Directive has been amended three times and was consolidated into the existing Directive on machinery (Directive 98/37/EC) which came into force in August 1998.

In 2001, the European Commission proposed a new Machinery Directive with the aim of modernizing, clarifying and simplifying the text of the existing Directive.

The resulting new Machinery Directive was adopted on 29 June 2006, and member states are required to transpose its requirements into national law by 29 June 2008, with the provisions of the Directive being applicable in member states from 29 December 2009. The 1992 Regulations and amendments are revoked when these Regulations come into force.

(b) Objectives
The new Machinery Directive has, like the existing Machinery Directive, the following main objectives:

➤ to ensure the free movement of machinery falling within its scope across the member states of the EU (an 'internal market objective');

➤ to provide for the appropriate level of health and safety for persons in the EU using and coming into contact with machinery (a 'health and safety objective');

➤ to provide for the appropriate level of environmental protection, and protection of animals, in situations where machinery is involved (an 'environmental and animal protection objective').

(c) Main requirements
The new Machinery Directive, and therefore the new Supply of Machinery Regulations, requires manufacturers of machinery (or their 'authorized representatives') to undertake the following activities before they can place machinery on the European market or put it into service:

(i) ensure the machinery satisfies relevant essential health and safety requirements;

(ii) ensure a technical file is made available, demonstrating that the design, manufacture and operation of machinery complies with certain essential health and safety requirements;

(iii) provide necessary information, relating to the machinery, such as instructions for use;

(iv) carry out appropriate procedures for assessing conformity, draw up a declaration of conformity and ensure this accompanies the machinery;

(v) affix a CE marking to the machinery.

In addition, before placing 'partly completed machinery' on the market, a manufacturer (or his 'authorized representative') is required to: prepare technical documentation covering the design, manufacture and operation of the partly completed machinery; prepare and provide assembly instructions for the partly completed machinery; and provide a 'declaration of incorporation' to show that partly completed machinery conforms to the necessary essential requirements.

The requirements on manufacturers (or their 'authorised representatives') in relation to machinery contained within the new Machinery Directive are broadly the same as those contained in the existing Machinery Directive, except where the two new types of machinery have been bought into scope, and where there are more specific requirements for partly completed machinery.

While machinery placed on the market before 29 December 2009 must continue to comply with Directive 98/37/EC, it can be assumed that a product that complies with the essential requirements of the new Machinery Directive continues to comply with the current Directive.

A manufacturer shall establish an EC Declaration of conformity according to Directive 2006/42/EC for products first placed on the market as from 29 December 2009.

The European Commission is issuing a mandate to CEN and Cenelec to develop the necessary new standards and ensure that the current standards are checked against Directive 2006/42/EC and adapted as necessary. Furthermore, all harmonized standards must include a reference to the new Directive. The Commission intends to publish a list of harmonized standards supporting Directive 2006/42/EC before the Directive becomes applicable.

(d) Principal changes
In brief, the principal changes:

➤ provide a new definition of the core term 'machinery', 'safety components' and 'lifting accessories', to make them easier to understand;

➤ introduce the concept of 'partly completed machinery', in order to address uncertainties about the respective duties of operators at various stages along complex production chains;

➤ introduce definitions of 'manufacturer' and 'placing on the market' which the previous Machinery Directive used but did not explicitly define (and which apply generally to 'New Approach' Directives);

➤ bring builders' goods hoists and cartridge-operated tools into the scope of the Regulations;

➤ clarify the interface with the UK Lifts Regulations;

➤ include a new voluntary full quality assurance module that will allow certain manufacturers (or 'responsible persons' in the terminology of the Regulations) of Annex IV products to carry out conformity assessment more efficiently than has been possible previously, which should be a particular benefit to those manufacturing prototype and bespoke machinery;

➤ introduce the new, lighter procedure for manufacturers using harmonized standards to produce hazardous products (as listed in Annex IV) whereby they are no longer required to deposit technical documentation for safe keeping by a notified body;

➤ make some small changes to the contents of Declarations of Conformity which will equip the enforcement authorities (sometimes referred to as the market surveillance authorities) with better means of tracing products. This will, in turn, make it easier for them to monitor the market and take action against non-compliant machinery and those supplying it and thus to ensure fair competition across the sector;

➤ introduce a new set of Essential Safety Requirements which are in a clearer format and a more logical sequence than in the existing Regulations. Manufacturers who take on board the new requirements, ahead of their officially coming into force, will find that they automatically fulfil the old set of requirements.

17.30.15 Sources of references

Buying New Machinery – A Short Guide to the Law INDG 271, 1998, HSE Books, ISBN 978 0 7176 1559 9.

Supplying New Machinery Advice to Suppliers INDG270, HSE 1998, Web only.

The Supply of Machinery (Safety) Regulations 2008, OPSI: http://www.opsi.gov.uk/si/si2008/uksi_20081597_en_1 Printed copies are available from the TSO bookshop.

17.31 Control of Vibration at Work Regulations 2005

17.31.1 Introduction

These Regulations implement European Directive Vibration Directive 2002/44/EC. They came into force in July 2005 with some transitional arrangements for the exposure limits until 2010 (2014 for WBV the whole-body vibration exposure limit value for agriculture and forestry sectors).

The Regulations impose duties on employers to protect employees who may be exposed to risk from vibration at work, and other persons who may be affected by the work.

17.31.2 Interpretation – Regulation 2

'Daily exposure' means the quantity of mechanical vibration to which a worker is exposed during a working day, normalized to an 8-hour reference period, which takes account of the magnitude and duration of the vibration.

'Hand–arm vibration' (HAV) means mechanical vibration which is transmitted into the hands and arms during a work activity.

'Whole-body vibration' (WBV) means mechanical vibration which is transmitted into the body when seated or standing through the supporting surfaces, during a work activity or as described in Regulation 5(3)(f).

17.31.3 Application – Regulation 3

For work equipment first provided to employees for use prior to 6 July 2007 and where compliances with the exposure limit values is not possible, employers have until 2010 to comply and in the case of agriculture and forestry 2014 (for WBV).

However, action must be taken to use the latest technical advances and the organizational measures in accordance with Regulation 6(2).

17.31.4 Exposure limit values and action values – Regulation 4

HAV

(a) 8-hour daily exposure limit value is 5 m/s^2 A(8)

(b) 8-hour daily exposure action value is 2.5 m/s^2 A(8)

(c) daily exposure ascertained as set out in Schedule 2 Part I.

WBV

(a) 8-hour daily exposure limit value is 1.15 m/s^2 A(8)

(b) 8-hour daily exposure action value is 0.5 m/s^2 A(8)

(c) daily exposure ascertained as set out in Schedule 2 Part I.

17.31.5 Assessment of risk to health created by vibration at the workplace – Regulation 5

A suitable and sufficient risk assessment must be made where work liable to expose employees is carried on. The risk assessment must identify the measures which need to be taken to comply with these Regulations.

Assessment of daily exposure should be by means of:

➤ observation of specific working practices;

➤ reference to relevant work equipment vibration data;

➤ if necessary, measurement of the magnitude of vibration to which employees are exposed;

➤ likelihood of exposure at or above an exposure action value or above an exposure limit value.

The risk assessment shall include consideration of:

➤ the magnitude, type and duration of exposure, including any exposure to intermittent vibration or repeated shocks;
➤ the effects of exposure to vibration on employees whose health is at particular risk from such exposure;
➤ any effects of vibration on the workplace and work equipment, including the proper handling of controls, the reading of indicators, the stability of structures and the security of joints;
➤ any information provided by the manufacturers of work equipment;
➤ the availability of replacement equipment designed to reduce exposure to vibration;
➤ any extension of exposure at the workplace to WBV beyond normal working hours, including exposure in rest facilities supervised by the employer;
➤ specific working conditions such as low temperatures;
➤ appropriate information obtained from health surveillance including, where possible, published information.

The risk assessment shall be reviewed regularly and forthwith if there:

➤ is reason to suspect that the risk assessment is no longer valid; or
➤ has been a significant change in the work to which the assessment relates;
and where, as a result of the review, changes to the risk assessment are required, those changes shall be made.
The employer shall record:

➤ the significant findings of the risk assessment as soon as is practicable after the risk assessment is made or changed;
➤ the measures which they have taken and which they intend to take to meet the requirements of Regulation 6 and 8.

17.31.6 *Elimination or control of exposure to vibration at the workplace – Regulation 6*

Following the general principles of prevention in the Management Regulations, the employer shall ensure that risk from the exposure of his or her employees to vibration is either eliminated at source or reduced as low as reasonably practicable.

Where this is not reasonably practicable and the risk assessment indicates that an exposure action value is likely to be reached or exceeded, the employer shall reduce exposure to as low as reasonably practicable by establishing and implementing a programme of organizational and technical measures which is appropriate.

Consideration must be given to:

➤ other working methods which eliminate or reduce exposure to vibration;
➤ a choice of work equipment of appropriate ergonomic design which, taking account of the work to be done, produces the least possible vibration;
➤ the provision of auxiliary equipment which reduces the risk of injuries caused by vibration;
➤ appropriate maintenance programmes for work equipment, the workplace and workplace systems;
➤ the design and layout of workplaces, work stations and rest facilities;
➤ suitable and sufficient information and training for employees, such that work equipment may be used correctly and safely, in order to minimize their exposure to vibration;
➤ limitation of the duration and intensity of exposure to vibration;
➤ appropriate work schedules with adequate rest periods;
➤ the provision of clothing to protect employees from cold and damp.

Subject to implementation dates, the employer shall:

➤ ensure that no employees are exposed to vibration above an exposure limit value; or
➤ ensure if an exposure limit value is exceeded, the employer shall forthwith:
  • take action to reduce exposure to vibration below the limit value;
  • identify the reason for that limit being exceeded;
  • modify the organizational and technical measures taken to prevent it being exceeded again.

This shall not apply where the exposure of an employee to vibration is usually below the exposure action value but varies markedly from time to time and may occasionally exceed the exposure limit value, provided that:

➤ any exposure to vibration averaged over 1 week is less than the exposure limit value;
➤ there is evidence to show that the risk from the actual pattern of exposure is less than the corresponding risk from constant exposure at the exposure limit value;
➤ risk is reduced to as low as reasonably practicable, taking into account the special circumstances;
➤ the employees concerned are subject to increased health surveillance, where such surveillance is appropriate.

Account must be taken of any employee whose health is likely to be particularly at risk from vibration.

17.31.7 Health surveillance – Regulation 7

If:

➤ the risk assessment indicates that there is a risk to the health of any employees who are, or are liable to be, exposed to vibration; or

➤ employees are exposed to vibration at or above an exposure action value

the employer shall ensure that such employees are under suitable health surveillance. Records must be kept, and employees given access to their own records and the enforcing authorities provided with copies, as may be required. A range of specified action is required if problems are found with the health surveillance results.

17.31.8 Information, instruction and training – Regulation 8

If:

➤ the risk assessment indicates that there is a risk to the health of employees who are, or who are liable to be, exposed to vibration; or

➤ employees are exposed to vibration at or above an exposure action value

the employer shall provide those employees and their representatives with suitable and sufficient information, instruction and training.

The information, instruction and training provided shall include:

➤ the organizational and technical measures taken in order to comply with the requirements of Regulation 6;

➤ the exposure limit values and action values;

➤ the significant findings of the risk assessment, including any measurements taken;

➤ the why and how to detect and report signs of injury;

➤ entitlement to appropriate health surveillance;

➤ safe working practices to minimize exposure to vibration;

➤ the collective results of any health surveillance undertaken in accordance with Regulation 7 in a form calculated to prevent those results from being identified as relating to a particular person.

The information, instruction and training required must be adapted to take account of significant changes in the type of work carried out or methods of work used by the employer, and cover all persons who carry out work in connection with the employer's duties under these Regulations.

17.31.9 Sources of references

Control Back-Pain Risks from Whole-Body Vibration Advice for Employers on the Control of Vibration at Work Regulations 2005 INDG242 (rev 1), HSE Books, 2005, ISBN 978 0 7176 6119 0.

Control the Risks from Hand-Arm Vibration Advice for Employers on the Control of Vibration at Work Regulations 2005, SI 2005 No. 1093, OPSI: http://www.opsi.gov.uk/si/si2005/uksi_20051093_en.pdf

Drive Away Bad Backs Advice for Mobile Machine Operators and Drivers INDG404, 2005, HSE Books, ISBN 9780 0 7176 6120 6.

Hand-Arm Vibration Advice for Employees Pocket Card INDG296 (rev 1), 2005, HSE Books, ISBN 978 0 7176 6118 3.

Hand-Arm Vibration: Control of Vibration at Work Regulations L140, HSE Books, 2005, ISBN 978 0 7176 6125 1.

Vibration Solutions Practical Way to Reduce the Risk of Hand-Arm Vibration HSG170, HSE Books, 1997, ISBN 978 0 7176 0954 3.

Whole Body Vibration: The Control of Vibration at Work Regulations Guidance on Regulations L141, HSE Books, 2005, ISBN 987 0 7176 6126 8.

17.32 Workplace (Health, Safety and Welfare) Regulations 1992 as amended in 2002

17.32.1 General

These Regulations were made to implement the European Directive on the minimum safety and health requirements for the workplace. A workplace for these purposes is defined very widely to include any part of non-domestic premises to which people have access while at work and any room, lobby, corridor, staircase or other means of access to or exit from them. The main exceptions to these rules are constructions sites, means of transport, mines and quarries and other mineral extraction sites.

Employers have a duty to ensure that workplaces under their control comply with the requirements of these Regulations. Any workplace and relevant equipment, devices and systems must be properly maintained. A maintenance system must be set up where appropriate, that is where a fault could result in a failure to comply and/or mechanical ventilation systems provided under Regulation 6.

The main requirements are summarized below. They are expressed in very general terms, and in each case it will be necessary to turn to the Approved Code or Practice associated with these Regulations for clarification of what is necessary to meet the objectives set.

17.32.2 Health: The Working Environment Regulations 6–10

Ventilation
Ventilation must be effective in enclosed areas, and any plant used for this purpose must incorporate warning devices to signal breakdowns which might endanger health or safety.

A reasonable temperature
A reasonable temperature must be maintained during working hours and sufficient thermometers must be provided to enable people at work to determine the temperature in any workroom. The temperature should be comfortable without the need for special clothing. Special guidance is available for areas like food processing where it could be very hot or very cold.

Temperature should be at least 16°C, or where strenuous effort is involved, 13°C.

Lighting
Lighting must be suitable and sufficient. This should be natural lighting sfarp.

Emergency lighting
Emergency lighting shall be provided where persons are especially exposed to danger if artificial light fails. Lights should avoid glare and dazzle and should not themselves cause a hazard. They should not be obscured, and be properly maintained.

Workplaces, furniture and fittings
Workplaces furniture and fittings should be kept sufficiently clean. Surfaces inside buildings shall be capable of being kept sufficiently clean.

Floors
Floors should not be slippery and wall surfaces should not increase fire risks.

Wastes
Wastes should not be allowed to accumulate, except in suitable receptacles and should be kept free from offensive waste products and discharges.

Room dimensions
Room dimensions have to allow adequate unoccupied space to work in and to move freely; 11 m³ minimum per person is required, excluding anything over 3 m high and furniture, etc.

Workstations
Workstations shall be suitable for any person in the workplace who is likely to work at that workstation and for any work that is likely to be done there.

Outside workstations
Outside workstations shall provide, sfarp, protection from adverse weather; provide adequate means of escape in emergencies; and ensure that no person is likely to slip or fall.

Seating
Seating shall be provided where work can or must be done sitting and shall be suitable for the person as well as the work. A footrest shall be provided where necessary.

17.32.3 Safety: Accident Prevention Regulations 5, 12–19

Maintenance
The workplace and equipment, devices and systems shall be maintained (including cleaned as appropriate) in an efficient state, efficient working order and in good repair, and where appropriate subject to a system of maintenance. This generally means planned rather than breakdown maintenance. Systems include ventilation, emergency lighting, safety fences, window cleaning devices, and moving walkways.

Floors and traffic routes
Floors and the surface of traffic routes shall be suitably constructed for their intended purpose and free of slope and holes (unless fenced). They should not be uneven or slippery. The traffic routes should be of adequate width and height to allow people and vehicles to circulate safely with ease and they should be kept free of obstructions.

Additional precautions are necessary where pedestrians have to cross or share vehicle routes. Open sides of staircases should be fenced with an upper rail 900 mm or higher and a lower rail. Loading bays should have exits or refuges to avoid people getting crushed by vehicles.

Falls and falling objects
Now covered by Work at Height Regulations 2005 (see Section 17.33).

Tanks and pits

Where there is a risk of falling into a tank, pit or structure containing a dangerous substance that is likely to:

➤ scald or burn;

➤ be poisonous or corrosive;

➤ have an asphyxiating gas fume or vapour;

➤ have any granular or free-flowing substance likely to cause harm,

measures must be taken to securely fence or cover the tank, pit or structure.

Ladders and roofs

Now covered by Work at Height Regulations 2005 (see Section 17.33).

Glazing

Windows and transparent doors and partitions must be appropriately marked and protected against breakage.

Windows

Windows and skylights must open and close safely, and be arranged so that people may not fall out of them. They must be capable of being cleaned safely.

Traffic routes

Pedestrians and vehicles must be able to circulate safely. Separation should be provided between vehicle and people at doors, gateways and common routes. Workplaces should have protection from vehicles.

Doors and gates

Doors, gates and moving walkways have to be of sound construction and fitted with appropriate safety devices.

17.32.4 Welfare: provision of facilities – Regulations 20–25

Sanitary conveniences and washing facilities

Suitable and sufficient sanitary conveniences and washing facilities should be provided at readily accessible places. The facilities must be kept clean, adequately ventilated and lit. Washing facilities should have running hot and cold or warm water, soap and clean towels or other method of cleaning or drying. If necessary, showers should be provided. Men and women should have separate facilities unless each facility is in a separate room with a lockable door and is for use by only one person at a time.

Drinking water

An adequate supply of wholesome drinking water, with an upward drinking jet or suitable cups, should be provided. Water should only be supplied in refillable enclosed containers where it cannot be obtained directly from a mains supply.

Accommodation for clothing and facilities for changing

Adequate, suitable and secure space should be provided to store workers' own clothing and special clothing. The facilities should allow for drying clothing. Changing facilities should also be provided for workers who change into special work clothing.

Facilities to rest and to eat meals

Suitable and sufficient, readily accessible, rest facilities should be provided. Arrangements should include suitable facilities to eat meals; adequate seats with backrests and tables; and means of heating food (unless hot food is available nearby) and making hot drinks.

Canteens or restaurants

Canteens and restaurants may be used as rest facilities, provided there is no obligation to purchase food.

Suitable rest facilities

Suitable rest facilities should be provided for pregnant women and nursing mothers. They should be near sanitary facilities and, where necessary, include the facility to lie down.

Non-smokers

This is now covered by the smoke-free legislation which bans smoking in largely enclosed or enclosed work or public places. See Section 17.34.13.

17.32.5 Sources of references

General Ventilation in the Workplace: Guidance for Employers HSG202, HSE Books, 2000, ISBN978 0 7176 1793 7.

How to Deal with Sick Building Syndrome: Guidance for Employers, Building Owners and Building Managers HSG132, HSE Books, 1995, ISBN 978 0 7176 0861 4.

Lighting at Work HSG38, HSE Books, 1998, ISBN 978 0 7176 1232 1.

Seating at Work HSG57, HSE Books, 1998, ISBN 978 0 7176 1231 4.

The Workplace (Health, Safety and Welfare) Regulations 1992, ISBN 978 0 11 025804 5, OPSI: http://www.opsi.gov.uk/si/si1992/Uksi_19923004_en_1.htm

Workplace Health, Safety and Welfare. Workplace (Health, Safety and Welfare) Regulations 1992 Approved Code of Practice and Guidance L24, 1992, HSE Books ISBN 978 0 7176 0413 5.

Workplace Health, Safety and Welfare A Short Guide, INDG244 (rev 20), 2007, HSE Books, ISBN 978 0 7176 6277 7.

17.33 Work at Height Regulations 2005 as amended in 2007

17.33.1 Introduction

These Regulations bring together all current requirements on work at height into one goal-based set of Regulations. They implement the requirements of the 2nd Amending Directive (2001/45/EC) to the Use of Work Equipment Directive (89/955/EEC) which sets out requirements for work at height. The 2nd Amending Directive has become known as the Temporary Work at Height Directive.

The Work at Height (Amendment) Regulations 2007, which came into force on 6 April 2007, apply to those who work at height providing instruction or leadership to one or more people engaged in caving or climbing by way of sport, recreation, team building or similar activities in Great Britain.

The Regulations require a risk assessment for all work conducted at height and arrangements to be put in place for:

➤ eliminating or minimizing risks from working at height;
➤ safe systems of work for organizing and performing work at height;
➤ safe systems for selecting suitable work equipment to perform work at height;
➤ safe systems for protecting people from the consequences of work at height.

17.33.2 Definitions – Regulation 2

'Work at height' means:

(a) work in any place, including a place at or below ground level;
(b) obtaining access to or egress from such place while at work except by a staircase in a permanent workplace where, if measures required by these Regulations were not taken, a person could fall a distance liable to cause personal injury.

'Working platform':

(a) means any platform used as a place of work or as a means of access to or egress from a place of work;
(b) includes any scaffold, suspended scaffold, cradle, mobile platform, trestle, gangway, gantry and stairway which is so used.

17.33.3 Organization, planning and competence – Regulations 4 and 5

Work at height must be properly planned, appropriately supervised and carried out in a manner which is, sfarp, safe. The selection of appropriate work equipment is included in the planning. Work must not be carried out if the weather conditions would jeopardize safety or health (this does not apply where members of the police, fire, ambulance or other emergency services are acting in an emergency).

All people involved in work at height activity including planning, organizing and supervising must be competent for such work, or if being trained, under competent supervision.

17.33.4 Avoidance of Risk – Regulation 6

A risk assessment carried out under the Management Regulations must be taken into account when identifying the measures required by these Regulations. Work at height should be avoided if there are reasonably practicable alternatives.

Where work at height is carried out employers must take suitable and sufficient measures to prevent persons falling a distance liable to cause personal injury. The measures include:

(a) ensuring work is carried out:
   (i) from an existing workplace;
   (ii) using existing means of access and egress that comply with Schedule 1 of the Regulations (assuming it is safe and ergonomic to do so).
(b) where this is not reasonably practicable, providing work equipment (sfarp) for preventing a fall occurring.

Employers must take steps to minimize the distance and the consequences of a fall, if it is not prevented. Where the distance cannot be minimized (sfarp) the consequence of a fall must be minimized and additional training, instruction and other additional suitable and sufficient measures must be adopted to prevent a person falling a distance liable to cause personal injury.

See Figure 17.10, which shows a flow chart for the work at height risk assessment.

Work at height – flow chart

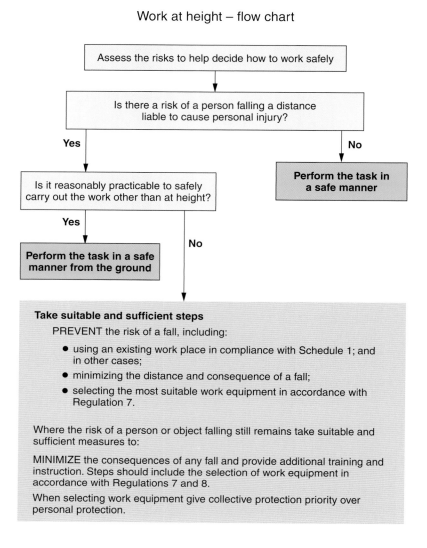

Figure 17.10 Work at height – flow chart.

17.33.5 General principles for selection of work equipment – Regulation 7

Collective protection measures must have priority over personal protection measures.

Employers must take account of:

➤ working conditions and the risks to the safety of persons at the place where the work equipment is to be used;
➤ the distance to be negotiated for access and egress;
➤ distance and consequences of a potential fall;
➤ duration and frequency of use;
➤ need for evacuation and rescue in an emergency;
➤ additional risks of using, installing and removing the work equipment used or evacuation and rescue from it;
➤ other provisions of the Regulations.

Work equipment must:

➤ be appropriate for the nature of the work and the foreseeable loadings;
➤ allow passage without risk;
➤ be the most suitable work equipment having regard in particular to Regulation 6.

17.33.6 Requirements for particular work equipment – Regulation 8

Regulation 8 requires that various pieces of work equipment comply with the schedules to the Regulations as follows:

(a) A guard-rail, toe-board, barrier or similar means of protection comply with Schedule 2.

Schedule 2 in summary:
- they must be suitable, of sufficient dimensions, of sufficient strength and rigidity;
- be so placed, secured and used to prevent accidental displacement;
- prevent fall of persons and materials;
- supporting structure to be of sufficient strength and suitable for the purpose;
- no lateral opening save at point of access to a ladder or stairway where an opening is necessary and:
  - must be in place except for a time to gain access perform a particular task and then replaced;
  - compensatory safety measures must be provided if protection removed temporarily.

(b) A working platform, comply with Schedule 3 Part 1 and, in addition, where scaffolding is provided, Schedule 3 Part 2.

Schedule 3 Part 1 in summary
Supporting structure must:
- be suitable, of sufficient strength and rigidity;
- if wheeled, be prevented, by appropriate device, from moving during work;
- not slip and have secure attachment;
- be stable while being erected, used modified or dismantled.

Working platform to:
- be suitable and strong enough;
- have no accidental displacement of components;
- remain stable during dismantling;
- be dismantled so as to prevent accidental displacement;
- be of sufficient dimensions for safe passage and use;
- have a suitable surface and be constructed to prevent people falling through;
- have a suitable surface and be constructed to prevent material or objects falling through unless measures have been taken to protect other persons from falling objects;
- be erected, used and maintained so that risk of slipping and tripping is prevented and no person can

be caught between working platform and adjacent structure;
- be not loaded so as to give risk of collapse or deformation.

Schedule 3 Part 2 additional for scaffolds:
- strength and stability calculations required unless they are available already or scaffold is assembled in conformity with a generally recognized standard configuration;
- an assembly, use and dismantling plan must be drawn up. Could be standard plan with supplements;
- copy of the plan and instructions to be available for persons doing the work;
- dimensions, form and layout to be suitable;
- when not available for use to be marked with warning signs and physical barrier preventing access;
- must be assembled, dismantled or significantly altered under supervision of competent person.

(c) a net, airbag or other collective safeguard for arresting falls which is not part of a personal fall protection system, comply with Schedule 4.

(d) any personal fall protection system to comply with Part 1 of Schedule 5.
- in the case of a work positioning system, comply with Part 2 of Schedule 5;
- in the case of rope access and positioning techniques, comply with Part 3 of Schedule 5;
- in the case of a fall arrest system, comply with Part 4 of Schedule 5;
- in the case of a work restraint system, comply with Part 5 of Schedule 5.

(e) a ladder to comply with Schedule 6.
Schedule 6 in summary:
- only to be used for work at height if the risk assessment demonstrates that the use of more suitable equipment is not justified because of the low risk and the short duration of use or existing features on site that cannot be altered;
- surface on which ladder rests must be stable, firm, of sufficient strength and suitable composition so its rungs remain horizontal;
- must be positioned to ensure stability;
- suspended ladder attached firmly without swing (except flexible ladder);
- portable ladder to be prevented from slipping by securing the stiles near their upper and lower ends and using an anti-slip device or by any other equivalent measures;
- when used for access must protrude above place of landing unless other firm handholds provided;

➤ interlocking or extension pieces must not move, relative to each other, during use;

➤ mobile ladder to be prevented from moving before being stepped on;

➤ where a ladder or run of ladders rises a vertical distance of 9 metres or more above its base, there shall, where reasonably practicable, be provided at suitable intervals sufficient safe landing areas or rest platforms;

➤ must be used so that a secure foothold and handhold are always available;

➤ must be used so that user can maintain a secure handhold while carrying a load (exceptions for using stepladders for low-risk, short-duration work).

17.33.7 Fragile surfaces – Regulation 9

Employers must take steps to prevent people falling through any fragile surface. Steps to be taken include:

➤ avoid, sfarp, passing across or near, or working on or near, a fragile surface;

➤ where this is not reasonably practicable:
  • provide suitable and sufficient platforms, covering, guard-rails or other similar means of support or protection and use them;
  • provide suitable and sufficient guard-rails to prevent persons falling through fragile surfaces;
  • where a risk of falling remains, take suitable and sufficient measures to minimize the distances and consequences of a fall.

➤ provide prominent warning signs at approach to the fragile surface, or make people aware of the fragile surface if not reasonably practicable;

➤ where the risk of falling remains despite the precautions a suitable and sufficient, fall arrest system must be provided.

17.33.8 Falling objects and danger areas – Regulations 10 and 11

Suitable and sufficient steps must be taken:

➤ to prevent the fall of any material or object;

➤ if reasonably practicable, to prevent the fall of any material or object, prevent persons being struck by falling objects or material (if liable to cause personal injury) by for example providing covered walkways or fan scaffolds;

➤ to prevent material or objects being thrown or tipped from height where it is likely to cause injury;

➤ to store materials and objects in such a way as to prevent risk to any person arising from collapse, overturning or unintended movement;

➤ where danger of being struck exists, unauthorized persons must be kept out of the area by suitable devices and the area clearly indicated;

➤ to store materials and objects so as to prevent them collapsing, overturning or move in a way that could be a risk to people.

17.33.9 Inspection of work equipment – Regulation 12

This Regulation applies only to work equipment to which Regulation 8 and Schedules 2–6 apply. The requirements include the following:

➤ where work equipment used for work at height depends for safety on how it is installed or assembled, the employer must ensure that it is not used after installation or assembly in any position unless it has been inspected in that place;

➤ where work equipment is exposed to conditions causing deterioration which is liable to result in dangerous situations, it must also be inspected at suitable intervals and each time that exceptional circumstances which are liable to jeopardize the safety of the work equipment have occurred, for example a severe storm;

➤ in addition, a working platform used for construction and from which a person could fall more than 2 m has to be inspected in that position (mobile equipment on the site) before use and within the last 7 days. The particulars for this inspection are set out in Schedule 7;

➤ no work equipment (lifting equipment is covered under LOLER) may either leave the undertaking or be obtained from another undertaking without evidence of in-date inspection;

➤ results of inspections must be recorded and retained until the next due inspection is recorded. Reports must be provided within 24 hours and kept at the site until construction work is complete and thereafter at the office for 3 months.

17.33.10 Inspection of places of work at height – Regulation 13

The surface and every parapet, permanent rail or other such fall protection measure of every place of work at height must be checked on each occasion before the place is used.

17.33.11 Duties of persons at work – Regulation 12

Every person must, where working under the control of another person, report to that person any activity or defect relating to work at height which they know is likely to endanger themselves or others.

Work equipment must be used in accordance with training and instructions.

17.33.12 Sources of references

A Head for Heights Guidance for Working at Height in Construction. Video. 2003, HSE Books, ISBN 978 0 7176 2217 7.

A Tool Box Talk on Leaning Ladders and Stepladders Safety INDG403, HSE Books, 2005, ISBN 978 0 7176 6106 0.

Health and Safety in Roof Work HSG 33, 1998, HSE Books.

Height Safe Essential Health and Safety Information for People who Work at Height. HSE's website: http://www.hse.gov.uk. ISBN 0 7176 1425 5.

Preventing Falls from Ladders and Stepladders: An Employer's Guide INDG402, HSE Books, 2005, ISBN 0 7176 6105 9.

Safe Use of Ladders and Stepladders: An Employers' Guide INDG402, 2005, HSE Books, ISBN 0 7176 6105 9.

The Work at Height (Amendment) Regulations 2007, SI 2007 No. 114, OPSI: http://www.opsi.gov.uk/si/si2007/20070114.htm

The Work at Height Regulations 2005, SI 2005/735, OPSI http://www.opsi.gov.uk/si/si2005/20050735

The Work at Height Regulations 2005 (as amended). A Brief Guide INDG401 (rev 1), 2007, HSE Books, ISBN 9780 7176 6231 9.

Top Tips for Ladders and Stepladder Safety INDG405, HSE Books, 2006, ISBN 9780 7176 6127 5.

Tower Scaffolds CIS10, HSE Books, 2005.

The Selection and Management of Mobile Elevating Work Platforms CIS 58, HSE 2008, Web only.

Useful website – HSE's Falls from Height: http://www.hse.gov.uk/falls

Using Access Equipment Safely in Building Maintenance Work C100, HSE 2008, website.

17.34 Other relevant Regulations in brief

The following Regulations are included with very brief summaries.

17.34.1 Control of Asbestos Regulations 2006

The Control of Asbestos Regulations 2006 came into force on 13 November 2006 (Asbestos Regulations – SI 2006/2739).

These Regulations bring together the three previous sets of Regulations covering the prohibition of asbestos, the control of asbestos at work and asbestos licencing.

The Regulations prohibit the importation, supply and use of all forms of asbestos. They continue the ban introduced for blue and brown asbestos 1985 and for white asbestos in 1999. They also continue the ban on the secondhand use of asbestos products such as asbestos cement sheets and asbestos boards and tiles, including panels which have been covered with paint or textured plaster containing asbestos.

The ban applies to new use of asbestos. If existing asbestos-containing materials are in good condition, they may be left in place, their condition monitored and managed to ensure they are not disturbed.

The Asbestos Regulations also include the 'duty to manage asbestos' in non-domestic premises. Guidance on the duty to manage asbestos can be found in the 'Approved Code of Practice The Management of Asbestos in Non-Domestic Premises', L127, ISBN 0 7176 6209 8 and on the duty to manage area of this website.

The Regulations require mandatory training for anyone liable to be exposed to asbestos fibres at work (see Regulation 10). This includes maintenance workers and others who may come into contact with or who may disturb asbestos (e.g. cable installers) as well as those involved in asbestos removal work.

When work with asbestos or which may disturb asbestos is being carried out, the Asbestos Regulations require employers and the self-employed to prevent exposure to asbestos fibres. Where this is not reasonably practicable, they must make sure that exposure is kept as low as reasonably practicable by measures other than the use of respiratory protective equipment. The spread of asbestos must be prevented. The Regulations specify the work methods and controls that should be used to prevent exposure and spread.

Worker exposure must be below the airborne exposure limit (control limit). The Asbestos Regulations have a single control limit for all types of asbestos of 0.1 fibres per cm^3. A control limit is a maximum concentration of asbestos fibres in the air (averaged over any continuous 4-hour period) that must not be exceeded.

In addition, short-term exposures must be strictly controlled and worker exposure should not exceed 0.6 fibres per cm^3 of air averaged over any continuous

10-minute period using respiratory protective equipment if exposure cannot be reduced sufficiently using other means.

Respiratory protective equipment is an important part of the control regime but it must not be the sole measure used to reduce exposure and should only be used to supplement other measures. Work methods that control the release of fibres such as those detailed in the Asbestos Essentials Task Sheet (http://www.hse.gov.uk/asbestos/essentials/index.htm) for non-licenced work should be used. Respiratory protective equipment must be suitable, must fit properly and must ensure that worker exposure is reduced as low as is reasonably practicable.

Most asbestos removal work must be undertaken by a licensed contractor but any decision on whether particular work is licensable is based on the risk. Work is only exempt from licencing if:

➤ the exposure of employees to asbestos fibres is sporadic and of low intensity (but exposure cannot be considered to be sporadic and of low intensity if the concentration of asbestos in the air is liable to exceed 0.6 fibres per cm^3 measured over 10 minutes);

➤ it is clear from the risk assessment that the exposure of any employee to asbestos will not exceed the control limit;

➤ the work involves:
  • short, non-continuous maintenance activities. Work can only be considered as short, non-continuous maintenance activities if any one person carries out work with these materials for less than 1 hour in a 7-day period. The total time spent by all workers on the work should not exceed a total of 2 hours;
  • removal of materials in which the asbestos fibres are firmly linked in a matrix. Such materials include: asbestos cement; textured decorative coatings and paints which contain asbestos; articles of bitumen, plastic, resin or rubber which contain asbestos where their thermal or acoustic properties are incidental to their main purpose (e.g. vinyl floor tiles, electric cables, roofing felt) and other insulation products which may be used at high temperatures but have no insulation purposes, for example gaskets, washers, ropes and seals;
  • encapsulation or sealing of asbestos-containing materials which are in good condition; or
  • air monitoring and control, and the collection and analysis of samples to find out if a specific material contains asbestos.

Under the Asbestos Regulations, anyone carrying out work on asbestos insulation, asbestos coating or asbestos insulating board (AIB) needs a licence issued by HSE unless they meet one of the exemptions above.

Although a licence may not be needed to carry out a particular job, there is still a need to comply with the rest of the requirements of the Asbestos Regulations.

If the work is licensable there are a number of additional duties, to:

➤ notify the enforcing authority responsible for the site where the work is being done (for example, HSE or the local authority);

➤ designate the work area (see Regulation 18 for details);

➤ prepare specific asbestos emergency procedures;

➤ pay for employees to undergo medical surveillance.

The Asbestos Regulations require any analysis of the concentration of asbestos in the air to be measured in accordance with the 1997 WHO recommended method.

From 6 April 2007, a clearance certificate for re-occupation may only be issued by a body accredited to do so. At the moment, such accreditation can only be provided by the United Kingdom Accreditation Service (http://www.ukas.com/).

Sources of references

More details of how to undertake work with asbestos-containing materials, the type of controls necessary, what training is required and analytical methods can be found in the following HSE publications.

A Comprehensive Guide to Managing Asbestos in Premises HSG 227, HSE Books, 2002, ISBN 978 0 7176 2381 5.

Approved Code of Practice Work with Materials containing Asbestos (http://www.hse.gov.uk/pubns/books/l143.htm) L143, HSE Books, 2006, ISBN 978 0 7176 6206 0.

Asbestos Essentials HSG 210 (second edition), HSE Books, 2008, ISBN 978 0 7176 6263 0 (Asbestos Essentials task sheets are available on the Asbestos Essentials area of the HSE website http://www.hse.gov.uk/asbestos/essentials/index.htm).

Asbestos Kills: A Quick Guide to Protecting Yourself INDG418, HSE Books, ISBN 978 0 7176 6271 5.

Asbestos: The Analysts' Guide for Sampling, Analysis and Clearance Procedures HSG 248, ISBN 978 0 7176 2875 9.

Asbestos: the Licensed Contractors Guide HSG 247, HSE Books, 2006, ISBN 978 0 7176 2874 2.

Introduction to Asbestos Essentials HSG 213, HSE Books, 2001, ISBN 978 0 7176 1901 6.

The Management of Asbestos in Non-Domestic Premises L127, HSE Books, 2006, ISBN 978 0 7176 6209 8.

17.34.2 Corporate Manslaughter and Corporate Homicide Act 2007

The Act creates a new statutory offence of corporate manslaughter which will replace the common law offence of manslaughter by gross negligence where corporations and similar entities are concerned. In Scotland the new offence will be called 'corporate homicide'. An organization will have committed the new offence if it:

- owes a duty of care to another person in defined circumstances;
- there is a management failure by its senior managers;
- it amounts to a gross breach of that duty resulting in a person's death.

On conviction the offence will be punishable by an unlimited fine and the courts will be able to make remedial orders requiring organizations to take steps to remedy the management failure concerned. It is important to note that the Act does not create a new individual liability. Individuals may still be charged with the existing offence of manslaughter by gross negligence. Crown immunity will not apply to the offence, although a number of public bodies and functions will be exempt from it (in defined circumstances).

The Act came into force on 6 April 2008. Available at: http://www.opsi.gov.uk/acts/acts2007/pdf/ukpga_20070019_en.pdf.

17.34.3 Disability Discrimination Act 1995 and 2005

The Disability Discrimination Act 1995 (DDA 1995)

The Disability Discrimination Act (DDA) 1995 aims to end the discrimination that many disabled people face. This Act gives disabled people rights in the areas of:

- employment;
- education;
- access to goods, facilities and services;
- buying or renting land or property.

The Act also allows the government to set minimum standards so that disabled people can use public transport easily.

From 1 October 2004, Part 3 of the DDA 1995 has required businesses and other organizations to take reasonable steps to tackle physical features that act as a barrier to disabled people who want to access their services.

This may mean to remove, alter or provide a reasonable means of avoiding physical features of a building which make access impossible or unreasonably difficult for disabled people. Examples include:

- putting in a ramp to replace steps;
- providing larger, well-defined signs for people with a visual impairment;
- improving access to toilet or washing facilities.

Businesses and organizations are called 'service provider[s]' and include shops, restaurants, leisure centres and places of worship.

The Disability Discrimination Act 2005 (DDA 2005)

In April 2005 a new Disability Discrimination Act was passed, which amends or extends existing provisions in the DDA 1995, including:

- making it unlawful for operators of transport vehicles to discriminate against disabled people;
- making it easier for disabled people to rent property and for tenants to make disability-related adaptations;
- making sure that private clubs with 25 or more members cannot keep disabled people out, just because they have a disability;
- extending protection to cover people who have HIV, cancer and multiple sclerosis from the moment they are diagnosed;
- ensuring that discrimination law covers all the activities of the public sector;
- requiring public bodies to promote equality of opportunity for disabled people.

Other changes came into force in December 2006 – the Disability Rights Commission (DRC) website has more details on these. Available at: http://www.opsi.gov.uk/acts/acts2005/ukpga_20050013_en_1.

17.34.4 Electrical Equipment (Safety) Regulations 1994

These Regulations came into force in January 1995 and relate to the supply of electrical equipment with a working voltage between 50 and 1000 volts and are made under the Consumer Protection Act 1987. They apply to suppliers, which include both landlords and letting agents.

The Regulations apply to all mains voltage household electrical goods and require them to be safe so that there is no risk of injury or death to humans or pets, or risk of damage to property. They do not apply to fixed electrical wiring and built-in appliances like central heating systems. The Regulations also require that instructions be provided where safety depends on the user being aware of certain issues and equipment should be labelled with the CE marking.

There are other electrical consumer Regulations, like The Plugs and Sockets etc. (Safety) Regulations 1994 and the Low Voltage Electrical Equipment (Safety) Regulations 1989.

Available at: http://www.opsi.gov.uk/si/si1994/Uksi_19943260_en_1.htm#tcon.

17.34.5 Gas Appliances (Safety) Regulations 1995

The Gas Appliance Regulations cover the safety standards on new gas appliances which have to:

➤ satisfy safety and efficiency standards;
➤ undergo type-examination and supervision during manufacture;
➤ carry the CE mark and specified information;
➤ be accompanied by instructions and warnings in the language of destination.

Available at: http://www.opsi.gov.uk/si/si1995/Uksi_19951629_en_1.htm.

17.34.6 Gas Safety (Installation and Use) Regulations 1998

The Installation and Use Regulations place duties on gas consumers, installers, suppliers and landlords to ensure that:

➤ only competent people work on gas installations;[3]
➤ no one is permitted to use suspect gas appliances;
➤ landlords are responsible, in certain cases, to make sure that fittings and flues are maintained;
➤ with the exception of room-sealed appliances there are restrictions on gas appliances in sleeping accommodation;

---

[3]From April 2009 a new gas installer registration scheme replaced the former CORGI gas scheme. The new 'Gas Safe Register' is run by the Capita Group Plc on behalf of the HSE. A Gas Safety engineer must be used for any type of gas work, including installation, maintenance and servicing. See: http://www.gassaferegister.co.uk/default.aspx.

➤ instantaneous gas water heaters must be room-sealed or fitted with appropriate safety devices.

Available at: http://www.opsi.gov.uk/si/si1998/19982451.htm.

17.34.7 Control of Lead at Work Regulations 2002

These Regulations came into force in November 2002 and impose requirements for the protection of employees who might be exposed to lead at work and others who might be affected by the work. The Regulations:

➤ require occupational exposure levels for lead and lead alkyls;
➤ require blood-lead action and suspension levels for women of reproductive capacity and others;
➤ re-impose a prohibition for women of reproductive capacity and young persons in specified activities;
➤ require an employer to carry out a risk assessment;
➤ require employers to restrict areas where exposures are likely to be significant if there is a failure of control measures;
➤ impose requirements for the examination and testing of engineering controls and RPE and the keeping of PPE;
➤ impose new sampling procedures for air monitoring;
➤ impose requirements in relation to medical surveillance;
➤ require information to be given to employees;
➤ require the keeping of records and identification of containers and pipes.

Available at: http://www.opsi.gov.uk/si/si1998/19982451.htm.

17.34.8 Occupiers Liability Acts 1957 and 1984 – Civil Law

The 1957 Act concerns the duty that the occupier of premises has towards visitors in relation to the condition of the premises and to things which have or have not been done to them. The Act imposes the following:

➤ there is a duty to take reasonable care to see that a visitor is reasonably safe in using the premises for the purpose for which they were invited or permitted by the occupier to be there;
➤ the common duty of care will differ depending on the visitor, so a greater duty is owed to children;
➤ an occupier can expect that a person in the exercise of their trade or profession will appreciate and guard against normal risks, for example, a window cleaner;

➤ no duty of care is owed to someone exercising a public right of way.

The 1984 Act extends the duty of care to persons other than visitors, i.e. trespassers. The occupier has to take reasonable care in all the circumstances to see that non-visitors do not get hurt on the premises because of its condition or the things done or not done to it. The occupier must cover perceived dangers and must have reasonable grounds to know that the trespassers may be in the vicinity.

Available at: http://www.opsi.gov.uk/acts/acts1984/PDF/ukpga_19840003_en.pdf.

17.34.9 Health and Safety (Offences) Act 2008

New legislation, the Health and Safety Offences Act 2008, increases penalties and provide courts with greater sentencing powers for those who flout health and safety legislation.

The Act raises the maximum penalties that can be imposed for breaching health and safety regulations in the lower courts from £5000 to £20 000 and the range of offences for which an individual can be imprisoned has also been broadened.

The Act amends section 33 of the HSW Act 1974, and raises the maximum penalties available to the courts in respect of certain health and safety offences. It received Royal Assent on 16 October 2008 and came into force in January 2009.

Available at: http://www.opsi.gov.uk/acts/acts2008/en/ukpgaen_20080020_en_1.

17.34.10 Control of Pesticides Regulations 1986

These Regulations made under the Food and Environment Protection Act 1985 all came into force by 1 January 1988. The Regulations apply to any pesticide or any substance used generally for plant control and protection against pests of all types, including anti-fouling paint used on boats. They do not apply to substances covered by other Acts like the Medicines Act 1968, The Food Safety Act 1990, those used in laboratories and a number of other specific applications.

No person may advertise, sell, supply, store, or use a pesticide unless it has received ministerial approval and the conditions of the approval have been complied with. The approval may be experimental, provisional or full, and the Minister has powers to impose conditions.

The Regulations also cover the need for users to be competent and have received adequate information and training. Certificates of competence (or working under the supervision of a person with a certificate) are required where pesticides approved for agricultural use are used commercially.

Defra issued a consultation document in August 2007 which covers the update of these Regulations as follows:

➤ At present there are four Regulations (plus their amendments) to control and monitor marketing and use of pesticides in England and Wales.
- The Control of Pesticides Regulations 1986 and its amendment in 1997;
- The Plant Protection Products Regulations 2005 and its amendment in 2006;
- The Plant Protection Products (Basic Conditions) Regulations 1997;
- The Plant Protection Products (Fees) Regulations 2007.
➤ Defra's proposal is to consolidate these four Regulations (plus amendments) into one Regulation.

Available at: http://www.pesticides.gov.uk/approvals.asp?id=2182.

17.34.11 Pressure Systems Safety Regulations 2000 (PSSR)

These Regulations came into effect in February 2000 and replace the Pressure Systems and Transportable Gas Containers Regulations 1989. Transportable gas containers are covered by the Use of Transportable Pressure Receptacles Regulations 1996 (SI 1996, No. 2092).

The aim of PSSR is to prevent serious injury from the hazard associated with stored energy as a result of a pressure system or one of its parts failing. The Regulations cover:

➤ steam at any pressure;
➤ gases which exert a pressure in excess of 0.5 bar above atmospheric pressure;
➤ fluids which may be mixtures of liquids, gases and vapours where the gas or vapour phase may exert a pressure in excess of 0.5 bar above atmospheric pressure.

With the exception of scalding from steam, the Regulations do not consider the effects of the hazardous contents being released following failure. The stored contents are of concern where they can accelerate wear and cause more rapid deterioration and an increased risk of failure.

Available at: http://www.opsi.gov.uk/si/si2000/uksi_20000128_en.pdf.

17.34.12 Personal Protective Equipment Regulations 2002

These Regulations relate to approximation of the laws of EU member states. They place duties on persons who place PPE on the market to comply with certain standards. These requirements are that PPE must satisfy the basic health and safety requirements which are applicable to that class or type of PPE, the appropriate conformity assessment procedures must be carried out, CE marking must be correctly affixed and the PPE must not compromise the safety of individuals, domestic animals or property when properly maintained and used.

Available at: http://www.opsi.gov.uk/si/si2002/uksi_20021144_en.pdf.

17.34.13 Smoke-free legislation

Five sets of smoke-free Regulations set out the detail of the smoke-free legislation. Within the UK smoking law is a devolved issue and therefore similar legislation has been enacted in Scotland, Wales and Northern Ireland.

1. The Smoke-free (Premises and Enforcement) Regulations set out definitions of 'enclosed' and 'substantially enclosed' and the bodies responsible for enforcing smoke-free legislation.
2. The Smoke-free (Exemptions and Vehicles) Regulations set out the exemptions to smoke-free legislation and vehicles required to be smoke free.
3. The Smoke-free (Penalties and Discounted Amounts) Regulations set out the levels of penalties for offences under smoke-free legislation.
4. The Smoke-free (Vehicle Operators and Penalty Notices) Regulations set out the responsibility on vehicle operators to prevent smoking in smoke-free vehicles and the form for fixed penalty notices.
5. The Smoke-free (Signs) Regulations set out the requirements for No-smoking signs required under smoke-free legislation.

The smoke-free law has been introduced to protect employees and the public from the harmful effects of secondhand smoke.

Key points are:

➤ On 1 July 2007 (Scotland, 26 March 2006; Wales, 2 April 2007), the smoke-free law was introduced. It is now against the law to smoke in virtually all 'enclosed' and 'substantially enclosed' public places and workplaces. See below for definitions.

➤ Public transport and work vehicles used by more than one person must be smoke free at all times.
➤ No-smoking signs must be displayed in all smoke-free premises and vehicles.
➤ Staff smoking rooms and indoor smoking areas are no longer allowed, so anyone who wants to smoke has to go outside.
➤ Managers of smoke-free premises and vehicles have legal responsibilities to prevent people from smoking.
➤ If anyone is uncertain where they can or cannot smoke, they need to look for the No-smoking signs or ask someone in charge.

Penalties and fines for breaking the smoke-free law

Local councils are responsible for enforcing the new law in England. If you do not comply with the smoke-free law, you will be committing a criminal offence. The fixed penalty notices and maximum fine for each offence are:

➤ **Smoking in smoke-free premises or work vehicles:** A fixed penalty notice of £50 (reduced to £30 if paid in 15 days) imposed on the person smoking. Or a maximum fine of £200 if prosecuted and convicted by a court.
➤ **Failure to display No-smoking signs:** A fixed penalty notice of £200 (reduced to £150 if paid in 15 days) imposed on whoever manages or occupies the smoke-free premises or vehicle. Or a maximum fine of £1000 if prosecuted and convicted by a court.
➤ **Failing to prevent smoking in a smoke-free place:** A maximum fine of £2500 imposed on whoever manages or controls the smoke-free premises or vehicle if prosecuted and convicted by a court. There is no fixed penalty notice for this offence.

Local councils are responsible for enforcing the new law in England. If someone is smoking in a smoke-free place or vehicle, people should alert the manager or the person in charge of the premises or vehicle in the first instance. Alternatively they can contact the relevant local council or phone the Smoke-free Compliance Line on 0800 587 1667 to make a report. This information will be passed to the relevant local authority to follow up as appropriate.

Definition of enclosed and substantially enclosed

Premises are considered '**enclosed**' if they have a ceiling or roof and (except for doors, windows or passageways) are wholly enclosed on either a permanent or temporary basis (Figure 17.11).

Premises are considered '**substantially enclosed**' if they have a ceiling or roof, but have an opening in the walls which is less than half the total area of the walls. The area of the opening does not include doors, windows or any other fittings that can be opened or shut.

Figure 17.11 (a) Example of substantially enclosed premises. (b) Example of non-substantially enclosed premises.

Businesses and organizations should contact their local council if they require further guidance on whether their premises are 'enclosed' or 'substantially enclosed'.

17.34.14 Working Time Regulations 1998 as amended by 2003 and 2007 Regulations

These Regulations came into force in October 1998 and, for specified workers, restrict the working week to 48 hours per 7-day period. Individuals can voluntarily agree to disapply the weekly working hours limit. Employers must keep a copy of all such individual agreements. Workers whose working time is not measured or predetermined, or who can themselves determine the duration of their working day, are excepted from weekly working time, night work, rest periods and breaks.

The Regulations were amended, with effect from 1 August 2003, to extend working time measures in full to all non-mobile workers in road, sea, inland waterways and lake transport, to all workers in the railway and offshore sectors, and to all workers in aviation who are not covered by the sectoral Aviation Directive. The Regulations applied to junior doctors from 1 August 2004.

Mobile workers in road transport have more limited protections. Those subject to European Drivers' hours rules 3820/85 are entitled to 4 weeks' paid annual leave and health assessments if night workers, with effect from 1 August 2003. Mobile workers not covered by European drivers' hours rules will be entitled to an average 48 hours per week, 4 weeks paid holiday, health assessments if night workers and adequate rest.

The Regulations were previously amended, with effect from 6 April 2003, to provide enhanced rights for adolescent workers.

The basic rights and protections that the Regulations provide are:

➤ a limit of an average of 48 hours a week which a worker can be required to work (though workers can choose to work more if they want to);
➤ a limit of an average of 8 hours work in 24 hours which night workers can be required to work;
➤ a right for night workers to receive free health assessments;
➤ a right to 11 hours' rest a day;
➤ a right to a day off each week;
➤ a right to an in-work rest break if the working day is longer than 6 hours;
➤ a right to 28 days' paid leave per year from April 2009.

1998 Regulations available at: http://www.opsi.gov.uk/si/si1998/19981833.htm

2003 amendment available at: http://www.opsi.gov.uk/si/si2003/20031684.htm

2007 amendment available at: http://www.opsi.gov.uk/si/si2007/20072079.htm

International aspects of health and safety

This chapter contains:

1. scope of the global occupational health and safety issues
2. the role and function of the International Labour Organization (ILO)
3. summary of major international health and safety management systems
4. key characteristics of health and safety management systems
5. the role of regulatory authorities
6. a comparison of fault and no-fault injury compensation schemes.

18.1 Introduction

In 2004, an International General Certificate was introduced by NEBOSH following requests from countries around the world and international companies with cross-border operations. The main difference between the National General Certificate and the International Certificate syllabuses is the exclusion of all UK legal issues and requirements.

NEBOSH allows, where appropriate, local occupational health and safety legal requirements and enforcement procedures to be covered during International General Certificate courses. In some parts of the world these requirements and procedures may well be based on UK law. There has also been a significant change in reference material in that ILO Conventions, Recommendations and Codes of Practice have now been included. This change is particularly significant for Chapter 11. NEBOSH has published a detailed syllabus for the International General Certificate which includes the recommended ILO references.

There have been several other minor changes in the International General Certificate, the majority of which affect Chapter 1 although there are some small changes to other chapters. Minor changes are discussed in Section 18.8 of this chapter.

Successful management of health and safety is a top priority throughout the world and for this reason a comparison of the three major occupational health and safety management systems is covered in this chapter.

The ILO estimates that 2 million women and men die each year as a result of occupational accidents and work-related diseases. Table 18.1 shows the global numbers in more detail.

In the USA in 2002, approximately 2 million workers were victims of workplace violence. In the UK, 1.7% of working adults (357 000 workers) were the victims of one or more incidents of workplace violence.

Ten per cent of all skin cancers are estimated to be attributable to workplace exposure to hazardous substances. Thirty seven per cent of miners in Latin America have silicosis, rising to 50% among miners over fifty. In India 54.6% of slate pencil workers and 36.2% of stone cutters have silicosis.

In the course of the 20th century, industrialized countries saw a clear decrease in serious injuries, not least

Event	Average (daily)	Annually
Work-related deaths	5000	2 000 000
Work-related deaths to children	60	22 000
Work-related accidents	740 000	270 000 000
Work-related disease	438 000	160 000 000
Hazardous substance deaths	1205	440 000
Asbestos-related deaths	274	100 000

Table 18.1 Numbers of global work-related adverse events

Figure 18.1 Good scaffold – Leaning tower, Pisa, Italy.

Figure 18.2 Local authority building site, Turkey.

because of real advances in making the workplace healthier and safer. The challenge is to extend the benefits of this experience to the whole working world. However, 1984 saw the worst chemical disaster ever when 2500 people were killed and over 200 000 injured in the space of a few hours at Bhopal. Not only were the workers affected but their families, their neighbours and whole communities. Twenty years later many people are still affected by the disaster and are dying as a result. The rusting remains of a once magnificent plant remain as a reminder of the disaster.

Experience has shown that a strong safety culture is beneficial for workers, employers and governments alike. Various prevention techniques have proved themselves effective, both in avoiding workplace accidents and illnesses and improving business performance. Today's high standards in some countries are a direct result of long-term policies encouraging tripartite social dialogue, collective bargaining between trade unions and employers, and effective health and safety legislation backed by potent labour inspection. The ILO believes that safety management systems like ILO-OSH 2001 provide a powerful tool for developing a sustainable safety and health culture at the enterprise level and mechanisms for the continual improvement of the working environment.

The size of the health and safety 'problem' in terms of numbers of work-related fatalities and injuries and incidence of ill-health will vary from country to country. However, these figures should be available from the statistics branch of the national regulator as they are available in the UK from the annual report on health and safety statistics from the Health and Safety Executive (HSE).

18.2 The role and function of the ILO

The ILO is a specialized agency of the United Nations that seeks to promote social justice through establishing and safeguarding internationally recognized human and labour rights. It was founded in 1919 by the Treaty of Versailles at the end of the First World War.

The motivation behind the creation of such an organization was primarily humanitarian. Working conditions at the time were becoming unacceptable to a civilized society. Long hours, unsafe, unhygienic and dangerous conditions were common in low-paid manufacturing careers. Indeed, in the wake of the Russian Revolution, there was concern that such working conditions could lead to social unrest and even other revolutions. The ILO was created as a tripartite organization with governments, employers and workers represented on its governing body.

The ILO formulates international labour standards and attempts to establish minimum rights including freedom of association, the right to organize, collective bargaining, abolition of forced labour, equality of opportunity and treatment and other standards that regulate conditions across all work-related activities.

Representatives of all ILO Member States meet annually in Geneva for the International Labour Conference, which acts as a forum where social and labour questions of importance to the entire world are discussed. At this conference, labour standards are adopted and decisions made on policy and future programmes of work.

The ILO has 178 Member States but if a country is not a member, the ILO still has influence as a source of guidance when social problems occur.

The main principles on which the ILO is based are:

1. labour is not a commodity;
2. freedom of expression and of association are essential to sustained progress;
3. poverty anywhere constitutes a danger to prosperity everywhere;
4. the 'war against want' requires to be carried on with unrelenting vigour within each nation, and by continuous and concerted international effort in which the representatives of workers and employers, enjoying

equal status with those of governments, join with them in free discussion and democratic decision with a view to the promotion of the common welfare.

A recent campaign launched by the ILO has been to seek to eliminate child labour throughout the world. In particular, the ILO is concerned about children who work in hazardous working conditions, bonded child labourers and extremely young working children. It is trying to create a worldwide movement to combat the problem by:

➤ implementing measures which will prevent child labour;
➤ withdrawing children from dangerous working conditions;
➤ providing alternatives;
➤ improving working conditions as a transitional measure towards the elimination of child labour.

18.2.1 ILO Conventions and Recommendations

The international labour standards were developed for four reasons. The main motivation was to improve working conditions with respect to health and safety and career advancement. The second motivation was to reduce the potential for social unrest as industrialization progressed. Thirdly, the Member States want common standards so that no single country has a competitive advantage over another due to poor working conditions. Finally, the union of these countries creates the possibility of a lasting peace based on social justice.

International labour standards are adopted by the International Labour Conference. They take the form of Conventions and Recommendations. At the present time, there are 187 Conventions and 198 Recommendations, some of which date back to 1919.

International labour standards contain flexibility measures to take into account the different conditions and levels of development among Member States. However, a government that ratifies a Convention must comply with all its articles. Standards reflect the different cultural and historical backgrounds of the Member States as well as their diverse legal systems and levels of economic development.

ILO occupational safety and health standards can be divided into four groups, and an example is given in each case:

1. Guiding policies for action

   The Occupational Safety and Health Convention, 1985 (No. 155) and its accompanying Recommendation

Figure 18.3 Poor standard of scaffold not to EU standards in Poland.

Figure 18.4 Mobile elevating work platform, France.

(No. 164) emphasize the need for preventative measures and a coherent national policy on occupational safety and health. They also stress employers' responsibilities and the rights and duties of workers.

2. Protection in given branches of economic activity

   The Safety and Health in Construction Convention, 1988 (No. 167) and its accompanying Recommendation (No. 175) stipulate the basic principles and measures to promote safety and health of workers in construction.

3. Protection against specific risks

   The Asbestos Convention, 1986 (No. 162) and its accompanying Recommendation (No. 172) gives

managerial, technical and medical measures to protect workers against asbestos dust.

4. Measures of protection

   Migrant Workers (Supplementary Provisions) Convention, 1975 (No. 143) aims to protect the safety and health of migrant workers.

ILO **Conventions** are international treaties signed by ILO Member States and each country has an obligation to comply with the standards that the Convention establishes.

In contrast, ILO **Recommendations** are non-binding instruments that often deal with the same topics as Conventions. Recommendations are adopted when the subject, or an aspect of it, is not considered suitable or appropriate at that time for a Convention. Recommendations guide the national policy of Member States so that a common international practice may develop and be followed by the adoption of a Convention.

ILO standards are the same for every Member State and the ILO has consistently opposed the concept of different standards for different regions of the world or groups of countries.

The standards are modified and modernized as needed. The Governing Body of the ILO periodically reviews individual standards to ensure their continuing relevance.

Supervision of international labour standards is conducted by requiring the countries that have ratified Conventions to periodically present a report with details of the measures that they have taken, in law and practice, to apply each ratified Convention. In parallel, employers' and workers' organizations can initiate contentious proceedings against a Member State for its alleged non-compliance with a convention it has ratified. In addition, any member country can lodge a complaint against another Member State which, in its opinion, has not ensured in a satisfactory manner the implementation of a Convention which both of them have ratified. Moreover, a special procedure exists in the field of freedom of association to deal with complaints submitted by governments or by employers' or workers' organizations against a Member State whether or not the country concerned has ratified the relevant Conventions. Finally, the ILO has systems in place to examine the enforcement of international labour standards in specific situations.

The ILO also publishes Codes of Practice, guidance and manuals on health and safety matters. These are often used as reference material by either those responsible for drafting detailed Regulations or those who have responsibility for health and safety within an organization. They are more detailed than either Conventions or Recommendations and suggest practical solutions for the application of ILO standards. Codes of Practice indicate 'what should be done'. They are developed by tripartite meetings of experts and the final publication is approved by the ILO Governing Body.

For example, the construction industry has a Safety and Health in Construction Convention, 1988 (No. 167) that obliges signatory ILO Member States to comply with the construction standards laid out in the Convention – the Convention is a relatively brief statement of those standards. The accompanying Recommendation (No. 175) gives additional information on the Convention statements. The Code of Practice gives more detailed information than the Recommendation. This can best be illustrated by contrasting the coverage of scaffolds and ladders by the three documents shown in Appendix 18.1.

The ILO Codes of Practice and guidelines on health and safety matters that are relevant to the International General Certificate are:

> Safety and Health in Construction (ILO Code of Practice);
> Ambient factors in the Workplace (ILO Code of Practice);
> Safety in the Use of Chemicals at Work (ILO Code of Practice);
> Recording and Notification of Occupational Accidents and Diseases;
> Ergonomic Checkpoints;
> Work Organization and Ergonomics;
> Occupational safety and health management systems (ILO Guidelines).

Important ILO Conventions (C) and Recommendations (R) in the field of occupational safety and health include:

> C 115 Radiation Protection and (R 114), 1960;
> C 120 Hygiene (Commerce and Offices) and (R 120), 1964;
> C 139 Occupational Cancer and (R 147), 1974;
> C 148 Working Environment (Air, Pollution, Noise and Vibration) and (R 156), 1977;
> C 155 Occupational Safety and Health and (R164), 1981;
> C 161 Occupational Health Services and (R 171), 1985;

Figure 18.5 Moving coal, Berlin waterway.

> C 162 Asbestos and (R 172), 1986;
> C 167 Safety and Health in Construction and (R 175), 1988;
> C 170 Chemicals and (R 177), 1990;
> C 174 Prevention of Major Industrial Accidents and (R 181), 1993;
> C 176 Safety and Health in Mines and (R 176), 1995;
> C 184 Safety and Health in Agriculture and (R 192), 2001;
> C 187 Promotional Framework for Occupational Safety and Health and (R 197), 2006;
> R 97 Protection of Workers' Health Recommendation, 1953;
> R 102 Welfare Facilities Recommendation, 1956;
> R 31 List of Occupational Diseases Recommendation, 2002.

Copies of Conventions, Recommendations and Codes of Practice can be obtained from the ILO website www.ilo.org/safework.

18.3 Major occupational health and safety management systems

There are three major occupational health and safety management systems that are in use globally. These are as follows:

➤ HSG65, which has been developed by the UK HSE and is described in Chapter 1.16. The syllabuses for both Certificate courses and the management chapter headings in this textbook have followed the elements of HSG65.

➤ BS OHSAS 18001 – 2007 has been developed by the British Standards Institute (BSI) from its recommended occupational health and safety management system BS 8800. It has been developed in conjunction with the ISO 9000 series for quality management and the ISO 14000 series for environmental management.

➤ ILO-OSH 2001 was developed by the ILO after an extensive study of many occupational health and safety management systems used across the world. It was established as an international system following the publication of 'Guidelines on occupational safety and health management systems' in 2001. It is very similar to OHSAS 18001.

18.3.1 Basic elements of all health and safety management systems

All recognized occupational health and safety management systems have some basic and common elements. These are:

➤ a planning phase;
➤ a performance phase;
➤ a performance assessment phase;
➤ a performance improvement phase.

The planning phase

The planning phase always includes a **policy statement** which outlines the health and safety aims, objectives and commitment of the **organization** and lines of responsibility. **Hazard identification and risk assessment** takes place during this phase and the significant hazards may well be included in the policy statement. It is important to note that in some reference texts, in particular those in languages other than English, the whole process of hazard identification, risk determination and the selection of risk reduction or control measures is termed 'risk assessment'.

However, all three occupational health and safety management systems described in this chapter refer to the individual elements of the process separately and use the term 'risk assessment' for the determination of risk only.

At the **planning stage**, emergency procedures should be developed and relevant health and safety legal requirements and other standards identified together with appropriate benchmarks from similar industries. An **organizational** structure must be defined so that health and safety responsibilities are allocated at all levels of the organization and issues such as competent persons and health and safety training are addressed. Realistic targets should be agreed within the organization and be published as part of the policy.

The performance phase

The performance phase will only be successful if there is good **communication** at and between all levels of the organization. This implies employee participation as both **worker representatives** and on **safety committees**. Effective communication with the workforce, for example with clear **safe systems of work** and other health and safety procedures, will not only aid the implementation and operation of the plan but also produce continual improvement of performance – a key requirement of all occupational health and safety, quality and environmental management systems. There should also be effective communication with other stakeholders, such as regulators, contractors, customers and trade unions. The performance phase must be **monitored** on a regular basis since this will indicate whether there is an effective occupational health and safety management system and a good **health and safety culture** within the organization.

The performance assessment phase

The performance assessment phase may be either **active or reactive** or, ideally, a mixture of both. Active assessment includes work-based **inspections and audits**, regular health and safety committee meetings, feedback from training sessions and a constant review of risk assessments. Reactive assessment relies on records of accident, work-related injuries and ill-health as well as near miss and any enforcement notices. Any recommended remedial or preventative actions, following an investigation, must be implemented immediately and monitored regularly.

The performance improvement phase

The performance improvement phase involves a **review** of the effectiveness of the health and safety management system and the identification of any weaknesses. The review, which should be undertaken by the management of the organization, will assess whether targets have been met and the reasons for any under-performance. Issues such as the level of resources made available, the vigilance of supervisors and the level of co-operation of the workforce should be considered at the review stage. When recommendations are made, the review process must define a timescale by which any improvements are implemented and this part of the process must also be monitored. **Continual improvement** implies a commitment to improve performance on a proactive continuous basis without waiting for a formal review to take place. Most management systems include an **audit** requirement, which may be either internal or external. The audit process examines the effectiveness of the whole management process and may act as a control on the review process. Many inquiry reports into health and safety management issues have asserted that health and safety performance should be subject to audit in the same way that financial performance must be audited.

In the publication 'Successful health and safety management – HSG 65', the HSE recommend a similar four-step approach to occupational health and safety management known as:

Plan – establish standards for health and safety management based on risk assessment and legal requirements

Do – implement plans to achieve objectives and standards

Check – measure progress with plans and compliance with standards

Act – review against objectives and standards and take appropriate action.

The **Plan-Do-Check-Act** for occupational health and safety management forms the basis of the three occupational health and safety management systems HSG65, OHSAS 18001 and ILO-OSH 2001.

18.3.2 HSG65

The UK HSE first published 'Successful health and safety management – HSG65 in 1991 and has amended it since then so that it is now more concerned with continual improvement and less with the attainment of minimum health and safety standards. Unlike the other two systems under consideration, HSG65 was developed by a regulator (HSE) and, therefore, is much more widely recognized in the UK. HSE inspectors use the framework of HSG65 when auditing the health and safety management arrangements of an organization, however it would be wrong to assume that it is the only system recognized by the HSE.

HSG65 is shown in Figure 18.6. An outline of the five key elements of HSG65 was given in Chapter 1 and is given in more detail here. The key elements are as follows:

1. **A clear health and safety policy** – Evidence shows that a well-considered policy contributes to business efficiency and continual improvement throughout the organization. It helps to minimize financial losses arising from avoidable accidents and demonstrates to the workforce that accidents are not necessarily the fault of any individual member of the workforce. Such a management attitude could lead to an increase in workforce co-operation, job satisfaction and productivity. This demonstration of senior management involvement offers evidence to all stakeholders that responsibilities to people and the environment are taken seriously by the organization. A good health and safety policy helps to ensure that there is a systematic approach to risk assessment and sufficient resources, in terms of people and money, have been allocated to protect the health and safety and welfare of the workforce. It can also support quality improvement programmes which are aimed at continual improvement.

2. **A well-defined health and safety organization** – The shared understanding of the organization's values and beliefs, at all levels of the organization, is an essential component of a positive health and safety culture. For a positive health and safety culture to be achieved, an organization must have clearly defined health and safety responsibilities so that there is always management control of health and safety throughout the organization. The formal organizational structure should be such that the promotion of health and safety becomes a collaborative activity between the workforce, safety representatives and the managers. An effective organization will be noted for good staff involvement and participation, high-quality communications, the promotion of competency and the empowerment of all employees to make informed contributions to the work of the organization.

3. **A clear health and safety plan** – This involves the setting and implementation of performance standards and procedures using an effective occupational health and safety management system. The plan is based on risk assessment methods to decide on priorities and set objectives for controlling or eliminating hazards and reducing risks. Measuring success requires the establishing of performance standards against which achievements can be identified.

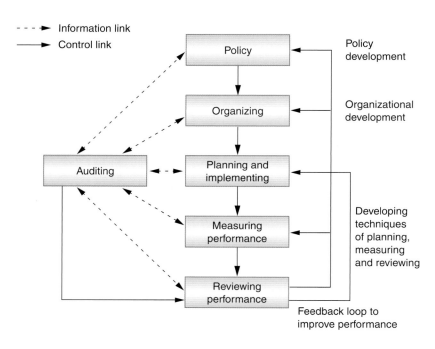

Figure 18.6 Key elements of HSG65.

4. **The measurement of health and safety performance** – This includes both active and reactive monitoring to see how effectively the occupational health and safety management system is working. Active monitoring involves looking at the premises, plant and substances plus the people, procedures and systems. Reactive monitoring discovers through investigation of accidents and incidents why controls have failed. It is also important to measure the organization against its own long-term goals and objectives.

5. **The audit and review of health and safety performance** – The results of monitoring and independent audits should be systematically reviewed to see if the management system is achieving the right results. This is not only required by the Health and Safety at Work (HSW) Act, but is part of any company's commitment to continual improvement. Comparisons should be made with internal performance indicators and the performance of external organizations with exemplary practices and high standards.

18.3.3 OHSAS 18001 – 2007

OHSAS 18001 and ILO-OSH 2001 have many similarities in the details of the two systems. One of the biggest differences

Figure 18.7 Sugar factory, Poland, moving to 18001 accreditation.

is that OHSAS 18001 is certifiable whereas ILO-OSH 2001 is not. OHSAS 18001, which is shown in Figure 18.8, was developed from the British Standard 8800 and was designed to be integrated with the two British Standards for quality and environmental management. Both occupational health and safety management systems have been

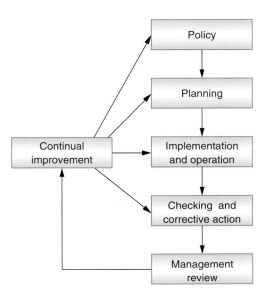

Figure 18.8 Key elements of OHSAS 18001.

shaped by internationally agreed occupational health and safety principles defined in international labour standards and the European Union Framework Directive 89/391/EEC. OHSAS 18001:2007 requires an external audit as part of the certification process. OHSAS 18002:2008 has also been produced by BSI as a guideline for implementation of OHSAS 18001. 18002 will be withdrawn when its contents are published as an International Standard.

Some of the principal differences between the 1999 and 2007 standards are:

➤ Health has been given a greater emphasis.
➤ It is now referred to as a 'Standard' which reflects the increasing adoption of OHSAS 18001 as the basis for national standards on occupational health and safety management systems.
➤ It has been more closely aligned with ISO 14001:2004 and ISO 9001:2000.
➤ 'Tolerable risk' has been replaced by 'Acceptable risk'.
➤ 'Accident' now is included in the term 'incident', and 'hazard' no longer includes damage to property or the workplace environment.

There are five elements in the OHSAS 18001:2007 occupational health and safety management system:

1. **Policy** – The general requirements are similar to HSG 65. It requires top management to define and authorize the organization's Occupational health and safety (OH&S) policy and ensure its OH&S management system:

   (a) is appropriate to the nature and scale of the organization's risks;
   (b) includes a commitment to the prevention of injury and ill-health and continual improvement;
   (c) includes a commitment to at least comply with applicable legal requirements and other requirements to which the organization subscribes;
   (d) provides a framework for setting and reviewing OH&S objectives;
   (e) is documented, implemented and maintained;
   (f) is communicated to all persons working under the control of the organization;
   (g) is available to interested parties;
   (h) is reviewed periodically.

2. **Planning** – The organization must establish, implement and maintain effective arrangements for the following:

   (a) Ongoing hazard identification, risk assessment and the establishment of the necessary control measures must take into account:
   (i) both routine and non-routine activities of all persons having access to the workplace including contractors and visitors;
   (ii) human factors;
   (iii) hazards which originate outside the workplace which may affect those inside;
   (iv) hazards created in the vicinity by work-related activities;
   (v) infrastructure, equipment and materials at the workplace;
   (vi) changes or proposed changes;
   (vii) modifications to the OH&S management system;
   (viii) any applicable legal requirements for risk assessment and control measures;
   (ix) the design of the work areas, processes, installations, machinery, equipment, procedures and work organization.

   (b) The results of the risk assessment must be considered when determining controls which shall be determined according to the following hierarchy:
   (i) elimination;
   (ii) substitution;
   (iii) engineering controls;
   (iv) signage/warnings and/or administrative controls;
   (v) personal protective equipment.

(c) The results must be documented and kept up to date.

(d) The organization must establish, implement and maintain a procedure for identifying and accessing the legal and other OH&S requirements that are applicable. This information must be communicated and kept up to date.

(e) The organization must establish, implement and maintain OH&S objectives which must be measurable (where practicable) and consistent with the policy and legal requirements. An effective programme and action plan for achieving the objectives must be established and implemented.

3. **Implementing and operation** – As with HSG65, there is a stress on the responsibility of top management for the effective implementation and operation of the system. Top management must demonstrate their commitment:

(a) Provide adequate resources.

(b) Define roles, allocate responsibilities and accountabilities, and delegate authorities.

(c) Appoint a member of top management with specific responsibilities for OH&S irrespective of their other responsibilities to ensure:
   (i) OH&S management system is established, implemented and maintained;
   (ii) reports on performance are presented to top management for review and used for improvement.

(d) The organization must ensure that all people under its control who perform tasks that could affect OH&S are competent on the basis of appropriate education, training or experience and must keep records.

(e) The organization must establish, implement and maintain procedures for effective internal communication with employees, contractors and other visitors. They must receive and appropriately respond to relevant external communications.

(f) The organization must establish, implement and maintain procedures for involvement of and consultation with employees and, as necessary, contractors, relating to hazard identification, risk assessment determination of suitable controls, incident investigation and the development and review of policies and procedures.

(g) Documents proportional to the risks and scale of the organization must be prepared and maintained with a proper system of control.

(h) Operational controls must be introduced to ensure that arrangements are in place for:
   (i) purchasing goods, equipment and services;
   (ii) contractors and other visitors;
   (iii) other situations as necessary;
   (iv) procedures for emergency preparedness and response must be prepared, tested and maintained. The organization must respond to actual emergency situations and mitigate adverse consequences.

See Chapters 3 and 6 for more information.

4. **Checking and corrective action** – This is similar to the performance measurement phase of HSG65. The organization has to establish, implement and maintain:

(a) procedures to monitor and measure OH&S performance on a regular basis. The procedures must cover:
   (i) both qualitative and quantitative measures;
   (ii) extent to which objectives have been met;
   (iii) effectiveness of controls;
   (iv) proactive measures of performance that measure conformance with OH&S programmes and controls;
   (v) reactive measures that monitor ill-health, incidents including accidents and near misses and any other historical evidence of poor performance;
   (vi) recording of data and results which can facilitate corrective action and preventative analysis.

(b) procedures for evaluating compliance with applicable legal and other requirements to which it subscribes and keep records of the evaluations.

(c) procedures to record, investigate and analyse incidents in a timely manner so that it can:
   (i) determine underlying deficiencies;
   (ii) identify corrective action;
   (iii) identify opportunities for prevention and continual improvement;
   (iv) communicate the results.

(d) procedures for dealing with actual and potential non-conformity and for taking corrective and/or preventative action.

(e) records to demonstrate conformity with the OH&S management system and OHSAS 18001.

(f) internal audits of the OH&S management system to ensure that it is up to date, properly implemented and effective. The audit programme must be properly planned, objective and impartial.

5. Management review – Top management are required to review the organization's OH&S management system at planned intervals to ensure its continued suitability and effectiveness. Reviews must assess opportunities for improvements and the need for change. They must be properly documented since they will form the base line from which continual improvement will be measured. Input to the reviews shall include:

(a) results of internal audits and evaluations of compliance;

(b) results of consultation;

(c) communications from external parties including complaints;

(d) OH&S performance;

(e) extent to which objectives have been met;

(f) status of investigations and corrective actions;

(g) follow up from previous reviews;

(h) changed circumstances, processes or local legislation;

(i) recommendations for improvement.

The output from the reviews must be consistent with the organization's policy, performance, resources and objectives.

The OHSAS 18001 management system recognizes that the system is bound to require modification and change as the activities of the organization change and national regulatory requirements change.

18.3.4 ILO-OSH 2001

The ILO has a considerable influence on the development of employment law in many countries across the world. As companies have become more international in terms of both markets and production bases, this influence of the ILO has increased and its working standards have become accepted in many parts of the world. Many of the ILO standards cover health and safety issues and are used in the International General Certificate course. As mentioned earlier, the ILO-OSH 2001 guidelines offer a recommended occupational health and safety management system based on an ILO survey of several contemporary schemes including HSG65 and OHSAS 18001. There are, therefore, many common elements between the three schemes. The guidelines are not legally binding and are not intended to replace national laws, Regulations or accepted standards.

At the national level, the guidelines should be used to establish a national framework for occupational health and safety management systems, preferably supported by national laws and Regulations. They should also provide guidance for the development of voluntary arrangements to strengthen compliance with Regulations and standards leading to continual improvement in occupational health and safety performance.

The ILO recognizes that a management system can usually only be successful in a country if there is some form of national policy on health and safety and occupational health and safety management systems. The ILO recommends, therefore, that the following general principles and procedures be established:

1. the implementation and integration of occupational health and safety management systems as part of the overall management of an organization;

2. the introduction and improvement of voluntary arrangements for the systematic identification, planning, implementation and improvement of occupational health and safety activities at both national and organizational levels;

3. the promotion of worker participation in occupational health and safety management at organizational level;

4. the implementation of continual improvement without unnecessary bureaucracy and cost;

5. the encouragement of national labour inspectors to support the arrangements at organizational level for a health and safety culture within the framework of an occupational health and safety management system;

6. the evaluation of the effectiveness of the national policies at regular intervals;

7. the evaluation of the effectiveness of occupational health and safety management systems within those organizations which are operating within the country;

8. the inclusion of all those affected by the organization, such as contractors, members of the public and temporary workers, at the same level of health and safety provision as employees.

Figure 18.9 shows the key elements of the ILO-OSH 2001 occupational health and safety management system. Each element will be discussed in turn but since it has been developed from HSG65 and OHSAS 18001 among other systems, there are many similarities. The elements are:

1. Policy – There is a more specific emphasis on worker participation, which is seen as an essential element of the management system and should be referenced in the policy statement. It is expected that workers and their health and safety representatives should have sufficient time and resources allocated to them so that they can participate actively in each element of the management system. The formation of a health

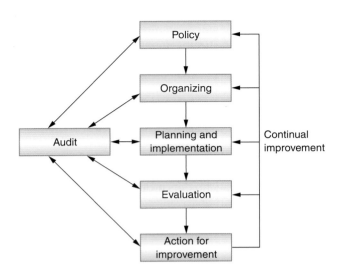

Figure 18.9 Key elements of ILO-OSH 2001.

and safety committee is also a recommended part of the system. The occupational health and safety management system should be compatible with or integrated with other management systems operating in the organization.

2. **Organizing** – There is much in common with both HSG65 and OHSAS 18001, with responsibility, accountability, competence, training and communication being key parts of this element. There is a specific responsibility to provide effective supervision to ensure the protection of the health and safety of workers and to establish prevention and health promotion programmes. Workers should have access to any records, such as accident and monitoring records, which are relevant to their occupational health and safety while respecting the need for confidentially. Health and safety training must be available to all members of the organization and be provided during normal working hours at no cost to employees.

3. **Planning and implementation** – Unlike other systems, the ILO system combines planning and implementation. Following an initial review of any existing health and safety management system, a plan should be developed to remedy any deficiencies found. The plan should support compliance with national laws and Regulations and include continual improvement of health and safety performance. It should contain measurable objectives which are realistic and achievable and, as with the other occupational health and safety management systems, hazard identification

and risk assessment. There should also be an adequate provision of resources and technical support. The plan must be capable of accommodating the impact on health and safety of any internal changes in the organization, such as new processes, new technologies and amalgamations with other organizations, or external changes due, for example, to changes in national laws or Regulations. As with OHSAS 18001, emergency and procurement arrangements and detailed arrangements for the selection and supervision of contractors must be included in the health and safety plan.

4. **Evaluation** – This is very similar to the performance measurement phase of HSG65 with a greater emphasis on the health and welfare of the worker. The recommendations concerning the investigation of work-related injuries, ill-health, diseases and incidents are identical to those for the management review element of OHSAS 18001.

5. **Action for improvement** – Arrangements should be introduced and maintained for any preventative and corrective action to be undertaken identified by performance monitoring, audits and management reviews of the health and safety management system. Arrangements should also be in place for the continual improvement of the management system. More details on the factors to be considered for continual improvement are given later in this chapter.

6. **Audit** – The ILO recommends that an audit should be performed by competent and trained personnel at agreed and regular intervals. It should cover all elements of the management system including worker participation, communication, procurement, contracting and continual improvement. The audit conclusions must state whether the health and safety management system is effective in meeting the organizational health and safety policy and objectives and promotes full worker participation. The audit should also check that there is compliance with national laws and Regulations and that there has been a satisfactory response to earlier audit findings.

It is clear that there is much that is common between the three management systems discussed and any differences are differences in emphasis. HSG65 was the first of the three systems and the emphasis was on legal compliance. OHSAS 18001 introduced the concept of continual improvement, integration with other management systems and certification. ILO-OSH 2001 combines the features of HSG65 and OHSAS 18001 and stresses the importance of worker participation for the system to be effective.

18.4 Other key characteristics of a health and safety management system

The four basic elements common to all occupational health and safety management systems, as described earlier in this chapter, contain the different activities of the system together with the detailed arrangements and activities required to deliver those activities. However, there are four key characteristics of a **successful** occupational health and safety management system:

➤ a positive health and safety culture;
➤ the involvement of all stakeholders;
➤ an effective audit;
➤ continual improvement.

18.4.1 A positive health and safety culture

In Chapter 4, the essential elements for a successful health and safety culture were detailed and discussed. In summary, they were:

➤ leadership and commitment to health and safety throughout the organization;
➤ an acceptance that high standards of health and safety are achievable;
➤ the identification of all significant hazards facing the workforce and others;
➤ a detailed assessment of health and safety risks in the organization and the development of appropriate control and monitoring systems;
➤ a health and safety policy statement outlining short- and long-term health and safety objectives. Such a policy should also include national codes of practice and health and safety standards;
➤ relevant communication and consultation procedures and training programmes for employees at all levels of the organization;
➤ systems for monitoring equipment, processes and procedures and the prompt rectification of any defects found;
➤ the prompt investigation of all incidents and accidents and reports made detailing any necessary remedial actions.

Some of these essential elements form part of the health and safety management system but unless all are present within the organization, it is unlikely that occupational health and safety will be managed successfully no matter which system is introduced. The chosen management system must be effective in reducing risks in the workplace else it will be nothing more than a paper exercise.

18.4.2 The involvement of stakeholders

There are a number of internal and external stakeholders of the organization who will have an interest and influence on the introduction and development of the occupational health and safety management system.

The internal stakeholders include:

➤ **Directors and trustees of the organization** – Following several reports on corporate governance in recent years culminating in the Combined Code of Corporate Governance 2003, the measurement of occupational health and safety performance and the attainment of health and safety targets have been recognized as being as important as other measures of business performance and targets. A report on health and safety performance should be presented at each board meeting and be a periodic agenda item for sub-committees of the board such as audit and risk management.
➤ **The workforce** – Without the full co-operation of the workforce, including contractors and temporary employees, the management of health and safety will not be successful. The workforce is best qualified to ensure and provide evidence that health and safety procedures and arrangements are actually being implemented at the workplace. So often this is not the case even though the occupational health and safety managements system is well designed and documented. Worker representatives can provide useful evidence on the effectiveness of the management system at shop floor level and they can also provide a useful channel of communication between the senior management and the workforce. One useful measure of the health and safety culture is the enthusiasm with which workers volunteer to become and continue to be representatives. For a representative to be really effective, some training is essential. Organizations which have a recognized trade union structure will have fewer problems with finding and training worker representatives. More information on the duties of worker representatives is given in Chapter 3.
➤ **Health and safety professionals** – Such professionals will often be appointed by the organization to manage

531

the occupational health and safety management system and monitor its implementation. They will, therefore, have a particular interest in the development of the system, the design of objectives and the definition of targets or goals. Unless they have a direct-line management responsibility, which is not very common, they can only act as advisers, but still have a positive influence on the health and safety culture of the organization. The appointed health and safety professional also liaises with other associated competent persons, such as for electrical appliance and local exhaust ventilation testing.

The external stakeholders include:

➤ **Insurance companies** – As compensation claims increase, insurance companies are requiring more and more evidence that health and safety is being effectively managed. There is increasing evidence that insurance companies are becoming less prepared to offer cover to organizations with a poor health and safety record and/or occupational health and safety management system.
➤ **Investors** – Health and safety risks will, with other risks, have an effect on investment decisions. Increasingly, investment organizations require evidence that these risks are being addressed before investment decisions are made.
➤ **Regulators** – In many parts of the world, particularly in the Far East, national regulators and legislation require certification to a recognized international occupational health and safety management system standard. In many countries, regulators use such a standard to measure the awareness of a particular organization of health and safety issues.
➤ **Customers** – Customers and others within the supply chain are increasingly insisting on some form of formal occupational health and safety management system to exist within the organization. The construction industry is a good example of this trend. Much of this demand is linked to the need for corporate social responsibility and its associated guidelines on global best practice.
➤ **Neighbours** – The extent of the interest of neighbours will depend on the nature of the activities of the organization and the effect that these activities have on them. The control of noise, and dust and other atmospheric contaminants are examples of common problem areas which can only be addressed on a continuing basis using a health, safety and environmental management system.

Figure 18.10 Construction site – Sienna.

➤ **International organizations** – The United Nations, the ILO, the International Monetary Fund, the World Bank and the World Health Organization are all examples of international bodies which have shown a direct or indirect interest in the management of occupational health and safety. In particular, the ILO is keen to see minimum standards of health and safety established around the world. The ILO works to ensure for everyone the right to work in freedom, dignity and security – which includes the right to a safe and healthy working environment. More than 70 ILO Conventions and Recommendations relate to questions of safety and health. In addition the ILO has issued more than 30 Codes of Practice on Occupational Health and Safety. For more information see their website www.ilo.org/safework.
➤ There is a concern of many of these international organizations that as production costs are reduced by relocating operations from one country to another, there is also a lowering in occupational health and safety standards. The introduction of internationally recognized occupational health and safety management systems will help to alleviate such fears.

18.4.3 An effective audit

An effective audit is the final step in the occupational health and safety management system control cycle. The 'feedback loop' produced by audit enables the reduction of risk levels and the effectiveness of the occupational health and safety management system to be improved.

Audit is a business discipline which is frequently used in finance, environmental matters and quality and can equally well be applied to health and safety. It will check on the implementation of occupational health and safety management systems and the adequacy and effectiveness of the management arrangements and risk control systems. Audit is critical to a health and safety management system but is not a substitute for the essential day-to-day management of health and safety.

The audit aims to establish that the three major components of any occupational health and safety management system are in place and operating effectively. It should show that:

➤ appropriate management arrangements are in place;
➤ adequate risk control systems exist, and are implemented and consistent with the hazard profile of the organization;
➤ appropriate workplace precautions are in place.

Where the organization is spread over a number of separate sites, the management arrangements linking the centre with these sites should be examined by the audit.

Some elements of the occupational health and safety management system do not need to be audited as often as others. For example, an audit to verify the implementation of critical risk control systems should be made more frequently than an audit of the management arrangements for health and safety of the whole organization. Where there are complex workplace precautions in place, such as in the chemical industry, it may also be necessary to undertake technical audits.

The audit programme should produce a comprehensive picture of the effectiveness of the health and safety management system in controlling risks. The programme must indicate when and how each component part will be audited. The audit team should include managers, safety representatives and workers. Such inclusiveness will help during the implementation of any audit recommendations.

More detailed information on the auditing process is given in Chapter 7.

When planning an audit, a decision has to be made as to whether to use internal or external auditors. When formal certification is required either by the organization or by the client, competent external auditors are essential. However, it is likely that, in such cases, internal auditors will also be used.

The advantage of using internal auditors is that they know the critical areas to monitor and will help to spread good practices around the organization. The disadvantages are that clients may question their independence and they may well be unaware of external benchmarks, unless they are auditing against a standard. These two disadvantages are the advantages of using external auditors. External auditors have the disadvantages that they do not know the organization, require much more documentation and can offer bland reports. It is also usually more expensive to use external rather than internal auditors. Auditor competence is an important issue and it is important that any auditors used, whether internal or external, are properly trained.

18.4.4 Continual improvement

Continual improvement has been mentioned several times in this chapter and is recognized as a vital element of all occupational health and safety management systems if they are to remain effective and efficient as internal and external changes affect the organization. Internal changes may be caused by business reorganization, such as a merger, new branches and changes in products, or by new technologies, employees, suppliers or contractors. External changes could include new or revised legislation, guidance or industrial standards, new information regarding hazards or campaigns by regulators.

Continual improvement need not necessarily be done at high cost or add to the complexity of the management system. Its benefits include:

➤ a decrease in the rate of injuries, ill-health and damage;
➤ a possible reduction in the resources required to manage the system;
➤ an acceptability of higher standards and an improved health and safety culture;
➤ overall improvements in the management system itself.

The simplest way to achieve continual improvement is to implement the recommendations of audits and management reviews, and use benchmarks from similar organizations and any revised national or industrial guidelines. The workforce, managers and supervisors often have very good suggestions on ways to improve processes and procedures. Such suggestions will lead to improvements in the management system. Finally the health and safety committee can be a very effective vehicle for continual improvement particularly if working parties are used to investigate specific issues within an agreed timeframe.

Figure 18.11 Have to consider different solutions in different countries – delivering furniture in Certaldo, Italy.

Figure 18.12 Typical health and safety legal framework.

18.5 The role of the regulatory authorities

18.5.1 The legal framework

The framework for regulating health and safety will vary across the world, for example European countries use the EU framework, and the Pacific Rim countries tend to use a USA framework whereas the Caribbean countries follow the UK framework. The course provider should be able to describe the legal and regulatory framework appertaining to any particular country.

Most legislation is driven by a framework of Acts, Regulations and support material including Codes of Practice and Standards, as illustrated in Figure 18.12. Within Europe there is another layer of legislation known as Directives, above the Member States' own legislation. These are legally binding on each state; for more information see Chapter 1. The US system of federal and individual state legislation is very similar.

18.5.2 Regulatory authorities and safety management systems

The role of the national regulatory authority is crucial to the successful implementation of an occupational health and safety management system. In many parts of the world, such as South East Asia, formal adoption of a recognized management system is required with third party auditing by government-approved auditors. In the USA, organizations with approved management systems may be exempted from normal inspections by the Occupational Safety and Health Administration.

In the UK, there has been a movement from prescriptive legislation to risk assessment by the employer and this movement is now occurring in many parts of the world including the EU. A management system is an essential tool to achieve this movement and such a system is implied in the UK Management of Health and Safety at Work Regulations. Countries such as Canada, Australia, New Zealand and Norway have developed occupational health and safety management systems as an encouragement for such self-regulation. The ILO-OSH 2001 system has been adopted by Germany, Sweden, Japan, Finland, Korea, China, Mexico, Costa Rica, Brazil, Indonesia, Vietnam, Malaysia, India, Thailand, the Czech Republic, Poland and Russia.

It is likely that more countries and multinational companies will expect occupational health and safety management systems to be adopted either as a legal duty or an implied duty following rulings from local courts of law.

18.6 The benefits and problems associated with occupational health and safety management systems

Occupational health and safety management systems have many benefits, of which the principal ones are:

> It is much easier to achieve and demonstrate legal compliance. Enforcement authorities have more confidence in organizations which have a health and safety management system in place.
> They ensure that health and safety is given the same emphasis as other business objectives, such as quality and finance. They will also aid integration, where appropriate, with other management systems.
> They enable significant health and safety risks to be addressed in a systematic manner.
> They can be used to show legal compliance with terms such as 'practicable' and 'so far as is reasonably practicable'.
> They indicate that the organization is prepared for an emergency.
> They illustrate that there is a genuine commitment to health and safety throughout the organization.

There are, however, several problems associated with occupational health and safety management systems, although most of them are soluble because they are caused by poor implementation of the system. The main problems are:

> The arrangements and procedures are not apparent at the workplace level and the audit process is only concerned with a desktop review of procedures.
> The documentation is excessive and not totally related to the organization due to the use of generic procedures.
> Other business objectives, such as production targets, lead to ad hoc changes in procedures.
> Integration, which should really be a benefit, can lead to a reduction in the resources and effort applied to health and safety.
> A lack of understanding by supervisors and the workforce leads to poor system implementation.

> The performance review is not implemented seriously thus causing cynicism throughout the organization.

18.7 Conclusions on the three health and safety management systems

Three occupational health and safety management systems have been described in some detail in this chapter and it has been shown that the similarities between them outweigh their differences. Any occupational health and safety management system will fail unless there is a positive health and safety culture within the organization and the active involvement of internal and external stakeholders.

A structured and well-organized occupational health and safety management system is essential for the maintenance of high health and safety standards within all organizations and countries. Some systems, such as OHSAS 18001, offer the opportunity for integration with quality and environmental management systems. This enables a sharing of resources although it is important that technical activities, such as health and safety risk assessment, are only undertaken by persons trained and competent in that area.

For an occupational health and safety management system to be successful, it must address workplace risks and be 'owned' by the workforce. It is, therefore, essential that the audit process examines shop floor health and safety behaviour to check that it mirrors that required by the health and safety management system.

Finally, whichever system is adopted, there must be continual improvement in health and safety performance if the application of the occupational health and safety management system is to succeed in the long term.

18.8 Other minor additions to the International General Certificate

The sources of health and safety information have been extended to include information provided by the websites or publicity offices of national or international agencies [e.g. ILO, Occupational Safety and Health Administration (USA), European Agency for Safety and Health at Work (EU) and Worksafe (Western Australia)]. The sources of information covered in Chapter 1 are also included.

Many countries have either fault or no-fault compensation schemes for workers involved in accidents. Knowledge of these schemes is important for those who work in several different countries.

18.8.1 Fault and no-fault injury compensation

In the UK, compensation for an injury following an accident is achieved by means of a successful legal action in a civil court (as discussed in Chapter 1). In such cases, injured employees sue their employer for negligence and the employer is found liable or at fault. This approach to compensation is adversarial, costly and can deter injured individuals of limited means from pursuing their claim. In a recent medical negligence claim in Ireland, costs were awarded against a couple who were acting on behalf of their disabled son, and they were faced with a bill for £3m.

The spiralling cost of insurance premiums to cover the increasing level and number of compensation awards, despite the Woolfe reforms (see Chapter 15), has led to another debate on the introduction of a no-fault compensation system. It has been estimated that in medical negligence cases, it takes on average 6 years to settle a claim and only 10% of claimants ever see any compensation.

No-fault compensation systems are available in many parts of the world, in particular New Zealand and several states in the USA. In these systems, amounts of compensation are agreed centrally at a national or state level according to the type and severity of the injury. The compensation is often in the form of a structured continuous award rather than a lump sum and may be awarded in the form of a service, such as nursing care, rather than cash.

The no-fault concept was first examined in the 1930s in the USA to achieve the award of compensation quickly in motor car accident claims without the need for litigation. It was introduced first in the State of Massachusetts in 1971 and is now mandatory in nine other states although several other states have either repealed or modified no-fault schemes.

In 1978, the Pearson Commission in the UK rejected a no-fault system for dealing with clinical negligence even though it acknowledged that the existing tort system was costly, cumbersome and prone to delay. Its principal reasons for rejection were given as the difficulties in reviewing the existing tort liability system and in determining the causes of injuries.

In New Zealand, there was a general dissatisfaction with the workers' compensation scheme, which was similar to the adversarial fault-based system used in Australia

Figure 18.13 Mobile tower crane – Cinque Terre, Italy.

The hazards of working in unfamiliar countries and/or climates are important health and safety issues – snake bites, diseases (such as malaria and yellow fever) and sunstroke. The course provider will relate these and any other particular hazards to the country for which the course is being provided.

and the UK. In 1974, a no-fault accident compensation system was introduced and administered by the Accident Compensation Corporation (ACC). This followed the publication of the Woodhouse Report in 1966 which advocated 24-hour accident cover for everybody in New Zealand.

The Woodhouse Report suggested the following five principles for any national compensation system:

➤ community responsibility;
➤ comprehensive entitlement irrespective of income or job status;
➤ complete rehabilitation for the injured party;
➤ real compensation for the injured party;
➤ administrative efficiency of the compensation scheme.

The proposed scheme was to be financed by channelling all accident insurance premiums to one national organization (the ACC).

The advantages of a no-fault compensation scheme include:

1. Accident claims are settled much quicker than in fault schemes.
2. Accident reporting rates will improve.
3. Accidents become much easier to investigate because blame is no longer an issue.
4. Normal disciplinary procedures within an organization or through a professional body are unaffected and can be used if the accident resulted from negligence on the part of an individual.
5. More funds are available from insurance premiums for the injured party and less used in the judicial and administrative process.

The possible disadvantages of a no-fault compensation system are:

1. There is often an increase in the number of claims, some of which may not be justified.
2. There is a lack of direct accountability of managers and employers for accidents.
3. Mental injury and trauma are often excluded from no-fault schemes because of the difficulty in measuring these conditions.
4. There is more difficulty in defining the causes of many injuries and industrial diseases than in a fault scheme.
5. The monetary value of compensation awards tend to be considerably lower than those in fault schemes (although this can be seen as an advantage).

No-fault compensation schemes exist in many countries including Canada and the Scandinavian countries. However, a recent attempt to introduce such a scheme in New South Wales, Australia, was defeated in the legislative assembly.

18.9 Practice NEBOSH questions for the International General Certificate

The International General Certificate is a new award and all the examination papers to date have used many questions taken from the question bank for the National General Certificate other than those questions directly related to UK law. Therefore, all practice questions given at the end of each chapter in this book can be used by students on the International General Certificate course except those at the end of Chapter 1 that relate to UK law. Typical questions that have been asked on the International General Certificate are given below:

1. **Outline** the economic benefits that an organization may obtain by implementing a successful health and safety management system.

2. (i) **Identify** the possible consequences of an accident to:
   (a) the injured workers
   (b) their employer.
   (ii) **Identify** the direct and indirect costs to the employer of accidents at work.
   (iii) **Outline** the actions an enforcement agency may take following an accident at work.

3. **Outline** the main features of:
   (i) a health and safety inspection
   (ii) a health and safety audit.

4. **Outline** the key elements of a health and safety management system.

5. **List** the ways in which a manager could involve workers in the improvement of health and safety in the workplace.

6. (i) **Outline** the factors that may affect the risk from manual handling in relation to:
   (a) the task
   (b) the load
   (c) the individual
   (d) the working environment.
   (ii) **Outline** a good manual handling technique that could be used when lifting a box weighing 12.5 kg.

7. (i) **List** possible health effects of working outside in extreme heat created by the sun.
   (ii) **Outline** control measures that could be used to minimize the effects of the sun on construction workers.

Appendix 18.1 Scaffolds and ladders

1.1 Convention (Safety and Health in Construction) (167)

Article 14

1. Where work cannot safely be done on or from the ground or from part of a building or other permanent structure, a safe and suitable scaffold shall be provided and maintained, or other equally safe and suitable provision shall be made.

2. In the absence of alternative safe means of access to elevated working places, suitable and sound ladders shall be provided. They shall be properly secured against inadvertent movement.

3. All scaffolds and ladders shall be constructed and used in accordance with national laws and Regulations.

4. Scaffolds shall be inspected by a competent person in such cases and at such times as shall be prescribed by national laws or Regulations.

1.2 Recommendation (Safety and Health in Construction) (175)

Scaffolds

16. Every scaffold and part thereof should be of suitable and sound material and of adequate size and strength for the purpose for which it is used and be maintained in a proper condition.

17. Every scaffold should be properly designed, erected and maintained so as to prevent collapse or accidental displacement when properly used.

18. The working platforms, gangways and stairways of scaffolds should be of such dimensions and so constructed and guarded as to protect persons against falling or being endangered by falling objects.

19. No scaffold should be overloaded or otherwise misused.

20. A scaffold should not be erected, substantially altered or dismantled except by or under the supervision of a competent person.

21. Scaffolds as prescribed by national laws or Regulations should be inspected, and the results recorded, by a competent person:
    (i) before being taken into use
    (ii) at periodic intervals thereafter
    (iii) after any alteration, interruption in use, exposure to weather or seismic condition or any other occurrence likely to have affected their strength or stability.

1.3 Code of Practice – Safety and Health in Construction

The Code of Practice covers scaffolds and ladders under the following topics over five pages:

1. general provisions
2. materials
3. design and construction
4. inspection and maintenance
5. lifting appliances on scaffolds
6. prefabricated scaffolds
7. use of scaffolds
8. suspended scaffolds.

Study skills

19

After reading this chapter you should be able to:

1. plan and organize self study work
2. organize revision
3. understand and organize information for an examination
4. understand the concept of memory maps
5. understand how to tackle examinations in the exam room and afterwards
6. understand about examiners' reports and what some of the latest reports are saying
7. understand the marks allocated to each NEBOSH question.

19.1 Introduction

The NEBOSH Certificate is like any other examination. To pass it there are several things a student needs to do. The best results can be achieved by planning in the following way:

➤ having clear and realistic goals, in both the short and the long term;
➤ having knowledge of some techniques for studying and for passing exams;
➤ having a well-organized approach;
➤ being motivated to learn.

It is often said that genius is 99% application and only 1% inspiration. Good organization and exam technique are secondary to academic ability when it comes to passing exams. Studying is an activity in its own right and there are a number of ways in which students can make their study much more effective to give themselves the greatest possible chance of success. The place where a person studies; planning of study time; adapting the type of study to suit the individual; styles of note-taking; improving memory and concentration skills; revision techniques; and being clear about the contents of the syllabus and about what is expected by the examiner – attention to all these details will maximize a student's available time and mental ability.

19.2 Finding a place to study

Studying does not always come naturally to people and it often has to be learned, so it is worth paying attention to the basics of study. It is easier to acquire the habit of study if the student works in a place that has been designated for the purpose. Choose somewhere reasonably quiet and free from distractions, with good ventilation (essential for alertness) and a comfortable temperature. Make sure there is enough light – a reading lamp helps to prevent eyestrain and tiredness.

Choose an upright chair rather than an armchair and make sure that the workspace is large enough. The size recommended by Warwick University, for example, is a minimum of 60 cm × 1.5 m. If possible, find a place where study materials do not have to be cleared away as it is very easy to put off starting work if they are not instantly accessible.

19.3 Planning for study

Start by making a study timetable. Planned study periods need to be established so that they do not get squeezed out of the busy working week. Depending on the type of syllabus being followed, the timetable will need to include time for carrying out assignments set by tutors, time for going over lecture notes and materials, any further reading rated as 'essential' and, if possible, reading rated as 'desirable'. Time should also be allocated for revision as the course progresses because the more regularly the information is revised, the more firmly it will become fixed in 'long-term memory'. This makes it very much easier to recall information for an exam.

After every hour of study, take a short break and, if possible, have a change of atmosphere. Physical exercise is very helpful in increasing the ability to concentrate and even a brisk walk round the block can do wonders for the attention span.

Variation is another way of holding the attention, so it helps to alter the type of activity. For example, time could be spent gathering information from books, the internet and so on, followed by drawing a diagram or a graph, writing, reading and so on.

19.4 Blocked thinking

Sometimes a student becomes 'blocked' in an area of study. If this happens, try leaving it alone for a few days and tackle a different area of the subject, since concentrating too hard on something that is too difficult is likely to produce loss of confidence and can be very de-motivating. It usually happens that, having had a break, the difficulty vanishes and the problem clears. This is either because the solution has emerged in the mind from another perspective, or because the student has learnt something new that has supplied the answer. Psychologists consider that it is possible that some problems can be solved during sleep when the mind has a chance to wander and apply lateral thinking.

19.5 Taking notes

The most efficient way to store notes is in a loose-leaf folder, because items can so easily be added. Write only on one side of the sheet and use margined paper to make it easier to add information. The facing pages can be used to make summaries or extra points. If notes are clearly written and well spaced out, they are much more straightforward

to work from and more attractive to return to later. Use colour, highlighting and underlining too; notes that look good are less daunting when the time comes to revise.

During revision it will be important to be able to identify subjects quickly, so use headings, numbering, lettering, bullet points, indention and so on. It is worth spending a bit of extra time if it will make revision easier.

Key words and phrases are better than continuous prose when note-taking. While writing down information it is easy to miss essential points that are being made by the tutor. The notes should be read through within 36 hours to make sure that they are completely clear when the time comes to revise. At this point it is convenient to add in anything that has been missed, while it is fresh in the mind. Reading through notes in this way also helps to fix the information in the memory and the level of recall will be further improved by reading them again a week later. Although it seems time-consuming, this technique will save a lot of time in the long term.

Many people find that they can understand and remember more effectively by making 'mind maps' (also known as 'pattern notes'). These will be discussed in more detail in the revision section.

19.6 Reading for study

Most students will be working while they are studying, so they will need techniques to help them make the best use of their time. It is not always necessary to read a whole book for study purposes, apart from books written specifically to a syllabus, such as this one. Use the index and be selective. In this book an example would be Chapter 17 on the law.

Where a book is not specifically written to a syllabus time saving techniques such as skim reading can be used. First of all, flip through the book to see what it contains, looking at the contents, summaries, introduction, tables, diagrams and so on. It should be possible to identify areas that are necessary to the syllabus. These can be marked with removable adhesive strips or 'post-it' notes.

Secondly, quickly read through the identified parts of the book. Two ways of doing this are:

(a) run a finger slowly down the centre of a page, watching it as it moves. The eye will pick up relevant words and phrases as the finger moves down;

(b) read in phrases, rather than word by word, which will increase the reading speed.

Both these methods need practice but they save a lot of time and effort, and it is better to read through a piece two or three times, quickly, than to read it once slowly. However, if there is a really difficult piece of information that is absolutely essential to understand, reading it aloud, slowly, will help.

➤ Finally, at the end of each section or chapter, make a few brief notes giving the essential points.

It is well worth learning to speed-read if a course involves a large amount of reading. Tony Buzan's series *The Mind Set* explains the technique in simple terms.

19.7 Organizing for revision

As the exam approaches, all the information from the course needs to be organized in preparation for revision. This should, ideally, have been an ongoing process throughout the course, but if not there will be extra work to be undertaken at this point. Information from lectures, reading, practical experience and assignments needs to be organized into a form that makes it readily accessible to the memory. In addition there will be the student's existing knowledge that is relevant to the syllabus. It is worth spending some time considering how this store of information relates to and can be used to expand and back up the new information from the course. December 2005 Examiners' Reports refer specifically to the advantages to be gained from this:

Paper A1 Question 4
'Some candidates, perhaps from their work experience, showed a good knowledge of a permit-to-work system and produced reasonable answers'.

They also warn of the dangers:

'Answers based solely on "what we do in our organization" ... do not earn high marks'.

19.8 Organizing information

There are a number of ways to organize information for an exam:

➤ reading back through the notes on a regular basis while the course is still in progress will help to fix the information in long-term memory;

➤ make revision cards: condensing the information down so that it fits on a set of postcards makes it possible to carry a lot of information around in a pocket. The activity of condensing the information is revision in itself. The revision material is then available for spare moments in the day, such as queuing in the canteen, sitting in a traffic jam on the motorway, waiting for the computer to function and so on.

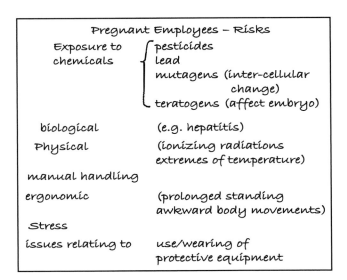

Figure 19.1 Revision notes.

For example: 'Outline the factors that may increase risks to pregnant employees'. On a card, the information could be condensed as shown in Figure 19.1.

➤ Make mind maps (also known as pattern notes): like revision cards, the act of making them is revision in itself. Because they are based on visual images they are easier for many people to remember and they can be made as complex or as simple as the individual requires. Using colour and imagery makes it easier to follow and to recall the information in a mind map (Buzan 2003). Figure 19.2 shows a mind map based on the report writing section of this book.

Use key words as an aid to memory. For example. 'ways of reducing the risk of a fire starting in the work place' list could be reduced to an eight-letter nonsense word, to jog the memory, as follows:

FLICSHEV:
Flammable liquids – provide proper storage facilities
Lubrication – regular lubrication of machinery
Incompatible chemicals need segregation
Control of hot work
Smoking – of cigarettes and materials, to be controlled
Housekeeping (good) – prevent accumulation of waste
Electrical equipment – needs frequent inspection for damage
Ventilation – outlets should not be obstructed

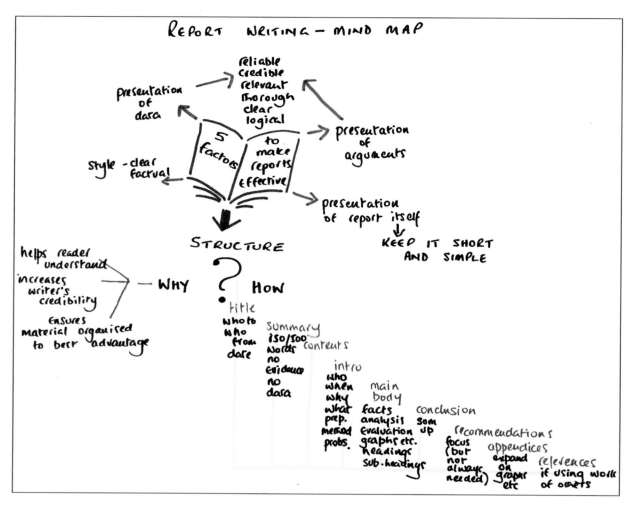

Figure 19.2 Mind map report writing.

The effort involved in constructing this list and inventing the word helps to fix the information in memory.

There are other methods to aid memory but these four are probably the most practical aids to exam technique.

19.9 How does memory work?

Understanding how memory works will help the student to use it more effectively. As in any scientific subject, there is still a lot to be discovered about the workings of the mind, and a considerable amount of disagreement about what has been discovered so far; but psychologists generally seem to agree about the way that memory works.

The process of remembering is divided, roughly, into two sections – short-term and long-term memory. Items that go into short-term memory, if no further attention is paid to them, will fade away and be forgotten. If they are rehearsed they will stay in short-term memory for a while, as, for example, when trying to remember a phone number to make a call. To put something into long-term memory demands a greater 'depth of processing', that is to say, more mental activity is required so that

(a) information stays in the memory (storage); and
(b) information can be found when it is needed (retrieval).

The techniques described in the section entitled 'Thinking about revision' give some ways in which storage and retrieval systems in the brain can be made to work effectively.

Contrary to popular opinion, mature students have many assets when it comes to learning. Because their life and work experience is almost always more comprehensive and of a higher complexity than that of younger students, they normally possess more 'schemas' (areas of knowledge) to which new information can be attached. In addition to this there will be more experiences already stored in the brain that can provide an explanation for the new pieces of knowledge that are being acquired. Add to this a very high level of motivation, a stronger level of incentive towards success and more determined application, and it becomes clear that the mature student has many learning advantages.

Research shows that, as learners, we take in 10% of what we read; 20% of what we hear; 30% of what we see; 50% of what we see and hear; 70% of what we ourselves say; and 90% of what we ourselves do (Northedge 2005). It is easy to see that people who are working at and studying a subject have an advantage.

It cannot be denied, however, that mature students tend to be very nervous about exams and often need a good deal of reassurance and support to enable them to realize their full potential.

19.10 How to deal with exams

There are three stages to taking an exam:

➤ planning and revision
➤ the exam room
➤ after the exam.

For people who require **special consideration**, there are provisions within NEBOSH to allow extra time or the use of special equipment. Students who think they may be eligible for reasonable adjustments/special consideration should apply to NEBOSH several weeks before the date of the exam. Reference should be made to the NEBOSH Guide.

19.10.1 Planning and revision

1. It is absolutely essential for students to know what they are going to be examined on and what form the exam will take. Students should read through the syllabus and if they are concerned about any area of it, this should be raised with the course tutor well before the date of the exam.

2. Read the examiners' reports for the subject. After the exams every year, the examiners highlight the most common mistakes made by students (see Section 19.11 on examiners' reports later in this chapter). They also provide useful information about, for example, pass rates, levels attained (distinction, credit, pass, fail), time management and other hints on exam technique.

3. It is very useful to work through some recent past papers, against the clock, to get used to the 'feel' of the exam. If possible, tutors should set up mock exams and make sure that the papers are marked. Some of the shorter courses will not be able to provide this service, and if this is the case, students should try to do at least one or two questions in their own time. The examiners' reports will give an indication of what could have gone into the answers. Past papers, examiners' reports and syllabuses are available from NEBOSH. The website is at www.nebosh.org.uk.

4. It is vital to know where the exam is to be held and the date and time. If possible, visit the building beforehand to help build confidence about the location, availability of parking and so on.

5. Make a chart of the time leading up to the exam. Include all activities, work, leisure, and social, as well as the time to be used for revision, so that the schedule is realistic.

6. Try to eat and sleep well and take some exercise.

7. Revision techniques are covered in the sections on revision and memory earlier in this chapter.

SET REALISTIC TARGETS, THEN ACHIEVE THEM. MAKE PLANS AND STICK TO THEM.

19.10.2 In the exam room

➤ Read through the exam paper very carefully.
➤ Check the instructions – how many questions have to be answered? From which section?
➤ Make a time plan.
➤ Underline command words, e.g. 'define', 'describe', 'explain', 'identify' and so on. NEBOSH is currently reviewing the command verbs and new guidance will be issued during 2009. Using the words in the question when writing the answers will help to keep the answer on track.
➤ Stick to the instructions given in the question. If the question says 'describe', then describe. If it says 'briefly' make sure the answer is brief.
➤ Write clearly. Illegible answers don't get marked.

➤ Look at the number of marks allocated to a question to pick up clues as to how much time should be spent on it.

➤ Mark questions which look possible and identify any that look impossible.

➤ It is rarely necessary to answer exam questions in a particular order. Start with the question that you feel most comfortable with since it will help to boost confidence. Make sure it is clearly identified by number for the examiner.

➤ Answer the question that is set, not the one you wish was on the paper.

➤ If ideas for other answers spring to mind while writing, jot down a reminder on a separate piece of paper. It is easy to forget that bit of information in the heat of the moment.

➤ Plan the use of time and plan the answers. Include some time to check over each answer.

➤ Stick to the time plan; stick to the point; make points quickly and clearly.

Terminology used in NEBOSH exams (taken from the guide to the NEBOSH National General Certificate of 2006 Appendix 3. At April 2009 this is under review and revised guidance will be issued during 2009)

Action verb	Meaning
define	provide a generally recognized or accepted definition
describe	give a word picture
explain	give a clear account of, or reasons for
give	provide without explanation (used normally with the instruction to 'give an example [or examples] of…)
identify	select and name
outline	give the most important features of (less depth than either 'explain' or 'describe' but more depth than 'list')
sketch	provide a simple line drawing using labels to identify specific features
state	a less demanding form of 'define', or where there is no generally recognized definition

Early marks in an exam question are easier to pick up than the last one or two, so make sure that all the questions are attempted within the time plan. No marks are given for correct information that is not relevant to the question. Examiners are only human and usually have to work under pressure; it is possible despite careful marking, that some vital and correct point may be missed if there is a mass of irrelevant information from which the point has to be extracted.

Don't be distracted by the behaviour of other students. Someone who is requesting more paper has not necessarily written a better answer; they may simply have larger handwriting. People who start to scribble madly as soon as they turn over the question sheet are not in possession of some extra ability – they simply haven't planned their exam paper properly.

Keep calm, plan carefully, don't panic.

19.10.3 After the exam

If there are several exams to be taken, it is important to keep confidence levels high. It is not a good idea to get into discussion about other people's experiences of the exam. After one exam, focus on the next. If something went wrong during the exam (for example, illness or severe family problems), the tutor and the examining board should be alerted immediately.

19.11 The examiners' reports

19.11.1 A few points from the examiners' reports

The latest reports available at the time of going to press are for March 2007 and June 2008. As in previous years, the examiners acknowledge the achievements and hard work of candidates, but they point out that avoidable mistakes are still being made and marks lost, simply through lack of examination technique.

The examiners advise that while acquisition of knowledge and understanding across the syllabus are clearly a prerequisite for success, examination technique is an essential. They define this as 'the skill of reading a question, identifying the breadth of issues relevant to define that question and putting them down on paper in a logical and coherent way and to the depth required'. Effective time planning is also emphasized.

The section on 'Recurrent Problems' is a clearly worded guide to the expectations of NEBOSH examiners. Examples of these problems include:

➤ not attempting all the required questions;
➤ failing to provide complete answers;
➤ not answering the question that is set;
➤ not applying the command words, for example 'describe', 'outline';
➤ not separating answers into different sub-sections;
➤ lack of time planning;
➤ illegible handwriting.

Here are some examples from the reports that show how good examination technique could have improved candidates' chances of gaining higher marks.

Paper NGC1 March 2007: there were 'many unstructured responses' to Q2, with evidence that some candidates would have gained more marks if they had taken time to plan their answers. In Q3, it was clear that some candidates had not read the question, as they provided a discussion of the '5 steps' instead of the principles of a safe system of work. Failure to read the question occurred again in Q7 and Q8, and also in Q10 where a reference to 'accidents' was missed in a question about an induction training programme.

In Paper NGC2 March 2007, Q5 asked for 'specific' measures in fire evacuation procedure, but several candidates provided only general measures.

Paper NGC1 June 2008 shows again that questions have not always been read carefully. For example in Q8, candidates were asked what might cause a health and safety culture to decline, but several supplied factors that would promote it. In Q9, some provided a list when the question asked for an 'outline'.

Paper NGC2 June 2008: where an 'outline' was requested, the instruction was not always followed (Q2). In Q6 candidates were asked to 'outline the factors' but some produced an outline of the hierarchy of control measures. In Q7 the term 'unsafe acts' was confused with 'unsafe conditions' in a question about a carpenter, and some candidates ignored the reference to tools in the second part of the question. In Q11, on reducing fire risk, some candidates did not read the question and gave methods for dealing with a fire once it had started.

The examiners also pointed out that in two instances the questions had appeared on previous papers. Studying past papers, together with the examiners' reports, is a useful exam technique that helps to clarify what is expected by the exam board.

19.11.2 Marks for NEBOSH questions

We have been asked about the allocation of marks to each question shown throughout the book. However, since the marks awarded to each part of a question can vary, only general guidance can be given.

All one- or two-part questions are given 8 marks with a minimum of 2 marks awarded to any part.

Questions with three or more parts are given 20 marks with a minimum of 4 marks for each part.

19.12 Conclusion

Passing health and safety exams and assessments has a lot in common with any other subject being examined. If candidates do not apply themselves to study effectively, read the questions carefully and plan their answers, success will be very limited. As once said by an old carpenter:

'Measure twice, think twice and cut once'
Applied to exams it can be changed to:

'Read twice, think twice and write once'

19.13 References

Buzan, T. and Buzan, B. *The Mind Map Book*. 3rd Edition Revised 2003, BBC active, ISBN 978 0 5634 8705 0.
Buzan, T. (2000) *The Mind Set*. BBC Worldwide (includes *Use Your Head, Master Your Memory, the Speed Reading Book* and *Use Your Memory*) ISBN 978 1 4066 1018 6.
Guide to the NEBOSH National General Certificate.
Leicester, Coventry and Nottingham Universities Study Skills booklets.
NEBOSH Examiners' Reports March 2007 and June 2008.
Northedge, Andrew: The Good Study Guide. The Open University 2005 ISBN 978 0 7492 5974 7

Specimen answers to NEBOSH examinations

20.1 Introduction

The NEBOSH National General Certificate is assessed by two written examination papers (NGC1 – The management of health and safety and NGC2 – Controlling workplace hazards) and a practical assessment (NGC3 – The practical application). Both examination papers and the practical application must be passed so that the NEBOSH National General Certificate may be awarded.

In this chapter, specimen answers are given to one long and two short questions for each of the two written papers. A specimen practical assessment and management report is also given for the practical application. It is important to stress that there are no unique answers to these questions but the answers provided should provide a useful guide to the depth and breadth expected by the NEBOSH Examiners. Candidates are strongly advised to read past examiners' reports which are published by

NEBOSH. These are very useful documents because they not only give an indication of the expected answers to each question of a particular paper but they also indicate some common errors made by candidates.

20.2 The written examinations

The previous chapter on study skills gives some useful advice on tackling examinations and should be read in conjunction with this chapter. NEBOSH has a commendably thorough system for question paper preparation to ensure that no candidates are disadvantaged by question ambiguity. Candidates should pay particular attention to the meaning of action verbs, such as 'outline', used in the questions.

Candidates have always had most difficulty with the action verb '**outline**' and, for this reason, several of the specimen questions chosen use '**outline**' so that some guidance on the depth of answer expected by the examiners can be given.

It is difficult to give a definitive guide on the exact length of answer to the examination questions because some expected answers will be longer than others and candidates answer in different ways. As a general guide, for the long answer question on the examination paper, it should take about 25 minutes to write about one and a half pages (550–620 words). Each of the 10 short answer questions require about half a page of writing (170–210 words).

Although candidates may not be guaranteed 100% by following the guidance in this chapter, they should obtain a comfortable pass.

20.2.1 NGC1 – The management of health and safety

Paper 1 – Question 1

(a) **Outline** the factors that should be considered when selecting individuals to assist in carrying out risk assessments in the workplace. **(5)**

(b) **Describe** the key stages of a general risk assessment. **(5)**

(c) **Outline** a hierarchy of measures for controlling exposures to hazardous substances. **(10)**

Answer:

(a) The most important factor is the competence and experience of the individuals in hazard identification and risk assessment. Some training in these areas should offer evidence of the required competence. They should be experienced in the process or activity under assessment and have technical knowledge of any plant or equipment used. They should have knowledge of any relevant standards, Health and Safety Executive (HSE) guidance and Regulations relating to the activity or process.

They must be keen and committed but also aware of their own limitations. They need good communication skills and be able to write interesting and accurate reports based on evidence and the detail found in health and safety standards, codes of practice, Regulations and guidance. Some IT skills would also be advantageous. Finally, the views of their immediate supervisor should be sought before they are selected as team members.

(b) There are five key stages to a risk assessment suggested by the HSE as follows:

The first stage is hazard identification, which involves looking at significant hazards which could result in serious harm to people. Trivial hazards should be ignored. This will involve touring the workplace concerned looking for the hazards in consultation with workers themselves and also reviewing any accidents, ill health or incidents that have occurred.

Stage 2 is to identify the person who could be harmed – this may be employees, visitors, contractors, neighbours or even the general public. Special groups at risk, like young persons, nursing or expectant mothers and people with a disability, should also be identified.

Figure 20.1 Ladders should be correctly angled.

Stage 3 is the evaluation of the risks and deciding if existing precautions or control measures are adequate. The purpose is to reduce all residual risks after controls have been put in to as low as is reasonably practicable. It is usual to have a qualitative approach and rank risks as high, medium or low after looking at the severity of likely harm and the likelihood of it happening. A simple risk matrix can be used to get a level of risk.

The team should then consider whether the existing controls are adequate and meet any guidance or legal standards using the hierarchy of controls and the General Principles of Prevention set out in the Management Regulations.

Stage 4 of the risk assessment is to record the significant findings which must be done if there are five or more people employed. The findings should include any action that is necessary to reduce risks and improve existing controls – preferably set against a timescale. The information contained in the risk assessment must be disseminated to employees and discussed at the next health and safety committee meeting.

Stage 5 is a timescale set to review and possibly revise the assessment which must also be done if there are significant changes in the workplace or the equipment and materials being used.

(c) The various stages of the usual hierarchy of risk controls are underlined in this answer.

Elimination or **substitution** is the best and most effective way of avoiding a severe hazard and its associated risks. Elimination occurs when a process or activity is totally abandoned because the associated risk is too high. Substitution describes the use of a less hazardous form of the substance. There are many examples of substitution, such as the use of water-based rather than oil-based paints and the use of asbestos substitutes.

In some cases it is possible to **change the method of working** so that exposures are reduced, such as the use of rods to clear drains instead of strong chemicals. It may be possible to use the substance in a safer form; for example, in liquid or pellets to prevent dust from powders. Sometimes the pattern of work can be changed so that people can do things in a more natural way; for example, by encouraging people in offices to take breaks from computer screens by getting up to photocopy or fetch documents.

Reduced or limited time exposure involves reducing the time that the employee is exposed to the hazardous substance by giving the employee either other work or rest periods.

If the above measures cannot be applied, then the next stage in the hierarchy is the introduction of **engineering controls**, such as isolation (using an enclosure, a barrier or guard), insulation (used on any electrical or temperature hazard) or ventilation (exhausting any hazardous fumes or gases either naturally or by the use of extractor fans and hoods). If ventilation is to be used, it must reduce the exposure level for employees to below the workplace exposure limit (WEL).

Housekeeping is a very cheap and effective means of controlling risks. It involves keeping the workplace clean and tidy at all times and maintaining good storage systems for hazardous substances.

A **safe system of work** is a requirement of the Health and Safety at Work (HSW) Act and describes the safe method of performing the job.

Training and information are important but should not be used in isolation. Information includes such items as signs, posters, systems of work and general health and safety arrangements.

Personal protective equipment (PPE) should only be used as a last resort. There are many reasons for this. It relies on people wearing the equipment at all times and it must be used properly.

Where necessary **health surveillance** should be introduced to monitor the effects on people and air quality may need to be monitored to check exposure.

Welfare facilities, which include general workplace ventilation, lighting and heating and the provision of drinking water, sanitation and washing facilities, are the next stage in the hierarchy.

All risk control measures, including **training and supervision** must be **monitored** by competent people to check on their continuing effectiveness. Periodically the risk control measures should be **reviewed**. Monitoring and other reports are crucial for the review to be useful. Reviews often take place at safety committee and/or at management meetings. A serious accident or incident should lead to an immediate review of the risk control measures in place.

Finally, special control requirements are needed for carcinogens.

Paper 1 – Question 2

Outline ways in which employers may motivate their employees to comply with health and safety procedures. **(8)**

Answer:

Motivation is the driving force behind the way a person acts or the way in which people are stimulated to act . The best way to motivate employees to comply with health and safety procedures is to improve their understanding of the consequences of not working safely, their knowledge of good safety practices and the promotion of their ownership of health and safety. This can be done by effective training (induction, refresher and continuous) and the provision of information showing the commitment of the organization to safety and by the encouragement of a positive health and safety culture with good communications systems. Managers should set a good example by encouraging safe behaviour and obeying all the health and safety rules themselves even when there is a difficult conflict between production schedules and health and safety standards. A good working environment

Figure 20.2 Motivating staff (NEBOSH).

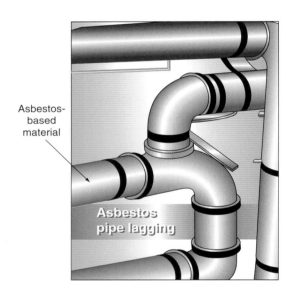

Asbestos-based material

Asbestos pipe lagging

Figure 20.3 Asbestos pipe lagging.

and welfare facilities will also encourage motivation. Involvement in the decision-making process in a meaningful way, such as regular team briefings, the development of risk assessments and safe systems of work, health and safety meetings and effective joint consultation

arrangements, will also improve motivation as will the use of incentive schemes. However, there are other important influences on motivation such as recognition and promotion opportunities, job security and job satisfaction. Self-interest, in all its forms, is a significant motivator.

Although somewhat negative, it is necessary sometimes to resort to disciplinary procedures to get people to behave in a safe way. This is rather like speed cameras on roads with the potential for fines and points on your licence.

Paper 1 – Question 3
(a) **Explain** why young persons may be at a greater risk from accidents at work. **(4)**
(b) **Outline** the measures that could be taken to minimize the risks to young employees. **(4)**

Answer:
(a) Young workers have a lack of experience, knowledge and awareness of risks in the workplace. They tend to be subject to peer pressure and behave in a boisterous manner. They are often willing to work hard and want to please their supervisor and can become over-enthusiastic. This can lead to the taking of risks without the realization of the consequences. Some younger workers have

underdeveloped communication skills and a limited attention span. Their physical strength and capabilities may not be fully developed and so they may be more vulnerable to injury when manually handling equipment and materials. They are also more susceptible to physical agents, biological and chemical agents such as temperature extremes, noise, vibration, radiation and hazardous substances.

(b) The Management of Health and Safety at Work Regulations require that a special risk assessment must be made before a young person is employed. This should help to identify the measures which should be taken to minimize the risks to young people.

Measures should include:

➤ additional supervision to ensure that they are closely looked after, particularly in the early stages of their employment;
➤ induction and other training to help them understand the hazards and risk at their work place;
➤ not allowing them to be exposed to extremes of temperature, noise or vibration;
➤ not allowing them to be exposed to radiation, or compressed air and diving work;
➤ carefully controlling levels of exposure to hazardous materials so that exposure to carcinogens is as near zero as possible and other exposure is below the WEL limits which are set for adults;
➤ not allowing them to use highly dangerous machinery like power presses and circular saws, explosives and mechanical lifting equipment such as fork-lift trucks;
➤ restricting the weight that young persons lift manually to well below any weights permitted for adults.

There should be clear lines of communication and regular appraisals. A health surveillance programme should also be in place.

20.2.2 NGC2 – Controlling workplace hazards

Paper 2 – Question 1

A glassworks produces covers for streetlights and industrial lighting. The process involves molten glass being blown by hand and shaped in moulds.

(i) **Identify FOUR** health effects that may be caused by working in the hot conditions of the glass factory. **(4)**

(ii) **Describe** measures that could be taken in order to minimize the health effects of working in such hot environments. **(6)**

(iii) **Outline** the factors relating to the *task* and the *load* that may affect the risk of injury to an employee engaged in stacking the finished product onto racking. **(10)**

Answer:

(i) 1. heat exhaustion due to high ambient temperature;
2. dehydration due to excessive sweating;
3. heart stress and, in extreme cases, heat stroke due to prolonged exposure to high ambient temperatures;
4. burns from handling hot molten glass;
5. the eyes can also be affected by high-intensity light from looking at molten glass (additional answer).

(ii) The health effects of working in a hot environment can be reduced by the gradual acclimatization of new workers. Even after the initial acclimatization, frequent rest periods will be necessary to allow the body to acclimatize to the hot conditions on a daily basis. Rest should be in cool areas which in summer may need to be artificially cooled. If, in addition, the humidity is high, a good supply of ventilation air will be needed to help control sweating. Adequate supply of cold drinking water is essential to avoid dehydration.

Workers in hot conditions should wear appropriate clothes, which must be a compromise between lighter garments to promote evaporation of perspiration, and protective clothes to prevent burns. It will be necessary to provide protective leather or fire-resistant aprons and gloves, and appropriate eye and face protection such as eye visors. Visors may need to be supplied with cooling air to keep people cool and permit proper vision. Screens could also be provided to protect workers from radiant heat. Periodic health surveillance should be provided.

Finally, workers should be trained to recognize ill-health effects on others.

(iii) The **task** should be analysed in detail so that all aspects of manual handling are covered including the use of mechanical assistance. This will involve a manual handling risk assessment. The number of people involved and personal factors, such as age and health, should also be considered. A satisfactory body posture must be adopted with the feet firmly on the ground and slightly apart. To avoid work-related upper limb disorders (WRULDs) there should be no stooping or twisting of the trunk; it should not be necessary to reach upwards as this will place additional stresses on the arms, back and shoulders. The further the load is held or manipulated from the trunk, the more difficult it is to control and the greater the stress imposed on the back. These risk factors are significantly increased if several of them are present at the same time. The load should not be carried over excessive distances (greater than 10 m). The frequency of lifting, and the vertical and horizontal distances the load needs to be carried (particularly if it has to be lifted from the ground and/or placed on a high shelf) can lead to fatigue and a greater risk of injury. If the loads are handled whilst the individual is seated, the legs are not used during lifting and stress is placed on the arms and back.

There should not be excessive pulling, pushing or sudden movements of the load. The state of floor surfaces and the footwear of the individual should ensure that slips and trips are avoided.

There should be sufficient rest or recovery periods and/or the changing of tasks particularly in the hot ambient temperatures of the glassworks. This enables the body to recover more easily from strenuous activity.

The imposition of a high rate of work is a particular problem with some automated production lines and can be addressed by spells on other work away from the line.

The handling capability of an individual is approximately halved when he/she becomes a member of a team. Visibility, obstructions and the roughness of the ground must all be considered when team handling takes place.

The **load** must also be carefully considered during the assessment – it may be too heavy. The maximum load that an individual can lift will depend on the capability of the individual and the position of the load relative to the body. There is therefore no safe load but guidance is available from HSE literature, which does give some advice on loading levels. If the load is too bulky or unwieldy, its handling is likely to pose a risk of injury. Visibility around the load is important, as is awareness that it may hit obstructions or become unstable in windy conditions. The position of the centre of gravity is important for stable lifting – it should be as close to the body as possible; however, this may be difficult if the load is hot, such as in boxes or trays of recently blown glass. They should be allowed to cool sufficiently.

The load becomes difficult to grasp when it is carried over slippery surfaces, has rounded corners or there is a lack of foot room. Sometimes the contents of the load are likely to shift. This is a particular problem when the load is a container full of smaller items, such as small glass covers. These are glass components which may shatter if dropped and leave shards of glass to be carefully cleared up.

The load is likely to be hot and could be sharp as well in places or when broken so that PPE, such as leather gloves and aprons along with eye protection, may be required.

Paper 2 – Question 2

Outline the precautions that may be needed when carrying out repairs to the flat roof of a building. **(8)**

(a)

(b)

Figure 20.4 Flat roof.

Answer:

Roof work is hazardous and requires a specific risk assessment and method statement prior to the commencement of work so that the required precautions may be identified. The particular hazards are fragile roofing materials, including those materials which deteriorate and become more brittle with age and exposure to sunlight, exposed edges, unsafe access equipment and falls.

There must be suitable means of access such as scaffolding, ladders and crawling boards. Suitable edge protection will be needed in the form of guard rails to prevent the fall of people or materials, and access must be restricted to the area below the work using visible barriers. Warning signs indicating that the roof is fragile, should be displayed at ground level. Protection should be provided in the form of covers where people work near to fragile materials and roof lights. The means of transporting materials to and from the roof may require netting under the roof and even weather protection.

Precautions will be required for other hazards associated with roof work, such as overhead services and obstructions, the use of equipment such as gas cylinders and bitumen boilers and manual handling.

Finally, only trained and competent persons must be allowed to work on roofs and they must wear footwear having a good grip. It is good practice to ensure that a person does not work alone on a roof.

Paper 2 – Question 3

For **EACH** of the following agents, **outline** the principal health effects **AND identify** a typical workplace situation in which a person might be exposed:

(i)	carbon monoxide	**(2)**
(ii)	asbestos	**(2)**
(iii)	legionella bacteria	**(2)**
(iv)	hepatitis virus.	**(2)**

Answer:

(i) Carbon monoxide is a colourless, tasteless and odourless gas. It causes headaches and breathlessness and, at higher concentrations, unconsciousness and death. The most common occurrence of carbon monoxide is in exhaust gas from a vehicle engine. Working in a vehicle repair garage without proper ventilation to exhaust gases would expose a person to carbon monoxide fumes.

(ii) Asbestos produces fine fibres which can become lodged in the lungs. This can lead to asbestosis (scaring of the lungs), lung cancer or mesothelioma – cancer of the lining of the lung. Asbestos can be found in buildings, in ceiling tiles and as lagging around heating pipes (see Figure 20.3). When these sites are disturbed, the asbestos fibres become airborne and inhalable affecting those engaged in maintenance or demolition work.

(iii) Legionella is an airborne bacterium and is found in a variety of warm water sources between 20°C and 45°C. It produces a form of pneumonia caused by the bacteria penetrating the alveoli in the lungs. The disease is known as Legionnaires' disease and has symptoms similar to influenza. The three most common systems at risk from the bacteria are water systems that incorporate a cooling tower, air conditioning units, and showers. People working on these systems or working in the area of infected systems are at risk, particularly if they are over 45 years of age and it affects men more than women.

(iv) Hepatitis is a disease of the liver and can cause high temperatures, nausea, jaundice and liver failure. The virus can be transmitted from infected faeces (Hepatitis A) or by infected blood (Hepatitis B and C). Hospital workers and first aiders who come into contact with blood products are at risk of hepatitis.

20.3 Hand drawn sketches

Here are examples of hand drawn sketches where NEBOSH have used the command 'sketch.'

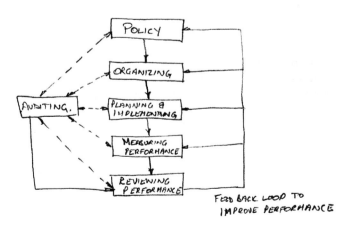

Figure 20.5 Flowchart for six elements of HSG65.

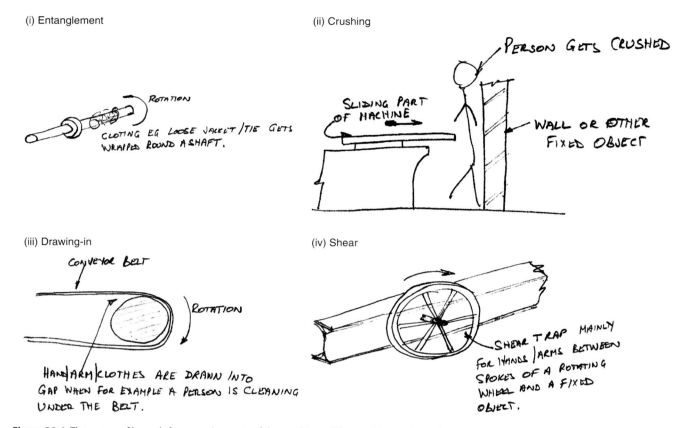

(i) Entanglement

ROTATION

CLOTHING EG LOOSE JACKET/TIE GETS WRAPPED ROUND A SHAFT.

(ii) Crushing

PERSON GETS CRUSHED

SLIDING PART OF MACHINE

WALL OR OTHER FIXED OBJECT

(iii) Drawing-in

CONVEYOR BELT

ROTATION

HAND/ARM/CLOTHES ARE DRAWN INTO GAP WHEN FOR EXAMPLE A PERSON IS CLEANING UNDER THE BELT.

(iv) Shear

SHEAR TRAP MAINLY FOR HANDS/ARMS BETWEEN SPOKES OF A ROTATING WHEEL AND A FIXED OBJECT.

Figure 20.6 The nature of hazards from moving parts of the machinery (Chapter 11, question 12).

20.4 NGC3 – the practical application

The guide to the NEBOSH National General Certificate is essential reading for all candidates before they attempt the practical application. The guide contains details on the aims of the application, the rules to be followed, the breadth and depth of observations and management report expected and the mark scheme.

In this section, the NEBOSH Guide is summarized in Section 20.4.1 and a specimen practical application, with associated observations and management report, is given in Section 20.4.2.

20.4.1 Requirements of the practical application

The practical application aims to test the application of basic health and safety knowledge to a workplace inspection. The test assesses the understanding of key issues and the ability to communicate the findings of the inspection in an effective way to a member of the senior management team. The formal aim of the assessment is to prove a candidate's ability to successfully complete two activities:

➤ to carry out unaided a safety inspection of a workplace, identifying the more common hazards, deciding whether they are adequately controlled and, where necessary, suggesting appropriate and cost-effective remedial action;

➤ to prepare a report that persuasively urges management to take appropriate action, explaining why such action is needed (including reference to possible breaches of legislation) and identifying, with due consideration of reasonable practicability, the remedial measures that should be implemented.

The inspection should take between 30 and 45 minutes and cover as wide a range of hazards as possible.

Observations and comments should be noted on the observation sheets provided by NEBOSH (see Appendix 20.1).

Candidates are expected to recognize actual and potential hazards and good – as well as bad – work practices and welfare provisions. While only brief details of each hazard are required, it is important that the assessor can subsequently identify:

➤ where the hazard was located;
➤ the nature of the hazard;
➤ the degree of risk associated with the hazard;
➤ the remedial actions, where appropriate, with relevant prioritization.

On completion of the inspection, candidates are allowed about 1 hour to produce the report, which must normally be in their own handwriting. Candidates are given a summary of the headings under which marks are allocated both for the report and observation sheets.

The whole assessment must be carried out under as near examination conditions as possible and should not normally take more than 2 hours. Candidates are not allowed to use previously prepared or organizational checklists or any other reference material.

For the inspection of the workplace, candidates are expected to do more than simply identify physical hazards such as unsafe machinery although this is important. In most workplaces they should find examples of chemical, fire, ergonomic, electrical, manual handling, slip, trip and fall (people and objects) hazards; consideration should also be given to health and physical hazards (e.g. dermatitis, noise and vibrations), access, emergency arrangements and welfare and environmental problems, such as heating or lighting. They need to comment when there is adequate control of hazards and where safe working practices are being observed, as well as when the opposite is the case.

Medium- and long-term actions must be noted as well as immediate actions to control any dangerous hazard. The distinction between immediate and longer-term action must be clear. The removal of a hazard is an example of immediate action, whereas the provision of information and training is longer term. The proposed remedial actions must not only remove or control the hazard but be realistic and cost-effective and the associated timescales realistic and appropriate.

Many candidates confuse comments with actions. The examiner looks for clear action proposals to remedy any observed weakness. The hazard recorded must be specific so that *examples* of poor housekeeping are recorded rather than a single statement of 'poor housekeeping'. For

some assessments, issues of employee awareness, supervision, maintenance, inspection and testing procedure may need to be considered. There is always a requirement for a monitoring system even if no action is required for a particular hazard at the time of the inspection. For most practical assessments, the examiner will be looking for between 20 and 30 observations.

For the **management report**, candidates are marked against the following five criteria:

1. *Selection of topics for urgent management action (0 to 10 marks)*

This requires candidates to emphasize those items on their observation sheets that they consider require urgent attention by management and to present them, together with suggested remedial actions (both short and longer term), in a logical and coherent manner. Those issues that are high risk and high priority, such as electrical safety, emergency exits and poor work practices, must be emphasized and highlighted. The report must differentiate between urgent and routine matters with simple cross-referencing to the observation sheets.

2. *Consideration of cost implications (0 to 5 marks)*

Candidates must demonstrate that they are aware of cost implications although they will not be judged on the value of the actual costs of a given measure.

If training is recommended as a solution to a problem, candidates should indicate if this is likely to require a few hours of work-based instruction or several days of more costly off-the-job training. It is the assessment of magnitude of the cost that is important, rather than precise figures. Words, such as 'cheap' and 'expensive', are better than giving no estimate. There should be an emphasis on the costs of taking no action (e.g. lower productivity, more compensation claims and possible enforcement action).

3. *Identification of breaches of legislation (0 to 5 marks)*

As reference material is not allowed during the application, candidates are only expected to indicate knowledge of the relevant legislation rather than the exact title and date. For example, Health and Safety at Work Act, First-Aid Regulations, PUWER and the Workplace Regulations are acceptable abbreviations on the observation sheets and in the management report.

The following list of Acts and Regulations would be relevant to all inspections:

➤ health and Safety at Work etc. Act;
➤ management of Health and Safety at Work Regulations;
➤ regulatory Reform (Fire Safety) Order;
➤ workplace (Health, Safety and Welfare) Regulations;

Most inspections would also include some of the following Regulations:

- manual Handling Operations Regulations;
- provision and Use of Work Equipment Regulations;
- personal Protective Equipment at Work Regulations;
- electricity at Work Regulations;
- control of Substances Hazardous to Health Regulations;
- health and Safety (Display Screen Equipment) Regulations;
- health and Safety First-Aid Regulations;
- control of Noise at Work Regulations;
- control of Vibration at Work Regulations;
- health and Safety (Safety Signs and Signals) Regulations;
- work at Height Regulations;
- health and Safety (Consultation with Employees) Regulations.

4. *Presentation of information (0 to 10 marks)*
A good report should normally comprise about three sides of handwritten A4 (i.e. about 500–750 words). It should cover the following points in a logical sequence:

- where and when the inspection took place;
- a brief summary of what was found;
- a short list of issues requiring urgent action by management with convincing arguments why such action is needed and calling attention to possible breaches of health and safety legislation;
- reference to the list of observations and recommended actions (which should be attached to the report), calling particular attention to any

recommendations which could have a high cost in terms of finance, inconvenience or time;
- a short conclusion.

5. *Effectiveness in convincing management to take action (0 to 15 marks)*
Managers are unlikely to have time to plough through lengthy reports. The report should be written in such terms that a manager would be able to take reasonable action based on facts relatively quickly. Therefore, it should be concise, readable and highly selective in terms of the action required by management. It should contain balanced arguments on why action is needed and explain the effect it would have on the standards of health and safety at the workplace.

20.4.2 Specimen practical application

The specimen practical assessment took place in a small joinery (woodworking) facility which is situated in a self-contained building. The facility is used to train 20 carpentry and joinery trainees and produce window frames for use on various building projects. The average age of the trainees is 24, although a minority of them are under 18 years. There are five employees, two supervisors, two technicians and an administrator, who administers the training scheme and liaises with the building projects.

Within the building are two workshops – a bench and a machine workshop – and an administration office, welfare facilities, a rest room, two lecture rooms and a storeroom. The assessment is concerned only with the workshops, the administration office and the welfare facilities.

The following management report was submitted to the Chief Executive of S & J Joinery Training Workshops.

The management report

1. Introduction
This report follows a health and safety inspection of the workshops, the administration office and welfare facilities at S & J Joinery Training Workshops.

The inspection took place on 7 June 2007. The report's aim is to highlight good practice and areas of concern requiring urgent action. The observations are listed in Appendix 20.1 attached to this report.

2. Summary
There have been no serious accidents since records began some five years ago. This good record appears to have led to a relaxation in the monitoring of standards. Several inexpensive improvements in the health and safety management systems are recommended.

There are some serious concerns involving fire precautions, the machine shop, manual handling and electrical equipment leading to breaches of the Health and Safety at Work Act (HSWA), the Management of Health and Safety at Work Regulations (MHSWR), the Regulatory Reform (Fire Safety) Order (Fire Regs.), the Provision and Use of Work Equipment Regulations (PUWER), the Personal Protective Equipment at Work Regulations (PPE Regs.), the Electricity at Work Regulations and the Workplace Regulations. Any of these breaches could attract enforcement action from the HSE.

All risk assessments must be reviewed as a matter of urgency.

A more detailed description of the topics requiring immediate attention is given below, with reference to the observation sheets given in brackets.

3. Issues requiring attention

(a) Management of health and safety

Many of the observed weaknesses result from a poor management system which could be improved if the technicians and administrator were given specific health and safety duties and, if necessary, relevant training of up to half-a-day in duration. At least one member of staff should be trained in first aid at a cost of approx £200.

In particular, the dangerous electrical equipment (20 and 25), the inappropriately guarded machines (22 and 23), the PPE problems (11 and 24) and the need to check that all equipment is switched off at the end of the day (35) can be addressed by assigning specific health and safety duties to staff. Supervisors also need to be more proactive in ensuring that benches are kept tidy (11, 15, 19 and 21) and trainees adhere to health and safety rules (15 and 24). COSHH assessments may need to be revised (19).

Many observations indicate that risk assessments, if they have been made, are no longer valid and need to be reviewed urgently. Finally, the MHSWR require a specific risk assessment to be made for persons under the age of 18 years (1).

The total cost of these recommendations will be between £200 and £500.

The health and safety committee must assess the quality and contents of the induction course given to trainees because of the large number of hazards found.

(b) Fire precautions

The Fire Regs require that fire risk assessments, fire-fighting arrangements, fire information and fire training be provided at the workshops. Therefore, there is an urgent need to install 'Fire Exit' signs (2), fire drills (4) and two fire extinguishers in the machine shop (25). This last point should have been noted by the extinguisher service contractor.

The emergency fire exit door should remain unobstructed at all times and unlocked whenever the building is in use (3). Security could be improved by fitting an alarm system which is activated whenever the door is opened or by the use of a quick release mechanism.

The total cost of all these improvements should not exceed £300.

The fire risk assessment must be reviewed.

(c) The machine shop

There are serious concerns regarding the guarding of the jig saw and the circular saw (22 and 23). The operation of these machines **must** cease immediately until they are properly guarded. The local exhaust ventilation (LEV) system must be examined and maintained on a regular basis with records being kept as required by PUWER and the broom (27) replaced by an industrial-grade vacuum cleaner (approximately £200). Wood dust causes health problems which can be very serious.

The rules concerning the use of the machines by trainees are commendable.

(d) Manual handling concerns

A mechanical lift truck should be purchased and used for the movement of window frames and timber (8). A manually operated truck should be sufficient at about £200 and much cheaper than a fork-lift truck with its considerable training costs.

Manual handling and trip hazards could be eased considerably by the erection of suitable shelving on the walls of the two workshops. Only one day's supply of timber should be kept in the workshops. Any excess timber stocks should remain in the storeroom. The use of shelving is a very cheap option (£50).

(e) Other issues

The PPE Regulations require that all personal protective equipment must be properly maintained and stored (11).

The heating system should be checked by a competent person because the workshops felt cold on the day of the inspection (below 16°C – the minimum recommended by the Workplace Regs Code of Practice) and there was an electric fire in the administration office (29). An improvement in housekeeping would ensure that floors and benches are kept clear and unobstructed.

4. Conclusions

The workshops were found to be clean, well maintained and well lit. The trainees clearly enjoyed working there. There were, however, a number of issues which need to be addressed urgently to avoid possible legal action or compensation claims following injuries. All these problems, other than the heating system, should be rectifiable within a total cost range of £1000–£1500.

Finally, the Health and Safety policy should be revisited and prominently displayed in both workshops.

Signed...J Brown.........   Date...7 June 2007...

Enclosures: [**Appendix 20.1**] **The practical assessment**

Appendix 20.1 The practical assessment

NATIONAL GENERAL CERTIFICATE

Unit NGC3 – Health and safety practical application

nebosh

Candidate's observation sheet

Sheet number __1__ of __6__

Candidate name	J Brown	Candidate number **C** 9887
Place inspected	S & J Joinery Training Workshops	Date of inspection _7_ / _6_ / _2007_

Observations List hazards, unsafe practices and good practices	Priority/risk (H, M, L)	Actions to be taken (if any) List all immediate **and** longer-term actions required	Time scale (immediate, 1 week, etc)
GENERAL OBSERVATIONS			
1. Several risk and COSHH assessments have been completed but the special risk assessment required by the Management Regs for young workers has not been done.	M –	Complete specific risk assessment for young persons. Establish a review timetable for all risk and COSHH assessments. Discuss all risk assessments at the Health and Safety committee meeting.	Immediate 1 month Each year
2. No Fire Exit signs are posted in the building. (Fire Regs)	H	Inform employees and trainees of fire escape routes. Obtain and install fire exit signs and review fire risk assessment. Organise some basic fire training.	Immediate 1 week 1 month
3. Emergency exit door is locked to protect building against unauthorised entry. Also the door was obstructed by several lengths of timber. (Fire Regs)	H –	Remove timber and any other obstacles and unlock door. Look for an alternative security arrangement and install. Monitor doors regularly and include this problem in the fire training.	Immediate 1 month Monthly
4. No fire drills have been undertaken for some time. (Fire Regs)	H	Undertake an immediate trial evacuation of the building to designated assembly point. Arrange 1 fire drill for all new trainees. Arrange at least 2 fire drills for employees per year and record evacuation time.	1 week 1 week after enrolment 6 months
5. All electrical equipment, other than the electric fire in the office and the sander in the workshop (see later), has been PAT tested.	–	Arrange for the fire and the sander to be PAT tested when repairs completed. All hand-held tools should be checked each week and PAT tested every 6 months. Ensure that records are kept and equipment re-tested at regular intervals.	1 week after repairs. Weekly 6 months Each year

NATIONAL GENERAL CERTIFICATE

Unit NGC3 – Health and safety practical application

nebosh

Candidate's observation sheet

Sheet number ___2___ of ___6___

Candidate name ___J Brown___

Candidate number **C** 9887

Place inspected ___S & J Joinery Training Workshops___

Date of inspection _7_ / _6_ / _2007_

Observations List hazards, unsafe practices and good practices	Priority/risk (H, M, L)	Actions to be taken (if any) List all immediate *and* longer-term actions required	Time scale (immediate, 1 week, etc)
GENERAL OBSERVATIONS (Cont.)			
6. Risk of slips, trips and falls due to trailing cables in both workshops. (Electricity Regs)	H	Fix covers over cables and re-route away from busy walkways. Reorganise workshop layout and/or work activities to reduce risk of tripping. Install additional power sockets to reduce the need for long trailing cables.	Immediate 1 week 6 months
7. Timber left lying on foor of both workshops – slip hazard.	H	Stack all unused timber in corner of workshop or adjacent free room. Fix up suitable shelving to store timber safely. Only store a day's requirement in the workshops. Include this arrangement in future training and monitor. Decide on a future strategy for timber storage and implement.	Immediate 2 weeks 6 months
8. Risk of back injuries from lifting heavy lengths of timber and manufactured window frames. (Manual Handling Regs)	H	Ensure that timber lengths are always lifted by at least 2 persons and window frames by 4 persons. Undertake a manual handling risk assessment to see whether mechanical assistance can be used. Give everyone a basic manual handling training course including supervisors. Monitor activities regularly.	Immediate 2 weeks 1 month Every 3 months
9. First aid box is fully and properly stocked and appropriate notices displayed in each workshop. (First Aid Regs)	L	No action required other than monitoring the contents of the first aid box to ensure that it is refilled.	Each week.
10. No trained first aider in the building. If required, help is obtained from next building.	M	Arrange for at least 1 member of staff (preferably 2) to be trained as a first aider.	3 months

NATIONAL GENERAL CERTIFICATE

Unit NGC3 – Health and safety practical application

nebosh

Candidate's observation sheet

Sheet number __3__ of __6__

Candidate name   J Brown

Candidate number **C** 9887

Place inspected   S & J Joinery Training Workshops

Date of inspection 7 / 6 / 2007

Observations List hazards, unsafe practices and good practices	Priority/risk (H, M, L)	Actions to be taken (if any) List all immediate *and* longer-term actions required	Time scale (immediate, 1 week, etc)
GENERAL OBSERVATIONS (Cont.)			
11. There several cases of torn gloves and scratched safety glasses on the benches in the 2 workshops. (PPE Regs)	H	Replace gloves and glasses. Ensure that all PPE is properly stored and maintained. Prominent notices should be displayed and all infringements recorded. Review all PPE risk assessments and monitor. Ensure that trainees receive instruction on the correct use of PPE.	Immediate 1 month 1 month
12. HSW Act poster and several other useful posters on manual handling, noise and the correct use of tools and machines, displayed in both workshops.	–	This is to be commended. May need updating in the longer term.	At induction
13. Lighting levels throughout were good.	L	No action required but keep under review.	
14. Toilet and washing facilities were satisfactory and cleaned on a regular basis. Barrier cream was provided and trainees instructed to use it. 1 toilet door lock was broken. (Workplace Regs)	L	Repair toilet door lock. Monitor cleanliness of toilets.	Immediate Daily.
15. Despite several notices forbidding the consumption of food and drink, several trainees had bottles of drinking water on their work benches. (Workplace Regs)	M	Supervisors must enforce ban on food and drink. Install a fresh drinking water facility in the trainee rest room with cups provided.	Immediate 1 month
16. Heating levels were low throughout the building and the workshops were cold. (Workplace Regs)	M	Check all room temperatures. Have boiler and heating system checked and, if necessary, provide auxiliary heating. Monitor to ensure that a reasonable temperature is maintained. This could be a costly problem in terms of finance, time and inconvenience.	Immediate 1 week Weekly from Sept to March
17. All fire extinguishers regularly serviced.	–	Commendable.	

NATIONAL GENERAL CERTIFICATE

Unit NGC3 – Health and safety practical application

nebosh

Candidate's observation sheet

Sheet number   4   of   6

Candidate name   J Brown

Candidate number **C** 9887

Place inspected   S & J Joinery Training Workshops

Date of inspection 7 / 6 / 2007

Observations List hazards, unsafe practices and good practices	Priority/risk (H, M, L)	Actions to be taken (if any) List all immediate **and** longer-term actions required	Time scale (immediate, 1 week, etc)
BENCH WORKSHOP 18. Several cupboard doors under the benches were left open leading to risk of minor injury.	L	Inform staff and trainees to keep all cupboards shut and post appropriate notice on notice board. Monitor compliance.	Immediate On-going
19. There was an unmarked jar containing a liquid (solvent?) on one of the benches. (COSHH)	H	Identify and label the jar or dispose of it. If it is identiÞed, ensure that a COSHH assessment has been completed. Supervisor to inform trainees of the unsafe practice. Ensure that the use of hazardous substances is covered during induction training.	Immediate Immediate Immediate Each induction
20. Frayed lead to hand-held sander – risk of electric shock. (Electricity Regs)	H	Remove from use. Replace lead, PAT test sander and place it on the PAT register. Include in regular PAT inspection schedule. Include electrical hazards and precautions in induction training for trainees	Immediate 1/2 weeks Every 3 months Each induction
21. Chisels and a Stanley knife left on benches. (PUWER)	H	Remove from benches and place in tool rack. Check to ascertain whether Stanley knife necessary. If it is, use one with a retractable blade. Report finding to supervisors and inform trainees of dangers. Monitor.	Immediate On-going

NATIONAL GENERAL CERTIFICATE

Unit NGC3 – Health and safety practical application

nebosh

Candidate's observation sheet

Sheet number 5 of 6

Candidate name   J Brown

Candidate number **C** 9887

Place inspected   S & J Joinery Training Workshops

Date of inspection 7 / 6 / 2007

Observations List hazards, unsafe practices and good practices	Priority/risk (H, M, L)	Actions to be taken (if any) List all immediate **and** longer-term actions required	Time scale (immediate, 1 week, etc)
MACHINE WORKSHOP			
22. Jig saw unguarded – guard has been removed. (PUWER)	H	Prohibit use of machine until guard installed. Monitor future use of jig saw. Include need for correct guards on machines in trainee training sessions.	Immediate On-going Each induction
23. Top guard on circular saw set too high. (PUWER)	H	Inform supervisor of dangerous hazard. Set guard to correct gap. Ensure that inspection of all machine guards is part of maintenance schedule and monitor compliance.	Immediate Immediate Each week
24. Noise assessment has been done but several persons not wearing ear defenders even though noise levels are in excess of the upper exposure action level and mandatory signs are in place. (Noise Regs)	H	Noise assessments are made every year and available in the machine shop. This is commendable. All persons in the machine shop must wear ear defenders when any machine is running. Include this in the trainee induction training.	Immediate Each induction
25. Only a water fire extinguisher available in the machine shop. All machines are electrically powered.	H	Replace with 2 fire extinguishers – a carbon dioxide and a powder one. Investigate the reasons for this occurrence with the servicing contractor. Again review fire risk assessment.	Immediate 3 months 1 week
26. LEV not been inspected for 3 years. Possible reasons for dust?	H	Arrange for immediate inspection by a competent person. Set up an on-going contract for future inspections. Train staff and monitor compliance.	Immediate 3 months
27. Only a broom available for sweeping up excess wood dust. Inhalation hazard.	H	Remove broom and replace with an industrial grade vacuum cleaner. Check efficiency of LEV (see 26 above). Train and monitor.	Immediate 3 months
28. Only trainees over the age of 18 years are allowed to use the machines and then only with close supervision.	L	Commendable.	

NATIONAL GENERAL CERTIFICATE

Unit NGC3 – Health and safety practical application

nebosh

Candidate's observation sheet

Sheet number   6   of   6

Candidate name   J Brown

Candidate number **C** 9887

Place inspected   S & J Joinery Training Workshops

Date of inspection 7 / 6 / 2007

Observations List hazards, unsafe practices and good practices	Priority/risk (H, M, L)	Actions to be taken (if any) List all immediate **and** longer-term actions required	Time scale (immediate, 1 week, etc)
ADMINISTRATION OFFICE			
29. Electric fire in the office has no guard and is not included on PAT register. (Electricity Regs)	H	Remove fire and fit a guard and undertake a PAT test. Place fire on PAT register if it is to be kept and monitor.	Immediate After remedial action
30. Health and safety policy statement completed but kept in office drawer unsigned and undated. (HASW Act)	–	Display policy statement on the main notice board in the workshops and communicate effectively to staff and trainees (at induction) after it has been signed by the chief executive of the company and dated.	Immediate
31. Employers' liability insurance certificate is up-to-date but not displayed on a notice board.	–	Display the Employers' liability insurance certificate on the main notice board in the workshops.	
32. The accident records are kept up-to-date with no serious accidents reported.	–	Commendable.	
33. All trainees and employees are given an induction training session that includes health and safety. Detailed records are kept of this and all other training.	–	This is commendable but given some of the problems found during this inspection, a review of the contents of the induction training should be undertaken.	
34. DSE assessment completed even though VDU only used for limited time. (DSE Regs)	L	Good.	
35. Risk of fire due to VDU being left on all the time.	M	Inform staff of the need to ensure that all power supplies to equipment are turned off when the equipment is not in use. Arrange a rota of staff to check all equipment at the end of the day.	Immediate 1 week

Index